MECHANISM
OF MYOFILAMENT
SLIDING IN MUSCLE
CONTRACTION

ADVANCES IN EXPERIMENTAL MEDICINE AND BIOLOGY

MECHANISM OF MYOFILAMENT SLIDING IN MUSCLE CONTRACTION

Edited by
Haruo Sugi
Teikyo University
Tokyo, Japan

and

Gerald H. Pollack
University of Washington
Seattle, Washington

Volume I

SPRINGER SCIENCE+BUSINESS MEDIA, LLC

Library of Congress Cataloging-in-Publication Data

Mechanism of myofilament sliding in muscle contraction / edited by
  Haruo Sugi and Gerald H. Pollack.
       p.   cm. -- (Advances in experimental medicine and biology ; v.
  332)
    Includes bibliographical references and index.
    ISBN 978-1-4613-6245-6   ISBN 978-1-4615-2872-2 (eBook)
    DOI 10.1007/978-1-4615-2872-2
    1. Muscle contraction--Congresses. 2. Cytoplasmic filaments-
  -Congresses. 3. Actin--Congresses. 4. Myosin--Congresses.
  I. Sugi, Haruo, 1933-   . II. Pollack, Gerald H. III. Series.
  QP321.M3385 1993
  591.1'852--dc20   93-22795
                                                          CIP

Proceedings of a symposium on Mechanism of Myofilament Sliding in Muscle Contraction, held
November 11–15, 1991, in Hakone, Japan

ISBN 978-1-4613-6245-6

© 1993 Springer Science+Business Media New York
Originally published by Plenum Press, New York in 1993
Softcover reprint of the hardcover 1st edition 1993

PREFACE

This volume presents the entire proceedings of the symposium organized by one of us (H.S.) on November 11 to 15, 1991 at Hakone, Japan, under the title of "Mechanism of Myofilament Sliding in Muscle Contraction."

Among various kinds of energy transduction mechanisms in biological systems, the mechanism of muscle contraction has been studied most intensively and extensively over many years. Since the monumental discovery by the two Huxleys and coworkers that muscle contraction results from relative sliding between the thick and thin myofilaments, attention of muscle investigators has been focused on the question, what makes the filaments slide past one another. In response to the above question, A.F. Huxley and Simmons put forward a contraction model in 1971, in which globular heads of myosin (cross-bridges) extending from the thick filament first attach to actin on the thin filament, and then change their angle of attachment to actin (power stroke) leading to force generation or myofilament sliding until they detach from the thin filament. The rocking cross-bridge contraction model seemed to be entirely consistent with the kinetic scheme of actomyosin ATPase published by Lymn and Taylor at the same time, thus giving a strong impression to the people concerned that the muscle contraction mechanism would soon be sorted out. In his review lecture in 1974, however, A.F. Huxley warned against such optimism by stating "The whole history of theories of muscular contraction during the last half century shows that even when a set of ideas seems to be well established, there is a large chance that it will be overthrown by some unexpected discovery."

Nearly twenty years has passed since Huxley gave the above review lecture. Despite the accumulation of our knowledge about structural changes during muscle contraction and the relation between the actomyosin ATPase cycle and mechanical events, derived from a number of new experimental techniques, we are still unable to answer the question, what makes the myofilaments slide past one another. In our feeling, it is uncertain whether the mechanism of muscle contraction will eventually be solved along the lines drawn in the early 1970's, or in a completely different way based on unexpected discoveries in the future.

Whichever the truth may be, the editors agree in that, for uncovering the mechanism of muscle contraction, it is necessary to pay attention to phenomena that appear inconsistent with the general lines of thinking, and to place more emphasis on experiments with intact or demembranated muscle fibers in which three-dimensional myofilament-lattice structures are preserved. Based on the above ideas, we have already published proceedings of three muscle symposia organized by us (in

1984) or one of us (H.S.) (in 1979 and 1988). It is our great pleasure that the above proceedings have been highly evaluated by readers, not only for the quality of papers contained, but also for the extensive discussions which constituted a most stimulating and informative part of the proceedings. This volume also contains a lot of such discussions which are, regrettably, always extremely time-consuming for us to edit. Thus, as with our previous publications, this volume is published after a delay of more than one year. Nevertheless, we believe that this volume will be useful and stimulating for readers, if the recent rather slow progress in the field of muscle research is taken into consideration.

Finally, we would like to express our hearty thanks to President Okinaga of Teikyo University, without whose generous support and encouragement the present symposium as well as the past three symposia could not have been held.

Haruo Sugi
Gerald H. Pollack

OBITUARY

William F. Harrington (1920 - 1992)

William F. Harrington, Professor of Biology at the Johns Hopkins University, died on October 31, 1992 after a heart attack at the age of 72. He was born in Seattle, earned his undergraduate and graduate degrees at the University of California at Berkeley, and finished his postgraduate work at Cambridge University and at the Carlsberg Laboratory in Copenhagen. After working at Iowa State University and at the National Heart Institute in Bethesda, Professor Harrington came to the Johns Hopkins University in 1960 and was chairman of the Biology department and director of the McCollum-Pratt Institute from 1973 to 1983.

As a distinguished muscle biochemist, he published more than 125 scientific papers, and had received a National Institute of Health Merit award for 10 years of research. He was visiting professor at Oxford University and at the University of Washington, and was director of the Institute for Biophysical Research on Macromolecular Assemblies at the Johns Hopkins University. As a member of the National Academy of Sciences and a fellow of the American Academy of Arts and Sciences, he served on the editorial boards of scientific journals.

Based on experimental results obtained in his laboratory, Professor Harrington has raised the possibility that the length of the subfragment-2 moiety of the myosin molecule shortens when it moves away from the thick filament backbone on muscle activation to contribute to muscle force generation. This unique hypothesis, quite different from the general view that the myosin subfragment-2 only serves as an elastic link connecting the myosin head (subfragment 1) to the thick filament, has stimulated the interest of muscle investigators and has evoked a great deal of discussion at many scientific meetings. Personally, I worked with him recently on the above topic, and was very much impressed with his enthusiasm in research work as well as his modest and friendly character. Besides his scientific activities, he loved chamber music and was a founding member and on the board of the concert series at the Johns Hopkins University.

As with everyone who knew him, I shall gravely miss Professor Harrington, a great investigator and a true gentleman in every sense.

Haruo Sugi

ACKNOWLEDGEMENT

The editors express their sincere thanks to Dr. Shoichi Okinaga, President of Teikyo University, for his interest, encouragement and generous financial support, which made this symposium possible as well as the previous three symposia held in 1978 (Tokyo), 1982 (Seattle) and 1986 (Hakone).

Our thanks are also due to Mr. Takashi Takeuchi, President of Japan Electron Optics Laboratory (JEOL) Co., Ltd. for their generous additional financial support.

We owe a debt of gratitude to Ms. Therese Grisham and Ms. Tiffany Jones at the University of Washington, who made enormous efforts in transcribing the tapes of the discussions, and to Dr. Hiroyuki Iwamoto and Miss Naoko Itagaki at Teikyo University for their help in preparing the text for camera-ready duplication, and to Drs. Teizo Tsuchiya, Takenori Yamada, Suechika Suzuki, Takakazu Kobayashi, Hiroyuki Iwamoto and Kazuhiro Oiwa in compiling the indices.

CONTENTS

II. REGULATORY MECHANISMS OF CONTRACTION

III. BIOCHEMICAL ASPECTS OF ACTIN-MYOSIN INTERACTION

IV. PROPERTIES OF ACTIN-MYOSIN SLIDING STUDIED BY *IN VITRO* MOVEMENT ASSAY SYSTEMS

V. PROPERTIES OF ACTIN-MYOSIN SLIDING AND FORCE GENERATION STUDIED BY *IN VITRO* FORCE-MOVEMENT ASSAY SYSTEMS

VI. STRACTURAL CHANGES DURING CONTRATON STUDIED BY X-RAY DIFFRACTION

VII. KINETIC PROPERTIES OF ACTIN-MYOSIN SLIDING IN MUSCLE STUDIED BY FLASH PHOTOLYSIS OF CAGED SUBSTANCES AND TEMPERATURE JUMP

VIII. KINETIC PROPERTIES OF ACTIN-MYOSIN SLIDING STUDIED WITH DEMEMBRANATED SYSTEMS

IX. KINETIC PROPERTIES OF ACTIN-MYOSIN SLIDING STUDIED WITH INTACT MUSCLE FIBERS

X. KINETIC PROPERTIES OF ACTIN-MYOSIN SLIDING STUDIED BY ENERGETICS EXPERIMENTS

SUMMARY AND DISCUSSION OF POSTER PRESENTATIONS

CONCLUDING DISCUSSION

INTRODUCTORY LECTURE ON MUSCLE CONTRACTION

A.F. Huxley:

Dr. Okinaga and Professor Sugi, most of all: thank you for bringing us all together. I congratulate Professor Sugi on the way in which he has collected such a large proportion of the people who are contributing to the present-day work in muscle and breaking new ground into the future.

I myself am just returning into the laboratory, having escaped from ten years in administrative posts. As a result, I have not got anything new of my own to present, so I intend to make a virtue of necessity and speak about the past. I shall discuss some of the many occasions during the last hundred years when muscle research went in a direction that turned out later to be wrong. I don't make any apology for this: we learn through mistakes. We learn most effectively from our own mistakes, but in relation to mistakes we have not yet made—or at any rate not yet recognized—we can still learn, even if not quite so well, from mistakes made by our predecessors. Of the wrong turnings about which I shall be speaking, some—a few—have been due to holding on too long to ideas that were superseded; more often, paradoxically, people have gone off in wrong directions through taking up new and substantially correct concepts or observations with too much enthusiasm, or sometimes prematurely

History is instructive in many ways: it can help us to understand what we are doing ourselves, and, I hope, help us to avoid mistakes. I have found reading in the old literature of muscle to be full of interest: I have been more and more impressed by the ability of our predecessors, including the ones who have in a sense taken the wrong tack. There is a tendency to speak of them using the words "nineteenth-century science" as a phrase of disparagement, but I have been repeatedly impressed by the ability and versatility and, particularly, the modernity of outlook of our predecessors a hundred years ago. Perhaps another warning about using this phrase is that the end of the twentieth century is very near, and soon our successors will be speaking about twentieth-century science the way we sometimes speak about nineteenth-century science.

Microscopy in the 19th Century

I have written and spoken often about the way in which much that was known from the use of the light microscope in the nineteenth century was lost[1-3]. I will first go through that again—probably some of you haven't heard me before. In the 1870s and '80s, it was common knowledge that the high refractive index and the birefringence of the A-band were due to rodlets of myosin that were present there. This belief was based on good experimental grounds. First of all, a remarkable observation was made by Ernst von Brücke[4], a most interesting and attractive person who wrote, among other things, on the theory of color in relation to painting. He measured the birefringence of muscle and found that when muscle fibers shortened or were stretched, the strength of their birefringence hardly altered, in contrast to the behaviour of most fibrous materials, whose birefringence increases greatly on stretch. He concluded that the birefringence was due to structures that he called disdiaclasts, little rodlets scattered about that changed their relative positions when the muscle shortened or lengthened but were not individually stretched. That was published in 1858. In 1869, Krause[5] published the observation that the A-band stays at constant length when the fiber shortens or is stretched, and he concluded that the disdiaclasts were in fact rodlets running from one end of the A-band to the other. This was confirmed by Engelmann[6] and by many other nineteenth-century observers, though not quite all. There was argument, but a strong majority held that the A-band stayed at more or less constant width. In the same book of 1869 in which Krause[5] published this observation, he also stated that applying a solution that was known to dissolve myosin took out the high refractive index material that makes the A-band visible.

Engelmann too was a very interesting person apart from his muscle work. He was versatile: he was originally a pupil of Ehrenberg and worked on infusoria. He was a considerable musician—Brahms dedicated a quartet to him—and he came from a family of publishers. In addition to establishing more firmly the constancy of A-band width, he gave correct descriptions of the other features of the band pattern—Z, H, M and N - and of the formation of "contraction bands" on extreme shortening—regions of high refractive index (but low birefringence) that appeared where the I-bands had been; since the I-bands at slack length have a lower refractive index than A this constituted a "reversal of striations" (this was the phenomenon that Niedergerke and I were setting out to reinvestigate in 1953 with my interference microscope when we noticed that the A bands stayed at constant width). Engelmann also established, by wide-ranging comparative observations on various types of muscle and on other intracellular contractile structures, that all "formed contractile elements" are birefringent with the slow axis in the direction of shortening.

The Decline of Microscopy

By 1950, however, much of this was forgotten and most of the rest was back to front in almost all the textbooks. The accepted belief was that shortening of a muscle fiber took place by narrowing of the A bands, and it was universally assumed that the myosin was present in continuous filaments running through both types of bands—largely, no doubt, because that was the only way in which it was possible to imagine that a muscle could shorten or produce tension.

How did it come about that this 19th century knowledge was lost? Here is one of the examples where it was progress that caused things to go off in a wrong direction. When I am emphasizing, as Professor Sugi did, that I am a physiologist and not a biochemist, I say that it was the rise of biochemistry that caused things to go wrong. Biochemistry really made its start in the 1890s, when it was discovered that cell extracts would do many of the things 'hat whole cells do in the way of chemical transformations. And of course this was the start of one of the most important approaches, not only to muscle research but to all kinds of biology. But with this recognition of the importance of chemical events, it became customary to say that contraction is a molecular process; you can't see molecules with a light microscope; therefore, it's no good expecting to learn much by looking at the striations down a microscope. The other, equally important advance that helped to put muscle off the rails was the general acceptance of the idea of evolution: recognition that all living things have a common origin, leading to the idea of the uniformity of Nature. You find repeatedly the statement that all kinds of contractility must be essentially the same, reinforced by Engelmann's observation of the association of birefringence with contractility.

Both of these propositions—the molecular nature of contraction and the uniformity of the nature of contraction wherever you find it—are stated explicitly in a paper of 1901 by Bernstein[7], one of my heroes as the founder of the ionic theory of excitation, who was upholding a surface-tension theory of muscle contraction. Both propositions were arguments for not paying attention to what you can see with the microscope. Smooth muscle contracts, and therefore if all kinds of contractility are essentially the same, the striations cannot be of fundamental significance.

Another factor that caused microscopy to go out of fashion was over-emphasis on the many kinds of artifact that can affect microscopic observations, due both to fixation and to the limitations imposed by the wavelength of visible light. In Britain, this distrust of the microscope was largely due to the emphasis on fixation artifacts by W.B. Hardy[8]; he is another of my heroes for having taken much of the mystery out of the behaviour of colloidal solutions. All methods have their artifacts and the right thing is not to give up a powerful method but to reduce the artifacts and to keep them in mind when interpreting its results.

Yet another of my heroes is of course A.V. Hill[9], particularly on account of his measurement and analysis of heat production by muscle, but one has to admit that he had a blind spot in relation to microscopy. Even in the 1950's, in one of his lectures he lists the methods by which muscle can be investigated—mechanical methods, chemical, thermal, and electrical—but he does not mention microscopy. Looking down a microscope was to him a useless occupation.

New Techniques in Microscopy

Another different kind of factor that led to the loss of these nineteenth-century observations is illustrated by two papers published in the first decade of this century, employing impressive new techniques. The first (Meigs, ref. 10) is by an American who spent some time working at Jena in the laboratory of Biedermann. Jena was of course the home of the Zeiss works and Köhler there had recently invented an ultra-violet microscope. Meigs used this microscope to obtain photographs of myofibrils. Z-lines were beautifully seen (genuinely by absorption), but where the image was in correct focus there was no trace whatever of an I-band. However, where the microscope was focussed a little high, a light zone resembling an I-band appeared flanking the Z-line, and Meigs claimed on the strength of this observation by an impressive new technique that the I-band was an optical artifact. The mistake that he made was to assume that his material, the flight muscle of a fly, was the same as what one might think of as "ordinary" striated muscle. As we now know, those asynchronous muscles undergo extremely small length changes and when they are slack, the length of the I-band is virtually zero. That was a mistake made by putting too much reliance on the proposition of the uniformity of Nature. Evolution gives rise to uniformity through common ancestry, but it is also the process by which the amazing diversity of living things has arisen.

The other paper in the first decade of this century that led people astray was published in 1909 by Hürthle[11]) in Germany. He made cinemicrograph records of muscle fibers from the legs of insects, principally with polarized light. Hürthle claimed that when these fibers shortened, it was at the expense of the width of the A-bands. It is a massive paper, more than 160 pages of *Pflügers Archiv*, and I think there cannot have been many people who read it carefully from beginning to end. I've read quite a lot of it. He went wrong through paying no attention to places where the fibers were somewhat stretched. In his published photographs, you can see that the A-bands then stay at about the width that they have in the shortened fibers, but he describes these fibers as atypical and dismisses them, persuading himself that the normal situation is a shortening of the A-band. When a fiber shortens so much that the thick filaments come into contact with the Z line, the structure of their ends is disrupted so that they lose their birefringence while the accumulation of the material of which they are composed causes an increase in mean refractive index. This shows up in ordinary light as a "contraction band", while in polarized light the width of the birefringent bands decreases.and the contraction bands have the same appearance as I-bands. Hürthle was using polarized light, and therefore concluded that shortening of the fiber as a whole was brought about by shortening within each A band.

Another factor that I think was important in the acceptance of the conclusions of these two papers was that some of the key people who might have picked up the origin of these misconceptions died at just about this time, Kölliker in 1905 arid Engelmann in 1909; I am sure either of them would have kept muscle on the right path. But another factor is that this work was followed by really rather bad microscopy by several workers which seemed to confirm Hürthle's conclusion. As I said microscopy went out of fashion—bright young people went into biochemistry and biophysics, and did not think it worth paying attention to microscopy. The textbook I used when I was an undergraduate (1935-39) quoted work by Holz[12]) in Germany using ordinary light microscopy on the relatively thick muscle fibers of frogs. Engelmann had insisted on using thin (10 - 15 μm) fibers from leg muscles of insects with ordinary (synchronous) striated muscle, where the striations are much broader than in vertebrate muscles. But with a fiber from a frog, perhaps 100 μm thick with striation spacing between 2 and 3 μm, it is almost impossible to get a reliable image with an ordinary microscope.

The Lactic Acid Theory

Another aspect of the rise of biochemistry was over-emphasis on lactic acid. The only chemical change known in muscle at the turn of the century was that when active, muscle produces lactic acid, an observation that was in fact first made by Berzelius, the Swedish chemist, in about 1810—an extraordinary feat. But as this was the only chemical change known, it was assumed that it must be the essential process that causes protein filaments to fold up, no doubt because hydrogen ions from the lactic acid neutralised negative charges on the filaments. This theory began to experience difficulties about 1924. Embden[13] found that much of the lactic acid was produced after the actual contraction. Tiegs[14] found another chemical change—creatine was liberated from muscle when it was active. It had been one of the mainstays of the lactic acid theory that there was a definite limit to the amount of lactic acid that a muscle could produce in one contraction, and this was taken as evidence that a precursor, present in limited quantity, was converted to lactic acid; then it was found that buffering the solution increased the amount of lactic acid formed, so it was apparently the acidification that caused the limitation of the amount of lactic acid produced, not exhaustion of a precursor. The liberation of phosphate from phosphocreatine was discovered in 1927. But Meyerhof, the greatest of the muscle biochemists of that era, who made many important discoveries, and many others, persisted in believing the lactic acid theory until it was finally demolished by Lundsgaard[15], who in 1930 did the notable experiment of poisoning a muscle with iodoacetic acid and finding that it would perform a substantial number of contractions without producing any lactic acid. That observation, by the way, was not part of an investigation of muscle contraction, but a by-product of investigating the cause of the increase in basal metabolism that takes place when your diet contains a large amount of protein.

The Eclipse of Darwinism

Leaving muscle for a moment, it seems to me there is a rather close parallel between the loss of nineteenth-century ideas on muscle around 1900 and the eclipse of Darwinian selection in relation to evolution. Mendelian genetics was rediscovered in 1900 and geneticists went overboard in attributing the direction of evolution to the appearance of new mutations. Natural selection, the process that Darwin had emphasized as the principal—though not the only—cause of evolutionary change, was relegated to an entirely secondary position, despite the fact that Mendelian heredity was exactly what Darwin's theory needed in order to avoid the difficulties created by the assumption that variability would tend to die out through blending. For the first thirty years of this century, it was quite unorthodox to lay much emphasis on natural selection as being important in evolution. Of course, this situation was reversed through the work of R.A. Fisher[16], J.B.S. Haldane, Chetverikov, Sewall Wright, and others, and neo-Darwinism took the stage in the 1930s, with natural selection playing an even more important part than Darwin had believed.

X-ray Diffraction

Another example of an impressive new technique that led people astray was the use of X-ray diffraction. In the 1930s, this was wide-angle X-ray diffraction, not the low-angle diffraction from which we have learned so much in the last forty years. Astbury and others recorded X-ray diagrams from resting and active muscle. These wide-angle diagrams showed remarkably little change in fact, but the authors somehow convinced themselves that contraction took place by an alteration in the folding in the proteins of which these supposed filaments were made. Indeed, I was taught that contraction was a switch from the alpha to the beta state

Tonic muscle and the All-or-None Law

Another major advance that led to a misconception was the all-or-none law of excitation. Of course, in relation to the heart, that was known from Bowditch's[17] work carried out in Ludwig's laboratory. Around 1900, it was an open question whether the gradation of force developed by skeletal muscle, stimulated either directly or through the nerve, was due to graded activity of individual fibers, or whether it was purely a matter of the number of fibers active. The crucial observations were made by Keith Lucas, who proved by separate but rather similar experiments on frog muscle that both muscle fiber excitation[18] and nerve conduction[19] are all-or-none processes in the individual fibers. In each of these experiments, Lucas demonstrated steps in the size of the mechanical twitch when plotted against strength of stimulus, which was increased smoothly. In 1922, E.D. Adrian[20], a pupil of Keith Lucas, showed that the action potential of individual muscle fibers is itself an all-or-none event. The most conclusive work on the all-or-none law was done on isolated fibers in the 1920s by G. Kato of Keio University in Tokyo.

Around 1920, there were many observations suggesting that some frog muscles had an alternative mode of contraction, a tonic mode that was not even twitch-like. This appeared to be contradicted by the experiments of Adrian & Bronk[21][22], recording electrical activity from living animals with a needle electrode that they could poke into the muscles of either cats or human beings. They found that even tonic activity of muscles consisted of asynchronous, twitch-like activity and unfused tetani. On this basis, most or all electrophysiologists (probably not all pharmacologists) dismissed the earlier ideas of a tonic mode of contraction, which were regarded with contempt, despite another piece of evidence that there was a second type of skeletal muscle fiber. Krüger in the 1930s, by ordinary light microscopy of stained transverse sections of frog muscle, showed that muscles like the familiar sartorius contain only fibers with very thin myofibrils while muscles that were known to give long-lasting contractures to KCl contained also fibers with great ribbon-like fibrils to which he gave the name Felderstruktur fibers. Perhaps another reason why the evidence for a tonic mode was lost was another piece of progress—the switch from using the frog gastrocnemius, an easy muscle to dissect but unsuitable for quantitative experimental work because it is so thick, to the sartorius. The sartorius does not contain Felderstruktur fibers, and gives only twitch-like activity. Finally, a conclusive demonstration of this other type of contraction was made by Tasaki, a pupil of Kato, in 1942, using again the gastrocnemius and isolating motor nerve fibers in the sciatic nerve. Stimulation of large-diameter motor nerve fibers would give twitch or tetanic responses, but a single stimulus to a small fiber caused no response while repetitive stimulation gave smooth contractions graded according to the frequency of the stimuli. It was later shown by Kuffler & Vaughan Williams[23)24)] that these two types of contraction were performed by different muscle fibers.

It was soon accepted that this was the situation in amphibia, but it was generally believed, on the basis of Adrian & Bronk's experiments, that special tonic fibers did not exist in mammals, and it was supposed that the small motor nerve fibers in mammals were innervating intrafusal muscle fibers. Another surprise came in 1961 when Matiushkin[25)] in the USSR published records of mechanical and electrical activity in what were well known to be the fastest muscles that anyone had studied, namely the external eye muscles of mammals, showing that they (especially the obliques) contain a small proportion of these tonic, non-spiking fibers. An equally great surprise is that fibers of the fast swimming muscles of bony fishes do not give all-or-none electrical responses.

Adrian himself, having shown that the muscle action potential is all-or-none, realized that this was a *sufficient* explanation for the all-or-none nature of the twitch, and that it was perfectly possible that the mechanical response was itself graded but the stimulus to the contractile material was the electrical event which he had found to be all-or-none. In the 1930's, Gelfan, from the U.S.A. but working partly in Adrian's laboratory, demonstrated graded behaviour of small regions of muscle fibers within the retrolingual membrane of the frog. He obtained graded responses by using very small electrodes. But his results were not widely accepted. I met Gelfan, I think in 1964, and he was justifiably bitter about the fact that he had been unable to get his results accepted in the United States because of the apparent contradiction of the well-established all-or-none law. In 1950, at the Physiological Congress in Copenhagen, I met Alex Forbes, who had been Professor of Physiology at Harvard and an outstanding figure in electrophysiology in the 1920s, and he boasted of having been the "watchdog of the all-or-none law". He was not the only one to give too much weight to the all-or-none law. Hodgkin records that when he visited the United States in 1937, with good evidence of graded sub-threshold activity in nerve fibers (a slow potential rise before it goes off explosively into a spike) he could not get this accepted by even such famous and important figures as Erlanger and Gasser. The early theories of sub-threshold electrical activity, notably by Rushton, one of my teachers, was that the gradation took place not by graded activity of any one small area of membrane, but by variation of the area that was in a fully active state at any one moment.

The Electron Microscope

Another impressive advance that caused a delay was the use of the electron microscope. The first useful electron micrographs of muscle to be published were by Hall, Jakus & Schmitt[26)], who claimed that they could see continuous filaments running through the A- and I-bands of the myofibrils that they studied. Another paper in 1950, one of whose authors was Albert Szent-Györgyi, again dealing with isolated fibrils, also claimed that continuous filaments could be seen (Rozsa et al., ref. 27). These claims were widely accepted as evidence for the theory then current that shortening was due to the folding up of continuous protein filaments. Hasselbach told me that H.H. Weber, who in 1930 had fresh evidence for myosin being in the A-band, had dropped this idea on account of this impressive evidence from electron microscopy.

Conclusion

Well, all this makes one wonder what misconceptions we may be harboring nowadays. It is easy to recognize misconceptions in the past—nothing is so useful as hindsight. History does repeat itself, but the trouble is that at any one moment, one doesn't know which bit of history it is that is going to be repeated: one can draw different lessons from different episodes in the past. As I have said, some of these sidetracks have been due to too much adherence to past ideas, as with this question of continuity of filaments; others have been due to too-active acceptance of correct ideas, as with the all-or-none law. At the present time we are experiencing a switch to the intra-molecular level, which may have analogues to the switch to the molecular level that took place nearly a hundred years ago, Will our successors find that important work of the last forty years is now being put aside because it doesn't tell us about what is happening inside the protein molecules that are clearly doing all the interesting things? Or are we clinging too hard to ideas that have been around for many years, like swinging cross-bridges? How are we to know?

I am sure that this meeting is going to help us on the path to discover what is right and what is wrong. The difficulty is that usually it is not as clear-cut as that: most things are partly right and partly wrong. There are very few clear-cut "yes" or "no" answers in biology. When I give a lecture on a biological topic, I find myself hesitating when I am about to say, "Such-and-such is always the case". I usually correct myself and have to say, "Nearly always ...".

REFERENCES

1. Huxley, A.F. in *The Pursuit of Nature* (eds. Hodgkin, A.L. and others.), 23-64 (Cambridge: University Press, 1977).
2. Huxley, A.F. *Reflections on muscle (Sherrington Lectures, 1977)*. (Liverpool: University Press, 1980).
3. Huxley, A.F. *Proc. Amer. philos. Soc.* **130**, 475-481 (1986).
4. Brücke, E. von *Denkschr. Akad. Wiss. Wien, math.-naturwiss. Kl.* 15, 69-84 (1858).
5. Krause, W. *Die motorischen Endplatten der quergestreiften Muskelfasern.* (Hannover: Hahn, 1869).
6. Engelmann, T.W. *Pflügers Archiv,* **7**, *Erster Artikel,* 33-71, *Zweiter Artikel,* 155-188 (1873).
7. Berstein, J. *Pflügers Arch.* **85**, 271-312 (1901).
8. Hardy, W.B. *J. Physiol. (Lond.)* **24**, 158-210 (1899).

9. Hill, A.V. *Proc. R. Soc. B* **126**, 136-195 (1938).

10. Meigs, E.B. *Z. Allg. Physiol.* **8**, 81-120 (1908).

11. Hürthle, K. *Pflügers Arch.* **126**, 1-164 (1909).

12. Holz, B. *Pflügers Arch.* **230**, 246-254 (1932).

13. Embden, G. *Klin. Wochenschr.* **3**, 1393-1396 (1924).

14. Tiegs, O.W. *Austr. J. Exp. Biol. Med. Sci.* **2**, 1 (1925).

15. Lundsgaard, E. *Biochem. Z.* **217**, 162-177 (1930).

16. Fisher, R.A. *The genetical theory of natural selection.* (Oxford: Clarendon Press, 1930). .

17. Bowditch, H.P. *Ber. Sächs. Ges.* **23**, 652-689 (1871).

18. Lucas, K. *J. Physiol. (Lond.)* **33**, 125-137 (1905).

19. Lucas, K. *J. Physiol. (Lond.)* **38**, 113-133 (1909).

20. Adrian, E.D. *Arch. Neerl. Physiol.* **7**, 330-332 (1922).

21. Adrian, E.D. & Bronk, D.W. *J. Physiol. (Lond.)* **66**, 81-101(1928).

22. Adrian, E.D. & Bronk, D.W. *J. Physiol. (Lond.)* **67**, 119-151 (1929).

23. Kuffler, S.W. & Vaughan Williams, E.M. *J. Physiol. (Lond.)* **121**, 289-317 (1953).

24. Kuffler, S.W. & Vaughan Williams, E.M. *J. Physiol. (Lond.)* **121**, 318-340 (1953).

25. Matiushkin, D.P. *Fiziol. Zh. SSSR.* **47**, 878-883 (in Russian). Translated in *Sechenov Physiol. J. USSR* **Jan. 1962**, 960-965 (1961).

26. Hall, C.E., Jakus, M.A. & Schmitt, F.O. *Biol. Bull., Woods Hole*, **90**, 32-50 (1946).

27. Rozsa, G., Szent-Györgyi, A. & Wyckoff, R.W.G. *Expt. Cell Res.* **1**, 194-205 (1950).

I. STRUCTURAL BASIS OF MYOFILAMENT SLIDING

INTRODUCTION

To clarify the mechanism of muscle contraction at the molecular level, detailed structural studies on the actin and myosin molecules and the myofilaments are of fundamental importance. A typical example of structural studies that led to a monumental finding is the establishment of the sliding filament mechanism in muscle contraction.

This chapter consists of seven papers dealing with structural features of muscle proteins and the myofilaments, attention being mainly focused on the mode of actin-myosin interaction and the possible conformational changes of myosin heads leading to muscle contraction. Holmes and others compared the atomic mode of F-actin structure reconstituted from cryo-electron microscopy of F-actin. They showed that the two structures are identical, and further found that, in the S-1 decorated filaments, each myosin head binds to two actin molecules on two distinct sites. Maedá and others describe a new crystal form of tropomyosin obtained from lobster tail muscle.

The next three papers are concerned with the actin-myosin linkages within muscle fibers. Reedy and coworkers performed detailed electron microscopic studies on the interesting features of rigor actin-myosin linkages in insect flight muscle, Using the quick-freezing and deep-etch techniques, Katayama reported distinct structural changes of myosin heads when they interact with actin, suggesting that they actually tilt while sliding past F-actin. Using the same techniques, Suzuki, Oshimi and Sugi examined the angular distribution of cross-bridges in muscle fibers, and showed that the cross-bridges tend to take angles around 90° in both contracting and rigor states. They also estimated the thin filament stiffness times unit length to be about 1.8×10^4 pN.

Trombitás and Pollack made elegant electron microscopic studies on the elastic properties of connecting filaments anchoring the thick filaments to the Z-line. They demonstrated the spring-like nature of connecting filaments in both insect flight and vertebrate skeletal muscles as well as other interesting features of connecting filament network along the sarcomere. Finally, Faruqi, Cross and Kendrick-Jones reported their interesting attempt to study the structure of myosin with the scanning tunneling microscope recently developed.

A COMPARISON OF THE ATOMIC MODEL OF F-ACTIN WITH CRYO-ELECTRON MICROGRAPHS OF ACTIN AND DECORATED ACTIN

K.C. Holmes, M. Tirion, D. Popp, M. Lorenz, W. Kabsch and R.A. Milligan*

Department of Biophysics
Max Planck Institute for Medical Research
Postfach 103820, Heidelberg, Germany
Department of Cell Biology
The Scripps Research Institute
10666 North Torrey Pines Road
La Jolla, California 92037, USA

ABSTRACT

We compare the atomic model calculated from the crystal structure and the X-ray fiber diagram of orientated F-actin[1] with the 3-D reconstructions produced from cryo-electron microscopy of actin[2]. Out to 30Å resolution the two structures are essentially identical. Furthermore, by combining the atomic model with the reconstruction of S1-decorated actin filaments[2] one can establish the nature of the actin binding site for myosin in the rigor complex. Each myosin head binds to two actin molecules on two distinct sites. Some of the actin residues involved in each of these binding sites can be identified. Furthermore, the atomic model of actin may be combined with the reconstruction of the S1 decorated thin filament to establish the tropomysosin binding site in the rigor complex. This result is compared with the model of tropomyosin-actin derived from an analysis of the X-ray fibre diagram of a reconstituted thin filament and are shown to be very similar.

Comparison of Actin Structures

By combining the atomic coordinates of the actin monomer derived from the X-ray crystallographic structure of the actin-DNase I complex solved by Kabsch et al[3] with the X-ray fibre diffraction pattern, Holmes et al[1] have produces an atomic model of F-actin (Fig. 1). Using the atomic coodinates from this model the cylindrical components of the Fourier-Bessel transform (the G_{nl}'s)[4] permitted by the helix selection rule for actin ($l = -6n + 13m$) have been calculated for the first 10 layer-lines out to a reciprocal space radial resolution of $1/20$Å^{-1}. The atomic scattering factors were corrected for water (solvent) scattering[1][5][6]. This is an

Mechanism of Myofilament Sliding in Muscle Contraction, Edited by
H. Sugi and G.H Pollack, Plenum Press, New York, 1993

15

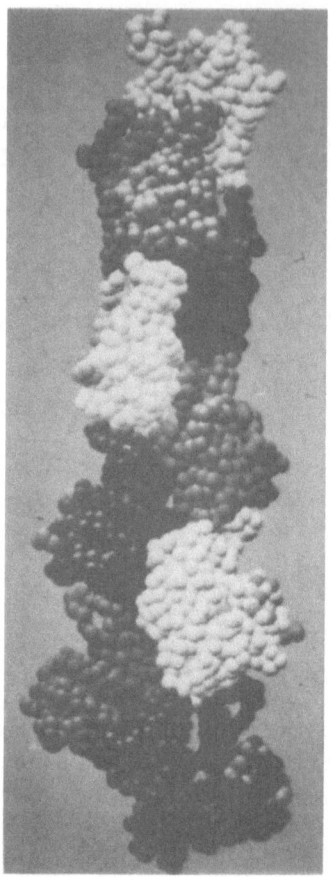

Fig. 1. Shown is a representation of the atomic model of F-actin[1] in which each sphere represents an amino-acid residue. The different shades serve to demarcate the subunits. The grey spheres show residues known from chemical cross-linking and other studies to be involved in myosin binding (see refs. 1 and 3 for references). In this and all subsequent diagrams the barbed end is at the bottom of the diagram.

important factor in the comparison since the the cryo-electron micrograph images are contrasted against vitreous ice, which has about the same density as water. The calculated G_{nl}'s were compared with the contrast-transfer-function-corrected G_{nl}'s derived from cryo-electron microscope images of actin filaments obtained by Milligan et al.[2] which extend to a resolution of $1/30Å^{-1}$. A search over all actin filament orientations was used to establish the best correspondence between the two data sets. The sum of squares of the algebraic differences between the two (complex) data sets was used as an index of goodness of fit. The atomic model was rotated by 5° intervals about the helix axis and translated in 5Å steps along the helix axis over one actin repeat. Since the structure is polar, both orientations of the model were considered. The best fit was unambiguous with a (sum of squares) R-factor of 20%.

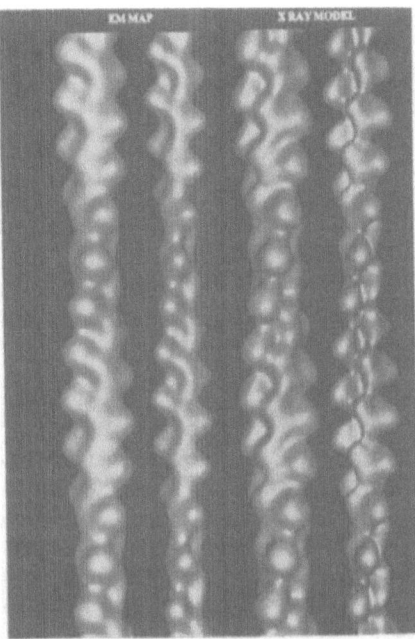

Fig. 2. Shown (left) is the reconstruction (at 20Å resolution) of F-actin at two cut levels calculated from G_{nl}'s derived from the atomic model of Holmes et al[1]. Right is shown the reconstruction (at 30Å resolution) at the corresponding cut levels calculated from cryo-electron micrographs of F-actin[2]. The G_{nl}'s have been corrected for the contrast transfer function.

Bearing in mind that the R-factor was based on the *algebraic* comparison of complex numbers and not, as in crystallography, on comparing the *modulus* of complex numbers, this result is as good as can be expected. The goodness of fit may be appreciated from Fig. 2 which shows reconstructions calculated at the same resolving power and two different cut levels from the experimental G_{nl}'s and those calculated from the atomic model. The agreement is striking.

Comparison of the Atomic Model with Decorated Actin

Having established the feasibility of a quantitative comparison between the atomic model and electron micrograph images obtained from filaments embedded in vitreous ice we were encouraged to combine the atomic model with images of S1-decorated actin and decorated thin filaments obtained by Milligan et al[2].

The result for decorated actin is shown in Fig. 3. The positioning of the actin is achieved by maximising the overlap between the electron-micrograph electron density and the atomic model for all electron density at a radius from the helix axis of less that 35Å. In the figure this electron density has been omitted for clarity and the atomic model of actin, represented as a Ca plot, has been inserted. The precision

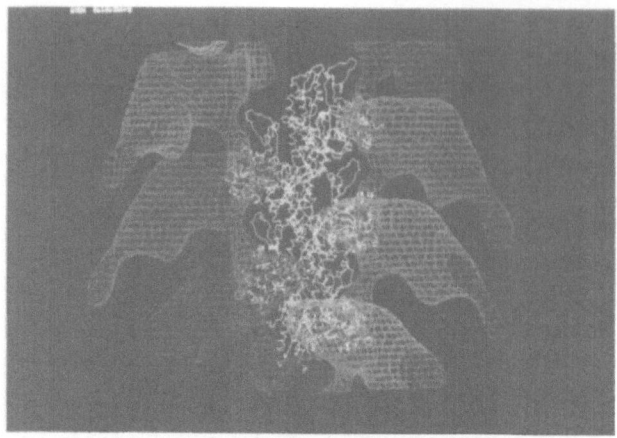

Fig. 3. The electron density calculated from electron-micrographs obtained from vitreous ice embedded S1-decorated-actin is shown (squirrel cage) together with the atomic model of F-actin (Ca representation). The electron density corresponding to actin has been omitted for clarity. Note the two binding sites[2]: a major binding site corresponding to the rigor binding site which encompasses the residues 2-4,20-25, 334-354; and a second binding site made by a finger-like extension from the myosin head which contacts actin on the back surface of the α-helix 79-93 and loop 94-99 (encompassing residues Arg37, His87, Arg95, Glu99, Glu100, and possibly Glu83).

of the actin position is estimated at about ± 3° in azimuth and ± 3Å along the helix axis.

The main actomyosin rigor interaction corresponds very well with the area expected (shown as white spheres in Fig. 1) on the basis of chemical cross-linking and chemical modification (see ref 3). However, Milligan et al[2] have pointed out that one S1 interacts with two actins. Some properties of the second site can also be ascertained (see below).

The major binding site can be recognised as involving the N-terminus of actin (negatively charged), the 20-25 loop (positively and negatively charged), and the hydrophobic helix (340-350). Kabsch et al[3] pointed out that this helix is unusual since it has three hydrophobic groups (Ile341, Ile345, Leu349) pointing out into the solvent. The present comparison shows that these hydrophobic residues occur in the middle of the rigor binding site so that they appear to contribute in an important way to the actomyosin interaction. Presumably they do not shown up in cross-linking experiments because of the low reactivity of hydrophobic side chains.

The second binding site can be seen to involve the finger-like extension of myosin S1 described by Milligan et al. This finger extends from the S1 attached to the neighboring actin along the long-pitch helix and makes contact with the the helix 79-93 and the following loop. The actin side chains taking part in the interaction appear to include His87, Arg95, Glu99, Glu100. The interaction apparently involves only charged residues.

Significance of the Second Binding Site

Residues between 90 and 100 of actin have already been proposed as part of the myosin binding site[7]. The present study suggests, however, that they are not part of the major myosin binding site but rather are part of a second binding site which is dominated by salt bridges. The existence of two actin binding sites on myosin has long been suspected on the basis of cross-linking experiments[6] (for review, see refs 8 and 10). In refs 8 and 9 it was argued that the two binding sites may be of significance in muscle contraction: if the binding to the two sites were to be sequential so that one bond had to be formed before the other, (an energy-requiring process) and if myosin had a hinge, one would have a natural basis for the cross-bridge power-stroke.

Building Tropomyosin into the Decorated Thin Filament Image

Tropomyosin can readily be built into the cryo-electron microscopic reconstructions of the decorated thin filament. Fig. 4 shows a Ca representation of a two-stranded coiled-coil of tropomyosin supercoiling around actin. By analysis of the repeating motifs in the primary sequence of tropomyosin it has been shown[11][12] that tropomyosin has a pseudo-repeat of 19 2/3 residues which permits it to supercoil on F-actin so that the backbone takes on *the symmetry of actin*. X-ray crystallographic analysis[13] of crystalline tropomyosin show that the tendency to supercoil is inherent in the tropomyosin structure and that the coiled-coil indeed has about the correct pitch to supercoil on actin. Therefore, a model was built on this basis (see ref 13) with a supercoil radius of 38Å (Fig. 4). This radius gives 39.6

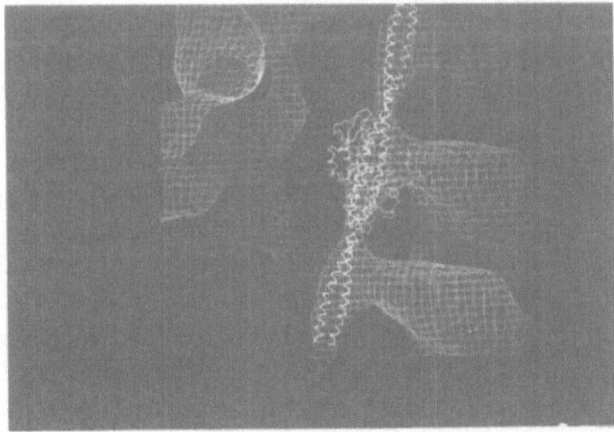

Fig. 4. The electron density calculated from cryo-electron micrographs of S1-decorated-thin-filaments is shown (squirrel cage) together with the atomic model of F-actin (Ca representation) and an atomic model of tropomyosin. The electron density corresponding to actin has been omitted for clarity. The radius of the tropomyosin is 38Å.

residues per actin for each tropomyosin chain which is very close to twice the pseudo repeat discovered by McLachlan & Stewart[11]. Moreover, as can be seen, it gives an excellent fit to the electron microscopic reconstructions. The relative position with respect to actin is depicted in Fig. 5b. The apparent interaction sites between tropomyosin and actin involve the outside strand of the β-sheet of the large domain and the neighboring α-helix (308-333) and seem in particular to involve the following hydrophilic residues: Asp311, Asn314, Lys326, Lys328.

The results are in general agreement with the pioneering determination of Spudich et al [14] although the radius obtained by these authors was rather smaller than the present determination.

The Position of Tropomyosin deduced from X-ray fibre Diagrams

Orientated gels of actin + tropomyosin give x-ray fibre diagrams of somewhat inferior quality to those of pure actin. However, they can be analysed to give the position of tropomyosin. The coordinates of super-coiled coiled-coil tropomyosin were generated as in ref 13. A 'typical' sequence of 39 residues (91-130 from rabbit tropomyosin A) was used. The radius, azimuth, displacement and origin of the tropomyosin were allowed to refine. The result is depicted in Fig. 5a. The radius of the tropomyosin is 38Å, in good agreement with the electron microscope result. As can be seen, the tropomyosin appears to be displaced about 10Å to the right compared with the electron microscope result. The difference suggests that the tropomyosin is forced to the left by the binding of myosin S1 in the rigor complex. However, since the azimuthal position of the tropomyosin in both determinations is subject to error this suggestion remains preliminary.

Implications for the Steric Blocking Hypothesis

At about the same time a number of groups[14-18] noted that the characteristic wide-angle diffraction peak on the second actin layer-line in muscle fibres altered its intensity depending on the physiological state of the thin filament. This intensity maxima arises from a Bessel function of order 4 (J_4) and reflects the degree of four-foldedness of the the thin filament. Typically in the permissive 'on' state (high Ca^{2+} and troponin, or actin + tropomyosin without troponin) the J_4 is strong. In the 'off' state (troponin without Ca^{2+}) the J_4 is weak implying that the tropomyosin moves so as to destroy the four-foldedness. This result was the bulwark of the so called 'steric blocking model' for explaining the control exercised by troponin and Ca^{2+} in the presence of tropomyosin over the acto-myosin interaction (the steric blocking model proposes that in the 'off' state tropomyosin is induced by an interaction with troponin to block the myosin binding site on actin : in the 'on' state it moves out of the way).

It has been widely assumed that tropomyosin would not take up the full symmetry of the actin genetic helix but would rather merely follow the long-pitch

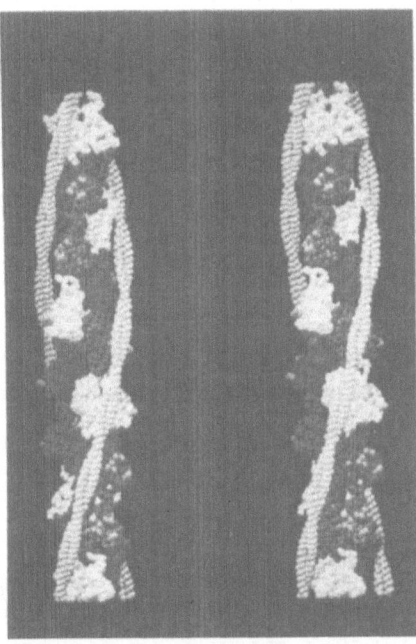

Fig. 5. A representation of an atomic model of actin and tropomyosin in which each sphere represents an amino-acid residue: (left) as deduced from analysis of x-ray fiber diffraction patterns of orientated gels of actin + tropomyosin; (right) as deduced from the reconstruction of decorated thin-filaments. The small azimuthal difference in position between the tropomyosin in the two results may be significant and could reflect the fact that myosin in the actomyosin rigor complex pushes the tropomyosin more to the left than the position it would take without myosin. In both cases the tropomyosin lies close to the β-strand and α-helix between 308-333. The interaction involves the residues Asp311, Asn314, Lys315, Lys326, Lys328.

helix. This would imply that changes in the diffraction pattern attibutable to a putative tropomyosin movement would be limited to the low-order layer-lines (the m = 0 branch of the transform) and in practice would only be observable on layer-lines 0-3. As was discussed above, the backbone of tropomyosin is rather likely to adopt the full actin symmetry. Thus a movement of tropomyosin will cause changes to *all* parts of the the observable fibre diffraction pattern.

This suggestion has important ramifications for the interpretation of x-ray fibre diagrams of orientated actin-tropomyosin gels or of diffraction from muscle fibres It is not necesary to invoke conformational changes of actin to explain changes in (say) the intensity of the 59Å layer-line on activating muscle. Model calculations show that such changes can be explained by an azimuthal movement of tropomyosin.

The steric blocking hypothesis maintains that the tropomyosin should be able to move so as to block the myosin binding site. The atomic models presented above show that this is indeed possible. If the tropomyosin is to retain its pseudo-symmetrical relationship to the actin helix then its radius remains constant. By moving the tropomyosin at a constant radius (rolling on the surface of actin would also be possible) one readily brings it into contact with the major myosin binding site. However, it is very hard to see how the tropomyosin can have any great influence on the second binding site, which is on the other surface of the actin and therefore not available to tropomyosin. It is thus tempting to identify the second binding site with the weak binding site[19] since not only is it not-tropomyosin-controllable but would also be sensitive to salt concentration.

ACKNOWLEDGEMENTS

This work was supported in part by research grant ASR39155 from NIH (to R.A.M.). R.A.M. is a Pew Scholar in the Biomedical Sciences

REFERENCES

1. Holmes, K.C., Popp. D., Gebhard, W. & Kabsch, W. *Nature* **347**, 44-49 (1990)
2. Milligan, R.A., Whittaker, M. & Safer, D. *Nature* **348**, 217-221. (1990)
3. Kabsch, W., Mannherz, H.-G., Suck, D., Pai, E. & Holmes K.C. *Nature* **347**, 37-44. (1990)
4. Klug, A., Crick, F.H.C., & Wyckoff, H.W. *Acta Cryst.* **11**, 273-283 (1958)
5. Langridge, R., Marvin, D.A., Seeds, W.E., Wilson, H., Cooper, R., Wilkins, M.H.F & Hamilton, L.D. *J. Mol. Biol.* **2**, 38-64 (1960)
6. Fraser, R.D.B., MacRae, T.P. & Suzuki, E. *J. Appl. Cryst.* **11**, 693-694 (1978)
7. Bertrand, R., Chaussepied, P., Kassab, R., Boyer, M., Roustan, C. & Benyamin, Y. *Biochemistry* **27**, 5728-5736 (1988)
8. Goody, R.S. & Holmes, K.C. *Biochim. Biophys. Acta* **726**, 13-39 (1984)
9. Holmes, K.C. & Goody, R.S. in *Contractile Mechanisms in Muscle* (eds. Pollack, G.H. and Sugi H.) Plenum N.Y. pp 373-384 (1984)
10. Carlier, M.F. *Int. Rev. Cytol.* **115**, 139-179 (1989)
11. McLachlan, A.D. & Stewart, M. *J. Mol. Biol.* **103**, 271-298 (1976)
12. Parry, D.A.D. *J. Mol. Biol.* **98**, 519-535 (1975)
13. Phillips, G.N.Jr., Fillers, J.P. & Cohen C. *J. Mol. Biol.* **192**, 111-131 (1986)
14. Spudich, J.A., Huxley, H.E. & Finch, J.T. *J. Mol. Biol.* **72**, 619-632 (1972)
15. Huxley, H.E. *Cold Spring Habor Symp. Quant. Biol.* **37**, 361-376 (1972)
16. Haselgrove, J.C. *Cold Spring Habor Symp. Quant. Biol.* **37**, 341-352 (1972)
17. Parry, D.A.D. & Squire J. *J. Mol. Biol.* **75**, 33-55 (1973)
18. Wakabayashi T., Huxley, H.E., Amos, L.D. & Klug, A. *J. Mol. Biol.* **93**, 477-497 (1975)
19. Brenner, B., Schoenberg, M., Chalovich, J.M., Greene, L.E. & Eisenberg, E. *Proc. Natl. Acad. Sci. USA* **79**, 7288-7291 (1982).

Discussion

Pollack: Now that it is clear that nebulin appears co-localized with the thin filaments, do you have any indication where nebulin might lie specifically?

Holmes: Unfortunately, I don't know.

Reedy: That's a good subject for the next meeting.

Gillis: Ken, could you tell us which site is blocked by tropomyosin? Is it the one with the finger, or is it the one with the big bulge of the myosin? I didn' t understand what you said.

Holmes: No, it' s the main rigor-binding site. I call it the "main rigor-binding site," because it is the biggest and firmest. All the cross-linking stuff indicates that it is the rigor-binding site—the one that can be blocked by tropomyosin quite easily. The other one, that Milligan has turned up, is on the other surface of actin (Milligan, R.A., Whittaker, M. and Safer, D. *Nature* **348**, 217-221, 1990). The tropomyosin really can' t get over there, so I don' t see any way it can block it.

Squire: When David Parry and I were doing modelling on the steric blocking mechanism, we tried giving a 50 Å repeat to the tropomyosin to see how much effect it would have on the higher layer lines (Parry, D.A.D. and Squire, J.M. *J. Mol. Biol.* **75**, 33-55, 1973). It was actually very hard to produce a big change there. Can you comment on that in light of what you just said?

Holmes: I' m not saying there is a big change; there are simply quite measurable changes. I have done atomic modelling on it, and I know exactly what happens. The changes are not wild—about 20%.

Kawai: How close do the myosin heads come to the helix of the actin?

Holmes: It is at high radius. I think you can divide actin from myosin roughly by putting a 40 Å cylinder through the whole thing.

Gergely: Can the second new site give an indication that the binding occurs only in the presence of ATP, or is it also in rigor?

Holmes: We only know what it looks like in the rigor complex. There is no other information.

Tregear: Ken, the cross-linking experiments only showed the binding of the myosin head to one actin, even in rigor, when we had hoped they would show two. Do you know whether this second site ought to show cross-linking?

Holmes: Well, there is much evidence that points to S-1s binding to two actins. Valentin-Ranc et al. just published a study in which they modified the actin-actin affinity by using pyridoxal phosphate to block one of the lysines (Valentin-Ranc, C., Combeau, C., Carlier, M.F., Panteloni, D. *J. Biol. Chem.* **266**, 17872-17879, 1991). This makes the affinities of actin for itself somewhat lower. When you have done that, you can actually study how S-1 influences the assembly of actin. It does so by interacting in such a way that the stoichiometry is two-to-one.

Gillis: I would like to return to the question of the two sites for binding actin, the big one and the "finger" one. If you compare that to the kinetics of movement of myosin toward actin as revealed by fast X-ray diffraction pattern, the first thing that occurs is that myosin moves toward actin before any force has developed. This is well known. So, the concept of the weak-binding site has been proposed. Your model suggests that the two bindings would occur in sequence. The first one, the weak one, would be some sort of prerequisite for the second one. Would you speculate about that, or is it too far-fetched?

Holmes: I think I have already speculated far too much. I shouldn' t have said it at all. The facts suggest that Milligan has seen these two sites and there is a fair amount of evidence now in the literature that points to the existence of two sites (Milligan, R.A., Whittaker, M., and Safer, D. *Nature* **348**, 217-221, 1990). That is, one S-1 seems to interact with two actins. So far, this is a quasi-fact. I think it is becoming a rather firm fact. Anything beyond that is pure speculation. All I am prepared to say is that if you build our atomic model into the Milligan results, what you find is that his finger comes close to the following residues: Arg 37; His 87; Arg 95; Glu 99; Glu 100, and possibly, Glu 83. The point is that there are a number of positively and negatively-charged residues there, but no hydrophobic residues. That is just a statement of what I see on that surface of actin. That is firm, but anything you might infer from that I think you had better do at your own risk.

A NEW CRYSTAL FORM OF TROPOMYOSIN

Andrea Miegel, Lan Lee, Zbigniew Dauter and Yuichiro Maéda

European Molecular Biology Laboratory at DESY
Notkestraße 85
D-2000 Hamburg 52, Germany

ABSTRACT

Tropomyosin crystals with a new morphology have been obtained from lobster tail muscle tropomyosin from which 11 residues at the carboxyl-terminus have been proteolytically removed to avoid head-to-tail polymerization. In contrast to the conventional Bailey crystal form in which the elongated tropomyosin molecules form a mesh, in the present crystals the molecules are packed side-to-side with the long axes parallel to the c-axis of the crystal. The unit cell is tetragonal with a = b = 109 Å, c = 509 Å, and the symmetry is either $P4_12_12$ or $P4_32_12$, with $4_1(4_3)$ helical axes parallel to the c-axis. This suggests that a group of molecules surrounding a local $4_1(4_3)$ axis is regarded as the building unit of the crystal. It is likely that the unit cell contains eight molecules with one molecule per asymmetric unit.

1. INTRODUCTION

Detailed three dimensional structural information on individual muscle proteins is essential for understanding the mechanism of muscle contraction and its regulation.

Our recent NMR study[1][2] of the carboxyl-terminus of the myosin rod, a member of the α-helical coiled-coil proteins, has suggested that the two chains of the molecule are not exactly in register. The NMR spectra indicated that the C-terminus of the molecule is unfolded and mobile, 9 residues from one chain and 12 residues from the other. The simplest interpretation of the asymmetric unfolding is that the two chains are staggered relative to one another by 3 residues. This is not consistent with Crick's model[3] for the packing of two α-helical peptides into a coiled-coil protein. The stagger would result in a weaker interface, making the molecule more flexible. Although the structure of the leucine zipper of GCN4 has recently been solved, indicating two identical chains exactly in register[4], it is still of great importance to solve the structure of other large α-helical coiled-coil molecules,

Mechanism of Myofilament Sliding in Muscle Contraction, Edited by
H. Sugi and G.H Pollack, Plenum Press, New York, 1993

especially molecules the function of which might be associated with alteration of its flexibility.

We have therefore carried out systematic trials to obtain crystals of coiled-coil proteins, including segments of myosin rods and tropomyosin. We have to date obtained crystals with a new morphology from the tropomyosin of lobster tail muscle.

Tropomyosin plays a key role in calcium regulation; the regulatory complex, one tropomyosin plus one troponin molecule, spans seven actin monomers on the actin filament and the signal of calcium binding to troponin is transmitted to actin via tropomyosin. Tropomyosin is an elongated molecule about 400 Å long. Two identical (or nearly identical) peptides, each forming an α-helix over almost its entire length, form a coiled-coil. In the coiled-coil the two chains are parallel (not anti-parallel) and almost in register. In solution, especially at low ionic strength, tropomyosin molecules polymerize through a head-to-tail interaction.

In spite of the substantial knowledge about the molecule, many questions are still left open, among others the nature and details of the intra-molecular interaction between the two chains, the nature of the head-to-tail (inter-molecular) interaction, and the nature and sites of interaction with troponin and actin. To answer these questions the structure of the molecule at an atomic resolution is essential.

Bailey[5] crystallized rabbit skeletal muscle tropomyosin when he first isolated the protein. Since then, the Bailey crystal form has been the object of intensive study[6-11]. However, because of the mesh-like packing of the molecules with a high solvent content (more than 95 %), the Bailey crystal form is not suitable for high resolution crystallography. We aimed at obtaining new crystal forms in which the molecules are packed side-to-side, not in a mesh.

2. RESULTS AND DISCUSSIONS

2.1. Strategy

The poor diffraction pattern from the Bailey crystal form arises from the high solvent content and/or intrinsic flexibility of the molecule. The solvent content is high as the elongated molecules form a kite-shaped mesh. The mesh is formed because (1) the building block of the crystal is not a single molecule but a chain of molecules which are polymerized in a head-to-tail manner, and (2) hydrophobic interactions are emphasized by adding ammonium sulfate as the precipitant.

Our intention is to find a way to make the molecules pack side-to-side, to reduce the solvent content of crystals and to minimize the thermal vibration of the molecules perpendicular to the molecular axis. To this end, (1) non-polymerizable tropomyosin has been prepared since the head-to-tail link would impose extra restrictions on molecular packing, and (2) organic solvents have been used as precipitants to reduce the dielectric constant of the solvent, emphasizing ionic interactions and suppressing hydrophobic interactions.

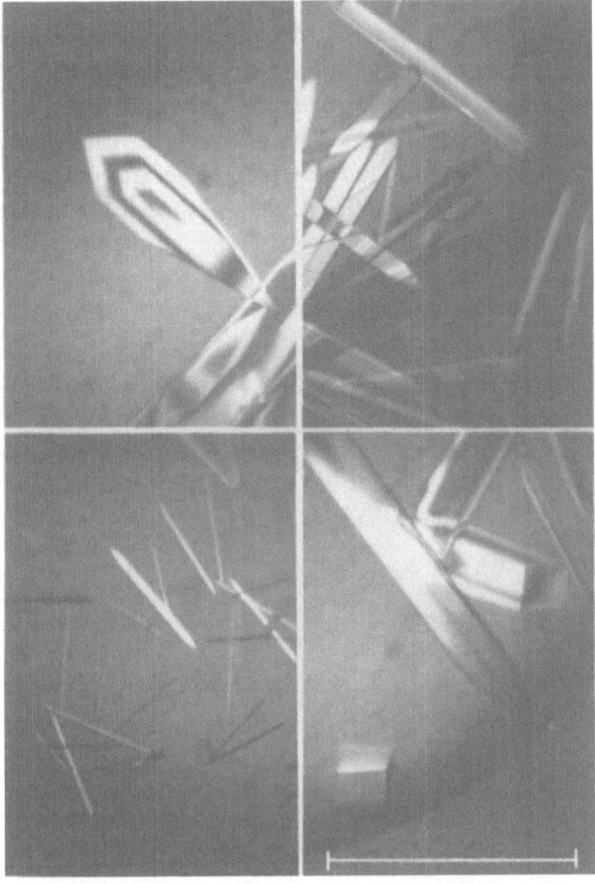

Fig. 1. The tropomyosin crystals photographed under a polarizing microscope. The bar represents a length of 1 mm.

2.2. Protein Preparation

In order to obtain non-polymerizable tropomyosin in large quantities and at high purity, rabbit skeletal muscle is not a suitable source. Firstly, the rabbit preparation consists of two isomers, α- and β-tropomyosin, which can not be easily separated from one another. Although rabbit heart muscle contains exclusively α-tropomyosin, this is not an ideal source for crystallization trials, which require substantial amounts of protein. Secondly, digestion of rabbit α-tropomyosin by carboxypeptidase A results in a heterogeneous population of molecules differing in the number of C-terminal residues removed by the enzyme. Even by employing an improved protocol for the proteolysis[12] and by taking precautions to dephosphorylate serine-283[13], the truncated tropomyosin is still substantially heterogeneous[14], as confirmed by our own results.

In the present study tropomyosin from lobster tail muscle has been prepared[15), and then subjected to carboxypeptidase A treatment to remove 11 residues from its C-terminus. After the enzyme digestion, the polypeptide chain has an apparent molecular weight of 32,000 kDa. Two identical polypeptide chains form a molecule. Lobster tail muscle tropomyosin consists of a single species, and homogeneous preparations of truncated and non-polymerizable tropomyosin can be easily obtained in large quantities (300 mg from 300 g of muscle). The carboxypeptidase A treatment gives rise to homogeneous preparations, probably due to the C-terminal

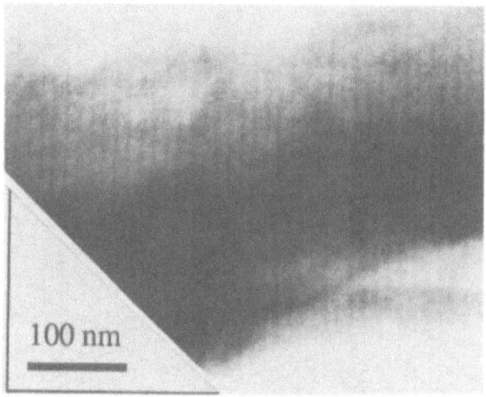

Fig. 2. An electron micrograph of a crushed tropomyosin crystal negatively stained. The striation spacing is about 130 Å. The bar represents 1000 Å.

peptide sequence -LDQTFSEKSGY, which differs substantially from rabbit α-tropomyosin (-LDHALAMTSI). The amino acid composition is also different from the rabbit counterpart[15). In spite of the chemical differences, the physical properties are not distinguishable from rabbit α-tropomyosin: the truncated lobster tropomyosin showed low viscosity at low salt concentrations, and a reduced affinity for F-actin (data not presented).

2.3. Crystallization

In the present study, it has been found that dimethyl sulfoxide (DMSO) is useful in crystallizing tropomyosin with side-to-side packing. Crystals begin to grow as needles (Fig. 1). After one to three weeks, the crystals grow into tetragonal rods with or without pointed ends, longer than 1 mm and 0.2 to 0.3 mm thick. Some are nearly isometric. Crystals grown under less favorable conditions tend to be broken into fibrous fragments.

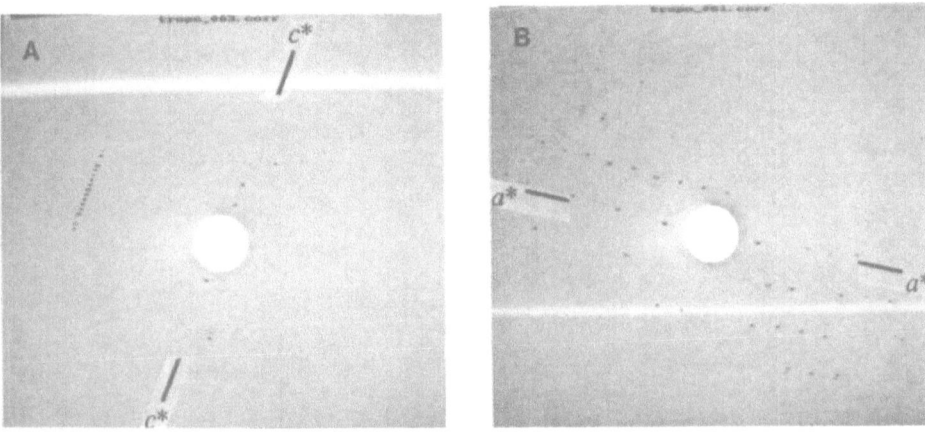

Fig. 3. X-ray diffraction patterns from tropomyosin crystals. The crystal was rotated by 2° around the axis perpendicular to the incident X-ray. In A, diffraction spots only with $l = 4n$ are observed on the c^*-axis (the arrow). In B, spots with $h = 2n + l$ are not seen on the a^*-axis (the arrow).

2.4. EM Observation of Crushed Crystals

Under the electron microscope, employing negative staining techniques with 2 % uranyl acetate, crushed crystals have shown striations with a regular spacing of about 130 Å (Fig. 2). The spacing corresponds to 1/3 of the molecular length and to 1/4 of the c-axis of the unit cell (see below). This indicates that the elongated molecules are packed side-to-side, parallel to the c-axis of the crystal, and within the crystal there is a stagger of 130 Å between molecules. It is very unlikely that two chains are separated, not forming a coiled-coil, since separated α-helices could not give rise to a stagger of 130 Å which is about the cross-over repeat of two α-helices[9] .

2.5. Diffraction Patterns

Diffraction patterns were recorded at beam-line X 31 of EMBL which receives synchrotron radiation from the DORIS ring at DESY. The ring was operated at 4.45 GeV with electrons currents of 15 to 40 mA. The wavelength was set at 1.28 Å, and patterns were recorded at 70 cm from the crystal on an imaging-plate scanner (J. Hendrix & A. Lentfer, unpublished).

The unit cell is tetragonal with a = b = 109 Å, c = 509 Å and the symmetry is either $P4_12_12$ or $P4_32_12$. Along the c^*-axis, only 001 reflexions with $l = 4n$ (n is an integer) are observed (Fig. 3A), indicating a $4_1(4_3)$ screw axis. This suggests that a group of molecules surrounding a 4_1 (4_3) axis may be regarded as the building unit of the crystal. Since the molecule may be at least 30 Å thick[4], it is likely that the unit cell contains eight molecules, one molecule per asymmetric unit. Since the c-axis of

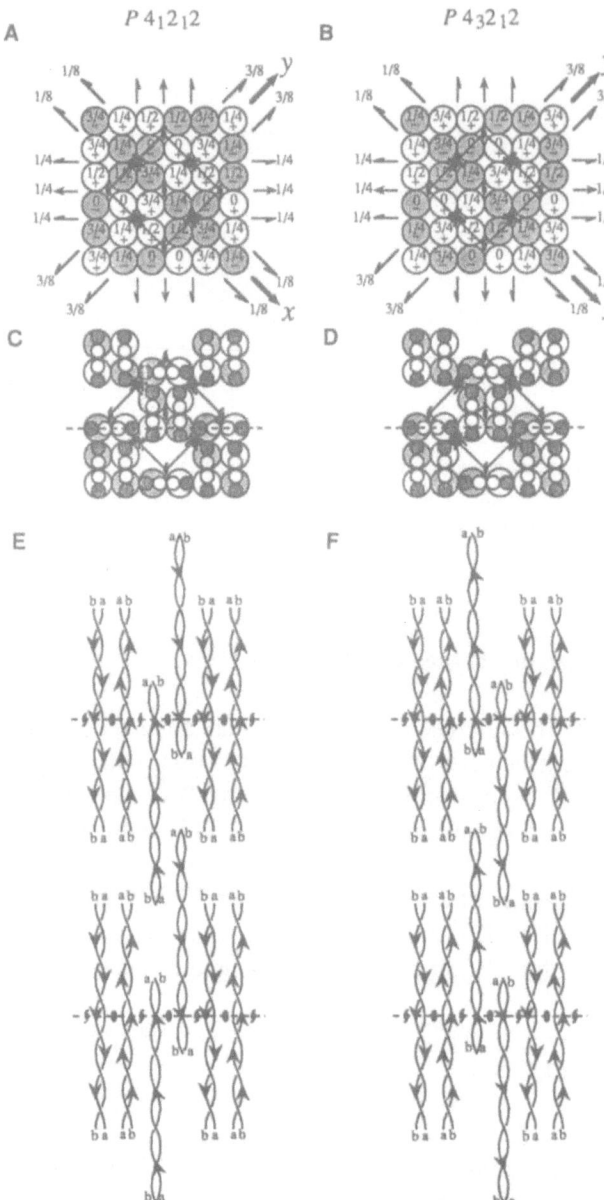

Fig. 4. Possible packing modes in the tropomyosin crystal. A, C and E, for $P4_12_12$ symmetry; B, D and E, for $P4_32_12$ symmetry. In A and B, symmetry elements and asymmetric units are represented. The square indicates the unit cell. Other symbols are after the International Tables for X-ray Crystallography, Vol. I[16]. C and D, a cross-sectional view of the x-y plane which is indicated as broken line in E and F. E and F, a side view of a plane parallel to the z-axis and diagonal to the x- and y-axes. The top view of the plane is indicated as the broken lines in B and D. In these figures, two chains (a and b) form a straight coiled-coil molecule, and the polarities (up + and down -) are indicated.

the crystal is longer than the length of the molecule by about 1/3 of the molecular length, there is a gap of this size between two molecules adjacent to each other in the direction of c-axis. This implies that molecules are axially too far apart to interact in a head-to-tail manner.

Diffraction spots have been observed to 8 Å resolution in the a*, b* plane, but along the c*-axis only to 15 Å resolution.

2.6. Packing

Based on the results described above, we propose two possible molecular packings within the crystal (Fig. 4), which we can not distinguish at present. The two packings differ from each other in the sense of the screw symmetry parallel to the c-axis. In either case, the crystal is bipolar, i.e. a half of the molecules point up and the rest down, due to the presence of two fold symmetry axis parallel to the a- and b-axes. Although the direction of the molecules must be on average parallel to the c-axis, and therefore the molecules are simplified as a straight coiled-coil in the side view in Fig. 4, the actual molecules in the crystal could be curved or even super-coiled as in the Bailey crystal form[9]. The stagger between adjacent molecules is 127 Å (1/4 x 509 Å), significantly shorter than 135 - 145 Å which has been deduced from the Bailey crystal form as the average pitch of the coiled-coil[9]. If the coiled-coil forms right-handed super-coils as it lies on the actin filament, then the pitch of the coiled-coil would be 128 Å. Therefore a stagger of 127 Å would enable adjacent molecules around the screw axis to interact in a broad edge to narrow edge manner. In this case 4_3 screw symmetry would be more plausible rather than 4_1 symmetry, since the super-coil is right-handed and the right handed 4_1 translation would result in steric conflict between molecules.

Although the present crystals diffract X-ray better than the Bailey crystal form, the crystals are not yet suitable for high resolution crystallography. We are now searching for a better crystal morphology and/or a way to reduce possible flexibility of the elongated molecule within the crystal.

ACKNOWLEDGEMENTS

We thank K.S.Wilson for critical reading of the manuscript and E.Mandelkow for letting us use the electron microscope in his laboratory. L.L. is an Alexander von Humboldt fellowship holder.

Note added in proof: Another crystal form of tropomyosin has been reported in Stewart, M., *J. Mol. Biol.*, **174**, 231-238 (1984) and Phillips, G.N.Jr., Cohen, C. & Stewart, M., *J. Mol. Biol.*, **195**, 219-233. (1987).

REFERENCES

1. Maeda, K., Kalbitzer, H.R. Rösch, A. Beneicke, W. Wittinghofer, A. & Maéda, Y. *FEBS Letters* **281**, 23-26 (1991).

2. Kalbitzer, H.R., Maeda, K., Rösch, A., Maéda, Y., Geyer, M., Beneicke, Neidig, K.-P. & Wittinghofer, A. *Biochemistry* **30**, 8083-8091 (1991).
3. Crick, F.H. *Acta Cryst.* **6**, 689-697 (1953).
4. O'Shea, E.K, Klemm, J.D, Kim, P.S. & Alber, T. *Science* **254**, 539-544 (1991).
5. Bailey, K. *Biochem.J.* **43**, 271-279 (1948).
6. Higashi, S. & Ooi, T. *J. Mol. Biol.* **34**, 699-701 (1968).
7. Caspar, D.L.D, Cohen, C. & Longley, W. *J. Mol. Biol.* **41**, 87-107 (1969).
8. Phillips, Jr.G.N, Lattman, E.E., Cummins, P, Lee, K.Y. & Cohen, C. *Nature* **278**, 413-417 (1979).
9. Phillips, Jr.G.N, Fillers, J.P. & Cohen, C. *J. Mol. Biol.* **192**, 111-131 (1986).
10. White, S.P, Cohen, C. & Phillips, Jr.G.N. *Nature* **325**, 826-828 (1987).
11. Cabral-Lilly, D. Phillips, Jr. G.N. Sosinsky, G.E., Melanson, L., Chacko, S. & Cohen, C. *Biophys. J.* **59**, 805-814 (1991).
12. Mak, A.S. & Smillie, L.B. *Biochem. Biophys. Res. Commun.* **101**, 208-214 (1981).
13. Mak, A.S., Smillie, L.B. & Bárány, M. *Proc. Natl. Acad. Sci. USA*. **75**, 3588-3592 (1978).
14. Walsh, T.P., Trueblood, C.E., Evans, R. & Weber, A. *J. Mol. Biol.* **182**, 265-269 (1984).
15. Miegel, A., Kobayashi, T. & Maéda, Y. *J. Muscle Res. Cell Motility* in press (1992).
16. Intenational tables for X-ray crystallography, Vol. I (eds. Henry, N.F.M. & Lonsdale, K., The Kynoch Press, Birmingham, England, 1969).

Discussion

Holmes: Am I right in thinking that the crystal symmetry you now see also refers to the molecule, so the molecule has a two-fold axis?

Maéda: Not necessarily. At the moment, we do not know if there is a non-crystallographic symmetry axis in the molecule.

Davis: Is it possible to take fragments of an alpha-helical coiled coil and see whether they have staggered ends that are not formed into an intercoiled coil? Do you detect anything with NMR there?

Maéda: We are now in the process of studying this. Our preliminary results are the following: We synthesized a peptide 40 amino-acid residues long. We looked at the details of the NMR pattern, and if the two chains were in complete symmetry, we would be able to look at the signals of only one chain. The second chain would not be visible under NMR. We saw signals from both chains. This means that the two chains, even in the folded part, are asymmetrical, which is completely consistent with our results from the C-terminus part.

EXPERIMENTS ON RIGOR CROSSBRIDGE ACTION AND FILAMENT SLIDING IN INSECT FLIGHT MUSCLE

M. K. Reedy, C. Lucaveche, M. C. Reedy and B. Somasundaram*

Duke University, Durham USA
*AFRC, Babraham, Cambridge, UK

ABSTRACT

We have explored three aspects of rigor crossbridge action:

1. Under rigor conditions, slow stretching (2% per hour) of insect flight muscle (IFM) from *Lethocerus* causes sarcomere ruptures but never filament sliding. However, in 1 mM AMPPNP, slow stretching (5%/h) causes filament sliding but no sarcomere ruptures, although stiffness equals rigor values. Thus loaded rigor attachments in IFM show no strain relief over several hours, but near-rigor states that allow short-term strain relief indicate different grades of strongly bound bridges, and suggest approaches to annealing the rigor lattice.

2. Sarcomeres of *Lethocerus* flight muscle, stretched 20-60% and then rigorized, show "hybrid" crossbridge patterns, with overlap zones in rigor, but H-bands relaxed and revealing four-stranded R-hand helical thick filament structure. The sharp boundary exhibits precise phasing between relaxed and rigor arrays along each thick filament. Extrapolating one lattice into the other should allow detailed modeling of the action of each myosin head as it enters rigor.

3. The "A-(bee)-Z problem" exposes a conflict about actin rotational alignment between A-bands and Z-bands of bee IFM, raising the possibility that rigor induction might rotate actins forcefully from one pattern to the other. As Squire[21] noted, 3-D reconstructions of Z-bands in relaxed bee IFM[2] imply A-bands where actin target zones form rings rather than helices around thick filaments. However, we confirm Trombitás *et al.*[23][24] that rigor crossbridges in bee IFM mark helically arrayed target zones. Moreover, we find that loose crossbridge interactions in relaxed bee IFM mark the same helical pattern. Thus no change of actin rotational alignment by rigor crossbridges seems necessary, but 3-D structure of IFM Z-bands should be re-evaluated regarding the apparent contradiction with A-band symmetry.

INTRODUCTION

The highly regular crossbridge array in insect flight muscles (IFM) of the asynchronous indirect type offers special advantages for electron microscope studies

Mechanism of Myofilament Sliding in Muscle Contraction, Edited by
H. Sugi and G.H Pollack, Plenum Press, New York, 1993

of structural changes associated with contraction. We have recently developed some evidence that throws light on the properties and actions of rigor crossbridges in IFM, whose structure is thought to exhibit the end-state of the power stroke. Here we group together three different sets of experiments, which show respectively how tightly rigor bridges bind actin, point toward a detailed description of the relax-rigor transition, and address the possibility that rigor induction forces actin filaments to rotate from one symmetry pattern to another.

1) In the first experiments, we tested whether rigor crossbridges would permit actin filament slippage in very slowly stretched fibers of insect flight muscle. If rigor sarcomeres elongated up to 6% as reported for rabbit psoas fibers[20], the change from rest length EM appearance of IFM would be unmistakable. The rate of filament sliding allowed by forced slippage should give some measure of how rapidly the insect rigor crossbridge lattice might "anneal" by strain-driven detachment-reattachment processes[18][19]. If it annealed fairly rapidly, then the typical highly ordered crossbridge array might be a secondary pattern, derived by annealing from a more disordered primary lattice state. However, if annealing is very slow, then the degree of order found in fibers quickly fixed after being rigorized[15] must represent non-annealed primary rigor. In this case the details of crossbridge structure, individually and in ensemble, would signify the order and timing of the final force-generating events, since every crossbridge would still be attached to the same actin site where ATP depletion first stranded it in rigor.

2) In the second set of experiments, we evaluated the unexpected preservation of relaxed thick filament structure in the H-band of IFM fibers which were put into rigor only after being stretched to ~150% of resting sarcomere length. The sharp transition from unbound relaxed heads to actin-bound rigor crossbridges at the boundary where each thick filament enters the overlap zone could open the way to detailed modeling of the structural change undergone by each myosin head during the rigor induction process, throwing more light on the origins of two-headed, single-headed and no-headed crossbridge attachments.

3) In the third set of experiments, we took up the challenge inherent in the 3-D reconstruction by Cheng & Deatherage[2], overlooked until Squire[21] drew attention to it, that bee Z-band structure implies quite a different arrangement of actin targets and crossbridge attachments in the A-band than the one worked out in *Lethocerus* IFM[3][12][17][22] and expected in bee IFM on the basis of published micrographs by Trombitás *et al.*[23][24].

METHODS

Dorsal longitudinal flight muscle in *Lethocerus* and bumblebee was exposed ventrally by dissection, glycerol-detergent permeabilized in the whole thorax or hemithorax and then stored 1-12 months at -80°C in relaxing buffer made up in 75% glycerol, as described previously[9][14]. For slow stretching of waterbug fibers at rates from 2%/hr to 1%/min, bundles of 2 fibers were mounted, end-glued to leave 4 mm free, briefly deglycerinated and slowly stretched on the same mechanical apparatus used in previous work for rabbit[20]. Length change was monitored as a function of applied DC voltage after newly recalibrating the displacement-to-voltage response of the length transducer, since optical position monitoring was not consistently reliable at the smallest displacements used. Sarcomere length was monitored by direct observation and manual measurement of the laser diffraction pattern, because laser diffraction lines from IFM were too faint and noisy for position readout on the photodiode array previously used for rabbit fibers. To achieve ATP depletion that was positive and complete, rigorization was accomplished at 23°C in rigor solution supplemented with hexokinase (Sigma; 20 units/ml), 1 mM d-glucose and 100 µM of myokinase inhibitor AP5A. The hexokinase tested "protease-free", in that 12-24 hr incubation of IFM and rabbit psoas myofibrils at 23°C removed no Z-band density by phase contrast microscopy. AMPPNP (Boehringer-Mannheim) was made up to 1 mM in a rigor solution identically supplemented with hexokinase-glucose and AP5A.

Bee IFM was fixed *in situ* in the hemithorax, either relaxed (after glycerol washout with relaxing solution) or in rigor (after a further washout with hexokinase-glucose rigor solution).

For EM waterbug fibers were fixed on the transducer and bee fibers *in situ* in the hemithorax by fixative made up in the final physiological solution, using 0.2% tannic acid to start, then after 10 min adding glutaraldehyde to 2.5% for 20-50 min more before rinsing three times in distilled water and postfixing 30 min in 1% uranyl acetate. Some bee IFM was fixed without adding aldehyde, using just TAURAC fixation[16]. Fibers were ethanol dehydrated and embedded, sectioned and stained as described[10].

RESULTS AND DISCUSSION

1. Can Waterbug IFM Filaments Slide in Rigor?

Control experiments showed that for *Lethocerus* flight muscle fibers, moderate stretching rates (5% per hr and faster) always produced normal filament sliding when the rigor solution was augmented with 5 mM MgATP, or 1 mM MgAMPPNP, or was made up nucleotide-free in 75% glycerol. Laser diffraction showed that sarcomere elongation was generalized, and coupled widening of H-bands and I-bands demonstrated filament sliding (Figs. 2, 4). Indeed, with a 6% increase of the ~2.65 µm rest-length sarcomeres by filament sliding, H-band width doubles. Measured by the convenient 39 nm crossbridge repeat, such doubling unmistakably widens the H-band from 4 to 8 repeats. In contrast, no signs of filament sliding were ever produced by slow stretching (2%/h for 3-6 h) of IFM in any aqueous rigor solution following the rigor contraction and 5 min soak in hexokinase-glucose rigor buffer that preceded every trial stretch. Laser diffraction showed no sarcomere elongation; diffraction maxima grew noisy and indistinct during stretch. EMs

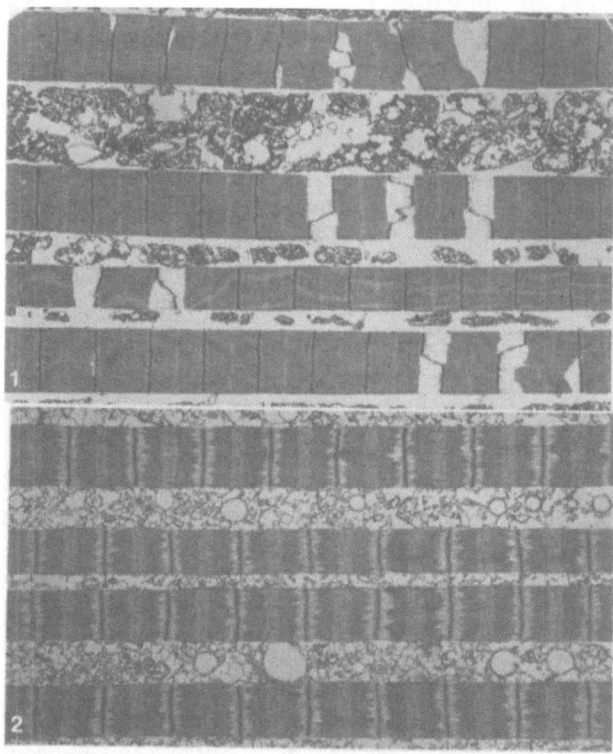

Fig. 1. Stretched 2%/hr (0.033%/min) for 3.3 hr, *Lethocerus* rigor fiber shows ruptured sarcomeres, but no widening of H-bands or I-bands.

Fig. 2. Stretched 5%/hr (0.083%/min) in 1 mM AMPPNP at 23°C, sarcomeres of *Lethocerus* fiber all show widening of H-bands and I-bands, indicating filament sliding. No ruptured sarcomeres were found.

showed that fibers stretched 6-8% in 3-4 hours always elongated only by rupturing of scattered sarcomeres, leaving all other sarcomeres at original length (Figs. 1, 3). This lack of any filament slippage, confirmed in all 8 rigor experiments, shows that rigor crossbridges bind too tightly in IFM to permit any rapid rate of strain relief according to the mechanisms considered in Schoenberg's analysis[18)19)]. We draw three lessons from this:

   i) The initial rigor attachment of each crossbridge must persist for several minutes or even hours at the same actin site on which it became "stranded" when it entered the rigor state (used up its last ATP). Thus, in fibers fixed immediately after rigorization, both the averaged order and the micro-disordered details of rigor crossbridge shape and angle are probably expressive of the final actions of each bridge as it generates and then sustains force upon entering rigor, justifying close correlation between structure and mechanics of rigorization[15)].

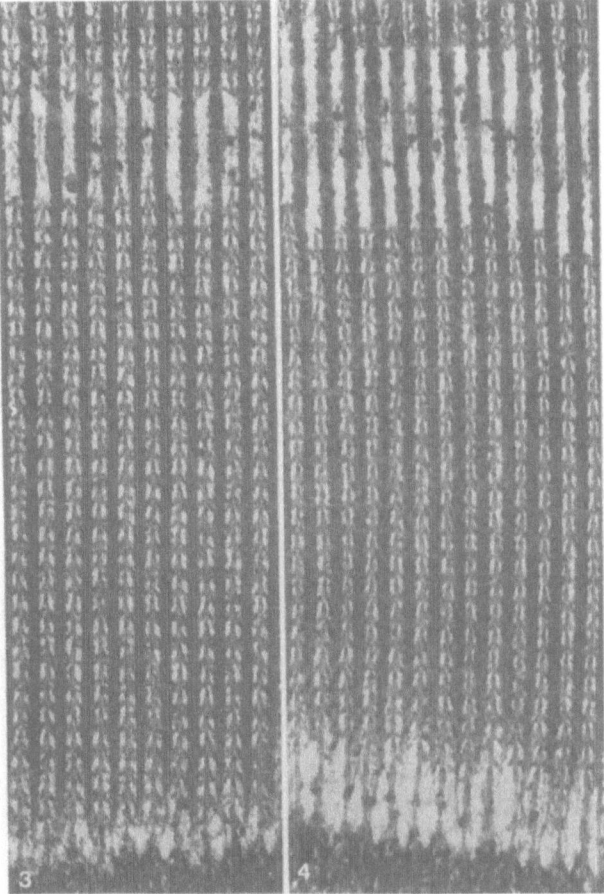

Fig. 3. After slow 7% fiber stretch, rigor sarcomeres keep regular crossbridge lattice and H-band width of about 0.15 μm, unchanged from H-band width typical of unstretched fibers.

Fig. 4. After slow 8-9% stretch in 1 mM AMPPNP, obvious widening of H-band to 0.3 μm or more indicates filament sliding.

ii) On the other hand, easy slippage of filaments in MgAMPPNP (both in the cold and at room temperature) indicate that the shape and attachment position of actin-bound crossbridges in AMPPNP need not be directly derived from the preceding rigor state, but may have been altered by strain-relieving detachment-reattachment events.

iii) Finally, these experiments suggest that a more perfectly crystalline rigor lattice might be "annealed" by slow withdrawal of AMPPNP or glycerol, thus giving a better reference standard by which to measure experimental perturbations of rigor structure.

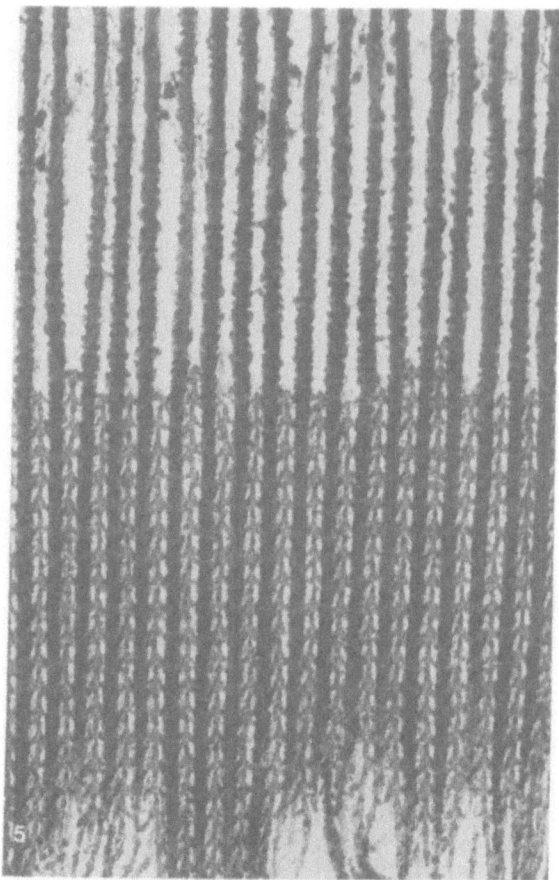

Fig. 5. Fibers stretched 30-50% at 1%/min in 5 mM MgATP relaxing buffer, then fixed after 5-30 min wash in rigor buffer show hybrid structure, with rigor chevrons in overlap zone, relaxed myosin head pattern in H-band.

2. Hybrid Relaxed/Rigor Sarcomeres in Waterbug IFM

The control experiments mentioned above first made us aware that stretching of relaxed IFM followed by 5-10 min hexokinase-glucose rigorization commonly rigorizes only the overlap zone, while retaining a relaxed arrangement of myosin heads in the non-overlap zone, or H-band (Fig. 5). Although this contradicts previously reported disordering of H-band heads on rigorization of rabbit muscle[5] and IFM[7], it has proven quite reproducible. Remarkably, in some experiments even overnight rigorization left the sarcomeres in this hybrid state (Figs. 6,7). In some regions excellent lateral register (Figs. 5,6) and even helical register (Fig. 7) seems to be preserved across the fibril by both rigor chevrons and 90° X 14.5 nm relaxed heads. The H-band shows good preservation of a helical arrangement of myosin

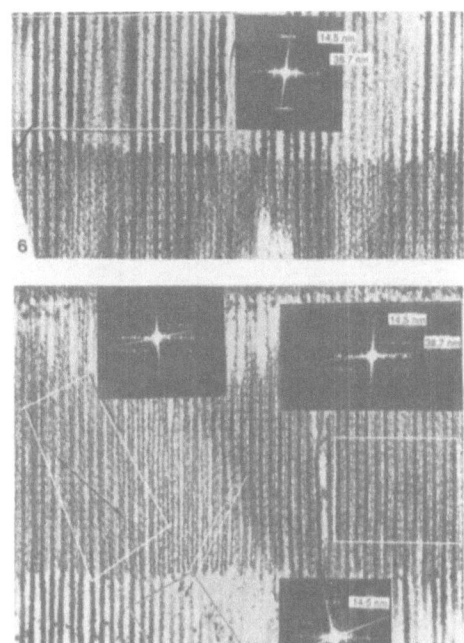

Fig. 6. Optical transform of actin-free relaxed H-band in hybrid sarcomere shows 14.5 nm meridional with 38.7 layer line well off the meridian in a position consistent with 4-stranded thick filament helical structure[7].

Fig. 7. Optical transforms, each from nearest framed region, show rigor pattern from myac layer (upper right), 1-sided helical patterns from rigor labeling of actin target zones (middle left) and myosin surface lattice (lower area). Section grazes near side of thick filaments in both rigor and relaxed regions, shows sloping helical tracks from left-handed helix in rigor zone (large black arrow) and right handed helix (white arrows with black heads) in relaxed H-band.

heads that is 4-stranded, as inferred from longitudinal sections (Figs. 6 & 7; and M.C. Reedy et al.[7]), and shown directly by the tetragonal profiles of single crowns on thick filaments in ultrathin transverse sections (Fig. 8). This symmetry is confirmed by the beautiful 3-D reconstruction from negative-stained thick filaments just published by Squire's group[4]. The thick filament reconstruction offers higher resolution, but the sectioned material has its own advantages. In order of increasing importance, these include:

i) It can display lattice features like lateral periodic register despite perturbed whole-filament register across the fibrillar ensemble (Figs. 5,6,7,8).

ii) As in Fig. 7, optical diffraction can show in side-by-side regions that the helical repeat period on thick filaments (in the H-band) is identical at 38.7 nm with

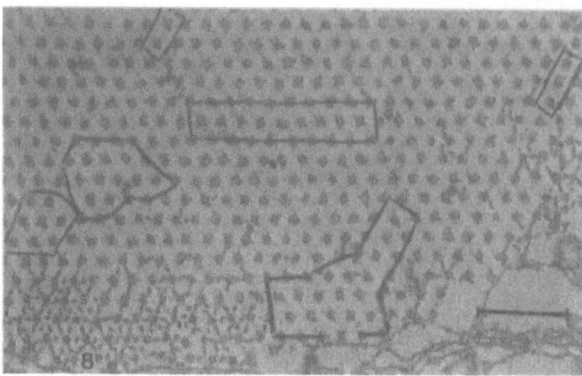

Fig. 8. 15-18 nm transverse sections of hybrid sarcomeres, sampling single levels of 14.5 nm repeat, confirm four-fold symmetry of thick filaments by frequency of square profiles. Square crowns are obviously larger than backbone diameter of rigorized thick filaments in overlap zone sampled along lower edge of fibril.

the helical repeat of thin filaments that is labeled by rigor crossbridges in the overlap zone.

iii) The hand of the thick filament 38.7 nm helix can be directly read out as right-handed, because it is obviously opposite in screw sense from the 2-stranded helix of rigor crossbridges (white-black *versus* black-on-white diagonal arrows in Fig. 7) that labels the actin target zones around and along the same thick filaments. This target zone helix is well-known to be left-handed[11][12].

iv) The abrupt transition at the boundary between overlap and non-overlap zones exhibits the phase relationships between relaxed and rigor lattices of myosin heads. Thus one lattice can be accurately extrapolated into superposition with the other along the same thick filament. This should allow detailed 3-D modeling of the structural changes undergone by each myosin head during the force-generating rigorizing power stroke, which involves the transition from relaxed (pre-power stroke) to rigor (end of power-stroke) conformation.

3. The "A-(Bee)-Z Problem": a Conflict of Actin Rotations

The Cheng & Deatherage[2] computer 3-D reconstruction of the intricate hexagonal lattice of Z-bands in honeybee (*Apis*) flight muscle was featured on the cover of J. Cell Biology a couple of years ago. Recently Squire[21] pointed out that extending this structure from Z-band to A-band gives actin filament rotational symmetry (Fig. 9 "Z") different from the A-band pattern known from giant waterbug (*Lethocerus*) IFM (Fig. 9 "A") [3][12][17][22] and expected in bee IFM on the basis of EMs from Trombitás et al.[23][24]. So we took a look at thin-section EMs from the A-band of bumblebee (*Bombus*) IFM. (Our bumblebee EMs and others' honeybee EMS indicate A-bands identical in structure.) First we looked at rigor, where we found the same pattern as in waterbug muscle. Fig. 9 shows how the center

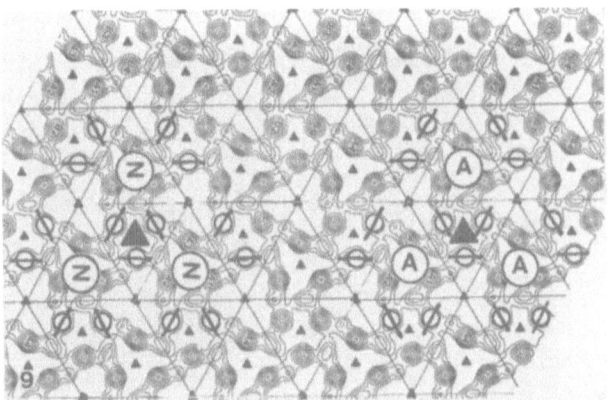

Fig. 9. Z-band and A-band inferences about actin orientation are shown, in overlays of A-band structure on a projection of the bee flight muscle Z-band lattice. Triangles, large and small, indicate threefold positions in the projected Z lattice; large circles represent thick filaments of one sarcomere projected onto the lattice. Where thick filaments are labeled "Z", the actin rotations around large triangles keep three-fold symmetry, in accord with 3-D reconstruction of bee Z-band. This implies that all six actin target zones are presented at the same level around each thick filament labeled "Z". Thus crossbridge labeling of actin targets should form rings around thick filaments in A-band, repeating axially every 38.7 nm. However, the bridge labeling actually observed follows a 2-stranded helix of target zones around each thick filament, indicating three different levels every 38.7 nm. This is diagrammed where thick filaments are labeled "A". Note that here the three actins around the large triangle do not exhibit the 3-fold symmetry shown by corresponding actins in "Z" group. (Modified from Cheng & Deatherage[2]).

of three-fold symmetry in the 3-D of bee Z-line implies six-fold actin symmetry around the thick filaments (labeled "Z" in Fig. 9) in the A-band, so that the 6 actins around each thick filament have their helices arranged to present actin "target zones" (segments of optimum subunit azimuth for crossbridge attachment) all at the same level, an annular pattern which repeats axially every 38.7 nm. However, our thin sections of *Bombus* in Figs. 10 and 12 show, just as in waterbug and as indicated by EMs of *Apis* IFM[23][24], that target zones in rigor IFM of bee follow a two-stranded helix along the cylindrical cage of six thin filaments surrounding each thick filament. Diagonal arrows in Fig. 10 show this helix directly where the section grazes near or far side of the myac layer. We have shown it to be left handed, as in waterbug (Fig. 12; other data not shown). This helix brings one opposed pair of the six actins into presenting optimum target zone orientation every 39/3 (= 12.9) nm.

In order to accommodate the evidence from Trombitás et al. yet preserve the otherwise devastating argument posed by annular target zone arrays against the match-mismatch theory[25] for stretch-activation in IFM, Squire has proposed that the Z-line actin rotational pattern in Fig. 9Z may prevail in the A-band of relaxed and stretch-activated contracting bee IFM, yet be labile enough to transform into the helical target-zone arrangement as an artifact of rigor cross bridge attachment. The

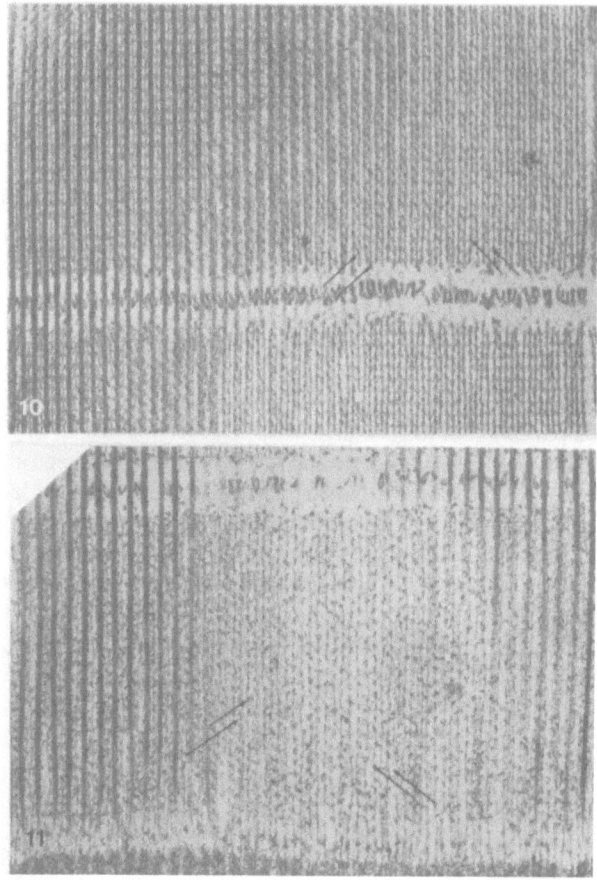

Fig. 10. Single filament layers from rigor bumblebee flight muscle, differing from similar EMs of waterbug flight muscle only in M-line and Z-line appearances. Helical pattern of actin target labeling by rigor crossbridges is clear from dashed diagonal lines found where section grazes thick filaments along one side as it passes from myac layer to actin layer. Fainter diagonal lines in actin layer (arrows) reflect rigor labeling from myac layers both above and below. Similar details are visible in Trombitás' EMs[23)24)] of honeybee flight muscle in rigor.

Fig. 11. Single filament layers from relaxed bumblebee flight muscle, fixed in tannic-acid-uranyl sequence[16)] to trap and intensify relaxed crossbridge contacts with thin filaments. Diagonal lines in lattice pattern of beading in actin layer (arrows) indicate same pattern of actin labeling by crossbridges seen in rigor.

action of rigor crossbridges implied by this proposal is that during rigor induction or subsequent annealing they would forcefully rotate the thin filaments from the pattern in Fig. 9Z to that in Fig. 9A, meanwhile obscuring or destroying the original Z-band symmetry. A full answer to this interesting idea will require good evidence from bee IFM about the rotational pattern of actin filaments in both relaxed A-band

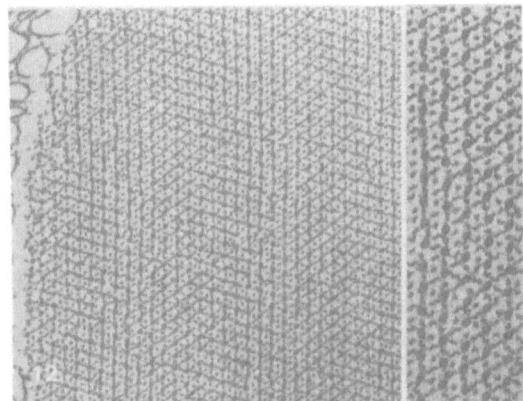

Fig. 12. 15-20 nm cross-section of rigor bumblebee flight muscle. Higher magnification inset shows flared Xes exactly like those seen in waterbug flight muscle. The progressive rotation of flared Xes across the fibril, due to slight obliquity of section, expresses how flared Xes advance along the helical array of target zones in 3 dimensions. Serial sections have verified a left-hand sense to this helix. Modeling has indicated that flared X structure alone is probably sufficient to verify a left- versus right-handed target zone helix, since the RH helix should show unflared, rather gaping Xes[3].

and rigor Z-band. We have begun with relaxed A-bands. Our initial evidence in Fig. 11 does not support Squire's proposal. This EM shows the A-band of glycerinated bee IFM which has never been rigorized because it was never depleted of 5 mM MgATP until after fixation. The diagonal arrows point out thin filament marking by densities that follow the same lattice pattern seen in similar views of rigor (Fig. 10), indicating material that is arrayed in helical, not annular fashion, around thick filaments. This same appearance is found in thin sections of relaxed *Lethocerus* IFM, where X-ray and EM evidence indicate that it represents fixative trapping and accumulation of relaxed crossbridges that are weakly associated with actin target zones[8],[13],[16]. Accordingly, Fig. 11 indicates that the A-band of bee IFM shows the same pattern of actin rotations and actin target zone presentations in relaxed muscle as in rigor. Alternatively, it might be argued that this is IFM troponin[6] rather than the marking of target zones by crossbridges, but it is difficult to see why troponin and target zones should agree in symmetry in waterbug A-bands but disagree in bee.

There is striking preliminary evidence from Auber[1] that thick filaments may in bee IFM be arranged with more perfect rotational and helical register than in waterbug IFM[21]. This could make bee even more useful than waterbug flight muscle for X-ray diffraction and EM studies of contraction. Thus it is important to start with a correct scheme of actin rotations. The evidence from Trombitás and ourselves about bee A-band structure does not support the scheme implied by the 3-D reconstruction of bee Z-band, so bee Z-band structure needs to be re-examined.

Supported in part by grants to MKR from NIH (AR 14317) and MDA, and to Richard Tregear from MRC and AFRC. We thank Richard Tregear for splendid lab hospitality and numerous helpful discussions.

REFERENCES

1. Auber, J. *C. R. Acad. Sci. (Paris)* **264**, 2916-2918 (1967).
2. Cheng, N.Q. & Deatherage, J.F. *J. Cell Biol.* **108**, 1761-1774 (1989).
3. Haselgrove, J.C. & Reedy, M K. *Biophys. J.* **24**, 713-728 (1978).
4. Morris, E.P., Squire, J.M. & Fuller, G.W. *J. Struct. Biol.* **107**, 237-249 (1991).
5. Padrón, R. & Craig, R. *Biophys. J.* **56**, 927-933 (1989).
6. Reedy, M.C., Bullard, B., Leonard, K. & Reedy, M.K. *J. Muscle Res. Cell Motility* **12**, 112 (1991).(Abstract)
7. Reedy, M.C., Magid, A.D. & Reedy, M.K. *Biophys. J.* **51**, 220a (1987).(Abstract)
8. Reedy, M.C., Reedy, M.K. & Goody, R.S. *J. Muscle Res. Cell Motility* **4**, 55-81 (1983).
9. Reedy, M C., Reedy, M.K. & Goody, R.S. *J. Muscle Res. Cell Motility* **8**, 473-503 (1987).
10. Reedy, M.C., Reedy, M.K. & Tregear, R.T. *J. Mol. Biol.* **204**, 357-383 (1988).
11. Reedy, M.K. *Am. Zool.* **7**, 465-481 (1967).
12. Reedy, M.K. *J. Mol. Biol.* **31**, 155-176 (1968).
13. Reedy, M.K., Goody, R.S., Hofmann, W. & Rosenbaum, G. *J. Muscle Res. Cell Motility* **4**, 25-53 (1983).
14. Reedy, M.K., Leonard, K.R., Freeman, R. & Arad, T. *J. Muscle Res. Cell Motility* **2**, 45-64 (1981).
15. Reedy, M.K. & Longley, W. *Biophys. J.* **51**, 220a (1987).
16. Reedy, M.K., Lucaveche, C. & Popp, D. *Biophys. J.* **59**, 579a (1991).
17. Reedy, M.K. & Reedy, M.C. *J. Mol. Biol.* **185**, 145-176 (1985).
18. Schoenberg, M. *Biophys. J.* **48**, 467-475 (1985).
19. Schoenberg, M. *Biophys. J.* **60**, 679-689 (1991).
20. Somasundaram, B., Newport, A. & Tregear, R.T. *J. Muscle Res. Cell Motility* **10**, 360-368 (1989).
21. Squire, J.M. *J. Muscle Res. Cell Motility* **13**, 183-189 (1992).
22. Taylor, K.A., Reedy, M.C., Reedy, M.K. & Crowther, R.A. *J. Mol. Biol.* (in press)
23. Trombitás, K., Baatsen, P.H. & Pollack, G.H. *Adv. Exp. Med. Biol.* **226**, 17-30 (1988).
24. Trombitás, K., Baatsen, P.H.W.W. & Pollack, G.H. *J. Ultrastruct. Mol. Struct. Res.* **100**, 13-30 (1988).
25. Wray, J.S. *Nature* **280**, 325-326 (1979).

Discussion

Pollack: This question has to do with the rigor cross-bridges at the end of the power stroke. As you know, at the previous Hakone meeting, Charles Trombitás and I published something about the anti-rigor cross-bridges (Sugi, H. and Pollack, G. H. *Molecular Mechanism of Contraction*, Plenum Press, 1988). Later, Mary Reedy also did some work on that (*Nature* **339**, 481-483, 1989). More recently, we found that if you allow the fiber to shorten during the induction of rigor, up to 50% of the cross-bridges are tilted backwards in the anti-rigor position. Given these observations, do you think it is still tenable that the rigor angle—the 45° angle—be associated with the end of a power stroke?

Reedy: You are describing a situation in which, after shortening, when rigor is induced by zero-force shortening, you end up with all the bridges in an anti-rigor position, which is totally the wrong orientation if the bridges are supposed to have produced the shortening. It is a confounding observation that I cannot explain at this point. The explanation simplest to me—but one that I can't readily apply to your finding—is that when you have arrowheads on actin in a rigor state it indicates actin polarity, no matter where you find the actin filaments Now, you are talking about bee muscle, aren't you?

Pollack: Yes.

Reedy: I haven't ever been able to see that in *Lethocerus* muscle. I have let Lethocerus go into rigor many times with no load on it, and I have never seen reverse chevrons. So, that's a puzzle arising in different specimens, but one I think worth careful comparative investigation.

Tregear: Regarding the inability to pull out sarcomeres in rigor, I would like to point out that in parallel with those experiments—in fact in the same lab by the same chap (Somasundaram, B. et al. *J. Muscle Res. Cell Motility* **10**, 360-368,1989)—they were pulling out rabbit fibers as well. The rabbit fibers would slip in rigor. So, the statement that rigor bridges won't slip is true for insect muscle, but apparently not true for rabbit muscle. Isn't that so?

Reedy: Absolutely right. It has to be limited in its application to insect muscle at this point, because the vertebrate results are different.

ter Keurs: Dr. Reedy, you commented on the potential rotation of the actin filament during interaction with the cross-bridge. I would like you to comment on the observations by Dr. Goldstein's group regarding the shape of the Z-band, which in the relaxed state in skeletal muscle of mammalia shows a small square configuration and shows what they call a basket-weave configuration during activity and force development of the muscle (Goldstein et al. *El. Micr. Rev.* **3**, 220-248, 1990). The simplest way to explain that observation would be by invoking rotation of the actin filaments during interaction with myosin. If the rotation is identical at the two sides of the Z-line, one would immediately conclude that the small squares should go into a basket-weave. Two questions: Do you find such a transition of the Z-structure in an intact muscle, or don't you? And do you have a comment regarding the rotation of actin filaments?

Reedy: No such observation has been made in the Z-band of insect muscle. It awaits further work of the kind that Dethridge reported (Cheng, N.Q. & Deatherage, J.F. *J. Cell Biol.* **108**, 1761-1774, 1989). No differences between rigor and relaxed, or relaxed and contracting Z-bands, have been developed as far as I know—nothing that corresponds to the small square/basket-weave transition is known from insect Z-bands. I think that deals with your first question. As for your second question, I meant to say that during the transition from relaxation to rigor, the limited amount of dynamic interaction required to develop that transition did not seem to produce any detectable rotational reorientation of actin. Rigor inductions could be applied some torque to actin, but the unchanging pitch of the actin helix and the apparently unchanged lattice arrangement of actin target zones give no sign of any such torque.

Squire: I wanted to go back to the comment about the Z-band, because I think the small square to basket-weave transition that has been seen implies a change in spacing of the Z-band. The Z-band also gives an X-ray reflection which does not change in spacing when you go from resting to active muscle.

Reedy: I think that's very true. The experiment has not been done under conditions where the lattice-spacing has been prevented by osmotic lock. It's the kind of experiment that is being contemplated in a couple of laboratories, but I don't know if it's really underway.

Holmes: The message I've taken home from this is that evidence is appearing that myosin alone, or myosin filaments at least, can be seen in different shapes and forms. What we are still a bit short of is evidence that actomyosin can be seen in different shapes and forms. I'm not saying it is impossible, but this is the more difficult end, and probably the end we all think the most interesting. What we have to worry about are things you have also discussed: for example, if you freeze it fast, or you fasten it down on a substrate, does that do any damage? These are things we have to worry about.

GROSS STRUCTURAL FEATURES OF MYOSIN HEAD DURING SLIDING MOVEMENT OF ACTIN AS STUDIED BY QUICK-FREEZE DEEP-ETCH ELECTRON MICROSCOPY

Eisaku Katayama

Department of Fine Morphology
Institute of Medical Science
University of Tokyo, Minato-ku, Tokyo 108 JAPAN

ABSTRACT

With quick-freeze deep-etch electron microscopy coupled with mica-flake technique, I showed previously that myosin subfragment-1 (S1) attached to F-actin in the presence of ATP is short and rounded, in contrast to its elongated and tilted appearance under rigor condition [*J. Biochem.* **106**, 751-770 (1989)]. I further indicated that each head of heavy meromyosin (HMM) changes its configuration in a likely manner as above by the addition of various nucleotides, i.e. heads were pear-shaped in the absence of nucleotide, in a ball-on-a-stick appearance when complexed with ADP and strongly kinked to the particular direction in the presence of ATP or ADP•Vi [*J. Muscle Res. Cell Motility* **12**, 313 (1991)]. Such morphological data not only corroborates the independent biophysical evidences suggesting gross conformational changes of myosin head upon binding ATP or ADP•Vi, but also provide strong evidence for the distinct polarity in the structure of each myosin head. Negatively stained image of chemically cross-linked acto-S1 also included cross-bridges sharply kinked to the same direction, confirming the above observation. Attempts were made to examine if such conformational change of myosin cross-bridge occurs during actomyosin superprecipitation. Samples were quick-frozen during rapid turbidity-increasing phase where actin filaments actively slide past myosin heads. The resultant image included actin-attached myosin heads all in a kinked configuration with the same polarity as observed for HMM. Several heads associated with a single actin filament were bent to the same direction suggesting that myosin heads might be in a kinked configuration with distinct polarity during contraction.

Mechanism of Myofilament Sliding in Muscle Contraction, Edited by
H. Sugi and G.H Pollack, Plenum Press, New York, 1993

47

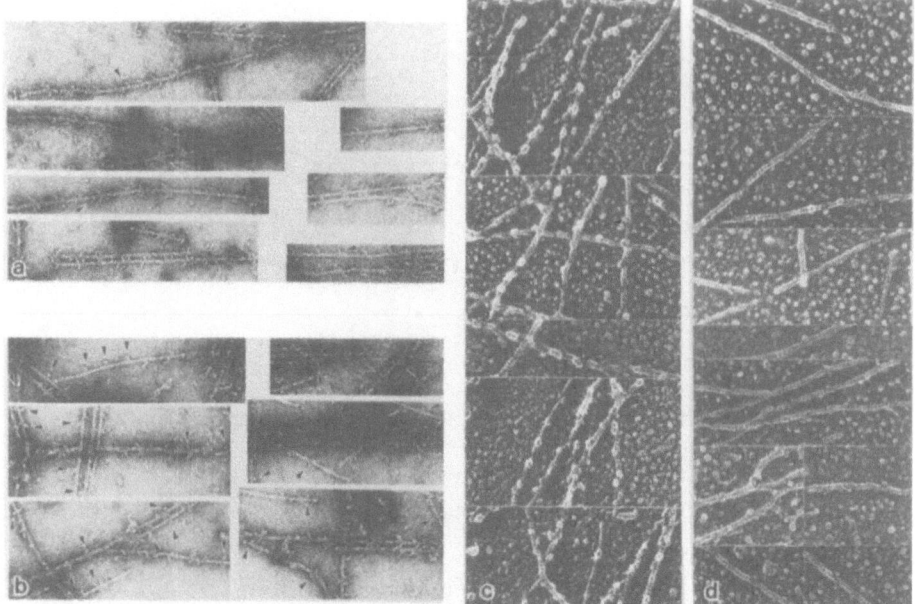

Fig. 1. a),b) Gallery of negaively stained images of S1 chemically cross-linked to F-actin. a) in the absence of ATP; b) in the presence of ATP [reproduced from reference 11) with permission]. Arrowheads indicate a) S1 heads whose substrucrue is well-observed; b) S1 heads with rounded (actually kinked) structure. c),d) Gallery of freeze-replicated S1 attached to actin; c) in the absence of ATP; d) in the presence of ATP [a) and b) were reproduced from reference 20); and c), d) were reproduced from reference 11) with permission].

1. Previous Studies and Methodological Considerations

Contraction of muscle and a variety of cell motility utilize the repetitive interaction between myosin and actin as a motor with concomitant hydrolysis of ATP as an energy source. It has long been accepted that such motor function would naturally involve some cyclic conformational change of myosin heads forming cross-bridge structure between thick and thin filaments (i.e. tilting cross-bridge hypothesis as the simplest model)[1)2)]. Among several postulated conformational states of myosin head during cross-bridge cycle, rigor state had been almost the only one which was subject to extensive analyses by high-resolution electron microscopy[3-7)]. The difficulty in visualizing the confirmational states other than rigor was derived from the extremely lowered affinity of myosin head to actin in the presence of ATP[8)] or analogous nucleotides. Conventional transmission electron microscopic methods such as negative staining or cryotransfer microscopy which are used to obtain high resolution image of protein molecules were not applicable for that purpose, because the specimens for these methods must be prepared under very low protein concentration which is not compatible with the requirement to keep sufficient fraction of myosin heads associated to actin filaments. Such difficulty could be avoided by the use of chemically cross-linked acto-myosin subfragment-1 (S1)[9] as a specimen and it was observed by negative staining that actin-attached S1 molecules were in a variety of shapes in the presence of ATP in contrast to uniformly elongated and tilted conformation under rigor condition[10][11] (Fig. 1a). Among various shapes observed with ATP was a notable population of images showing conspicuously small and rounded S1 particles[10][11] (Fig. 1b). A similar shape change by the addition of ATP also occurred in the cross-linked samples embedded in vitrified ice[12], confirming that the observed morphological change could not have been a staining artifact. A serious argument for the interpretation of such images[10-12] was, however, the possibility the some S1 particles could be simply "tethered" to actin, in the presence of ATP, by the connection through one covalent bond[13]. Thus, different strategy was necessary to observe uncross-lined S1 associated with actin even in the presence of ATP under physiological conditions.

Quick-freeze deep-etch electron microscopy coupled with mica-flake technique[14] seems to be an ideal methodology to be applied to such situation, both in theory and practice. In this technique, protein samples were adsorbed for a short time to the slurry of fine mica-flakes. They are then instantly frozen, by slamming onto copper block which are cooled close to liquid helium temperature, before water molecules in the sample rearrange to form crystalline ice. Frozen samples are fractured, etched for a short time followed by low-angle rotary-shadowing and replication. It should be noted that this etching process removes only a superficial layer of free water while protein-bound water molecules still remain preserving native structure of the protein[15]. Smooth mica surface not only supports the protein molecules *in situ* but also prevents the shadowing material from coating the rearside of the sample molecules. Hence, the resultant image could reveal a faithful surface replica of the functioning molecules in solution.

By exploiting superior freezing technique as such as the mica-flakes whose surface was appropriately modified for the specific purpose[11], I could successfully visualize the structure of uncross-linked acto-S1 samples under physiological conditions in the absence or presence of various nucleotides[11]. Actin-attached particles observed in this way appeared all short and rounded in the presence of ATP (Fig. 1d), whereas they were elongated and protruded obliquely from F-actin (Fig. 1c) exactly in the same manner as observed by negative staining. Both the result of parallel pelletting experiment and the number of actin-attached S1 calculated from the dissociation constant under such conditions[8] were in a reasonable agreement with the number of short and rounded particles seen in the presence of ATP, indicating that observed particles were certainly S1 molecules. The image of acto-S1 in the presence of ADP plus inorganic vanadate (Vi) was indistinguishable from that with ATP, in accordance with the biochemical data that S1•ADP•Vi forms a stable analogue[16] of S1•ADP•Pi. Such rounding of S1 also accounts well for the data from independent biophysical studies[17][18] that the distance between actin and alkali light-chain of S1 as estimated by fluorescence energy transfer was less than half when S1 was in an active or weakly-bound conformation compared with the value for rigor conformation.

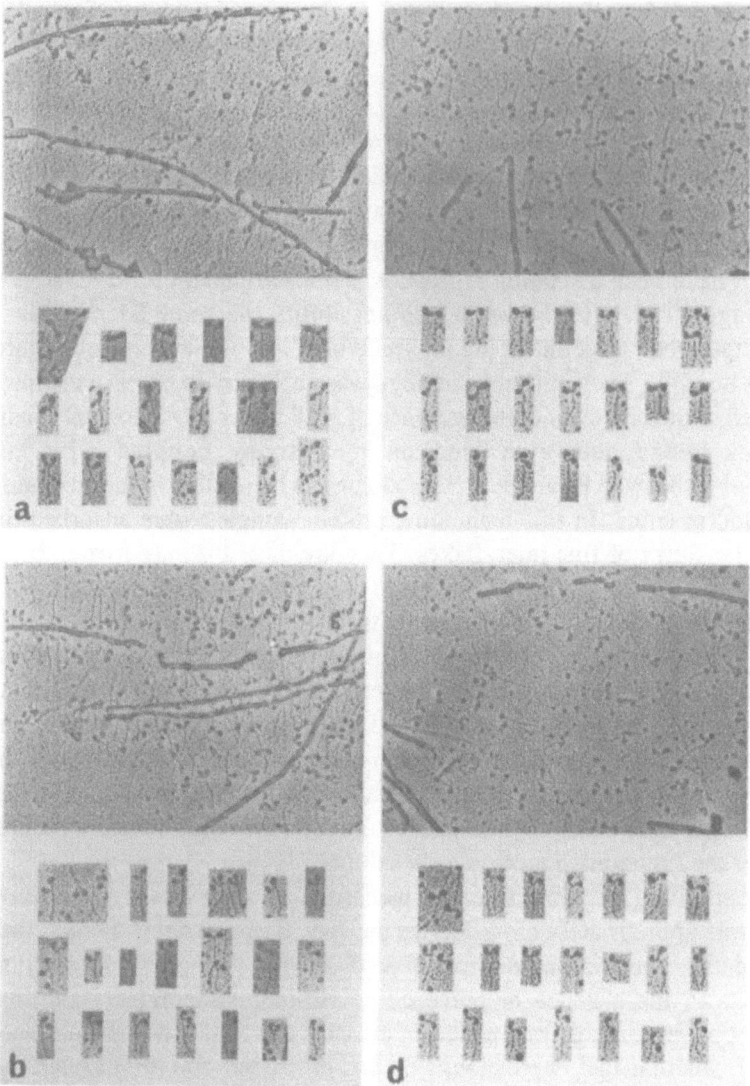

Fig. 2. General view of low concentration acto-HMM and gallery of individual HMM particles in the absence or presence of various nucleotides; a) no nucleotide; b) ADP; c) ATP; d) ADP•Vi [reproduced form reference 20)].

2. Origin of Observed Conformational Change of Myosin Head

Actin-attached S1 was thus short and rounded in the presence of ATP. Then, the question arose how elongated myosin head could become short and rounded. Since heavy meromyosin (HMM) consists of two heads and subfragment-2 (S2) moiety as a tail, it must be a much easier sample than a simply ellipsoidal S1 to understand how myosin head changes its shape as observed when it binds ATP. HMM molecules in

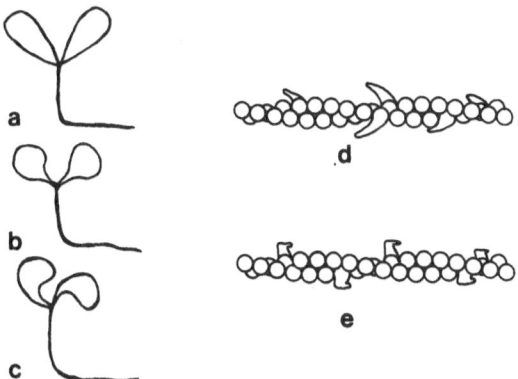

Fig. 3. Schematic drawings of specific nucleotide-induced conformation of myosin head. Freeze-replica images of a) HMM in the absence of nucleotide; b) HMM in the presence of ADP; c) HMM in the presence of ATP; d) HMM in the presence of ADP•Vi. Negatively stained images of S1 chemically cross-linked to F-actin; d) in the absence of ATP, e) in the presence of ATP [reproduced from reference 20)].

the absence or presence of various nucleotides, were visualized by the same electron microscopic technique as before. In the absence of nucleotide, HMM appeared very similar to conventionally rotary-shadowed molecules with two pear-shaped heads and S2 tail connecting them. Elongated heads were straight or only slightly curved in the absence of nucleotide[19)20)] (Fig. 2a) and presented virtually the same configuration whether bound or not bound to F-actin. when ADP was added, each head of HMM became seemingly fatter at the tip and narrowed at the neck giving rise to an "ball-on-a-stick" appearance (Fig. 2b). The effect of the addition of ATP to HMM sample was striking. Many heads in the field were sharply kinked at the middle of the particle and showed a distinct habit of adsorption in terms of the direction of bending (clockwise if followed form the neck to the tip of the head)(Fig. 2c). A fraction of HMM molecules had one of two heads which was classified to a straight rigor-type or a fatter ADP-induced type configuration. Inclusion of ADP•Vi, instead of ATP, made the field of HMM even more abundant in kinked conformer population, again with the absolute habit in polarity (Fig. 2d). There were some HMM particles which did not at first look kinked in a single view but proved, by stereo examination, being really kinked and standing up from the mica surface. All of these observations strongly indicate that myosin head sharply kinks while it binds and hydrolyses ATP to ADP•Pi[19)20)]. I reexamined carefully the negatively stained image of chemically cross-linked acto-S1 in the presence of ATP[11)] and noticed that S1 particles which were previously recognized as rounded were actually the molecules strongly kinked to the same direction throughout the same actin filament (Fig. 1b), confirming the above interpretation. The rounding of S1 described in the preceding section would be thus reasonably attributable to the strong flection within myosin head (Fig. 3). The results from the measurement of rotational relaxation coefficient of S1 by Agguire et al.[21)] and Highsmith & Eden[22)]

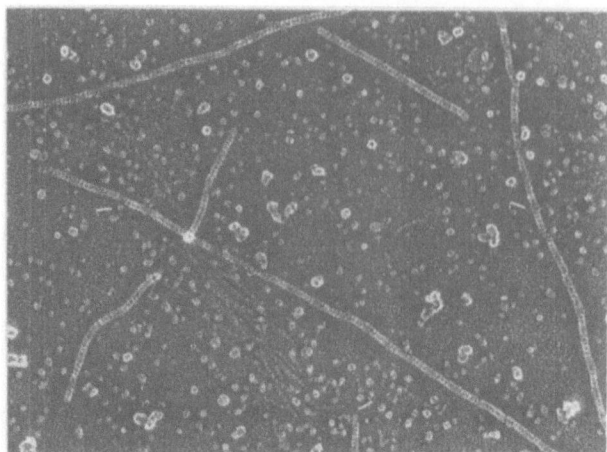

Fig. 4. Freeze-replicated image of actomyosin during superprecipitation. Three myosin heads are attached to actin filaments (indicated by arrowheads). All of them are kinked to a clockwise direction.

are notable. Despite the methods they used were quite different from each other, they reached to the same conclusion that the radius of gyration of S1 becomes smaller when it is complexed with ADP•Vi, providing the supporting evidence for the present electron microscopic observation. Further interesting possibility might be addressed that the three kinds of configurations observed in this study for rigor, ADP-bound and ADP•Pi-bound myosin head could possibly correspond to each of three spectroscopic conformers which were first reported by Morita and her colleagues[23].

3. Structure of Myosin Head during Active Sliding Movement of Actin

Using S1 and HMM as test samples, I showed that myosin head changes from straight pear-shape to rounded or kinked conformation upon binding of ATP or ADP•Pi. The next and more interesting question is whether such conformational change really occurs in intact myosin heads on which actin filament slides unidirectionally to develop force. Actomyosin superprecipitation has been used as a good *in vitro* model of muscle contraction. Following the addition of ATP, visual appearance of actomyosin solution gradually changes from initial semitransparent phase to final ground-glass-like turbidity[24]. This process involves active sliding movement of actin filaments past myosin heads, especially during rapid turbidity increasing phase[25]. According to Harada et al.[26], mica surface works as a good alternative of appropriately treated cover-glass to support myosin molecules in *in vitro* actin motility assay. Utilizing this convenient property of mica substrate, I attempted to observe the structure of myosin heads on which actin filaments are actively sliding. While the turbidity of actomyosin mixture was continuously monitored, a fraction of the sample was taken up from the cuvette at certain time-points and was mixed with mica slurry for a short time followed by quick-freezing on liquid-helium-operated machine. As expected, freeze-replica of these samples gave clear views of interacting actomyosin (Fig. 4). The cores of thick filaments were often disassembled by the strong adsorption onto flat mica surface and were observed as parallel rows of myosin molecules. Myosin heads were splayed around the core and showed either ADP-induced or ATP-induced kinked configuration with the polarity according to the habit of adsorption onto mica surface. Among myosin molecules which were in close contact with actin filaments, some of the heads showed a clearly definable configuration. The heads of such molecules observed were, so far, all in a kinked configuration. This would suggest the possibility that myosin head might take a kinked configuration during sliding movement of actin filaments, at least for most of the fraction of ATPase cycle time. In a few images, multiple myosin heads attaching to a single actin filament were viewed as kinked to the same direction, in a similar manner as observed in negatively stained cross-linked acto-S1. During superprecipitation, actin filaments slide freely on myosin filaments without load. This *in vitro* phenomenon is, hence in more strict sense, regarded as an appropriate model of zero-load contraction. According to Ishijima et al.[17], measurement of force developed by actomyosin *in vitro* presented a quite different fluctuation profile under isometric conditions compared with that under isotonic conditions. Distribution among conformers of actin-attached myosin heads could considerably differ under different mode of contraction. More elaborate work is necessary to further clarify this point.

REFERENCES

1. Huxley, A.F. *J. Physiol. (Lond.)* **243**, 1-43 (1974).
2. Huxley, H.E. *Science* **164**, 1356-1366 (1969).
3. Wakabayashi, T. & Toyoshima, C. *J. Biochem.* **90**, 683-701 (1981).
4. Katayama, E. & Wakabayashi, T. *J. Biochem.* **90**, 703-714 (1981).
5. Taylor, K.A. & Amos, L.A. *J. Mol. Biol.* **147**, 297-324 (1981).
6. Vibert, P. & Craig, R. *J. Mol. Biol.* **157**, 299-319 (1982).
7. Milligan, R.A. & Flicker, P. *J. Cell Biol.* **105**, 29-39 (1987).
8. Stein, L.A., Schwarz, R.P. Jr., Chock, P.B. & Eisenberg, E. *Proc. Natl Acad. Sci. USA* **82**, 3895-3909 (1979).
9. Mornet, D., Pantel, P., Audemard, E. & Kassab, R. *Biochem. Biophys. Res. Commun.* **89**, 925-932 (1979).
10. Craig, R., Greene, L.E. & Eisenberg, E. *Proc. Natl. Acad. Sci. USA* **82**, 3247-3251 (1985).
11. Katayama, E. *J. Biochem.* **106**, 751-770 (1989).
12. Applegate, D. & Flicker, P. *J. Biol. Chem.* **262**, 6856-6863 (1987).
13. Sutoh, K. *Biochemistry* **22**, 1579-1585 (1983).
14. Heuser, J.E. *J. Mol. Biol.* **169**, 155-195 (1983).
15. Gross, H. in *Cryotechniques in Biological Electron Microscopy* (eds. Steinbrecht, R.A. & Zierold, K.) 205-215 (Springer-Verlag, Berlin, Heidelberg, 1987).
16. Goodno, C.C. & Taylor, E.W. *Proc. Natl. Acad. Sci. USA* **79**, 21-25 (1982).
17. Bhandari, D.C. Trayer, H.R. & Trayer, I.P. *FEBS Lett.* **187**, 160-166 (1985).
18. Trayer, H.R. & Trayer, I.P. *Biochemistry* **27**, 5718-5727 (1988).

19. Katayama, E. *J. Musc. Res. Cell Motility* **12**, 313 (1991).
20. Katayama, E.(submitted).
21. Aguirre, R., Lin, S-H., Gonsoulin, F., Wang, C.-K. & Cheung, H. *Biochemistry* **28**, 799-807 (1989).
22. Highsmith, S. & Eden, D. *Biochemistry* **29**, 4087-4093 (1990).
23. Morita, F. & Ishigami, F. *J. Biochem.* **81**, 305-312 (1977).
24. Ebashi, S. Kodama, A. *J. Biochem.* **60**, 733-734 (1966).
25. Higashi-Fujime, S. *Cold Spring Harbor Symp. Quant Biol.* **46**, 69-75 (1982).
26. Harada, Y., Sakurada, K., Aoki, T., Thomas, D.D. & Yanagida, T. *J. Mol. Biol.* **216**, 49-68 (1990).
27. Ishijima, A., Doi, T., Sakurada, K. & Yanagida, T. *Nature* **352**, 301-306 (1991).

Discussion

Morales: Dr. Katayama, I would like to mention two points that you might consider. Recently, there were some measurements involving the cross-linking pattern in S-1, and it was shown that there were significant differences when the cross-linking was done at 298° Kelvin or 273° Kelvin. But you are going to 4° Kelvin. Is it possible that the structure of the protein itself might be temperature-sensitive, and especially that its response to ATP may be very different? The second question is: do you see some serious conflict with the results of experiments with the deuterated actin and heavy water mixture, in which Curmi and Mendelson felt they were showing no shape change (*J. Mol. Biol.* **203**, 781-798, 1988)?

Katayama: To the first question: I haven't actually done the ATP experiment by comparing the temperature effects, but I think I have to do it, because the three structures are comparable to the three kinds of conformers that were found spectrophotometrically (Morita et al. *J. Biochem.* **81**, 305-312, 1977).

Morales: I think the FRET measurement you mentioned that was done by Trayer could be accounted for by a rigid movement of the whole S-1 relative to the actin.

Katayama: One fluorescence was put on the alkali light chain, the other on actin, and the distance between them decreased to less than half by adding ATP. If the fluorescence was put on SH-1 instead of the light chain, the distance did not change much, or rather, increased slightly (Arata, T. *J. Mol. Biol.* **191**, 107-116, 1986). As to the other question, Mendelson's group has been studying the structure of S-1 and heavy meromyosin extensively, and they could not find a structural change (e.g., *J. Mol. Biol.* **177**, 153-191, 1984). But I think Dr. Wakabayashi may have some comments. I believe that previous studies may have missed it, just because of the too-low resolution of X-ray technique.

Morales: I would think they could detect what you showed.

Edman: I just noticed that only one of the two heads was attached to the filaments. Is that genuine?

Katayama: Under rigor conditions, sometimes HMM can even cross-link the actin filaments, as shown by negative staining (*J. Mol. Biol.* **133**, 569-556, 1979).

Edman: So, it differs between rigor and when you add ATP?

Katayama: Yes. Under rigor conditions, there are two or three possibilities—one-head binding or two-head binding to a single actin filament or to two actin filaments.

Kawai: What is the state of phosphate?

Katayama: That would be ADP•P_i, I believe.

Kawai: What happens if you add phosphate to it?

Katayama: Nothing.

Reedy: I would like to comment that just as your very interesting results do not find an echo in the work of Mendelson, likewise many times with insect muscle we made X-ray patterns in ADP and never saw any indication of a departure from rigor structure. So, that's another puzzle. This is rabbit, not insect, so maybe this is one of those cases of a species difference, but that would be surprising. I think these are very interesting results that need to be examined further.

CROSS-BRIDGE ANGLE DISTRIBUTION AND THIN FILAMENT STIFFNESS IN FROG SKELETAL MUSCLE FIBERS AS STUDIED BY QUICK-FREEZE DEEP-ETCH ELECTRON MICROSCOPY

Suechika Suzuki, Yoko Oshimi and Haruo Sugi

Department of Physiology
School of Medicine
Teikyo University
Itabashi-ku, Tokyo 173, Japan

ABSTRACT

To give information about changes in orientation of myosin heads (cross-bridges) during contraction, mechanically skinned frog muscle fibers were rapidly frozen in various states, and cross-bridge angles were measured on the freeze-etch replicas. Histograms of cross-bridge angle distribution showed a peak around 90° in relaxed, contracting and rigor states. The proportion of cross-bridges taking angles around 90° decreased when rigor fibers were stretched or released before freezing. These results are explained by assuming the stretch-induced tilting of cross-bridges due to elastic recoil of the thin filaments in the I-band.

As a matter of fact, the axial spacing of actin monomers in the thin filament increased with increasing rigor force before freezing. The stiffness times unit length of the thin filament was estimated to be about 1.8×10^4 pN.

INTRODUCTION

In the contraction model of Huxley and Simmons[1], the myosin heads (cross-bridges) change their orientation relative to the thin filaments, and pulls the elastic link connecting them to the thick filaments, thus generating force or muscle shortening. The above mechanism implies that externally applied muscle length changes may also change the cross-bridge orientation.

As an attempt to detect the changes in cross-bridge orientation in muscle, we performed electron microscopic studies, in which mechanically skinned frog muscle fibers were rapidly frozen in relaxed, contracting and rigor states, and the cross-bridge angles to the thin filament axis were measured on the freeze-etch replicas with a digital image processor. The opportunity was also taken to measure the axial spacing of actin monomers in the thin filament when rigor fibers were frozen with and without imposed length changes.

Mechanism of Myofilament Sliding in Muscle Contraction, Edited by
H. Sugi and G.H Pollack, Plenum Press, New York, 1993

57

S. Suzuki et al.

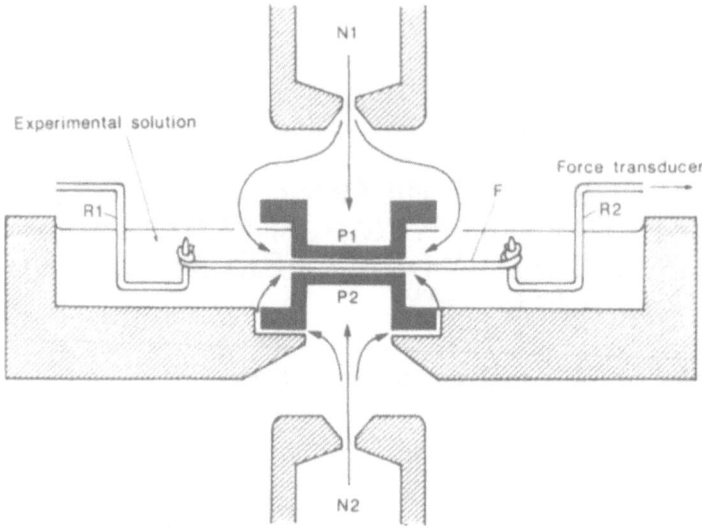

Fig. 1. Schematic drawing of experimental arrangement. The skinned fiber preparation F was mounted horizontally between two rods R1 and R2, its middle portion being placed between two hat-shaped plates P1 and P2. R1 was fixed in position, while R2 was connected to the force transducer. After the experimental solution in the experimental chamber was drained, the fiber was rapidly frozen by jetting liquid propane from two nozzles N1 and N2 to the plates P1 and P2 and the fibre F as indicated by arrows. From reference (4).

METHODS

As shown in Fig. 1, a single mechanically skinned muscle fiber (sarcomere length 2.3-2.4 μm) prepared from the semitendinosus muscle of the frog *Rana catesbeiana* was mounted horizontally between two stainless-steel rods in an experimental chamber filled with relaxing solution. The fiber was made to contract isometrically with contracting solution, or put into rigor state with rigor solution. The temperature of the solutions was kept at 4°C. The middle portion of the fiber was placed between two hat-shaped gold plate (Balzers). Immediately before freezing, the solution in the chamber was drained, and the fiber was rapidly frozen by jetting liquid propane (about -190°C) from two nozzles to the gold plates and the fiber[2)3)]. After the completion of freezing, one of the two gold plates with middle portion of the frozen fiber attached was removed from the chamber in liquid nitrogen, and put on a freeze-fracture machine (Balzers BAF 301 or 400D) to fracture the superficial layer of the fiber at 10^{-6} torr. After etching (4 min, -95°C), the specimen was rotary shadowed with Pt/C (thickness 2 nm), and was dissolved in household bleach, The replica images were observed using a JEOL JEM100CX electron microscope.

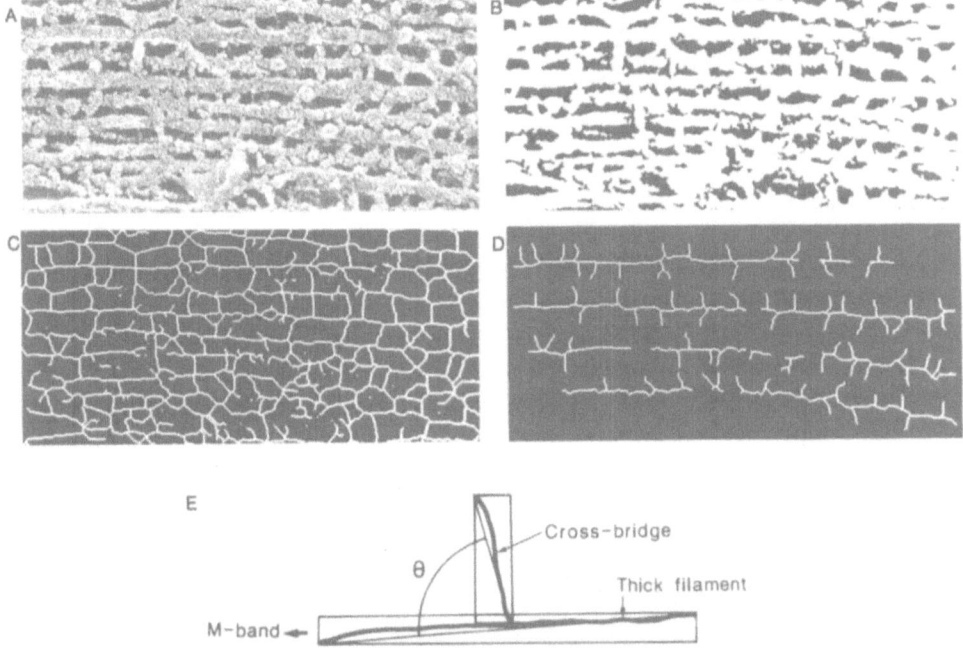

Fig. 2. Diagram illustrating the method of measurement of the cross-bridge angles to the filaments on freeze-etch replica images. The original replica image of the filaments and cross-bridges (A) was first converted into the binary image (B) and then further converted into the picture (C) with thinning process. After erasing the lines of the thin (or thick) filaments and other unidentified structures, the picture (D) only containing lines of thick (or thin) filaments and cross-bridges. The cross-bridge angle to the thick (or thin) filament was measured as the angle (q) between two diagonal lines as shown in E. For further explanation, see text. Modified from reference (4).

The measurement of angles between the cross-bridges and the thick and thin filaments with a digital image processor (Toshiba Tospix II) is illustrated in Fig. 2. Enlarged eléctron micrographs (A) were first processed to produce binary images showing contours of the cross-bridges and the filaments on dark background (B). The binary images were subjected to thinning process to obtain lines corresponding to the central axis of the cross-bridges and the filaments (C). After erasing lines of the thin (thick) filaments and other unidentified structures (D), the angle of each cross-bridge to the thick filament was measured as the angle between the diagonal lines of two rectangles, consisting of vertical and horizontal lines drawn to just surround the central axis of the cross-bridge and that of a thick (thin) filament segment (50-100 nm) around the cross-bridge (E). The angles were measured from the M-band side in each sarcomere; when a cross-bridge inclined toward the M-line, its angle to the thick filament was acute and that to the thin filament was obtuse.

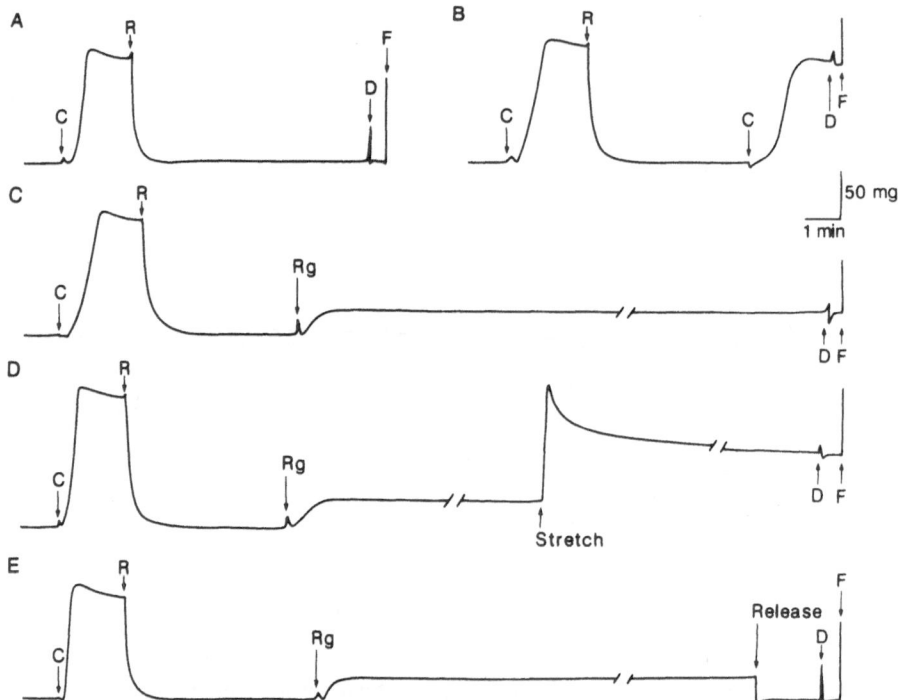

Fig. 3. Examples of force records when the fiber was rapidly frozen in resting state (A), in contracting state (B), in standard rigor state (C), in stretched rigor state (D), and in released rigor state (E). In each record, the time of application of contracting solution (C), relaxing solution (R) or rigor solution (Rg) is indicated as well as the time of draining experimental solution in the chamber (D) and the time of subsequent rapid freezing of the fiber (F). From reference (4).

The fibers mounted in the experimental chamber were first made to contract isometrically with contracting solution to record their maximum Ca^{2+}-activated isometric force P_0 and were relaxed in relaxing solution. Then the fibers were frozen rapidly (1) in relaxed state in relaxing solution (Fig. 3A), (2) in contracting state in contracting solution (Fig. 3B), and in rigor state in Ca-free rigor solution after steady rigor force (about 0.2 P_0) was attained (standard rigor state, Fig. 3C). The fibers in rigor state were also frozen rapidly (4) after a stretch (about 5%) with which the steady force rose to about 0.6 P_0 (stretched rigor state, Fig. 3D), and (5) after a release to reduce the force just to zero (released rigor state, Fig. 3E). Further details of the methods have been described elsewhere[4].

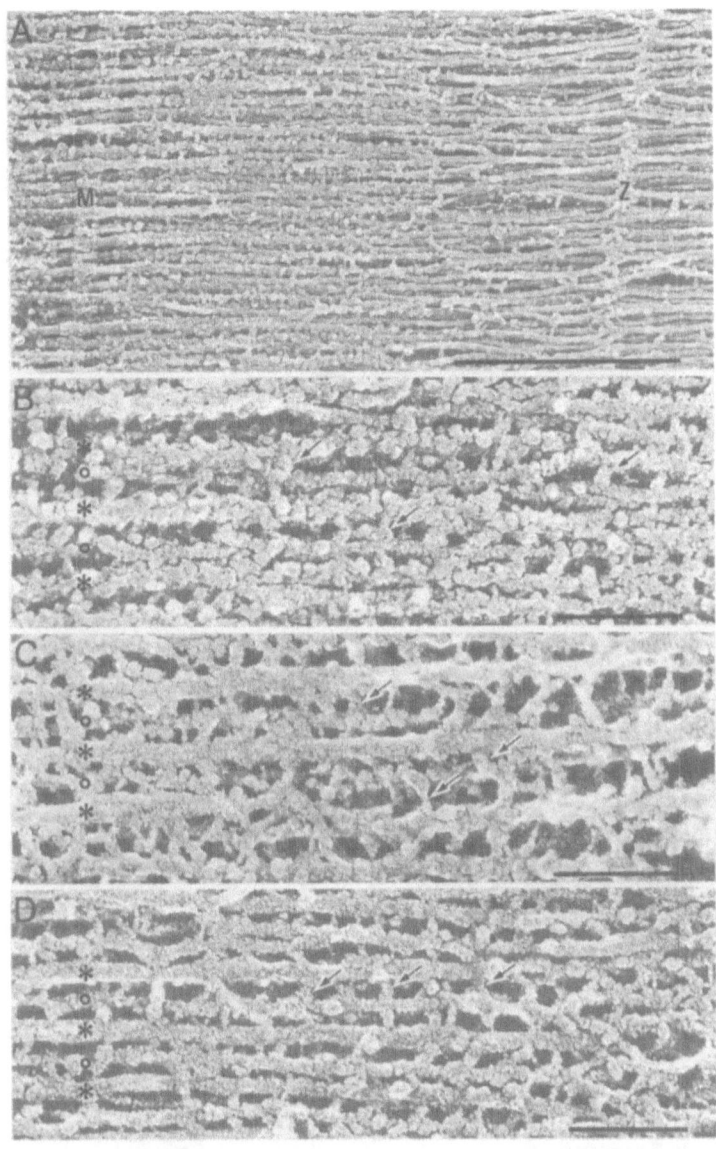

Fig. 4. Freeze-etch replica images of mechanically skinned muscle fibers in various physiological states. A, low magnification view of half sarcomere in a relaxed fiber, in which the M- band (M) and Z- band (Z) structures as well as the thick and thin filaments are clearly visible. Bar, 0.5 μm. x111,800. B-D, high magnification view of the thick and thin filaments in the filament overlap region in relaxed (B), contracting (C) and standard rigor (D) states. Note that the thick (*) and thin (o) filaments appear alternately and run in parallel with one another. Some of the rod-like cross-bridges between adjacent thick and thin filaments are indicated by arrows. Bar, 0.1 μm. B, x312,700; C, x321,800; D, x310,500. Modified from reference (4).

RESULTS

Appearance and Distribution of Cross-bridges along Thick Filament

Fig. 4 shows a low magnification view of a sarcomere (A) and high magnification views of the myofilament overlap region in relaxed (B). Contracting (C) and rigor (D) states in the freeze-etch replica images of frozen fibers. In all the states examined, one thin filament was frequently seen between two thick filaments with dark background between them, indicating that only a single layer of thick filaments and a single layer of neighbouring thin filaments were exposed for microscopic observation (see Fig. 9).

The number of cross-bridges per 0.5 μm per one side of thick filament was 4.9 ± 2.5 (S.D., n = 100) in relaxed state, 17.8 ± 2.7 (n = 100) in contracting state, and 16.9 ± 4.1 (n = 100) in standard rigor state. The value in relaxed state was less than one third of the values in contracting and standard rigor states.

Fig. 5 shows histograms of interval between the two adjacent cross-bridges along each side of the thick filament. In relaxed state, the cross-bridge interval showed an extremely wide variation with a peak at 32-34 nm (A), while in

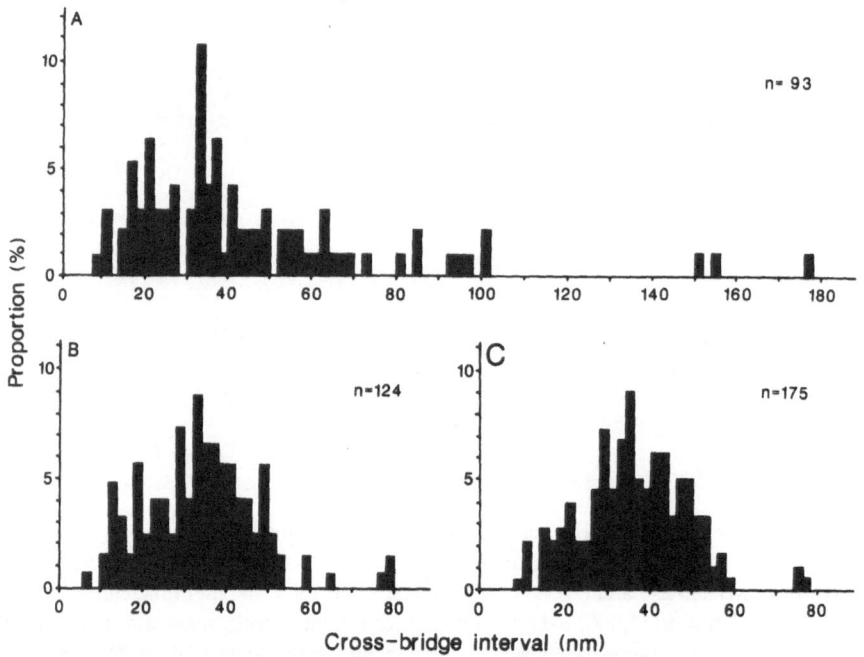

Fig. 5. Histograms showing distribution of interval between two adjacent cross-bridges along each side of the thick filament in relaxed state (A), in contracting state (B) and in standard rigor state (C). Total number of measurements is given alongside each histogram. From reference (4).

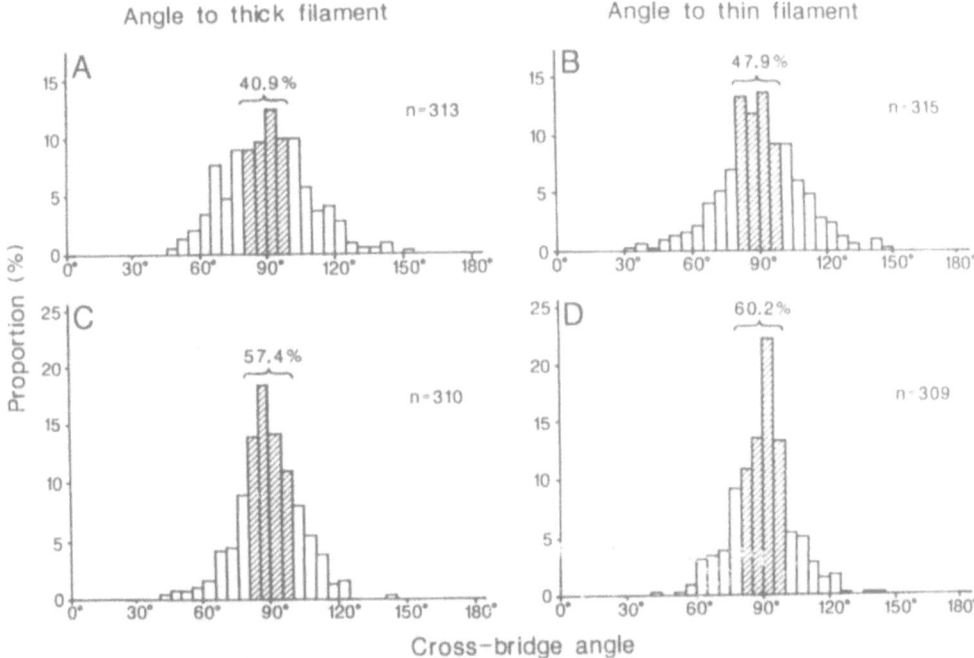

Fig. 6. Histograms showing distribution of cross-bridge angles to the thick and thin filaments in relaxed state (A and B) and in contracting state (C and D). Total number of measurements is given alongside each histogram. In Figs. 6 and 7, histograms on the left side and those on the right side are concerned with cross-bridge angles to the thick filament and those to the thin filament, respectively. From reference (4).

contracting and standard rigor states the variation was much smaller with peaks at 34-36 and 32-34 nm (B, C).

Cross-bridge Angle Distribution in Relaxed, Contracting and Rigor States

Histograms of cross-bridge angle distribution in relaxed and contracting states are presented in Fig. 6. In both cases, the cross-bridge angles to the thick and thin filaments showed a peak around 90°. In relaxed state, the proportion of cross-bridges taking angles of 90 ± 10° was 40.9% for angles to the thick filament and 47.9% for angles to the thin filament (A, B). In contracting state, this value increased to 57.4% for angles to the thick filament and to 60.2% for angles to the thin filament (C, D).

Histograms of cross-bridge angle distribution in various rigor states are shown in Fig. 7. In standard rigor state, the cross-bridge angle distribution was analogous to that in contracting state; the proportion of cross-bridges taking angles of 90 ± 10° was 50.3% for angles to the thick filament and 55.5% for angles to the thin filament (A, B). In stretched rigor state, the value for angles to the thick filament was 52.4% and did not differ appreciably from the corresponding value in standard rigor state (C). However, the value for angles to the thin filament was decreased to 46.3%

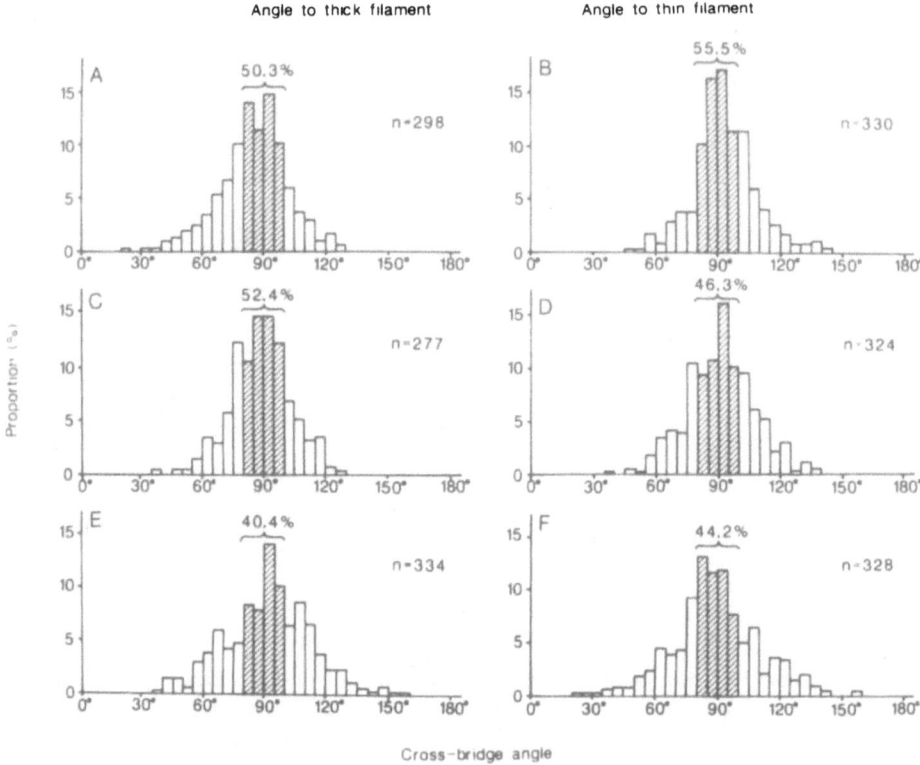

Fig. 7. Histograms showing distribution of cross-bridge angles to the thick and thin filaments in standard rigor state (A and B), in stretched rigor states (C and D), and in released rigor state (E and F). From reference (4).

compared to the corresponding value in standard rigor state (D). In released rigor state, on the other hand, the value decreased for angles to both the thick and thin filaments to 40.4% and 44.2% respectively (E, F).

Axial Spacing of Actin Monomers at Various Rigor Forces in the Thin Filament

As actin monomers constituting the thin filament were clearly observed in the electron micrographs in the myofilament non-overlap region (I-band) of frozen fibers, the opportunity was taken to estimate the thin filament stiffness by measuring the axial spacing of actin monomers on the thin filament segment image using the technique of optical diffraction. Fig. 8 shows a typical thin filament segment image (A) and its diffraction pattern (B). The results obtained are summarized in Fig. 8C. It is assumed that each thin filament supports an equal force of 210 pN at P_O, and this force is scaled with the force in the fiber relative to P_O[5]. The axial spacing of actin monomers increased significantly ($P < 0.01$) with increasing force in the fiber, From the slope of the line relating the actin monomer axial spacing to force in the

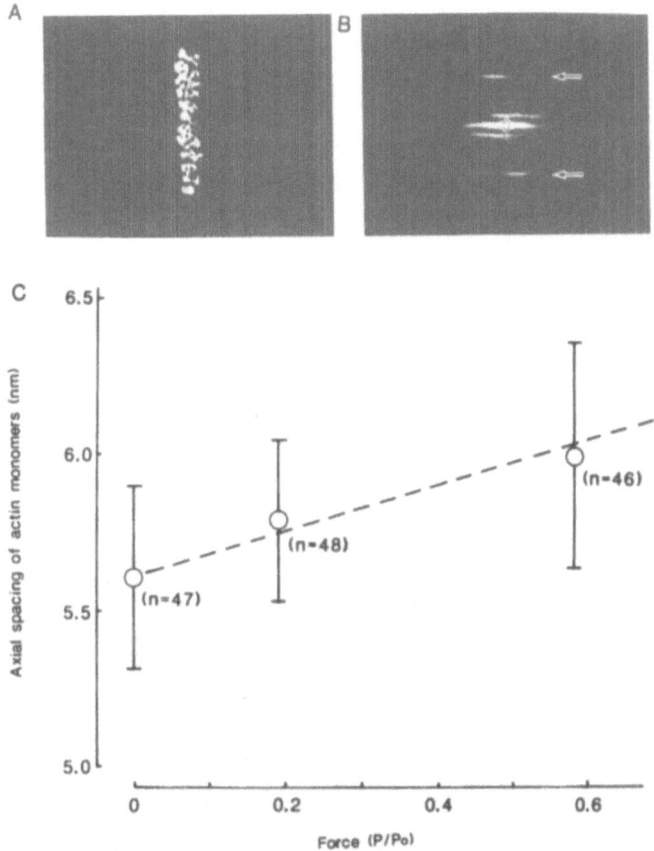

Fig. 8. Measurement of axial translation of actin monomers in the thin filament. A, example of thin filament segment replica images obtained from the filament nonoverlap region. B, optical diffraction pattern from the image shown in A. Layer line based on axial spacing of actin monomers is indicated by arrow. C, relation between the rigor force in the fibre immediately before rapid freezing and the axial spacing of actin monomers in the thin filament. Each data point with vertical bar represents mean ± S.D. obtained from 46-48 thin filament segments of two to three different fibres. From reference (4).

fiber, the thin filament stiffness times unit length was estimated to be about 1.8×10^4 pN, a value not much different from that estimated by measuring the thin filament length in stretched rabbit psoas fibers in rigor state[5].

DISCUSSION

Implications of Cross-bridge Distribution along the Thick Filament

The present measurement of cross-bridge angles were performed only with the replica images in which the thick and thin filaments appeared alternately and ran

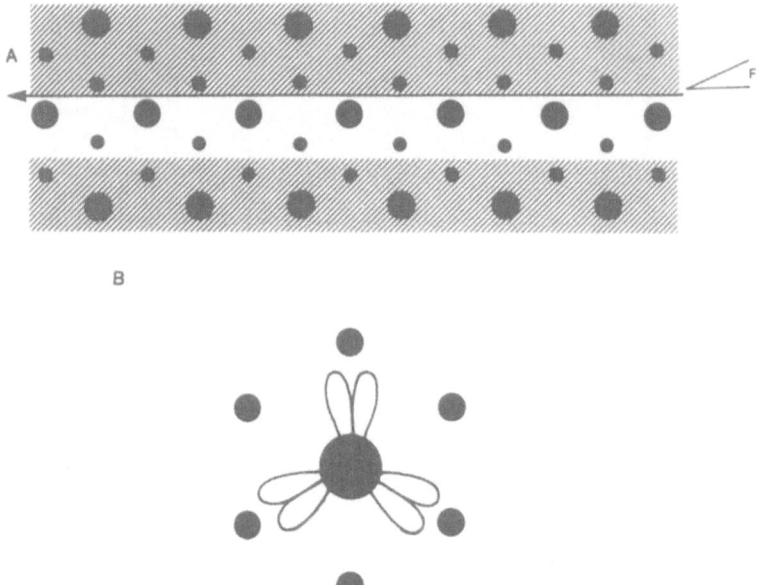

Fig. 9. A, rationale of the freeze-etch replica images in which the thick and thin filaments appear alternately. The freeze-fractures (F) in the present study travel through the fiber cross-section as indicated by arrow to remove the filament lattice in the upper shaded area. The subsequent deep-etch process exposes a layer of the thick filaments (large filled circles) and a layer of the thin filaments (small filled circles), while the filament lattice in the lower shaded area remains embedded in ice. B, cross-sectional view of a thick filament (large filled circle) at a 14.3 nm spaced axial level where three pairs of myosin heads are located at angular intervals of 120°. The thick filament is surrounded by six neighbouring thin filaments (small filled circles). From reference (4).

parallel with one another (Fig. 4). Stereo view observations indicated that the thin filaments were located at the level slightly underneath the thick filament plane. As the thin filaments are located in trigonal positions in the hexagonally packed thick filaments, we consider that the filament lattice is fractured along the line shown in Fig. 9A.

It is generally agreed that, in the thick filament of vertebrate skeletal muscle, three pairs of myosin heads are located in each 14.3 nm-spaced axial level at angular intervals of 120°[6] (Fig. 9B). On this basis, there should be about 12 pairs of myosin heads per 0.5 μm per one specific side of the thick filament, since the myosin head arrangement repeats at 42.9 nm intervals (500/42.9 = 11.7). Assuming that each observed cross-bridge corresponds to one myosin head, the number of observed cross-bridges along one side of the thick filament indicates that the proportion of myosin heads attached to the thin filaments is about 20, 74 and 70% in relaxed, contracting and rigor states respectively. It should be kept in mind, however, that the present method can not distinguish attached cross-bridges from apparent ones.

In contracting and standard rigor states, the cross-bridge interval showed a peak around 35 nm (Fig. 5B, C), which is equal to the crossover repeat of actin helix (35.5 nm)[7]. This suggests that the probability of attachment of myosin heads to the thin filament is largest when the distance between the thick and thin filaments is shortest. As the axial repeat of myosin head arrays on the thick filament is largest when the distance between the thick and thin filaments is shortest. As axial repeat of myosin head arrays on the thick filament (42.9 nm) does not match the actin helix in the thin filament, the present results imply that myosin heads can conform their interval to the axial repeat of actin helix, possibly by the flexibility of their S-2 region.

Implications of Cross-bridge Angle Distribution in Relaxed, Contracting and Rigor States

The cross-bridge angle distribution exhibited a peak around 90° in all the states examined (Figs. 6 and 7). As the cross-bridges in relaxed state may be mostly apparent ones, their angle distribution (Fig. 6A, B) seems to be consistent with the view that, in relaxed state, myosin heads are swinging freely around the equilibrium position where the force in the S-2 link is zero[8].

In contracting and standard rigor states, the proportion of cross-bridges taking angles around 90° increased compared to that in relaxed state (Figs. 6C, D and 7A, B), indicating the tendency of myosin heads to take angles around 90° to both the thick and thin filaments when they generate active or rigor isometric forces, though the present results give no information about whether force generation is coupled with changes in cross-bridge angle.

In the stretched rigor state, on the other hand, the proportion of myosin heads taking angles around 90° to the thin filament decreased by about 10% compared to that in standard rigor state (Fig. 7B, D), while the corresponding values to the thick filaments did not differ appreciably between standard and stretched rigor states (Fig.

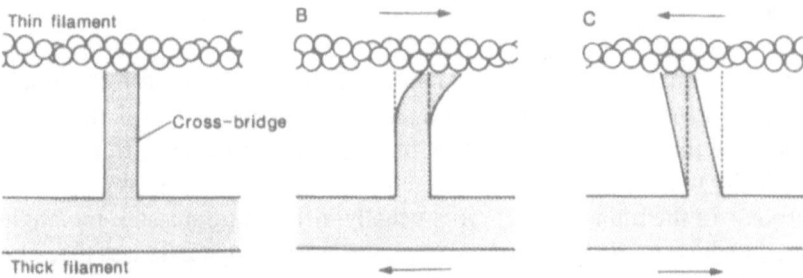

Fig. 10. Possible changes in cross-bridge configuration which accounts for the results shown in Fig. 7. A, a cross-bridge with an angle of 90° to both thick and thin filaments in standard rigor state. B, bending at the distal portion of a cross-bridge in stretched rigor state. C, tilting of the whole cross-bridge toward M-band in released rigor state. Arrows in B and C indicate directions of filament sliding caused by stretch and release. From reference (4).

7A, C). This seems to suggest the mechanism illustrated in Fig. 10A, B; if a rigor cross-bridge is displaced in the direction opposite to that of fiber shortening, its flexible distal portion is distorted to change its angle to the thin filament, while its proximal portion is rigid (Fig. 10A, B). As a result, the proportion of cross-bridges taking angles around 90° decreases with corresponding increase in the proportion of cross-bridges taking acute angles to the thin filament.

In released rigor state, the proportion of cross-bridges with angles around 90° to both the thick and thin filaments decreased by about 10% compared to the corresponding values in standard rigor state (Fig. 7A, B, E, F). This might result from tilting of myosin heads caused by elastic recoil of thin filament in the filament nonoverlap region (I-band) when rigor force is reduced from 0.2 P_0 to zero (Fig. 3E). In the present study, the thin filament stiffness times unit length was estimated to be about 1.8×10^4 pN (Fig. 8). On this basis, the elastic recoil of the thin filament in the I-band during transition from standard rigor to released rigor state would be about 1 nm/half sarcomere. This would in turn cause tilt of each attached myosin head toward the M-band to result in a decrease in the proportion of myosin heads with angles about 90° to both the thick and thin filaments by about 10%, as illustrated in Fig. 10A, C.

REFERENCES

1. Huxley, A.F. & Simmons, R.M. *Nature* **233**, 533-538 (1971).
2. Müller, M., Meister, N. & Moore, H. *Mikroskopie* **36**, 129-140 (1980).
3. Suzuki, S. & Pollack, G.H. *J. Cell Biol.* **102**, 1093-1098 (1986).
4. Suzuki, S., Oshimi, Y. & Sugi, H. *J. Electron Microsc.* in press (1993).
5. Suzuki, S. & Sugi, H. *J. Gen. Physiol.* **81**, 531-546 (1983).
6. Squire, J.M. *J. Mol. Biol.* **72**, 125-138 (1972).
7. Parry, D.A.D. & Squire, J.M. *J. Mol. Biol.* **75**, 33-35 (1973).
8. Huxley, A.F. *Prog. Biophys. Biophys. Chem.* **7**, 255-318 (1957).

Discussion

Morales: I have the same temperature comment as I did on Katayama's paper. With regard to the distribution of the angle, since the S-1 is a rigid body, it is positioned by three angles. Are you assuming that, just happily, it is only the angle perpendicular to the beam of the electron microscope that changes?

Suzuki: In the replica image, it is usually difficult to measure the angle that is not perpendicular to the beam. In our case, the cross-bridge axis was measured as an average direction in the projection image, because we used the digital image processor.

Huxley: Was it Giovanni Cecchi who made stiffness measurements at different lengths with varying amounts of H-zone and I-band, but with full overlap of thin filament over the cross-bridges, and I think obtained substantially less stiffness than

Ford, Simmons, and I had obtained? I think that was a much better measurement to compare to this kind of thing. Can you quote a figure that would be related to the figures we have just been given?

Cecchi: Our figure for the compliance was measured as compared to the total compliance of the muscle fiber. It was about 15%. In other words, 15% of the compliance of the total muscle fiber will be attributed to actin filament compliance. It is not completely clear to me how the figure you have given is expressed in this way, because if I understood it correctly, I see you have about an 8% increase in length when the force goes from zero to P_0. This is quite a large compliance, I think.

Suzuki: That's right. Our result of rigor stretch indicates this large compliance. In addition, when the fiber is released, the actin length changes by about 1.5%.

Reedy: That's going from 57 Å back to 56 Å spacing?

Suzuki: Yes.

Huxley: It seemed to me that the changed spacing on the thin filament that Dr. Suzuki observed was of a scale that ought to be visible with X-ray. I have questioned Hugh Huxley several times about changes of thin filaments' spacing according to tension, and he says there is no change within 0.1%. I want to hear what other X-ray people say about this.

Holmes: Actually, Richard Tregear did these experiments (unpublished). He did careful experiments on this using the synchotron in the old days of doing stretched rigor muscle and relaxed rigor muscle and looked at the actin spacing. I don't think we could observe a difference, and this was with a tension that Richard might like to comment on; it was quite a lot. But you could argue that this was cross-linked rigor muscle. We still don't know what part of that was contributed to by the actin. But, in that situation, there wasn't very much to be seen. I personally have never seen anything, but it could be that we have not done the right experiment.

Tregear: We got quite good measurements on myosin spacing, but I don't think we really had good measurements on actin spacing.

Holmes: In the actin spacings we have looked at, there was nothing much happening (Huxley, H.E. et al. *Nature* **206**, 1358). The X-ray results say that there is nothing observable so far above half a per cent.

Sugi: Many times, our group intended to look at actin-based layer line spacing changes. But actin layer line measurement is very difficult, because its shape is round and somewhat broad.

Cecchi: We found (Bagni et al. *J. Muscle Res. Cell Motility* **11**, 371-377, 1990) that the stiffness of the actin filament was a bit smaller than the value given by Ford, Huxley, and Simmons [Ford, L.E. et al. *J. Physiol. (London)* **378**, 175-94, 1981]. However, it was much higher than the value given by Dr. Suzuki. If I understood the diagram, a decrease of about 1.5% in length will be required to discharge the tension completely on the actin filament. This would be more than 20 times higher than the amount we know we have to apply to the whole fiber to decrease the tension to zero. Is that correct?

Suzuki: In our case, we used mechanically skinned fibers, so that the elasticity might be a little different from that of the intact fiber.

Yanagida: We are now directly measuring the stiffness of single actin filaments by using microneedles and single actin filaments. Preliminary results indicate that actin filaments one micron long must be stretched by at least 0.5 %. In that sense, my result supports Dr. Suzuki's.

Holmes: We were talking about 3.5% and you say 0.5%. Is that correct? The X-ray work would agree with 0.5%. I would like to suggest, Dr. Yanagida, that you have measured the breaking strain of actin. And your result, if I remember, is about 100 pN, isn't it?

Yanagida: In my previous *Nature* paper I made a mistake in the calculation of the needle, so the present corrected result is that breakage occurs at 400 pN.

Holmes: Okay, but this is quite stiff, which it ought to be, otherwise we wouldn't be able to use our muscles. But doesn't this put a limit on the elasticity of actin? The breakage rates of actin filaments can be related to a work function ΔW by the Arrhenius equation

$$k_{break} = A \exp (-\Delta W/RT)$$

Taking a simple model for the mechanism of fiber breakage, the work function ΔW will be equal to the breaking strain multiplied by the distance one has to go through before the fiber breaks, which in turn is inversely proportional to the stiffness of the fiber. Thus, if the fiber is stiff, the rate of breakage will be fast for a given breaking strain, whereas if the fiber is compliant, the rate will be slow. My suggestion is that since you have measured the breaking strain, such arguments might be used to put bounds on the stiffness. For example, since the measured breaking strain is quite high if the fiber is very compliant, the estimated breakage rates could be much slower than is actually observed.

Yanagida: We determine the stiffness of the microneedle directly. We did not determine the stiffness from the breaking measurement. In my case, both ends of the actin filament were coated by two kinds of microneedles, and we applied a sinusoidal length change to one end and measured the movement of another needle. From this very simple measurement we could easily determine the stiffness of the microneedle. Then we obtained that result, showing that the actin filament must be stretched by 0.5% during isometric contraction.

Edman: You have found a change in the bridge-spacing when you released the rigor muscle, but how is that explained? Is the compliance used up during stretch?

Sugi: It originates from the characteristic properties of the head. It's apparently not symmetric.

Edman: It shouldn't depend on the head I think, according to your explanation.

Sugi: It depends on the molecular structure around the area the force is applied. My present feeling is that it somehow behaves non-symmetrically.

ELASTIC PROPERTIES OF CONNECTING FILAMENTS ALONG THE SARCOMERE

Károly Trombitás and Gerald H. Pollack

Bioengineering WD-12
University of Washington
Seattle, WA 98195

ABSTRACT

The elasticity of the connecting filament—the filament that anchors the thick filament to the Z-line—has been investigated using rigor release, freeze-break and immunolabelling techniques. When relaxed insect flight muscle was stretched and then allowed to go into rigor, then released, the recoil forces of the connecting filaments caused sarcomeres to shorten. Thin filaments, prevented from sliding by rigor links, were found crumpled against the Z-line. Thus, rigor release experiments demonstrate the spring-like nature of the connecting filaments in insect flight muscle.

In vertebrate skeletal muscle, however, the same protocol did not result in sarcomere shortening. Absence of shortening was due to either smaller stiffness of connecting filaments and/or higher stiffness of the thin filaments relative to insect flight muscle. The spring-like nature of the connecting filament was confirmed with the freeze break technique. When the frozen sarcomeres were broken along the A-I junction, the broken connecting filaments retracted to the N_1-line level, independently of the thin filaments, demonstrating the basic elastic nature of these filaments.

To study the elastic properties of the connecting filaments along the sarcomere, the muscle was labelled with monoclonal antibodies against a titin epitope near the N_1-line, and another very near the A-I junction in the I-band. Before labelling, fibers were pre-stretched to varying extents. Based on filament retraction and epitope translation with stretch, we could conclude: (1) the A-band domain of the connecting filament is ordinarily bound to the thick filaments; (2) at higher degrees of stretch, connecting filaments become free of the thick filaments, and the freed segments are intrinsically elastic; (3) between the A-I junction and the N_1-line, connecting filaments behave independently of thin filaments; between N_1- and Z-lines, however, they are firmly associated with the thin filaments.

Mechanism of Myofilament Sliding in Muscle Contraction, Edited by
H. Sugi and G.H Pollack, Plenum Press, New York, 1993

71

INTRODUCTION

Recent results of biochemical analyses indicate that giant structural proteins can be found in sarcomeres of vertebrate skeletal[1][2], heart[1][2] and insect flight muscle[3-6]. These proteins are thought to comprise a third, partly extensible filament system over and above the thick and thin filaments[1][2][7], and have been considered to account for the support of passive tension in stretched muscle[8-10]. The I-band domain of these filaments (the connecting filaments) was shown to be stretchable[7][11-13], whereas the A-band domain appeared to be bound to the thick filaments[11-15]. Nevertheless, there has been no direct evidence to indicate whether this stretchable element can exert recoil force at physiological sarcomere length.

We therefore performed experiments to investigate elastic properties of the connecting filaments in both the insect flight muscle and vertebrate skeletal muscle. Based on data obtained from "rigor-release" and "freeze-break" experiments, it appears that the connecting filament behaves as an independent spring: namely, the stretched filament can retract. Furthermore, the elastic properties of the connecting filaments along the I-band are not uniform. Between the A-I junction and N_1-line they are free and elastic, but between the N_1-line and Z-line they are either inelastic or, more likely, they are tightly bound to the thin filaments.

Connecting Filaments in Insect Flight Muscle

Evidence for the presence of connecting filaments between thick filaments and Z-line has been obtained from insect flight muscle, both at resting length (Fig. 1a), and after having been stretched in rigor. In the rigor condition cross-bridges are rigidly bound to the thin filaments, so the applied external stress breaks the thin filaments along the Z-line. Thin filaments remain completely in the A-band, leaving a gap between A-band and the Z-line. The elongated connecting filaments protrude from the ends of the thick filaments, crossing the gap and running to the Z-line[16-18] (Fig. 1b). In transverse sections the connecting filament lattice in the gap has the same organization and dimensions as the thick filament lattice in the A-band[17,18].

The nature of these connecting filaments was characterized with antibody-labeling experiments. Antibodies prepared against connecting filament protein[4] (projectin) labelled specifically the connecting filaments in the gap (Fig. 1c) (Trombitás, unpublished).

The most powerful evidence for the spring-like nature of connecting filaments comes from rigor-release experiments[18-20]. When stretched muscle (Fig. 2a) was transferred into rigor solution and released, the recoil forces of the connecting filaments caused the sarcomere to shorten. Thin filaments, prevented from sliding by rigor links, were found crumpled against the Z-line (Fig. 2b, c).

Although the connecting filaments are highly elastic, sudden length changes can cause breakage. Rigor-release experiments reveal the site of breakage. The muscle in Fig. 2d was suddenly stretched, transferred into rigor solution, and released after rigor had developed. In half-sarcomeres in which the connecting filaments remained

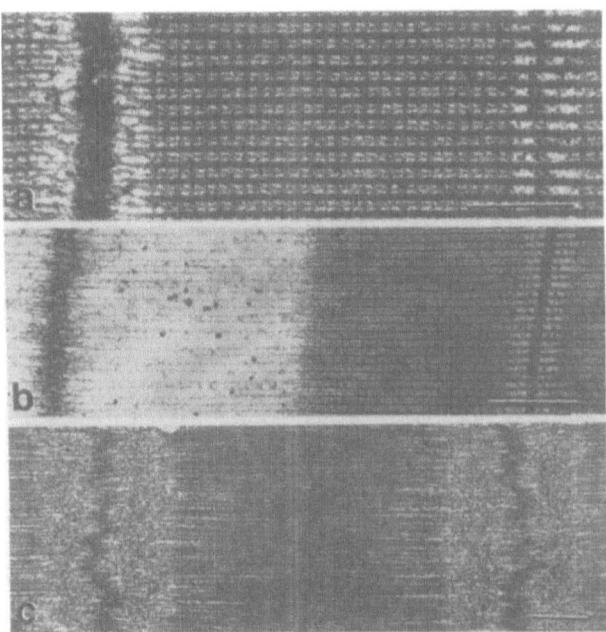

Fig. 1. (a): Bee flight muscle at resting length. Section contains single layer of filaments. Connecting filaments interconnect ends of thick filaments with Z-line. (b): Bee flight muscle stretched in rigor. Thin filaments that were broken along the Z-line remained in the A-band because of strong rigor links. The connecting filaments, attached between the Z-line and the end of the thick filaments, were elongated by the stretch, and filled the gap between the A-band and the Z-line. (c): Antibodies against connecting filament protein (projectin) specifically labelled the connecting filaments in the gap. The Z-line was not labelled. Bar, 0.5μm.

intact, the I-band shortened the same way as in Fig. 2b,c. In half-sarcomeres in which the connecting filaments were broken as a result of the stretch, the I-band remained elongated. The broken connecting filaments retract to a site near the Z-line (Fig. 2d, arrows). Note that the retracted filaments did not reach the Z-line.

According to Fig. 2, the considerable recoil force of the connecting filament is responsible for the high resting tension found in insect flight muscle, and can return the sarcomere to rest length after stretch.

Connecting Filaments in Vertebrate Striated Muscle

The discovery of titin[1][2], a giant structural protein, led eventually to the conclusion that the sarcomere contained, in addition to the thick and thin filaments, a third longitudinal filament system composed of titin. Titin epitopes have been demonstrated to change their position in the I-band as a function of the I-band width. For this reason, it has been hypothesized that titin forms an extensible filament system[1][7][11][12]. But individual titin filaments have never been visualized in the physiologically intact sarcomere; the only visible part of the extensible filament

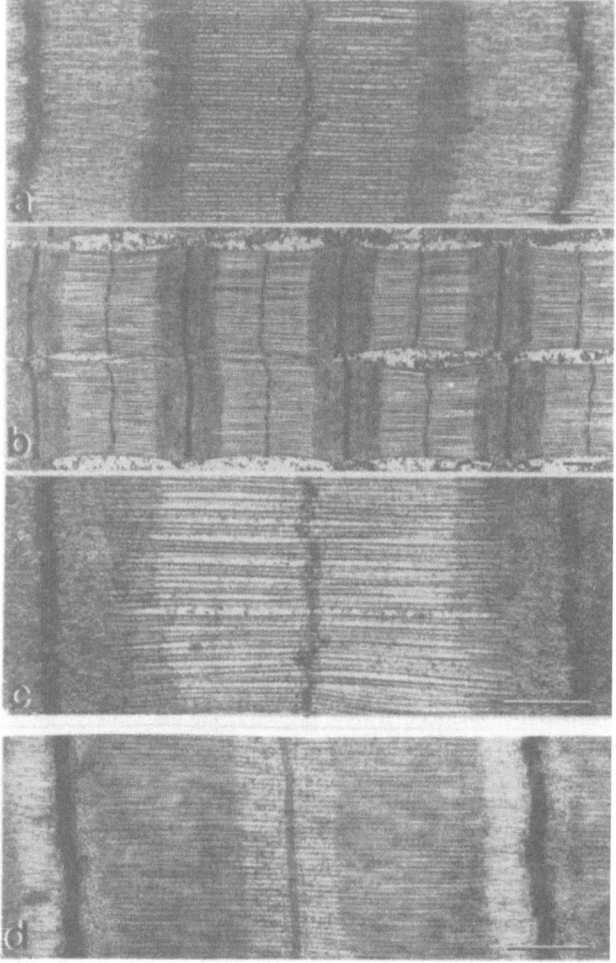

Fig. 2. Rigor release experiment: (a): Elongated sarcomere from bee flight muscle fiber, stretched in the relaxed state. The H-zone is twice as long as the I-band. (b): Stretched muscle released in rigor. The restoring force of the connecting filaments caused the I-band to shorten, thereby crumpling the thin filaments along the Z-line. (c): Higher magnification image of sarcomere in (b). (d): Bee flight muscle, stretched abruptly in the relaxed state, and released in rigor. Where the connecting filaments remained intact, the I-band disappeared. Where connecting filaments had broken, the I-band remained extended. Bar, 0.5µm.

system is the gap filament—the region of the connecting filament between thin filament tips and A-band—seen in highly stretched sarcomeres. Therefore, identification of the gap-filament protein is the key needed to unlock the existence of an independent elastic filament system: if gap filaments could be labelled with an antibody against one of the titin epitopes ordinarily present in the I-band of unstretched muscle, then one could say that the gap filament is a segment of the connecting filament. Although a few reports have been presented about the titin

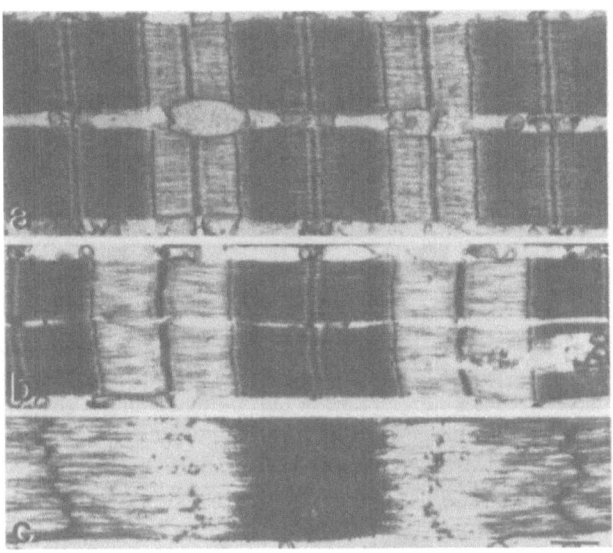

Fig. 3. Frog semitendinosus muscle labelled with monoclonal anti-titin (T-11, Sigma). (a): At resting sarcomere length the titin epitopes are found in the I-band near the A-I junction. (b): Stretched sarcomere. The epitopes remain in the same position relative to the A-band. (c): Very highly stretched sarcomere (sarcomere length 5.4 μm). Titin epitopes are found in the middle of the gap. Bar, 0.5μm.

content of the gap filament[1)12)21)], experiments of the kind mentioned above have not yielded conclusive results.

In recent experiments using monoclonal anti-titin[7)] (T-11 Sigma), we succeeded in demonstrating that one of the titin epitopes normally present in the unstretched or moderately stretched I-band (Fig 3a,b) labels the gap filaments in a regular manner[22)] (Fig. 3c). This finding proves that the gap filament in vertebrate muscle is the overstretched segment of the connecting filament. Furthermore, the antibody deposits separate A-band and I-band titin domains in the gap, showing that titin's A-band domain, which is functionally inextensible at physiological sarcomere length, is intrinsically elastic[12)13)23)].

When overstretched rigor muscle was released, the connecting filaments shortened the sarcomeres, thereby eliminating the gap. But the thin filaments were not crumpled as they were in the insect flight muscle; sarcomeres remained 3.6 μm long. Similar results were reported using relaxed muscle[24)]. It seems either that the thin filaments are stiffer and/or the connecting filaments are less stiff than in insect flight muscle. Similar results were achieved when the rigor-release experiment was carried out on stretched muscle that had a small overlap zone. The connecting filaments failed to cause sarcomere shortening.

Because these experiments provided no direct evidence concerning the connecting filament's spring-like nature at physiological sarcomere length, we pursued a second approach. We developed the "freeze-break" method[25)]. Small

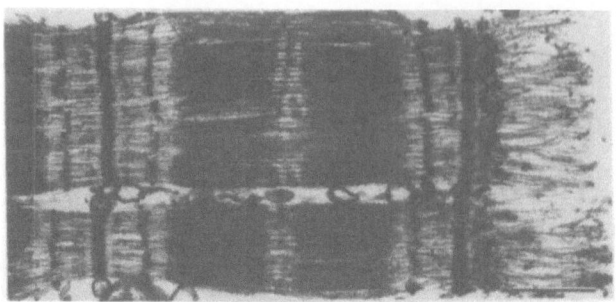

Fig. 4. Freeze-break experiment: The broken fiber was labelled with monoclonal antibody to titin (RT-13). In intact sarcomeres, the titin epitopes are symmetrical relative to the Z-line. In regions where the sarcomeres were broken at the A-I junction, the epitopes are no longer symmetrical. The broken thin filaments remained in their original position, but the titin epitope in the broken half-sarcomere retracted independently of the thin filaments, to very near the Z-line. The retracted titin epitope has not reached the Z-line; retraction stops at the N_1-line level. Bar, 0.5μm.

bundles of freshly prepared rabbit psoas muscle fibers were quickly frozen and broken under liquid nitrogen to fracture sarcomeres in planes perpendicular to the filament axis. The still-frozen specimens were thawed during fixation to allow elastic filaments to retract. The broken specimens were then labelled with monoclonal anti-titin antibodies (RT 13) against an unique epitope in the I-band. In the broken sarcomeres, the connecting filaments retracted independently of the thin filaments, forming a dense band just near the Z-line (Fig. 4). The antibodies labelled

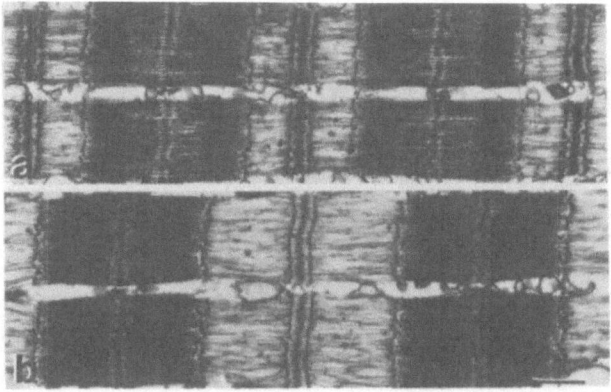

Fig. 5. Frog tibialis anterior muscle labelled with anti-titin antibodies (T-11 and T-12). T-11 epitope is located near the A-I junction, T-12 epitope near the N_1-line. Although the distance between the two T-11 epitopes is twice as great in stretched sarcomeres (b) as in short sarcomeres (a), showing that the length of the connecting filaments is twice as great in the elongated sarcomeres as in short sarcomeres, the T-12 epitope positions did not change relative to the Z-line. Bar, 0.5μm.

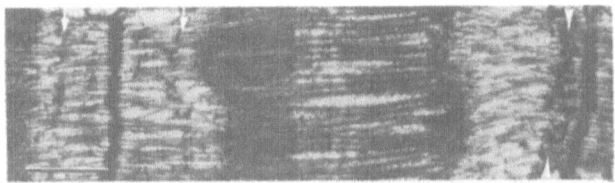

Fig. 6. Highly stretched muscle (sarcomere length 3.4 μm) released to near resting length. Released muscle was labelled with anti-titin, as above. Generally, titin filaments remained intact (arrows denote epitopes). In some instances, the titin-filament set was broken, and the titin filaments retracted (arrowhead). Where the titin filaments had broken, the half-sarcomere remained elongated, independently of the release. Bar, 0.5μm.

the retracted filaments, confirming the titin composition of these filaments. The retracted filament apparently never reached the Z-line; retraction stopped at the N_1-line. It appears that the connecting filaments (composed of titin) are independent of the thin filaments, and that they have highly elastic spring-like properties, just as in insect flight muscle. Furthermore, they seem either to associate firmly with the thin filaments only at the N_1-line level, or to have an inelastic domain near the Z-line[24].

To distinguish between these two options, frog semitendinosus fibers were labelled with monoclonal antibodies against a titin epitope near the N_1-line[7][15] (T-12, Boehringer), and another one near the A-I junction in the I-band[7][15] (T-11, Sigma). Before labelling, fibers were pre-stretched to varying extents. Fig. 5 shows that although the distance between the T-11 epitope and the Z-line increased with increasing sarcomere length (namely, the connecting filaments were elongated), the T-12 epitope remained in the same position. Since in cross section, we could not find extra filaments between the N_1-line and the Z-line over and above "thin" filaments, it seems that the connecting filaments are firmly associated in this region with the thin filaments.

In immunoelectron microscopic experiments, we could obtain data concerning the physiological significance of the connecting filaments. When the stretch was performed quickly, as the sarcomere length approached 3.6 μm the connecting filaments apparently broke and retracted in some half-sarcomeres. Anti-titin labelling revealed a density similar to the one seen with the freeze-break procedure. When such stretched muscle was released, the intact half-sarcomere returned approximately to its resting length. The half-sarcomere with broken connecting filaments did not return, however, but remained stretched[25] (Fig. 6). This confirms that it is the connecting filament that is responsible for returning the stretched sarcomere to its resting length.

CONCLUSION

Since titin has been found in skeletal and heart muscles of a wide range of vertebrate and invertebrate species[1], and mini-titin has been found in the insect-flight and leg muscles and other invertebrate species[5)6)], the connecting filament would seem to be a universal elastic element of the sarcomere. The common feature of the connecting filament is its physical connection between the end of the thick filament and the Z-line (or near the Z-line) as an independent, elastic element of the sarcomere.

Although the physiological role of the connecting filament is not yet fully revealed, we can draw some conclusions from these experimental results. As Fig. 4 implies, one of the important physiological roles of the connecting filaments is to return the sarcomere to its resting length.

A further implicative feature of the results is the degree of connecting filament retraction. Retraction occurs all the way to a point near the Z-line in both insect-flight and vertebrate muscle. The natural (unstrained) length of the elastic connecting filament is thus extremely short, and implies that even at resting sarcomere length, the (longer) connecting filament could be under tension. Thus, the connecting filament could exert tensile stress on the thick filament even in the rest-length sarcomere, balanced by some compressive force—perhaps a weak interaction between actin and myosin or a stiff additional component of the connecting filament[20)]. Thus, the thick-filament-centering action of the connecting filaments[10)] may exist not only in stretched sarcomeres, but in resting-length sarcomeres as well.

ACKNOWLEDGEMENTS

We thank Dr. Judy Saide for her generous donation of projectin antibodies, and Dr. Kuan Wang for his generosity in allowing us to make use of a titin antibody (RT-13) developed in his laboratory.

REFERENCES

1. Wang, K. in *Cell and Muscle Motility* (ed. Shay, J. W.) 312-369 (Plenum Press, New York, 1985).
2. Maruyama, K. *Int. Rev. Cytol.* **104**, 81-114 (1986).
3. Bullard, B., Hammond, K.A. & Luke, B.M. *J. Mol. Biol.* **115**, 417-440 (1977).
4. Saide, J.D. *J. Mol. Biol.* **153**, 661-679 (1981).
5. Nave, R.& Weber, K. *J. Cell Sci.* **95**, 535-544 (1990).
6. Hu, D.H., Matsuno, A., Terakado, K., Matsuura, T., Kimura, S. & Maruyama, K. *J. Muscle Res. Cell Motility* **11**, 497-511 (1990).
7. Fürst, D.O., Osborn, M., Nave, R.& Weber, K. *J. Cell Biol.* **106**, 1563-1572 (1988).
8. Yoshioka, T., Higuchi, H., Kimura, S., Ohashi, K., Umazume, Y. & Maruyama, K. *Biomed. Res.* **7**, 181-186 (1986).
9. Funatsu, T., Higuchi, H. & Ishiwata S. *J. Cell. Biol.* **110**, 53-62 (1990).
10. Horowits, R. & Podolsky, R.J. *J. Cell Biol.* **105**, 2217-2223 (1987).
11. Wang, K., Wright, J. & Ramirez-Mitchell, R. *J. Cell Biol.* **99**, 435a. (Abstract) (1984)..
12. Itoh, Y., Suzuki, T., Kimura, S., Ohashi, K., Higuchi, H., Sawada, H., Shimizu, T., Shibata M. & Maruyama K. *J. Biochem. (Tokyo)* **104**, 504-508 (1988).
13. Pierobon-Bormioli, S., Betto, R. & Salviati G. *J. Muscle Res. Cell Motility* **10**, 446-56 (1990).
14. Whiting, A., Wardale, J. & Trinick, J. *J. Mol. Biol.* **205**, 263-268 (1988).

15. Fürst, D.O., Nave, R., Osborn M. & Weber K. *J. Cell Sci.* **94**, 119-125 (1989).
16. Reedy, M.K. in *Contractility of Muscle Cell and Related Process* (ed. Podolsky, R. J.) 229-246 (Prentice-Hall, Inc., New York, 1971).
17. White, D.C.S. & Thorson, J. *Prog. Biophys.* **27**, 173-255.
18. Trombitás, K. & Tigyi-Sebes, A. in *Insect Flight Muscle* (ed. Tregear, R. T.) 79-90 (Elsevier, 1977).
19. Trombitás, K. & Pollack, G.H. in *Molecular Mechanism of Muscle Contraction* (eds. Sugi, H. and Pollack, G. H.) 17-30 (Plenum Press, New York, 1988).
20. Pollack, G.H. in *Muscles and Molecules*, 61-81 (Ebner and Sons, Seattle, 1990).
21. La Salle, F., Robson, R.M., Lusby, M.L., Parrish, F.C., Stromer, H.M. & Huiatt, T.W. *J. Cell Biol.* **97**, 258a (1983).
22. Trombitás, K., Baatsen, P.H.W.W., Kellermayer, M.S.Z. & Pollack, G.H. *J. Cell Sci.*, **100**, 809-814 (1992).
23. Wang, K. & Wright, J. *Biophys. J.* **53**, 25a (1988).
24. Maruyama, K., Matsuno, A., Higuchi, H., Shimaoka, S., Kimura S. & Shimizu T. *J. Muscle Res. Cell Motility* **10**, 350-9 (1989).
25. Trombitás, K., Pollack, G.H., Wright, J. & Wang, K. (submitted).

Discussion

Huxley: Do you know the papers by Lucy Brown, Hill, and myself? These concerned frog fibers that had been longitudinally compressed and seen in the state when they were wavy. But the sarcomere length at which individual fibers became wavy varied quite considerably. I don't know why, but there's certainly no sharp and uniform length at which passive shortening ceases.

Pollack: Is your point that the natural length of the titin filament is perhaps not so well defined?

Huxley: Not necessarily. It might be something quite different. It might be circumferential forces or osmotic forces affecting the length and acting against titin. I think there are all kinds of possibilities.

Pollack: Yes, I think so too.

Edman: When a fiber is shortened and below slack length, it returns automatically to rest length. Do you have any comment on that?

Pollack: Yes, I think what happens is that the thick filament shortens as the fiber shortens, and when the fiber "wishes" to return to its natural length, it is simply because the thick filament reverts to its natural length of 1.6 μm.

Edman: So you don't think that passive elements are involved?

Pollack: There may be passive elements. The evidence I have seen in the literature implies that during contraction, the thick filament shortens. The shortening can be considerable, especially at the very short length. So, I would guess that during the period of relaxation, the thick filament would simply return to its natural length by a mechanism that is not yet clear. There certainly could be other possibilities, like some sort of radial force that would perhaps compress the myofibril and prompt it to return to its natural length.

Edman: Or perhaps the sarcotubular system?

Pollack: Sure, that's another possibility.

STRUCTURAL STUDIES ON THE CONFORMATIONS OF MYOSIN

A.R. Faruqi, R.A. Cross* and J. Kendrick-Jones

MRC Laboratory of Molecular Biology
Hills Road
Cambridge CB2 2QH, U.K.
* Marie Curie Research Institute
The Chart, Oxted
Surrey RH8 OTL, U.K.

ABSTRACT

Myosin is the major motor protein found in vertebrate striated and smooth muscle and in non-muscle cells where, in association with actin, its main role is to convert chemical energy into mechanical work. Smooth muscle and non-muscle myosin adopts a number of different conformations : for example an unfolded (6S) form which is capable of forming filaments and generating force and a 'folded' (10S) form, which is most probably a storage form incapable of forming filaments. In the 10S form the products of ATP cleavage are trapped by the folded tails and the ATPase activity of myosin is greatly reduced. It is believed that a transition to the unfolded 6S form is necessary prior to filament formation. We report here on two relatively low resolution structural techniques for studying hydrated myosin. We have used a relatively recent development in microscopy, the scanning tunneling microscope, to image a series of biologically interesting specimen, mainly to evaluate the potential of the technique. There are significant potential advantages for imaging biological specimen with the STM as the imaging is done in air and the specimen can be imaged without a metal coating. Our experience with imaging myosin suggests that good images of hydrated myosin can be obtained but with poor reproducibilty. We have also carried out small angle solution x-ray scattering studies on the two myosin conformations to explore the possibilities of doing kinetic measurements on the transition between the two states. Small angle scattering from the S1 fragment and re-constituted parts of the rod have also been carried out and the data is compared with expected scattering from model structures.

Mechanism of Myofilament Sliding in Muscle Contraction, Edited by
H. Sugi and G.H Pollack, Plenum Press, New York, 1993

81

INTRODUCTION

Myosin is the major motor protein found in all types of muscle and many non-muscle cells. The main role of myosin in muscle cells is to generate force, which it does by association with actin using ATP hydrolysis as the chemical fuel which is converted into mechanical work. Among the functions of myosin in non-muscle cells is in cytokinesis, cell motility and various aspects of cell regulation including cell signalling and in the control of cell shape[1].

Myosin has an extremely asymmetric structure consisting of two 'pear-shaped' heads, known as the subfragment 1 (S1 subunit), which contain the binding sites for both actin and for ATP. The heads are attached to a long rod-like tail, 1500 Å long and 20 Å in diameter, as shown in schematic form in Fig. 1[2]. There are, in addition, two light chains situated near the 'neck' region of the head which are involved in the regulation of certain types of muscle. According to a recently determined electron microscope structure of S1 crystals[3], there are three domains in the structure. The largest domain, consisting of about 60% of the S1 volume contains the actin binding site, the intermediate domain contains about 30% of the volume and the smallest

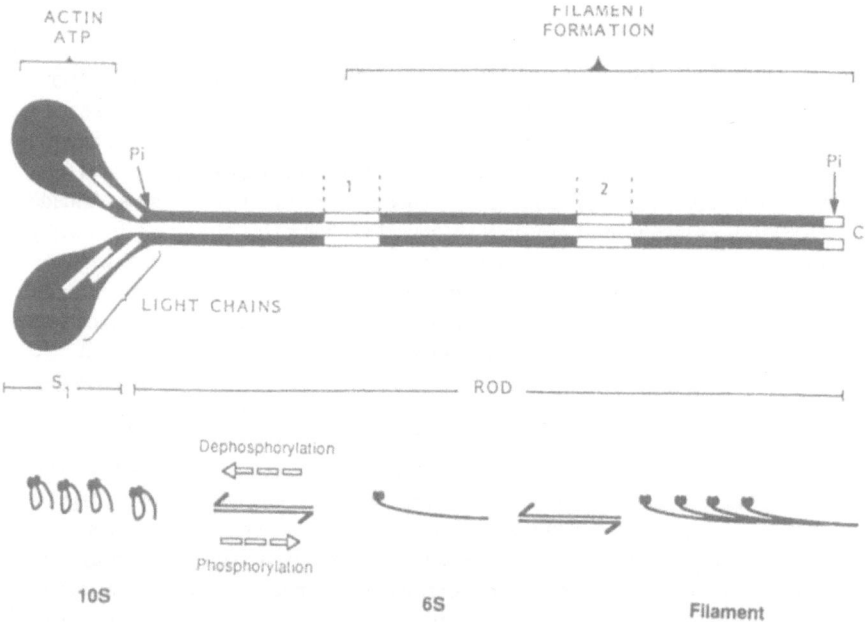

Fig. 1. Schematic diagram of the myosin molecule showing the highly asymmetric structure consisting of two globular heads, containing the binding sites for both actin and ATP, attached to a long rod-like tail. The rod is involved in filament assembly and the two regions are shown, where the rod bends to form the loop for the 'folded' 10S form of the monomer. Although the location of the light chains is not known precisely, the site for light chain phosphorylation, which leads to the 10S to 6S transition is believed to be near the 'neck' region of the S1[2].

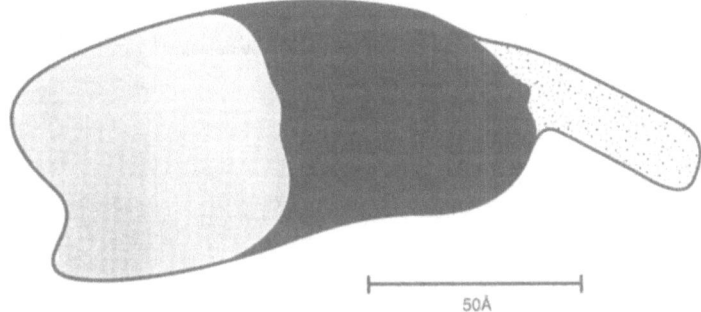

50Å

Fig. 2. Model of the myosin S1 approximated from a recent structure determination using electron microscopy of thin sectioned S1 crystals[3]. The structure consists of three unequal domains: a large domain with about 60% of the total S1 volume which also contains the actin binding site, a central domain with about 30% of the volume and a short stalk-like domain with about 10% of the volume. The ATPase site is probably at the junction between the large and central domains.

domain about 10%. An approximation to the model is shown in Fig. 2 where the three domains are shaded differently for convenience; this structure has been used in the modelling of the S1 experimental data. Although myosin containing filaments in striated muscle are stable structures this does not appear to be true for non-muscle cells where myosin filaments are probably only assembled when they are needed. There are at least two conformations of myosin: a 'straight', unfolded form of the myosin monomer, capable of forming filaments and generating force in association with actin filaments which has a sedimentation coefficient of 6S and a 'folded' form in which the rod part appears to loop around and attach to the head; the folded form of myosin monomer has a sedimentation coefficient of 10S[4-6]. The 10S form contains the product of ATP hydrolysis on the heads. An increased calcium level, which leads to phosphorylation of the light chain by a calmodulin dependant light chain kinase, appears to be the cellular trigger for force generation and leads to a myosin transition from 10S to 6S and filament formation; the pathway is shown partly in Fig. 1. Lowering of the calcium levels result in dephosphorylation of the myosin light chains and filament disassembly.

Solution X-ray Scattering Studies on Myosin

Solution x-ray scattering has been carried out on unfolded myosin molecule, e.g. the subfragment 1 (S1), the rod part (LMM) and smaller 'expressed' portions of the rod. Although it is possible to get only relatively low resolution structural information, the specimen is kept in a solution at the appropriate ionic strength and it is possible to study changes in molecular conformation in response to chemical stimuli[7]. Our main interest in this technique has been focused on obtaining scattering 'signatures' from both the unfolded (6S) and the folded (10S) monomers(6S), folded myosin monomers (10S) and on various parts of the

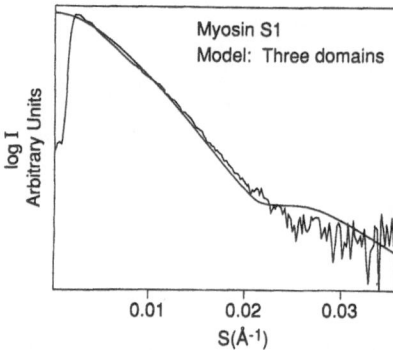

Fig. 3. Comparison of scattering by myosin S1 fragment with scattering from the 'model' S1 shown in Fig. 2. Data extends from s =. 0028 to .037 and yields a radius of gyration of 37 Å, calculated from the inner part of the scattering curve. The model curve is smooth.

conformations of the myosin monomer to assess the feasibility of monitoring dynamic transitions between the 10S and 6S states which lead, in the next step, to filament formation[8]. Preliminary results on scattering from the 6S/10S conformations have been reported previously[7,9].

The small angle scattering measurements were made on a double (point) focusing camera using two mirrors set on a rotating anode x-ray generator with a multiwire area detector to record the scattering pattern[9][10]. After circular averaging patterns were converted to a format suitable as input to a general purpose data reduction and analysis program, OTOKO[11], which was used for plotting and part of the analysis of the data, e.g. radius of gyration, etc. Scattering from 'model' structures were generated using ATOMIN[12], a program which uses Debye's formula for computing the scattering from a structure of arbitrary shape and size. All scattering curves are plotted as relative intensities with the model curves normalised over the whole scattering range accessible to the instrument(with the camera geometry used), which is from s =.0028 to s =.037. The small angle limit is imposed by camera resolution and the upper limit by the size of the detector. All the scattering data presented spans this range.

Myosin S1 with intact light chains(molecular weight 128 kDa) was prepared by papain digestion of chicken skeletal muscle myosin[13]. The scattering curve from the myosin S1 fragment after buffer subtraction, and plotted on an arbitrary intensity scale is shown in Fig. 3. Data was summed from several separate experiments and from two concentrations: 16 mg/ml and 32 mg/ml, to obtain adequate counts in the outer region. The inner part of the scattering pattern, i.e. the low angle region reflects the large scale structure of the S1, giving a radius of gyration of 42.5Å, in agreement with other recent measurements[14]. The outer region of the scattering curve corresponds to the finer features in the structure of the S1; the agreement between data and model is not as good as for the inner part, perhaps due to the fact that though the model assumes a 'hard' and well defined structure, in reality it may be more 'fuzzy' and may have an ill-defined boundary. The scattering from a

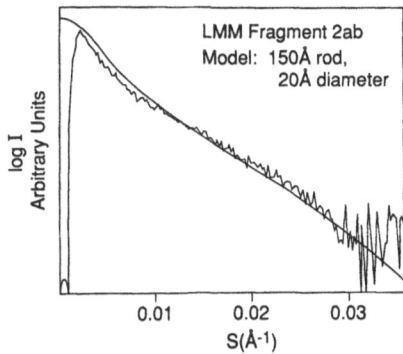

Fig. 4. Comparison of scattering by part of the rod fragment with a model scattering from a rod 150 Å long and 20 Å in diameter. The model curve is smooth.

'model' structure, closely based on recently published electron microscope data[3] of S1 crystals is also shown in Fig. 3 for comparison with the S1 data. Scattering data has also been obtained from S1 in the presence of ATP in order to detect the presence of conformational changes within the domains[18], if any, on binding ATP. The model S1 curve does not fit the data very well, though there are suggestions that the 'rod-like' component may have increased somewhat. However, it has not been possible so far to model the three-domain structure with a plausible new conformation to give a good fit to the experimental data.

Scattering from the straight tail form of myosin (6S), is shown in Fig. 4. The myosin was purified from scallop muscle and the scattering was obtained from a specimen with a concentration of 13.8 mg/ml in a buffer consisting of 0.6 M NaCl, 2mM $MgCl_2$, 10 Mm NaPi and 1 mM DTT at pH 7.0. There is good agreement with a model which assumes a 1500 Å rod with a diameter of 20 Å and a pair of heads 160 Å long with an opening of 35 degrees between the long axes of the heads. The model curves are sensitive to the opening angle between the heads and the agreement between data and model is much worse for heads at 180 degrees.

Scattering from bacterially expressed rod fragments[16] from a non-muscle myosin are shown in Fig. 5. The length of the fragment (from the genetic sequence) is expected to be about 150 Å and the data fits very well with a model scattering curve from a 150 Å rod with 20 Å diameter. If the reconstituted object were not in the form of a rod, or if the diameter of the rod were different, it would be evident from the data.

Scanning Tunneling Microscopy

We have used the scanning tunneling microscope (abbreviated as STM), in collaboration with J. Finch (MRC) and M. Walls (Cavendish Laboratory, Cambridge), primarily to assess the feasibilty of using it as an imaging device in structural and cell biology. The technique of STM was started by the work of Binnig

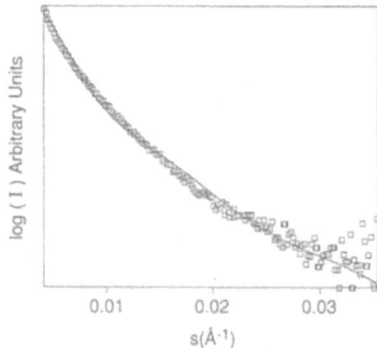

Fig. 5. Comparison of scattering from the unfolded (6S) form of myosin monomers (obtained from scallop muscle; concentration:13.8 mg/ml) and a 6S model described in the text[7].

and Rohrer (many references to biological applications of STM and scanning force microscopy are contained in a recent review by Engel[17])), who produced the first microscope and demonstrated its power with high resolution images. The great advantage of this technique was apparent to those involved in the study of surface sciences immediately as the STM produces extremely high resolution surface topographs to a level which even showed individual atoms. The first images were obtained under ultra-high vacuum conditions but were soon follwed by images obtained in air and even under liquids.

The principle of operation of the STM is discussed briefly to highlight the extreme sensitivity of the device but also to bring out the drawbacks. The STM operates by measuring the tunneling current between two conductors (or semi-conductors) separated by a vacuum. Even though classical electrodynamics would not permit such a phenomena, according to quantum mechanics there is a small probability of the electrons leaking through the vacuum gap. If one of the conductors is made into a very fine tip the tunneling is confined to a very small area. Further, if the tip is scanned across the second conductor and if the current is kept constant by moving the tip vertically to keep the gap constant between the conductors, one can obtain a 'topograph' of the surface by plotting the vertical movements of the scanning tip. There is an acute dependance of the tunneling current on the gap; typically a 1 Å change in the tunneling gap affects the tunneling current by an order of magnitude.

The heart of the STM is a very sharp tip made of a heavy metal, such as platinum, which scans in a raster fashion very close to the surface of the specimen under observation.The scan is made in three dimensions with the help of a very sensitive piezo-electric translator, which can typically produce a 10 Å movement for 1 volt applied to the translator. The tip is kept within a few angstrom from the surface and the tunelling current is measured in the height servo and it is kept constant by varying the gap between the tip and the surface. As discussed above one can keep this distance constant to ~0.1 Å. An image is built up by making a series of scans which are usually displayed as topographs on a television monitor.

Scanning tunneling microscopy is a new form of imaging, specially for biological specimen and thus involves different procedures to those used in conventional electron microscopy. Some of these points are given below along with problems encountered based on our (limited) experience with the STM.

1. As mentioned earlier, STM imaging is carried out by controlling the tunneling current through the specimen. In order to obtain good spatial contrast it is essential to deposit the molecule on a surface which is flat to atomic dimensions otherwise the spatial height variations in the substrate are likely to mask the molecular 'shape'. Ideally one needs a crystal lattice plane and highly ordered pyrolytic graphite (HOPG), with a freshly cleaved surface, has been found to be a suitable substrate. The other requirements of the substrate, viz. electrical conductivity and mechanical rigidity are also satisfied by HOPG. An excellent side benefit of using HOPG is that it is a good test specimen and the satisfactory operation of the STM can be checked by imaging the carbon atoms in the lattice. There have been mixed results in imaging biological molecules deposited on a HOPG substrate for a number of reasons. Firstly, the adhesion of the molecules is not very strong to the substrate and this often results in a shift of the molecule between scans due to the pressure exerted by the scanning tip. This leads to non-reproducible scans and is one of the main hurdles which need to be overcome if STM imaging of biological molecules is to become a routine exercise. Secondly, imaging artifacts from the graphite surface need to be carefully eliminated as they can appear very similar to 'DNA' type structures. An alternative substrate, coated mica, also has drawbacks due to the size of the coating particle, usually 20-50 Å, which restricts resolution. Clearly, considerable effort is needed to develop specialised substrates for the specimen under study.

A possible application of the improved resolution STM could be to image the folded (10S) monomer. Electron microscopy of rotary shadowed 10S monomers demonstrate that it is folded at the two 'bend' regions in the tail(see Fig. l) and suggest that a portion of the tail region near the distal (from the S1) end may be stabilising the folded conformation[2]. Higher resolution images would be extremely useful in elucidating rod-head contacts. Another possible application of higher resolution imaging would be to find out whether phosphorylation/dephosphorylation of the light chains affect the S1 by locking or unlocking it in a given conformation.

2. The lateral resolution is related to the tip radius which should ideally be the radius of one or a few atoms, an ideal difficult to attain. More effort is needed in producing adequate tips.

3. The mechanism of contrast formation in STM imaging is not properly understood. The magnitide of the tunneling currents are such that one requires 10^{10} electrons/sec conducted through an insulator. There have been some possible explanations to explain this effect based on the modification of the substrate work function by the adsorbed molecule, and this might be satisfactory for relatively thin molecules like DNA (~25Å), but more difficult to imagine for much thicker molecules such as the S1 part of the myosin molecule.

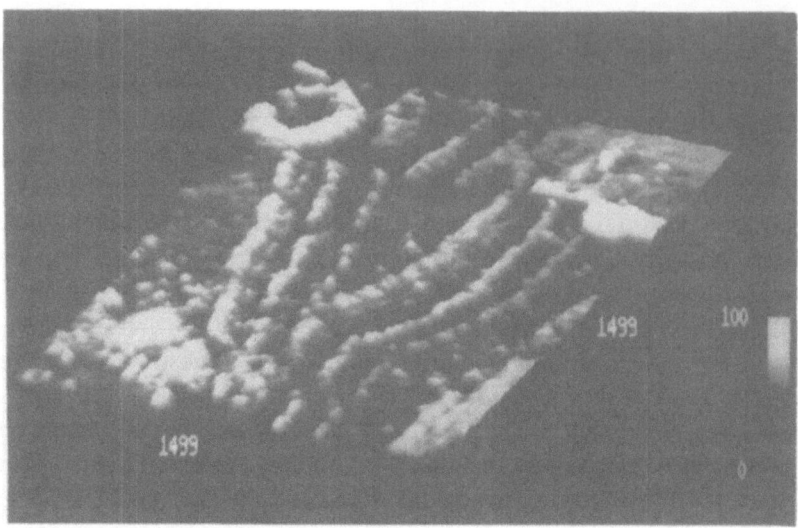

Fig. 6. Scanning Tunneling Microscope generated image of myosin monomers in the unfolded 6S state. Myosin from vertebrate smooth muscle at a concentration of 5 μg/ml was deposited on a HOPG substrate and dried briefly prior to imaging. The myosin monomer on the left side of the view shows up with good resolution. However, several artifacts are also present and these are discussed in the text.

We have imaged the 6S (unfolded) form of vertebrate smooth muscle myosin (concentration 5 μg/ml) deposited on HOPG, after drying in air for a few minutes, i.e. still partially hydrated. One of the great attractions of myosin as a test specimen is that because of the very distinctive and asymmetric shape, there is less likelihood of scanning artifacts simulating the molecule. An image, obtained with a WA Technology instrument, is shown in Fig. 6. The image has several molecules of myosin but only one appears to have both its heads intact, and can be seen on the left hand side. The dimensions of the molecule are consistent with accepted values; i.e. the length is approximately 1500 Å and the S1 heads are between 150 and 200 Å. The thickness of the molecule is less than expected but this appears to be a general problem with STM measurements. Such preliminary results, on myosin and on other molecules of biological interest, are very encouraging, but suggest that if further effort in the following areas can be invested, it might lead to dramatic improvements in the technique.

1. Specimen Preparation. Much better adherence is required between the molecule being imaged and the substrate to prevent shifts in the position and to allow repeated scans of the same molecule. Aggregation of molecules makes it difficult to image individual molecules and it would be very desirable to prevent it from happening.

2. The mechanism of contrast formation is ill-understood, specially for 'large' objects which are insulators. It is important for understanding this phenomena in the hope that suggestions for improvements will emerge from this study.

3. It is very difficult to change the area of scan from a large field of view to a small one; this technical facility would be enormously helpful in imaging a molecule at high resolution after initial location in a low magnification field of view.

4. As stated earlier, STM is a relatively new and thus untried technique. It is very important for the success of the technique to gain as wide an experience in imaging different molecules as possible and compare them with their 'accepted' structure to find out if there are any systematic differences.

REFERENCES

1. Warrick, H.M. & Spudich, J.A. *Ann. Rev. Cell Biol.* **3**, 379-421 (1987)
2. Citi, S. & Kendrick-Jones, J. *Bioassays*, **7**, 155-159 (1987)
3. Winkelman, D.A., Baker, T.S. & Rayment, I. *J. Cell Biol.* **114**, 701-713 (1991)
4. Onishi, H. & Wakabayashi, T. *J. Biochem.* **92**, 871-879 (1982)
5. Cross, R.A., Cross, K.E. & Sobieszek, A. *EMBO J.* **5**, 2637-2641 (1986)
6. Trybus, K.M. *Curr. Opin. Cell Biol.* **3**, 105-111 (1991)
7. Faruqi, A.R., Cross, R.A. & Kendrick-Jones, J. *J. Cell Sci. Supplement* **14**, 23-26 (1991)
8. Cross, R.A. *J. Muscle Res. Cell Motility* **9**, 108-110(1988)
9. Faruqi, A.R., Cross, R.A. & Kendrick-Jones, J. *J. Appl. Cryst.* **24**, 852-856 (1991)
10. Faruqi, A.R. & Andrews, H. *Nucl. Instr. Meth.* **A283**, 445-447 (1989)
11. Koch, M.H.J. & Bendall, P. *Proc. Dig. Equip. Users Soc.* 13-16(1981)
12. Bordas, J. Mant, G. & Nave, C. *"ATOMIN"* Unpublished Modelling Program, Daresbury Laboratory.
13. Margossian, S.S. & Lowey, S. *Methods Enzymol.* **85**, 55-71 (1982)
14. Gorrigos, M. & Vachette, P. *Biophys. J.*,**55**, 80a (1989)
15. Winkelman, D.A., Mckeel, H. & Rayment, I. *J. Mol. Biol* **181**, 487-501 (1985)
16. Cross, R.A., Hodge, T.P. & Kendrick-Jones, J. *J. Cell Sci. Supplement* **14**, 17-21 (1991)
17. Engel, A. *Ann. Rev. Biophys. Biophys. Chem.* **20**, 79-108 (1991)
18. Huxley, H.E. & Kress, M. *J. Muscle Res. Cell Motility* **6**, 153-161 (1985)

Discussion

Gillis: Have you tried coating your supporting EM grid with polylysine so your specimens stick better?

Faruqi: No, we haven't tried polylysine.

Gillis: I suppose your myosin molecules are in a salt solution and not in pure water. Are the ions of the solution the electrical conductors?

Faruqi: They could be.

Reedy: Recently, there was a serious re-examination of DNA images on graphite, because it was found that the control preparations contained many DNA images (Clennen, C.R. & Beebe, T.P., Jr. *Science* **251**, 640-642, 1991). Did you look at empty graphite to see if you could find your specimen?

Faruqi: Yes. It is easy to find DNA on graphite, whether you put it there or not. Actually, we used myosin to evaluate the technique for the simple reason that it is asymmetric, and it is difficult even for graphite to simulate myosin. But it's true, the field is rich with artifacts.

Reedy: But you are saying you do not find myosin-like artifacts?

Faruqi: Not whole myosin-like artifacts. Nevertheless, you can certainly find blobs that may look like heads, or bits that look like rods.

Brenner: Did you try imaging with the atomic force microscope? That might get you around some of the problems you have here.

Faruqi: Yes, but you have different problems. The atomic force microscope certainly does not need conducting molecules; however, its resolution is not as good as that of the tunneling microscope, so you may find that the molecule is just a blur. The force-microscope pictures of F-actin I have seen are not quite as impressive as the tunneling microscope ones. But you are right, of course. If the resolution could be improved, it would be the obvious solution. The AFM certainly has great potential for imaging molecular assemblies.

Holmes: May I ask a leading question? Given that vitreous ice microscopy is so good and so quantitative, why do you bother?

Faruqi: Because the specimen is in solution and can be kept at near-physiological conditions during imaging.

Holmes: Something that is frozen fast is essentially in solution as well.

Faruqi: Well, this new technique belongs to a subset of a more general technique known as "scanned probe microscopy," and it might work. It might produce better results; we don't know, as yet.

Holmes: Well, it can't give you an electron density, because it only looks at the surface. You don't know what it's telling you because it's telling you about some conductivity you don't understand.

Faruqi: When you start with a new technique, I'm afraid you either have to pursue it to understand it better or give up. As I mentioned in my talk, it is certainly essential to improve our understanding of contrast in the image.

Holmes: Yes, it is worth choosing a technique that tells you something you want to know.

Brenner: I think at least it would help—in reference to the comments of Manuel Morales—not to go down to very low temperatures where you don't know what is happening to the proteins.

Holmes: Yes, but if you freeze fast, I don't think that is a real problem either. Dr. Faruqi, when you did this modelling—when you took the Winkelman shape and plotted it against your low-angle scattering data—how sensitive was it? If you had done it as a Guinier plot, you would have seen that the low-angle region is insensitive to shape. How much chance does one have of seeing something?

Faruqi: The model we used was published about three or four weeks ago, and so I can't answer your question, because we haven't done anything else (Winkelman, D.A., Baker, T.S. & Rayment, I. *J. Cell Biol.* **114**, 701-713, 1991). We haven't changed the model to see how sensitive it is compared to our data.

Holmes: Yes, but how far do your data deviate from a Gaussian curve? If it were a Gaussian curve, you have no information about shape. The information lies in the difference between what you observe and a Gaussian curve.

Faruqi: The modelling we did was done by filling the structure of the "Winkelman S-1" with 10 Å spheres and using the Debye scattering formula, written into a program by E. Mandelkow, and C. Nave et al. and obtained from Daresbury Laboratory. The data and model extend to ca. 25 Å, well beyond the "Guinier" region.

Tregear: I would be interested to know from Dr. Faruqi what the potential resolution of the STM is. Is this a technique that is potentially going to give enormous resolution? Is that the answer to Ken's previous criticism?

Faruqi: It is not really clear what it will be able to produce, because so little has been done. However, despite the paper that Mike Reedy referred to, which considers only artifacts produced with the STM, there are some DNA pictures that have come out of Cal Tech, for instance, that show base pairs with only a modicum of image processing. That is a special case, because DNA is a very thin molecule. It is only 20 Å. The sort of resolution they can get from it is tremendous. I don't think I have seen anything like that from any other type of microscopy. Whether that will be the case for a much thicker molecule like the S-1 is an open question.

Tregear: So, there is the potential that you might not have to use crystallography in order to get atomic resolution.

Faruqi: No, that's not true. STM can only give you topographs of molecules. It's just a surface technique. It will never tell you anything about molecular connections or anything that lies below the surface. So, no, you will always have to do crystallography, I'm afraid.

II. REGULATORY MECHANISMS OF CONTRACTION

INTRODUCTION

The contraction-relaxation cycle in various kinds of muscle is primarily regulated by the change in the intracellular Ca^{2+} concentration, but is also influenced by a number of factors other than calcium. The mechanism of muscle contraction can be approached from studies on how the actin-myosin interaction is switched on and off or modified by these factors in various kinds of muscle.

Ashley and others used the technique of laser flash photolysis of caged ATP and caged Ca^{2+} Chelator (Diazo-2) to study how the rate of contraction and relaxation is regulated in frog muscle fibers (actin regulated) and scallop muscle (myosin regulated). Gergely, Grabareck and Tao investigated conformational changes of TnC associated with Ca^{2+}-induced triggering of muscle contraction using mutants of TnC and presented evidence for movement of helical segments upon Ca^{2+}-binding to TnC. Babu, Su and Gulati studied the mechanism of Ca^{2+}-binding in the EF-hand in TnC by its genetic engineering with interesting results.

To study the weak actin-myosin binding state during muscle contraction, Barnett and Schoenberg modified the cross-bridges with pPDM or NPM in muscle fibers, and showed almost no Ca^{2+}-sensitivity of their weak binding to actin. Hou, Johnson and Rall studied the role of parvalbumin in relaxation of frog skeletal muscle with the conclusion that parvalbumin facilitates relaxation in the 0 to 20°C temperature range. Winegrad made histochemical measurement of Ca- and actin-activated myosin ATPase in cryostatic sections of rapidly frozen rat ventricular traveculae, and the results obtained were consistent with the production of a regulatory factor within the fiber bundle. Maughan and coworkers made interesting studies on the effect of deficiency of tropomyosin in the flight muscle of *Drosopila* by genetic cross of a flightless mutant with wild type flies. Both flight ability and wing beat frequency were found to be dependent on the gene dosage. The ultrastructure and mechanochemistry of isolated fibers were also affected by the gene dosage.

Rüegg and others examined the effect of peptides derived from the sequence of S-1 domain of myosin heavy chain on the regulation of cardiac contractility, and found that the peptides derived from the sequence around SH thiol groups (cyc 707) have a "Ca-sensitizing" effect on demembranated ventricular fiber bundles. The use of such peptides (peptide mimetics) seem very promising in the future research work. Gordon and Rigway summarize the experimental results supporting the following idea. On muscle activation, Ca^{2+} first binds to TnC in the thin filament to initiate strong myosin binding to the thin filament. Then, the strong myosin binding in turn causes additional activation either by increasing Ca^{2+}-binding to TnC or by changing the thin filament structure.

Pfitzer, Fischer and Chalovich studied the role of caldesmon in contraction of chicken gizzard smooth muscle, suggesting that it modulates contraction but has no effect on passive tension maintenance. Finally, Gailly, Gillis and Capony examined the effect of brevin, which is known to sever actin filaments into pieces in a Ca^{2+}-dependent manner, on the mechanical properties of skinned muscle fibers isolated from guinea-pig taenia coli, and found that it accelerates unloaded shortening velocity probably by affecting the cytoplasm viscosity.

ACTIVATION AND RELAXATION MECHANISMS IN SINGLE MUSCLE FIBRES

C.C. Ashley, T.J. Lea, I.P. Mulligan, R.E. Palmer and S.J. Simnett

University Laboratory of Physiology
Parks Road, Oxford, OX1 3PT. England

ABSTRACT

The effect of Ca^{2+} on the time course of force generation in frog skinned muscle fibres has been investigated using laser flash photolysis of the caged-calcium, either nitr-5 or DM-Nitrophen. Gradations in the rate and extent of contraction could be achieved by changing the energy of the laser pulse, which varied the amount of caged Ca^{2+} photolysed and hence the amount of calcium released. The half-time for force development at 12°C was noticeably calcium-sensitive when small amounts of calcium were released (low energy pulses) but did not change appreciably for calcium releases which produced a final tension of more than 50% of the maximal tension at pCa 4.5. This result is unlikely to be due to calcium binding to the regulatory sites of troponin C when on the thin filament, as this process is considered rapid (k_{on} 10^8 M^{-1} s^{-1}, k_{off} 100 s^{-1}). Our experimental results show that force develops relatively rapidly at intermediate Ca^{2+} which produce only partial activation (i.e. 50% Pmax or greater). This would not be the case if the affinity of the regulatory sites changes slowly with crossbridge attachment. The kinetics of calcium exchange with the regulatory sites may be much more rapid than crossbridge cycling, so that if calcium binding to a particular functional unit induces crossbridge attachment and force production, the force producing state may be maintained long after calcium has dissociated from that particular functional unit. The relaxation of skinned muscle fibres has also been successfully studied following the rapid uptake of Ca^{2+} by a photolabile chelator Diazo-2, a photolabile derivative of BAPTA, which is rapidly (> 2000 s^{-1}) converted from a chelator of low Ca^{2+} affinity (K_d 2.2 μM) to a high affinity chelator (K_d 0.073 μM). We have used single skinned muscle fibres from both frog (actin regulated) and scallop striated muscle (myosin regulated), to study the time course of muscle relaxation. This procedure has enabled us to examine the effects of the intracellular metabolites, ADP, P_i and H^+ upon the rate of relaxation. Single skinned muscle fibres from the semitendinosis muscle of the frog *Rana temporaria* to relax with a mean half-time of 56.0 ± 4.1 ms (range 30-100 ms, n = 18) at 12°C, which is faster than the relaxation observed in the intact muscles (half-time 133 ms at 14°C) and similar to the rate of the fast phase of tension decay in intact single fibres (20 s^{-1} at 10°C). The presence of 6.3 mM free ADP led to an increase in the mean half time of relaxation to 123 ± 8 ms (n = 7).

Mechanism of Myofilament Sliding in Muscle Contraction, Edited by
H. Sugi and G.H Pollack, Plenum Press, New York, 1993

97

In frog a pH 6.5 led to a slowing of the fast phase of relaxation, and a pH 7.5 an increase in the rate of the fast phase, compared to pH 7.0.

INTRODUCTION

As the site of release of calcium in vertebrate skeletal muscle, the SR, is spatially separated by some 2-4 μm from its site of action on the contractile proteins, methods have been developed to bypass this diffusion step and permit rapid activation and deactivation of permeabilized muscle by the use of photolabile caged compounds and laser flash photolysis methods. This has given insight into the kinetics of thin filament activation processes independent of the influence of the sarcoplasmic reticulum (SR) and the newer methods are able to bypass the major diffusional delays inherent in previous activation methods[1-3].

Evidence for a Transient Free Ca^{2+} Change

It was not until 1967 that the aequorin molecule[4] was used within a biological system, a striated muscle fibre, to detect a transient free Ca^{2+} response as a result of electrical activation[5][6] and provided vital kinetic information for modelling[7]. This initial observation made with aequorin indicated that the relations between free calcium and force were not simple ones. Thus, the free calcium and force did not rise and fall concomitantly; there was an appreciable delay between the maximum of the aequorin light emission, and hence the free calcium in the cell and peak force output. This unexpected finding led to the immediate suggestion that the free calcium change within the muscle cell may be controlling the rate of force development, so that the reactions which produced force were rate controlling steps and were out of equilibrium with the free calcium concentration, at least following a brief electrical stimulation[7]. However the more detailed investigation of these reactions required the use of the skinned fibre preparation where the influence of the sarcoplasmic reticulum and surface membrane events have been removed. In addition, the details of the cuvette kinetics of the thin filament calcium binding proteins also needed to be known.

Kinetics of Ca^{2+} Binding to TnC.

In order to understand the detailed mechanism of how thin filament calcium activation leads to force production, knowledge of the apparent rate of calcium binding to TnC and displacement from TnC are important. Knowledge of the equilibrium-binding constants alone is insufficient. It has been possible to measure some of these rate constants by using the fluorescent probe dansylaziridine (DANZ). DANZ-labelled troponin C exhibits a strong fluorescence upon excitation at 350 nm, when calcium occupies the calcium-specific regulatory sites[8]; whilst calcium occupancy of the high-affinity sites causes a slight decrease in fluorescence. The change in fluorescence was measured when DANZ-labelled STnC or STn were rapidly mixed with Ca^{2+} [9]. Significantly both STn and isolated STnC both bound Ca^{2+} rapidly within the mixing time of the instrument (2-3 ms). When the converse experiment was performed that is either calcium-saturated STn or STnC were mixed rapidly with the chelator EGTA, the DANZ fluorescence intensity decreased in a quasi-exponential manner within about 40 ms for STn and even more rapidly for STnC. The half-time of the fluorescence decay suggested a rate constant (k_{off} Ca^{2+}) of approximately 23 s^{-1} for STn, > 300 s^{-1} for STnC at $20°C$ [10]. These measurements imply that calcium binding to these low affinity (Ca^{2+} specific (T)) sites is fast, approximately (k_{on} Ca^{2+}) $> 10^8 M^{-1}s^{-1}$ and that these two low-affinity sites are equivalent and independent in their Ca^{2+} binding [10]. The rate of calcium dissociation from the higher affinity Ca^{2+}-Mg^{2+} (P) sites is slower, (k_{off} Ca^{2+}) 3 s^{-1}. Although the calcium on-rate to these high-affinity sites is also fast in the absence of magnesium; in the presence of a physiological free Mg^{2+} concentration (1-3 mM [11]) these sites are largely occupied by magnesium and therefore the apparent rate of calcium binding is limited by the need to displace bound magnesium; a relatively slow process (k_{off} Mg^{2+}) 2 s^{-1} [12][13].

Thus these kinetic considerations [9] suggested that the low-affinity (T) calcium-binding sites are the major regulatory sites of STnC in skeletal muscle, as calcium binding to the high-affinity sites (P) (at physiological magnesium concentrations) would be too slow to account for the observed rate of force development during a twitch, although a more prolonged elevated free Ca^{2+}, characteristic of a tetanic response, may well displace bound Mg^{2+} from these sites as well as from the Ca^{2+}-Mg^{2+} sites of the soluble calcium-binding proteins (pavalbumins) present in millimolor concentrations in many muscle cells [12][13].

It is, therefore, likely that only the two low-affinity regulatory sites play the major role in contraction.

Skinned Fibre Experiments: Activation Kinetics

The development of skinned fibre from which the surface membrane has either been removed [14] or made highly permeable, whilst the contractile system remains intact, has proved a very important preparation in the study of contractile mechanisms [15][16]. This allows the internal milieu of the fibre to be readily controlled. However the use of this preparation to examine the kinetics of muscle contraction, in a way analogous to the stopped-flow or quenched-flow experiments conducted with isolated proteins in solution, is limited by the rate at which the concentration of materials within the muscle fibre can be altered.

A procedure for minimizing this diffusion-induced equilibration time for changes in Ca^{2+} concentration (pCa, where pCa = $-log_{10}[Ca^{2+}]$) within skinned muscle fibres was described by Ashley and Moisescu [1][2][17][18]: the 'pCa clamp' method. Using this technique Ashley and Moisescu [17][19] were able to study the relation between Ca^{2+}, steady-state force and the rate of force production. However these experiments were still undoubtedly heavily influenced by diffusion delays

Table 1.

	Ca^{2+}			Mg^{2+}		
	K_d (μM)	k_{on} (M^{-1} s^{-1})	k_{off} (s^{-1})	K_d (mM)	k_{on} (M^{-1} s^{-1})	k_{off} (s^{-1})
Unphotolysed nitr-5	0·27 [a]	2·2 × 10^8 [b]	60 [c]	6·0 [d]	4 × 10^4	240
Photolysed nitr-5	12·0 [e]	2·8 × 10^8	3300 [d]	6·0 [d]	4 × 10^4	240
TNC (T site) [f,g,h]	1·3	1·0 × 10^8	130	5·0	1 × 10^4	50

[a] Determined in the LR muscle solutions (0·2 M ionic strength) containing 3·0 mM-nitr-5 and different total Ca concentrations by measuring Ca^{2+} with the Ca microelectrode. [Bound Ca]/[free Ca^{2+}] was plotted against [bound Ca] in a Scatchard plot, in which the slope = $-K = 1/K_d$.

[b] $k_{on} = k_{off}/K_d$.

[c] As for BAPTA (Tsien, 1980), since BAPTA is the parent Ca^{2+} chelator.

[d] Adams, Kao, Grynkiewicz, Minta & Tsien (1988).

[e] Mean of 6 μM at 0·1 M ionic strength and 18 μM at 0·3 M ionic strength (Tsien & Zucker, 1986).

[f] Gillis, Thomason, Lefevre & Kretsinger (1982).

[g] Robertson, Johnson & Potter (1981).

[h] The total concentration of TNC 'T' sites was taken as 140 μM, the values reported for vertebrate muscle (Gillis et al. 1982).

The model assumes nitr-5 is distributed homogeneously within the myofibrillar bundle, although there is no direct evidence for this. The myofibrillar space appears as accessible to unphotolysed nitr-5 as to EGTA (Ashley, 1983), if one compares the rate of force development in activating solutions containing either of the two chelators.

associated with the entry of both Ca-EGTA^{2-} and Ca^{2+} into the skinned muscle fibre[19)20)].

Caged Compounds

One important method of examining in detail the activation processes associated with skinned fibres, as well as circumventing the major problem of the diffusion delay, is to allow the compounds to diffuse into the fibres in an inactive or relatively inactive form and then to convert them to an active form in situ. Such a technique became possible with the introduction of photolabile 'caged' compounds[21)22)]. The first photolabile compound to be used extensively in muscle research was caged ATP devised by Kaplan and this has been used extensively by Goldman and colleagues to probe the kinetics of relations between force and the acto-myosin ATPase cycle[23)].

Caged Calcium

The experiments with caged ATP have demonstrated the way in which the technique of flash photolysis of caged compounds can be applied to study the reaction rates of processes. The application of similar techniques to the study of the activation of muscle by Ca^{2+} had to await the development of a suitable compound. This required the synthesis of a calcium chelator whose affinity for calcium would fall upon photolysis. Several such photosensitive chelators have now been synthesized[24-27)]; the nitr series devised by R.Y. Tsien is based upon the parent chelator BAPTA[28)] and retains its selectivity for Ca^{2+} over Mg^{2+} and its relative pH insensitivity, while the compound DM-nitrophen is based upon the chelator EDTA.

The first compound synthesized by Tsien and Adams, nitr-1, demonstrated the feasibility of this approach[24)25)]. Unfortunately this particular compound had a low

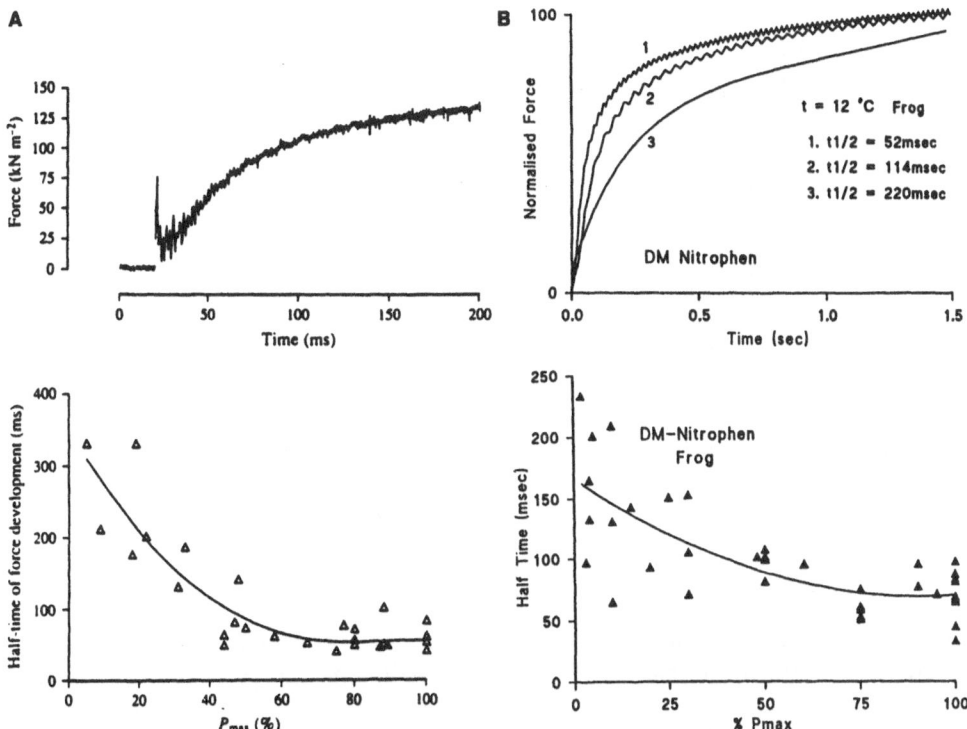

Fig. 1. Laser flash photolysis of nitr-5 in chemical skinned single fibres from frog semitindinosus muscle by a 50 ns pulse 1(A) or DM-nitrophen (1B) of light at 347 nm from a frequency doubled ruby laser. Tension transient (upper) from a single fibre showing force rising from zero to maximum following photolysis. Temp., 12°C; ionic strength, 0.2 M; pH, 7.0; SL = 2.1 μm, air flash, d = 100 μm. Results from contractions in single permeabilized fibres from frog semitendinosus muscle showing (lower) the half-time for force development plotted against the final force developed after photolysis (% P_{max}). Least squares third-degree polynomial regression line. (from Ashley et al., 1991; S.J. Simnett, I.P. Mulligan & C.C. Ashley, unpublished).

extinction coefficient at wavelengths above 300 nm and this limited its usefulness in biological experiments.

The compound, nitr-5, now meets however most of the criteria required for the successful use in both neurophysiological studies[29] as well as in studies of SR Ca^{2+} release and in muscle kinetics[3][30-35]. Its K_d prior to photolysis is 270 nM (t = 12°C, I = 0.2 M) (Table 1) which changes to 12 μM after photolysis. The quantum efficiency has been measured[25] as 0.012 for calcium-free nitr-5 and 0.035 for calcium-saturated nitr-5 (ε = 5500 M^{-1} cm^{-1} at 355-365 nm prior to photolysis). The only photolysis by-product is water, rather than the methanol produced on photolysis of nitr-2.

By rapidly (< 5 ms) increasing the free-calcium concentration within the myofibrillar space of skinned muscle (Fig. 1) using laser flash photolysis of the

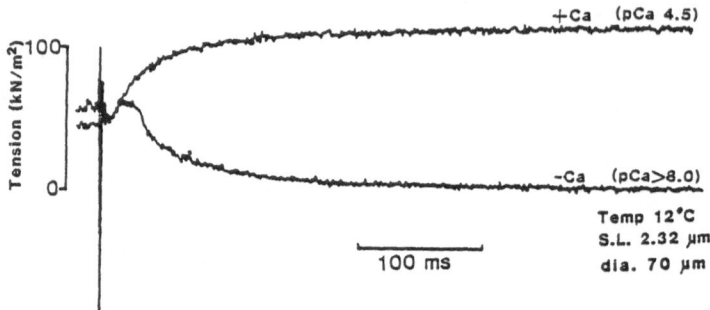

Fig. 2. Time course of force development in chemically skinned single frog fibre following photolysis of caged-ATP at different pCa's. Single skinned fibre from frog semitendinosus muscle at 12°C. The fibre is initially in rigor prior to the laser flash, which liberates approximately 1 mM Mg-ATP within the myofibrillar matrix. This causes the fibre to contract (upper trace) or to relax (lower trace) depending on the pCa, 4.5 and 8.0 respectively. (from Mulligan, I.P. & Ashley, C.C., unpublished)

photolabile calcium chelator, nitr-5 (Fig. 1A) or DM-nitrophen (Fig. 1B), the delay imposed by diffusion in earlier pCa-clamp experiments can be overcome. Single fibres from the semitendinosus muscle of the frog, *Rana temporaria*, permeabilized with Triton X-100 (1% w/v) could be fully activated using this technique. The half-time for force development was 40 ± 2 ms S.E.M. (n = 14) for nitr-5[30-33] at 12°C which is similar to that found in electrically-stimulated intact frog sartorius muscles (36 ms for tetanized frog muscle at 12°C[36]) and in single tibialis anterior fibres (36.9 ms at 12°C[37]) and is considerably faster than the half-time of force development seen in experiments using the pCa clamp method which at pCa 5.88 was 1.6 s^{-1} at 4°C[2].

Gradations in the rate and extent of contraction could also be achieved by changing the energy of the laser pulse which illuminated the fibre and which varied the amount of nitr-5 or DM-nitrophen photolysed and hence the amount of calcium released. These single skinned fibres could therefore be rapidly activated in a graded manner, a procedure which is not possible with intact electrically stimulated vertebrate muscle fibres and the kinetics of activation examined at different free Ca^{2+} concentrations. The half-time for force development at 12°C was noticeably calcium-sensitive when small amounts of calcium were released (low energy pulses) but did not change appreciably for calcium releases which produced a final tension of more than 50% of the maximal tension of pCa 4.5 (Fig. 1). This result is unlikely to be due to calcium binding to the regulatory sites of skeletal troponin C (STnC) when on the thin filament, as this process, as has already been discussed, is considered rapid (k_{on} 10^8 M^{-1} s^{-1}, k_{off} 100 s^{-1}; 20°C[3][38][39]) at least when STnC is measured in cuvette experiments. However the relation observed between %P_{max} and the half-time of force development may be best explained, assuming that shortening is not making a major contribution to events, by cooperativity in the rate constant for tension development, where this rate constant is dependent on the state of adjacent regulatory units (see later).

Comparison with the Rate of Tension Development in Experiments Using Caged-ATP: Double Cage Experiments.

When muscle fibres were incubated in 5 mM caged-ATP in the absence of other nucleotides, the muscle fibres were the rigor state. In the presence of calcium ions (pCa 4.5), laser pulse photolysis of caged-ATP resulted in an increase in the MgATP concentration to 1 mM or more at 50 s^{-1} and tension rose in an approximately exponential fashion with a rate constant of 25 ± 8 s^{-1} at 12°C (n = 3) using the same exponential fitting procedure as above, and on average tension rose more rapidly in caged-ATP experiments than in nitr-5 experiments (Fig. 1a and 2).

In experiments with cATP in the absence of Ca^{2+} (pCa > 8.0), tension first rose and then declined to its pre-photolysis level with a rate constant of 34 sec^{-1} at 12°C. This result suggests that protein cooperativity (S1 binding) can transiently switch on the thin filament and permit force development (Fig. 2).

In the presence of both cATP and caged calcium (nitr-5), the fibre, initially in rigor, upon photolysis produced a force transient which had a rise time indistinguishable from that of cATP alone (+ Ca^{2+}).

Monte Carlo Model: 3-state Model with Cooperativity and Enhanced Ca^{2+} binding

As suggested by Hill[40], the Monte Carlo technique (for reviews see 41-43) can be applied to models of muscle activation by calcium and provides the ability to consider explicitly the cooperative interactions between regulatory units along the thin filament. However it is essentially a statistical technique which only samples a relatively small number of the possible configurations of the model. It is therefore possible for the sample it takes to be unrepresentative of the general population of states. This may occur if it becomes trapped in a pseudo-steady state. These problems may be overcome to some extent by sampling large populations over long periods of time. We have developed a Monte Carlo model of the activation of muscle contraction by calcium from the equilibrium-binding model[40]. This model is composed of functional units each of which may be one of three states:

State	Calcium bound	Myosin bound
0	No	No
1	Yes	No
2	Yes/No	Yes

These functional units are arranged in linear chains and the state of each unit is influenced by the states of the two adjacent units. The stochastic history of each functional unit may be followed through a large number of transitions (see ref. 44). The model follows an ensemble of 3000 regulatory units, grouped into 100 chains of

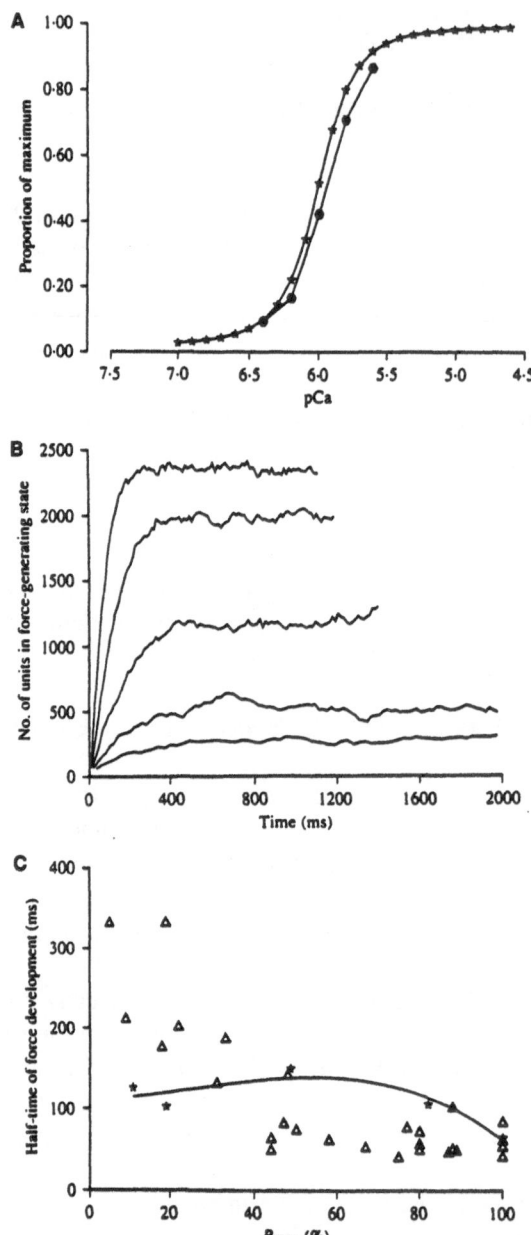

Fig. 3. (A) pCa-force relation derived from the 3 state model using the matrix method of
Hill[40] (✳) or from a Monte Carlo simulation (●). The difference between the lines is due to
the influence of the ends of the linear chains in the Monte Carlo simulation (which are only 30
units long), whilst in the matrix method the thin filaments are assumed to be infinitely long.
(B) Monte Carlo simulation of the development of force in a frog muscle fibre following a step
change in the free Ca^{2+} concentration, based on the model of Hill[40]. Final pCa values after the
step change are 5.6, 5.8, 6.0, 6.2 and 6.4.

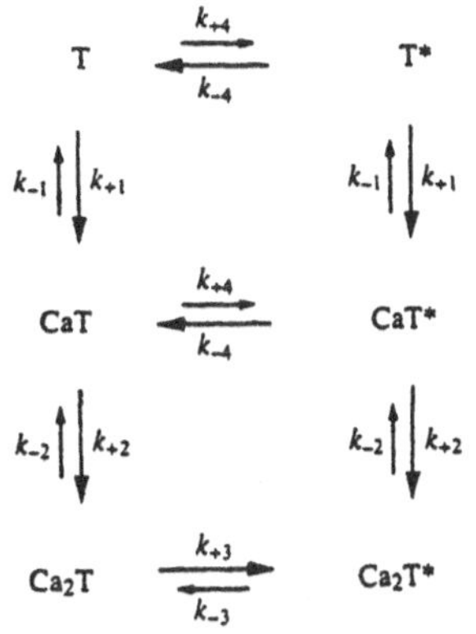

$$T \underset{k_{-4}}{\overset{k_{+4}}{\rightleftharpoons}} T^*$$

$$k_{-1} \Big\updownarrow k_{+1} \qquad\qquad k_{-1} \Big\updownarrow k_{+1}$$

$$CaT \underset{k_{-4}}{\overset{k_{+4}}{\rightleftharpoons}} CaT^*$$

$$k_{-2} \Big\updownarrow k_{+2} \qquad\qquad k_{-2} \Big\updownarrow k_{+2}$$

$$Ca_2T \underset{k_{-3}}{\overset{k_{+3}}{\rightleftharpoons}} Ca_2T^*$$

Fig .4. Six-state reaction scheme for the model in which the affinity of troponin C for calcium does not change with the formation of force generating cross-bridges. The rates of calcium binding to the two regulatory sites remains constant K_{+1} 2 x 10^8 M^{-1} s^{-1} and k_{+2} 10^8 M^{-1} s^{-1} as do the rates of dissociastion k_{-1} 100 s^{-1} and k_{-2} 200 s^{-1}. The rates of transition between the two states: non-force generating T and force generating T* vary in two subsequent models. In the first, these rates depend only on calcium binding to the individual functional unit, functional unit comprises one troponin-tropomyosin complex and its associated actin monomers (~7) and myosin cross-bridges (~2)), whilst in the second model cooperative interactions along the thin filament modulate these rates, dependent on the calcium occupancy of the regulatory sites on adjacent functional units. (from Ref : 3).

30 units each, as 30 is approximately the number of troponin-tropomyosin units on a 1 μm actin strand[45].

The results of this simulation are shown in Figure 3. The steady state relation is founded on the equilibrium constants and interaction factors used by Hill[40] and it is therefore reassuring to see that the steady-state pCa-force relation is virtually identical to that derived by the matrix method of solving the partition functions of a long linear chain of interacting units. Figure 3 shows the development of force predicted by this model following a step change in the calcium. As this particular model relies on an increase in the affinity of TnC for calcium as cross-bridges bind, the rise of force following step changes in the free calcium ion concentration, which result in intermediate force levels (> 50% and less than 100% P_{max}), is relatively slow. This property is an inherent feature of models which have a change in the calcium-binding induced by the relatively slow formation of force-generating cross-bridges and this prediction is at odds with our experimental observations (Fig. 1a, b). It is possible that the change in affinity for calcium is produced by a stage earlier than cross-bridge attachment, but this possibility seems less plausible.

(C) Relation between final tension developed (%P_{max}) and the half-time of tension development. Results from Monte Carlo simulation (✳) (least squares third-degree polynomial regression line) and experimental results (Δ). (from Ref : 3).

Monte Carlo Model: 6-state Model with Constant Ca^{2+} affinity, with and without Cooperativity

Although it has been suggested by Bremel & Weber and others that rigor cross-bridges enhance the affinity of the regulatory sites on the thin filament for calcium[46-48], it is not clear that cycling cross-bridges produce a similar effect. Pan and Solaro[47] did not find any increase in the affinity for calcium of cardiac myofibrils, measured by ^{45}Ca binding, as force developed. However Güth and Potter[49] found an increase with force development in the apparent affinity of the regulatory sites for calcium, as measured by the fluorescence changes in rabbit psoas muscle fibres, whose TnC had been replaced with TnC$_{DANZ}$. Therefore, in view of this ambiguity, a second Monte Carlo model of the activation of muscle contraction by calcium has been developed. This does not require a change in the affinity of the thin filament for calcium as force is generated.

This version has two calcium-binding sites per functional unit and thus is a closer replica of vertebrate skeletal muscle. This means that each functional unit can be in any one of six different states (Fig. 4):

State	No. of calcium ions bound	Myosin bound
0	0	No
1	1	No
2	2	No
3	0	Yes
4	1	Yes
5	2	Yes

The rates of calcium binding to and dissociation from the regulatory sites on these functional units are independent of both calcium binding to the other site on the functional unit and the state of the adjacent units. However the rates of transition from the inactive non-force generating states to the active force generating states depends not only on the number of calcium ions bound to that particular unit but also on the number of calcium ions bound to adjacent functional units. In this fashion it is possible to generate a calcium binding-pCa relation which is non-cooperative, whilst the force-pCa relation is cooperative.

One feature that becomes obvious from the Monte Carlo simulations is that the relation between the lifetimes of the calcium-troponin complex and the force generating state may play an important part in determining both the force-pCa relation and the kinetics of force development[50]. Thus if the time during which calcium remains attached to troponin is short compared with the lifetime of the force generating state, virtually maximal force may be produced by relatively low levels of calcium occupancy of the regulatory sites on TnC.

These considerations lead to the model shown in the figure (Fig. 4). As the transition to the force-generating state does not change the affinity of the regulatory sites for calcium, it follows that under some conditions the transition to the force generating state followed by calcium binding is less energetically favourable than calcium binding followed by the transition to the force-generating state. Thus under these conditions the system will reach a steady state in which the proportion of the units in each of the states is constant, but individual units tend to move round the sequence of states in the reaction scheme in an anti-clockwise direction. The transition to the force-generating state occurring most often when calcium is bound and the reverse transition occurring most often when calcium is not bound. The energy needed for this could be derived from the hydrolysis of ATP. Figure 5A shows the results of this simultation in the absence of cooperative interactions along the thin filament. Figure 5B shows the results when cooperative kinetic interactions along the length of the thin filament are included. In the latter case, the steady-state pCa-force relation is steeper than the pCa-calcium binding relation and the two curves intersect at approximately 50% P_{max} and 50% calcium occupancy (Fig. 5B). This pCa-force relation differs from that obtained from models in which the amount of force produced is proportional to the number of TnC molecules with calcium ions bound to both their regulatory sites. This latter relation is less steep and is shifted towards a lower pCa, that is a higher free Ca^{2+} value. The development of force following a step change in the free calcium ion concentration is shown in Figure 5 panels b for a series of free Ca^{2+} increments. The larger step changes in pCa are followed by larger and faster changes in force. The relation between the half-time for force development and the final force level is displayed in Figure 5 panels c and compared with the experimental results, where the prediction is close to that observed experimentally, with either nitr-5 or DM-nitrophen.

Finally, this result would both be in accord with both the Ca^{45} equilibrium binding experiments of Pan and Solaro[47] and Fuchs[51] in rabbit where no significant increase in ^{45}Ca binding was observed in actively cycling cross-bridges compared to the rigor case (-ATP). Also mechanical perturbation designed to break cross-bridges in frog led to no significant change in the free Ca^{2+} level as detected with aequorin[52].

Caged Chelators : Diazo-2

Recently diazo-2[53][54] a photolabile derivative of BAPTA, has been described which upon photolysis is rapidly (> 2000 s^{-1}) converted from a chelator with a low affinity for Ca^{2+} ($K_d = 2.2$ μM) to one with a high affinity for Ca^{2+} ($K_d = 0.073$ μM).

Rates of relaxation in single frog skinned fibres can be achieved with a caged chelator, which are at least as fast as those seen *in vivo* whilst the fall in free calcium is likely to be much faster ($t_{1/2} < 1$ ms)[55-59] than even the fast phase of the relaxation *in vivo* at 12°C ($t_{1/2} = 70$-80 ms), while in myosin-regulated scallop striated muscle the rates achieved are far faster than intact muscle[58]. It is now possible to use this method to study how rapidly the TnC bound Ca^{2+} falls, using

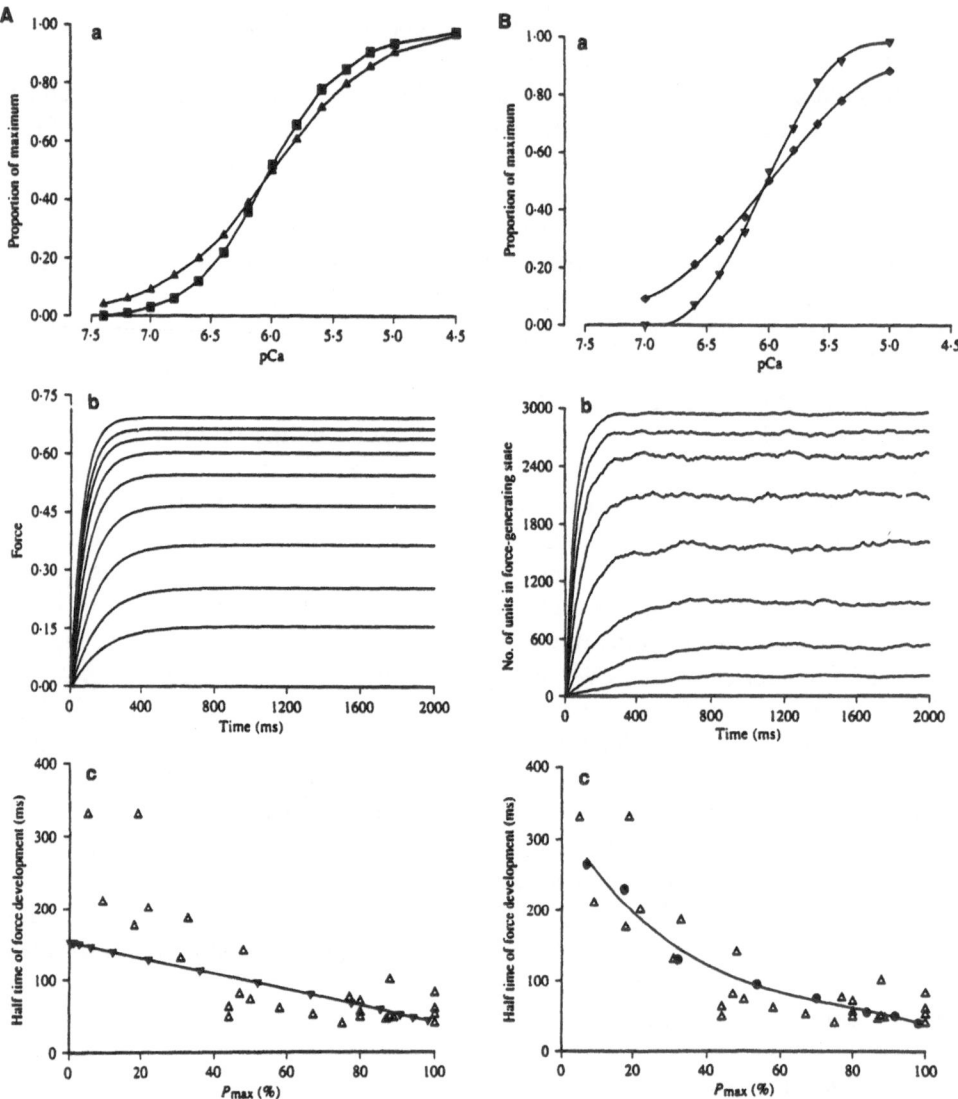

Fig. 5. (A) Model based on the six-state reaction scheme in which the rates k_{+3}, k_{-3}, k_{+4} and k_{-4} depend only on calcium binding to the regulatory sites of an individual troponin k_{+1} 2 x 10^8 M^{-1} s^{-1}, k_{-1} 10^8 M^{-1} s^{-1}, k_{-2} 200 s^{-1}, k_{+3} 5 s^{-1}, k_{+4} o s^{-1}, k_{-4} 5 s^{-1}. *No cooperative interactions.* (a) Steady state reactions: Force-pCa (■); calcium binding to regulatory sites - pCa (▲). (b) Force transients following a step change in free calcium to pCa 4.5, 5.0, 5.2, 5.4, 5.6, 5.8, 6.0, 6.2 and 6.4. (c) Relation between the half time for force development and the final force level reached. Model (●); experimental results (Δ). (from Ref : 3). (B) Metropolis Monte Carlo simulation of force development in a frog muscle fibre following a step change in the free Ca^{2+} concentration, based on a model with a constant affinity of troponin C for calcium: *Positive cooperativity.* (a) Steady-state relations of this model: pCa-force (▼); pCa-calcium binding to regulatory sites (◆). 6-state reaction scheme for the model in which the affinity of troponin C for calcium does not change with the formation of force-generating cross-bridges. k_{+1} 2 x 10^8 M^{-1} s^{-1}; k_{-1} 100 s^{-1}; k_{+2} 10^8 M^{-1} s^{-1}, k_{-2} 200 s^{-1}. The other rate constants

fluorescently labelled derivatives such as TnC_{DANZ} and to determine which are the steps which limit relaxation *in vivo*.

Effects of H+ and ADP on Relaxation with Diazo-2.

In a fatigued muscle fibre, the concentrations of ADP, P_i and H+ are all increased and relaxation is slowed. We have used the technique of laser flash photolysis of the caged calcium-chelator, diazo-2, to investigate the direct effect of changes in pH (pH 6.5, 7.0, 7.5) upon tension during relaxation of single chemically skinned frog fibres, when the effects of the sarcoplasmic reticulum are absent. The relaxation transients were closely fitted with two exponentials, a fast (42.3 ± 1.4; pH 7.0) and a slow process (12.0 ± 0.7; pH 7.0). The fast phase of relaxation was pH and ADP sensitive; lowering pH or increasing ADP led to a slowing of the rate of force decline and raising pH leading to an increase of the rate. The rate of the slow phase was unaltered by changing pH over the range investigated. Thus the slowing of relaxation in fatigued muscle may be due, in part, to the direct action of protons on the myofilaments (Ref : 59).

SR Release Mechanisms Studied with Caged Ca^{2+}(nitr-5): Barnacle Muscle

Ca^{2+}-induced Ca^{2+} release (CICR) is thought to play an important role in excitation-contraction coupling in cardiac muscle[60)61)]. Although CICR was first described in skinned fibres of frog skeletal muscle[62)63)] its importance in the generation of the skeletal muscle contraction has been doubted because rates of release from the SR in skinned fibres were found to be much slower than in intact fibres during a twitch[64)65)]. In addition the source of trigger Ca^{2+} is unclear; although there is a small Ca^{2+} inward current during the action potential it does not seem to be necessary for excitation-contraction coupling[66)67)].

In order to examine CICR using the laser-induced photolysis of nitr-5 to produce a jump in free Ca^{2+}, the barnacle myofibrillar preparation was first equilibrated in a muscle solution containing 0.1 mM nitr-5 as the predominant Ca^{2+} buffer at a pCa of 6.8-6.6. Figure 6 B shows the ryanodine sensitivity of the responses, giving the Ca^{2+} fraction produced by SR release (CICR) (Ref : 68).

Direct activation of Triton-treated myofibrils by photolysis of 2.0 mM-nitr-5 (initial pCa 6.4) gave contractions of up to 100% P_0 and a mean rise half-time of 164

(k_{+3}, k_{-3}, k_{+4} and k_{-4}, depend on the calcium occupancy of the neighbouring functional units. (b) Force transients following a step change in the free Ca^{2+} concentration based on a model with a constant affinity of troponin C for calcium. Final pCa values after the step change in free calcium concentration are 5.0, 5.4, 5.6, 5.8, 6.0, 6.2, 6.4, 6.6 and 7.0. (c) Plot of the half time for force development against %P_{max}. Values derived from the Metropolis Monte Carlo simulation (●); experimental data points from single semitendinosus muscle fibres of frog (Δ). Least squares regression line to points from the Monte Carlo simulation. (see Ref : 3 for details).

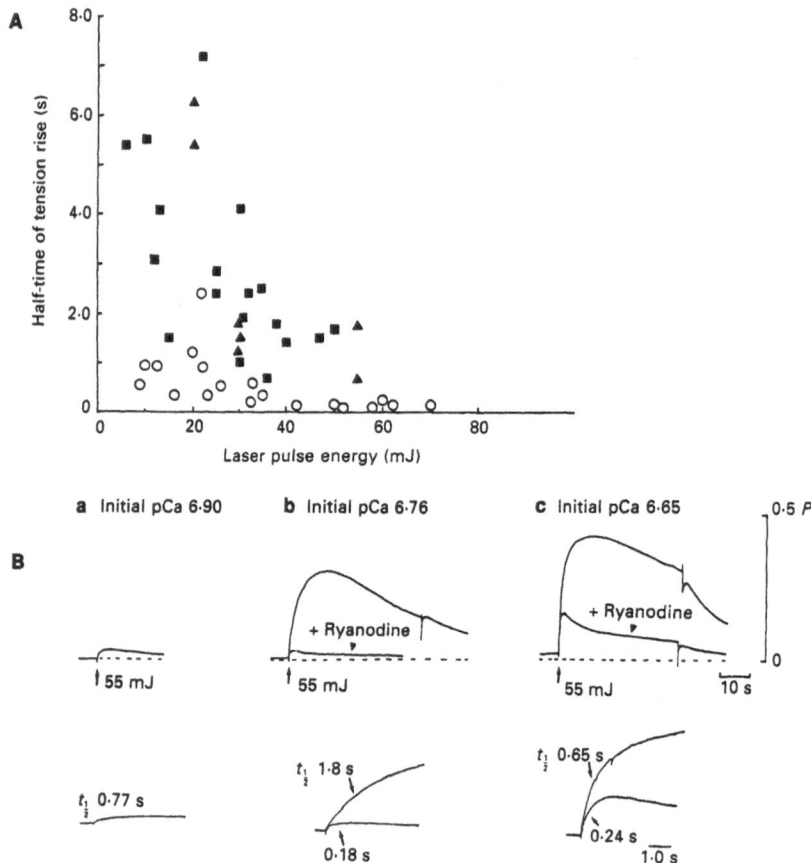

Fig. 6. (A) The half-times for the rise in tension in barnacle myofibrillar bundles following photolysis of 0.1 mM-nitr-5 at initial pCa 6.8-6.6 (■ and ▲) and of 2.0 mM- nitr-5 at initial pCa 6.4 (O), plotted against laser pulse energy (mJ). Combined data from twenty myofibrillar bundles: each point represents a single measurement. For the 0.1 mM-nitr-5 data, the bundle (with intact SR) was either in air (▲) or in solution in the trough (■) at the time of the laser pulse. The 2.0 mM-nitr-5 data are from either bundles with intact SR or bundles after treatment with Triton. (from Ref : 68). (B) Isometric tension records from a single barnacle myofibrillar bundle (diameter 140 µm) showing the effect of ryanodine on responses to photolysis of 0.1 mM-nitr-5 in air at three initial pCa values. Responses to a 55 mJ laser pulse were obtained after equilibration at pCa 6.90 (a), 6.76 (b) and 6.65 (c). The bundle was then transferred to a relaxing solution containing 10^{-4} M- ryanodine for 20 min and then a 'low relaxing' solution. The protocol with a 55 mJ pulse was then repeated in b and c. Upper traces, pen recordings; lower traces, oscilloscope traces at a faster time calibration. Records before and after ryanodine are superimposed. After response b, the SR was reloaded with Ca^{2+} using a pCa of 6.4 (5 mM-EGTA solution) for 5 min. (from Ref : 68).

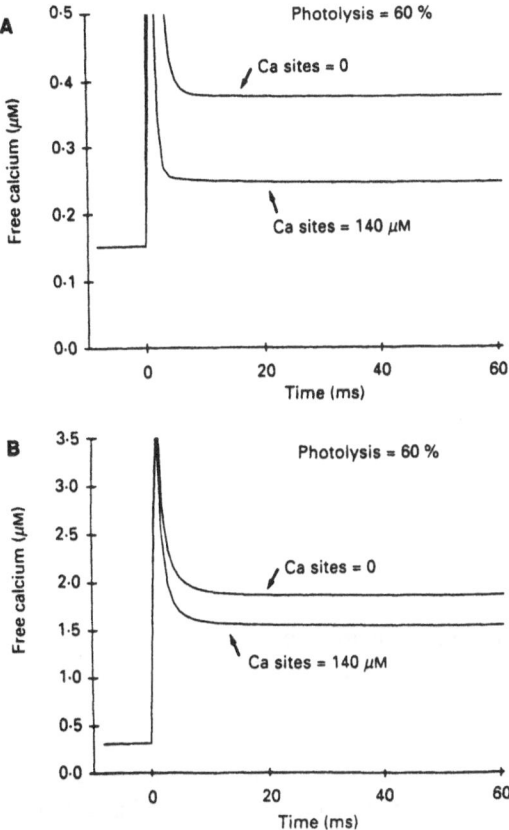

Fig. 7. The time courses of the rise in free Ca^{2+} concentration following partial photolysis of: (A), 0.1 mM-nitr-5 (initial pCa 6.8); and (B), 2.0 mM-nitr-5 (initial pCa 6.5), as predicted by the computer model (see Table 1 for details). At t = 0 ms the effect of the laser pulse is simulated to produce 60% photolysis of Ca^{2+}-nitr-5 and Mg^{2+}-nitr-5 and 30% of free nitr-5. Myofibrillar binding sites are either ignored or included at 140 μM with a dissociation constant of 1.3 μM. The nitr-5 is assumed to be homogeneously distributed in the myofibrillar bundle. The effects of the SR on the free Ca^{2+} are ignored. (from ref : 68)

ms at 12°C (n = 9 for contractions greater than 40% P_0)(Fig. 6A). Both the amplitude and the rate of these contractions were dependent on the laser pulse energy.

The Ca^{2+}-induced Ca^{2+} release responses obtained with photolysis (0.1mM nitr-5 experiments) were significantly slower than the fastest rate of tetanus development which has been recorded from intact fibres of barnacle muscle (mean half-time = 177 ms at 12°C) or when directly activated (2mM nitr-5 experiments)(Fig. 6A). This could mean that either Ca^{2+}-induced Ca^{2+} release is less efficient in isolated myofibrillar bundles than in intact fibres or that Ca^{2+}-induced Ca^{2+} release is not the primary Ca^{2+} releasing mechanism in excitation-contraction coupling in barnacle muscle (Ref : 68).

Time Course of Ca^{2+} Jump.

A separate computer program was written using a NAG library routine to predict the time course of the jumps in free Ca^{2+} following nitr-5 photolysis; this used published 'on' and 'off' kinetic constants for Ca^{2+} and Mg^{2+} binding to photolysed and unphotolysed nitr-5 and the Ca^{2+}-specific sites of TnC (Table 1). The program did not take into account any changes in Ca^{2+} due to SR activity. The shape of the Ca^{2+} jump for 60% photolysis of 0.1 mM-nitr-5 at an initial pCa of 6.8 shows an immediate rapid spike followed within about 10 ms by a steady plateau level (Fig. 7B). Presumably the spike is predicted by the model because of rebinding of the Ca^{2+}, which is released from photolysis of Ca-nitr-5, to the remaining unphotolysed nitr-5 and to the TnC sites. The prediction also shows that at 0.1 mM-nitr-5 the size of the Ca^{2+} jump will depend critically on the extent of Ca^{2+} binding by the TnC sites; this dependence is much less evident in the predicted Ca^{2+} jump for 2.0 mM-nitr-5 (Ref : 68) (Fig. 7B).

Conclusions

The exact reaction mechanisms which underlie the response of the contractile syswtem to small changes in free calcium are still not clearly understood, nor are the steps which lead to the release of this calcium from the sarcoplasmic reticulum as a result of surface membrane depolarization. This article has described some of the recent developments in both of these areas, particularly the progress that has been made through the use of photolabile chelators which have permitted Ca^{2+} kinetics to be studied in virtually intact muscle systems free of the major diffusion problems which plagued earlier attempts. However the exact time-course of calcium binding to the thin filament calcium sensor TnC during activation is still unclear; are the reaction kinetics the same as in the cuvette? Equally important, what is the relation between calcium binding and force generation, for as this article indicates, it is possible that not all the regulatory sites on TnC need to be occupied by calcium to support force? How cooperative is the system and what influence do cross-bridges have on this cooperativity and on TnC calcium binding when actively cycling; if feedback occurs which cross-bridge state is important?

ACKNOWLEDGEMENTS

I.P.M. wishes to thank the British Heart Foundation for an Intermediate Fellowship and T.J.L. and C.C.A. the MDA and MRC for financial support. R.E.P. and S.J.S. thank the SERC for Studentships. Special thanks are due to Drs. Roger Tsien and Steven Adams who provided valuable early samples of the photolabile chelators. Finally, our thanks to Miss Christine Lake for her excellent typing and to the authors and editors of journals cited in the text for permission to reproduce published illustrations.

REFERENCES

1. Ashley, C.C. & Moisescu, D.G. *J. Physiol. (Lond.)* **233**, 8-9P (1973).
2. Moisescu, D.G. *Nature* **262**, 610-613 (1976).
3. Ashley, C.C., Mulligan, I.P. & Lea, T.J. *Q. Rev. Biophys.* **24**, 1-73 (1991).
4. Shimomura, O., Johnson, F.H. & Saiga, Y. *J. Cell. Comp. Physiol.* **59**, 223-239 (1962).
5. Ridgway, E.B. & Ashley, C.C. *Biochem. Biophys. Res. Commun.* **29**, 229-233 (1967).
6. Ashley, C.C. & Ridgway, E.B. *J. Physiol. (Lond.)* **209**, 105-130 (1970).
7. Ashley, C.C. & Moisescu, D.G. *Nature (New Biol.)* **237**, 208-211. (1972).
8. Johnson, J.D., Collins, J.H. & Potter, J.D. *J. Biol. Chem.* **253**, 6451-6458 (1978).
9. Johnson, J.D., Charlton, S.C. & Potter, J.D. *J. Biol. Chem.* **254**, 3497-3502 (1979).
10. Rosenfeld, S.S. & Taylor, E.W. *J. Biol. Chem.* **260**, 242-251 (1985).
11. Hess, P., Metzger, P. & Weingart, R. *J. Physiol. (Lond.)* **329**, 173-188 (1982).
12. Robertson, S.P., Johnson, J.D. & Potter, J.D. *Biophys. J.* **34**, 559-569 (1981).
13. Gillis, J.M., Thomason, D., Lefevre, J. & Kretsinger, R.H. *J. Muscle Res. Cell Motility* **3**, 377-398 (1982).
14. Natori, R. *Jikeikai Med. J.* **1**, 119-126 (1954).
15. Hellam, D.C. & Podolsky, R.J. *J. Physiol. (Lond.)* **200**, 807-819 (1969).
16. Julian, F.J. *J. Physiol. (Lond.)* **218**, 117-145 (1971).
17. Ashley, C.C. & Moisescu, D.G. in *Calcium Transport in Contraction and Secretion* (ed. Carafoli, E. et al.) 517-525 (North Holland, Amsterdam, 1975).
18. Endo, M. *Cold Spring Harbor Symp. Quant. Biol.* **37**, 505-510 (1973).
19. Ashley, C.C. *Ann. N.Y. Acad. Sci.* **307**, 308-329 (1978).
20. Ashley, C.C. in *Calcium in Biology* (ed. Spiro, T.) 109-173 (Wiley, New York, 1983).
21. Engels, J. & Schlaeger, E.J. *J. Med. Chem.* **20**, 907-911 (1977).
22. Kaplan, J.H., Forbush, B. & Hoffman, J.F. *Biochemistry* **17**, 1929-1935 (1978).
23. Goldman, Y.E., Hibberd, M.G., McCray, J. & Trentham, D.R. *Nature* **300**, 701-705 (1982).
24. Adams, S.J., Kao, J.P.Y. & Tsien, R.Y. *J. Gen. Physiol.* **88**, 9-10a (1986).
25. Adams, S.J., Kao, J.P.Y., Grynkiewicz, G. Minta, A. & Tsien, R.Y. *J. Am. Chem. Soc.* **110**, 3212-3220 (1988).
26. Kaplan, J.H. & Ellis-Davies, G.C.R. *Proc. Natl. Acad. Sci. USA.* **85**, 6571-6575 (1988).
27. Ellis-Davies, G.C.R. & Kaplan, J.H. *J. Org. Chem.* **53**, 1966-1969 (1988).
28. Tsien, R.Y. *Biochemistry* **19**, 2396-2404 (1980).
29. Gurney, A.M., Tsien, R.Y. & Lester, H.A. *Proc. Natl. Acad. Sci. USA.* **84**, 3496-3500 (1987).
30. Ashley, C.C., Barsotti, R.J., Ferenczi, M.A., Lea, T.J. & Mulligan, I.P. *J. Physiol. (Lond.)* **394**, 24P (1987a).
31. Ashley, C.C., Barsotti, R.J., Ferenczi, M.A., Lea, T.J., Mulligan, I.P. & Tsien, R.Y. *J. Physiol. (Lond.)* **390**, 144P (1987b).
32. Ashley, C.C., Barsotti, R.J., Ferenczi, M.A., Lea, T.J. & Mulligan, I.P. *J. Physiol. (Lond.)* **398**, 71P (1988).
33. Ashley, C.C., Barsotti, R.J., Ferenczi, M.A., Lea, T.J. & Mulligan, I.P. in *Biochemical Approaches to Cellular Calcium* (ed. Reid, E.) **19**, 131-132 (Royal Society of Chemistry, London, 1989).
34. Ashley, C.C., Lea, T.J., Mulligan, I.P. & Timmerman, M.P. *J. Physiol. (Lond.)* **414**, 50P.
35. Lea, T.J., Fenton, M.J., Potter, J.D. & Ashley, C.C. *Biochim. Biophys. Acta.* **1034**, 186-194 (1990).
36. Kress, M., Huxley, H.E., Faruqi, A.R. & Hendrix, J. *J. Mol. Biol.* **188**, 325-342 (1986).
37. Griffiths, P.J., Duchateau, J.J., Maeda, Y., Potter, J.D. & Ashley, C.C. *Pflügers Arch.* **415**, 554-565 (1990).

38. Johnson, J.D., Robinson, D.E., Robertson, J.D., Schwartz, A. & Potter, J.D. in *The Regulation of Muscle Contraction : E-C coupling* (ed. Grinnel, A.D.) 241-259 (Academic Press, New York, 1981).

39. Griffiths, P.J., Potter, J.D., Coles, B., Strang, P. & Ashley, C.C. *FEBS Lett.* **176**, 144-150 (1984).

40. Hill, T.L. *Biophys. J.* **44**, 383-396 (1983).

41. Binder, K. in *Topics in Current Physics* Vol.11. (Springer Verlag, Berlin, 1983).

42. Binder, K. in *Topics in Current Physics* Vol.7 (2nd edition). (Springer Verlag, Berlin, 1986)

43. Kalos, M.H. & Whitlock, P.A. *Monte Carlo Methods.* (Wiley, New York, 1986).

44. Chen, Y.D. & Hill, T.L. *Proc. Natl. Acad. Sci. USA.* **80**, 7520-7523 (1983).

45. Brandt, P.W., Diamond, M.S. & Rutchik, J.S. *J. Mol. Biol.* **195**, 885-896 (1987).

46. Bremel, R.D. & Weber, A. *Nature (New Biol.)* **238**, 97-101 (1972).

47. Pan, B.S. & Solaro, R.J. *J. Biol. Chem.* **262**, 7839-7849 (1987).

48. Wnuk, W., Schoechlin, M. & Stein, E.A. *J. Biol. Chem.* **259**, 9017-9023 (1984).

49. Güth, K. & Potter, J.D. *J. Biol. Chem.* **262**, 13627-13655 (1987).

50. Brandt, P.W., Cox, R.N., Kawai, M. & Robinson, T. *J. Gen. Physiol.* **79**, 997-1016 (1982).

51. Fuchs, F. *J. Muscle Res. Cell Motility* **6**, 477-486 (1985).

52. Cecchi, G., Griffiths, P.J. & Taylor, S.R. *Adv. Exp. Biol. Med.* **170**, 455-466 (1984).

53. Harootunian, A.T., Kao, J.Y.P. & Tsien, R.Y. *Cold Spring Harbor Symp. Quant.Biol.* **53**, 935-943 (1988).

54. Adams, S.R., Kao, J.P.Y. & Tsien, R.Y. *J. Am. Chem. Soc.* **111**, 7957-7968 (1989).

55. Mulligan, I.P. & Ashley, C.C. *FEBS Lett.* **255**, 196-200 (1989).

56. Ashley, C.C., Mulligan, I.P. & Palmer, R.E. *J. Physiol. (Lond.)* **426**, 31P (1990).

57. Mulligan, I.P., Adams, S.R., Tsien, R.Y., Potter, J.D. & Ashley, C.C. *Biophys. J.* **57**, 541a (1990).

58. Palmer, R.E., Mulligan, I.P., Nunn, C. & Ashley, C.C. *Biochem. Biophys. Res. Commun.* **168**, 295-300 (1990).

59. Palmer, R.E., Simnett, S.J., Mulligan, I.P. & Ashley, C.C. *Biochem. Biophys. Res. Commun.* **181**, 1337-1342 (1992)

60. Fabiato, A. *Am. J. Physiol.* **245**, C1-C4 (1983).

61. Fabiato, A. *Biophys. J.* **97**, 195a (1985).

62. Endo, M., Tanaka, M. & Ogawa, Y. *Nature* **228**, 34-36 (1970).

63. Ford, L.E. & Podolsky, R.J. *Science* **167**, 58-59.

64. Endo, M. *Physiol. Rev.* **57**, 71-108 (1977).

65. Baylor, S.M., Hollingworth, S. & Marshall, M.W. *J. Physiol. (Lond.)* **344**, 625-666 (1983).

66. McClesky, E.W. *J. Physiol. (Lond.)* **361**, 231-249 (1985).

67. Brum, G., Stephani, E. & Rios, E. *Can. J. Physiol. Pharmacol.* **65**, 681-685 (1986).

68. Lea, T.J. & Ashley, C.C. *J. Physiol. (Lond.)* **427**, 435-453 (1990).

Discussion

Huxley: You gave the figure for rate of relaxation *in vivo* of frog fibers as 66 milliseconds. Was that a half-time? Anyhow, relaxation in a live frog fiber goes with a very slow phase until there is a shoulder, after which elongation occurs locally within the fiber. Does the figure that you took, which seemed to agree reasonably well with your *in vitro* one, correspond to the slow phase or to this artifactual rapid phase, as I call it?

Ashley: As you say, the rapid exponential phase *in vivo* is due to sarcomeric disorganization. We have deliberately tried not to make any assumptions in our caged diazo experiments about what the relaxation phase actually does. It is a tricky point. Nevertheless, for the overall fast relaxation phase with intact frog—we took the figures from Paul Edman's paper—our diazo 2 values ($T_{1/2}$) seem to be within the same range, but I haven't deliberately pushed that analogy [Edman, P. and Flitney, F.W. *J. Physiol. (Lond.)* **329**, 1-20, 1982]. I think it would be dangerous to do so.

Gulati: With regard to your model of tropomyosin-tropomyosin interaction for the cooperativity in partially TnC-extracted fibers, a number of people have shown that the cooperativity decreases [Brandt et al. *J. Mol. Biol.* **195**, 885, 1987; Gulati, J. *J . Physiol. (Lond)* **420**, 1398, 1989]. In the model you were using, if the units were acting independently, I would expect that the cooperativity would not be affected by partial TnC extraction unless you made additional assumptions. Have you modelled that?

Ashley: No, we haven't tried anything with partial extraction.

THE MOLECULAR SWITCH IN TROPONIN C

John Gergely, Zenon Grabarek and Terence Tao

Department of Muscle Research
Boston Biomedical Research Institute
Dept. of Neurology
Massachusetts General Hospital
Department of Biological Chemistry and Molecular Pharmacology
Harvard Medical School
Boston, Massachusetts, USA

ABSTRACT

Conformational changes in troponin C (TnC) associated with Ca^{2+}-induced triggering of muscle contraction are discussed in light of the model proposed by Herzberg, Moult and James (*J. Biol. Chem.* **261**, 2638, 1986) and of our recent work on mutants of troponin C. The model involves a Ca^{2+}-induced angular movement of one pair of α-helical segments relative to another pair of helices in the N-terminal domain. A disulfide bridge introduced into the N-terminal domain reversibly blocks the key conformational transition and the Ca^{2+}-regulatory activity. Binding of troponin I (TnI) to the disulfide form of TnC is weakened owing to the blocking of its interaction with the N-terminal domain; however incorporation of the mutant into TnC-extracted myofibrils is not abolished. Introduction of a Cys residue in the C-terminal domain of TnC leads to disulfide formation between it and the indigenous Cys-98, with accompanying inhibition of regulatory activity attributable to interference with binding to TnI and, consequently, incorporation into the thin filaments. Evidence for movement of helical segments upon Ca^{2+}-binding to TnC was obtained by measurements of excimer fluorescence and of resonance energy transfer with probes attached to Cys residues introduced by site-directed mutagenesis at suitable locations. Introduction of a disulfide bridge into calmodulin, another member of the super-family of Ca^{2+}-binding proteins to which TnC belongs, abolishes its interaction with target enzymes. This suggests that the type of conformational change involving angular movement of helical segments that takes place in TnC is also involved in signal transmission in other Ca^{2+}-dependent regulatory proteins.

Mechanism of Myofilament Sliding in Muscle Contraction, Edited by
H. Sugi and G.H Pollack, Plenum Press, New York, 1993

INTRODUCTION

It is now well established that muscle contraction is initiated by the release of Ca^{2+} ions from the sarcoplasmic reticulum and relaxation occurs as calcium is reaccumulated by the calcium pump in the membrane of the sarcoplasmic reticulum[1]. The work of Ebashi and his colleagues[2] brought to light the existence of a regulatory protein complex in thin filaments of striated muscle which eventually was shown to consist of tropomyosin and three additional subunits forming the troponin complex[3-8]. Troponin C (TnC) is the receptor for the Ca^{2+} that triggers activation of contraction. The other two components were named troponin I (TnI) and troponin T (TnT), based on the assumption that their primary roles were the inhibition of actin-myosin interaction and binding to tropomyosin, respectively. Early structural studies suggested that activation of the actin-myosin system involves a movement of tropomyosin along the surface of the actin filament towards the long pitch groove leading to uncovering the—or part of the—myosin binding site and thereby permitting the binding of myosin to actin[9-11]. Studies of the effect of Ca^{2+} on the binding of myosin or its active subfragment (S-1) to a regulated, viz. tropomyosin-troponin containing, F-actin, and of the kinetics of the ATPase cycle provided evidence against the view that regulation by Ca^{2+} primarily involves facilitation of binding to actin but suggested that a kinetic step in the crossbridge cycle would be accelerated[12]. Recent x-ray diffraction and electron microscopic data[13][14] still support the view that some structural change in the thin filaments may be the first step that initiates activation. A Ca^{2+}-induced movement of the TnI is indicated by resonance energy transfer and crosslinking experiments[15]. The recent elucidation of the actin structure at the atomic level[16][17] will help to clarify unresolved questions concerning details of the interaction of myosin and tropomyosin with actin and the mechanism of regulation of contraction. In this paper we shall review recent work dealing with the changes that take place in troponin C upon calcium binding.

The Structure of TnC

Troponin C belongs to a super-family of calcium binding proteins, the first member of which was parvalbumin, a protein originally found in fish muscle. Parvalbumin contains the prototype of the Ca^{2+} binding site present in all members of this super-family. Each Ca^{2+} binding site contains a helix-loop-helix structure (HLH) often referred to as the EF-hand on the basis of the nomenclature applied to parvalbumin[18]. The loop provides coordinating groups for the bound Ca^{2+} at the vertices of a pentagonal bipyramid[19]. In TnC there are eight α-helical segments (designated A-H) associated with the four Ca^{2+} binding HLH sites and an additional short N terminal helix. The two C-terminal sites in TnC have high Ca^{2+} affinity and also bind Mg^{2+}, whereas the two N-terminal ones are specific for Ca^{2+} and constitute the trigger for activation. An early structural model proposed by

Kretsinger and Barry[20] for troponin C suggested that two parvalbumin-like domains each containing two HLH units form a compact, more or less globular, structure. The structure derived from x-ray diffraction studies[21][22] on troponin C crystals turned out to be different from that postulated earlier; two distinct globular domains, each containing pairs of Ca^{2+} binding sites were not in contact with each other but joined by a long central α-helical stretch containing two helices, D and E, joined by a short linker, whose α-helical character had previously not been recognized.

The Herzberg-Moult-James (HMJ Model)

James and his colleagues[23], having noted that under conditions of crystallization only the C-terminal domain contained bound Ca^{2+}, suggested that the structural differences between the two domains would furnish a clue to the changes that should occur on Ca^{2+}-binding to the triggering sites in the N-terminal domain. They showed that by a joint movement of helices B and C, changing the angle between them and helices A and D but leaving other structural components essentially undisturbed, a structure quite similar to that found in the Ca^{2+}-bound C-terminus can be produced. They also showed that the postulated movement of helices B and C would expose a hydrophobic patch that could serve as a binding site for troponin I.

The cloning of the cDNA encoding troponin C and its expression with a high yield in *E. coli*[24], made it possible to explore the consequences of the suggested structural change. It seemed that by replacing the appropriate residues by Cys a disulfide bridge could form in the Ca^{2+}-free state which then would prevent the expected Ca^{2+}-induced movement and interfere with TnI binding and regulation. A review of interatomic distances in turkey TnC (kindly supplied by Dr. O. Herzberg) suggested that optimal sites for replacement by Cys were Gln 51 in the B-C linker and Gln 85 in the D helix, or the homologous residues in rabbit TnC, Gln 48 and Gln 82, respectively to form an -S-S- bond which, when formed, would preclude the about 10 Å separation of these residues predicted by the HMJ model in the Ca^{2+}-bound form.

We, therefore, produced a mutant of rabbit TnC with Cys residues in positions 48 and 82 having also replaced the native Cys 98 by Leu[25]. The latter modification was made to avoid dimerization of TnC via disulfide bonds or possible additional internal disulfide bond formation. The ready formation of a disulfide bridge between the introduced Cys residues was apparent from the changed electrophoretic mobility of the purified mutant TnC; the mobility change was reversed to that corresponding to wild type TnC upon reduction by DTT. The oxidized mutant showed reduced Ca^{2+} binding affinity and reduced TnI binding, the latter being attributed to the inability of the N-terminal domain to interact with TnI[26]. More important was the inability of the oxidized mutant TnC to mediate Ca^{2+}-activation of ATPase activity of TnC-free myofibrils. In all these respects the reduced form of the mutant exhibited normal behavior. It should be noted that the -SH groups of the reduced form were protected against reoxidation to -S-S- bonds by carboxamidomethylation or carboxymethylation; the Ca^{2+} binding affinity was actually enhanced by carboxymethylation, presumably owing to the presence of additional charges favoring separation of the α-helices. Direct modification of charged residues by mutagenesis in the putative area where movement of the helices should take place was also shown to affect the physiological properties of troponin C[27].

Evidence for Ca^{2+}-induced Movement in TnC

While these experiments were clearly consistent with the postulated mechanism[23], and indicated its physiological relevance, additional more recent work has provided direct evidence for the actual movement taking place in troponin C upon Ca^{2+} binding to the regulatory sites. For these experiments other mutants were produced with Cys residues introduced at sites permitting attachment of optical probes. The sites had to be chosen so that on the one hand the 3-D structure would accommodate the probe without distortion and, on the other hand, the distance change would be sufficiently large for detection. Changes in the distance between two probes would be indicated by a change in the excimer fluorescence of two vicinal pyrene probes or, quantitatively, by changes in the resonance energy transfer efficiency between a suitably placed donor and acceptor. A decrease in the excimer emission of Cys mutants at positions 12, in helix A, and 49, in the linker between helices B and C, labelled with a pyrene-maleimide probe indicated a separation of helices A and B. The actually measured change on Ca^{2+} binding was an increase of 13 Å which is in excellent agreement with the 11 Å increase predicted by the HMJ model[28].

Interactions with TnI

While the combination of structural information and results derived from experiments with engineered proteins has given considerable insight into the events that occur in TnC upon Ca^{2+} binding, more work will be needed to understand all the details of the way in which the two domains of TnC interact with TnI and how this interaction eventually leads to the movement of TnI resulting in an increased tightening of the interaction between TnI and TnC and loosening of the interaction with actin. While in earlier work considerable emphasis has been placed on the interaction of the so-called inhibitory segment of troponin I, residues 96-116, with a region containing the sulfhydryl group (Cys 98) at the C-terminal end of the central helix[29], recent crosslinking work has shown that the same region of TnI can also interact with residues in helix C in the N-terminal domain of TnC[30)31]. This would be consistent with the regulatory function of the N-terminal domain supported by the results discussed above.

Further insights into the role of the C-terminal domain in the interaction with TnI are provided by a TnC mutant containing an additional Cys residue, Cys-122[32]; the latter is able to crosslink with the indigenous Cys 98, the disulfide being reducible with DTT. The oxidized form of the mutant is characterized by a greatly reduced affinity for TnI and reduced ability to confer Ca^{2+} sensitivity on TnC-extracted myofibrils. Significant regulatory activity was obtained at higher concentrations of the mutant. The reduction in affinity for TnI on disulfide formation is about ten times greater than that occurring on disulfide formation in the N-terminal domain. This is consistent with the conclusions reached in earlier work based on Ca^{2+} binding equilibrium and kinetic studies that the primary regulatory events take place in the N-terminal domain while the C-terminal domain plays chiefly a structural—or in the light of the results with the mutant—TnI anchoring role.

The apparent ability of a relatively short segment of TnI to interact with these two regions of TnC is consistent with a two-prong interaction between TnC and TnI. However, the fact that, on the basis of the crystal structure the two TnI binding regions would be at some distance raises some questions concerning the physiological relevance of the extended structure revealed by x-ray crystallography. There is a body of evidence suggesting that the actual solution structure of TnC and possibly the structure in situ may involve some bending of the central helix[33]. Most recent molecular dynamic simulations also point to a more compact structure with a bent central helix in solution[34]. Persechini and Kretsinger[35] have suggested a bending in the analagous region of calmodulin—the ubiquitous Ca^{2+} binding protein present in most cells which also belongs to the same family possessing four Ca^{2+} binding helix-loop-helix motifs—permitting the interaction of an α-helical segment of myosin light chain kinase with both globular domains of it. Whether the inhibitory TnI segment could be involved in similar interactions with the two domains of TnC will have to await further studies. Similarly, the functional significance of the N-terminal region of TnI, whose interaction with TnC has recently been documented[36], requires clarification. It should be noted that when calmodulin was subjected to site directed mutagenesis and introduction of disulfide bonds in positions analogous to those in TnC its ability to activate target enzymes became inhibited. This suggests that the opening up of hydrophobic region upon movement of a pair of helices is an essential step in the mechanism of action of that protein too. It seems likely that it represents a general mechanism of signal transduction in Ca^{2+}-mediated regulatory proteins[37].

Conclusions

There is now a large body of evidence in support of the view that Ca^{2+}-binding to the low affinity sites of TnC induces a movement of helices B and C away from helices A and D, thus opening a hydrophobic cavity—a site of interaction with TnI. Another site of similar structure is formed by the helical segments in the C-terminal domain. Both sites appear to interact with the inhibitory segment of TnI. Whereas the interactions at both sites are necessary for the full regulatory activity of TnC, the interaction at the C-terminal domain stabilizes the complex and that involving the N-terminal domain is directly linked to the release of inhibition. In the absence of Ca^{2+}

the inhibitory region of TnI would preferentially bind to actin and upon Ca^{2+} binding to sites I and II it would switch to the site in the N-terminal domain of TnC. Detachment of TnI from actin gives rise to further events in thin filament regulation.

ACKNOWLEDGEMENTS

The authors' work was supported by grants from NIH, MDA, and AHA.

REFERENCES

1. Ashley, C.D., Mulligan, I.P. & Lea, T.J. *Quart. Rev. Biophys.* **24**, 1-77 (1991).
2. Ebashi, S. & Endo, M. *Prog. Biophys. Mol. Biol.* **18**, 123-183 (1969).
3. Schaub, M.C. & Perry, S.V. *Biochem. J.* **115**, 993-1004 (1969).
4. Greaser, M. & Gergely, J. *J. Biol. Chem.* **246**, 4226-4233 (1971).
5. Leavis, P.C. & Gergely, J. *CRC Crit. Rev. Biochem.* **16**, 235-305 (1984).
6. Ohtsuki, I., Maruyama, K. & Ebashi, S. *Adv. Prot. Chem.* **38**, 1-67 (1986).
7. Zot, H.G. & Potter, J.D. *Ann. Rev. Biophys. Biophys. Chem.* **16**, 535-559 (1987).
8. Grabarek, Z., Tao, T. & Gergely, J. *J. Muscle Res. Cell Motility* **13**,383-393, (1992).
9. Huxley, H.E. *Cold Spring Harbor Symp. Quant. Biol.* **37**, 361-376 (1972).
10. Haselgrove, J. *Cold Spring Harbor Symp. Quant. Biol.* **37**, 341-352 (1972).
11. Parry, D.A.D. & Squire, J.M. *J. Mol. Biol.* **75**, 33-55.
12. Chalovich, J.M. & Eisenberg, E. *J. Biol. Chem.* **257**, 2432-2437 (1982).
13. Kress, M., Huxley, H.E, Faruqi, A.R. & Hendrix, J. *J. Mol. Biol.* **188**, 325-342 (1986).
14. Milligan, R.A., Whittaker, M. & Safer, D. *Nature* **348**, 217-221 (1990)
15. Tao, T., Gong, B.-J. & Leavis, P.C. *Science* **247**, 1339-1341.
16. Holmes, K.C., Popp, D., Gebhard, W. & Kabsch, W. *Nature* **347**, 44-49 (1990).
17. Kabsch, W., Mannherz, H.G., Suck, D., Pai, E.F. & Holmes, K.C. *Nature* **347**, 37-44 (1990).
18. Kretsinger, R.H. & Nockolds, C.E. *J. Biol. Chem.* **248**, 3313-3326 (1973).
19. Strynadka, N.C.J. & James, M.N.G. *Ann. Rev. Biochem.* **58**, 951-998 (1989).
20. Kretsinger, R.H. & Barry, C.D. *Biochim. Biophys. Acta* **405**, 40-52 (1975).
21. Sundaralingam, M., Bergstrom, R., Strasburg, G., Rao, S.T., Roychoudhury, P., Greaser, M. & Wang, D.C. *Science* **227**, 945-948 (1985).
22. Herzberg, O. & James, M.N.G. *Nature* **313**, 653-659 (1985).
23. Herzberg, O., Moult, J. & James, M.N.G. *J. Biol. Chem.* **261**, 2638-2644 (1986).
24. Wang, C.L.A., Sarkar, S., Gergely, J. & Tao, T. *J. Biol. Chem.* **265**, 4953-4957 (1990)
25. Grabarek, Z., Tan, R.-Y., Wang, J., Tao. T. & Gergely, J. *Nature* **345**, 132-135 (1990).
26. Gusev, N.B., Grabarek, Z. & Gergely, J. *J. Biol. Chem.* **266**, 16622-16626 (1991).
27. Fujimori, K., Sorenson, M., Herzberg, O, Moult, J. & Reinach, F.C. *Nature* **344**, 182-184 (1990).
28. Wang, C.L.A., Gergely, J. & Tao, T. *Biophys. J.* **59**, 218a (1991).
29. Leszyk, J., Collins, J.H., Leavis, P.C. & Tao, T. *Biochemistry* **27**, 6983-6987 (1988).
30. Leszyk, J., Grabarek, Z., Gergely, J. & Collins, J. *Biochemistry* **29**, 299-304 (1990).
31. Kobayashi, T., Tao, T., Grabarek, Z., Gergely, J. & Collins, J.H. *J. Biol. Chem.* **266**, 13746-13751 (1991).

32. Grabarek, Z., Gusev, N., Tan, R.-Y. & Gergely, J. *Biophys. J.* **59**, 582a, 1991.
33. Heidorn, D.B. & Trewhella, J. *Biochemistry* **27**, 909-914 (1988).
34. Mehler, E.L., Pascual-Ahuir, J.L. & Weinstein, H. *Protein Engineering* **4**, 625-637 (1991).
35. Kretsinger, R.H. & Nockolds, C.E. *J. Biol. Chem.* **248**, 3313-3326 (1973).
36. Ngai, S.M. & Hodges, R.S. *Biophys. J.* **59**, 586a, (1991).
37. Grabarek, Z., Tan, R.-Y. & Gergely, J. *Biophys. J.* **59**, 23a (1991).

Discussion

Maéda: If I follow correctly the measurements of distances between helices A, B, and D with troponin-C and -I alone, do you think the same thing happens within the whole complex of troponin?

Gergely: The distances and the changes induced by Ca^{2+} are the same in the whole complex.

Gulati: Do you envision an interaction between C-terminal sites and N-terminal sites? In other words, do you see a role for the central helix to transmit a message from one terminus to the other?

Gergely: In the work I spoke of today there was no indication of that, but there are other types of experiments where some interaction can be detected.

Gordon: Do you think that plays a role in activation?

Gergely: That is hard to say, particularly because we don't know really what the disposition of the two domains is *in situ*.

Gulati: Could I come back to that? One reason I was asking you was because we made a mutant in which we deleted the KGKSEEE, the seven residues right in the middle of the helix, and we find remarkably that it increases the calcium sensitivity; this presumably suggests that the helix length may have an effect under those conditions.

THE MECHANISM OF Ca²⁺-COORDINATION IN THE EF-HAND OF TnC, BY CASSETTE MUTAGENESIS

Árvind Babu, Hong Su and Jagdish Gulati

The Molecular Physiology Laboratory
Division of Cardiology
Departments of Medicine & Physiology/Biophysics
Albert Einstein College of Medicine
Bronx, NY 10461

ABSTRACT

Genetic engineering of TnC and skinned fiber physiology on rabbit psoas muscle are combined to study the mechanisms of Ca²⁺-binding in the EF-hand in TnC. Of the six coordinating positions (X,Y,Z,-Y,-X & -Z) for Ca²⁺-binding in the loop, the X position is invariably occupied by an aspartate, and the -Z position by a glutamate. X-ray analysis has indicated that both oxygen atoms of the β-carboxylate in aspartate (in X) are extensively hydrogen bonded to other residues in the loop. When this aspartate in site II was replaced by a glutamate (γ-carboxylate), Ca²⁺-binding was annihilated, and the mutant was unable to regulate force development in the fiber. Similarly, glutamate for aspartate exchange in the -Z position of site I also inactivated the site as well as its function in skinned fiber. Mutations in the Y position indicated that a glutamate was unacceptable in place of aspartate but that an asparagine was acceptable. The Ca²⁺-sensitivity with asparagine was also similar to that of the wild type. The study indicates a powerful approach for defining the physicochemical principles governing Ca-coordination and sensitivity in Ca-binding proteins. Furthermore, by comparison with findings on chemically synthesized peptides, the results show that behavior of the EF-hand in TnC is modified by quaternary structure of the molecule.

1. Introduction

Ca²⁺ binding to TnC modulates the interaction of TnC with TnI and triggers the onset of contraction in heart and muscle by releasing the cross-bridge cycle. TnC has four potential Ca²⁺-binding sites; these sites share an helix-loop-helix motif (the EF-hand) where Ca ion is coordinated within the loop[1]. Fig. 1 indicates the sequence of rabbit fast skeletal TnC with its four EF hands: I & II are the Ca²⁺-specific sites and III & IV the Ca²⁺-Mg²⁺ sites. These latter also have nearly 100-fold greater affinity

Mechanism of Myofilament Sliding in Muscle Contraction, Edited by
H. Sugi and G.H Pollack, Plenum Press, New York, 1993

125

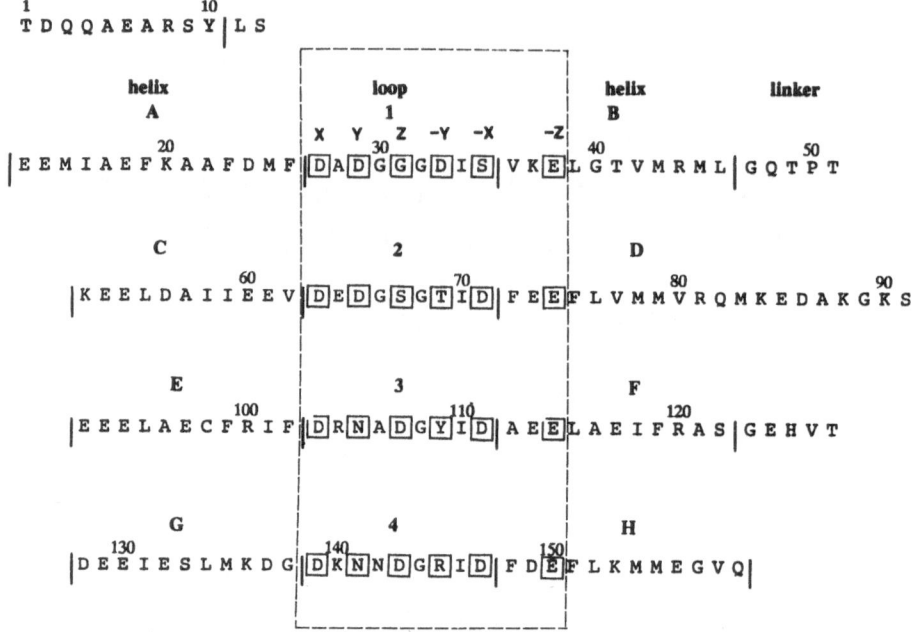

Fig. 1. Sequence of rabbit fast skeletal muscle TnC.

for Ca^{2+} than that of sites I & II. On the basis of x-ray crystallography, it could be defined that six amino acid residues are involved in Ca^{2+}-coordination; five are in the loop domain (X,Y,Z,-Y,-X positions) and the sixth one (-Z coordinate) is in the F-helix of the putative EF hand[2)3)]. The present study is part of an effort towards structure-function analysis of the EF-hand, to delineate the role(s) of the residues in the various coordinating positions in determining the properties of the Ca^{2+}-binding sites of TnC in physiological milieu.

The structure-function studies of TnC are helped by combining genetic engineering with skinned fiber physiology as a result of the recent demonstrations that TnC mutants can be functionally incorporated in the skinned fibers of skeletal and cardiac muscles[4)5)]. To facilitate the present investigations, we have synthesized a unique TnC encoding gene (from the amino acid sequence of rabbit fast skeletal muscle) with several restriction sites, by utilizing the concept developed in the Khorana lab[6)]. These restriction sites are helpful in making the mutants by vastly simplifying the process due to the ability to utilize cassette mutagenesis on a routine basis. Our fiber results with these mutants extend the previous findings based on x-ray structural analysis for Ca^{2+}-coordination in the EF-hand and provide new insights into the mechanism of Ca^{2+}-sensitivity. The present results with genetic engineering are also compared with those in the literature on chemically synthesized peptides.

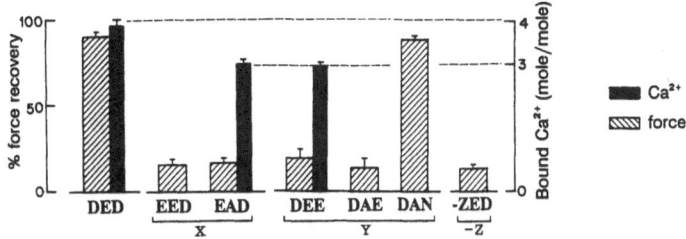

Fig. 2. Force regulation and Ca^{2+}-binding properties of TnC mutants. Each bar indicates the mean ± S.E.M. of at least 3 experiments.

2. Rationale for Mutagenesis

Comparisons of the loop sequences in Fig. 1 indicate that the X coordinate is invariably occupied by an aspartate. The crystal structure of TnC, solved with 2Å resolution, had indicated that aspartate in the X position at the helix-loop juncture has both oxygen atoms of the carboxylate involved in extensive hydrogen bonding to other residues in the loop[3]. Thus it was of interest to know whether the γ-carboxylate of a glutamate would substitute for β-carboxylate of aspartate in this position.

Next, the Y coordinate was of interest in the present study because while it too is an aspartate in sites I and II, the corresponding coordinate is an asparagine in sites III and IV. Since the former sites are Ca^{2+}-specific and have affinities close to two orders of magnitude lower than those of the Ca^{2+}-Mg^{2+} sites III and IV[7], we sought to determine whether an asparagine would be accepted in the Y-coordinate of a Ca-specific site and whether that would influence the overall metal-ion sensitivity.

Finally, a glutamate appears to be essential in the -Z position, from x-ray structural analysis, since both oxygen atoms of the carboxylate group are invariably involved in coordinating the Ca-ion in a bidentate manner[3]. With mutagenesis, we aim to gain direct experimental test of this mechanism as well.

3.0. Performance of TnC Mutants in Skinned Fibers

The results of the various mutants are summarized in Fig. 2. Two types of measurements are given: Cross-lined bars show the force development in skinned muscle fibers with maximal Ca^{2+}-activation, and the black bars show the results of total Ca^{2+}-binding of the purified mutants in solution. The first set of bars are the results with wild type TnC: The force in skinned fibers containing the TnC is observed to be close to normal (85-100%), and the wild type protein binds 4 Ca^{2+} ions mole per mole of protein, which is also the expected result[7].

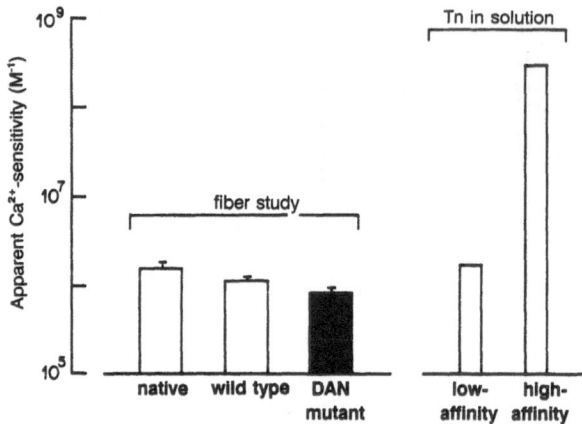

Fig. 3. Ca-affinity of the DAN mutant. Also shown is the Ca-sensitivity of the low- and high-affinity sites in the whole troponin. (Tn data taken from Potter and Johnson[7]) and corrected for physiological ionic strength using the results from Ogawa[13]) .

3.1. D →E Mutations in the X and Y Coordinates in Site II

The mutants in which the aspartate in either the X or the Y coordinates was replaced with glutamate (EED & DEE mutants) failed to regulate force development with Ca^{2+}. We next tested whether this failure might be the result of the close proximity of the side chain carboxyl groups in EE pairs rather than that of D→E replacement per se. Such close proximity in other proteins is known to favor hydrogen bonding between the carboxylates and modify their pKa's[8]), which would have potentially annihilated Ca^{2+}-coordination in the site. Thus, new mutants were made by replacing the middle E with alanine (EAD & DAE) that would eliminate the possible EE-pair effect. As seen in Fig. 2, these proteins remained ineffective in restoring the force regulation. The combined results indicate that close proximity of the carboxyl groups in the EE paired configuration was not a factor in the inactivation of the mutated site in TnC.

Ca^{2+}-binding Measurements. The Ca^{2+} binding was also checked in the EAD and DEE mutants, to investigate whether the loss of force regulation following E for D exchange in the X or Y coordinates was accompanied by alteration of the Ca-binding properties of the site. Compared with the wild type, Ca^{2+} binding in each case was reduced by one unit, indicating that the loss of function in these mutants, in skinned fibers, was associated with annihilation of metal-ion coordination presumably in mutated site II.

3.2 D →N Mutations in the Y Coordinates in Site II

With the substitution of asparagine for aspartate in the Y coordinate, the mutant TnC (DAN mutant) was found to be active in force regulating function of the

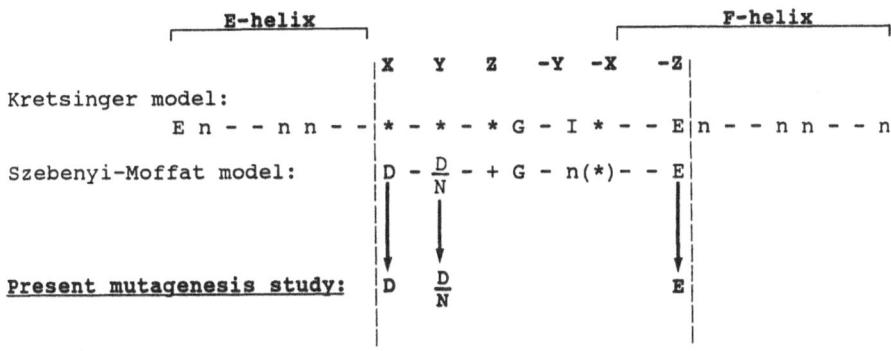

Fig. 4. Primary sequence predictions and results for a putative EF-hand. n indicates preference for a non-polar residue; * any residue with an oxygen donating atom; - no specific residue; + Asp, Asn or Ser.

protein. This indicates that both oxygens in aspartate are non-essential in this instance in the Y coordinate.

We also checked the Ca^{2+}-sensitivity of the fiber with DAN mutant. These results are shown in Fig. 3. The pCa-force relations were determined on fibers with the mutant and compared with wild type. The apparent Ca^{2+}-sensitivity for force regulation is slightly lower for the mutant (pCa50 = 6.05 ± 0.05 for wild type, 5.85 ± 0.02 for DAN). Also shown are data from literature comparing the Ca^{2+}-sensitivity of Tn in solution for both the high affinity (Ca^{2+}-Mg^{2+} sites) and the low affinity sites. We draw attention to the fact that the Ca^{2+}-sensitivity for force regulation with DAN mutant remains close to that of the low affinity sites of the protein dissolved in solution.

3.3. E→D Mutation in the -Z Coordinate

To address the question whether glutamate was critical in -Z coordinate, we substituted an aspartate in this position in site I (putative -ZED mutant). Aspartate was selected for its β-carboxylate, which should still be able to contribute both oxygen atoms for bidentate coordination of Ca^{2+} as predicted for glutamate by x-ray crystallography. The results shown in Fig. 2 indicate that the new mutant was inactive in the skinned fiber. The findings indicate that requirement for glutamate per s is critical in the -Z coordinate and that the mere availability of two oxygen atoms for a bidentate coordination of the Ca-ion is insufficient. Thus stereochemical factors are dominant in the mechanism of Ca-coordination.

4. Mechanisms of Ca-Coordination & Sensitivity

The results of this study can be nicely summarized in the context of the proposed models for a generic EF-hand, and they support the main features of the revised scheme of Szebenyi and Moffat[9] (Fig. 4). We now provide direct experimental proof that the first residue of the loop (X position) requires an aspartate; a glutamate (γ-carboxylate) was not tolerated. These solution studies extend the results of x-ray crystal analysis, which indicated a prominent role for this aspartate at the helix-loop-helix juncture, due to extensive hydrogen bonding of the carboxylate oxygen atoms with other residues in the loop[3].

In the Y coordinate too, we find that a glutamate is unable to replace aspartate but, in this case we show further that an asparagine can replace aspartate. This result was obviously not completely unexpected since an asparagine exists in the same position in loops III and IV of TnC (Fig. 1), but we were also interested in the finding that asparagine replacement in our experiments was unaccompanied by any increase in the overall Ca^{2+}-sensitivity. Therefore, we conclude that difference in the sensitivities of Ca^{2+}-specific and Ca^{2+}-Mg^{2+} sites is either unrelated to asparagine in the Y position or requires other residues in addition to asparagine. Since this obviously is an important question in the modulation of Ca-sensitivity, it would be worthwhile to examine whether other residues in the loops are more effective in modulating the sensitivity of the EF-hand in TnC. The helices flanking the loops could be important in this mechanism as well. For instance, curiously, compared to about 50% homology between the various loops, there is relatively little overall sequence homology between the flanking helices. Another possibility to be kept in mind, in interpreting the present results, is that the quaternary structure of TnC may also participate in the mechanisms of Ca^{2+}-Mg^{2+} and Ca^{2+}-specific sites. This would be interesting to check with chimeras: chimeras, one in which the entire loop III has substituted for loop II, and another in which the complete EF hand for loop III has been duplicated in the position of site II, could be particularly insightful.

Finally, on the residue specificity for Ca^{2+}-binding in the -Z coordinate, the results of protein in solution support the idea that glutamate is essential in this position for a possible bidentate coordination[3]. Further, we now find that aspartate was unsuitable in this position in TnC. Aspartate may be unsuitable because of its shorter length[10]. However, the results leave open the possibility that a side chain of similar length to glutamate even without a carboxylate (e.g., glutamine) may be able to function in some manner. But, it is interesting that the variant loop in ICaBP also shows a glutamate in the -Z coordinate[9]. On the other hand, active sites in the regulatory light chains (RLC1) from several species do have an aspartate in -Z coordinate (see, ref.11). This again raises the possibility that the function of a residue in any given coordinating position in the loop depends on physicochemical characteristics of the particular residue as well as on the overall structure of the rest of the loop and/or quaternary structure of the protein.

5. Relation to Other Studies

It is also instructive to compare the present results with those from another recent approach in which peptides of 34 residues—equalling the unitary EF-hand—are synthesized. The substitution of asparagine for aspartate in the putative Y coordinate of such a 34-residue synthetic peptide (unitary EF-hand) caused a nearly

3-fold decrease in the Ca^{2+}-dissociation constant (i.e., increased Ca^{2+}-sensitivity)[12], but in contrast no such effect was seen with the DA$\underline{N}$ mutant of TnC in skinned fibers. This suggests that the principles governing the properties of the "isolated" unitary EF-hand may differ from those governing the intact protein. This is also a further indication that the quaternary structure of TnC, and possibly of other Ca-binding proteins as well, affects the sensitivity mechanism in the binding of metal-ion in the EF-hand.

ACKNOWLEDGEMENTS

We thank Dr. Christopher Miller (Howard Hughes Medical Institute, Brandeis University) for helping us with synthesis of the TnC encoding gene. Dr. John Russell was helpful throughout the study, and Youngju Ryu helped with some of the experiments. The grant support was from NIH AR33736 & HL37412, NY Heart Association & the Blumkin Fund.

REFERENCES

1. Herzberg, O., & James, M.N.G. *J. Mol. Biol.* **203**, 761-779 (1988).
2. Kretsinger, R.H. *CRC Crit. Rev. Biochem.* **8**, 119-174 (1980).
3. Strynadka, N.C.J., & James, M.N.G. *Ann. Rev. Biochem.* **58**, 951-998 (1989).
4. Putkey, J.A., Sweeney, H.L., & Campbell, S.T. *J. Biol. Chem.* **264**, 12370-12378 (1989).
5. Gulati, J., Sonnenblick,E & Babu, A. *J. Physiol (Lond)* **441**, 305-324 (1991).
6. Ferretti, L., Karnik, S.S., Khorana, H.G., Nassal, M., & Oprian, D. *Proc. Natl. Acad. Sci. USA* **83**, 599-603 (1986).
7. Potter, J.D., & Johnson, J.D. in *Calcium and Cell Function* (Cheung, W.Y., ed) Vol. 2, pp. 145-173 (Academic Press, New York, 1982).
8. Caspar, D.L.D. *Adv. Prot. Chem.* **18**, 37-121 (1963)
9. Szebenyi, D.M.E. & Moffat, K. *J. Biol. Chem.* **261**, 8761-8777 (1986).
10. Gariepy, J., & Hodges, R.S. *FEBS Lett.* **160**, 1-6 (1983).
11. Collins, J.H. *J. Muscle Res. Cell Motility* **12**, 3-25 (1991).
12. Shaw, G.S., Hodges, R.S., & Sykes, B.D. *Biochemistry* **30**, 8339-8347 (1991).
13. Ogawa,Y. *J. Biochem. (Tokyo)* **97**, 1011-1023 (1985).

Discussion

Ashley: In the mutants that don't restore force, how do you know they actually bind to the myofilaments?

Gulati: We do two tests to show that binding has occurred. The first test is to actually take those fibers and run them on the gel. Another quick test that is done in every experiment is after you have loaded the bacterially-synthesized mutant into the fiber and recorded the force response, the fiber is loaded further with the wild type. In none of the instances that I have described do you get extra force.

THE STRENGTH OF BINDING OF THE WEAKLY-BINDING CROSSBRIDGE CREATED BY SULFHYDRYL MODIFICATION HAS VERY LOW CALCIUM SENSITIVITY

Vincent A. Barnett and Mark Schoenberg

Laboratory of Physical Biology
National Institute of Arthritis and Musculoskeletal and Skin Diseases
National Institutes of Health
Bethesda, MD 20892

ABSTRACT

The acto-subfragment-1•ATP state is an important intermediate in the Ca-activated acto-S1 ATPase reaction, suggesting that the myosin•ATP crossbridge seen in muscle fibers similarly may be an important intermediate in the contractile cycle. Treatment of muscle fibers with either para-phenylenedimaleimide (pPDM) or N-phenylmaleimide (NPM) alters the myosin crossbridges so that they bind to the actin filament with about the same affinity as the myosin•ATP crossbridge. Additionally, the treated crossbridges and the myosin•ATP crossbridge have virtually identical attachment and detachment rate constants. Thus the treated crossbridges appear to be reasonable analogues of the weakly-binding myosin•ATP crossbridges of relaxed fibers and studies of the treated fibers may shed some light on the behavior of the physiologically important myosin•ATP crossbridge. We have examined the influence of Ca^{2+} on the binding and rate constants of pPDM- and NPM-treated weakly-binding crossbridges. In agreement with earlier solution studies, we found almost no Ca-sensitivity of the binding of pPDM- or NPM-treated crossbridges.

INTRODUCTION

Since the discovery of the roles of both calcium and crossbridges in muscle contraction, the action of calcium on the crossbridge cycle has been of considerable interest. In solution[1], and in the fiber[2] calcium binds to troponin-tropomyosin and increases the affinity of the strongly-binding myosin subfragment-1 states for actin. It has also been shown, at least in solution, that calcium binding accelerates a step in the acto-myosin ATPase cycle near the phosphate release step[3].

Mechanism of Myofilament Sliding in Muscle Contraction, Edited by
H. Sugi and G.H Pollack, Plenum Press, New York, 1993

133

In solution, studies have shown that the strength of binding of the weakly-binding states of myosin subfragment-1, S1•ATP and S1•ADP•Pi, to actin in the presence of troponin and tropomyosin is relatively insensitive to calcium[3]. These experiments are easy to carry out in solution since both in the presence and absence of calcium, the strongly-binding S1•ADP state is short-lived and the predominant states are S1•ATP and S1•ADP•Pi.

In a fiber, the experimental situation is more complicated. In the absence of calcium, the predominant states are the myosin•ATP and myosin•ADP•Pi states[4]. In the presence of calcium, however, the predominant state is an altogether different one, the strongly-binding myosin•ADP state[5]. For this reason, experiments in fibers to measure the Ca^{2+}-sensitivity of the myosin•ATP crossbridges have not been carried out.

To circumvent this problem, attempts have been made to examine the question of calcium-sensitivity of the weakly-binding myosin•ATP crossbridges by examining the calcium-sensitivity of the binding of other weakly-binding crossbridges, particularly those that do not appear to undergo a major state change when calcium is added. Dantzig et al. have examined the strength of binding of the ATP-γ-S crossbridge. They found that binding of the ATP-γ-S crossbridge was very weak in the absence of calcium and increased significantly in the presence of calcium. A possible difficulty with using the ATP-γ-S crossbridge as an analogue of the myosin•ATP crossbridge is that the crossbridge detachment kinetics for the ATP-γ-S crossbridge, as estimated from the chord stiffness - duration of stretch relationship, more resemble those of the more strongly-binding myosin.PPi crossbridge than those of the weakly-binding myosin•ATP crossbridge[2,6-8].

Another analogue of the weakly-binding myosin•ATP state[9-13], the sulfhydryl-modified crossbridge, has attachment and detachment rate constants, and strength of binding over a range of ionic strengths, that are virtually identical to those of the myosin•ATP crossbridge[13]. The sulfhydryl-modified crossbridge appears locked in a weakly-binding state, apparently incapable of undergoing a transition to a strongly-binding state[13]. The current paper, therefore, examines the calcium sensitivity of the strength of binding of this crossbridge. It is found that the strength of binding of the weakly-binding crossbridge created by sulfhydryl modification is hardly sensitive to calcium over the physiological concentration range.

METHODS

The method of dissecting single rabbit psoas fibers and measuring their chord stiffness - duration of stretch relationship has been reported previously[14]. The technique for locking crossbridges in their weakly-binding configuration by alkylation of their sulfhydryls also has been reported previously[13].

RESULTS

Fig. 1 shows the response of a skinned rabbit psoas fiber at 5°C to exposure to pCa 6.2 and pCa 4.5 contracting solution. It also shows the response of the same

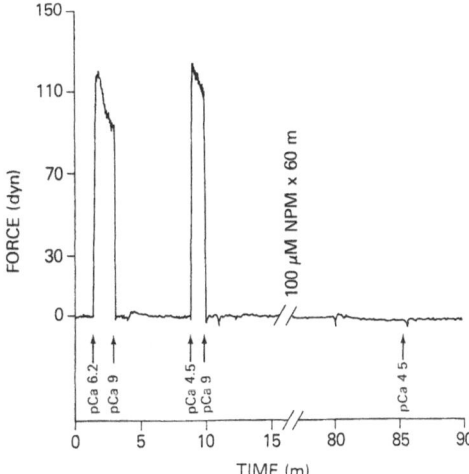

Fig. 1. Effect of 100 μM NPM treatment for 1 hr. on calcium-activated force. Vertical arrows show times of change in calcium. Note that before treatment with NPM, the skinned rabbit psoas fiber produced approximately 100 dyn of force in response to either pCa 6.2 or 4.5, but after treatment for 60 min with NPM, the fiber produced no detectable force in response to calcium.

fiber to pCa 4.5 solution after the fiber was treated for 1 hr. with 0.1 mM NPM. It is clear that treatment of a fiber for 1 hr. with NPM produces crossbridges that generate no active force. Treatment of fibers for 1 hr with 0.2 mM pPDM produces a similar result (data not shown). Previously it was shown that 1 hr. of treatment with either of these agents appears to put all of the crossbridges in a weakly-binding state having kinetics similar to those of the myosin•ATP crossbridge[13]. The data of Fig. 1 are consistent with this. They furthermore show that the weakly-binding crossbridges produced by NPM treatment do not produce force in the presence of ATP and calcium. A similar result was previously shown by Chaen et al. for pPDM treatment[12].

A useful way of examining crossbridge kinetics is to measure a muscle fiber's chord stiffness - duration of stretch relationship. If chord stiffness is plotted versus the logarithm of the duration of stretch used to measure the stiffness, the position of the curve along the logarithmic axis is related to the crossbridge detachment rate constants and the amplitude of the curve is related to the number of attached crossbridges, with the latter being related to the relative magnitude of the attachment and detachment rate constants[15].

Fig. 2 shows a number of experiments designed to examine the calcium-sensitivity of the kinetics of the weakly-binding crossbridges created by treatment with alkylating agents. Each frame, representing one full experiment, shows the chord stiffness - duration of stretch relationship for a normal control fiber in ATP-containing relaxing solution (squares), the same fiber in relaxing solution after it was treated for 1 hr. with an alkylating agent (triangles), and the treated fiber after

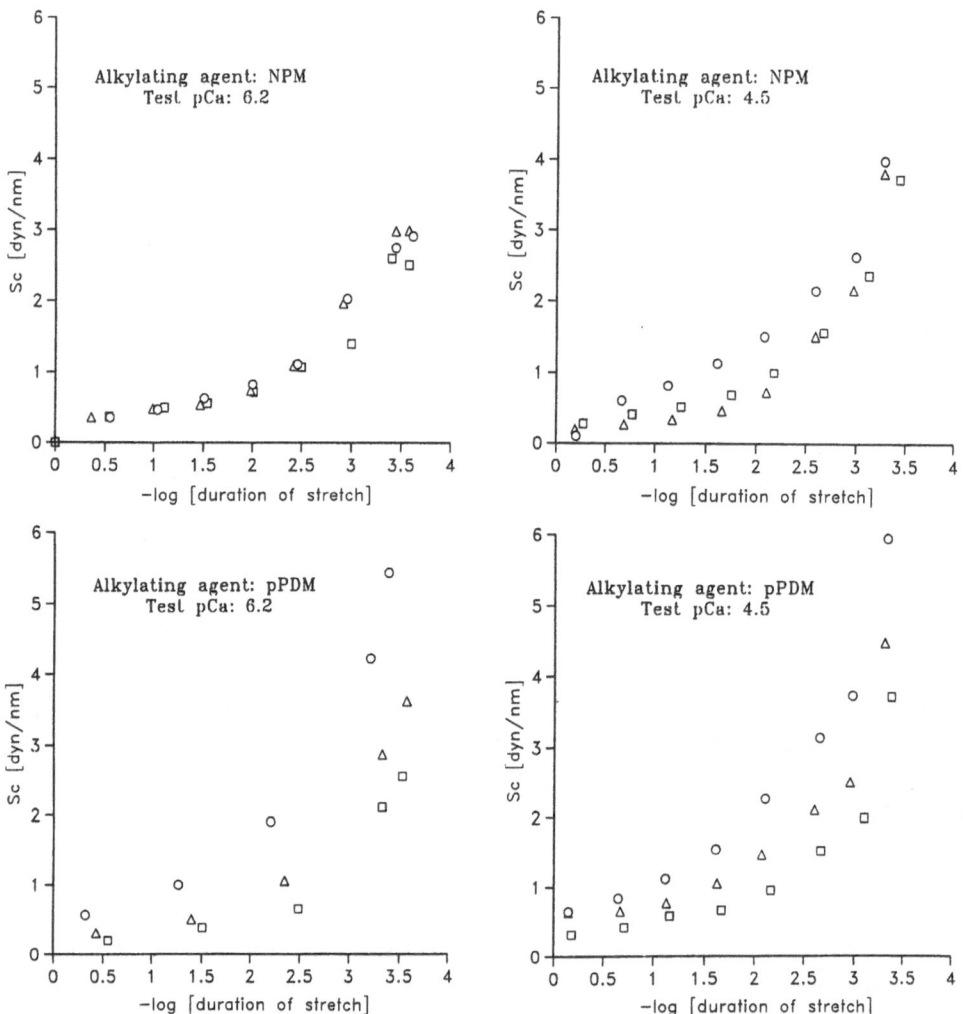

Fig. 2. A montage of experimental results showing effect of calcium on the attachment and detachment rate constants of NPM- and pPDM-treated crossbridges, as measured from the chord stiffness (S_C) - duration of stretch (t_d) relationship. Data obtained from single skinned rabbit psoas fibers bathed in solutions of ionic strength 40 mM at 5°C. Squares (□), untreated fiber under relaxing conditions; Triangles (Δ), same fiber under relaxing conditions after 1 hr of treatment with alkylating agent; Circles (O), treated fiber under activating conditions. Change of detachment rate constant is manifest as a horizontal shift of S_C - log (t_d) relationship; change of attachment rate constant is manifest as a change in vertical scaling of the relationship. It is seen that changes in calcium over the physiological range have very little effect upon these crossbridge properties.

being placed in contracting solution (circles). The alkylating agent was either 0.1 M NPM or 0.2 mM pPDM and the pCa of the contracting solution was either 6.2 or 4.5. The experiments shown in Fig. 2 are representative examples from a much larger number of experiments performed. Two things are clear. Firstly, the treated fibers have kinetics identical to those of the normal relaxed myosin•ATP crossbridge. Secondly, the kinetics of the NPM- or pPDM-treated crossbridges (attachment and detachment rate constants) are more or less independent of calcium concentration.

DISCUSSION

Because of its importance with regard to muscle regulation, the calcium sensitivity of weakly-binding myosin states has been of considerable interest. In solution, where the weakly-binding myosin•ATP and myosin•ADP•Pi states predominate both in the presence and absence of calcium, there has been good evidence that the strength of binding of the weakly-binding states is relatively independent of calcium. One possible exception to this is the case of heavy meromyosin where calcium may have as much as a 5 - 10-fold effect. The pPDM-treated myosin state also has been studied in solution. Solution studies have shown that the two sulfhydryl-reactive groups of pPDM cross-link myosin's SH1 and SH2 sulfhydryls. They show also that the strength of binding to actin of myosin subfragment-1 treated with pPDM exhibits very little calcium sensitivity (< a factor of 2)[11].

The first suggestion that one could react the myosin in fibers with pPDM came from the work of Chaen et al.[12]. They showed that treatment of a muscle fiber with pPDM inhibits the ability of the muscle fiber to make force. Fig. 1 shows the same result for NPM that Chaen et al. showed for pPDM. The data are consistent with previous results that treatment of muscle fibers with pPDM or NPM also inhibits a muscle fiber's rigor stiffness[13]. These results, along with the non-calcium data in Fig. 2, are most easily interpreted in terms of the hypothesis that alkylating agents react with the myosin in skeletal muscle fibers and modify the myosin in such a way that it is irreversibly locked in a weakly-binding crossbridge state.

Fig. 2 confirms that after treatment for 1 hr with either NPM or pPDM, the myosin crossbridges appear to be in a state whose attachment and detachment rate constants are virtually identical to those of the myosin•ATP crossbridges of normal fibers[13]. Additional experiments show this to be true not only for the low ionic strength condition shown in Fig. 2, but also at normal ionic strength (data not shown). Thus the crossbridges created by reaction with the alkylating agents NPM and pPDM have kinetics virtually identical to those of normal weakly-binding myosin•ATP crossbridges. The data of Fig. 2 also show that the kinetics, and therefore the strength of binding, of the crossbridges created by alkylation of myosin, are affected hardly at all by changes in calcium concentration over the physiological range.

While the calcium sensitivity results strictly apply only to the weakly-binding crossbridges produced by reaction with NPM or pPDM, it is none-the-less tempting to speculate about the calcium sensitivity of the normal, weakly-binding, myosin•ATP crossbridges of untreated fibers. For reasons discussed in the Introduction, experiments to measure directly the calcium sensitivity of myosin•ATP crossbridges in fibers have not been carried out. The finding that for NPM- and pPDM-treated myosin the calcium sensitivity exhibited in the fiber is similar in magnitude to that exhibited in solution, and the fact that the kinetics of the NPM-treated, pPDM-treated, and normal myosin•ATP crossbridges are so similar, all strongly suggest that binding of the myosin•ATP crossbridge in the fiber is not calcium-sensitive, just as the binding of myosin subfragment-1•ATP to actin in solution[3] is not calcium sensitive.

REFERENCES

1. Greene, L.E. & Eisenberg, E. *Proc. Natl. Acad. Sci. USA*. **77**, 2616-2620 (1980).
2. Brenner, B., Yu, L.C., Greene, L.E., Eisenberg, E. & Schoenberg, M. *Biophys. J.* **50**, 1101-1108 (1986).
3. Chalovich, J. M. & Eisenberg, E. *J. Biol. Chem.* **252**, 2432-2437 (1982).
4. Marston, S.B. & Tregear, R.T. *Nature* **235**, 23-24 (1972).
5. Siemankowski, R.F., Wiseman, M.D. & White, H.D.. *Proc. Natl. Acad. Sci. USA*. **82**, 658-662 (1985).
6. Dantzig, J.A., Walker, J.W., Trentham, D.R. & Goldman, Y.E. *Proc. Natl. Acad. Sci. USA* **82**, 658-662 (1988).
7. Schoenberg, M. *Adv. Exp. Med. Biol.* **226**, 189-202 (1988).
8. Schoenberg, M. *Biophys. J.* **60**, 690-696 (1991).
9. Wells, J.A. & Yount, Y.G.. *Proc. Natl. Acad. Sci. USA*. **76**, 4966-4970 (1979).
10. Duong, A. & Reisler, E.. *Biochemistry* **28**, 3502-3509 (1989).
11. Chalovich, J.M., Greene, L.E. & Eisenberg, E. *Proc. Natl. Acad. Sci. USA*. **80**, 4909-4913 (1983).
12. Chaen, S., Shimada, M. & Sugi, H. *J. Biol. Chem.* **261**, 13632-13636 (1986).
13. Barnett, V.A., Ehrlich, A. & Schoenberg, M. *Biophys. J.* **61**, 358-367 (1992).
14. Schoenberg, M. *Biophys. J.* **54**, 135-148 (1988).
15. Schoenberg, M. *Biophys. J.* **48**, 467-475 (1985).

Discussion

Morales: Do you know where in the myosin NPM is binding? pPDM cross-links SH1 and SH2, but little is known about NPM.

Schoenberg: We have started to work on that.

Brenner: Was the modification done under relaxing conditions? What happens if you do it under rigor conditions?

Schoenberg: In order to convert the cross-bridges to the weakly binding form, you need to treat under relaxing conditions. In solution, rigor conditions prevent the conversion to weakly binding myosins. The same is true in the fiber. If you treat under rigor conditions, you don't create these weakly binding cross-bridges.

Huxley: I understand this was mainly done at 40 mM ionic strength. One might expect quite non-specific binding between protein molecules, simply because it's at low ionic strength. Is there evidence that the binding you get is in any way related to the binding we are interested in as a preliminary to contraction?

Schoenberg: Yes, there is. We have done a number of experiments to try to relate the binding in the fiber to what you would expect the binding to be between actin and S1-ATP in solution. We changed the ionic strength to see if the binding in the fiber corresponds to the known change in binding strength between actin and myosin, which is measured in solution. The other modification we made to make certain that in the fiber we looked at the kind of binding that one sees between actin and myosin in solution is to change the major anion. We switched between chloride and proprionate. Again, the change you see in the apparent binding in the fiber corresponds to exactly the kind of change you would expect from the change in binding constant measured in solution.

Huxley: Is there any reason for thinking what you see in solution is what happens *in vivo?*

Schoenberg: I'll give a rather short answer to that question, an answer that summarizes about 15 years of research. We have developed a model of the fiber, and using that model, we can accurately predict what happens in the fiber solely from the binding and detachment rate constants measured in solution. That has encouraged us to believe that there is a very strong correlation between what happens in solution and what happens in the fiber.

Cecchi: When you show displacement against force, that is a stiffness measurement only if you are sure that there are no velocity changes during the stretch and there is no viscosity in the fiber. Is that correct?

Schoenberg: If you slide the filaments past each other, and the cross-bridges don't detach during the period of stretch, you are going to measure the force-displacement relationship of the attached cross-bridges. The measurements that Professor Huxley has done, as well as a number of other measurements, suggest that that relationship is approximately linear and should correspond to the cross-bridge stiffness. What my lab has generally measured is a parameter I call the "apparent stiffness" or more formally, the "chord stiffness." This is not a true stiffness in the sense of your correct formal definition. It is a quantity I defined that has the dimensions of a true stiffness, but, unlike a true stiffness, its magnitude depends upon the duration and extent of the stretch used to measure it. It is a useful quantity because if you keep the extent of the stretch at some fixed value, say 2 nm/half-sarcomere, which we usually use, the dependence of the chord stiffness on the duration of the stretch used to measure it provides information about the cross-bridge detachment rate constants (Schoenberg, M. *Biophys. J.* **48**, 467-475, 1985).

Griffiths: When you were showing stiffness traces you were saying that the fall-off of stiffness at longer durations was due to detachment of bridges. I was wondering what the transient looked like when you actually terminated the ramp. Do you have a mono-exponential decline back to zero?

Schoenberg: It's not mono-exponential; it does decay back to zero, though.

Tawada: You said your modified myosin lost its sensitivity to calcium in skinned fibers. But you modified whole skinned fibers with an SH-modifying reagent. So, I'm wondering whether the loss of sensitivity is produced because the tropomyosin-troponin regulatory system was impaired by this SH-modification in your system. Is there any biochemical experiment that indicates this possibility?

Schoenberg: We used both radioactive pPDM and also radio-active N-phenyl maleimide to see where these reagents were binding. The pPDM binds almost exclusively to myosin and maybe titin. The N-phenylmaleimide is somewhat more promiscuous, but it binds in significant amounts only to the myosin heavy chain (Barnett et al. *Biophys. J.* **61**, 358-367, 1992).

ROLE OF PARVALBUMIN IN RELAXATION OF FROG SKELETAL MUSCLE

Tien-tzu Hou, J. David Johnson* and Jack A. Rall

*Departments of Physiology and *Medical Biochemistry*
Ohio State University
Columbus, OH 43210 USA

ABSTRACT

Experiments were done to test the hypothesis that parvalbumin (PA) promotes relaxation in frog skeletal muscle. Single fibers and purified PA from *R. temporaria* skeletal muscle were used to determine the relationship between Ca^{2+} and Mg^{2+} dissociation rates from PA and changes in relaxation rate as a function of isometric tetanus duration at 0°C. Relaxation rate slows as a function of tetanus duration with a rate constant of 1.18 s^{-1}. Recovery of relaxation rate after a prolonged tetanus exhibits a rate constant of 0.12 s^{-1}. Dissociation rate constants for Mg^{2+} and Ca^{2+} from purified PA are 0.93 s^{-1} and 0.19 s^{-1}, respectively. Thus rates of slowing and recovery of relaxation rate may be controlled by Mg^{2+} and Ca^{2+} dissociation from PA, respectively. The influence of temperature on relaxation rate and on Ca^{2+} and Mg^{2+} dissociation rates from purified PA also was examined. The magnitude of slowing of relaxation rate with increasing tetanus duration, relative to the final, steady value of relaxation rate, is greater at 0 than at 10°C. In the 0 to 10°C range, the Q_{10} for relaxation rate increases with increasing tetanus duration. Both of these observations can be explained if the Q_{10} for Ca^{2+} uptake by the sarcoplasmic reticulum is greater than the Q_{10} for Ca^{2+} sequestration by PA during relaxation. When Ca^{2+} and Mg^{2+} dissociation rates from PA at various temperatures are compared to other proposed indicators of PA function, it is concluded that PA facilitates relaxation of frog skeletal muscle in the 0 to 20°C range.

INTRODUCTION

It has been proposed that parvalbumin (PA) promotes relaxation in fast contracting skeletal muscle (see Fig. 1)[1-3]. PA is a soluble Ca^{2+}-binding protein found in high concentrations in the sarcoplasm of vertebrate fast contracting skeletal muscle (for review see Heizmann[4]). In frog skeletal muscle the [PA] is 0.5 to 0.76 mM[5)6] compared to a troponin (TN) concentration of 0.09 mM[7]. PA binds two Ca^{2+} per mol with high affinity (K_d of 10 nM) in competition with Mg^{2+} (K_d of

Mechanism of Myofilament Sliding in Muscle Contraction, Edited by
H. Sugi and G.H Pollack, Plenum Press, New York, 1993

141

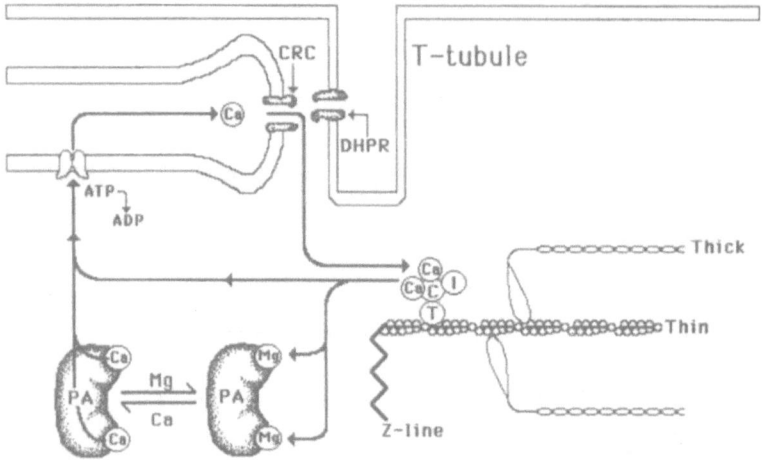

Fig. 1. Illustration of the proposed role of PA in promotion of skeletal muscle relaxation.

100 μM)[8]. Under resting conditions, PA is thought to be loaded predominantly with Mg^{2+} (ref. 9). Thus the net rate of Ca^{2+} uptake by PA during contraction would be determined by the rate of dissociation of Mg^{2+} from PA. This rate is much slower than the rate of Ca^{2+} binding to Ca^{2+}-specific sites of troponin[10]. Thus Ca^{2+} would preferentially bind to TN during muscle activation. In contrast, PA can deplete isolated myofibrils of Ca^{2+} and isolated sarcoplasmic reticulum (SR) can deplete PA of Ca^{2+} (ref. 2). Computer simulation studies indicate that, in theory, Ca^{2+} exchange for Mg^{2+} on PA during relaxation can occur rapidly enough for PA to promote relaxation in frog skeletal muscle by acting in parallel with the SR[11-13]. Furthermore, the content of PA in various mammalian skeletal muscles correlates positively with relaxation speed[14].

If PA promotes relaxation, its effect would be expected to diminish as tetanus duration increases since PA would progressively fill with Ca^{2+}. The time course of slowing of relaxation rate with increasing tetanus duration should be related closely to the rate of dissociation of Mg^{2+} from PA. This result is expected because [PA·Mg], which is directly related to the capacity of PA to buffer Ca^{2+} during relaxation and thus promote relaxation, decreases as Ca^{2+} exchanges for Mg^{2+} during a tetanus. Furthermore since PA is freely soluble in the sarcoplasm[15], relaxation rate should decrease with increased tetanus duration with a time course determined by the rate of Mg^{2+} dissociation from purified PA. Evidence that PA binds Ca^{2+} during a maintained tetanus is derived from electron probe studies which measure total $[Ca^{2+}]$ and studies which measure free $[Ca^{2+}]$. The total increase in $[Ca^{2+}]$ in the sarcoplasm during a tetanus is about 1 mM[16]. This increase is much greater than the concentration of the Ca^{2+}-specific sites on TN (0.18 mM)[7] but is similar to the total Ca^{2+}-binding sites on TN and PA. Relaxation rate does slow progressively with increased tetanus duration in frog skeletal muscles[17-19]. Also the rate of decay of the Ca^{2+} transient slows with increasing tetanus duration[20][21].

There is a direct correlation between the rate of decay of the Ca^{2+} transient and rate of relaxation[20]. This result suggests that relaxation rate is controlled by the rate of Ca^{2+} uptake by and/or release from intracellular Ca^{2+} buffers. Mg^{2+} dissociation from PA during contraction should result in a transient increase in the free $[Mg^{2+}]$ during contraction. A Mg^{2+} transient has been observed in frog skeletal muscle[22].

If PA promotes relaxation, the time course of recovery of relaxation rate after a prolonged tetanus should be closely related to the re-formation of PA•Mg after a tetanus. The rate of re-formation of PA•Mg after a prolonged tetanus is expected to be limited by and thus similar to the rate of Ca^{2+} dissociation from PA[10)22]. Recovery of relaxation rate should be slower than that for progressive slowing of relaxation rate because Ca^{2+} dissociates from PA more slowly than does Mg^{2+} (ref. 23). The observed time course of recovery of relaxation rate in frog skeletal muscle is slower than the time course of progressive slowing of relaxation rate[18)19]. This interpretation suggests that sarcoplasmic Ca^{2+} should not return to the SR until after relaxation is complete. Electron probe studies[24] verify this prediction and indicate that Ca^{2+} does not completely return to the SR until after relaxation. Also this interpretation suggests that the free $[Mg^{2+}]$ should decrease after a tetanus to the pre-tetanus value with a time course similar to that for Ca^{2+} dissociation from PA.

It has been predicted that in skeletal muscle of poikilotherms PA would have its greatest relative effect at low temperatures[11]. According to this prediction, the Q_{10} for the rate of Ca^{2+} uptake by the SR is greater than the Q_{10} for the rate of Ca^{2+} binding to PA•Mg. The Q_{10} for the rate of Ca^{2+} binding to PA•Mg is controlled by the Q_{10} of Mg^{2+} dissociation from PA. This argument leads to two testable predictions. First, the relative magnitude of slowing of relaxation rate with increasing tetanus duration should be greater at lower temperatures. Second, the Q_{10} for relaxation rate from an isometric tetanus should increase with increasing tetanus duration. This last result would be expected because relaxation rate would be controlled by two processes, operating in parallel, with different Q_{10}s, one of which declines exponentially with increasing tetanus duration.

Experiments were designed to quantitatively test the hypothesis that PA•Mg promotes relaxation in fast contracting skeletal muscle of the frog[6)25].

RESULTS AND DISCUSSION

Intact single fibers from the tibialis anterior muscle and purified PA from skeletal muscle of *R. temporaria* were utilized. Mechanical experiments were conducted at 0 and 10°C because the effect of PA is predicted to be greatest at low temperatures where Ca^{2+} uptake by the SR is depressed[11]. Fibers were set to a resting sarcomere length of 2.2 μm. Mg^{2+} and Ca^{2+} dissociation rates from purified PA were measured in the temperature range of 0 to 20°C so that these results could be compared to the mechanical studies and to other data in the literature. The following sections are organized around predictions of this hypothesis that have been tested.

Table 1. Influence of temperature on Mg^{2+} dissociation rate constants from purified PA IVb of *R. temporaria*[6)25)]

Temperature °C	Mg^{2+} dissociation rate s^{-1}	Q_{10}	N
0	0.93 ± 0.02*		5
		1.9 ± 0.1	
10	1.76 ± 0.07		5
		1.9 ± 0.1	
20	3.42 ± 0.14		5

*Mean ± S.D.

Prediction #1: Time Course of Slowing of Relaxation Rate with Increasing Tetanus Duration is Limited by Mg^{2+} Dissociation Rate from Purified PA.

Isometric tetani of various durations were produced in random order. Relaxation rate was calculated as the inverse of the time required for force to fall from 95% to 80% ($1/t_{95-80\%}$) of the peak value after the last stimulus. The 80% point occurred before the "shoulder" of relaxation. This measure of relaxation rate was used because: a) it occurs during the phase of relaxation where sarcomere length is constant and thus the fiber is truly isometric[26)] and b) it significantly correlates with changes in rate of fall of the calcium transient during relaxation[20)].

At 0°C relaxation rate slowed as tetanus duration increased and data was fitted to an exponential equation of the form: $R = R_0 e^{-t\tau_r} + R_s$ where R is relaxation rate, R_0 is initial "extra" rate of relaxation at t = 0, τ_r is rate constant for the effect, and R_s is steady value of relaxation rate[19)]. Mean values (± S.D.) (n = 17) are: $\tau_r = 1.18 ± 0.35$ s^{-1}; $R_0 = 2.89 ± 0.79$ s^{-1}; $R_s = 2.22 ± 0.33$ s^{-1} (ref 6). Thus relaxation rate decreases exponentially with increasing tetanus duration at a rate of $1.18 ± 0.35$ s^{-1} at 0°C. Relaxation rate for a 0.3 s tetanus (shortest tetanus in which peak force is reached) is about 2 times faster than the value of relaxation rate for a 4 s tetanus. Slowing of relaxation rate with increasing tetanus duration probably is not associated with the fall in peak tetanus force since greater than 99% of the slowing of relaxation rate occurs with < 5% decrease in peak tetanus force in a 4 s tetanus. After 4 s of stimulation, there was no further slowing of relaxation rate in tetani as long as 15 s in duration even though peak tetanus force continues to decline slowly.

Terbium (Tb^{3+}) can substitute for Ca^{2+} in binding to Ca^{2+}-binding proteins[27)]. Fluorescence intensity of Tb^{3+} increases upon binding to Ca^{2+} binding sites and it can therefore act as a reporter group for the study of the characteristics of Ca^{2+} binding sites. Tb^{3+} binding to apo-PA occurred within the 2 ms mixing time of the stopped-flow instrument. Thus the rate of fluorescence change upon Tb^{3+} binding to PA saturated with Mg^{2+} provided an indirect measure of the dissociation rate of Mg^{2+} from PA. Experiments were conducted under conditions similar to those in fibers, i.e., 150 mM ionic strength, pH 7.0 and 0°C (for further details, see Hou et

Table 2. Influence of temperature on Ca^{2+} dissociation rate constants from purified PA IVb of *R. temporaria*[6)25)]

Temperature °C	Ca^{2+} dissociation rate s^{-1}	Q_{10}	N
0	0.19 ± 0.01*		5
		2.4 ± 0.1	
10	0.46 ± 0.02		5
		2.2 ± 0.1	
20	1.03 ± 0.03		5

*Mean ± S.D.

al.[6)]). Data were fit with a single exponential with a rate constant of 0.93 ± 0.02 s^{-1} (Table 1)[6)25)]. Thus the Mg^{2+} dissociation rate from purified PA is 0 93 s^{-1} at 0°C. Mg^{2+} dissociation rate from PA IVb was not dependent on [Mg^{2+}] in the range of 10 to 100 mM Mg^{2+} and was not significantly different from that observed for PA IVa. *Conclusion*: Rate of slowing of relaxation rate in frog fibers is similar to Mg^{2+} dissociation rate from purified PA at 0°C. The same conclusion is reached from results at 10°C[25)].

Prediction #2: Time Course of Recovery of Relaxation Rate after a Prolonged Tetanus is Limited by Ca^{2+} Dissociation Rate from Purified PA.

Fibers were given a 4 s conditioning tetanus at 0°C to produce maximum slowing of relaxation rate and then after a variable rest interval were given a 0.5 s test tetanus to measure recovery of relaxation rate. Conditioning and test tetani were produced in pairs. The duration of the test tetanus was selected to just produce peak force. Rest interval between pairs was 30 min. Five to seven pairs of tetani were given in random order. Rest intervals between conditioning and test tetani varied from 0.8 to 30 s. Relaxation rate of the test tetanus was divided by relaxation rate of the conditioning tetanus of its pair. This procedure corrected for any variation in fiber performance throughout an experiment. Time course of recovery of this ratio of relaxation rates with increasing rest interval was fitted with an exponential equation: $R_n = R_i + R_e (1-e^{-t\kappa_r})$, where R_n is the ratio of relaxation rate in a test tetanus divided by relaxation rate in the conditioning tetanus, R_i is the ratio of relaxation rates at zero time interval between tetani which is taken as t = 0, R_e is the magnitude of the effect and κ_r is the rate constant. Mean values (n = 14) are: $\kappa_r = 0.12 ± 0.02$ s^{-1}; $R_e = 0.55 ± 0.17$ and $R_i = 0.95 ± 0.03$ (ref. 6). Since the ratio of R_n after a 30 min rest period to ($R_e + R_i$) (0.99 ± 0.09 [n = 12]) is not significantly different from 1.0, the recovery process is adequately explained by a single exponential function. Thus relaxation rate recovers at a rate of 0.12 s^{-1} at 0°C.

Rate of fluorescence change upon Tb^{3+} binding to PA saturated with Ca^{2+} provided an indirect measure of the dissociation rate of Ca^{2+} from PA. The data is fit with a single exponential with a rate constant of 0.19 s^{-1} $\pm$ 0.01 s^{-1} (Table 2)[6)25)]. Thus Ca^{2+} dissociation rate from purified PA is 0.19 s^{-1} at 0°C. The Ca^{2+} dissociation rate from PA IVb was not dependent on [Ca^{2+}] in the range of 0.5 to 2 mM and was not significantly different from that observed for PA IVa. *Conclusion*: Rate of recovery of relaxation rate after a prolonged tetanus in frog fibers is similar to Ca^{2+} dissociation rate from purified PA at 0°C. The same conclusion is reached from results at 10°C[25)].

Prediction #3: Time Course of the Increase in Free [Mg^{2+}] during a Contraction is Limited by Mg^{2+} Dissociation Rate from Purified PA.

With prolonged voltage-clamp depolarization, the maximum rate of increase in free [Mg^{2+}] as measured by the indicator antipyrylazo III in the sarcoplasm of *R. temporaria* skeletal muscle fibers is 3-4 s^{-1} at 18°C[22)]. Mg^{2+} dissociation rate from PA IVb saturated with Mg^{2+} was measured at 0, 10 and 20°C (Table 1)[6)25)]. Using Q$_{10}$s from Table 1, calculated dissociation rate for Mg^{2+} from PA at 18°C is 3 s^{-1}, *Conclusion:* Time course of increase in free [Mg^{2+}] in frog fibers at 18°C is precisely described by Mg^{2+} dissociation rate from purified PA.

Prediction #4: Time Course of Slowing of Rate of Decline of Free [Ca^{2+}] during Relaxation with Increasing Tetanus Duration is Limited by Mg^{2+} Dissociation Rate from Purified PA.

Rate of decline of the Ca^{2+} transient during relaxation slows with increasing tetanus duration in *R. temporaria* skeletal muscle[20)21)]. At 10°C rate of slowing of decline of the Ca^{2+} transient with increasing tetanus duration is 1.7 s^{-1} (ref. 20). Mg^{2+} dissociation rate from PA at 10°C is 1.76 ± 0.07 (Table 1)[6)25)]. *Conclusion:* Time course of slowing of rate of decline of free [Ca^{2+}] during relaxation with increasing tetanus duration in frog fibers at 10°C is precisely described by Mg^{2+} dissociation rate from purified PA.

Prediction #5: Return of Free [Mg^{2+}] to Baseline after a Contraction is Limited by the Ca^{2+} Dissociation Rate from Purified PA.

With repetitive action potentials or voltage-clamp depolarization, free [Mg^{2+}] in *R. temporaria* skeletal muscle recovered to the prestimulus baseline with an average rate constant between 0.5 and 1.0 s^{-1} at 18°C[22)]. Ca^{2+} dissociation rate from PA IVb saturated with Ca^{2+} was measured at 0, 10 and 20°C (Table 2)[6)25)]. Using Q$_{10}$s from Table 2, calculated dissociation rate for Ca^{2+} from PA at 18°C is 0.9 s^{-1} *Conclusion:* Return of free [Mg^{2+}] to baseline after a contraction in frog skeletal muscle at 18°C is precisely described by Ca^{2+} dissociation rate from purified PA.

Table 3. Time course of exponential slowing of relaxation rate $(1/t_{95-80})$* as a function of temperature in tibialis anterior fibers of *R. temporaria*[25]

	0°C	10°C	Q_{10}#
R_o, s^{-1}	5.72 ± 1.96@	6.88 ± 3.20	1.23 ± 0.41
τ_r, s^{-1}	1.75 ± 0.59	2.96 ± 2.00	1.88 ± 1.16
Rs, s^{-1}	2.75 ± 0.39	10.00 ± 1.62	3.66 ± 0.45
N	5	5	

*Data fit to equation: relaxation rate = $R_o e{-}t\tau_r + R_s$.
#Calculated on a pairwise basis
@Mean ± S.D.

Prediction #6: Return of Free [Ca^{2+}] after Relaxation from a Prolonged Tetanus to Resting Value is Limited by Ca^{2+} Dissociation Rate from Purified PA.

After relaxation is complete, free [Ca^{2+}] has not returned to the prestimulus value but only does so slowly[20]. Rate constant for this slow return of free [Ca^{2+}] is 0.33 s^{-1} in *R. temporaria* skeletal muscle fibers at 10°C[20]. Rate of Ca^{2+} dissociation from PA at 10°C is 0.46 ± 0.02 s^{-1} (Table 2)[6,25]. *Conclusion*: Slow return of free [Ca^{2+}] to the resting value after a prolonged tetanus of frog skeletal muscle at 10°C is similar to rate of Ca^{2+} dissociation from purified PA.

Prediction #7: Return of SR Ca^{2+} Content to Resting Value after a Tetanus is Limited by Ca^{2+} Dissociation Rate from Purified PA.

Electron probe analysis of SR Ca^{2+} content indicates that Ca^{2+} returns to the SR after a prolonged tetanus at 23°C in *R. pipiens* skeletal muscle with a half-time of 1.1s[24]. Using the Q_{10}s for Ca^{2+} dissociation from PA (Table 2)[6,25], calculated half-time for Ca^{2+} dissociation from PA at 23°C is 0.5 s. *Conclusion:* Time course of Ca^{2+} return to SR after relaxation in frog skeletal muscle is similar to Ca^{2+} dissociation from purified PA.

Prediction #8: Relative Magnitude of Slowing of Relaxation Rate is Greater at Lower Temperatures.

This prediction arises from the suggestion that Q_{10} for Ca^{2+} uptake by SR is greater than Q_{10} for Ca^{2+} binding to PA•Mg[11]. Thus at lower temperatures PA•Mg would play a relatively greater role in promoting relaxation. Slowing of relaxation rate with increasing tetanus duration was examined in the same fibers at 10 and 0°C (Table 3)[25]. Results were fit with an exponential equation (see Prediction #1). When analyzed on a pair-wise basis, magnitude of slowing of relaxation rate (R_o) relative to steady value of relaxation rate (R_s) is significantly greater at 0°C (R_o/R_s = 2.08 ± 0.34) than at 10°C (0.72 ± 0.19). Preliminary studies at 20°C also showed that relaxation rate slows exponentially with increasing tetanus duration. *Conclusion:* Relative magnitude of slowing of relaxation rate in frog skeletal muscle is greater at 0°C than at 10°C.

Prediction #9: Q_{10} for Relaxation Rate from an Isometric Tetanus Increases with Increasing Tetanus Duration.

If PA promotes relaxation and if Q_{10} for Ca^{2+} uptake by SR is greater than Q_{10} for Ca^{2+} binding to PA•Mg, then Q_{10} for relaxation should depend on tetanus duration. This is true because Q_{10} for relaxation from a brief tetanus would be a combination of Q_{10}s of SR and Ca^{2+} binding to PA•Mg whereas Q_{10} for relaxation from a prolonged tetanus would reflect Q_{10} of SR alone. Therefore Q_{10} for relaxation would be predicted to increase with increasing tetanus duration. Using data from Table 3 (ref. 25), relaxation rate can be calculated at 0 to 10°C for tetani of various durations. For example, Q_{10} for relaxation rate of 0.5 s tetanus is 2.3 whereas that for a tetanus of 4 s duration is 3.6. Further support for this interpretation of the results is derived from the observation that the Q_{10} for Mg^{2+} dissociation from PA (1.9, Table 1)[6)25)] is the same as the Q_{10} for the exponential rate of slowing of relaxation rate with increasing tetanus duration (τ_r, 1.88, Table 3)[25)]. *Conclusion:* Q_{10} for relaxation rate from an isometric tetanus of frog skeletal muscle increases with increasing tetanus duration.

Prediction #10: During Muscle Relaxation, PA•Mg Can Accumulate Ca^{2+} Rapidly Enough to Contribute to Absolute Rate of Relaxation.

It is important to consider whether Ca^{2+}/Mg^{2+} exchange on PA is rapid enough to contribute to absolute rate of relaxation in a tetanus in frog skeletal muscle. Consider relaxation from a 0.5 s tetanus at 0°C. For muscle to relax, Ca^{2+} bound to Ca^{2+}-specific sites of TN must be removed. Concentration of Ca^{2+}-specific sites of TN is 0.18 mM[7)]. [PA] in frog muscle is 0.76 mM[6)]. If about half of the 2 Ca^{2+}/Mg^{2+} binding sites on PA are occupied by Mg^{2+} at rest then [PA•Mg] is 0.76 mM[6)]. Based on Mg^{2+} dissociation rate from PA (~ 1 s[-1], prediction #1), during 0.5 s of stimulation about 40% of PA•Mg exchanges for PA•Ca. Thus about 60% of PA•Mg or 0.46 mM is present at the beginning of relaxation. Based on this information and the assumption that Ca^{2+} dissociation from Ca^{2+}-specific sites of TN (about 30 s[-1] at 22°C) is not rate limiting[28)], time course of Ca^{2+} removal from troponin by PA•Mg during relaxation can be estimated as: $Y = [0.18 - 0.46 (1-e^{-1t})]/0.18$ where Y is the fraction of Ca^{2+}-specific sites of TN occupied by Ca^{2+} at time t. Time courses of a 0.5 s tetanus at 0°C and predicted Ca^{2+} removal from TN by PA•Mg during relaxation are shown in Figure 2. The actual rate of decline of sarcoplasmic free [Ca^{2+}] would be more rapid still since SR is proposed to function in parallel with PA. An estimate of rate of decline of free [Ca^{2+}] at 0°C can be derived from observations that the rate of fall of free [Ca^{2+}] during relaxation from a 0.5 s tetanus at 5°C is about 9 s[-1] with a Q_{10} of 2.8 (ref. 29). Thus free [Ca^{2+}]

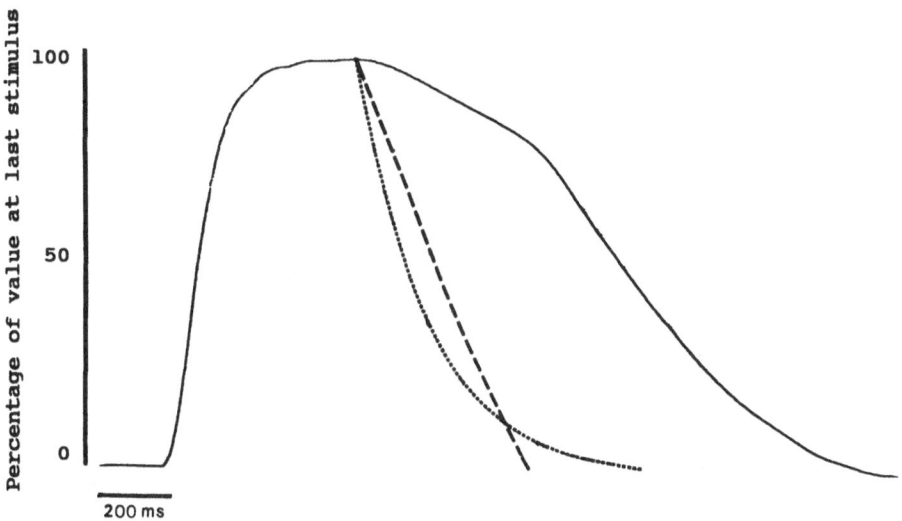

Fig. 2. Comparison of time courses of mechanical relaxation, calculated Ca^{2+} removal from TN by PA•Mg during relaxation and estimated Ca^{2+} transient during relaxation in frog skeletal muscle at 0 °C. Solid line: time course of an isometric tetanus of 0.5 s in a tibialis anterior fiber from *R. temporaria* . Dashed line: decline of [TN•Ca] complex by PA•Mg binding Ca^{2+} during relaxation expressed as a fraction of its maximal value (0.18 mM) just before force starts to decline (see text). Dotted line: relative decrease of sarcoplasmic [Ca^{2+}] from its maximal value (10 μM). Figure taken from ref. 6.

should decline during relaxation in a 0.5 s tetanus at 0°C with a rate constant of about 5 s^{-1}. This fall in free [Ca^{2+}] can be compared to Ca^{2+} removal from TN by PA•Mg in Figure 2. *Conclusion:* Based on this analysis, rate of Ca^{2+} binding by PA is fast enough to contribute to rate of relaxation from a tetanus at 0°C in frog skeletal muscle.

SUMMARY

Results from other studies combined with Ca^{2+} and Mg^{2+} dissociation rates from purified PA determined in this study provide convincing quantitative evidence that during contraction in frog skeletal muscle Ca^{2+} binds to PA•Mg in a reaction that is limited by Mg^{2+} dissociation rate from PA•Mg and that this process is reversed during and after relaxation in a reaction that is limited by Ca^{2+} dissociation rate from PA•Ca (see predictions #3 - #7). Whereas these results are consistent with the hypothesis that PA•Mg promotes muscle relaxation, they do not prove it. It is conceivable that Ca^{2+} binding to PA•Mg may occur too slowly or to too small an extent to contribute significantly to muscle relaxation.

The experiments correlating changes in rates of relaxation with Ca^{2+} binding to PA•Mg and Mg^{2+} binding to PA•Ca (predictions #1 and #2) are crucial in the argument because they represent the first quantitative studies that relate muscle relaxation to kinetics of Ca^{2+} and Mg^{2+} exchange on PA in the same species under the same conditions. Results from these studies provide strong support for the hypothesis that PA•Mg promotes relaxation in frog skeletal muscle at 0 and 10°C. Furthermore, considering the [PA•Mg] in a muscle, Ca^{2+}/Mg^{2+} exchange on PA appears to be rapid enough and extensive enough to contribute significantly to the absolute rate of relaxation from a tetanus (prediction #10). Taken together these results strongly suggest that PA•Mg promotes relaxation in frog skeletal muscle.

ACKNOWLEDGEMENTS

This work was supported by NIH grants AR20792 and DK33727, The Brenner Foundation of Ohio State University College of Medicine and The American Heart Association, Central Ohio Heart Chapter Inc.

REFERENCES

1. Briggs, N. *Fed. Proc.* **34**, 540 (1975).
2. Gerday, C. & Gillis, J.M. *J. Physiol. (Lond.)* **258**, 96-97P (1976).
3. Pechere, J.-F., Derancourt, J. & Haiech, J. *FEBS Lett.* **75**, 111-114 (1977).
4. Heizmann, C.W. *Experientia* **40**, 910-921 (1984).
5. Gosselin-Rey, C. & Gerday, C. *Biochim. Biophys. Acta* **492**, 53-63 (1977).
6. Hou, T.-T., Johnson, J.D. & Rall, J.A. *J. Physiol. (Lond.)* **441**, 285-304 (1991).
7. Yates, L.D. & Greaser, M.L. *J. Biol. Chem.* **258**, 5770-5774 (1983).
8. Potter, J.D., Johnson, J.D., Dedman, J.R., Schreiber, W.E., Mandel, F., Jackson, R.L. & Means, A.R. In *Calcium-binding Proteins and Calcium Function* (Wasserman, R.H., Corradino, R.A., Carafoli, E., Kretsinger, R.H., Maclennan, D.H. & Siegel, F.L.) 239-250 (North-Holland, New York, 1977)
9. Haiech, J., Derancourt, J., Pechere, J-F. & Demaille, J.G. *Biochem.* **18**, 2752-2758 (1979).
10. Robertson, S.P., Johnson, J.D. & Potter, J.D. *Biophys. J.* **34**, 559-569 (1981).
11. Gillis, J.M., Thomason, D., Lefevre, J. & Kretsinger, R.H. *J. Muscle Res. Cell Motility* **3**, 377-398 (1982).
12. Baylor, S.M., Chandler, W.K. & Marshall, M.W. *J. Physiol. (Lond.)* **344**, 625-666 (1983).
13. Cannell, M.B. & Allen, D.G. *Biophys. J.* **45**, 913-925 (1984).
14. Heizmann, C.W., Berchtold, M.W. & Rowlerson, A.M. *Proc. Nat. Acad. Sci.* **79**, 7243-7247 (1982).
15. Gillis, J.M., Piront, A. & Gosselin-Rey, C. *Biochim. Biophys. Acta* **585**, 444-450 (1979).
16. Somlyo, A.V., Gonzalez-Serratos, H., Shuman, H., McClellan, G. & Somlyo, A.P. *J. Cell Biol.* **90**, 577-594 (1981).
17. Abbott, B.C. *J. Physiol. (Lond.)* **112**, 438-445 (1951).
18. Curtin, N.A. *J. Muscle Res. Cell Motility* **7**, 269-275 (1986) .
19. Peckham, M. & Woledge, R.C. *J. Physiol. (Lond.)* **374**, 123-135 (1986).
20. Cannell, M.B. *J. Physiol. (Lond.)* **376**, 203-218 (1986).
21. Blinks, J.R., Rudel, R. & Taylor, S.R. *J. Physiol. (Lond.)* **277**, 291-323 (1978).
22. Irving, M., Maylie, J., Sizto, N.L. & Chandler, W.K. *J. Gen. Physiol.* **93**, 585-608 (1989).

23 . Ogawa, Y. & Tanokura, M. *J. Biochem.* **99**, 81-89 (1986) .

24. Somlyo, A.V., McClellan, G., Gonzalez-Serratos, H. & Somlyo, A.P. *J Biol Chem.* **260** 6801-6807 (1985).

25. Hou, T.-T., Johnson, J.D. & Rall, J.A. *J. Physiol. (Lond.)* **449**, 399-410 (1992).

26. Edman, K.A.P. & Flitney, F.W. *J. Physiol. (Lond.)* **329**, 1-20 (1982).

27. Dockter, M.E. *In Calcium and Cell Function,* vol. 4 (Cheung, W.Y.) 175-208 (Academic Press, New York, 1983).

28. Johnson, J.D., Robinson, D.E, Robertson, S.P. Schwartz, A. & Potter, J.D. In *The Regulation of Muscle Contraction* (Grinnell, A.D. & Brazier, M.B.) 241-257 (Academic Press, New York, 1981).

29. Cannell, M.B. *J. Physiol. (Lond.)* **329**, 44-45P (1982).

Discussion

Pollack: You found that the rate of relaxation is a function of the duration of the tetanus, but several other things go on as the tetanus duration changes. I therefore wonder whether you can be certain that the effect is due to parvalbumin. The first is that, as you know, the ends of the single fibers tend to shorten during contraction, and with a longer duration contraction the ends may shorten more than in the shorter duration contractions. Since the relaxation involves undoing these dynamics, I wonder whether the sarcomere dynamics may have something to do with the differences in the rate of relaxation. The second possibility is one that Stuart Taylor recently reported: that the single fibers get significantly fatter during contraction (Neering et al. *Biophys J.* **59**, 926-933, 1991). The process apparently involves some dynamics of water—either going from the ends of the cell to the middle or from the outside to the inside. Again, this would have to be reversed during relaxation. I wonder whether this could also play a role.

Rall: The best answer I can give is that we measured relaxation rate during the so-called isometric period. Edman and Flitney have shown that the striation spacing during that period of time is constant [*J. Physiol.* (Lond.) **329**, 1-20, 1982]. We did all these experiments at a sarcomere length of 2.2 μm, so we are not talking about phenomena associated with stretched fibers. That's the first point. The second point is that the data fit very well with these results, and they fit in a quantitative sense. So, while I don't know about the movements of water, I would argue that the data fit very well. The third point is that one of the reasons for developing a skinned muscle preparation is to be able to manipulate directly the things like the parvalbumin concentration. In principle, we should be able to recreate the relaxation rate one measures in the intact fiber.

Curtin: I wanted to ask about the isometric phase of relaxation and the exponential phase of relaxation. Do you ever see these two phases in your twitch-type responses?

Rall: We haven't yet seen those two phases. Sometimes you can see them in an isometric twitch in an intact fiber, but usually you don't. What we are really seeing is essentially a relaxation rate. With regard to the different phases, we have noticed in the intact fibers by prolonging tetanus duration, for example, that it's not just the isometric phase that increases but actually the exponential phase too.

Curtin: We saw that too, but I think that one wants to know whether what you're looking at is an isometric phase or one with movement.

Rall: Right. I would imagine that in the case of this twitch we have sarcomere shortening and sarcomere lengthening. So we don't have a truly isometric situation.

Kushmerick: During the 15 second tetanus, the rate of relaxation wasn't altered. Yet, there must have been a very large increase in the inorganic phosphate content and possibly a decrease in pH. Therefore, under those conditions, what is the explanation for the absence of effects on relaxation rate? Increase in Pi and decrease in pH predict a slowing.

Rall: Yes, one of the points I wanted to emphasize is that the complete effect of the slowing in relaxation at $0°$ occurs in four seconds, and that 10 further seconds of stimulation cause very little change. Of course, during those 10 seconds of stimulation, one would expect changes in ADP and inorganic phosphate. So I would conclude that this response is not associated with those components. On the other hand, this effect occurs with very little fatigue. If there is a large amount of fatigue in the fiber, maybe these other factors come into play.

Ashley: From what I remember, our modelling couldn't explain the prolonged elevated free calcium during relaxation by simply assuming a slow K_{off} for Ca^{2+} from either parvalbumin or whatever, because with a slow K_{off} the sarcoplasmic reticulum pulls the calcium down immediately (Ashley, Moisescu, and Rose, 1974). So there must be some other alteration that will produce this elevated free calcium.

Rall: All I showed was a correlation with the results from Cannell at $10°$ C in frog fibers [*J. Physiol. (Lond.)* **376**, 203-218, 1986]. He found that the free calcium concentration approached the baseline after relaxation· with a rate of 0.33 per second. I noted the fact that this rate is very similar to the calcium off rate from purified parvalbumin of 0.46 per second.

Edman: Jack, do you think that parvalbumin is uniformly distributed in the cell? The reason I am asking is that the non-uniform sarcomere behavior during relaxation disappears to a great extent after fatiguing the fiber when frog parvalbumin is no longer active.

Rall: We have done no experiments on that. Actually, the person to refer to is Jean-Marie Gillis. He has looked at the diffusibility of parvalbumin in fibers and is better able to comment on that point.

Gillis: Yes, we studied the diffusibility of parvalbumin (Gillis, J.M. et al. *Biochim. Biophys. Acta* **585**, 444-450, 1979). When you take a fiber, skin it mechanically, and keep it in oil, you can transfer it to a gel for studying immuno-diffusion. We found that the parvalbumin leaves the fiber very easily. So, that fact points out that the parvalbumin is not bound somewhere in the cell; it is readily diffusible. Also, this protein is reluctant to bind to any sort of protein. It has a very independent mind!

Edman: Does it disappear from chemically skinned fibers?

Gillis: Yes, of course, it disappears in chemically skinned fibers because you are in solution. It is just washed out.

Maughan: Since there are two isoforms of parvalbumin, both at approximately equal concentrations, have you measured any differential properties between these two? And, in fact, do you attach any physiological significance to the fact that there are two isoforms?

Rall: I didn't have time to mention it in the talk, but the magnesium and calcium disassociation rates are the same for each isoform. That was a little disappointing. So we have no evidence for any difference with regard to this parameter in the function of these isoforms.

ter Keurs: When calcium comes off TnC and binds to parvalbumin, it drives off a magnesium. Is it possible that that magnesium may drive off other calciums from other TnC at other sites, so that the process can become regenerative?

Rall: I don't know the answer to that, except to say that the calcium-magnesium binding sites on parvalbumin are thought to be very similar to the calcium-magnesium binding sites on troponin, and the calcium-specific sites are thought to have a much lower affinity for magnesium. So, I don't think something like that would happen.

T. Yamada: How does acidosis affect the slowing of relaxation? What about the intracellular pH change?

Rall: Under the circumstances we have studied, where there is little fatigue of force, there is likely to be a very small change in the intra-cellular pH. On the other hand, if we drive the muscle to a fatiguing type of contraction, there will be large changes in the products of ATP hydrolysis, including hydrogen ions. We did not study that effect. Nancy Curtin may know more about this than I do. Do you want to add something?

Curtin: It's clear that acidosis and fatigue produce effects that are quite similar to prolonged stimulation in the sense that the isometric part of the relaxation becomes slower. Certainly, the pH effects there are probably due to slowing of calcium pumping by the SR and not anything to do with parvalbumin. You can get things that look like the same effects by a different route.

Gillis: It has been known for a long time that reducing the pH has no effect on calcium binding to parvalbumin.

Homsher: A question for the parvalbumin people. Parvalbumin is going to cost the muscle a lot of energy to make and maintain. What is the survival advantage for the organism?

Rall: If I could paraphrase something that Wilfred Mommaerts said a long time ago, I'd say "relax now and pay later." What I mean by this statement is that parvalbumin, by acting in parallel with the sarcoplasmic reticulum, promotes relaxation "now" without directly using ATP. ATP is used "later" to return calcium slowly to the SR after relaxation is complete.

EVIDENCE FOR THE EXISTENCE OF ENDOTHELIAL FACTORS REGULATING CONTRACTILITY IN RAT HEART

Saul Winegrad

Department of Physiology
School of Medicine
University of Pennsylvania
Philadelphia, PA 19104-6085 U.S.A

ABSTRACT

Force developed by isolated papillary muscle decreases as the cross-sectional area increases. The basis for this decline in force is not clear in as much as theoretical considerations and experimental data have indicated that the rate of diffusion of oxygen into thin bundles should not be limiting. Decline of maximum Ca-activated force with increasing cross-sectional area of detergent skinned papillary muscle can be attributed to the accumulation of inorganic phosphate in the center of the bundle. In both cases, the bundle of intact cells with a possible limitation of diffusion of oxygen into the bundle and of skinned cells with a limitation of diffusion of P_i outward, the lowest level of activity should be in the center of the bundle. We have used quantitative histochemistry for measuring Ca- and actin-activated myosin ATPase activity in cryostatic sections of rapidly frozen isolated traveculae. The technique is very sensitive and has sufficient spatial resolution to resolve individual myofibrils. At different times after dissection, ventricular trabeculae were quickly frozen, transversely sectioned and Ca- and actin-activated myosin ATPase, measured in serial sections both without and with 1 μM cAMP in the assay solution. In none of over 40 trabeculae studied was there an inward gradient of actin-activated ATPase activity of myosin. The most superficial cells had very low enzymatic activity. Cyclic AMP decreased the gradient by raising the enzymatic activity of the less active cells more that the more active cells. Ca-activated myosin ATPase was always uniform across the transverse section. These observations are incompatible with a limitation of diffusion of oxygen inward or of an inhibitory factor produced by muscular activity outward. The results are, however, consistent with the production of a regulatory factor within the bundle.

The role of endothelial cells in the regulation of tone of vascular smooth muscle was first demonstrated by Furchgott and Zawadski[1]. In preparations with and without functioning endothelial cells, they found a difference in the response of

Mechanism of Myofilament Sliding in Muscle Contraction, Edited by
H. Sugi and G.H Pollack, Plenum Press, New York, 1993

155

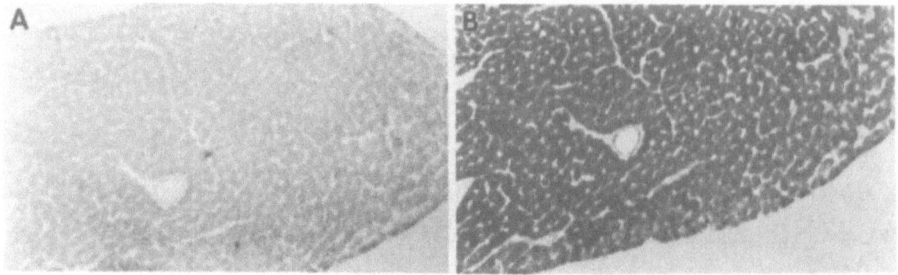

Fig. 1. The ATPase activity of actomyosin (left) and Ca-activated myosin (right) in cryostatic (serial) sections of a rat heart quickly frozen after prompt removal from the animal[8)9)]. The enzymatic activity is directly proportional to density. The Ca-activated ATPase activity is normally about 2.5 times that of actomyosin[11)].

vascular smooth muscle to agents known to alter vasomotor tone. Neurotransmitters such as acetylcholine and humoral agents such as bradykinin required the presence of the endothelial cells for their action on vascular tone to occur. Subsequently several substances with marked effect on vascular tone have been shown to be elaborated by the endothelial cells. These include nitric oxide, a potent vasodilator formed from arginine[2)], and endothelin, a peptide with very stong vasoconstricting effect[3)]. Endothelial cells also respond to changes in local oxygen tension and to shear force produced by fluid flowing past the cells[4)].

The possibility that endothelial cells may also influence the contractility of cardiac myocytes was raised by Brutsaert and his collaborators[5)6)]. They selectively damaged the endocardial endothelium of isolated cardiac trabeculae by a very brief exposure of the tissue to Triton X-100 and then observed both a decrease in peak force and an earlier onset of relaxation during the contraction of the tissue. Following up on these studies, Smith and coworkers[7)] showed that the medium incubating a culture of cardiac endothelial cells could reverse the changes produced in the contraction by damaging the endocardial endothelium.

We have investigated the contribution of endothelial cells to the modulation of cardiac contractility by two different approaches. One involved the measurement of actomyosin ATPase activity in myocardial cells and the relation of differences in enzymatic activity to the proximity to endothelial cells[8)9)] In the second the coronary venous effluent from an isolated perfused heart was assayed for activity in changing the contraction of heart muscle[10)].

A. Studies of Actomyosin ATPase Activity

When a heart is removed from an animal and suspended as an isolated, perfused working heart, the coronary perfusion is reestablished within one minute following removal of the organ from the animal. Within 10-30 minutes, the isolated heart has become very stable in its hemodynamic function as measured by systolic pressure,

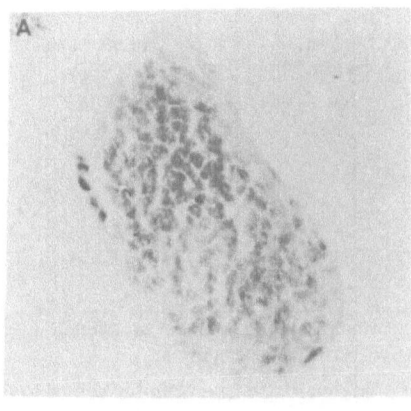

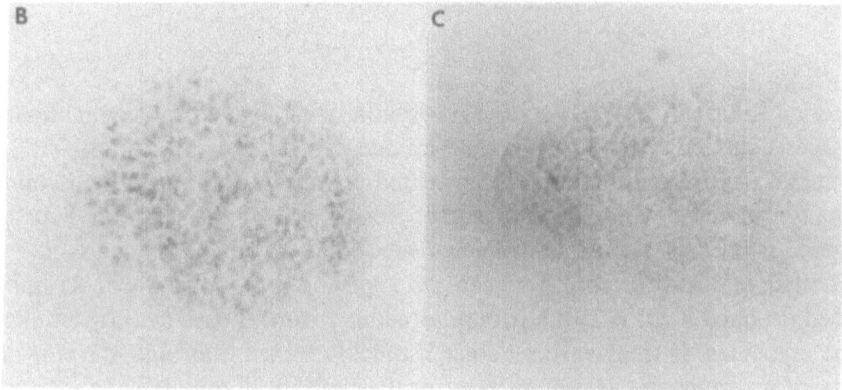

Fig. 2. The actomyosin ATPase activity of cells in transverse sections of quickly frozen isolated superfused cardiac trabeculae of different thicknesses[8][9]. The ATPase activity is directly proportional to the density, which is due to accumulation of reaction product.

cardiac output, cardiac power, cardiac efficiency and oxygen extraction. The level of function is then maintained for at least 2-3 hours.

The ATPase activity of the actomyosin in the cardiac cells can be measured with a spatial resolution of less than 1 micron by quantitative histochemistry applied to thin cryostatic sections of the isolated heart after it has been quickly frozen. In the sections of heart frozen immediately after removal from the animal and of hearts that had been perfused *in vitro* for periods as long as 120 minutes, the actomyosin ATPase activity was always uniform among the myocardial cells (Fig. 1). No gradients were observed near blood vessels or near the endocardium.

The results with isolated papillary muscles or isolated endocardial trabeculae were quite different (Fig. 2). A very long period of time is required before the mechanical performance of the isolated tissue becomes stable. In general at least 100-120 minutes of superfusion are necessary. During this period, the contraction generally declines continuously and becomes shorter. In a minority of cases, there is an early rise in developed force before the decline begins to occur. Once the

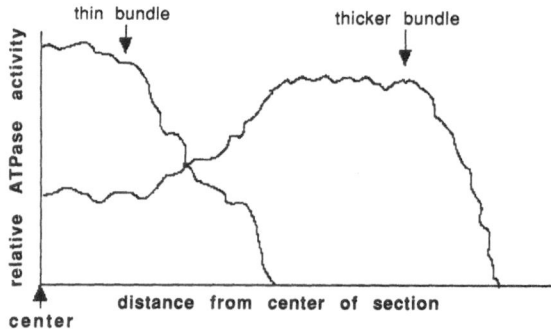

Fig. 3. Density of reaction product along a line from the center to the surface of a transverse section of tissue bundles. Two typical examples with bundles of different thicknesses are shown.

contraction has stabilized, it can remain essentially unchanged for several hours. If the tissue is chemically skinned by a 30 minute exposure to 1% Triton X-100, the maximum Ca-activiated force can be measured. Skinning the preparation terminates the change in the development of force. Regardless of the stage at which the detergent has been used, maximum Ca-activated force becomes very constant.

Another major difference between the superfused isolated trabecula and the perfused, isolated heart is in the pattern of actomyosin ATPase activity among the cardiac myocytes. In transverse sections of quickly frozen trabecula, the enzymatic activity of the myocytes is not uniform for the first 60-90 minutes of superfusion. During this time, which corresponds to most of the period of stabilization of mechanical function, there are gradients of ATPase activity among the myocytes, particularly in the vicinity of the small arteries and the endocardial endothelium. The first 1-2 cell layers next to a small blood vessel have a higher ATPase activity, and the cells immediately below the endocardial endothelium have a lower ATPase activity. This pattern of ATPase activity cannot be explained by a simple limitation of oxygen diffusion from the surface of the trabecula as the gradient is in the wrong direction, and the gradient exists around the blood vessels even though there is no perfusion of oxygenated fluid within them.

Another characteristic of the pattern of ATPase activity is that it varies with the transverse cross-sectional area. In thin bundles there are two zones of ATPase activity with higher enzymatic activity in the center (Fig. 2,3). In thicker bundles, a third zone of intermediate level of activity appears in the center, producing a donut-like pattern. As the bundles become thicker, the level of ATPase activity of the most active cells in the tissue section decreases. In serial sections of these tissues Ca-activated ATPase activity of myosin, which does not require interaction between actin and myosin, is uniform regardless of the thickness of the tissue bundle. Nonuniformity is specific for actomyosin ATPase activity and is not the result of some nonspecific effect such as sectioning.

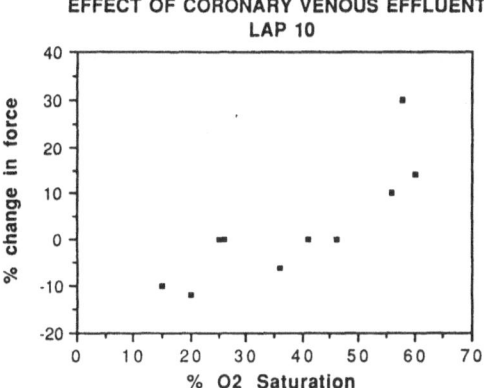

Fig. 4. The relation between the change in peak tension produced when the normal superfusion medium of an isolated trabecula is replaced by coronary venous effluent and the oxygen tension (given as % saturation) of the effluent before it has been reequilibrated with 95% O_2:5% O_2.

There are important differences between the conditions for endothelial cells in isolated perfused hearts and isolated superfused trabeculae. Since the perfusion of the coronary circulation is rapidly re-established after removal of the heart from the animal, endothelial cells in the isolated intact heart are still exposed to an environment similar to that of the heart in situ. In the case of the isolated trabecula, the endothelial cells lining the blood vessels no longer experience shear force or high oxygen tension, and the endocardial endothelium is exposed to a different kind of physical force from what is normally present in the in situ beating heart. Since differences in oxygen tension and in shear force are known to alter the release of endothelial derived factors operating on smooth muscle function, it is reasonable to consider that a similar mechanism might be involved in the changes in ATPase activity in myocardial cells.

B. Assay of Coronary Venous Effluent

In order to determine whether coronary vascular endothelial cells release substances that alter the contractiltiy of the heart, a new technique has been developed in which the effluent from the coronary sinus of an isolated perfused working heart is collected and, after reoxygenation, is used to superfuse a trabecula isolated from another heart. The systolic pressure and the filling pressure of the perfused heart are controlled. Cardiac output, coronary blood flow and both arterial and coronary sinus oxygen tensions are measured. The effect of the coronary venous effluent on the shape fo the isometric contraction of the isolated trabecula is measured after the mechanical performance of the trabecula has stabilized.

In a majority of experiments, the coronary venous effluent altered the peak tension and the duration of the contraction without changing the rate of rise of force. The direction and the amplitude of the change were related to the oxygen tension of the coronary effluent before it had been reoxygenated prior to its use as a superfusion medium (Fig. 4). When the oxygen tension had been high, the coronary effluent increased peak developed force, and when oxygen tension had been low, peak developed force declined. With the oxygen tension in the intermediate range, no change in contractility occurred. When these studies were repeated after the trabecula had been treated with 0.5% Triton X-100 in Krebs' solution for a sufficiently brief period to damage only the endocardial endothelium selectively, the response of the trabecula to coronary venous effluent was substantially altered. The positive effect of coronary effluent originally having a high oxygen tension essentially disappeared and the negative effect at low oxygen tension was reduced.

The amplitude but not the direction of the effect of coronary effluent was related to the rate of coronary flow in the perfused heart at the time the coronary effluent was collected. As the rate of coronary flow increased the amplitude of the change in contractility produced by the coronary effluent decreased regardless of whether the change was positive or negative. At the highest levels of coronary flow no change was produced by the coronary effluent. There are at least two explanations for the effect of coronary flow, a change in shear force on the endothelial cells and a simple dilution of the secretion of the endothelial cells by the larger volume of fluid flowing past the cells per unit time.

It is important to note, however, that the rates of coronary flow that occur in isolated perfused hearts are considerably greater than those that normally occur in the in situ heart. The lowest level of coronary flow observed with the isolated heart is similar to what occurs in the in situ heart in an animal exercising intensively. There are indications from other studies that the endothelial cells in the unperfused vessels may not produce cardioactive factors. If this is correct, then as coronary flow increases from zero, the rate of production of cardioactive factors rises to a maximum within the normal physiological range, peaking during normal intense exercise and declining at unphysiologically high rates of flow.

C. An *in vitro* Model

With the observed relations between the effect of endothelial derived factors and the levels of both oxygen tension and coronary flow, a model has been derived to explain the patterns of actomyosin ATPase activity in the transverse sections of superfused trabeculae. The properties of the model are: 1) release of an up regulating substance by endothelial cells in blood vessels at a rate that is proportional to the local oxygen tension and 2) inhibition of the release of cardioactive substances from the endocardial endothelium because of the abnormal shear force on the surface of the superfused bundle of cells.

The model (Fig. 5) reproduces the three major characteristics of the patterns of actomyosin ATPase activity in transverse sections of the trabeculae: 1) In thin trabeculae there are 2 zones of ATPase activity with the higher activity in the center; 2) in bundle of intermediate or greater thickness, there are 3 zones of activity with the highest in the middle and the lowest at the surface; and 3) as the thickness of bundles increases the level of maximum ATPase activity decreases. While this model is consistent with the data, it is important to recognize that the model is not unique,

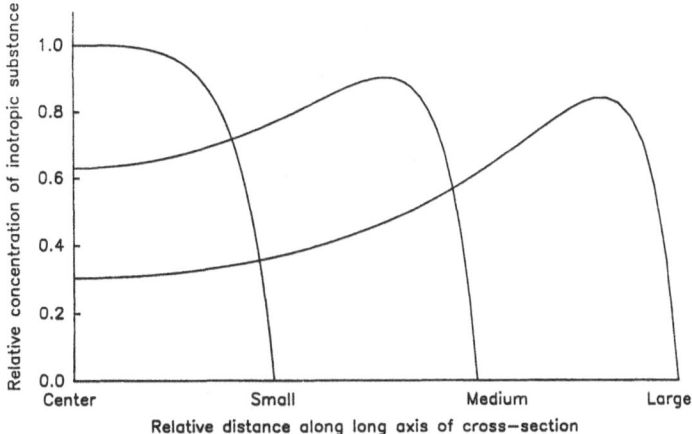

Fig. 5. The calculated actomyosin ATPase activity along a line from the center to the surface of a transverse section of each of three different hypothetical trabeculae of different thicknesses. Details of the model given in the text.

and other explanations such as the production of a negative or down regulating factor by the endocardial endothelium or the washout of the up regulating substance from the bundle surface by the high rates of superfusion could give the same hypothetical ATPase distribution.

DISCUSSION

The new experiments that have been described demonstrate several properties of myocardial tissue:

1) Under normal conditions of perfusion of blood vessels and shear force on endothelial cells, the ATPase activity of actomyosin is uniform. When the physical and chemical environment of endothelial cells is changed, the properties of contractile cells become nonuniform with gradients developing in the immediate vicinity of the endothelial cells.

2) Coronary blood vessels modify the medium perfusing them, presumably by adding active substances, and, as a result, the medium can alter the contractile properties of the myocardial cells exposed to the medium.

3) The direction of the change in contractility produced by the effluent from the coronary vascular system is related to the coronary venous oxygen tension and therefore presumably the tissue oxygen tension.

4) The amplitude of the change in contractility produced by coronary vascular effluent, regardless of direction, is related to the rate of coronary flow, presumably as a result of different shear forces on the vascular cells.

5) Damage to the endothelial cells in the assay bundle of cardiac muscle markedly changes the response of the tissue to coronary effluent.

These observations collectively suggest that after passage of the perfusion medium through the coronary vasculature, the medium contains four different kinds of factors representing two different kinds of regulatory systems, one up regulating and the other down regulating, as regards modification of the contractility of cardiac muscle. Among the properties of the contractile system that are regulated is control of the interaction between actin and myosin, as evidenced by the change in the ATPase activity of actin activated but not Ca-activated myosin. Two different components of each system are present in the effluent, a signal to the endothelial cells and a cardioactive substance produced by the endothelial cells in response to the signal. The presence of both components is shown by the change in response of the trabecula to the coronary effluent following by damage to the endocardial endothelium in the trabecula. The fact that although the response changes, it does not disappear indicates that a substance acting directly on the myocytes is present as well as a substance modulating endothelial cell function.

The presence of a mechanism for modulating cardiac contractility in response to oxygen tension and coronary flow would help balance the supply of energy to the myocardium with the work performed by the heart. As the oxygen tension in the tissues drops, the contractility decreases, lowering the work performed by the heart, and as the rate of coronary flow rises to levels near its normal peak, the contractility begins to decline, thus decreasing the likelihood that the work performed by the heart will outstrip its energy supply.

ACKNOWLEDGEMENTS

The research reported in this paper was supported by grants from The National Institutes of Health, U.S.P.H.S. and from the W.W. Smith Charitable Trust.

REFERENCES

1. Furchgott R. & Zawadski J. *Nature* **288**, 373-376, (1980)
2. Palmer R., Ferrigi A. & Moncada S. *Nature* **327**, 524-526, (1987).
3. Yanagisawa M., Kurihara H., Kimura S., Tomobe Y., Kobayashi M., Mitsui Y., Yasaki Y., Goto K. & Masaki T. *Nature* **332**, 411-415, (1988).
4. Rubanyi G., Romero J. & Vanhoutte P. *Am. J. Physiol.* **250**, H1145-1149, (1986).
5. Brutsaert D., Meulemans, Sipido K. & Sys. S. *Circ. Res.* **62**, 358-366, (1988).
6. Brutsaert D. *Ann. Rev. Physiol.* **51**, 263-273, (1989).
7. Smith J., Shah A. & Lewis M.J. *J. Physiol. (Lond.)* **439**, 1-14, (1991).
8. Lin L.-E., McClellan G., Weisberg A. & Winegrad, S. *J. Physiol. (Lond.)* **441**, 73-94, (1991).
9. McClellan G., Weisberg A., Kato N., Ramaciotti C., Sharkey A. & Winegrad S. *Circ.. Res.* **70**, 787-803 (1992).

10. Ramaciotti C., Sharkey A., McClellan G. & Winegrad S. *Proc. Natl. Acad. Sci. USA*, **849**, 4033-4036 (1992).
11. Winegrad S, Weisberg A., Lin L.-E. & McClellan G. *Circ. Res.* **58**, 83-95, (1986).

Discussion

ter Keurs: Saul, what evidence do you have that shear forces are involved, instead of simply stretch of endothelial cells of the coronary system? The second question is, which endothelial cells are involved in this process? Endothelial cells of arterioles, venules, or capillaries?

Winegrad: The answer to the first question is that we can't distinguish what kinds of physical distortion of the endothelial cell are involved. The evidence that we have is simply that it is related to the flow rate within the blood vessel wall. In response to the second question, we can say that the evidence favors small arteries. There is no evidence that endothelial cells in capillaries play any role at all. The sensitivity and the spatial resolution of the technique are sufficient that we would be able to detect gradients within individual cells. If capillary endothelial cells were important, we should have been able to see their effect. I should point out that this actually is consistent with some of the studies that have been done on the localization of receptors and localization of secretion sites for endothelin and some other endothelial factors. Capillaries are different from arterioles and venules.

Rüegg: Endothelin is supposed to have a calcium-sensitizing effect on cardiac muscle according to Kraemer and others (Kraemer, B.K., Smith, J.W.& Kelly, R.A. *Circ. Res.* **68**, 269-279, 1991) Do you have any evidence whether your factor also has such an effect?

Winegrad: We haven't looked because we've used ATPase activity under conditions where calcium is not limiting to make the evaluation. We plan to do these experiments, but they require a different procedure. The one we used was specifically designed to avoid changes in calcium sensitivity and to look for other changes in the contractile proteins.

EFFECTS OF TROPOMYOSIN DEFICIENCY IN FLIGHT MUSCLE OF *DROSOPHILA MELANOGASTER*

Justin Molloy, Andrew Kreuz*, Rehae Miller**, Terese Tansey**, and David Maughan†§

Department of Biology
University of York
Heslington, United Kingdom
*Department of Molecular Genetics
Ohio State University
Columbus OH 43210
**Department of Biology
Georgetown University
Washington DC 20057
†Department of Physiology and Biophysics
University of Vermont
Burlington VT 05405

ABSTRACT

We have studied the structure and function of muscle fibers in which tropomyosin stoichiometry has been reduced by genetic mutation. We used a *Drosophila melanogaster* flightless mutant *Ifm(3)3* and a genetic cross of this mutant with wild type flies to achieve a gradation of tropomyosin gene dosage. We measured the flight ability and wingbeat frequency of the live insects and the ultrastructure and mechanochemistry of isolated single flight muscle fibers.

Flight ability is impaired when tropomyosin gene dosage is reduced. Wingbeat frequency also depends upon gene dosage as well as the severity of myofilament lattice disruption and the number of myofilaments in the organized core of the myofibrils. A reduction in number of myofilaments appears to result in a reduction in active muscle stiffness without resulting in an appreciable change in kinetics of force production.

Ifm(3)3 is trapped in a relaxed state and cannot generate active force. However, tight-binding rigor cross-bridges are able to form; in the absence of ATP, *Ifm(3)3* muscle fibers have high stiffness and force.

---

§To whom correspondence should be addressed.

Mechanism of Myofilament Sliding in Muscle Contraction, Edited by
H. Sugi and G.H Pollack, Plenum Press, New York, 1993

INTRODUCTION

Muscle protein mutants of *Drosophila* have been selected for on the basis of flightless behavior[1][2]. Many of these mutants have indirect flight muscle (IFM) myofibrils which show peripheral disruption to the myofilament lattice or reduced number of myofilaments. Either of these ultrastructural changes would reduce the overall muscle stiffness and therefore the resonance frequency of the wings. There would also be a reduction in power output concomitant with loss of myofilaments.

Mogami and Hotta[1] isolated a flightless mutant, *Ifm(3)3*, later shown to have a defective TmI gene[3]. This mutation reduces the amount of an indirect flight muscle specific isoform of tropomyosin, which results in gross disruption of the myofilament lattice.

We have measured the wingbeat frequency of flies of wild type (+/+), homozygous *Ifm(3)3*, and heterozygous (*Ifm(3)3/+*) lines. We have also examined the ultrastructure and mechanical properties of chemically skinned single muscle fibers of these flies. Results are interpreted in terms of cross-bridge theory and the resonance properties of the wing.

METHODS

Mutant Lines

Wild type *Drosophila melanogaster* were of the Canton-Special strain; *Ifm(3)3* flies, isolated by Mogami and Hotta[1], were obtained from R. Storti.

Muscles and Test Apparatus

Dorsal longitudinal indirect flight muscle fibers were isolated from the thoraces of 3-5 day old females and chemically skinned for ~1 h (8°C) in relaxing solution containing either 50 µg/ml saponin or 0.5% w/v Triton X-100. Relaxing solution (pCa > 8) contained 18 mM MgATP, 1 mM free Mg^{2+}, 5 mM EGTA, 20 mM BES buffer (pH 7.0), 50 mM sucrose, at 150 mM ionic strength (adjusted with K methane sulfonate).

The skinned fiber was transferred to a 10 µl drop of relaxing solution in a temperature controlled oil-filled chamber. The fiber was attached via aluminum clips to an Akers force transducer and a piezoelectric length controller. Sinusoidal length perturbations of 0.5% fiber length (peak-to-peak) and 1-240 Hz were applied. Length and force transients were monitored by an oscilloscope and digitized by an analog-to-digital system. Fiber dimensions were measured with a filar micrometer.

To activate the skinned fibers, relaxing solution was replaced with activating solution (pCa 4) which, except for added calcium, had essentially the same ionic composition as relaxing solution. Rigor was induced by replacing the activating solution with an ATP-free solution similar in composition to relaxing solution except ATP was omitted and K methane sulfonate concentration increased to maintain 0.15 M ionic strength. Experiments were conducted at 12°C and 18 mM MgATP to avoid diffusion limitation of substrate from the bath into the core of the skinned fiber.

Muscle Stiffness

Complex stiffness was calculated by dividing ΔF by $\Delta L/L$, where ΔL and ΔF are the corresponding amplitudes of applied length change and resulting force response, respectively, and L is the fiber length. $\Delta L/L$ was expressed as a percentage. The fiber was initially stretched just taut. The elastic stiffness component (in phase with the applied length changes) consists of total myosin crossbridge stiffness plus parallel resting stiffness; therefore, we considered the in-phase stiffness an indicator of number of crossbridges. Fibers have in-phase stiffnesses that are greatest in rigor, less so in the active state (pCa 4), and least in the relaxed state (pCa > 8). In addition, we considered the viscous stiffness component (out-of-phase with the applied length change) an indicator of crossbridge cycling kinetics. The active fiber exhibited a marked increase in negative out-of-phase stiffness compared to its relaxed or rigor state, indicating that the skinned fiber performs mechanical work on the apparatus driving the oscillatory length changes. The frequency at which the lowest (largest negative) out-of-phase stiffness was recorded is a close measure of the rate at which the active process responsible for driving the wing beat occurs *in vitro*; it is also the frequency of maximum mechanical work. *In vivo* the flight muscles drive the wing beat at the resonant frequency of the flight system (see later discussion).

Wing Beat Measurements

Wing beat frequency was measured by tethering flies and recording their wing movements by an optical device[4]. Frequency components were extracted using a spectrum analysis routine.

RESULTS AND DISCUSSION

Ifm(3)3 Mutant Strain

In *Drosophila*, tropomyosin is encoded by two genes (*TmI* and *TmII*) both of whose primary transcripts are differentially spliced to produce several tropomyosin isoforms[6][7] *Ifm(3)3* is an insertion mutation in the *TmI* gene that greatly reduces the accumulation of IFM specific TmI (Ifm-TmI). *Ifm(3)3* is totally flightless (Table 1). The *Ifm(3)3/+* heterozygote, which has roughly half the accumulation of IFM-TmI, has a flight ability mid-way between that of the wild type and homozygote (Table 1).

Table 1. Flight abilities and wing beat frequencies of wild type flies and *Ifm(3)3* heterozygous and homozygous mutants with reduced tropomyosin I gene expression.

Genotype	Flight histogram				Flight score (arcsin tranform.)			Wingbeat freq.
	as	bs	b	s	as/total	as+bs/total	Sum	Hz
CS/CS	0.60	0.22	0.11	0.08	50.5	64.6	115.1	205 ± 10
CS/Ifm(3)3	0.00	0.55	0.42	0.03	0	47.7	47.7	145 ± 18
Ifm(3)3/Ifm(3)3	0.00	0.00	0.82	0.19	0	0	0	0

Flight indices (from Tansey et al.[13]) indicate what fraction of flies deposited on a swing in a box remain on the swing (s), jump or fall to the bottom of the chamber (b), jump or fly to the wall below the swing (bs), or fly to the wall above the swing (as) after a certain period of time. Flight score (sum) calculated from flight histogram by taking the arcsin of the square root of the fraction found either above or above and below the swing[14], and summing the results.

We attempt here to explain the graded flight ability of the Ifm-TmI deficient flies in terms of cross-bridge theory and the resonance properties of the flight system.

Mechanical Properties of IFM from *Ifm(3)3*

"Stretch activation" is common to many types of muscle, and in insects it is responsible for powering the oscillatory wing movement. Stretch activation in *Ifm(3)3* and wild type flies was examined here by analyzing force responses to sinusoidal length perturbations applied at different frequencies[8][9].

Skinned IFM from wild type flies showed a pronounced stretch activation at pCa 4, indicated by the large negative out-of-phase stiffness (Fig. 1A). In contrast, homozygous *Ifm(3)3* flies showed complete absence of stretch activation at pCa 4 (Fig. 1C, compare to 1A). Indeed, the in-phase stiffness of fibers at pCa 4 was similar to that at pCa 8, indicating complete absence of Ca^{2+} regulated contraction. Although homozygous *Ifm(3)3* fibers did not produce an active response, they were nevertheless able to generate both tension and a marked increase in stiffness in rigor solution.

We were surprised that *Ifm(3)3* lacked any detectable active response since the myofibrils, though grossly disrupted, do contain a core of well-ordered myofilament lattice[5]. In cross-section, the lattice is similar in appearance to that of wild type, and the number of filaments in the core lattice (224 ± 18 thick filaments/myofibril compared to 1072 ± 46 in wild type) sufficiently great to be detected mechanically if able to be activated. However, sarcomere lengths within the organized regions are ~0.7 μm shorter than the sarcomeres of wild type flies, and Z-bands are not as straight[5].

The jitter in axial alignment of filaments and shorter sarcomere length raises the possibility that stretch-activation is blocked because of misaligned filaments.

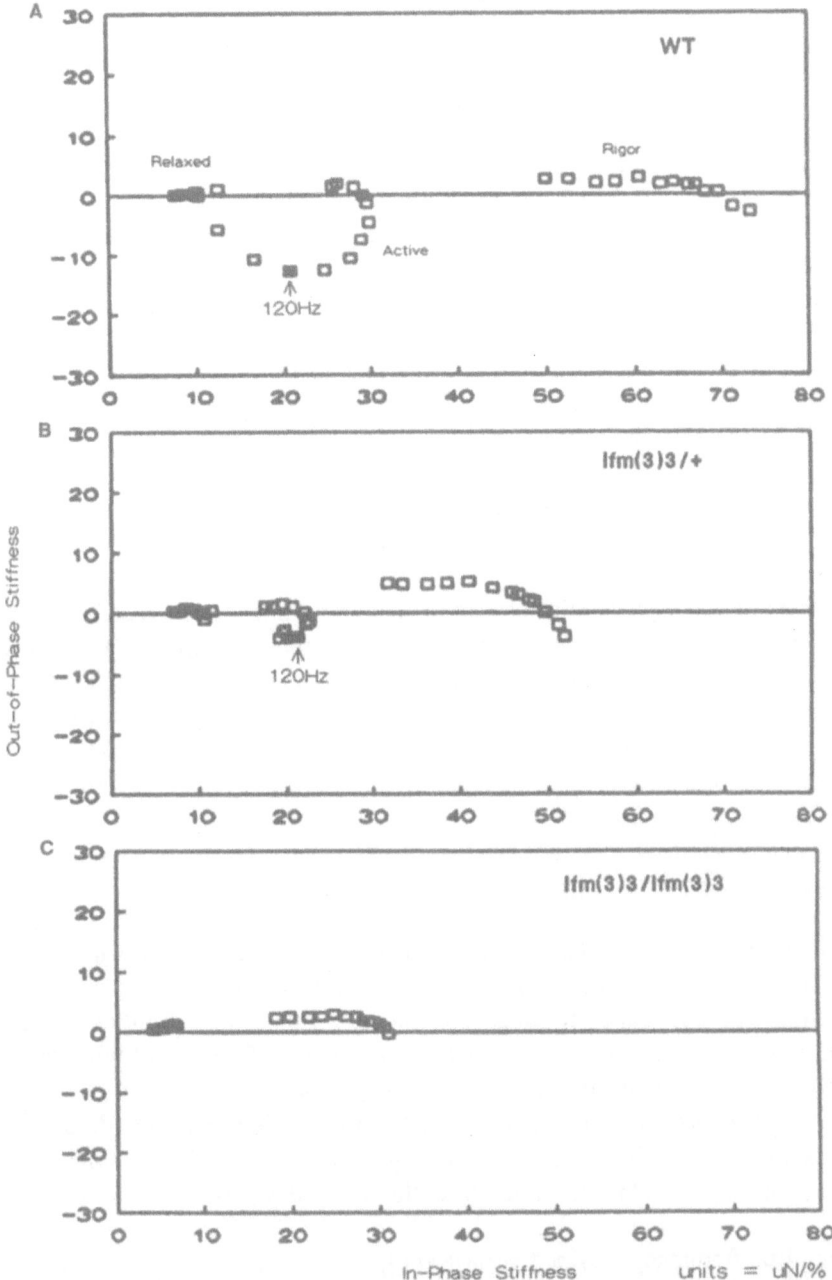

Fig. 1. Nyquist plots of complex stiffness in calcium-activated skinned indirect flight muscle fibers from wild type and tropomyosin deficient flies: A, wild type Canton-S; B, *Ifm(3)3* heterozygotes; C, *Ifm(3)3* homozygotes. Force measured in response to length perturbations applied at different frequencies (0.5 - 240 Hz; 120 Hz indicated by solid circle). Stiffness calculated by dividing change in force by percent change in muscle length (see Methods). Results for relaxed (pCa >8), active (pCa 4) and rigor (ATP free) states indicated. Fiber cross-sections ~10^{-8} m^{-2}; therefore, 1 μN/% (per fiber) ~10 kN m^{-2}.

Filament alignment is important to Wray's[10] model of stretch activation which depends upon exact registration of the thick and thin filaments. However, the lack of Ca^{2+} activation in *Ifm(3)3* skinned fibers indicates that the fiber has also lost its Ca^{2+} regulatory ability, perhaps through dissociation of troponin-T (otherwise bound to Ifm-TmI) and the rest of the troponin complex. Some Ifm-TmI specific isoform may be present in the core, but not enough to confer normal regulatory function. We also cannot rule out the possibility that the Ifm specific TnH (which over part of its length is homologous to Ifm-TmI) occupies the Ifm-TmI vacancy, but cannot fulfill the Ifm-TmI function. In any case, myosin is able to tightly bind actin in the absence of nucleotide in the mutant muscle (Fig. 1C, compare to 1A).

Heterozygous *Ifm(3)3* flies (progeny of *Ifm(3)3* crosses with wild type Canton-S flies) show active responses similar, but not identical, to that of the wild type flies (Fig. 1B, compare to 1A). As mentioned earlier, the frequency at which maximum negative out-of-phase stiffness occurs is an index of the rate of stretch activation, and it is also the frequency at which maximum mechanical work is produced. This frequency in heterozygous *Ifm(3)3* flies is similar to that of wild type flies (~120 Hz, solid squares). Thus the kinetics of delayed tension generation do not appear to be affected by gene dosage.

Active muscle in-phase stiffness, an indicator of number of myosin crossbridges, does tend to increase with gene dosage. In phase stiffness for *Ifm(3)3/+* fibers is less than that of homozygous wild type. The increase in number of myosin crossbridges correlates with an increase in the number of filaments that are part of a well-ordered core and with an increase in sarcomere length. *Ifm(3)3/+* flies contain 632 ± 67 thick filaments/myofibril and have sarcomere lengths intermediate between those of wild type and homozygous *Ifm(3)3*.

Wing Beat Frequencies

Wingbeat frequency of the mutant strains also depended on gene dosage (Table 1, last column) and number of filaments in the well-ordered core. *Ifm(3)3* homozygotes do not beat their wings, accounting for their total flightlessness, while *Ifm(3)3* heterozygotes do beat their wings but at a rate 70% of that of wild type. This reduced rate may account for their impaired flight ability.

The wingbeat is driven by the two sets of indirect flight muscles (dorsal-ventral and dorsal-longitudinal) whose resonant frequency is governed by the inertia of the wing and the stiffness of its mounting (muscles and cuticle)[11]:

$$\text{wing beat frequency} = 1/2\pi * \sqrt{(k_m + k_c)/I},$$

where k_m and k_c represent muscle and cuticle stiffnesses, and I represents the moment of inertia of the wings.

It is apparent from the above equation that if most of the mounting stiffness of the wing is in the flight muscle, then the resonance frequency of the wings will be determined primarily by the muscle stiffness. Together, our results (Fig. 1 and

Table 1) suggest that the stiffness determining wing beat frequency resides primarily in the flight muscle myofilaments, in both the high resting (parallel) stiffness and active actomyosin crossbridges formed in that part of the lattice that is well-ordered.

The results also suggest that the flight muscles are the major stores for elastic energy during the wingstroke. This is important because storage of wing inertial energy over each half wing-stroke cycle is essential for high efficiency, especially when the wingbeat frequency is high[12].

Several conclusions can thus be reached:

1) The family of genotypes tested here indicate that muscle stiffness is a major factor in determining wing resonance and as such is the major store for elastic energy during the wingstroke cycle.

2) Wing beat frequency appears to be a determinant of flight ability, with impairment of flight associated with reduced wing beat frequency.

3) In the flightless homozygous *Ifm(3)3* tropomyosin mutant, crossbridges are not able to bind and actively cycle in the tropomyosin deficient flight muscle, but in the absence of nucleotide strong-binding, rigor cross-bridges are able to form.

ACKNOWLEDGEMENTS

We thank Ric Schaaf and Kevin Bickford who contributed their technical help to this project. This work was supported by grants from the National Institutes of Health (D.M.) and National Science Foundation (T.T.). J.M. was a visiting NATO fellow at the University of Vermont.

REFERENCES

1. Mogami, K. & Hotta, Y. *Mol. Gen. Genet.* **183**, 409-417 (1981)
2. Sparrow, J., Drummond, D. Peckham, M. Hennessey, E. & White, D. *J. Cell Sci.* **14**, 73-78 (1991)
3. Karlik, C.C. & Fyrberg, E.A. *Cell* **41**, 57-66 (1985).
4. Unwin, D.M. & Ellington, C. P. *J. Exp. Biol.* **82**, 377-378 (1979).
5. Tansey, T., Schultz, J., Miller R., & Storti, R. *Mol. Cell. Biol.*(1991), in press.
6. Basi, G.S. & Storti, R.V. *J. Biol. Chem.* **261**, 819-827 (1986).
7. Karlik, C.C. & Fyrberg, E.A. *Mol. Cell. Biol.* **6**, 1965-1973 (1986).
8. Kawai, M. & Brandt, P.W. *J. Muscle Res. Cell Motility* **1**, 279-303 (1980).
9. Thorson, J. & White, D.C.S. *J. Physiol. (Lond.)* **343**, 59-84, (1983).
10. Wray, J. *Nature* **280**, 325-326 (1979)
11. Pringle, J.W.S. *Insect Flight.* Cambridge, Cambridge Univ. Press. (1957)
12. Ellington, C.P. *J. Exp. Biol.* **115**, 293-304 (1985).
13. Tansey, T., Mikus, M.D. Dumoulin, M. & Storti, R.V. *EMBO J.* **6** 1375-1383.
14. Sheppard, D.E. *Drosph. Inf. Serv.* **51**, 150 (1974)

Discussion

Sellers: We have some data that tropomyosin increases the sliding rate of actin filaments over phosphorylated *Lethocerus* myosin by a factor of 10 in the motility assays (unpublished). So, the tropomyosin may very well have important influence on the kinetics of actomyosin interactions.

Kawai: Dave, it's interesting to see whether the rigor stiffness of this homozygote mutant is similar to that of the control. Is the time course of tension the same as in the control?

Maughan: We haven't measured the time course of tension development. The steady-state tension is extremely low in these preparations. What we do measure is the complex stiffness of this preparation which is independent of baseline drift in the tension record.

Kawai: It's surprising that the 120 Hz frequency point shows no out-of-phase stiffness in the case of the mutant homozygote.

Maughan: There is no stretch activation in the homozygote. You saw that the locus of the active points was superimposable over the relaxed points, so there wasn't any out-of-phase stiffness. There was no work done on the system in that case.

Kawai: Your mutant strain, the tropomyosin *Ifm (3)3* homozygote, is not calcium activatable, but the system can be activated by myosin-head attachment in rigor.

Maughan: *Ifm (3)3* shows rigor tension and stiffness, but there is no stretch activation.

Kawai: Then you could perhaps bring it into the rigor condition, add a small amount of ATP, and then make the system work.

Maughan: A good idea. We tried that and even under those conditions the tropomyosin *Ifm(3)3* homozygote could not be activated. We could see no detectable negative out-of-phase stiffness over a range of ATP concentrations of 50 µM to 5 mM.

Pollack: The degree of structural impairment seems to be correlated with the amount of tropomyosin that is missing. Could you speculate on the reason why that might be the case?

Maughan: This is characteristic of many of these heterozygote mutants. The myosin light-chain 2 mutant heterozygote also shows a peripheral disruption. I think it's related to the myofilament lattice assembly program: if these contractile elements aren't present, the assembly just doesn't occur. That would imply that these elements are missing, that somehow cross-bridge interaction, or contraction, is essential in the assembly process at some point.

Pollack: Could it be that these elements all play some subtle structural role?

Maughan: Absolutely.

EFFECT OF MYOSIN HEAVY CHAIN PEPTIDES ON CONTRACTILE ACTIVATION OF SKINNED CARDIAC MUSCLE FIBRES

J.C. Rüegg, J.D. Strauss, C. Zeugner and I. Trayer*

II. Institute of Physiology
University of Heidelberg
W-6900 Heidelberg
Federal Republic of Germany
*School of Biochemistry
University of Birmingham
Edgbaston B15 2TT
United Kingdom

ABSTRACT

Peptides derived from the sequence of the S1 domain of the myosin heavy chain were tested for their effects on the regulation of cardiac contractility. Basal calcium responsiveness of the contractile apparatus in terms of isometric tension generation and ATPase was determined in chemically demembranated ventricular fibre bundles. Incubation with a series of peptides derived from the peptide sequence around SH thiol group (Cys 707) resulted in a measurable increase in isometric tension and ATPase activity at sub-maximal concentrations of calcium but not at saturating levels of calcium activity, thus demonstrating a "calcium-sensitizing" effect of these peptides. The effects of two of these peptides, S1 687-716 and S1 701-717, are demonstrated to mimic, but importantly were not additive with, the calcium sensitization induced by lowering ATP concentration to 10 μM from 10 mM. This suggests the possibility of a similar mechanism of action underlying both types of sensitization. Because these effects demonstrate tissue specificity, were sensitive with respect to potency to not only amino acid composition but also sequence, and could not be duplicated by a similarly charged, non-homologous peptide, we attribute the effects to be specific to the sequences of these peptides. These data provide further evidence that the sequence between residues 687 and 717 of the S1 domain of the myosin heavy chain influences the calcium responsiveness of the contractile apparatus.

Mechanism of Myofilament Sliding in Muscle Contraction, Edited by
H. Sugi and G.H Pollack, Plenum Press, New York, 1993

173

INTRODUCTION

In the past, we enjoyed a comparatively simple view of muscle contraction in which the generation of force was ascribable to one set of proteins, actin and myosin, and the regulation of calcium responsiveness was attributable to a different set of proteins. It is now clear that the interaction of actin and myosin is not only important in crossbridge cycling, that is in the process of force generation, but may also contribute to the regulation of the calcium-responsiveness of the contractile system. As described by Eisenberg and Greene[1], myosin subfragment 1 (S1) interacting with the thin filament potentiates the acto-S1 ATPase. Thus it was demonstrated that crossbridges interacting with actin may increase the calcium-responsiveness of the contractile apparatus. This latter consequence of actin-myosin interaction may be particularly important in cardiac muscle which is known to contract under ischemic conditions, that is at low ATP concentration, even at basal levels of free calcium[2]. Under these conditions, i.e. low ATP concentration, at least a portion of the crossbridges are in a nucleotide free rigor like state[3]. This portion of rigor bridges may be sufficient to permit activation of the thin filament even at low levels of calcium.

Thus, we may argue that the interaction of S1 with actin is not only responsible for activity of the myosin ATPase but also has a calcium-sensitizing effect on force production. Indeed, according to Güth and Potter[4], attached or cycling crossbridges reduce the calcium concentration required for half-maximal activation. We show here that the effect of attaching crossbridges on calcium-responsiveness may be mimicked by peptides that contain the region of the myosin heavy chain around the SH1 thiol group (Cys 707), which has been shown to be an actin binding domain[5][6]. These peptides increase both force and ATPase activity of skinned cardiac fibre at submaximal calcium activation, that is to say, these peptides sensitize the fibre to calcium.

METHODS

Fibre bundles of right pig ventricle of < 200 μm diameter were skinned with Triton X-100, attached to an AME-801 force transducer, and relaxed in a well stirred bathing solution (cf. Rüegg et al.[7]) containing (in mM): imidazole, 30; ATP, 10; creatine phosphate, 10; NaN_3, 5; calcium EGTA, 5; $MgCl_2$, 12.5; creatine kinase 380 U ml^{-1} ; dithiothreitol, 5; Pi 5. Ionic strength was adjusted to 100 mM with KCl, pH 6.7, 20°C. Contraction was induced by immersion into an similar solution in which EGTA was replaced by a calcium-EGTA buffer. The buffered calcium ion concentration was determined from the ratio of EGTA to calcium EGTA essentially according to the method of Fabiato and Fabiato[8], but using an apparent dissociation constant of 1.6 μM for the calcium EGTA buffer at pH 6.7 and 20°C. ATPase activity of skinned fibres was determined according to Güth and Wojciechowski[9]. The myosin peptides were prepared as described earlier[6] and stored as freeze-dried

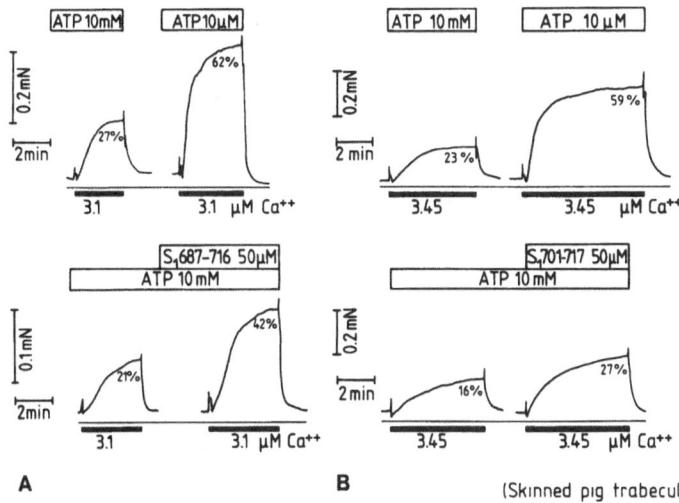

Fig. 1. A. Upper panel shows the increase in contractile force elicited by a fixed submaximal calcium concentration when the ATP concentration was lowered to 10 μM in presence of an ATP regenerating system. On the lower panel it is shown that a similar "potentiating" effect may be produced by the addition of the peptide S1 687-716 in a concentration of 50 μM. B. The upper panel again demonstrates the same sensitizing effect of reduced ATP in another fibre. The lower panel shows the effects of incubation of the smaller peptide, S1 701-717.

powders. Just prior to use, these were dissolved in 10 mM imidazole buffer, pH 6.7. MKI peptide was purchased from Peninsula Laboratories. This peptide was also dissolved in 10 mM imidazole and the pH of the solution adjusted to 6.7 prior to use.

RESULTS AND DISCUSSION

Lowering the ATP concentration increases the calcium-responsiveness, or rather, the apparent calcium sensitivity. Thus, a much lower Ca^{2+} concentration is required to induce a given submaximal force of skinned cardiac fibres when the ATP concentration is reduced to 10 μM (Fig. 1). At a concentration of 50 μM, one of the peptides (S1 687-716) mimicked the potentiation of the calcium induced contractile response observed when lowering the ATP concentration in skinned fibres. As can be seen in the upper panels of the first figure, contractile force elicited by a given submaximal calcium concentration can be increased by lowering the ATP concentration to 10 μM in presence of an ATP regenerating system. A similar "potentiating" effect may be produced by the addition of the peptide S1 687-716 in a concentration of 50 μM, as seen in the lower panel of Fig. 1A. As demonstrated in the adjacent panel (Fig. 1B), similar results were found when fibres were tested against a smaller peptide based on this same region of S1 (S1 701-717). We

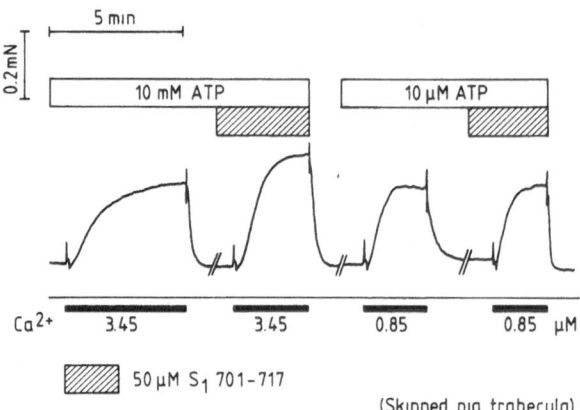

Fig. 2. The effect of the calcium-sensitizing peptide S1 701-717 in solutions containing 10 μM ATP. Note that at 10μM ATP, substantially less calcium (0.85 μM) was required to generate tension similar to that generated at 10mM ATP with 3.45 μM calcium. The peptide caused a sensitization to calcium only in the presence of 10 mM ATP, indicating that the calcium sensitization associated with reduced ATP and with 50μM peptide S1 701-717 are not additive.

hypothesized, based on the similarity between the effects of reduced ATP concentration and the peptides, that the mechanisms of action may be related.

We tested whether the two types of effects were additive, and thus dependent upon different mechanisms, by incubating fibres simultaneously with both low ATP concentration and peptide after calcium sensitization was induced by lowering ATP concentration. Addition of the peptide S1 701-717 had no additional force-enhancing effect on fibres which had already been sensitized by lowering the ATP concentration. However, in the same fibre bundle, this peptide did increase the calcium-response at high ATP concentration (Fig. 2.). If the effects were brought about by entirely different mechanisms, we might have expected the effects to be additive. Either incubation with the peptide or lowering the ATP concentration resulted in calcium sensitization, but the effects were not additive.

The magnitude of the effects are dependent upon peptide concentration, 50 μM producing about 50% of the maximal effect. Interestingly, this concentration corresponds approximately to the dissociation constant (~20 μM) of the peptide-actin complex as determined by Keane et al.[6]. Moreover, at maximal activation by Ca^{2+} (at a saturating concentration of Ca^{2+} greater than 10 μM) the force is not enhanced by the peptide, indicating an effect on calcium sensitivity per se and not on the calcium efficacy.

When relative force and ATPase activity of skinned fibres are plotted versus the pCa in the presence and absence of peptide S1 701-717, it can be seen that the peptide displaces the relation to the left, towards lower calcium ion concentrations. (Fig. 3) These results, therefore, suggest that the region from which these peptides are derived, the region around the SH1 thiol group, may have an important role in determining the calcium responsiveness of contractile structures in the heart. The

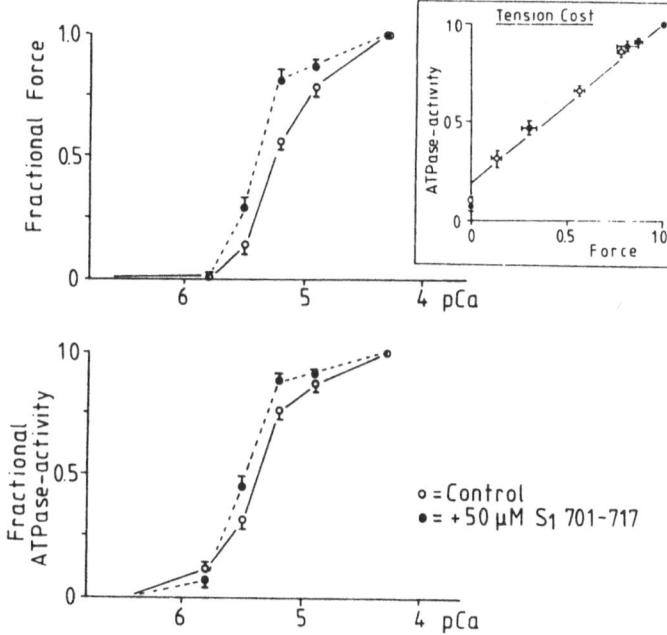

Fig. 3. Relative force-pCa (upper panel) and ATPase-pCa (lower panel) relations in the presence and absence of peptide S1 701-717. Force and ATPase were recorded simultaneously as a function of calcium concentration before and during incubation with 50 μM S1 701-717. Calcium sensitization was apparent in both cases and is manifested as a leftward shift in the relations. Inset: The relation between the relative force and ATPase activity, described as energetic cost of isometric tension. Note that tension cost is unchanged by sensitization of this peptide.

relation between ATPase activity and relative force, describing the energetic cost of isometric tension, is plotted as an inset in Fig. 3. The tension cost is unchanged by sensitization of the system by this peptide. This may be taken as evidence that crossbridge detachment rate constant remains unchanged and the effect is on the degree of activation.

By synthesizing a series of peptides whose sequences overlap and include both this specific region and flanking sequences, the specific important sequence may be mapped. As can be seen by the activity profile in Fig. 4, peptide mimetics of residues 687-716 or smaller peptides derived from within that region increase the calcium-responsiveness while peptides derived from the protein sequence flanking this region have no effect. Thus, by using different peptides as probes, the calcium sensitizing region of S1 could be delineated and isolated to the sequence 687-716. It is interesting to note that the calcium sensitizing effects of myosin peptides may be antagonized by a peptide from the inhibitory region of TnI, amino acid residues 104-115 of the native skeletal sequence[10]. This finding, taken with other evidence, suggests TnI and myosin may both bind to the same region of the N-terminal end of actin.

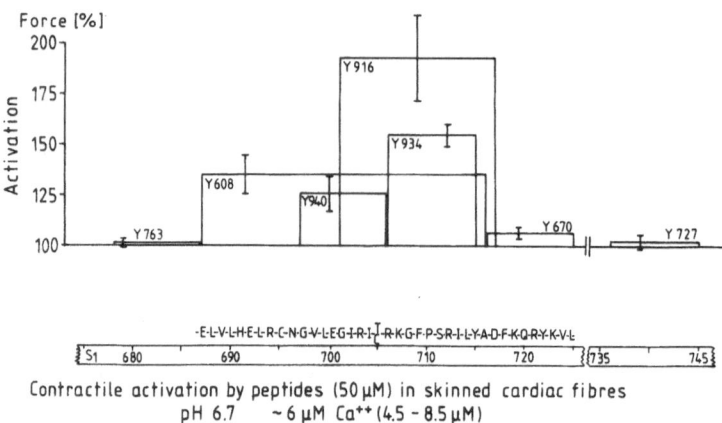

Fig. 4. Activity profile of several peptides derived from the native protein sequence of S1. Calcium responsiveness is on the ordinate and residue number on the abscissa. Activation is defined as percent increase in tension with addition of 50 μM peptide at conditions of constant calcium (~6 μM) and ATP (10 mM). Y916 is the peptide S1 701-717 and Y608 is S1 687-716. Note that peptides derived from the sequence flanking the 687-716 region have little or no activity while peptides derived from within this region all have substantial activity.

As a working hypothesis we therefore propose that:

1. The myosin peptide induces calcium sensitization by competing with TnI for actin, thereby depressing the inhibitory effect of the latter protein, TnI, on calcium-responsiveness.

2. The myosin peptides mimic the potentiation effect of attaching crossbridges and thus increase the calcium responsiveness in an ATP dependent manner. By weakening the TnI-actin interaction, the interaction between TnI and TnC should be increased and the conformation of TnC so altered as to increase the calcium sensitivity of the system.

It is interesting that the critical region of myosin which interacts with actin bears a net of four positive charges, whereas the N-terminal end of actin is negatively charged. Hence, the interaction may be of an electrostatic nature. In this respect, it is noteworthy that some other peptides with a similar composition but a scrambled sequence have similar although always weaker effects. Furthermore, the effects may be attenuated by raising the ionic strength, supporting the idea of an electrostatic mechanism. On the other hand the peptide effects show tissue specificity; there is little effect on skinned fibres of taenia coli of guinea pigs while the effects on skinned fibres from skeletal muscle are intermediate[6]. Furthermore, another peptide with a similar charge density but no sequence homology, known as MKI and derived from the sequence of the myosin 20 kD light chain, was also tested in these fibres. This peptide had virtually no effect on either the calcium sensitivity or maximal isometric tension in cardiac tissue (Fig. 5) despite the fact that this peptide has been shown to have significant effects on the contractility of skinned smooth muscle fibres[11].

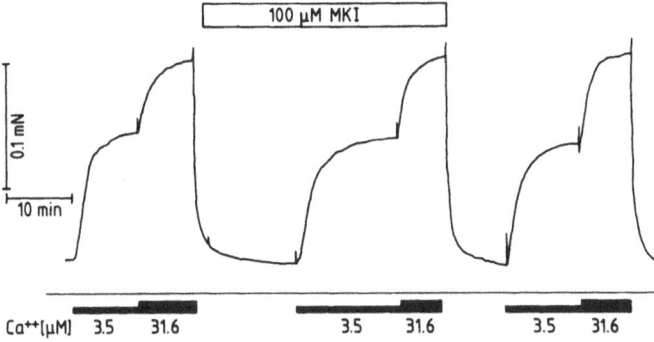

Fig. 5. Effect of 100 μM peptide MKI on isometric tension at submaximal and maximal calcium concentrations. Bracketing control contractions before and after peptide incubation demonstrate the stability of the preparation. Identical results were obtained in 5 different fibres. This peptide, lys-lys-arg-ala-arg-ala-ala-thre-ser, is based on the sequence around the phosphorylation site of smooth muscle myosin light chain (LC20 11-19) and has an especially dense positive charge on the N-terminus with an overall charge similar to the S1 peptides. It did not mimic the effects of the S1 peptides.

This region of S1 only represents one area on the molecule that is involved at the acto-myosin interface during the contractile process. Others must exist and be responsible for the energy transduction process. Nevertheless, it is interesting that a functional "sub-site" can be dissected from such a complex structure. It is not assumed that these peptides necessarily adopt the same structure free in solution as they do as part of the parent S1 molecule, but that they do possess sufficient information in their sequence to adopt the appropriate conformation in the presence of the partner molecular template, namely, actin. Furthermore, the results show that they are capable of inducing the same conformational change(s) in actin as does this same region in intact S1. From this, it might be attractive to hypothesize that a pharmaceutical agent to be used as a clinical calcium sensitizer might be designed after structure of the peptide. However, while affinity of the peptides for actin may be described as high (Kd = 20 μM)[6], the affinity is not nearly high enough for use as a pharmaceutical tool. Nevertheless, by using modern NMR techniques to determine the three-dimensional structure of the peptides when bound to actin, the requirements for recognition at this critical actin site may be determined precisely. It should then be possible to design and synthesize structurally stable peptide analogues that would bind more tightly and in a specific and tissue dependent manner, thus opening the way for selective intervention in the regulation of the contractile process.

ACKNOWLEDGEMENTS

This work was supported in part by a grant from the Deutsche Forschung Gemeinschaft. JDS is a fellow of the Alexander von Humboldt Foundation. This

180   J.C. Rüegg et al.

material is based upon work supported in part by the North Atlantic Treaty
Organization through a grant awarded March 1990 (JDS).

REFERENCES

1. Eisenberg, E. & Greene, L.E. *Ann. Rev. Physiol.* **42**, 293-309 (1980).
2. Allshire, A., Piper, M., Cuthbertson, K.S.R. & Cobbold, P.H. *Biochem. J.* **244**, 381-385
   (1987).
3. Ridgway, E.B., Gordon, A.M. & Martyn, D.A. *Science* **219**, 1075-1077 (1983).
4. Güth, K. & Potter J.D. *J. Biol. Chem.* **262**, 13627-13635 (1987).
5. Suzuki, R., Nishi, N, Tokura, S. & Morita, F. *J. Biol. Chem* **262**, 11410-11412 (1987).
6. Keane, A.M., Trayer, I.P., Levine, B.A., Zeugner, C. & Rüegg, J.C. et al. *Nature* **344**, 265-
   268 (1990)
7. Rüegg, J.C., Zeugner, C., Van Eyk, J., Kay, C.M. & Hodges, R.S. *Pflügers Arch.* **414**, 430-
   436 (1989).
8. Fabiato, A. & Fabiato, F. *J. Physiol. (Paris)* **75**, 463-505 (1979).
9. Güth, K. & Wojciechowski, R. *Pflügers Arch.* **407**, 552-557 (1986).
10. Rüegg, J.C., Zeugner, C., Van Eyk, J.E., Hodges R.S. & Trayer, I.P *Peptides as Probes in
    Muscle Research* (ed. Rüegg, J.C.), 95-110 (Springer Verlag, Heidelberg/Berlin, 1991).
11. Strauss, J.D., de Lanerolle, P. & Paul, R.J. *Am. J. Physiol.* **262**, C1437-C1445 (1992)

Discussion

Gillis: With the peptide, can you prevent the formation of rigor links when you
shift your skinned preparation from a relaxing medium to a low ATP rigor
medium?

Rüegg: We did not try to find out, but I assume that some rigor links will be
formed.

Goldman: Since the actin peptide activated the S1-ATPase, can you speculate on
why G-actin doesn't activate the S1-ATPase?

Rüegg: That's an important question. However, I think it was Gerald Offer and
colleagues who found that G-actin will activate the S1-ATPase by a factor of four
(Offer, G. et al. *J. Mol. Biol.* **66**, 435-444, 1972).

Goldman: Are the conditions similar to how you test your peptide?

Rüegg: Yes.

Holmes: I thought the work with the actin peptide looked very encouraging.
However, I might remind everybody that J. A. Borden determined the structure of
an ATP-binding peptide from actin, which is 18 residues long, and this comes out to
be a beta hairpin (*Biochemistry* **26**, 6023-30, 1987). Unfortunately, this peptide is
actually in an α-helix and is 30 Å away from the ATP. What I'm trying to say is: be
careful with peptides.

Rüegg: Yes, because they may have a different structure in solution than in the
bound state. That is, of course, an important point; however, some of the peptides in
solution may have structure. Heidi Hamm found that some have and some have not
(in: *Peptides as Probes in Muscle Research* J.C. Rüegg, ed., Springer Verlag,

Heidelberg, 1991). But when they bind, they obtain some kind of structure by "induced fit."

Holmes: They may or may not.

Rüegg: Yes, they may or may not, but interestingly, they have an effect on skinned fibers.

Tregear: But surely they may induce an effect that wasn't there originally. There may be a terribly irrelevant peptide which may not have been doing that *in vivo*.

Rüegg: Yes. It would be very important to find out what the structure of a peptide is when it binds. The idea that comes to mind is that this is the kind of structure the peptide ought to have in solution, too. Perhaps one could stabilize that structure by some kind of cross-link and then use that peptide, which has a stabilized structure, or a related bio-organic peptide mimetic molecule. This might have a much higher affinity for the binding site, since it would already be in the actual structure that binds.

Holmes: Could I just add that natural product chemistry shows us short peptides that are biologically active, are nearly always cyclized, and are also often cross-linked. One does all that one can to give them some kind of structure.

Rüegg: That is where we would quite clearly like to go in the future.

CROSS-BRIDGES AFFECT BOTH TnC STRUCTURE AND CALCIUM AFFINITY IN MUSCLE FIBERS

A.M. Gordon and E.B. Ridgway

Departments of Physiology and Biophysics
University of Washington
Seattle, WA 98195 and
Medical College of Virginia
Richmond, VA 23298

ABSTRACT

In vertebrate striated muscle, calcium binding to troponin initiates contraction, a strong interaction of actin and myosin. In isolated proteins and skinned fibers, the strong interaction of myosin with actin also affects troponin. Fluorescent labels attached to troponin C show structural changes in the TnC environment with cross-bridge attachment and also with calcium binding. Evidence that this effect of cross-bridges also occurs in intact striated muscle comes from studies in partially activated cardiac or skeletal muscle by others and in barnacle muscle by us. Length changes which detach myosin cross-bridges produce a brief burst of extra calcium that can be detected by aequorin in activated, voltage clamped single barnacle muscle fibers. That this calcium is coming from calcium bound to the activating site (troponin-C) is supported by several pieces of evidence. Studies on the dependence of the extra calcium on force and the time of the length change are consistent with the amplitude of the extra calcium being proportional to the bound calcium (CaTnC) and with increased cross-bridge attachment and force increasing calcium binding to troponin-C by up to a factor of 10. Importantly, stretch of active muscle (which first detaches cross-bridges and then enhances steady force) gives a biphasic response: first extra calcium (presumably due to cross-bridge detachment) and then, decreased calcium (presumably due to enhanced calcium binding to TnC). The enhanced calcium binding we see with elevated force (via strained cross-bridges) implies that calcium binding to TnC is enhanced not only be cross-bridge attachment but also by cross-bridge (or thin filament) strain.

This effect of cross-bridge attachment/force on calcium binding is consistent with a dual mechanism of calcium activation of contraction. First, calcium binds to troponin in the thin filament activating strong myosin binding to the thin filament. Then, strong myosin binding in turn provides additional activation either by increasing calcium binding or by changing the thin filament structure directly allowing additional cross-bridge attachment.

Mechanism of Myofilament Sliding in Muscle Contraction, Edited by
H. Sugi and G.H Pollack, Plenum Press, New York, 1993

It is well established that the initiating event in the activation of contraction in vertebrate striated muscle is the binding of calcium to troponin as first shown by Ebashi and coworkers[1]. Since that time evidence has accumulated from isolated protein studies (initiated by Bremel and Weber[2]) and skinned fiber studies that strongly attached cross-bridges also bring about structural changes in the thin filaments associated with activation of actin-myosin interaction and contraction. This evidence has been reviewed by us[3,4] and others[5-7].

In skinned muscle fibers, data from our lab[8] Potter's lab[9] and Rüegg's lab[10] all show that rigor and cycling cross-bridges both cause changes in the TnC structure independent of calcium binding, as measured by fluorescent changes of labeled TnC exchanged into skeletal muscle for the endogenous TnC. These data would be consistent with the results of Reuben et al[11] and Goldman et al[12] that rigor cross-bridges can activate contraction in skinned skeletal muscle fibers in the absence of calcium. Another indication that cross-bridges can activate the thin filament is the change in calcium binding to TnC brought on by cross-bridges. Fuchs and coworkers[13,14] have shown that rigor cross-bridges enhance ^{45}Ca binding in skinned skeletal and cardiac muscle, but that cycling cross-bridges only appeared to enhance ^{45}Ca binding in skinned cardiac muscle[14] but not skeletal muscle[15]. Thus there may be a difference in the effect between muscle types.

Is there evidence for a cross-bridge effect on activation in intact muscle? X-ray diffraction studies by Huxley and coworkers[16] have shown changes in the second actin layer line associated with tropomyosin movement occurring in fibers stimulated at sarcomere lengths beyond thick and thin filament overlap where there should be no cross-bridge attachment. Furthermore, they showed that these changes in the second actin layer line preceded the change in equatorial reflections associated with cross-bridge attachment to the thin filament and also the subsequent force development. Thus, in so far as tropomyosin movement signals thin filament activation, their data is consistent with calcium alone bringing about activation at the beginning of contraction, causing tropomyosin movement before cross-bridges attach. In addition, their data shows that in fibers stimulated at a sarcomere length where there is filament overlap (and thus cross-bridge attachment), the return of the second actin layer line (and presumably tropomyosin to its resting, inhibiting position) is greatly delayed from that seen in fibers stretched to sarcomere lengths beyond filament overlap. Thus their data supports the concept of cross-bridges prolonging activation in rabbit skeletal muscle fibers. To our knowledge this is the best direct evidence of a role for cross-bridges in the activation of skeletal muscle.

Strong circumstantial evidence for an effect of cross-bridges in enhancing calcium binding to TnC in intact muscle comes from studies from our lab on barnacle muscle[17,18], Allen's and Kurihara's labs on cardiac muscle[19], and Stephenson's lab[20] on skeletal muscle. These studies all show extra calcium when length is decreased sufficiently to cause cross-bridge detachment during the declining phase of the calcium transient in active muscle fibers. The rationale for these studies is that if calcium binding to TnC depends on cross-bridge attachment, then decreasing length to decrease cross-bridge attachment should decrease calcium

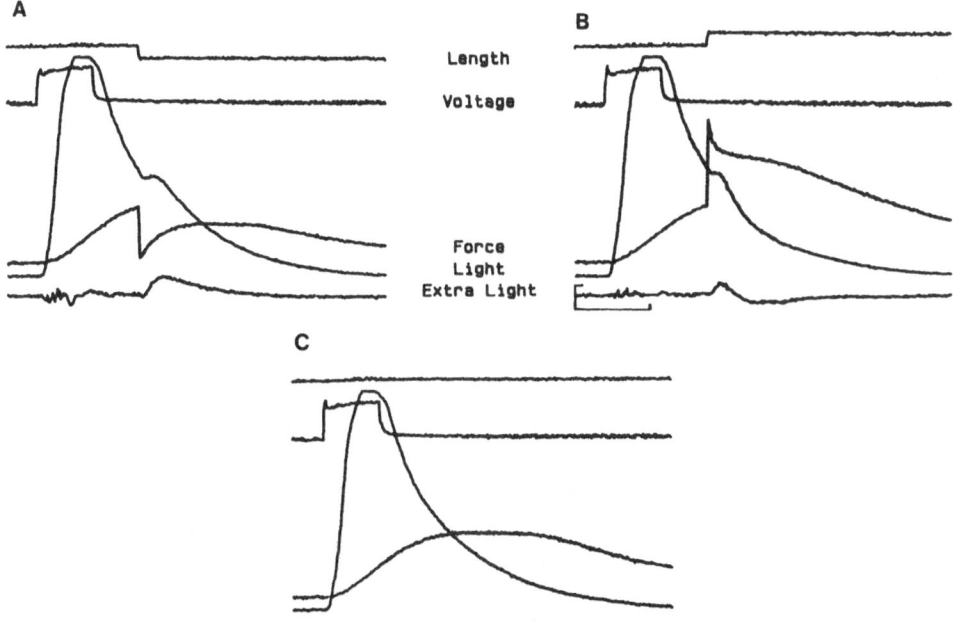

Length
Voltage

Force
Light
Extra Light

Fig. 1. A. Extra light on shortening: B. biphasic response to stretch in an aequorin-injected, voltage-clamped barnacle single muscle fiber. C. Control record with no length change. Simultaneous recordings of muscle length, membrane potential, light from bioluminescent aequorin, force, and extra light [subtracting the control light trace (C) from the light trace with the length change] during 3% length changes during the declining phase of the calcium transient demonstrate the extra light on shortening (A) and the complex biphasic response to stretch (B). Shortening produces a transient increase in intracellular calcium: stretch produces a transient increase in intracellular calcium followed by transient decrease in calcium. During shortening, the force falls and then redevelops: after stretch, the force falls somewhat but remains elevated. Note that in each light record, the top is clipped. Fiber length, 20 mm. Fiber weight, 36 mg. The calibration bar is 1.4 mm length, 50 mV voltage, 5.0 g force, 0.5 V for light and extra light, and 400 ms for time. Temperature, 8°C.

bound to TnC and elevate free calcium. One of the reasons why this is such a sensitive technique for measuring changes in bound calcium is that the total TnC is nearly 10^{-4} M and thus the bound calcium is in the same range. A change in bound calcium therefore produces a large relative change in free calcium which varies from 10^{-7} M at rest to only about 10^{-5} M during maximum activation. So it is actually easier to detect a change in free calcium than a change in bound calcium. These studies support the hypothesis that calcium binding to TnC is a function of cross-bridge attachment and force. A role for cross-bridges in thin filament activation thus seems likely.

Figure 1A shows the extra calcium seen with a shortening step during the declining phase of the calcium transient in voltage clamped barnacle single muscle fibers microinjected with the calcium sensitive photoprotein aequorin. Our arguments that this extra calcium is coming from calcium previously bound to TnC

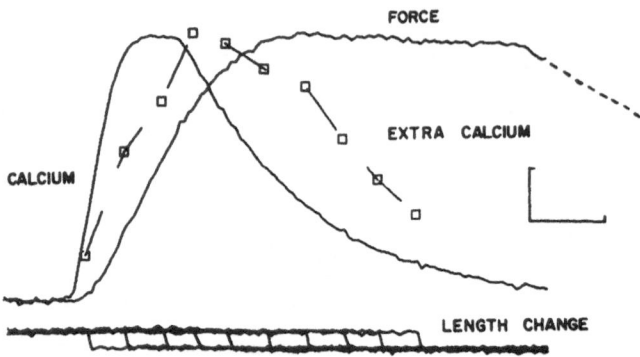

Fig. 2. Time course of the peak in extra calcium is intermediate to that of the free calcium and force. The free calcium is calculated from a control calcium transient. The peak of the extra calcium observed in response to shortening steps at the various times indicated by the length changes is calculated and plotted as occurring at the time of the length change. All records (except the length steps) have been scaled to the same maximum for comparison of the relative time courses. Extra calcium is scaled by 10.8X and plotted as squares connected by a dashed line. Force is displayed at 1.9 g cal with relaxation indicated by the dotted line taken from another record. Length steps are each 0.5 mm. Horizontal sweep, 400 ms/cal. Fiber length, 21 mm; weight, 45 mg; temperature, 7.5°C.

and not from the sarcoplasmic reticulum or other membrane sources were summarized in Ridgway and Gordon[18]. Additional evidence that extra calcium does not come from these other sources is that it is still observed in skinned fibers when both sarcoplasmic reticulum and surface membrane have been destroyed by treatment with detergents[20][21]. The idea that this extra calcium comes from the activating sites is supported by experiments showing that the amplitude of the extra calcium (for a shortening step at a particular time) increases with the peak force when the peak force is increased by increasing the stimulus intensity. Furthermore, for shortening steps at different times during the calcium transient, the peak amplitude of the extra light has a time course that is intermediate between that of free calcium and that of force (see Figure 2).

It is known that decreases in sarcomere length decrease the calcium sensitivity in skinned fibers[22]. From this one might surmise decreased calcium binding with decreased length. That the decrease in calcium affinity of TnC (causing the increase in free calcium with shortening steps) is due to a decrease in cross-bridge attachment or force and not to the above length effect was shown by our previous experiments[23]. Here using a two step protocol and varying the time between the first and second step, we showed that the amount of extra calcium on the second shortening step depends on the force that has redeveloped by the time of the second step rather than depending only on the length change which was the same in each case. Another demonstration that this extra calcium depends on cross-bridge attachment/force was provided by Kurihara et al.[24] who showed [by blocking cross-bridge attachment with 2,3-butanedione monoxime (BDM)] that the amount of extra calcium with a given length step decreased dramatically with decreased cross-bridge attachment/force even though the calcium transient did not. Thus both sets of experiments imply that there is an increase in calcium binding to TnC with increases in cross-bridge attachment/force.

Although TnC is known to be the calcium sensor in mammalian cardiac and skeletal muscle, we questioned whether TnC was responsible for thin filament activation in barnacle striated muscle. TnC has been extracted from barnacle muscle and sequenced[25]. It has just one functional calcium binding site at each end, site 2 in the N-terminus and site 4 in the C-terminus. Furthermore, Ashley et al.[26] and our lab[27] have shown that when TnC is extracted from skinned barnacle muscle fibers, they lose their calcium activated force which can be restored by adding back TnC purified from barnacle muscle. Thus there is now little doubt that barnacle muscle is a thin filament regulated system[28] with TnC as the calcium sensitizing protein.

We have further hypothesized that the shortening step (by changing the calcium affinity of TnC) effectively samples the calcium bound to TnC (CaTnC)[18]. We will show below (in experiments such as those shown in Figure 2, for shortening steps at different times during the calcium transient) that the assumption that the peak of the extra calcium is proportional to the CaTnC is justified. This will be done by deriving a mathematical function that relates the calcium affinity of TnC to the muscle force. Then (from a model of activation) we calculate the free calcium in response to the length change and show that the peak amplitude of the extra calcium is proportional to the calcium bound to TnC at the time of the length change. We can demonstrate that all the measurements are consistent with calcium binding to TnC depending on cross-bridge attachment/force.

If we assume (1) that the peak of the extra calcium seen on shortening is proportional to CaTnC and (2) that the aequorin signal gives a measure of free calcium, then we can calculate the calcium association (k_{on}) and dissociation (k_{off}) rates and thus the calcium affinity required to fit the data to the simple reaction:

$$Ca + TnC \underset{k_{off}}{\overset{k_{on}}{\rightleftharpoons}} CaTnC.$$

When this is done for kon rates of 10^5 M^{-1}s^{-1} and higher, it is impossible to find unique values for k_{off} that (given the free calcium measured with aequorin) can then fit calcium binding to TnC (CaTnC) over the whole time course of contraction. It is possible, however, to find values that can fit the bound calcium during the declining phase of the calcium transient. During this time the force is relatively constant (see Figure 3). This procedure gives a relative calcium affinity for that particular force level. If the force is increased by increasing the stimulus intensity, both the free and bound calcium increase. When these new values are fit with the same k_{on}, the k_{off} required to fit the data is decreased and the K_{aff} is thereby increased. Plotting the relative increase in affinity as a function of the fiber force (expressed as force/unit area), produces a curve such as that shown in Figure 4. It can be seen that the relative

A.M. Gordon and E.B. Ridgway

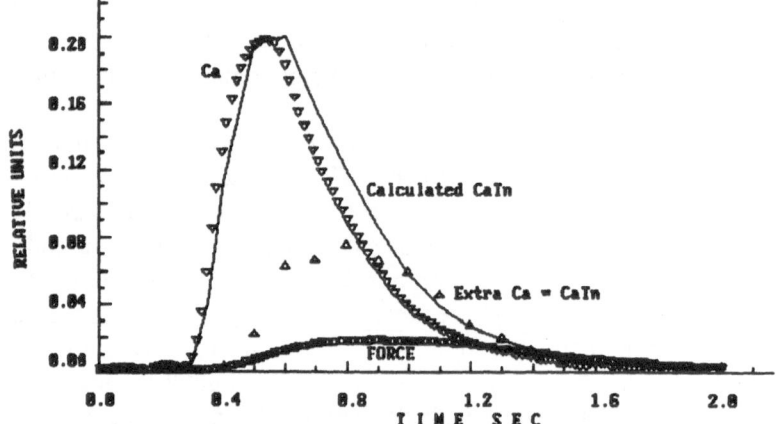

Fig. 3. Calculation of CaTn from measured free Ca using a constant affinity- comparison with the time course of the peak in extra Ca. CaTn (labelled calculated CaTn, the solid line) is calculated using the estimate of free Ca from the aequorin light record (labelled Ca, the inverted triangles) using a fixed k_{on} of 6×10^6 $M^{-1}s^{-1}$ and k_{off} of 30 s^{-1} and compared to the observed peak of the extra Ca (labelled Extra Ca = CaTn, the triangles) for length changes at different times during the Ca transient. The extra Ca is determined in the same manner as in Figure 2. The force is as labelled. Curves scaled to allow comparison of the relative time courses. Note that the calculated CaTn only fits the Extra Ca late in the Ca transient near the time when the force is at its peak and relatively constant. It does not fit the data during the rising phase of both the Ca transient and the twitch, greatly overestimating the amount of bound Ca. Other values of the on and off rates could fit the data as long as the affinity is similar. Since the absolute values of free Ca and CaTn are not known, the affinity is only an estimate. Note however that with the on rate in this range no one single affinity fits the data over the whole time period: a lower affinity is required early and a higher later. This suggests that there is an increase in affinity with the increase in force.

affinity of TnC for calcium increases dramatically as the force (cross-bridge attachment) is increased.

Under the assumption that the peak of the extra calcium seen on shortening is proportional to the calcium bound to TnC, the clear implication is that the calcium affinity of TnC increases dramatically with force. Since barnacle fibers can produce forces up to about 6 kg/cm^2 [29], much of the change in calcium affinity occurs at relatively low values of force (far below the maximum), and thus influences calcium activation of these muscles, particularly at low levels of activation, by enhancing calcium binding. It is, of course, because the calcium affinity of TnC so dramatically decreases with decreasing force that explains the extra calcium seen with shortening steps as in Figure 1A.

We have done preliminary modelling studies using (1) the above relationship between K_{aff} and force, (2) calculations of the free calcium from realistic sarcoplasmic reticulum calcium releasing and uptake functions, (3) activation of force through the rate constant of cross-bridge attachment being dependent on

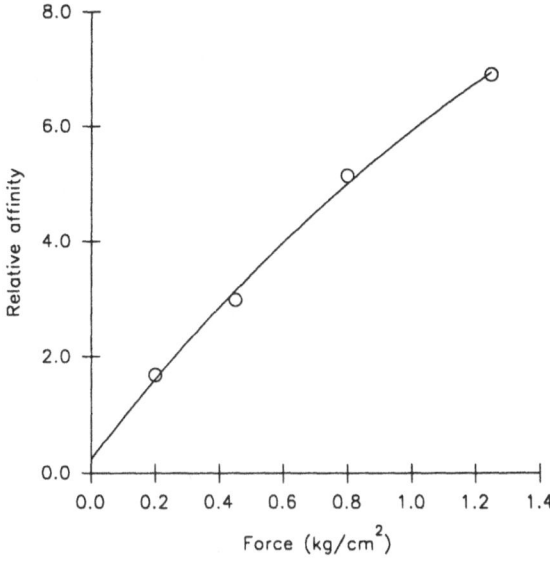

Fig. 4. Plot of calculated calcium affinity as a function of force/unit area from the calculated fit to free calcium and extra calcium (assumed equal to CaTn) during the time in the twitch when the force is relatively constant for records such as the one in Figure 3. Data from five fibers. To combine the data from different fibers with different force/unit area, the affinity for each fiber was expressed relative to the value at the force near 0.45 kg/cm². Since data for each fiber were taken at different force/unit area values, averages of the relative values were made with the following force/unit area ranges, 0.19-0.23 (0.2), 0.40-0.49 (0.45), 0.6-0.82 (0.8), and 1.25 (one fiber) expressed as kg/cm² and plotted at the indicated value.

CaTnC binding, and have been able to show that the peak of the extra calcium seen with shortening steps is proportional to the calcium bound to TnC. This shows that the data is consistent with the conclusions that at moderate levels of activation the peak of the extra calcium is a good relative measure of the calcium bound to TnC. In these modelling studies, single calcium binding to TnC was assumed sufficient to activate (consistent with the single active calcium binding site in the N-terminal region of TnC, Collins et al.[25]). Although this data and analysis imply that calcium binding to TnC does depend on muscle force, it can not easily distinguish between binding being a function of force and binding being a function of cross-bridge attachment *per se*. In fact, since these are often proportional, one needs to design specific experimental conditions to make such a distinction.

One way to separate force and attachment is to stretch the fiber during activation. With this mechanical manipulation, muscle force is increased greatly without a large increase in cross-bridge attachment[30]. When a voltage clamped, aequorin injected fiber is stretched during the declining phase of the calcium transient, the light response is biphasic (Figure 1B). Our interpretation of the biphasic response to stretch is that the stretch first detaches cross-bridges, which decreases calcium binding to TnC giving rise to an elevated free calcium, and then

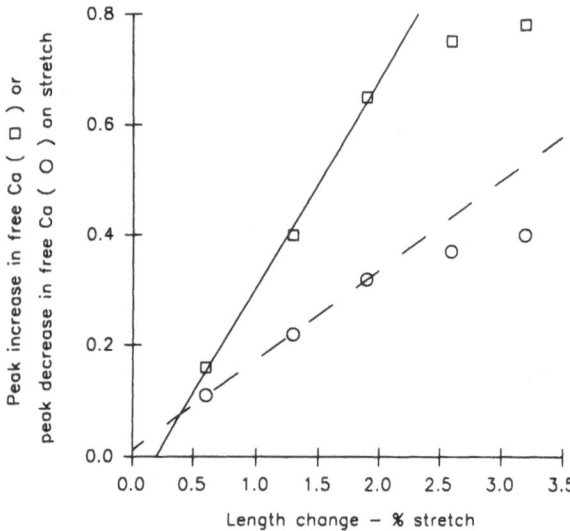

Fig. 5. The amplitude of both phases of the extra light in response to stretch of an aequorin injected, voltage clamped barnacle single muscle fiber as shown in Figure 1B depends on the amplitude of the stretch. The figure shows the amplitude of the peak increase of extra light (extra calcium) during the positive phase (squares) and the peak of the decrease in extra light (decrease in intracellular calcium) during the negative phase (circles) plotted as a function of the amplitude of the length change for a typical fiber. The curves are extrapolated using a least-squares linear fit to the first three points. Note that the intercept on the length axis is near the origin for the decrease in calcium during the negative phase, but that the increase in calcium in response to stretch has a threshold at about 0.2% of the fiber length. This threshold averaged 0.41% for all fibers. Thus fibers have to be stretched above about 0.4% of the fiber length to produce this extra calcium in response to stretch, yet the decrease in calcium with stretch occurs for any amplitude of stretch. Only stretches of greater than 0.4% of muscle length might be expected to detach cross-bridges, yet stretches of any amplitude would tend to elevate the force after stretch.

these cross-bridges reattach but in a more highly strained configuration to support the higher force. Accompanying this higher force, the calcium affinity of TnC is increased resulting in increased calcium binding and free calcium then of course falls to less than control.

We suggested this interpretation of the biphasic response to stretch in a recent paper[31] along with additional supporting evidence. We will summarize the supporting evidence here. Evidence that the increase in free calcium with stretch comes from cross-bridges detaching comes from experiments showing that the extra calcium is only seen with stretches greater than about 0.4% of the muscle length (see Figure 5). Smaller stretches, either steps or ramps, do not lead to increased free calcium. This is consistent with the length change needing to be large enough to detach cross-bridges in this long sarcomere length muscle. In contrast, the decrease in free calcium (the second phase in response to stretch) occurs at all length changes, consistent with it being due to the elevated force or cross-bridge strain after the stretch. Presumably this elevated force is somehow enhancing the calcium binding to TnC and thus reducing the free calcium. Additional support for this interpretation of the two phases comes from the experimental observation that the relative amplitudes of these two phases are a function of the time of the stretch. Stretches made earlier in the contraction enhance the decrease in free calcium with stretch; later stretches enhance the extra calcium on stretch. The relative time courses of these two effects fit with the idea that the decrease in free calcium phase is proportional to the free calcium and the increase in calcium phase is proportional to calcium bound to TnC (see Figure 7 in reference 31 for details). This result would be expected if stretch (breaking cross-bridges) leads to the increased free calcium by decreasing calcium bound to TnC, while stretch (increasing cross-bridge strain) leads to enhanced calcium binding to TnC, with the calcium to accomplish this coming from the free pool.

Thus our data on the effects of muscle stretch is consistent with cross-bridge strain per se having some effect on the thin filament to enhance calcium binding. Whether the strain-induced changes are the major cause of the increased calcium binding to TnC or whether the major effect is just through cross-bridge attachment per se awaits future experiments.

In conclusion, the studies done in intact muscle fibers support the role of cross-bridges in thin filament activation of muscle contraction. The evidence is consistent with a dual mechanism of activation whereby elevated calcium initially causes thin filament activation with attached cross-bridges (or force generation) helping to sustain activation through enhanced calcium binding or even possibly direct activation. The enhanced calcium binding to TnC with cross-bridge attachment enhances steady activation at lower levels of calcium and importantly steepens the relationship between calcium and muscle force.

ACKNOWLEDGEMENTS

We thank Dr. Dennis Willows and the staff at Friday Harbor Laboratories for their hospitality. We thank Dr. O. Shimomura for the gift of native aequorin and for the gift of several purified molecular species of aequorin. We greatly appreciate the assistance of Martha Mathiason in the preparation of the manuscript and figures and with writing computer programs used in the data analysis. This work was supported by NIH grants NS08384 and AM35597 and by a Grant-in-Aid from the Virginia Affiliate of the American Heart Association.

REFERENCES

1. Ebashi, S. & Endo, M. *Prog. Biophys. Mol. Biol.* **18**,123-183 (1968).
2. Bremel, R.D. & Weber, A. *Nature* **238**, 97-101 (1972).
3. Gordon, A.M. in *Muscular Contraction* (ed. Simmons, R.M.) 163-179 (Cambridge University Press, 1992).
4. Gordon, A.M. & Yates, L.D. in *Molecular and Cellular Aspects of Muscle Contraction and Cell Motility* (ed. Sugi, H.) 1-36 (Springer-Verlag, 1992).

5. Weber, A. & Murray, J.M. *Physiol. Rev.* **53**, 612-673 (1973).
6. Eisenberg, E. & Hill, T.L. *Science* **227**, 999-1006 (1985).
7. El-Saleh, S.C., Warber, K.D. & Potter, J.D. *J. Muscle Res. Cell Motility* **7**, 387-404 (1986).
8. Gordon, A.M., Ridgway, E.B., Yates, L.D. & Allen, T. *Adv. Exp. Med. Biol.* **226**, 89-98 (1988).
9. Güth, K. & Potter, J.D. *J. Biol. Chem.* **262**, 13627-13635 (1987).
10. Morano, I. & Rüegg, J.C. *Pflügers Arch.* **418**, 333-337 (1991).
11. Reuben, J.P., Brandt, P.W., Berman, M. & Grundfest, H. *J. Gen. Physiol.* **57**, 385-407 (1971).
12. Goldman, Y.E., Hibberd, M.G. & Trentham, D.R. *J. Physiol. (Lond.)* **354**, 605-624 (1984).
13. Fuchs, F. *Biochim. Biophys. Acta* **491**, 523-531(1977).
14. Hofmann, P.A. & Fuchs, F. *Am. J. Physiol.* **253**, C541-C546 (1987).
15. Fuchs, F. *J. Muscle Res. Cell Motility* **6**, 477-486 (1985).
16. Kress, M., Huxley, H.E., Faruqi, A.R. & Hendrix, J. *J. Mol. Biol.* **188**, 325-342 (1986).
17. Gordon, A.M. & Ridgway, E.B. *Eur. J. Cardiol.* **7**, 27-34 (1978).
18. Ridgway, E.B. & Gordon, A.M. *J. Gen. Physiol.* **83**, 75-103 (1984).
19. Allen, D.G. & Kurihara, S. *J. Physiol. (Lond.)* **327**, 79-94 (1982).
20. Stephenson, D.G. & Wendt, I.R. *J. Muscle Res. Cell Motility* **5**, 243-272 (1984).
21. Allen, D.G. & Kentish, J.C. *J. Physiol. (Lond.)* **407**, 489-503 (1988).
22. Endo, M. *Nature New Biol.* **237**, 211-213 (1972).
23. Gordon, A.M. & Ridgway, E.B. *J. Gen. Physiol.* **90**, 321-340 (1987).
24. Kurihara, S., Saeki, Y., Hongo, K., Tanaka, E. & Sudo, N. *Jpn. J. Physiol.* **40**, 915-920 (1990).
25. Collins, J.H., Theibert, J.L., Francois, J.-M., Ashley, C.C. & Potter, J.D. *Biochem.* **30**, 702-707 (1991).
26. Ashley, C.C., Kerrick, W.G., Lea, T.J., Khalil, R. & Potter, J.D. *Biophys. J.* **51**, 327a (1987).
27. Qian, Y., Gordon, A.M. & Luo, Z.X. *Biophys. J.* **59**, 584a (1991).
28. Dubyak, G.R. *J. Muscle Res. Cell Motility* **6**, 275-292 (1985).
29. Griffiths, P.J., Duchateau, J.J., Maéda, Y., Potter, J.D. & Ashley, C.C. *Pflügers Arch.* **415**, 554-565 (1990).
30. Sugi, H. & Tsuchiya, T. *J. Physiol. (Lond.)* **407**, 215-229 (1988).
31. Gordon, A.M. & Ridgway, E.B. *J. Gen. Physiol.* **96**, 1013-1035 (1990).

Discussion

Edman: Is there an increase in this extra calcium fraction when you increase the amount of the shortening?

Gordon: Yes.

Edman: Is it proportional to the length change?

Gordon: No, it's not. The extra calcium accompanying the length change saturates for length changes above about 10%. Those are large length changes, which are doing more than just detaching cross-bridges. Since barnacle muscle is a long sarcomere-length muscle with presumably a similar cross-bridge stroke to that in other muscle, length changes of only a few tenths of a percent should detach cross-bridges (Griffiths, P.J. et al. *Pflügers Arch.* **415**, 554-565, 1990).

Edman: Wouldn't this imply that the effect must be something other than merely a decrease in force in the bridges?

Gordon: Probably, but with these experiments we really can't distinguish between a decrease in force and a decrease in attachment. The experiments involving muscle stretch imply that cross-bridge strain may play a role as well.

ter Keurs: What are the implications of your observations and model for the force-velocity relation?

Gordon: For force-velocity relations at maximum activation, it should have very little effect, but it should affect the system at very low levels of activation, decreasing calcium binding with filament sliding.

Ashley: I have two quick questions. One relates to the observations that Dr. Sugi published recently on barnacle (Iwamoto et al. *J. Exp. Biol.* **148**, 281, 1990). He finds that in certain situations it behaves like a "catch muscle." Do you observe that, and do you think that affects what you are modelling?

Gordon: We haven't observed it looking like a "catch muscle," but there is a hysteresis in calcium sensitivity (Ridgway, E.B. et al. *Science* **219**, 1075-1077, 1983).

Ashley: Secondly, why is the extra light so delayed? Maybe I don't understand what's going on, but I would have thought it should have occurred much faster if it is associated with the rapid breakage of cross-bridges. You can't explain it by the delay in the Aequorin response which is small on this time scale. Why does the increase in Aequorin light occur after the bridges have been broken? It seems to follow force recovery, rather than the bridges breaking. We have seen the same phenomena, I might add, in intact fibers. I'm just curious about your interpretation.

Gordon: You are correct. It is delayed. The time course of the free calcium can be delayed somewhat by calcium binding at other sites, but the calculated record is still faster than the measured record. Also, the Aequorin does introduce a slight delay, but it is short compared to the observed delay. To date, our modelling can predict the peak height but not the time course of the extra light on shortening.

Gulati: I would expect that when you increase the sarcomere length, decreasing the overlap, there should be a change in the amount of calcium released.

Gordon: With the hypothesis that I proposed, there is not a direct length-dependent effect. Any length-dependent effect would operate through changes in attached cross-bridges or force.

Gulati: Let me try again. With increased sarcomere length, decreasing the overlap, you should get a decrease in the extra calcium released by the length step in your hypothesis.

Gordon: Yes, depending on where the muscle fiber is on the force-length curve. In this preparation as we set it up, as you decrease the length, you decrease the maximum force. With decreased initial length, you also see a decrease in the extra calcium observed for a given length change.

Gergely: When you refer to detachment, do you really mean detachment, or a shift from a strong-binding to a weak-binding state?

Gordon: In this model detachment and weak binding are equivalent, so, yes, it could mean a shift to a weak-binding state.

Griffiths: Dr. Gordon, you said that the feedback effect on calcium binding comes from force. But you seem to have a point on Figure 2 where there was no force and still you got an elevation of calcium through shortening. I wonder if you could clarify that.

Gordon: We never obtain extra calcium if there is no force. What may make it appear that there is little force when we observe extra calcium is that we plotted the extra calcium as occurring at the time that the length change is initiated. The peak in extra calcium occurs later. Neither stretch nor release of a relaxed fiber produces extra calcium.

Ishiwata: If the calcium binding affinity to troponin-C is increased by stretching the fiber, the information transfer should be mediated by the actin. So, I think that something is happening in the actin filament, a strain-induced change of the dipole moment or something that triggers the change of the binding of troponin-C. Do you have any evidence for the change in actin by stretching?

Gordon: We do not at this time, but we are investigating the mechanism of the strain-induced enhancement of troponin-C in binding. If there were strain—individual changes in thin filament structure—it could produce a delay between the time of the length or force change and the time when the affinity changes. That could contribute to the kinetics of the process. The real issue is what the molecular mechanism is of this cross-bridge attachment/force-induced change in Ca affinity. We can't answer that now.

PHOSPHORYLATION-CONTRACTION COUPLING IN SMOOTH MUSCLE: ROLE OF CALDESMON

Gabriele Pfitzer[†], Wolfgang Fischer and Joseph M. Chalovich[*]

II. Physiologisches Institut
Universität Heidelberg, INF 326
D-6900 Heidelberg, Germany
*Department of Biochemistry
East Carolina University School of Medicine
Greenville, North Carolina, 27858-4354, U.S.A.

ABSTRACT

In intact smooth muscle strips from chicken gizzard, carbachol elicited brief, phasic contractions which were associated with a very rapid, transient phosphorylation of the 20 kDa myosin light chains. Phosphorylation was not significantly different from basal levels after 30 s while force still amounted to 50% of the peak value. The rate of tension decline could be increased by addition of atropine, even at apparently basal phosphorylation levels suggesting a phosphorylation independent regulation. The force, at a given level of phosphorylation, could also be modulated by addition of the actin binding, putative regulatory protein, caldesmon. Caldesmon, inhibits phosphorylation dependent force in skinned fiber bundles of chicken gizzard without affecting myosin light chain phosphorylation. This suggests that caldesmon might modulate contraction in smooth muscle. Moreover our results suggest that caldesmon does not function to maintain passive tension.

INTRODUCTION

Myosin light chain phosphorylation (LC_{20}) plays a primary role in activating smooth muscle contraction. This has recently been emphasized by Itoh and coworkers[1] who showed that smooth muscle cells could be activated by injection of the Ca^{2+}-calmodulin independent myosin light chain kinase (MLCK) at basal levels of intracellular Ca^{2+}. However, under many conditions, a temporal dissociation between force and LC_{20} phosphorylation has been observed in both tonic[2] and

---

[†]to whom correspondence should be addressed

Mechanism of Myofilament Sliding in Muscle Contraction, Edited by
H. Sugi and G.H Pollack, Plenum Press, New York, 1993

phasic[3][4] smooth muscles. Moreover the coupling between force and LC_{20} phosphorylation may be variable and depend on the origin of the smooth muscle[5], and also on the mode of stimulation[6]. These results suggest that additional regulatory mechanisms may operate. However, at least the temporal dissociation between force and LC_{20} phosphorylation may be modelled on the basis of the phosphorylation theory[7]. According to this model ("latch"-theory), attached, dephosphorylated crossbridges which are generated by dephosphorylation of attached, phosphorylated crossbridges contribute to force maintenance. This implies, that under conditions where force is maintained at low levels of LC_{20} phosphorylation, the rate of relaxation is limited by the rate of detachment of latch-bridges. The model, therefore, does not account for regulation of relaxation at low levels of LC_{20} phosphorylation easily.[8]

In this study we show, 1) that the rate of relaxation of intact fiber bundles from chicken gizzard is regulated at apparently basal levels of LC_{20} phosphorylation, and 2) that the thin filament linked protein, caldesmon, which has been suggested to function to regulate the latch state inhibits contraction in skinned fibers from chicken gizzard.

METHODS

Chicken gizzards were collected at a local farm immediately after sacrifice of the animal and immersed in ice-cold physiological salt solution containing (in mM) NaCl 150, KCl 5, $MgCl_2$ 2, $CaCl_2$ 5, HEPES 24 (pH 7.4 at 37°C) and glucose 10, gassed with 100% oxygen. For studies using intact muscle, small strips of approximately 15 mm length and 0.4-0.5 mm diameter were cut out of the superficial circumferential muscle layer, mounted between a force transducer and a length step generator (Ling-Dynamics 101 Vibrator), and incubated in PSS maintained at 37°C as described previously[4]. For studies using skinned fibres, muscle strips of about 0.2 mm diameter were incubated in a skinning buffer containing 1% (vol/vol) Triton - X 100 for 4 hours[9]. Fibres were stored until use (maximally 4 days) at -20°C in 50% (vol/vol) glycerol and relaxing solution. For isometric force measurements, thin fibre bundles (100 - 150 μm diameter) were glued between an AME 801 force transducer and a glass rod extending from a micrometer drive[9].

The constitutively active fragment of myosin light chain kinase[10] (I-MLCK) as well as caldesmon or the fragments of caldesmon were allowed to diffuse into the skinned fibers for 30 min in rigor solution containing (in mM): KCl 50, imidazole 25 (pH 7.0), EGTA 4, $MgCl_2$ 3, DTE 2. Contraction was inititated by switching to the ATP containing relaxing solution consisting of (in mM) KCl 50, imidazole 25 (pH 7.0, 21-23°C), $MgCl_2$ 4, EGTA 4, creatine phosphate 4, DTE 2 ATP 1, phosphocreatine kinase (140 U/ml). Contraction solution had the same composition as relaxing solution exept for the addition of 4 mM $CaCl_2$ and 1 μM calmodulin.

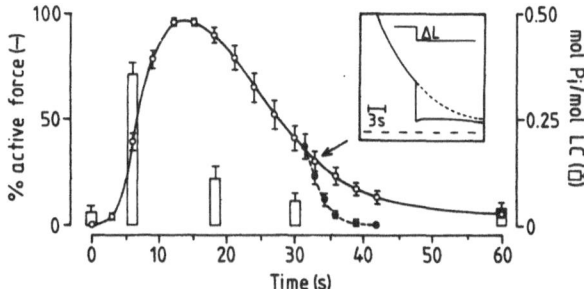

Fig. 1. Time course of force and LC_{20} phosphorylation in response to 3 μM carbachol (open circles) and following addition of atropine (closed circles). Note, the increase in the relaxation rate following addition of atropine. Force at the time of freezing was normalized in respect to the preceding contraction. Each point represents the mean value ± SEM from 5 fibres taken from different animals. Peak force was 21 ± 2 N/cm² (n = 49). Inset: tension response following a quick release of 5% L_0 applied at arrow.

Solutions containing tryptic fragments of proteins also contained 0.4 mg/ml soybean trypsin inhibitor.

Myosin light chain phosphorylation was determined as previously described[4]. The details of protein purification will be published elsewhere[11].

RESULTS AND DISCUSSION

I. Regulation of Relaxation in Intact Fibers of Chicken Gizzard

Intact strips of chicken gizzard responded to stimulation with carbachol with a phasic contraction lasting less than 60 seconds[4]. A smiliar contractile response has also been shown by Ozaki and coworkers[12]. The carbachol induced contraction was associated with a very rapid LC_{20} phosphorylation transient having its peak at 6 s, and which was basal after 30 s while force still amounted to 50% of the peak value (Fig. 1). At LC_{20} phosphorylation levels indistinguisable from basal, there was virtually no force recovery following a quick release (Fig. 1) suggesting that tension was maintained passively possibly by noncycling crossbridges. Addition of atropine at this point increased the rate of relaxation (Fig. 1), suggesting that the net detachment of dephosphorylated crossbridges may be regulated independent of changes in the apparent LC_{20} phosphorylation.

According to the "latch" model of Hai and Murphy[7], tension at low levels of LC_{20} phosphorylation is maintained by so-called latch bridges which are generated by dephosphorylation of attached crossbridges, and which have a low detachment rate compared to phosphorylated crossbridges. Thus at low or even basal levels of LC_{20} phosphorylation there has to be an elevated LC_{20} phosphorylation turnover which we cannot exclude in our experiments if, in the continued presence of carbachol, MLCK activity was suprabasal. Antagonizing the action of carbachol by

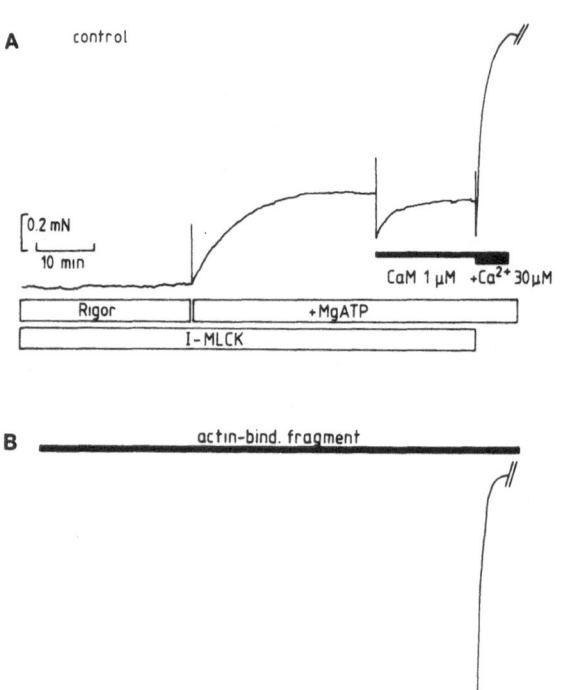

Fig. 2. Inhibition of phosphorylation dependent force by an actin-binding fragment of caldesmon. A: Following preincubation in rigor solution, the constitutively active fragment of MLCK (I-MLCK) elicited force in the presence of MgATP which was not enhanced by addition of exogenous calmodulin (CaM); Ca^{2+} (30 μM) further increased force. B: In the presence of the actin-binding fragment, force in response to I-MLCK was completely inhibited and was reversed in the presence of Ca^{2+}.

atropine would decrease the MLCK activity to basal levels thereby preventing the replacement of detaching latch bridges by new ones. The relaxation rate in the presence of atropine would then reflect the detachment rate of latch bridges which, however, would have to be much higher than that proposed by Murphy and co-workers. Based on the assumptions of Hai and Murphy[7], we modelled the force and LC_{20} phosphorylation transients. The force and LC_{20} phosphorylation transients observed in the presence of carbachol could be modelled according to the latch model whereby the rate constant of latch bridge detachment was higher (0.1 sec^{-1}) than that in tonic smooth muscles (0.01 sec^{-1})[7]. However, we could not find a set of parameters, with the activity of MLCK as the only regulated parameter, that would account simultaneously for the contraction elicited by carbachol and for the 5-fold faster relaxation in the presence of atropine. This suggests that a second regulatory mechanisms is operative regulating the net detachment of dephosphorylated crossbridges.

II. Inhibition of Force by Caldesmon in Skinned Fiber Bundles from Chicken Gizzard

The effect of caldesmon and fragments of caldesmon on tension development was examined in skinned fibers from chicken gizzard activated by the calmodulin independent, constitutively active fragment of MLCK[10] in the absence of Ca^{2+}. This approach allowed us to study the effect of caldesmon without any possible interference of caldesmon with the physiologcial Ca^{2+} and calmodulin dependent activation pathway of smooth muscle while maintaining the phosphorylation-dephosphorylation cycle of myosin light chains.

In the presence of a C-terminal, 20 kDa actin-binding fragment of caldesmon, tension elicited by I-MLCK in the absence of Ca^{2+} was significantly inhibited (Fig.

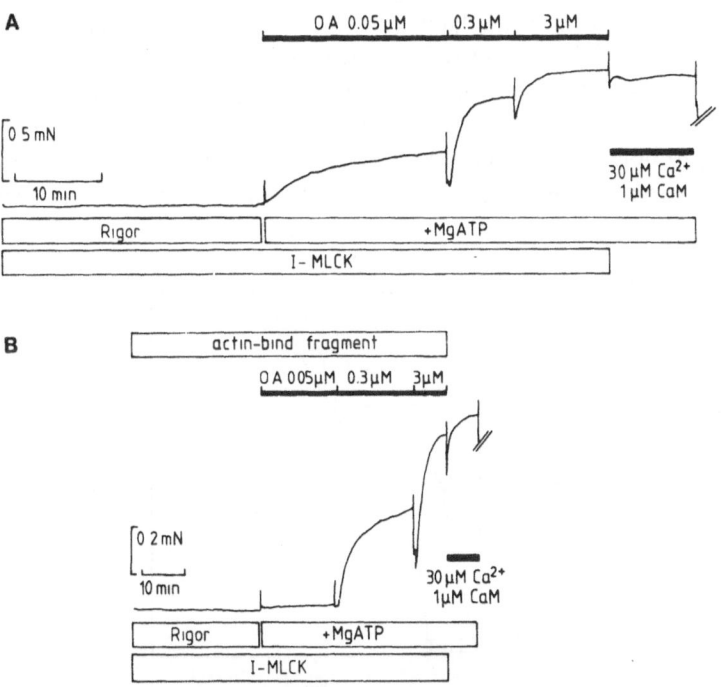

Fig. 3. Okadaic acid (O.A.) dose dependently increased force elicited by I-MLCK. Note, that LC_{20} phosphorylation was sufficient to induce maximal force as there was no further increase with Ca^{2+} and calmodulin (CaM). The actin-binding fragment of caldesmon shifted the dose response to higher concentrations of O.A. Note, that force was completely inhibited in the presence of 0.05 μM O.A., while 3 μM O.A. nearly reversed the effect of the fragment.

2) and amounted to 2.04 ± 0.4 N/cm^2 (n = 10) as compared to 5.3 ± 0.4 N/cm^2 in paired control fibers. Intact caldesmon inhibited force to a smaller extent (3.6 ± 0.3 N/cm^2) while a N-terminal myosin-binding fragment was without effect. Inhibition of force was reversed in the presence of 30 μM Ca^{2+}, and 1 μM calmodulin (Fig. 2) which weaken the binding of caldesmon to actin[13]).

The phosphatase inhibitor, okadaic acid, dose-dependently increased tension elicited by I-MLCK (Fig. 3a). In the presence of the actin-binding fragment, higher concentrations of okadaic acid were required to reach the same level of force (Fig. 3b). High concentrations of okadaic acid reversed the inhibitory effect of the actin-binding fragment of caldesmon. In the presence of okadaic acid, there was an increase in phosphorylation of myosin light chains. However, it is not clear at present whether high levels of LC$_{20}$ phosphorylation are related to the reversal of the inhibitory action of caldesmon. Caldesmon and the actin-binding fragment were also phosphorylated in the presence of okadaic acid, and phosphorylation of caldesmon has been suggested to relieve the inhibition of smooth muscle actomyosin ATPase[14)15].

The inhibition of force was observed under conditions where significant levels of force were maintained at low levels of LC$_{20}$ phosphorylation (Fig. 4). Thus, caldesmon and the actin-binding fragment inhibited contractions in skinned fibers which were reminiscent of the latch state in intact smooth muscle[7]). Moreover, the relationship between force and LC$_{20}$ phosphorylation was shifted to the right (Fig. 4) indicating that in the presence of the actin-binding fragment of caldesmon, higher levels of LC$_{20}$ phosphorylation were required to support a given level of force. It is interesting to note that the relation between force elicited by I-MLCK and LC$_{20}$ phosphorylation is curvilinear much in the same way as has been reported

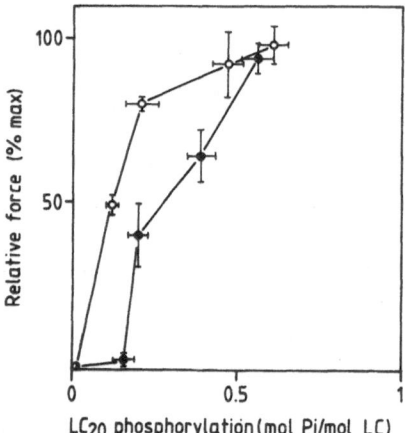

Fig. 4. Relation between force and LC$_{20}$ phosphorylation with (closed circles) and without (open circles) the actin-binding fragment of caldesmon. Experimental protocol as in Fig. 3. exept that only one dose of O.A. was applied in a single experiment. Values are mean ± SEM, n = 4-6 from at least 5 different gizzards.

previously when force was elicited by calcium[16][17]. This suggests that the curvilinear relation between force and LC_{20} phosphorylation is a property of the dependent regulatory mechanism.

It has been suggested that caldesmon may support tension during the latch state[18][19] by virtue of its ability to bind simultaneously to actin and myosin[20][21]. Then inhibition of force might be due to a competitive inhibition of the interaction of endogenous caldesmon with actin and myosin by the actin- and myosin-binding fragments of caldesmon. If this were true, inhibition of force would have been expected to occur in the presence of both the actin- and myosin-binding fragments. On the other hand, intact caldesmon should increase force by virtue of reconstitution of the fibers which are partially devoid of caldesmon due to skinning[22]. Clearly, our results do not support a role of caldesmon in maintaining passive force, because both intact caldesmon and the actin-binding fragment inhibited tension while the myosin-binding fragment was without effect. Rather, our results support a role of caldesmon in maintaining a relaxed state or regulation of relaxation, whereby caldesmon does not seem to act as a switch but rather as a modulator of LC_{20} phosphorylation dependent force. A similar conclusion has been reached by Hemric and Haeberle using *in vitro* motility assays[23][24]. Thus caldesmon appears to inhibit contractility of smooth muscle by virtue of its inhibition of binding of myosin directly to actin[19]. Whether the mechanism of inhibition of force in skinned smooth muscle is similar to that in striated muscle[25] remains to be seen.

In conclusion, we have shown that the level of force that a smooth muscle produces is not uniquely defined by the level of phosphorylation of the myosin light chains. This raises the possibility that other smooth muscle proteins, such as caldesmon, may modulate force production. Our results do not support the involvement of caldesmon in tension maintenance by crosslinking myosin with actin. However, caldesmon could be involved in the regulation of force at basal levels of LC_{20} phosphorylation observed in intact fibers from chicken gizzards.

ACKNOWLEDGEMENTS

We thank C. Zeugner and M. Troschka for expert technical assistance. This work was supported by grants from DFG (Pf 226/1-1 and Pf 226/1-2 to GP), NIH (AR-35216 and AR40540-01A1 to JMC), and NATO (900257 to JMC).

REFERENCES

1. Itoh, T., Ikebe M., Kargacin G.J., Hartshorne D.J., Kemp B.E. & Fay F.S. *Nature* **338**, 164-167 (1989)
2. Kamm, K.E. & Stull, J.T. *Ann. Rev. Pharmacol. Toxicol.* **25**, 593-620 (1985).
3. Himpens, B., Matthijs, G., Somlyo, A.V., Butler T.M. & Somlyo, A.P. *J. Gen. Physiol.* **92**, 713-729 (1988).

4. Fischer, W. & Pfitzer, G. *FEBS Letters* **258**, 59-62 (1989).

5. Gerthoffer, W.T., Murphey, K.A. & Mangini, J. *Biophys. J.* **61**, A16 (1992).

6. Suematsu, E., Resnick, M. & Morgan, K.G. *Am J. Physiol.* **261**, C253-C258 (1991).

7. Hai, C.-M. & Murphy, R.A. *Am. J. Physiol.* **254**, C99-C106 (1988).

8. Marston, S.B. *J. Musc. Res. Cell Motility* **10**, 97-100 (1989).

9. Pfitzer, G., Merkel, L., Rüegg, J.C. & Hofmann, F. *Pflügers Arch.* **407**, 87-91 (1986).

10. Ikebe, M., Stepinska, M., Kemp, B.E., Means, A.R. & Hartshorne, D.J. *J. Biol. Chem.* **260**, 13828-13834 (1987).

11. Pfitzer, G., Zeugner, C., Troschka, M. & Chalovich, J.M. *Proc. Natl. Acad. Sci. USA* (submitted).

12. Ozaki, H., Kasai, H., Hori, M., Sato, K., Ishihara H. & Karaki, H. *Naunyn Schmiedeberg's Arch. Pharmacol.* **341**, 262-267 (1990).

13. Smith, C.W.J., Prichard, K. & Marston, S.B. *J.Biol. Chem.* **262**, 116-122 (1987).

14. Ngai, P.K. & Walsh, M.P. *J. Biol. Chem.* **259**, 13656-13659 (1984).

15. Sutherland, C.& Walsh, M.P. *J. Biol. Chem.* **264**, 578-583 (1989).

16. Pfitzer, G., Schmidt, U., Troschka, M. & Gröschel-Stewart, U. *Pflügers Arch.* **420** (Suppl 1), R 101 (1992).

17. Kenney, R.E., Hoar, P.E. & Kerrick, W.G.L. *J. Biol. Chem.* **265**, 8642-8649 (1990).

18. Sutherland, C. & Walsh, M.P. *J. Biol. Chem.* **264**, 578-583 (1989).

19. Hemric, M. & Chalovich, J.M. *J. Biol. Chem.* **263**, 1878-1885 (1988).

20. Ikebe, M. & Reardon, S. *J. Biol. Chem.* **263**, 3055-3058 (1988).

21. Hemric, M.E. & Chalovich, J.M. *J. Biol. Chem.* **265**,19672-19678 (1990).

22. Kossmann, T., Fürst, D. & Small, J.V. *J. Muscle Res. Cell Motility* **8**, 135-144 (1987).

23. Hemric, M.E. & Haeberle, J.R. *Biophys. J.* **61**, A6 (1992).

24. Haeberle, J.R. & Hemric, M.E. *Biophys. J.* **61**, A6 (1992)

25. Brenner, B., Yu, L.C. & Chalovich, J.M. *Proc. Natl. Acad. Sci. USA* **88**, 5739-5743 (1991).

Discussion

Katayama: Are almost all the thin filaments saturated with caldesmon, or is there still some space?

Pfitzer: We do not think that all the thin filaments are saturated with caldesmon. In fact, the molar ratio of actin:caldesmon is lowest in phasic smooth muscles (Haeberle et al. *J. Muscle Res. Cell Motility* **13**, 81-89, 1992). Chicken gizzard is a phasic smooth muscle and, moreover, in the skinned fibers, some caldesmon may have been extracted. Thus we believe that there is some space.

Katayama: Do you have any assay; SDS gel estimation, for instance?

Pfitzer: No, we haven't done that yet.

Sugi: Dr. Ashley mentioned our previous experiment with barnacle fibers (Iwamoto et al. *J. Exp. Biol.* **198**, 281-291, 1990). Our main result was that during prolonged electrical stimulation, force rises slowly up to the peak, while unloaded shortening velocity, one measure of the ability to shorten, reaches a peak rapidly during the rising phase, and falls. The time course is similar to your phosphorylation curves, so I suggest you try to measure the ability to shorten under unloaded conditions.

Pfitzer: For technical reasons, it has not been possible to measure unloaded shortening velocity in intact fiber bundles from chicken gizzard. We estimated the active state by the quick release method. Tension recovery in the initial phase of the contraction is certainly much more complete and very rapid as compared with the declining phase. So we have some evidence that within the contraction there is a change in contractile kinetics, i.e., some force is maintained passively.

Sellers: Might an easier way to do the experiment and sort out the relative contributions of caldesmon and myosin phosphorylation be to thiophosphorylate the skinned fiber, and then add, in the presence of ATP, the caldesmon fragments to see whether this could fully inhibit the force?

Pfitzer: Yes, I purposely didn't do this kind of experiment yet. We plan to do that as well. But, for instance, in cardiac muscle it is known that thiophosphorylation has a different effect than phosphorylation. Using the constitutively active fragment of MLCK has the advantage that the phosphorylation/dephosphorylation cycle of myosin light chains is maintained, which is not the case when they are stably thiophosphorylated. In the light of current theories of the latch model (cf. Hai, C.M., & Murphy, R.A. *Am. J. Physiol.* **254**, C99, 1988; *Ann. Rev. Physiol.* **51**, 285-298, 1989), this may be important.

Huxley: A good many years ago, Professor Ebashi had evidence of calcium-sensitive force development in smooth muscle without phosphorylation of the myosin, and that this was dependent upon a protein that he called leiotonin. What is the present view of smooth muscle people about those observations?

Pfitzer: I think it is very difficult to say that you really have a contraction without any phosphorylation. I think I've shown this one experiment where we get tonic tension with the potassium channel blocker and we get very high forces at a very low level of phosphorylation also in skinned fibers. But the problem really comes in deciding whether it is really zero phosphorylation or zero increase in phosphorylation. At the moment, it is very difficult to decide from both skinned smooth muscle and intact smooth muscle.

Huxley: Those contractions that you showed were extremely slow, and without redevelopment. I don't remember the details of Ebashi's results, but I think they were at what you might call normal speed of force development.

INFLUENCE OF Ca-ACTIVATED BREVIN ON THE MECHANICAL PROPERTIES OF SKINNED SMOOTH MUSCLE

Ph. Gailly, J.M. Gillis and J.P. Capony*

Department of Physiology
University of Louvain
1200 Bruxelles, Belgium
* C.N.R.S., Route de Mende
Montpellier, France.

ABSTRACT

Solutions of purified brevin were applied to skinned thin bundles or isolated fibres of smooth muscle. This produced a sharp drop of isometric tension, an effect due to the severing effect of brevin on actin filaments, partially depleted from tropomyosin in skinned preparations. On skinned single fibres, brevin accelerates the speed of unloaded shortening. As no effect was detected on the myofibrillar ATPase turnover rate, brevin was thought to affect the viscosity of the cytoplasm. This was confirmed by analysis of the cytoplasm stiffness which decreased in the presence of brevin. It is proposed that Ca-activated brevin acts on actin-filamin gels, set in parallel to the contractile apparatus.

INTRODUCTION

Smooth muscle is one of the most important source of gelsolin and brevin, two closely related proteins which sever actin filaments into short pieces[1]. This effect depends on the presence of Ca, in micromolar concentration, thus occurs in conditions which prevail during muscle activation. This is a paradoxical situation as any severing effect on actin filaments could have disastrous effects on the mechanical output. We reported here our studies of various mechanical properties of skinned smooth muscle, with or without the presence of brevin.

Mechanism of Myofilament Sliding in Muscle Contraction, Edited by
H. Sugi and G.H Pollack, Plenum Press, New York, 1993

MATERIAL AND METHODS

Strips of taenia coli from decapitated guinea pigs were prepared according to Sparrow et al.[2], using Triton X100 to demembranate. Preparations were kept in relaxing solution, at pH 6.4, to reduce myosin loss during storage. Isolated cells from taenia coli were prepared by enzymic digestion, following Small's procedure[3], using a Ca-dependent collagenase preparation (from Boehringer).

Mechanical arrangements. Thin bundles of skinned taenia coli were clamped, at both ends, with light aluminum clips and attached, at one end to a force transducer (Akers, Norway) and at the other end to a device for generating sinusoidal length changes (peak to peak amplitude : 0.25-0.5 % length; frequency : from 1 to 100 Hz). In experiments with bundles, the mechanical parameter under study (e.g. isometric contraction or stiffness measurements) was first examined in the absence of brevin, as it was washed out during the skinning process. Then inactive brevin was let to diffuse into the bundle by prolonged soaking (2h) in 1μM brevin dissolved in a Ca-free, relaxing medium. This loading period was necessary given the high molecular weight (90 kDa) of brevin.

Isolated cells were observed by phase contrast microscopy and solutions were passed between slide and coverslip. Only well elongated and relaxed fibres, of about 400 μm long were selected. They were demembranated first by passing the Triton-relaxing solution for 0.5 min. This was sufficient to remove 90% of the brevin content. The cell could be re-loaded with inactive brevin when appropriate, and finally activated. Unloaded shortening velocity was observed, and the time needed to shorten by 50% of the resting length (see Results) recorded. Fibres twisting or contracting unevenly were not retained. Each preparation could only be used once.

Solutions (mM). Relaxing : KCl : 100; KH_2PO_4 : 6; Mg-ATP : 5; imidazole : 10; EGTA : 4; NaN_3 : 1; pH 6.7. Contracting: same as relaxing, without potassium phosphate, containing either $Ca^{2+} 10^{-4}$ M, or Ca-EGTA/EGTA mixture to obtained Ca^{2+} buffered concentration ranging from 10^{-9} M to 10^{-5} M, computed according to Fabiato & Fabiato[4]. Brevin was purified from bovine serum as described by Soua et al.[5] and lyophilized.

RESULTS

1. Isometric Tension

In brevin loaded bundles, the amount of isometric tension obtained was only 10 to 20 % of the tension developed by the same bundles before the introduction of brevin. We checked that this effect results genuinely from the activation of brevin by Ca^{2+} when the contracting solution was admitted and not from a proteolytic activity contaminating the brevin preparation.

This dramatic reduction of tension in the presence of brevin suggests that the transmission of force was, to a large extent, interrupted by cuts in the actin filaments. In our experimental conditions, the brevin/actin molar ratio was about

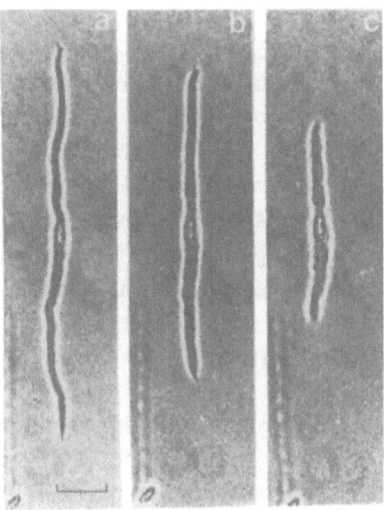

Fig. 1. Unloaded shortening of an isolated smooth muscle cell (taenia coli) observed in phase contrast microscopy. Times of contraction : (a): 0 s, (b) 6 s, (c) 12 s. In this example, a 50 % shortening was obtained in 12 s, corresponding to an average speed of 12 $\mu m.s^{-1}$. Scale bar : 35 μm. From ref. 13.

1/800, thus few cuts were expected, as one molecule of brevin is needed to obtain one cut in F-actin, by insertion of brevin between two actin monomers. As transmission of force requires uninterrupted structures in series, it is clear that tension will be extremely sensitive to the severing action of brevin : only few cuts can produce the dramatic loss of tension we observed. A priori, it is difficult to see any relevance of this effect to physiology as it would make the contractile machinery very inefficient. Moreover, it is well established that actin filaments, covered with tropomyosin are not severed by brevin[6]. We think that the effect we observed results from a very limited depletion of tropomyosin in skinned preparations, making some actin filaments accessible to brevin. One can thus expect that in the living cell, actin filaments interacting with myosin and transmitting force are fully protected by tropomyosin from the action of brevin.

2. Unloaded Shortening Velocity of Single Fibres

The speed of unloaded contraction was estimated as described in the Methods.

A typical example is illustrated at the Fig. 1. In the absence of brevin, the mean time needed to get a 50% shortening was 17.8 sec ($\pm$ 6.7, S.D. n = 30) corresponding to 11 $\mu m.sec^{-1}$. For cells loaded with brevin the mean shortening time dropped to 9.9 sec ($\pm$ 3.9, S.D. n = 22). Thus in that case, shortening speed increased by 1.7 times. For each of the seven taenia coli preparations, we compared the *fastest shortenings*, without and with brevin. Again a 1.7 increase of speed was noted in the presence of brevin.

To understand the origin of this effect, we measured the ATPase activity of isolated myofibrils from taenia coli, without and with brevin (in that case we maintained a brevin/actin molar ratio of about 1/800 as in our mechanical experiments). The ATPase activity was practically unaffected.

We hypothesize that a faster shortening reflects a reduction of the internal resistance to movement, resulting of the effect of brevin on non-contractile structures. This may be detected by analysis of the cytoplasmic stiffness.

3. Stiffness of the Smooth Muscle Cytoplasm

This parameter was studied by submitting thin bundles of skinned taenia coli to sinusoidal stretches and releases. To eliminate any interference from the actin-myosin interaction, we used inhibitors of the myosin ATPase (mersalyl[7]), an organic mercurial or vanadate ions[8])).

The stiffness of inhibited preparations, kept in Ca-free relaxing solutions, without brevin, was taken as reference. In the presence of Ca-activated brevin, this reference stiffness decreased by about 30 %, on the average. A 10% drop was detectable after 1-2 min. The full effect was obtained at 2×10^{-6} M Ca^{2+} and was half-maximal at 6×10^{-7} M Ca^{2+}.

The decrease of stiffness in vanadate-inhibited bundles was very variable : it ranged from 8 to 59 % . The smallest effect was obtained in preparations where the actin-myosin interaction was not 100 % inhibited : they developed a small increase of stiffness in the presence of calcium alone (no brevin). This suggests that the structures undergoing a decrease of stiffness were arranged in parallel with those developing an increase of stiffness.

The stiffness we measured had a 'complex' nature : it contained contributions from both viscous and elastic elements. These contributions can be separated by analyzing the phase delay between imposed sinusoidal length changes and the corresponding sinusoidal force response. At 50 Hz, this phase lag was 20.5° in reference conditions, it dropped to 16.5° in the presence of Ca-activated brevin. From the reduction of the amplitude of the complex stiffness (-30%) and of the phase lag (- 4°), it can be computed that the viscous component decreased by 43% and the elastic one, by 29% (Nyquist diagram analysis).

Finally, for the sake of comparison, we studied the stiffness in the presence of a concentrated (0.6 M) KI solution which depolymerizes F-actin. This produced a 75 % decrease of the bundle stiffness.

DISCUSSION

The increase of unloaded shortening velocity in the presence of brevin was quite unexpected. As a direct effect on the ATPase turnover rate could be excluded, we

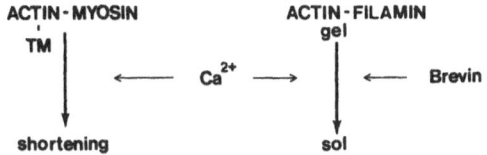

Fig. 2. Model of simultaneous activation of actin-myosin and of brevin by calcium. From ref.14

hypothesized that brevin acted by reducing some internal resistance to movement. Actin containing structures were certainly involved, as the action of brevin is very specific. For reasons developed above, actin filaments forming with myosin the contractile apparatus are most likely protected from the action of brevin and therefore cannot contribute to the observed effect. However it is well known that the actin content in smooth muscle greatly exceeds the amount of myosin.

V. Small and co-workers[9)10)] have provided some evidence that a large part of the muscle actin is organized into filamentary domains, free of myosin, but rich in filamin; they extend along the fibre axis, side by side with the contractile apparatus. The actin of these domains can be selectively isolated from that associated with myosin[11)]. Filamin crosslinks actin filaments to form a gelified network, where actin may not be totally protected by tropomyosin because binding of filamin displaces tropomyosin[12)].

The study of the cytoplasmic stiffness suggests that the target of brevin would be this actin-filamin gelified network.

As illustrated in Fig. 2, upon activation, the increased cytoplasmic Ca^{2+} concentration would, on the one hand, activate the contractile machinery, and, on the other hand, activate brevin, causing the solation of the actin-filamin gel. This would greatly reduced the viscosity of the cytoplasm and allow a faster shortening.

ACKNOWLEDGEMENTS

Figures 1 and 2 were reproduced with the permission of the editors of the Journal of Muscle Research and Cell Motility.

REFERENCES

1. Yin, H.L., Albrecht, J.H. & Fattoum, A. *J. Cell Biol.* **91**, 901-906 (1981).
2. Sparrow, M.P., Mrwa, U., Hoffmann, F. & Rüegg, J.C. *FEBS lett.* **125**, 141-145 (1981).
3. Small, J.V. *J. Cell Sci.* **24**, 327-349. (1977).
4. Fabiato, A. & Fabiato, F. *J. Physiol. (Paris)* **75**, 463-505 (1979).
5. Soua, Z., Porte, F., Harricane, M.-C., Feinberg, J. & Capony, J.-P. *Eur. J. Biochem.* **153**, 275-287 (1985).
6. Fattoum, A., Hartwig, J.H. & Stossel, T.P. *Biochem.* **22**, 1187-1193 (1983).
7. Weber, H.H. & Portzehl, H. *Prog. Biophys. Biophys. Chem.* **4**, 60-111 (1954).
8. Goodno, C.G. & Taylor, E.W. *Proc. Natl. Acad. Sci. USA* **79**, 21-25 (1982).

9. Small, J.V., Fürst, D.O. & De Mey, J. *J. Cell Biol.* **102**, 210-220. (1986).
10. Kossmann T., Fürst D. & Small J.V. *J. Muscle Res. Cell Motility* **8**, 135-144. (1987).
11. Lehman, W., Sheldon, A. & Madonia, W. *Biochim. Biophys. Acta* **914**, 35-39. (1987).
12. Zeece, M.G., Robson, R.M. & Bechtel, P.J *Biochim. Biophys. Acta* **581**, 365-370 (1979).
13. Gailly, Ph., Lejeune, Th., Capony, J.P. & Gillis, J.M. *J. Muscle Res. Cell Motility* **11**, 293-301 (1990).
14. Gailly, Ph., Gillis, J.M. & Capony, J.P. *J. Muscle Res. Cell Motility* **12**, 333-339 (1991).

Discussion

Yagi: In the equatorial X-ray diffraction pattern from smooth muscle, there is a spot showing the presence of a bundle of actin filaments, which was found by Jack Lowy many years ago (Elliot, G.F., & Lowy, J. *Nature* **219**, 156-7, 1963). It shows a center-to-center distance of 12 nm and the presence of very tightly packed gels of actin filaments. We found that this spot becomes broader during contraction. So, I wonder if brevin helps contraction by making the gel much softer? Is that possible?

Gillis: Yes, sure.

Sellers: What's the mechanism for re-annealing? Could you possibly incorporate into the mechanism you propose a method to account for the increase in force, or for the force maintenance without light-chain phosphorylation later in the contraction? The mechanism you propose would break down the cytoskeletal actin gel at the start of the contraction; can you reconstitute again in time to account for some of the peculiar mechanical properties of smooth muscle such as "latch," where tension is maintained even though the phosphorylation of myosin has decreased?

Gillis: F-actin can be quickly cut into pieces by brevin in the presence of calcium, and if you remove calcium by EGTA, long filaments reform. So the system is reversible. That's one point.

Sellers: But, could you reform the network quickly enough to account for the latch-like behavior?

Gillis: The only kinetic experiment I know about the reversibility is from people checking the length of the filament with electron microscopy. The shortest time for reversing is six seconds So possibly in a tonic smooth muscle it could work.

Rall: Does brevin have any effect on isometric force-generating capacity?

Gillis: Yes. We didn't study the isometric tension on single fibers, but on skinned bundles which were kept for at least a few days in relaxing solution. In that case, we observed a dramatic effect on isometric tension which was suppressed by about 90%. We were very worried about this effect. But, in fact, looking carefully through the literature, we found that Vic Small's group in Salzburg (*J. Muscle Res. Cell Motility* **8**, 135-44) reported that during the skinning procedure some tropomyosin is lost. You must be aware that the isometric tension test is extremely sensitive to any cut of the actin filament system. It is like having a big chain—you cut in the middle and nothing is transmitted. So, in spite of the fact that we have a molar ratio of 1 brevin:800 actin monomers, even very few cuts of the filament could have a dramatic effect on tension production. But this effect on tension, we think, is the result of a skinning artifact—the loss of tropomyosin, even 5% of the tropomyosin.

Goldman: In order to account for the 50% change in shortening velocity, I would think that the alteration you observe in the viscous modulus would have to be enough to provide the amount of force that would occur on the force-velocity curve at 50% velocity. I wonder if the change in the viscous modulus that you observe is quantitatively sufficient?

Gillis: I cannot produce an immediate answer, but that is a very important question.

Rüegg: What is the physiological function of gelsolin in smooth muscle?

Gillis: Well, that was the purpose of the experiment! We don't know yet, I'm afraid. The first hypothesis is to try and find a function for it.

Rüegg: I'm sure you know that at the beginning of atherosclerosis vascular smooth muscle cells can become very motile; they move like amoeboid cells. Then you have quite a different kind of contactility, maybe actin-based. One could have a brevin effect here.

Gillis: Yes, in macrophages, of course, it is possible for brevin and gelsolin to change the cytoskeleton and to allow a change of shape of the macrophage. But of course, in smooth muscle, we do not expect that the shape of the cell would change very much.

Rüegg: Under the conditions that I mentioned, yes. I think it might be very important in patho-physiological conditions.

Ishiwata: To my knowledge, tropomyosin cannot completely inhibit the function of gelsolin in the case of skeletal muscle, at least as reported last year by Funatsu, Higuchi and me (*J. Cell. Biol.* **110**, 53-62, 1990). Tropomyosin binding slows down the severing function of gelsolin, but it is not a complete inhibitor. Do you think that, in the case of smooth muscle, the apparent difference of the function of gelsolin is due to the difference of the species of tropomyosin?

Gillis: I cannot say "yes," our actin was perfectly protected by tropomyosin. The only thing I can say is that, in this case, the amount of gelsolin was extremely small compared with the amounts of tropomyosin and actin. So, we suspect that the actin covered by tropomyosin was, in fact, protected.

Ishiwata: We also found that rigor bridges were more effective as an inhibitor of the function of gelsolin.

Gillis: We found the same.

Ishiwata: What is your condition? Rigor?

Gillis: No, this is not rigor. But in another set of experiments, we found exactly what you said, that the rigor protects, somehow. But in the present case, the fibers were in relaxing solution and passed from relaxation to activation.

ter Keurs: Dr. Gillis, is the following scenario possible: If the smooth muscle cell is transiently activated by a calcium pulse, it will contract, brevin will be activated and will therefore lyse actin filaments. That will lead to a shortened cell. If the brevin effect is terminated and actin anneals again, do you end up with a stiffer and shorter cell, given constant physical properties of actin? And could this effect therefore be responsible for tonic effects on stiffness of smooth muscle cells?

Gillis: That's possible, due to the fact that the reversibility of the effect is rather fast once the calcium is removed. It is conceivable that at the shortened length, the actin-filamin gel re-forms, but at a shorter length, and therefore can maintain the shorter length passively.

III. BIOCHEMICAL ASPECTS OF ACTIN-MYOSIN INTERACTION

INTRODUCTION

The biochemical aspect of muscle contraction is the ATPase activity of myosin and actomyosin in solution, though the function of muscle to shorten and to do work is lost in the protein samples to be studied. The five papers in this chapter aim at giving information about the molecular mechanism of myofilament sliding with various techniques of biochemistry, physiochemistry and molecular genetics.

Onishi's work is concerned with an old and unsolved question, whether the two heads of a myosin molecule interact with actin independently of each other or interact with actin in some cooperative manner. He applied zero-length cross-linker to rigor complex formed by F-actin and chicken gizzard HMM, and found that the two myosin heads were cross-linked, suggesting that the two heads are in contact with each other when they attach to F-actin. In order to obtain information about the behaviour of myosin heads during contraction, it is important to find out suitable probes attached to myosin heads. Burghardt and Ajtai developed a new technique which made it possible to obtain fluorescence polarization and electron paramagnetic resonance signals arising from the same donor. As the two signals generate complementary information, this technique seems promising for future research work.

Morales, Ue and Bivin address a question how the effects of nucleotide binding to, and its hydrolysis at, a site on the myosin head are transmitted to the myosin head-actin interface to result in myofilament sliding. Based on the proximity map of amino acid residues in the myosin head, constructed by the method of FRET (fluorescent resonance energy transfer), they suggest that a strand of 50kDa domain (residues 510-540) and a strand of 20kDa domain (residues 697-719) are involved in the above transmission mechanism. Sutoh made an attempt to determine the area on actin surface interacting with myosin. He first constructed several mutant genes from *Dictyostelium* by the site-directed mutagenesis. The mutations were designed to change amino acid residues in actin subdomain 1 to histidine. The mutant genes were then expressed in *Dictyostelium* cells. *In vitro* motility assays showed that the N-terminal cluster of acidic residues is essential for the actin-myosin sliding, while the C-terminal cluster of acidic residues is not essential for the sliding. Okamoto and Cremo examined the properties of photochemically cleaved myosin in the presence of vanadate and ATP, and found that the photochemical cleavage was more specific than the proteolytic cleavage, providing a means of a single-site cleaved myosin sample for studying the contraction mechanism.

OBITUARY: Ken Hotta (1922-1990)

On December 15 1990 the international community of muscle researchers lost Ken Hotta, at the age of 68. Professor Hotta received his Doctorate in Physics from Nagoya University in 1958, but soon turned his attention to biological applications, particularly to the basic phenomena of muscle contraction and excitation-contraction coupling. From 1958 to 1982 he worked at major American universities (Maryland, Dartmouth, California), where he made significant contributions to knowledge of the actomyosin-ATP and the sarcoplasmic reticular systems. In 1982 he returned to Japan to found a new physiological institute at Nagoya City University, which he directed until his retirement in 1988. During these years he used his basic knowledge to explore a great variety of muscle-related problems, particularly in heart and other viscera. Professor Hotta's wisdom, love of life, and good humor will be sorely missed by his many friends in Japan and abroad.

Manuel F. Morales

INTERACTION BETWEEN TWO MYOSIN HEADS IN ACTO-SMOOTH MUSCLE HEAVY MEROMYOSIN RIGOR COMPLEX

Hirofumi Onishi

Department of Structural Analysis
National Cardiovascular Center Research Institute
Suita, Osaka 565, Japan

ABSTRACT

Muscle contraction occurs as a result of the cyclic interaction between actin and myosin, coupled with the hydrolysis of ATP by the myosin heads[1]. A myosin molecule consists of two globular heads and a rod-shaped tail[2]. It is thought that each myosin head functions as a contractile machinery unit, because each head shows ATPase activity and binds to F-actin. A still unsolved question is whether or not the two heads of a myosin molecule interact with each other during their cyclic interaction with F-actin. Here we report that when chicken gizzard heavy meromyosin (HMM) in its rigor complex with F-actin is reacted with the zero-length cross-linker 1-ethyl-3-[3-(dimethylamino)propyl] carbodiimide (EDC), the two heads of the HMM molecule are cross-linked. This result suggests that the two heads of smooth muscle myosin are in contact with each other when myosin is attached to F-actin. It is thus possible that the two myosin heads interact with each other when cross-bridges are formed.

When EDC cross-links two polypeptides, it activates a carboxyl group of one polypeptide, and catalyzes cross-linking of the activated carboxyl group with an amino group of another polypeptide only when they are very closely located. We have used this reagent in order to investigate proximity of two myosin heads in the acto-smooth muscle HMM rigor complex.

EDC was added to rigor complexes formed between rabbit skeletal muscle F-actin and tryptic chicken gizzard HMM in order to introduce cross-linking (Fig. 1)[3]. Experiments were performed by using the combination of either F-actin and fluorescent HMM (panels A and B) or fluorescent F-actin and HMM (panels C and D). The protein band profiles (panels A and C) in SDS electrophoresed gels indicated that, in both combinations, the reaction led to the formation of two new products with M_r 115K and 60K. As shown in lane 0, the tryptic HMM consists of

Mechanism of Myofilament Sliding in Muscle Contraction, Edited by
H. Sugi and G.H Pollack, Plenum Press, New York, 1993

217

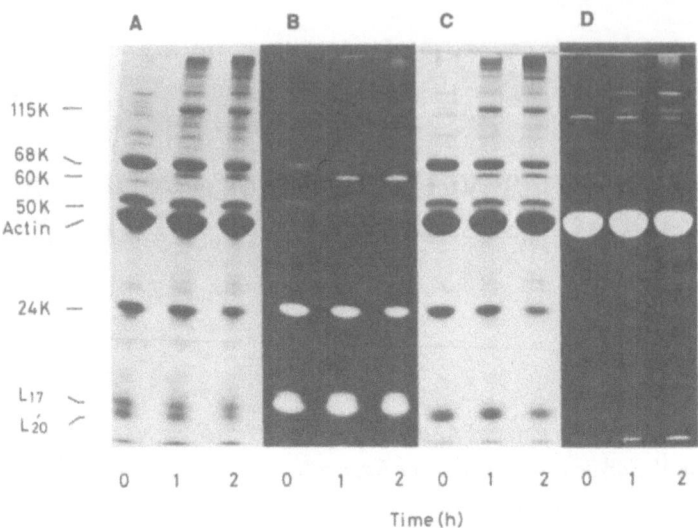

Fig. 1. EDC-Catalyzed cross-linking of the rigor complex formed between F-actin and gizzard HMM. The rigor complex (1 mg/ml) either between F-actin and AEDANS-modified fluorescent HMM (A and B) or between DACM-modified fluorescent F-actin and HMM (C and D) was reacted with 5 mM EDC in 0.1 M KCl, 10 mM $MgCl_2$, and 10 mM imidazole-hydrochloride (pH 7.0) at 24°C. At time 0 h and at 1 and 2 h following cross-linking, a protein aliquot was withdrawn and mixed with excess 2-mercaptoethanol to terminate the reaction. Protein samples (10 μg of HMM) was submitted to SDS gel electrophoresis. The Coomassie blue staining patterns (A and C) and the fluorescence profiles (B and D) of the same gels are shown. See ref. 3 for details of fluorescent labeling of F-actin and HMM.

three heavy chain fragments of 24K, 50K, and 68K, the essential (17K) light chain, and a fragment of the regulatory (20K) light chain. The fluorescent profiles illustrated in panel B indicates that fluorescent labels in HMM were incorporated into the newly formed 60K species. This result suggests that the 60K species contained the 24K fragment and/or the essential light chain as a component since the fluorescent HMM used in the experiment had fluorescent 17K and 24K bands (lane 0 of panel B). On the other hand, the combination of fluorescent F-actin and HMM resulted in the formation of the 115K fluorescent species (panel D). Therefore, we concluded that the 115K species was a cross-linked complex between actin and HMM components. This latter experiment also indicated that the 60K species did not contain actin. Moreover, we found that the amino acid compositions of the 24K and 60K peptides were very similar to each other. Since the composition of each HMM component, with the exception of the 24K component, differed from the observed composition of the 60K peptide, we concluded that the 60K species was a cross-linked complex between two 24K heavy chain fragments.

The 60K peptide was further fragmented with cyanogen bromide, and the resulting peptides were separated by reverse-phase HPLC and subsequent gel filtration[4]. A single cross-linked peptide was isolated. The amino acid composition

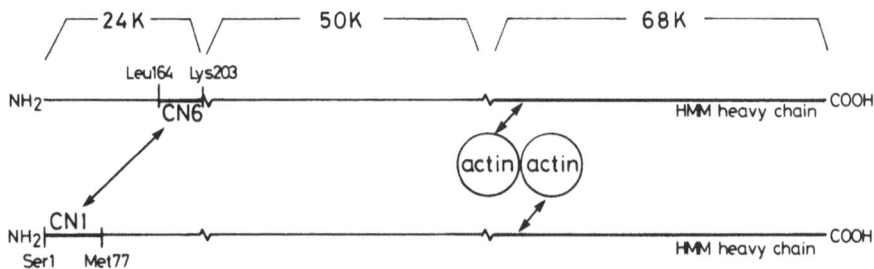

Fig. 2. Schematic representation of EDC cross-linking sites between two heavy chains (long double-headed arrow) and between actin and a myosin head (short double-headed arrows). Wavy lines indicate tryptic cleavage sites in the HMM heavy chain. CN1 and CN6 are fragments containing residues 1-77 and 164-203, respectively, produced from the NH_2-terminal 24K segment of the HMM heavy chain by CNBr cleavage. See ref. 4 for details.

of the peptide was identical with the value calculated for the equimolar mixture of CN1 (residues 1-77) and CN6 (residues 164-203). Since CN1 and CN6 were localized at different positions in the primary sequence of the 24K fragment as shown in Fig. 2, the two contact regions are different parts in the HMM head. In order to identify the cross-linking site within the sequence from residues 164 to 203, the cross-linked peptide (CN1•CN6) was analyzed by using a protein sequencer. Only one PTH-amino acid was produced in each cycle of Edman degradation and the amino acid sequence was identical to the sequence of uncross-linked CN6. When compared to CN6, a significantly low yield of PTH-glutamic acid was obtained at cycle 5 for CN1•CN6. This result suggests that this glutamic acid (at position 168) is a cross-linking site.

When the EDC cross-linked product of acto-gizzard HMM was digested with alpha-chymotrypsin in the presence of Mg-ATP, the 60K fluorescent band disappeared with a concomitant appearance of the 40K fluorescent band[5]. This result indicates that the 40K peptide was a degraded product of the 60K cross-linked complex. Judging from its M_r, we concluded that the 40K peptide was the dimer of the truncated COOH-terminal product (20K) from the 24K fragment. The 40K cross-linked peptide and the uncross-linked, 20K truncated product were separately fragmented with lysylendopeptidase, and peptide fragments were analyzed by reverse-phase HPLC. When we quantitatively compared the two elution profiles, a significant difference was observed for a peak containing residues 145-182. This peak was greatly reduced in the lysylendopeptidase sample of the 40K peptide. This result suggests that this peptide is involved in the cross-linking. Only one new major peak was identified in the fragmented 40K sample that eluted slightly earlier than the peak containing 145-182. Its amino acid composition was identical to the calculated composition for the equimolar mixture of residues 53-66 and 145-182. Since the peptide containing 145-182 was too long to identify the cross-linking sites, the peptides in this peak were further fragmented into smaller pieces with cyanogen bromide and then chromatographed on a reverse-phase HPLC column. When the first peak was analyzed using a protein sequencer, two PTH amino acids were

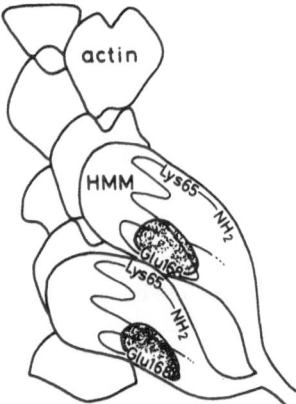

Fig. 3. Schematic drawing of EDC cross-linking sites on gizzard HMM heads attached to F-actin. In this model, the pointed end of F-actin is toward the top. Therefore, myosin heads slide upward along F-actin. The ATP-binding site is depicted by the speckled area. See ref. 5 for details.

produced for each cycle of Edman degradation. From already known sequence of the myosin heavy chain, we determined that the two peptides were residues 53-66 and 164-182 (Fig. 2). Two PTH amino acids of Glu-168 (cycle 5) and Lys-65 (cycle 13) were absent from the sequence of the cross-linked peptide when compared to the sequences of the two components. On the basis of these results, we concluded that cross-linking occurred between these two amino acids.

After acto-gizzard HMM was cross-linked with EDC, the mixture was separated into a supernatant and a pellet by ultracentrifugation in the presence of Mg-ATP[6]. The HMM irreversibly cross-linked with actin cosedimented with the actin during ultracentrifugation. HMM molecules contained in the supernatant were examined by electron microscopy after rotary shadowing with platinum/carbon. Rotary-shadowed images showed that about half of the HMM molecules had a contact between two heads. The percentage of head-to-head contact was very close to the densitometrically estimated amount of the 60K band formed by cross-linking between two 24K fragments. Most of the contacts were observed between two heads of the same molecule. Cross-linking between the two HMM heads occurred within the distal, more globular half of each head. This result suggests that the cross-linking sites are located within the distal half of HMM heads.

The present study allows us to predict the geometry of the myosin heads attached to F-actin. The sites involved in the cross-linking were identified as Glu-168 on one head of HMM and Lys-65 on the other. Since Glu-168 was just adjacent to the sequence location of the ATP-binding site suggested by Walker et al.[7], it appears that with the head-to-head contacted HMM molecule, one contact region is positioned near the ATP-binding site. As shown in Fig. 3, the two HMM heads are laterally in contact. As one contact region is close to the ATP-binding site, the location of this region within the HMM head can be estimated from the position of the ATP-binding site. Tokunaga et al.[8] have shown the gross position of the ATP-binding site on bound subfragment-1 by combining the use of three-dimensional image reconstruction and site-specific labeling with the avidin-biotin system. The position for ATP binding that they reported appears to be close to the contact region facing the lower surface of the HMM head than to that facing the upper surface. Therefore, we propose that one contact region which includes Glu-168 is exposed to the lower surface of the HMM head. Our model agrees well with the structure of acto-subfragment-1 reconstructed from the data of fluorescence resonance energy transfer studies[9]. Only one pair of amino acid residues was identified to be the cross-linking sites. This result suggests that the contact between two HMM heads occurs at specific affinity sites. The EDC cross-linking between two HMM heads resulted in an increase of the apparent affinity of HMM to actin[6]. We have not yet obtained evidence that direct contact between two myosin heads occurs in cross-bridges of the contracting muscle. Nevertheless, we propose a hypothesis that the two heads of a myosin molecule interact with each other in the cross-bridge.

ACKNOWLEDGEMENTS

This work has done in collaborations with Drs. K. Fujiwara, T. Maita, and G. Matsuda. Grant-in-Aids from the Ministry of Education, Science and Culture of Japan and Research Grants from the Ministry of Health and Welfare of Japan supported this work.

REFERENCES

1. Huxley, H.E. *Science* **164**, 1356-1366 (1969).
2. Lowey, S., Slayter, H. S., Weeds, A.G. & Baker, H. *J. Mol. Biol.* **42**, 1-29 (1969).
3. Onishi, H., Maita, T., Matsuda, G. & Fujiwara, K. *Biochemistry* **28**, 1898-1904 (1989).
4. Onishi, H., Maita, T., Matsuda, G. & Fujiwara, K. *Biochemistry* **28**, 1905-1912 (1989).
5. Onishi, H., Maita, T., Matsuda, G. & Fujiwara, K. *J. Biol. Chem.* **265**, 19362-19368 (1990).
6. Onishi, H. & Fujiwara, K. *Biochemistry* **29**, 3013-3023 (1990).
7. Walker, J.E., Saraste, M., Runswick, M. J. & Gay, N.J. *EMBO J.* **1**, 945-951 (1982).
8. Tokunaga, M., Sutoh, K., Toyoshima, C. & Wakabayashi, T. *Nature* **329**, 635-638 (1987).
9. Botts, J., Thomason, J.F. & Morales, M.F. *Proc. Natl. Acad. Sci. USA* **86**, 2204-2208 (1989)

Discussion

Sellers: Do you know anything about the activity of the cross-linked myosin? It has been proposed that smooth muscle HMM has an equivalent of the 10S form, which is, I think, 9S. Do you know whether the cross-linked material can be relaxed? Does it have a low-ATPase activity when it is phosphorylated?

Onishi: I haven't yet done the cross-linking study using phosphorylated HMM. Therefore, I cannot answer your final question. I have measured the activity of

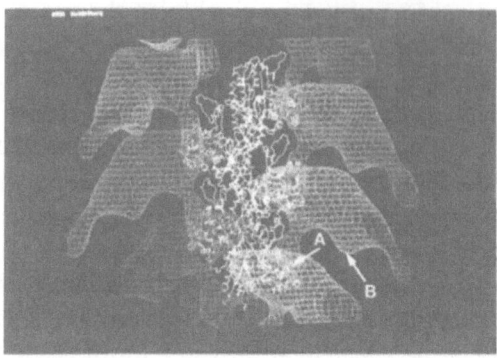

Fig. D1. A reconstruction of decorated actin from the cryoelectronmicroscopy data of R.A. Milligan et al. (*Nature* **348**, 217-221, 1990). The S1 molecules are shown as surfaces by a squirrel cage representation. The actin helix is represented by lines joining the C_α atoms. Note finger-like extension from one S1 to the neighboring actin underneath (A). At the bulge (B) neighboring S1 molecules are about 2.5 nm apart.

cross-linked HMM. The results indicate that the head-to-head cross-linking has no effect on the activity of HMM itself, whereas it strengthens the affinity of HMM and actin during the ATPase reaction.

Simmons: Do you know if both heads actually bind in the cross-linked myosin in the absence of ATP? Have you measured the binding constant?

Onishi: I have already mentioned the binding affinity between the myosin head and actin in the presence of ATP. However, I haven't measured the binding constant in the absence of ATP.

Holmes: I would like to show a result of Ron Milligan's (Fig. D1). If the HMM behaves like the S1 in Milligan's reconstruction the heads don't come close to each other except where the finger-like extension from the next S1 along the helix reaches down and touches both the actin and the S1 underneath (A in Fig. D1). This is the only place a short cross-linker would appear able to cross-link two S1s.

Onishi: I think there are two possible regions for the attachment between two myosin heads. One region is located in the middle of the myosin head, and the other region is located in the tip.

Holmes: Yes, but one of those is the point B in Fig. D1; the two molecules are here actually about 2.5 nm apart.

Sellers: What's the distance at the finger region?

Holmes: The S1s appear to touch.

Sellers: With HMM, though, the tails would have to converge back in again at the S2 region.

Holmes: That's certainly true. Therefore, it would be interesting to see if the cross-linking would work with S1.

Onishi: In Toyoshima and Wakabayashi's previous work, they also reported the three-dimensional reconstituted structure of the acto-S1 complex [Toyoshima, C., and Wakabayashi, T. *J. Biochem.* (*Tokyo*) **97**, 219-243, 1985]. In their study, the two myosin heads were closer to each other in the middle of the myosin head than in the tip.

Holmes: I can only say that the Milligan reconstruction is of very high quality.

Morales: Dr. Onishi, did I understand you correctly that the rigor affinity is hardly changed after you have cross-linked?

Onishi: No, I didn't say that. I have not yet done that experiment under rigor conditions.

Morales: But the ATPase was unaffected by the cross-linking?

Onishi: Of HMM itself, yes.

Morales: Even though the cross-linking site was very near the ATPase site?

Onishi: Yes.

Morales: It seems remarkable that although two HMM molecules have been cross-linked very near the ATPase site of each, ATP molecules can continue to bind and be processed at the same rate as before.

LUMINESCENT/PARAMAGNETIC PROBES OF ORDER AND ORIENTATION IN BIOLOGICAL ASSEMBLIES: THE TRANSFORMATION OF LUMINESCENT PROBES INTO π-RADICALS BY PHOTOCHEMICAL REDUCTION WITHIN THE CONTRACTILE APPARATUS

Thomas P. Burghardt and Katalin Ajtai

Department of Biochemistry and Molecular Biology
Mayo Foundation
Rochester, MN 55905 U.S.A

ABSTRACT

Several unsubstituted xanthene dyes (eosin, erythrosin, and fluorescein) were irradiated by laser light at their absorption maximum in the presence of different reducing agents. Due to photochemical reduction the quinoidal structure of the xanthene ring is transformed into a semiquinone and a π-radical is formed having a characteristic electron paramagnetic resonance (EPR) signal of an unpaired electron spin with proton hyperfine interactions. A strong EPR signal is observed from the dye in solution or when specifically attached to myosin following irradiation in the presence of dithiothreitol or cysteine. The spectroscopic methods of fluorescence polarization and EPR are useful in the study of ordered biological assemblies. These methods generate complementary information about the order of the system but a consistent quantitative interpretation of the related data is complicated because the signals arise from different donors. Our method allows us to detect both signals from the same donor. We applied our new technique to the study of skeletal muscle fibers. The fluorescent dye iodoacetamidofluorescein was covalently attached to the reactive thiol of the myosin molecule in muscle fibers. Fluorescence polarization and EPR spectroscopy were performed on the labeled fibers in rigor. Both signals indicate a highly ordered system characteristic of cross-bridges bound to actin.

INTRODUCTION

The molecular mechanism of muscle contraction involves the cyclical interaction of the proteins myosin and actin. One model of contraction has the enzymatic head region of myosin (the cross-bridge) rolling without slipping on the

Mechanism of Myofilament Sliding in Muscle Contraction, Edited by
H. Sugi and G.H Pollack, Plenum Press, New York, 1993

225

actin filament during ATP hydrolysis, to produce the relative sliding movement of myosin and actin filaments that is characteristic of muscle contraction[1][2]. Since this model has the cross-bridge rotating, experimental efforts were made to detect changes in cross-bridge orientation that accompany the physiological state changes in the muscle fibers. Luminescent and paramagnetic probes of myosin cross-bridges can indicate cross-bridge orientation and are used to study cross-bridge orientation changes during muscle contraction.

Luminescence polarization, and electron paramagnetic resonance (EPR), spectroscopy employing extrinsic probes may be used to quantitate orientation and order in biological assemblies because of their sensitivity to differences or changes in probe orientation[3-5]. Luminescent and paramagnetic probes of myosin cross-bridges can indicate cross-bridge orientation changes due to physiological state changes in muscle. In the skeletal muscle fiber the reactive thiols SH1 and SH2 are routinely specifically modified by covalent probes of this type[3][4][6][7]. Previously, probe orientation data collected from different probes of SH1 were interpreted as indicating contradictory extents of cross-bridge rotation in fibers. Some probes of SH1 were said to indicate a large cross-bridge rotation when the fiber changes its physiological state while others were said to indicate almost no change under identical conditions[8-11]. These contradictions arise from (i) the complexity of the 3-dimensional motion of the rotating cross-bridge and (ii) the difficulty in comparing data originating from different signal donors on SH1. These difficulties in interpreting probe data are common problems in probe studies of order and orientation in biological assemblies.

We address point (ii) by introducing probes that are both luminescent and paramagnetic. Luminescent unsubstituted xanthene dyes (e.g., fluorescein, eosin, or erythrosin) can be converted to a free radical when they absorb a photon in the presence of specific reducing agents[12-20]. The kinetics of this photochemical process are slow enough at low excitation light levels that luminescence polarization spectroscopy measurements can be made without appreciable probe photobleaching or photoreduction to the free radical. Free radical formation is appreciable at the higher excitation light levels such that a characteristic EPR signal is observed under these conditions. We modified the conditions of this reaction for use with proteins by introducing mild reductants such as dithiolthreitol (DTT) or cysteine. When labeling proteins with these dyes we are able to measure the fluorescence polarization and EPR spectra from the same probe under identical conditions.

Previously, we introduced a method to combine data from various (different) fluorescent and spin probes of myosin to enhance the resolution of the individual probe angular distributions[21][22]. The data from luminescent/paramagnetic probes are readily combined using this formalism and these probes also have a unique property that simplifies the use of our method for combining probe data sets. When the signals come from a spin and a different fluorescent probe we must search for the nonlinear parameters relating the probe-fixed reference frames. With luminescent/paramagnetic probes the probability of dipolar photon absorption directly modifies the lineshape of the EPR spectrum so that the relationship between spin and fluorescent probe frames may be determined experimentally. Thus the introduction of luminescent/paramagnetic probes reduces the number of unknowns in the equations defining the global solution for the probe distributions.

These xanthene dyes solve other difficult problems related to the nativity of the muscle when using spin probes and to spin probe specificity. First, conventional nitroxide spin labels require conditions that protect the radical from undergoing unwanted oxidation or reduction so that the routine use of antioxidants, such as DTT or mercaptoethanol, to protect the native structure of proteins by preventing disulfide bridge formation, is not permitted. Conditions consistent with fluorescence or the photoreduction of our xanthene dyes (involving the presence of antioxidants) are better for protecting the integrity of protein structure than the conventional nitroxide spin labels. Second, checking the specificity of nitroxide probes presented special problems due to the difficulty in detecting free radicals after prolonged biochemical manipulations associated with gel electrophoresis and the necessity of eliminating nonspecific spin signals with a ferricyanide wash that caused confusion in identifying the position of the active spin label[23]. Furthermore, the ferricyanide wash of spin labeled muscle fibers damages the fiber since treated fibers loose Ca^{2+} sensitivity and contract in relaxing conditions[24][25]. The iodoacetamido based xanthene dyes on fibers give EPR signals characteristic of a specific and immobilized probe and their detection in gels is routine and quantitative.

METHODS

EPR Measurements

EPR measurements were carried out at X-band on a Bruker Model ER200 series instrument using a TM_{110} cylindrical cavity. For these measurements we continuously illuminated samples inside the EPR cavity with monochromatic light from an argon ion laser propagating at right angles to the Zeeman field. The excitation wavelengths used were near to the absorption maximum of the dye, i.e., 488 nm with fluorescein or 514 nm with eosin or erythrosin. No EPR signal was observed without light excitation and the shape of the spectra were unaffected by the presence of oxygen.

RESULTS AND DISCUSSION

Light Induced EPR Studies of Xanthene Dyes

We report here studies of fluorescein and its halogenated derivatives eosin and erythrosin. Eosin and erythrosin have a high efficiency for excited singlet to triplet state conversion when in solution[26]. The lifetime of the triplet is many orders of magnitude longer than the singlet state and in these two probes the quantum efficiency for phosphorescent emission is significant. Because of these properties eosin and erythrosin, when modified to react with specific side chains of proteins, are used to detect rotational relaxation rates of their protein hosts using time-resolved optical techniques[27]. The longer lifetime of the triplet probes allows investigation of slower relaxation rates than is possible with singlet probes. Fluorescein has a lower efficiency for triplet conversion and in most applications is used as a fluorescent probe of proteins in biological assemblies (see Table 1).

The chemistry of the photoreduction of these dyes was studied by measuring steady state or time resolved absorption following flash or continuous illumination photolysis[28] as well as with EPR measurements of the radical intermediates[12]. In time-resolved experiments, studies of the kinetics of the dye reaction were conducted with phenol as the reducing agent[28]. From these studies it was proposed that the dye was reduced from its triplet state by electron transfer from phenol producing free radical forms of both the dye and the phenol.

Subsequent work using EPR to detect radical intermediates from eosin and phenol in methanol following illumination by a flash of visible light in the presence of oxygen, showed that not the phenoxy radical but its product from the further reaction with the reduced eosin was detected[20]. This product, the p-benzosemiquinone anion radical, has a characteristic five-line EPR spectrum due to an unpaired electron interacting with four equivalent protons. The relative intensity of the lines is given by the ratios 1:4:6:4:1[29]. In this case the reduced eosin free radical was not detected because of its rapid reaction with the phenoxy radical. We also observed the production of the p-benzosemiquinone with EPR from eosin and phenol in phosphate buffer (pH 7.0) under constant illumination from the 514 nm line from an argon ion laser.

In other previous work, the reduced dye free radical was detected using EPR from eosin and ascorbic acid in pyridine in the absence of oxygen and under constant illumination with visible light[12]. The ascorbic acid served as the reducing agent. The EPR spectrum obtained had three lines with hyperfine splitting of 4.6 Gauss. The ratio of the intensity of the components of the hyperfine structure was 1:2:1. The proposed detailed mechanism for the formation of the reduced erythrosin free radical is summarized in Fig. 1. The unpaired π-electron of the erythrosin must interact with two protons at equivalent positions in the molecule to give the relative intensity of the components of the hyperfine structure of 1:2:1. These protons are at positions 8 and 1 in rings a and c, respectively.

Table 1. Triplet quantum yields, Φ_T, at 23°C[26]

compound	solvent	concentration (M)	Φ_T
fluorescein (Fl)	aqueous, pH 9	8×10^{-7}	0.05 ± 0.02
Fl(Br$_2$)	aqueous, pH 9	8×10^{-7}	0.49 ± 0.07
eosin(Fl[Br$_4$])	aqueous, pH 9	6×10^{-7}	0.71 ± 0.10
erythrosin (FlI$_4$)	aqueous, pH 9	9×10^{-7}	1.07 ± 0.13

We performed experiments on fluorescein, eosin, and erythrosin. DTT or cysteine acted as reductants for the probes free in solution or when the probes were covalently bound to proteins or protein assemblies and did not disturb biological activity. EPR spectra were recorded from 0.2 mM fluorescein, eosin, and erythrosin in 4mM DTT, 5mM phosphate buffer pH 7.0. EPR experimental parameters were 1 mW microwave power and 0.5 G modulation amplitude in a TM_{110} cavity. Laser power was continuous at 100 mW. The EPR spectra from fluorescein freely moving in solution at room temperature and in 50% glycerol at -15°C are shown in Fig. 2.

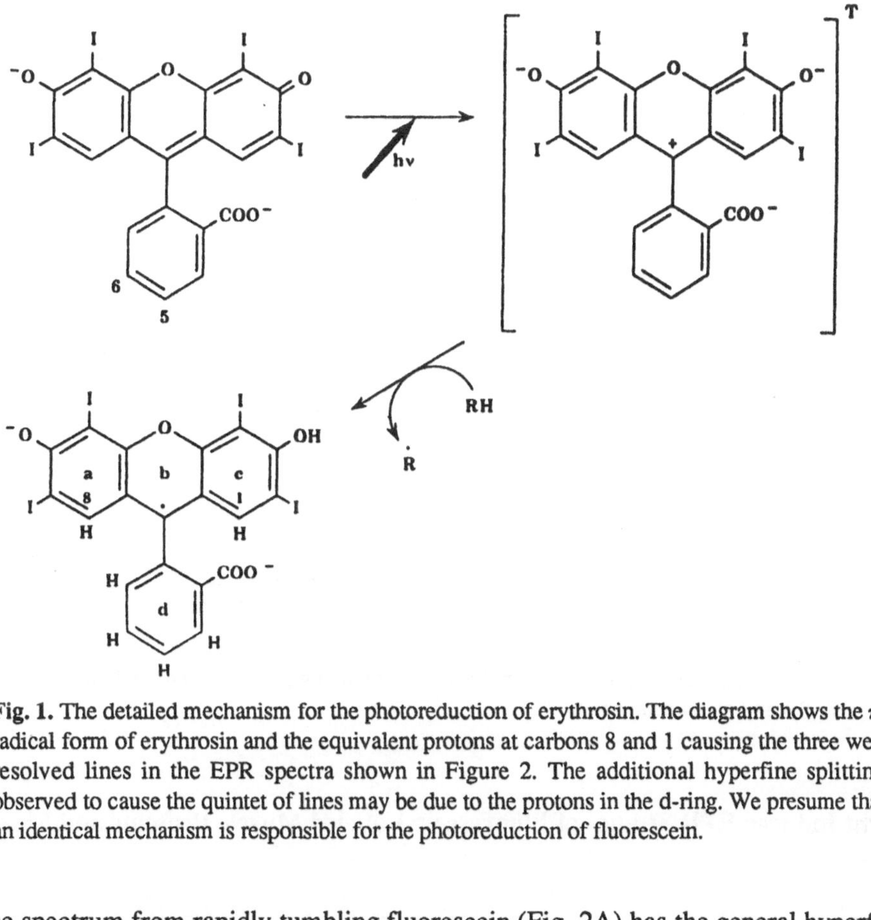

Fig. 1. The detailed mechanism for the photoreduction of erythrosin. The diagram shows the π-radical form of erythrosin and the equivalent protons at carbons 8 and 1 causing the three well resolved lines in the EPR spectra shown in Figure 2. The additional hyperfine splitting observed to cause the quintet of lines may be due to the protons in the d-ring. We presume that an identical mechanism is responsible for the photoreduction of fluorescein.

The spectrum from rapidly tumbling fluorescein (Fig. 2A) has the general hyperfine features of an unpaired electron interacting with two equivalent protons as observed previously[12]. Additional weaker hyperfine interactions, that are not fully resolved, split the original three lines suggesting an interaction with a different set of four approximately equivalent protons. These protons are possibly those in the d-ring of the molecule shown in Fig. 1. The spectrum from randomly oriented immobilized fluorescein (Fig. 2B) is broadened such that most of the hyperfine features are lost. EPR spectra from eosin and erythrosin are qualitatively very similar to that shown in Fig. 2 (data not shown).

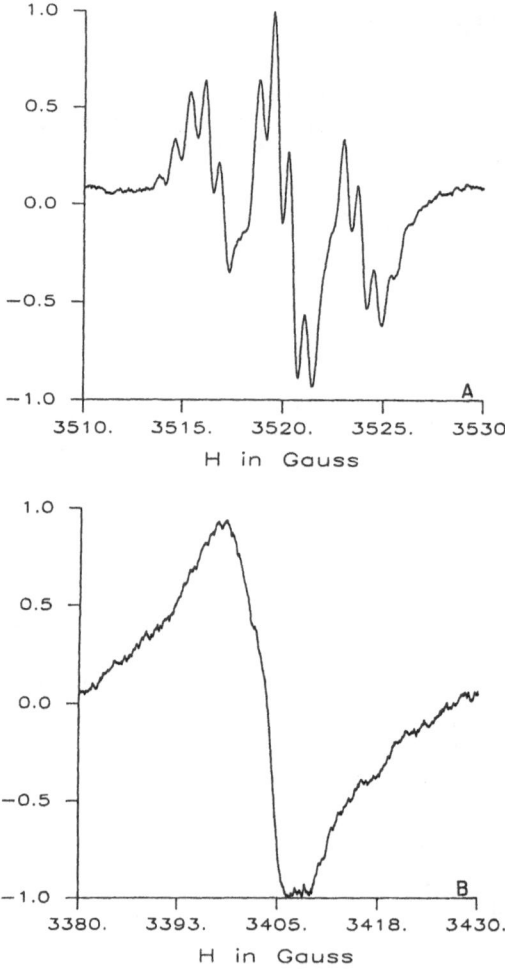

Fig. 2. EPR spectrum of the fluorescein radical freely tumbling in solution (A) and immobilized in 50% glycerol at low temperature (B).

Light Induced EPR Studies of Fluorescein Labeled Muscle Proteins and Muscle Fibers

We labeled the fast reacting thiol of myosin in glycerinated muscle fibers with iodoacetamidofluorescein (IAF) and performed experiments on fibers in rigor. The spectra, shown in Fig. 3, are from IAF labeled fibers in rigor that are oriented with fiber axis ∥ (top) or ⊥ (bottom) to the Zeeman field. Comparison of these spectra show the sensitivity of the shape of the signal to changes in probe orientation. The labeled fiber system, identical to that used in the EPR studies, was also used in fluorescence polarization studies. The fluorescence polarization excitation spectra from fluorescein labeled fibers in rigor and from fluorescein labeled myosin

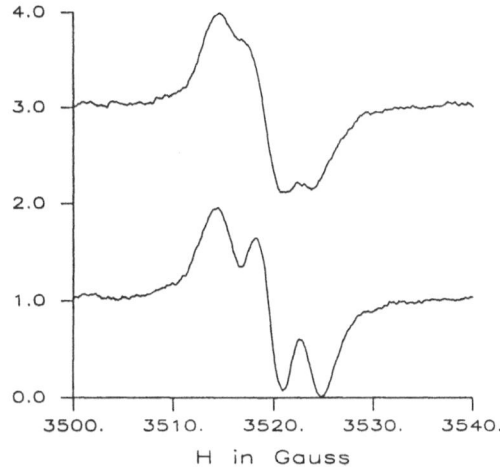

H in Gauss

Fig. 3. The EPR spectra from fluorescein labeled muscle fibers at 4°C in rigor oriented either parallel (top) or perpendicular (bottom) to the Zeeman field.

subfragment 1 (IAF-S1) in 50% glycerol at -15°C were also measured. Comparison of the polarization ratios from the labeled fibers with the spectrum from the immobilized randomly oriented IAF-S1 indicates that the probes labeling the muscle fiber are ordered as expected from the EPR results.

EPR spectra from immobilized IAF-S1 or IAF-S1 freely moving in solution, in the presence and absence of nucleotides, indicate that the probe is immobilized on the surface of S1 and not affected by nucleotide binding. Time-resolved fluorescence studies of IAF-S1 indicate that the fluorescence intensity changes when S1 binds ATP or actin[30-32]. Other studies indicate fluorescein inhibits ADP release during the ATPase cycle in fibers[33]. These data raise doubts about the suitability of IAF for studies of active fibers. Maleimido and iodoacetamido eosin probes have been used to modify fibers and isolated muscle proteins with varying degrees of success[34)35]. The eosin probe also has a phosphorescence signal in addition to fluorescence and EPR signals.

Our approach that utilizes one chemical group as a luminescent and paramagnetic probe permits us to detect the orientation changes of protein components of the biological assembly, to a degree exceeding that available by introducing two conventional probes, but with the advantage that only one protocol for the specific and unharmful modification of the protein need be found. This approach also allows us to experiment on a labeled biological sample that is closer to its native state than if nitroxide labeled samples were used. This is because the conventional nitroxide spin label requires conditions that protect the radical from undergoing unwanted oxidation or reduction so that the routine use of antioxidants, such as DTT, to protect the native structure of proteins by preventing disulfide bridge formation is not permitted. Thus, the conditions consistent with the measurement of fluorescence or the photoreduction of our xanthene dyes are better for protecting the integrity of the protein structure than the conventional nitroxide spin label methods.

The application of our new method to the study of the actomyosin interaction is very promising. Preliminary calculations suggest that the z-axis of the principal magnetic frame of the fluorescein probe from fibers in rigor is ~60° from the fiber axis. This is a new orientation from which to study cross-bridge rotational movement suggesting that this probe may provide independent information on cross-bridge rotational dynamics during contraction. Other untried luminescent/paramagnetic probes may provide additional unique orientations without the problems affecting IAF labeled fibers[30-33]. Additionally, the angular resolution enhancement provided by techniques for combining multiple probes of myosin are readily applied to luminescent/paramagnetic probes[21][22]. These methods benefit from the ability to experimentally determine of the relationship between spin and luminescent probe reference frames.

ACKNOWLEDGEMENTS

We thank N. Simmons and D. Toft for excellent technical assistance, and, E. H. Hellen, P. J. K. Ilich, and F. G. Prendergast for helpful discussions. This work was supported by the National Science Foundation (DMB 8819755), the National Institutes of Health (R01 AR 39288), the American Heart Association (Grant-in-Aid 900644), and by the Mayo Foundation. T. P. B. is an Established Investigator of the American Heart Association.

REFERENCES

1. Huxley, H.E. *Science* **164**, 1356-1366 (1969).
2. Huxley, A.F. & Simmons, R.M. *Nature* **233**, 533-538 (1971).
3. Morales, M.F., Borejdo, J., Botts, J., Cooke, R., Mendelson, R.A., & Takashi, R. *Ann. Rev. Phys. Chem.* **22**, 319-351 (1982).
4. Gergely, J. & Seidel, J.C. *Handbook of Physiology* Sec. **10**, 257-274 (1983).
5. Arata, T. & Shimizu, H. *J. Mol. Biol.* **151**, 411-437 (1981).
6. Ajtai, K. & Burghardt, T P. *Biochemistry* **28**, 2204-2210 (1989).
7. Burghardt, T.P. and Ajtai, K. in: *Molecular Mechanism in Muscle Contraction*: Vol **3**, (ed. Squire, J. M.) 211-239 (MacMillan, London, 1990).
8. Cooke, R., Crowder, M.S. & Thomas, D.D. *Nature* **300**, 776-778 (1982).
9. Ajtai, K. and Burghardt, T.P. (1986) *Biochemistry* **25**, 6203-6207 (1986).
10. Ajtai, K., French, A.R. & Burghardt, T.P. *Biophys. J.* **56**, 535-541 (1989).
11. Fajer, P.G., Fajer, E.A., Matta, J.J. & Thomas, D.D. *Biochemistry* **29**, 5865-5871 (1990).
12. Bubnov, N.N., Kibalko, L.A., Tsepalov, V.F. & Shliapintokh, V. IA. *Optics and Spectroscopy* **VII**, 71-72 (1959).
13. Oster, G. & Adelman, A.H. *J. Am. Chem. Soc.* **78**, 913-916 (1956) .
14. Adelman, A.H. & Oster, G. *J. Am. Chem. Soc.* **78**, 3977-3980 (1956).
15. Uchida, K., Kato, S. & Koizumi, M. *Nature* **184**, 1620-1621 (1959).

16. Grossweiner, L.I. & Zwicker, E.F. *J. Chem. Phys.* **31**, 1141-1142 (1959).
17. Grossweiner, L.I. & Zwicker, E.F. *J. Chem. Phys.* **34**, 1411-1417 (1961).
18. Kasche, V. & Lindquist, L. *Photochem. Photobiol.* **4**, 923-933 (1965).
19. Kasche, V. *Photochem. Photobiol.* **6**, 643-650 (1967).
20. Leaver, I.H. *Aust. J. Chem.* **24**, 891-894 (1971).
21. Burghardt, T.P. & Ajtai, K. *Biochemistry*, **31**, 200-206 (1992).
22. Ajtai, K., Ringler, A. & Burghardt, T.P. *Biochemistry*, **31**, 207-217 (1992).
23. Ajtai, K., Pótó, L. & Burghardt, T.P. *Biochemistry* **29**, 7733-7741 (1990).
24. Fajer, P.G., Fajer, E.A., Brunsvold, N. J. & Thomas, D.D. *Biophys. J.* **53**, 513-524.
25. Ajtai, K. & Burghardt, T.P. unpublished observation (1991).
26. Bowers, P.G. & Porter, G. *Proc. Roy. Soc. A* **299**, 348-353 (1967).
27. Ludescher, R.D., Eads, T.M. & Thomas, D.D. in: *Optical Studies of Muscle Cross-Bridges*
    (eds. Baskin, R.J. & Yeh, Y.) 33-65 (C.R.C. Press, Boca Raton, FL, 1987).
28. Zwicker, E.F. & Grossweiner, L.I. *J. Chem. Phys.* **67**, 549-555 (1963).
29. Venkataraman, B., Segal, B.G. & Fraenkl, G.K. *J. Chem. Phys.* **30**, 1006-1017 (1959).
30. Marsh, D.J., Stein, L.A., Eisenberg, E. & Lowey, S. *Biochemistry* **21**, 1925-1928 (1982).
31. Ando, T. *Biochemistry* **23**, 375-381 (1984).
32. Aguirre, R., Gonsoulin, F. & Cheung, H.C. *Biochemistry* **25**, 6827-6835 (1986).
33. Borejdo, J., Ando, T. & Burghardt, T.P. *Biochim. Biophys. Acta* **828**, 172-176 (1985).
34. Stein, R.A., Ludescher, R.D., Dahlberg, P.S., Fajer, P.G., Bennett, R.L. & Thomas, D.D.
    Biochemistry **29**, 10023-10031 (1990).
35. Eads, T.M., Thomas, D.D. & Austin, R.H. *J. Mol. Biol.* **179**, 55-81 (1984).

Discussion

Arata: How high is the angular resolution of π-electron EPR? Is it possible to distinguish two angles?

Burghardt: It's not clear yet. That work is going on right now. The only thing we have been able to show definitively is that the probe does sense the fact that you tilt the fiber in the spectrometer. But the resolution is definitely lower than the nitroxides, and right now we don't have a good handle on how it is quantitatively.

Morales: Yesterday there was a paper by Dr. Suzuki in which an angular distribution of cross-bridges was measured from EM pictures. I was wondering if the order of magnitude of the half-width somehow corresponded to it. You didn't have time to say what kind of distributions you get.

Burghardt: You mean from any probe? Well, the spin labels have roughly a 15° half-width. I don't quite remember what the resolution was on the EM pictures.

Sugi: In Suzuki's presentation in my laboratory the angle ranged from 30° to 120°. Is that consistent?

Morales: The probe should certainly pick that up.

Burghardt: I would also like to mention that you can enhance the angular resolution by combining the probes. The pictures show the ability to pick out two orientations.

THE REGION IN MYOSIN S-1 THAT MAY BE INVOLVED IN ENERGY TRANSDUCTION

Manuel F. Morales, Kathleen Ue and Donald B. Bivin

Dept. of Physiology and Biophysics
University of the Pacific
2155 Webster St.
San Francisco, CA, 94115, USA.

ABSTRACT

Newly-reported structural information about certain proximities between points on bound nucleotide and points on the heavy chain of myosin S-1 are incorporated into a previously-reported [Botts, J. Thomason, J.F. & Morales, M.F. *Proc. Nat. Acad. Sci. USA*, **86**, 2204-2208 (1989)] structure of S-1. The resulting, enhanced structure is then used to identify some functionalities (e.g., the ATP-perturbable tryptophans), and to explain certain observations (e.g., some concerning the role of bound Mg^{2+} in the spectral response of TNBS-labelled Lys-83, and some concerning the response of the S-1 CD signal to nucleotide binding and to temperature change). These considerations lead to the suggestion that a strand of the 50 kDa "domain" (residues 510 to 540), and a strand of the 20 kDa 'domain" (residues 697-719) are involved in transmitting the effects of nucleotide binding and hydrolysis to the loop (constituted from the same "domains") that reaches a major (S-1)-actin interface.

Progress in explaining contraction in physical terms has passed through various stages. The present stage could be said to consist of the "standard model", evolved from references (1), (2), and (3). The essence of this model is that myosin S-1 and actin (or an analogous pair of proteins in other contractile systems), while remaining bound to each other, change their relative spatial relations. If one combines this notion with the knowledge that ATPase on S-1 proceeds in steps [4)5)], and assumes a correspondence between steps in the ATPase cycle and steps in the relational cycle[6)], one arrives at the present day theory of the contractile event.

We have pointed out[7)] that important and satisfying as these ideas are, they are not the final solution to the problem of contraction. Specifically, they offer no explanation for the correlation between the ATPase states and the relational states. They therefore do not provide a means of calculating from first principles the

Mechanism of Myofilament Sliding in Muscle Contraction, Edited by
H. Sugi and G.H Pollack, Plenum Press, New York, 1993

235

contractile force from the ATPase events (the test to be passed by a complete theory). What has seemed to us the key to the final solution are the events that go on "inside" of S-1 and actin. Simplistically, we have suggested that deformation of the S-1 structure at the ATPase ("N-site") may be transmitted over a long distance to actin-binding sites ("A-sites"), where these transmitted effects alter the relations between S-1 and actin[8)9)]. So we say that energy transduction truly occurs in the acto-(S-1).

These suggestions cannot be tested without some knowledge of acto-S-1 structure, so there the matter might sit, waiting for a crystallographic study of S-1. Unwilling to wait for this cataclysmic advance, we have tried to deduce at least a crude 3-dimensional structure based on an oriented lattice of FRET distance estimates, improved by a knowledge of sequence homologies and by crosslinking proximities[10)]. With this structure in mind we tried to interpret experiments in which structural changes were induced by nucleotide binding. This has led us to the idea that these structural changes consist of relative movement between two strands of a loop[11)] connecting the N- and A-sites. Were this idea "hardened" by future research, especially crystallography, it would point the way to the final solution, as it purports to say *in chemical terms* what are the three essential elements of the transducer—ATPase, transmitter, and work performer—thus enabling in principle an *ab initio* calculation of the force.

In the present paper we first show how the original structure proposal[10)] can continue to assimilate information reported since its publication, and then we describe how certain experiments not previously interpreted can be rationalized by these ideas. Some of the newest data are in a paper currently in press[12)].

Until recently what was known about the N-site was that it resided mainly on the 25 kDa "domain" of S-1, probably between residues 168 (phosphate end) and 195 (adenine end). This assignment is consistent with homology dictates and with the results of photoaffinity experiments. Our proposed structure of course suggested proximities between the N-site and other points of S-1, but corroborating experimental evidence was lacking. An important advance since 1988-89 is the discovery by Burke, et al.[13)] that the "B-sequence" of Walker, et al.[14)] resides on the 20 kDa "domain", toward the C-terminal from Cys-707 ("SH-1"). A feature of the B-sequence is that it contains a terminal aspartyl which anchors to the protein the Mg^{2+} bridge from the gamma phosphate of bound ATP. According to Burke, et al. this is Asp-719 in the amino acid sequence of S-1. This identification is easy to accommodate into our proposed structure, so it has been drawn into an updated version of the structure (Fig. 1). This Mg^{2+} bridge is a transitory structure during the ATPase cycle, providing a mechanical connection between the 25 kDa and 20 kDa "domains". This connector is undoubtedly affected during the hydrolytic transition; since it stabilizes the ligand, it must have important kinetic consequences. Three further clues about the environment of bound nucleotide have recently been found. Grammer and Yount[15)] have shown that Ser-243 is near a phosphate of the bound nucleotide, and Mahmood, et al.[16)] have found that Ser-324 is 0.6-0.7 nm from the 3'(2') ribose oxygens of the bound nucleotide. These proximities give

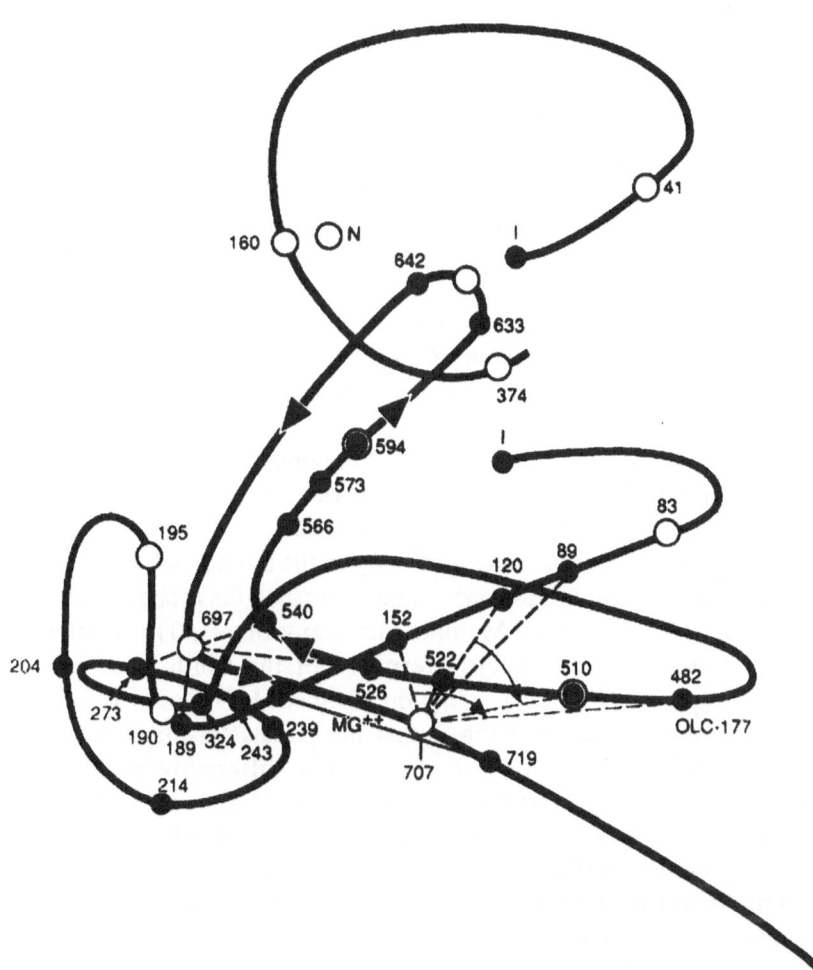

Fig. 1. Approximate structure of myosin S-1 attached to actin. The original version of this figure is Fig. 2B of reference (10), which was a composite of sequence and sequence homology information, interpoint and radial distance estimates based on fluorescence resonance energy transfer, photoaffinity, and crosslinking results. Since reference (10) was written, new proximities have been reported (see references cited in the text); the incorporation of this new information enhances and slightly alters the figure. Specifically, the topology of the N-terminal region of the 50 kDa "domain" is now better defined, the ATP-perturbable tryptophan residues (Trp-510, Trp-594) are tentatively identified, and the Mg^{2+} is portrayed as bridging Asp-719 of the 20 kDa "domain" to a phosphate oxygen of bound nucleotide (the bound nucleotide is not shown on either this or the earlier figure, but is thought to interact with the 25 kDa "domain" between residues 168 and 195)

definition to a region of the 50 kDa "domain" for which we had none before. They have been accommodated in Fig. 1 while respecting a third new constraint [Agarwal, et al.[17)]], which requires that Arg-239 be about 1 nm from Cys-697 ("SH-2"). The final new localization is that of the ATP-perturbable tryptophans discovered by Morita[18)], developed into fluorescent signals by Werber, et al.[19)], and made famous in the ATPase kinetic studies of Trentham, et al.[4)]. The reasons why we believe these tryptophans to be primarily Trp-510 and secondarily Trp-594) are detailed in reference (12); briefly, they are that: Torgerson[20)] showed in intact S-1 that, among the five tryptophans, the ATP-perturbable ones cannot be fluorescence-quenched by acrylamide. From studies of the isolated 50 kDa "domain" Muhlrad, et al.[21)] showed that it contains the protected residues. Our construction[10)] shows that Trp-510 and Trp-594 are on the moving elements, while Trp-440 is far away. While it is possible as in the past to use the emission of these tryptophans empirically, knowing something of their sequence and spatial location should enhance the value of their signals.

We turn finally to certain experiments not previously considered in terms of our structure. Firstly, Muhlrad[22)] studied the spectroscopic response of trinitrophenylated S-1 when nucleotides bind, with and without Mg^{2+}. As is well known, Tonomura and his associates[23)] pioneered the specific labelling of Lys-83 of the 25 kDa "domain" with trinitrobenzene sulfonate, and the probe signals have been widely studied and applied, by Muhlrad and many others. The special result of reference (22) was that the spectral response of the trinitrobenzene sulfonate "reporter" was quite different depending on whether Mg^{2+} was or was not present. It can now be appreciated from Fig. 1 that the effects (displacements, tensions) resulting from filling the N-site would propagate differently along the 25 kDa "domain" toward Lys-83, depending on whether the aforementioned Mg^{2+} was or wasn't in place. It should also be noted that the motion of this member of 25 kDa (which Burke and Lu have shown to near to Cys-707) in response to ATP-binding may be important for function, since Miyanishi, et al.[24)] showed that antibodies attached near to residue 80 block fiber contraction.

A second set of experiments that seems interpretable with the aid of Fig. 1 is one[12)] recently made with Dr. Curtis Johnson. In this work we studied, as a function of temperature, both the circular dichroism signal (negative intensity at 222 nm, the so-called "amide band") and the 335 nm tryptophan fluorescence emission not quenched by acrylamide (i.e. that part of the emission that arises mainly from the ATP-perturbable tryptophans; see above). In the temperature range, 0-16°C, both of these parameters suffer small, reversible, and very similar decreases. Also observed in the same temperature regime were very small, but definite, CD signal increases upon adding ATP, and of course also the well known large increases in fluorescence emission. We take these observations collectively to mean that probably on the 50 kDa "domain" strand bearing Trp-510 the temperature or ATP-addition influences formed structure. Other authors[25-27)] have given evidence that in the 20 kDa strand joining Cys-697 to Cys-707 deformation is also possible. Our structure suggests that these two strands are contiguous and roughly parallel; they are on the path between the N-site and the loop that extends to the A-site, and they seem to move relative to one another in transduction[10]. It would be of great interest to know if this transducing region is also a part of the "motional" region once suggested from NMR observations[28]. In sum, it appears at present that in the ATP-induced relative slide between strands the 50 kDa strand translates with local rotation of W_{510}; more remarkably, the movement of C_{697} toward C_{707} seems to be accomplished without extensive change in secondary structure.

ACKNOWLEDGEMENTS

This work was supported by USPHS Grant HL-44200, by NSF Grant DMB-9003692, and by a Fogarty Scholarship-in-Residence to M.F.M.

REFERENCES

1. Reedy, M.K., Holmes, K. & Tregear, R.T. *Nature* **207**, 1276-1280 (1965).
2. Huxley, H.E. *Science* **164**, 1356-1366 (1969).
3. Huxley, A.F. & Simmons, R.M. *Nature* **233**, 533-538 (1971).
4. Trentham, D.R., Eccleston, J.F. & Bagshaw, C.R. *Quart. Rev. Biophys.* **9**, 217-281 (1976).
5. Lymn, R.W. & Taylor, E.W. *Biochemistry* **10**, 4617-4624 (1971).
6. Botts, J., Cooke, R., Dos Remedios, C.G., Duke, J.A., Mendelson, R.A., Morales, M.F., Tokiwa, T., Viniegra, G. & Yount, R.G. *Cold Spring Harbor Symp. Quant. Biol.* **37**, 195-200 (1973)
7. Morales, M.F. & Botts, J. *Proc. Natl. Acad. Sci. USA* **76**, 3857-3859 (1979).
8. Morales, M.F. in *Perspectives on Biological Energy Transduction* (ed. Mukohata, Y.) 3-12 (Academic Press, Tokyo, 1987).
9. Botts, J., Takashi, R., Torgerson, P.M., Hozumi, T., Muhlrad, A., Mornet, D. & Morales, M.F. *Proc. Natl. Acad. Sci. USA* **81**, 2060-2064 (1984).
10. Botts, J., Thomason, J.F. & Morales, M.F. *Proc. Natl. Acad. Sci. USA* **86**, 2204-2208 (1989).
11. Mornet, D., Ue, K. & Morales, M.F. *Proc. Natl. Acad. Sci. USA* **82**, 1658-1662 (1985).
12. Johnson, W.C., Bivin, D.B., Ue, K. & Morales, M.F. *Proc. Natl. Acad. Sci. USA* **88**, 9748-9750 (1991).
13. Burke, M., Rajasekharan, K.N., Maruta, 5. & Ikebe, M. *FEBS Lett.* **262**, 185-188 (1990).
14. Walker, J.E., Saraste, M., Runswick, M.J. & Gay, M.J. *EMBO J.* **1**, 945-951 (1982).
15. Grammer, J. & Yount, R.G. *Biophys. J.* **59**, 226a (1991).
16. Mahmood, R., Elzinga, M. & Yount, R.G. *Biochemistry* **28**, 3989-3995 (1989).
17. Agarwal, R., Rajasekharan, K.N. & Burke, M. *Biochemistry* **266**, 2272-2275 (1991).
18. Morita, F. *J. Biol. Chem.* **242**, 4501-4506 (1967).
19. Werber, M., Szent-Györgyi, A.G. & Fasman, G.D. *Biochemistry* **21**, 1284-1294 (1972).
20. Torgerson, P.M. *Biochemistry* **23**, 3002-3007 (1984).
21. Muhlrad, A., Kasprzak, A.A., Ue, K., Ajtai, K. & Burghardt, T.P. *Biochim. Biophys. Acta* **869**, 128-140 (1986).
22. Muhlrad, A. *Biochim. Biophys. Acta* **493**, 154-166 (1977).
23. Kubo, A., Tokura, 5. & Tonomura, Y. *J. Biol. Chem.* **235**, 2835-2839 (1960).

24. Miyanishi, T., Horiuti, K., Endo, M., & Matsuda, G. *Eur. J. Biochem.* **171**, 31-35 (1988).
25. Burke, M., Reisler, E. & Harrington, W.F. *Biochemistry* **15**, 1923-1927.
26. Dalbey, R.E., Weiel, J. & Yount, R.G. *Biochemistry* **22**, 4696-4706 (1983).
27. Cheung, H.C., Gryczynski, I., Malak, H., Wiczk, W. & Lakowicz, J.R. *Biophys. Chem.* **40**, 1-17 (1991).
28. Highsmith, S., Akasaka, K., Konrad, M., Goody, R., Holmes, K., Jardetzky, N. & Jardetzky, O. *Biochemistry* **18**, 4238-4244 (1979)

IDENTIFICATION OF ACTIN SURFACE INTERACTING
WITH MYOSIN DURING THE ACTIN-MYOSIN SLIDING

Kazuo Sutoh

Department of Pure and Applied Sciences
College of Arts and Sciences
University of Tokyo
Komaba, Tokyo 153, Japan

ABSTRACT

We constructed several mutant actin genes from the Dictyostelium actin 15 gene by the site-directed mutagenesis. Mutations were designed to change acidic residues in actin subdomain 1 to histidine residues. Amino acid replacements were: D1H (single replacement of Asp1 to His), D4H, D1H/D4H (double replacements of Asp1 and Asp4 to histidine), D1H/E3H/D4H (triple replacements of Asp1, Glu3 and Asp4 to histidine), D24H/D25H, E99H/E100H, E360H/E361H, and D363H/E364H. Mutant genes were then expressed in Dictyostelium cells. *In vitro* motility assays were carried out for purified actins to see whether the mutations affect sliding motion of actin filaments driven by HMM. The assays showed that replacement of N-terminal acidic residues inhibited the sliding. Replacement of D24/D25 and E99/E100 also resulted in inhibition of the sliding motion. However, replacement of acidic residues at the C-terminal cluster E360/E361/D363/E363 did not resulted in loss of motility.

INTRODUCTION

The subdomain 1 of actin[1] has high density of acidic residues including a cluster at the N-terminus. Although amino acid sequences of actins from various sources show significant variation at the N-terminal region, it appears that the acidic nature of the N-terminal segment has been invariant during evolution. It suggests that the N-terminal cluster of acidic residues takes some essential roles in force generation. In fact our previous crosslinking experiments[2][3] showed that N-terminal acidic residues are in contact with myosin in the rigor complex. The crosslinking experiments also showed that a C-terminal cluster of acidic residues E360/E361/D363/E364 is also in contact with myosin in the rigor complex. The N- and C-terminal clusters are located closely with each other on the subdomain 1.

Mechanism of Myofilament Sliding in Muscle Contraction, Edited by
H. Sugi and G.H Pollack, Plenum Press, New York, 1993

241

Other acidic residues D24/D25, and E99/E100 are also closely located to the N- and C-terminal clusters, forming a surface area with high density of negative charges. In order to understand functions of these acidic residues, we replaced them with histidine residues by the site-directed mutagenesis. If these residues are functionally involved in force generation, we can expect that charge reversions by the mutations would result in loss of the actin-myosin interaction.

We expressed the mutant actins in Dictyostelium cells, purified them, and assayed their ability to activate HMM ATPase activity and to slide over HMM[4] .

Expression and Purification of Mutant Actins

Mutant actin genes were constructed from Dictyostelium actin 15 gene[5] by the oligonucleotide-directed mutagenesis. These actin genes, which has its own promoter and terminator regions, were then inserted into two types of transformation vectors for Dictyostelium cells, i.e. B10 (integration vector) and pnDeI (extrachromosomal vector). B10 vectors were introduced into Dictyostelium cells by the $CaPO_4$ method. For PnDeI vectors, electroporation was used to transform cells. Transformed cells were selected by culturing cells in culture medium HL5 with an antibiotics G418 (20 µg/mL). Selected cells were then cloned by picking up cells from a single colony. Two or three cycles of the cloning were carried out to obtain mutant strains. Expression of mutant actins were checked by two-dimensional gel electrophoresis since the amino acid replacements are reflected in shift of the isoelectric point of actin. Cloned cells expressing mutant actins were stocked in 10% DMSO or as spores.

Cells expressing a mutant actin were grown in HL5 containing G418 to a density of 10^7 cells/mL and harvested by centrifugation at 1,000 g. Cells were then suspended and disrupted in 30 mM MOPS, 2.5 mM EGTA, 5 mM $MgCl_2$, and 0.2 mM ATP (pH 7.0) containing protease inhibitors (1 mM PMSF, 20 µg/mL of pepstatin and chymostatin, and 100 µg/mL of leupeptin). After centrifugation at 10,000 g for 10 min to remove cell debris, supernatant was further centrifuged at 541,000 g for 1 h to precipitate cytoskeletons. Supernatant was then applied onto a DEAE HPLC column. Proteins were eluted by a linear gradient of NaCl up to 0.5 M. Mutant actins and wild type actins were eluted into different fractions. They were polymerized by addition of 2 mM $MgCl_2$, 0.2 mM ATP and 10 mM MOPS (final concentrations), and then dialyzed against 50 mM NaCl, 20 mM MOPS and 2 mM $MgCl_2$ (pH 7.0). Resulting F-actin solution was centrifuged at 541,000 g for 30 min. F-actin pellet was solubilized in 2 mM Tris-HCl, 0.2 mM ATP, and 0.2 mM $CaCl_2$ (pH 8.0) and dialyzed against the same buffer. The solution was finally cleared by centrifugation at 541,000 g for 1 h. G-actin was polymerized by addition of 50 mM NaCl, 20 mM MOPS, and 2 mM $MgCl_2$ (pH 7.0) (final concentrations). Resulting F-actin was centrifuged at 541,000 g for 10 min. Soft pellet was then washed in 2.5 mM KCl, 10 mM MOPS, 4 mM $MgCl_2$, 0.2 mM DTT (pH 7.0) containing

Table 1. *In vitro* motility assay of sliding of mutant actins on HMM.

	Polymerization	Rh- Phalloidin	Rigor Binding	Motility
Wild	Normal	+	+	++
D1H	Normal	+	+	+
D4H	Normal	+	+	+
D1H,D4H	Normal	+	+	±
D1H,E3H,D4H	Normal	+	+	−
D24H,D25H	Normal	+	+	−
E99H,E100H	Normal	+	+	−
E360H,E361H	Normal	+	+	++
D363H,E364H	Normal	+	+	++

phallacidin (10 μg/mL). Finally, the phallacidin-stabilized F-actin was suspended into the above buffer.

In Vitro Motility Assay

In vitro motility assay was carried out according to the method already published[6)7)] . Measurements were carried out at 22.5°C in 25 mM KCl, 25 mM imidazole, 4 mM MgCl$_2$, 1 mM DTT (pH 7.4). Wild type Dictyostelium actin filaments moved on HMM coated on a cover glass at the average rate of 2.0 μm/sec. The mutant actin D1H filaments moved at the average rate of 1.3 μm/sec. The double mutant D1H/D4H filaments did not slide continuously. They moved for some time, and stopped. After a while they resumed the sliding movement. As an average, 20% of filaments moved. Average velocity of these sliding filaments was 0.4 μm/sec. Thus, replacement of Asp residues one by one with His residues resulted in decrease of the sliding velocity.

Similar measurements were carried out for other mutants (Table 1). Mutants D1H/E3H/D4H, D24H/D25H, and E99H/E100H did not show the sliding motion. However, these mutants were normal in other properties; they polymerized normally, bound phalloidin, and bound myosin to form the rigor complex in an ATP-dependent way. Only the motility was lost by the mutations, indicating that the loss of motility resulted from selective disruption of the site recognized by myosin during the actin-myosin sliding, not from nonspecific conformational changes of actin induced by the mutations.

Quite interestingly, mutants E360H/E361H and D363H/E364H were motile. They moved with virtually the same velocity as that of the wild type actin.

Actin-Activated Myosin ATPase

Actin-activated ATPase activities of HMM were measured in the presence of various amount of the wild type actin and the N-terminal mutants, i.e. D1H and

Table 2. Sliding velocity and ATPase activity of mutant actins.

	Sliding velocity (µm/sec)	ATPase activity	
		Km (µM)	Vmax (1/sec)
Wild Actin	1.95	7	14.6
D1H	1.34	6	6.4
D1H,D4H	0.42	5	1.2
D1H,E3H,D4H	0	-	0

D1H/D4H. The activation of HMM ATPase activity was gradually lost by the D-H substitution, consistent with the motility assay. Km (apparent affinity of actin and ATP-saturated HMM) and Vmax (ATPase activity in the presence of saturating amount of actin) were calculated as shown in Table 2. Vmax gradually dropped by the D-H replacement while Km stayed virtually constant.

Conclusion

We can conclude from these results that the N-terminal cluster of acidic residues is essential for the actin-myosin sliding. Residues D24/D25 and residues E99/E100 are also essential for the sliding motion, while the C-terminal cluster of acidic residues, i.e. E360/E361/D363/E364 is not essential at least for the sliding motion.

REFERENCES

1. Kabsch, W., Mannhertz, H.G., Suck, D., Pai, E.F. & Holmes, K.C. *Nature* 347, 37-44 (1990).
2. Sutoh, K. *Biochemistry* 21, 3654-3661 (1982).
3. Sutoh, K. *Biochemistry* 22, 1579-1585 (1983).
4. Sutoh, K., Ando, M., Sutoh, K. & Toyoshima, Y. *Proc. Natl. Acad. Sci. USA* 88, 7711-7714 (1991).
5. Knecht, D., Cohen, S.M., Loomis, W.F. & Loodish, H.F. *Mol. Cell Biol.* 6, 3973-3983 (1986).
6. Kron, S.J. & Spudich, J.A. *Proc. Natl. Acad. Sci. USA* 83, 6272-6276 (1986).
7. Harada, Y., Noguchi, A., Kishino, A. & Yanagida, T. *Nature* 326, 805-808 (1987).

Discussion

Sellers: What happens if you make co-polymers of one of your mutants that does not move, and one of your mutants that slides—for example, your wild type. How does that behave?

Sutoh: If we mix wild-type actin and D1H:D4H, double mutant at the one-to-one ratio, its sliding velocity is almost half the velocity of the wild-type actin.

Simmons: I wonder at what conditions you measured the velocity in the motility assay? Was it at saturating head density?

Sutoh: Do you mean the concentration of HMM?

Simmons: Yes.

Sutoh: We used a nitrocellulose membrane and 20 µg/ml of HMM.

Simmons: Did you check whether that actually saturated the velocity? Because if it didn't, you have two possible interpretations. One is that the effective Michaelis-Menten dissociation rate constant changed. The other is that access to heads is reduced, so that at any one time fewer heads are able to interact in the motility assay.

Sutoh: What I can say is that when we increased the concentration of HMM further than that, we observed an accumulation of molecules as multiple layers. I suppose that the surface of the membrane is saturated with HMM at that concentration.

Morales: How does the motility assay depend on the ionic strength?

Sutoh: I'm not sure. We used only one condition—25 mM KCl and 25 mM Imidazole.

Morales: I was wondering, though: if this is a question of columbic interaction, the ionic strength might behave like reducing the charge. Dr. Sellers, do you know?

Sellers: The rate increases as you go up in ionic strength.

Morales: Indefinitely?

Homsher: No. It plateaus out at about 100 mM.

Holmes: The group of residues around 360 that don't seem to have too much effect on at least the velocity of movement have been shown by Milligan *et al.* to be near where one end of the A-1-A-2 light chain ends up (Milligan, R.A. et al. *Nature* **348**, 217-221, 1990). Moreover, Ian Trayer's group showed through NMR that the C-terminus of actin is also close to the N-terminus of the A-1 light chain (Trayer, I.P. et al. *Eur. J. Biochem.* **164**, 259-256, 1987). So, there are grounds for believing that this junction containing the C-terminal acidic residues is somehow involved with light-chain binding.

PHOTOCHEMICAL CLEAVAGE OF MYOSIN HEAVY CHAIN AND THE EFFECT ON THE INTERACTION WITH ACTIN

Yoh Okamoto and Christine Cremo*

Division of Bioengineering
Department of Applied Chemistry
Muroran Institute of Technology
Muroran 050 Japan
*Biochemistry and Biophysics Department
Washington State University
Pullman WA 99164-4660 U.S.A.

ABSTRACT

Myosin from rabbit skeletal muscle has been photochemically cleaved in the presence of vanadate ion and ATP. Two cleavage sites termed V1 and V2 within the S1 head region were studied. In the presence of magnesium ion both sites were cleaved, but in the absence of divalent cation cleavage only occurred at the V2 site. V2 cleaved myosin had higher K^+-EDTA-ATPase and actin activated Mg^{2+}-ATPase activity than V1,V2 cleaved myosin. Immunochemical characterization shows that the photochemical cleavage is more specific than that of proteolytic cleavage since breakdown of light chains was not observed for the photochemical method. This must be one of the best ways to prepare a single site cleaved myosin for the study of molecular mechanism of sliding.

Myosin plays a key role in muscle contraction and other forms of cell motility. Each myosin molecule contains two heavy chains of about 200 kDa and two pairs of light chains of about 20 kDa each. The amino terminal half of each of the heavy chains with one of each type of the light chains form the two heads of myosin that have the ATPase active site and actin binding sites. The carboxy terminal half of the two heavy chains fold together into α-helical coiled coil structure to form the rod-like tail of myosin. The motility function is localized in the head region of myosin since a new family of single headed myosins, called myosin I, without long tail has more recently been found and the myosin I does not assemble into filament[1][2]. The three dimensional structure of actin has been determined and the multi-domain structure was found[3]. Key binding sites on actin molecule for myosin are beginning

Mechanism of Myofilament Sliding in Muscle Contraction, Edited by
H. Sugi and G.H Pollack, Plenum Press, New York, 1993

247

uncovered. In contrast, our knowledge about the myosin head region where the principal energy transducing process takes place is far from atomic resolution. Under these circumstances we need to know more about the structural and functional organization of the myosin head. Previously, we have worked mainly on two subjects. Photoaffinity labeling techniques have been used to map key sites for nucleotide binding on the primary sequence[4][5]. Secondly, we have tried to isolate a smaller piece of the head fragment from myosin after making a specific proteolytic nick within the head[6]. The smaller fragment is then dissociated from the remaining molecule by a heat treatment. It has proved difficult to isolate a pure form of this smaller active fragment, because of incomplete inactivation of the protease activity during the heat treatment.

This is why we have to use a similar protocol to prepare a cleaved myosin but without the use of protease. As a first step to prepare such a myosin, we describe the biochemical properties of myosin that has been photochemically cleaved using vanadate ion.

MATERIALS AND METHODS

Myosin and actin were prepared from rabbit skeletal muscle by the methods of Kielley and Bradley[7] and Spudich and Watt[8]. ATP and ADP were purchased from Yamasa Co.. Orthovanadate was purchased from Wako Pure Chem. Co. and the stock solution (0.1M, pH10.0) was prepared according to the procedure of Goodno[9]. A super high pressure Hg lamp (USH-102D) and the power unit were the products of Ushio Co.. Myosin solutions were irradiated at no more than 2mm deep in a petri dish (Corning Pyrex covered) that was placed on half melted ice 5cm distant from the Hg lamp. At each time point aliquots were taken and vanadate and ADP were removed for ATPase assays and SDS-PAGE analysis. K^+-EDTA-ATPase and Ca^{2+}-ATPase activities were measured in 0.5M KCl, 20mM Tris-Cl(pH8.0), 2mM EDTA and 30mM KCl, 20mM Tris-Cl(pH8.0), 2mM $CaCl_2$, respectively. Actin-activated Mg^{2+}-ATPase activity were assayed in 30 mM KCl, 20mM HEPES-KOH(pH7.0), 2mM $MgCl_2$. These ATPase reactions were done at 25°C and the liberated Pi was determined by the Fiske-SubbaRow method[10]. SDS PAGE were done using the buffer system of Laemmli[11]. Western blotting was performed using nylon membrane (sartorius) on a Biorad Co. semi-dry blotter. Peroxidase conjugated secondary antibodies were purchased from Cappel Co. and Konica immunostain system was used. All other chemicals were of analytical grade.

RESULTS

Vanadate was initially known as a strong inhibitor of ATPase. The effective concentration for inhibition varies upon the kind of ATPase. In the case of myosin ATPase, orthovanadate can be tightly bound to the active site with ADP which

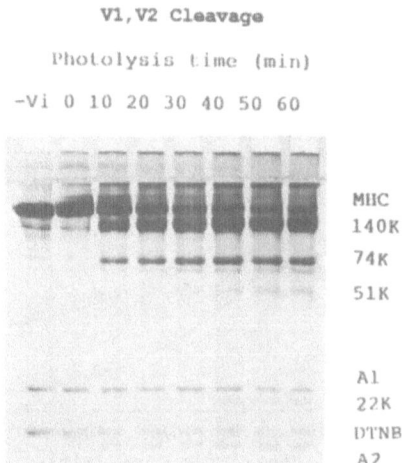

Fig. 1. Vanadate dependent photochemical cleavage in the presence of ATP and magnesium. During the photolysis aliquots (20 µg) were taken and applied to 12% SDS-PAGE.

results in proportional inactivation as described by Goodno[9]. The nucleotide-vanadate complex is extremely stable with a half life of many days. This unusual behavior has enabled us to "trap" the photoaffinity analog of ADP, NANDP, at the active site of gizzard myosin[12]. During the course of this study we noticed that both trapped vanadate and NANDP leaked out of the active site spontaneously when irradiated using a 300 nm cut-off filter. But the trapping can be maintained if irradiated using a 400 nm cut-off filter[12]. This was not a special property of NANDP since similar leakage was observed with vanadate trapped ADP. This light sensitive leakage of trapped nucleotide was not found when we used chemical crosslinking of SH1 and SH2 for the trapping[14]. Thus it appeared that vanadate trapping was sensitive to light of wavelengths between 300-400 nm. Lee Eiford et al. found a vanadate dependent photochemical cleavage of dynein during their photoaffinity labeling study[13]. The cleavage was site specific and was paralleled by complete loss of the ATPase activity. Myosin and subfragment 1 have also been cleaved photochemically using vanadate[14-18]. There are three sites for vanadate dependent photochemical cleavage. Two of them are in the S1 heavy chain and the third one is reported to be located near the carboxy terminus of the rod[16][17]. The specificity of the photochemical cleavage reactions appears to be much higher than that of proteolytic cleavage. This property may be useful for the study of internal structure and the function of myosin head in the sliding mechanism. We attempted to characterize the effect of cleavage at each of the two vanadate sites in the head region of the myosin molecule. In the presence of ATP the time course of the cleavage at two sites within the head region were monitored by SDS PAGE. In the presence of magnesium ion, cleavage was observed at 74 kDa(V2 site) and 22 kDa(V1 site) from the amino terminus. The rate of cleavage at V2 was higher than that of V1. Under these conditions, neither of the light chains were cleaved (Fig. 1). The ATPase

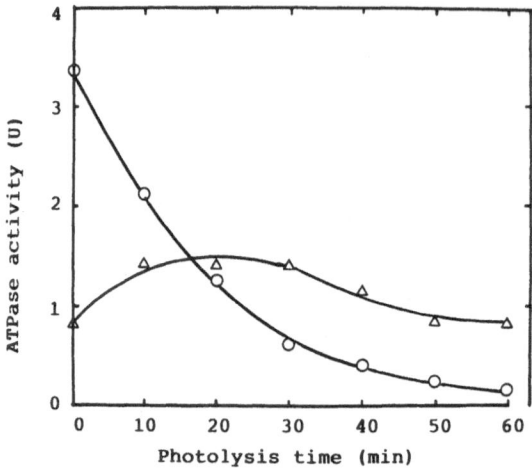

Fig. 2. K+-EDTA- (O) and Ca2+- ATPase (Δ) activities were assayed after taking aliquots as indicated in Fig. 1.

activity profile during The photochemical cleavage shows a rapid inactivation of K+-EDTA-ATPase activity and a rather constant Ca2+-ATPase activity (Fig. 2). After 60 min of irradiation about 90% of myosin heavy chain was cleaved. In order to determine the effect of each cleavage at either the V1 or the V2 site on the change in the biological functions, we tried to find more specific cleavage conditions. In contrast to previous cleavage condition (Figs. 1 & 2), the V2 site, 74 kDa from amino terminus was specifically cleaved in the presence of EDTA as shown in

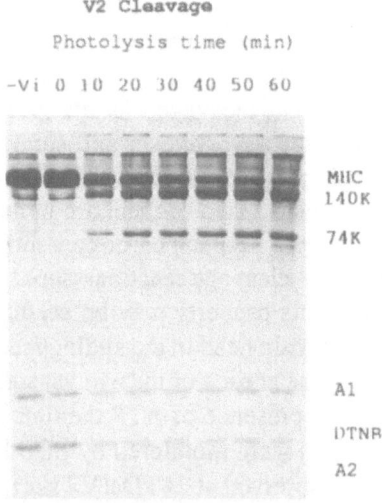

Fig. 3. Vanadate dependent photochemical cleavage in the presence of ATP and EDTA. During the photolysis aliquots (20 μg) were taken and applied to 12% SDS-PAGE.

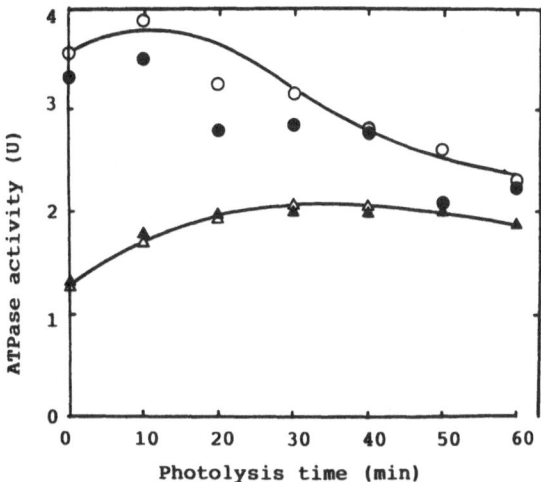

Fig. 4. K$^+$-EDTA- (O,●) and Ca^{2+} - ATPase (△,▲) activities were assayed after taking aliquots as indicated in Fig. 3. Open symbols were the same as in Fig, 2. Closed symbols indicate activities assayed after NaBH$_4$ treatment which may reduce aldehyde, if exists, to regenerate serine as reported[15].

Figure 3. After 60 min irradiation in the presence of EDTA as shown in Figure 4, only 30% of the K$^+$-EDTA-ATPase activity was lost, as compared to 94% in the presence of Mg^{2+} (Fig. 2). This is consistent with the identification of the V1 site as Ser180 in the consensus phosphate binding loop[15]. In contrast to the monotonous decrease of the K$^+$-EDTA-ATPase activity, the Ca^{2+}-ATPase activity shows biphasic changes as the photolysis proceeds. As a possible reason for this complexity, a photooxidation of Ser180 at the V1 site can be considered, which causes significant changes in both K$^+$-EDTA-and Ca^{2+}-ATPase activities[14]. The resultant serine aldehyde can be reduced using NaBH$_4$ to regenerate serine[15]. Neither type of ATPase activity was altered after treatment of V2 cleaved myosin with NaBH$_4$ as shown in Figure 4. Therefore, the possibility of the presence of serine aldehydes that affect activity was excluded. It is concluded that the V2 cleavage alone causes the Ca^{2+}-ATPase activity to increase by about 80% as shown in Figure 4.

As a first step to determine whether the vanadate cleavage altered the affinity for actin and the extent of maximum activation of the Mg^{2+}-ATPase activity by actin, the ATPase activities were measured as shown in Figure 5. In the presence of a seven fold molar excess of F-actin over myosin active sites, the activity increased to 120% after 10 min of irradiation in the presence and absence of Mg^{2+}. After this the activities decreased continuously. The significant loss of the actin-activated Mg^{2+}-ATPase activity of the V2 cleaved myosin suggests that V2 cleavage can alter the interaction with actin since both Ca^{2+}- and K$^+$-EDTA-ATPase activities stay at relatively high levels as shown in Figure 4. For a more extensive investigation of the changes in the interaction with actin we are going to determine the activity at various concentrations of F-actin and also the extent of cleavage at each site. In this way we

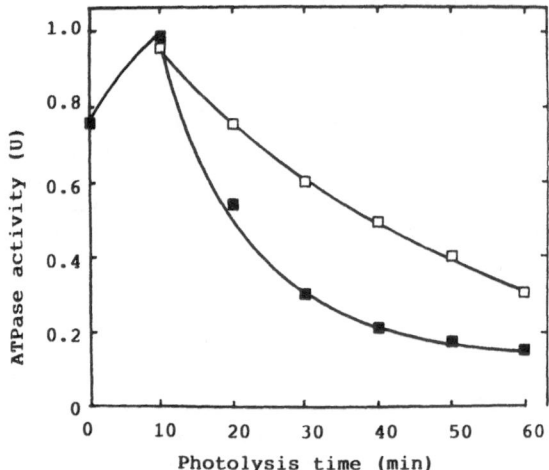

Fig. 5. Actin-activated Mg^{2+}-ATPase activities after vanadate dependent photochemical cleavage in the presence of magnesium (■) and EDTA (□)

will see the relationship between the changes in the affinity with actin with the maximum value of the activation and the extent of cleavage.

It is important to confirm the assignment of each of the cleaved products after V2 specific cleavage. We tested this by a western blot assay using antiserums raised against the 50 kDa and 20 kDa heavy chain fragments as shown in Figure 6. Anti 50 kDa antiserum bound to 74 kDa band and the uncleaved myosin heavy chain. In the case of the anti 20 kDa antiserum the 140 kDa fragment and the uncleaved myosin

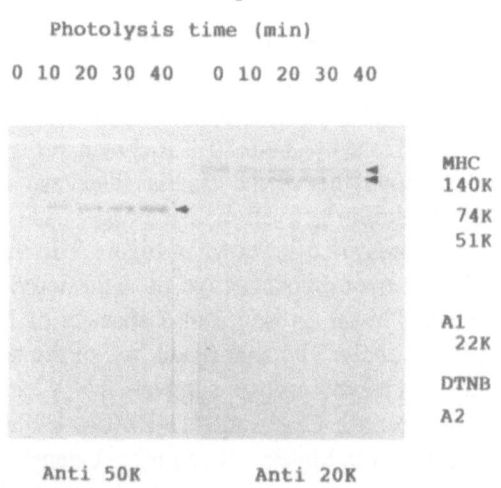

Fig. 6. Western blot assay of V2 cleaved myosin. Blotted membranes were incubated with serums raised against 50 kDa or 20 kDa heavy chain fragments and then immunostained.

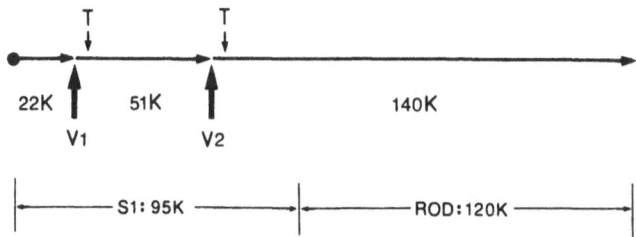

Fig. 7. Myosin heavy chain vanadate dependent photo-cleavable sites.

heavy chain were the positive bands. These data confirms the cleavage site model shown in Figure 7. The absence of any low molecular weight positive bands indicates that the cleavage at the V2 site was specific under these experimental conditions.

DISCUSSION

In attempting to find a specific vanadate dependent photochemical cleavage condition for rabbit skeletal muscle myosin, we found two types of intramolecularly cleaved myosins by using either magnesium or EDTA in the medium. From the previous work on nucleotide trapping in the active site of myosin, it is known that, in the presence of EDTA, nucleotide can not be stably trapped in the active site[19]. However, the added ATP appears to protect the V1 site in the active site against vanadate interaction while the V2 site can be cleaved exclusively. The resultant myosin appears to be more active than that produced in the presence of magnesium. From these observations the V1 site is confirmed to be the important site for nucleotide binding in the catalytic cycle of the ATPase reaction and also for the trapping with vanadate. This is consistent with the observation that Ser180, which is oxidized prior to V1, is in the consensus sequence among nucleotide binding proteins[15]. What is the relevance of the vanadate binding and cleavage at the V2 site? In a parallel experiment this cleavage doesn't require the presence of ATP[17]. Thus, the V2 site may be a nucleotide independent phosphate binding site. It would now be useful to test the V2 cleaved myosin using a more biological method such as an *in vitro* motility assay. In our preliminary motility assay, the affinity of V2 cleaved myosin to F-actin in the presence of magnesium and ATP appears to be reduced (data not shown). In conclusion the vanadate dependent photochemical cleavage reaction could be a powerful way to prepare a single site alteration in myosin which may be superior in specificity to that of proteolytic enzyme.

ACKNOWLEDGEMENTS

This research was supported in part by a Grant-in-Aid for Special research for the molecular mechanism of sliding from the Ministry of Education, Science and Culture of Japan. This work has been supported in part by a grant from the Tokyo Ohka Foundation for the Promotion of Science and Technology.

REFERENCES

1. Korn, E.D. & Hammer, J.A.III *Ann. Rev. Biophys. Biophys. Chem.* **17**, 23-45 (1988).
2. Pollard, T.D. et al. *Ann. Rev. Physiol* . **53**, 653-681 (1991).
3. Kabsh, W. et al. *Nature* **347**, 37-44 (1990).
4. Okamoto, Y. & Yount, R.G. *Proc. Natl. Acad. Sci. USA* **82**, 1575-1579 (1985).
5. Mahmood, R. & Yount, R.G. *J. Biol. Chem.* **259**, 12956-12959 (1984).
6. Okamoto, Y. & Sekine, T. *J. Biol. Chem.* **262**, 7851-7854 (1987).
7. Kielley, W.W. & Bradley, L.B. *J. Biol. Chem.* **218**, 653-659 (1956).
8. Spudich, J.A. & Watt, S. *J. Biol. Chem.* **246**, 4866-4871 (1971).
9. Goodno, C.C. *Methods Enzymol.* **85**, 116-123 (1982).
10. Fiske, C.H. & SubbaRow, Y. *J. Biol. Chem.* **66**, 375-400 (1925).
11. Laemmli, U.K. *Nature* **227**, 6 80-685 (1970).
12. Okamoto, Y . et al . *Nature* **324**, 78-80 (1986).
13. Lee-Eiford, A. et al. *J. Biol. Chem.* **261**, 2337-2342 (1986).
14. Grammer, J.C. et al. *Biochemistry* **27**, 8408-8415 (1988).
15. Cremo, C.R. et al. *Biochemistry* **27**, 8415-8420 (1988).
16. Mocz, G. *Eur. J. Biochem.* **179**, 373-378 (1989).
17. Cremo, C.R. et al . *Biochemlstry* **29**, 7982-7990 (1990).
18. Muhlrad, A. et al. *Biochemistry* **30**, 958-965 (1991).
19. Wells, J.A. & Yount, R.G. *Methods Enzymol.* **85**, 93-116 (1982).

Discussion

Sellers: In the ATPase, do you know whether this is a K_m or a V_{max} effect that you are seeing?

Okamoto: I cannot say yet. We are planning to check it.

IV. PROPERTIES OF ACTIN-MYOSIN SLIDING STUDIED BY *IN VITRO* MOVEMENT ASSAY SYSTEMS

INTRODUCTION

Recent development of *in vitro* movement (motility) assay systems has enabled us to study the ATP-dependent sliding between actin and myosin under microscopic observation. The assay systems are classified into two different types; in one type, fluorescently labeled F-actin is made to slide on myosin bound to a glass surface, while in the other type myosin-coated latex beads are made to slide on well-organized actin filament arrays (actin cables) occurring in the giant algal cells.

Toyoshima studied the modes of binding of S-1 or HMM to the nitrocellulose-coated glass surface by proteolytic digestion of the samples on the nitrocellulose film and analyzing the released proteolytic products and the resulting change in F-actin motility. She concludes that S-1 and HMM bind to the nitrocellulose film in different modes, but only the S-1 or the HMM bound to the film at its tail region may be responsible for F-actin sliding. Sellers, Umemoto and Cuda investigated the factors determining smooth muscle mechanical activity with special reference to phosphorylation to myosin light chain. They used both types of assay systems mentioned above as well as biochemical measurements, and showed that the velocity of actin-myosin sliding in smooth muscle can be modulated by the presence and absence of tropomyosin or caldesmon on the thin filament and the degree of myosin phosphrylation. Homsher, Wang and Sellers carefully compared the dependence of [ATP], [ADP], [Pi], pH, iotonic strength and temperature between the sliding velocity of F-actin on myosin(V_f) and literature values of the unloaded shortening velocity of muscle fibers (V_u), with the conclusion that V_f is a good analogue of V_u only under certain conditions, indicating that the results of motility assay should be cautiously interpreted.

Tawada presented a hypothesis in which viscous-like "friction" arising from weak-binding actin-myosin interaction is a determinant of "unloaded" actin-myosin sliding velocity in the *in vitro* motility assay systems. According to him, his hypothesis well predicts the experimental data with mixtures of different myosins and is also consistent with the properties of another protein motors such as dynein-microtubule system.

HOW ARE MYOSIN FRAGMENTS BOUND TO NITROCELLULOSE FILM ?

Yoko Y. Toyoshima

Department of Biology
Ochanomizu University
Ohtsuka, Tokyo 112, Japan

ABSTRACT

In an attempt to identify the binding sites of myosin fragments to the nitrocellulose film in a flow cell, proteolytic digestion was performed of myosin fragments bound to the substrate. Chymotryptic digestion of papain•Mg-subfragment-1 (S1) showed that a large proportion of S1 were bound near the tail and that the digestion site was protected for some S1. The binding of S1 near the tail was also confirmed by the production of extra long subfragment-2 (S2) by chymotryptic digestion of heavy meromyosin (HMM). Papain digestion of HMM released slightly smaller HMM from the substrate and impaired the motility at the same time. Hence, the motility of HMM is supported largely by those bound to the substrate near the C terminus. These results demonstrate the importance of flexible regions of myosin in generating active movements.

In vitro motility assay has been providing much insight into the molecular events involved in motility. We have developed a simple and reproducible assay system and demonstrated that head portion of myosin (subfragment-1) is sufficient for generating active movement[1]. In this system, purified proteins are loaded into a flow cell consisting of cover slip and slide glass; cover slip is coated by nitrocellulose film, on which motor proteins generate active movement on addition of ATP. Movements of single actin filaments are observed by fluorescence microscopy[2]. This system is now widely used combined with other techniques[3-6].

One of the important things to be clarified with this system is how myosin fragments are bound to the nitrocellulose film. Identification of the binding sites to the substrate may give us critical information on how myosin generates active movement. First, if the fragments were bound to the substrate at the middle or top of the head, the "swinging" cross-bridge model[7] will hardly be able to explain the step size measured[8-12]. Second, this information will restrict the flexible parts in myosin

Mechanism of Myofilament Sliding in Muscle Contraction, Edited by
H. Sugi and G.H Pollack, Plenum Press, New York, 1993

259

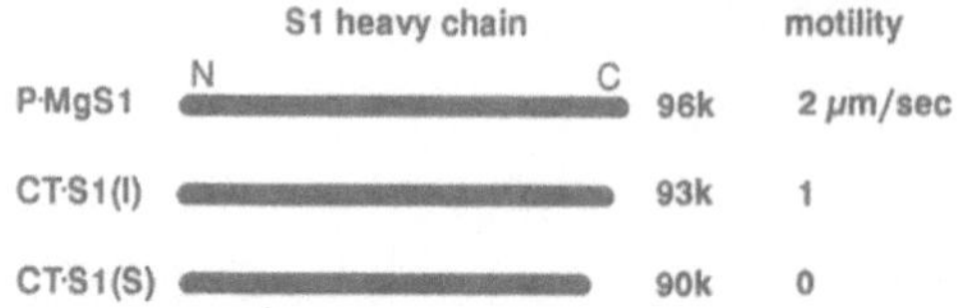

	S1 heavy chain		motility
P·MgS1	N ⬛⬛⬛⬛⬛⬛ C	96k	2 μm/sec
CT·S1(l)	⬛⬛⬛⬛⬛⬛	93k	1
CT·S1(S)	⬛⬛⬛⬛⬛	90k	0

Fig. 1. Size of the heavy chain and the motility of three kinds of myosin subfragment-1 (S1) used in this study. *P MgS1*: S1 obtained by papain digestion of myosin in the presence of Mg^{2+}, *CT S1 (l)* and *CT S1 (s)*: S1 obtained by brief and extensive chymotrypsin digestions, respectively. Motility refers to the movement of actin filaments *in vitro*[1].

heads responsible for generating force in any direction; the direction appears to be determined solely by actin filaments[13][14].

As the first attempt, myosin fragments bound to the substrate were digested by several proteases. The digestion products, those released from the substrate and those retained were examined by SDS gel electrophoresis. The motility of the digestion products were also monitored.

1. Binding of Subfragment 1 to the Nitrocellulose Film

Figure 1 illustrates three kinds of S1 used in this study. They have heavy chains of slightly different lengths and show different motility. Papain·Mg-S1 (P·Mg-S1) has the longest heavy chain (96 kd) and moves actin filaments at 2 μm/sec. S1 prepared by chymotrypsin digestion (CT-S1) contains at least two types with different heavy chain lengths. Extensive digestion yields only short S1, showing no motility. Brief digestion results in a mixture of short and long S1, which can partially be separated by gel filtration chromatography. The fractions containing long CT-S1 move actin filaments at 1 μm/sec.

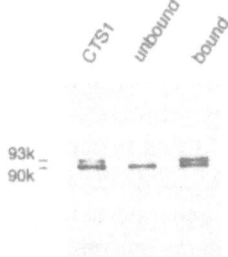

Fig. 2. Binding of chymotryptic S1 of different heavy chain length to the nitrocellulose film. The mixture of two kinds of chymotryptic S1 *(S1(CT))* was applied to the flow cell coated with nitrocellulose film, left to stand for 2 min, and unbound proteins *(unbound)* were collected by removing the solution from the flow cell. Bound proteins *(bound)* were recovered by washing the flow cell with 1% SDS. They were loaded on a 7.5% polyacrylamide gel and electrophoresed. Note that the bound fraction contained heavy chains of 93k at much higher ratio than in the original mixture.

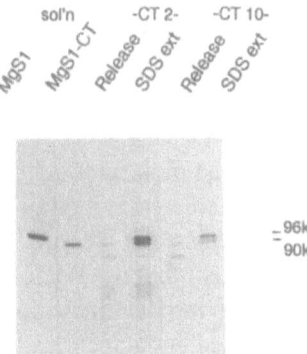

Fig. 3. Chymotryptic digestion of papainMg-S1 on the nitrocellulose film. S1 obtained by papain digestion (*MgS1*) was further digested by chymotrypsin in solution (*MgS1-CT*), or on nitrocellulose film (*CT2* and *CT10*). 50 µg of S1 was applied to a flow cell and digested for 5 min at 25°C at 2 µg/ml (*MgS1-CT* and *CT2*) or 10 µg/ml (*CT10*) of chymotrypsin in the presence of EDTA. *Release*, polypeptides released from the substrate by the digestion; *SDS ext*, polypeptides retained on the substrate and recovered by washing the flow cell by 1% SDS. Note that a large proportion of S1 was retained on the substrate keeping the original heavy chain length, although the same amount of chymotrypsin digested the heavy chain completely in solution.

The long CT-S1 showed higher ability in binding to the substrate. Fig. 2 compares the specimen loaded into the flow cell, the bound and the unbound fractions. The bound fraction was retrieved by washing the flow cell by 1% SDS after incubation for 10 minutes. This gel shows that long CT-S1 was bound to the substrate at much higher ratio than in the original mixture. Hence the C terminal region of S1 affects the binding ability to the nitrocellulose film.

Chymotryptic digestion was performed of papain Mg-S1 (P•Mg-S1) bound to the nitrocellulose film. The questions here are (i) whether chymotryptic S1 is generated and (ii) whether chymotryptic S1 is released from the substrate. The amounts of chymotrypsin (2 and 10 µg/ml) were chosen so that all P•Mg-S1 were cleaved into CT-S1 in solution. Fig. 3 shows that the amount of proteins released by digestion was not much (10% or less); even at 10 µg/ml of chymotrypsin, most of the

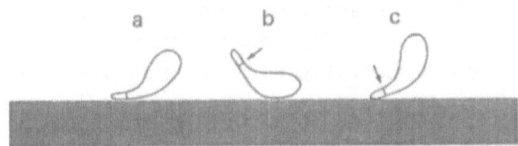

Fig. 4. Schematic diagram showing possible ways of binding of papain Mg-S1 to the nitrocellulose film. *a*; S1 bound to the nitrocellulose film around neck region, making the chymotryptic digestion site protected. *b*; S1 attached at its tip or middle part. *c*; S1 attached to nitrocellulose film at the C terminus end part absent in chymotryptic S1. Arrows indicate chymotryptic digestion sites.

proteins were still attached to the substrate, leaving about 2/3 of P•Mg-S1 intact. However, the digestion impaired the motility. The movement was similar to the one at a lower density of S1, suggesting that the number of heads contributing to the movement was much decreased.

These experiments suggest that S1 molecules are bound to the nitrocellulose film in several ways. Most of them are tethered at the tail part, having the digestion site protected (Fig. 4a). Some may be tethered at the middle of S1 (Fig. 4 b); this type of S1 will be digested with chymotrypsin but not released. Only about 10% of S1 are bound at the end of the tail part, having the digestion site exposed (Fig. 4c); in this case, the digested S1 will be released and appear as the 90k band (Fig. 3). The last population may be most responsible for the motility.

2. Binding of Heavy Meromyosin to the Nitrocellulose Film

In the presence of Mg^{2+}, chymotrypsin cleaves myosin to yield heavy meromyosin (HMM) and light meromyosin. In the presence of EDTA, chymotrypsin digests HMM into S1 and subfragment-2 (S2). S2 may have different lengths; short S2 lacks C terminus 20k of long S2[15]).

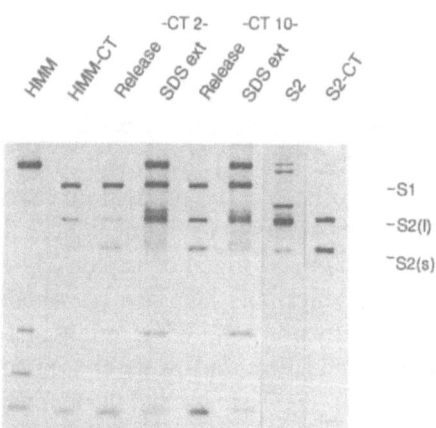

Fig. 5. Chymotryptic digestion of heavymeromyosin (HMM) on the nitrocellulose film. Chymotryptic HMM (*HMM*) bound to the nitrocellulose film was further digested by chymotrypsin in the presence of EDTA, and the digestion products (*CT2* and *CT10*) were examined on a 13% polyacrylamide SDS gel. Also shown are: HMM digested by chymotrypsin in solution (*HMM-CT*), crude chymotryptic subfragment-2 (S2) preparation (*S2*) and its chymotrypsin digests (*S2-CT*). The digestion conditions were the same as in Fig. 3. *Release*, polypeptides released from the substrate by the digestion; *SDS extract*, polypeptides retained on the substrate and extracted by washing the flow cell with 1% SDS. Due to the overlap of the band from bovine serum albumin used for blocking the substrate from nonspecific binding, the band labeled *S2(l)* may not reflect the actual amount of S2. Note that a large proportion of HMM on the substrate withstood the chymotrypsin attack, and that obscure bands appeared between *S1* and *S2(l)* in the lanes for the retained polypeptides.

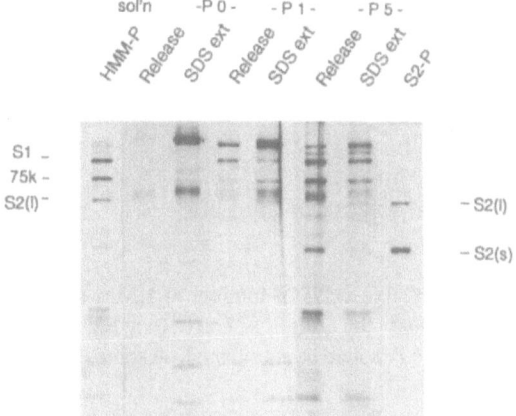

Fig. 6. Papain digestion of HMM on the nitrocellulose film. Chymotryptic HMM bound to the nitrocellulose film was digested further by papain in the presence of Mg^{2+}; the digestion products (*P1* and *P5*) were examined on a 13% polyacrylamide gel. 50 μg of HMM was applied to each flow cell and digested for 5 min at 25°C at 1 (*P1*) or 5 (*P5*) μg/ml of papain. *P0* shows the retained (*SDS ext*) and released (*Release*) polypeptides by incubation under the same conditions. Also shown are: digestion products of HMM (*HMM-P*), and of chymotryptic S2 (*S2-P*) in solution. Movement of actin filaments were observed with *P1* but not with *P5*.

Fig. 5 examines the chymotryptic digestion of HMM bound to the nitrocellulose film. The released polypeptides included S1, long and short S2; all the bands corresponding to these polypeptides were clearly defined. In contrast, the bands for S2 in the retained polypeptides were obscured, showing multiple weak bands. Especially interesting were those between S1 and long S2 bands. Since these bands reacted strongly with anti-S2 antibodies (kind gift from Dr. W.F. Harrington), they must represent extra long S2 (longer than the normal long S2). Considering the multiple chymotrypsin digestion sites at S1-S2 junction, this result indicates that some of the digestion sites were protected by binding to the substrate. The amount of HMM retained on the substrate was also much larger than expected from the digestion in solution.

Papain digestion of chymotryptic HMM in the presence of Mg^{2+} yielded polypeptides different from chymotryptic ones (Fig. 6). The striking thing here is that HMM, only slightly smaller than the original, was released by papain digestion. At the same time, the motility was much impaired. Actin filaments moved with *P1* but not with *P5* in Fig. 6. Even with *P1*, the movement of actin filaments suggested that the number of myosin heads contributing was much decreased. Again the bands for long S2 were multiple. Short S2 was found only in the released products, showing that the digestion site producing short S2 was protected by binding to the substrate.

These results indicate that the motility of HMM is supported largely by those bound to the substrate near the HMM-LMM junction (Fig. 7a). A large population of HMM are bound so that some of digestion sites are protected (Fig. 7b). The presence

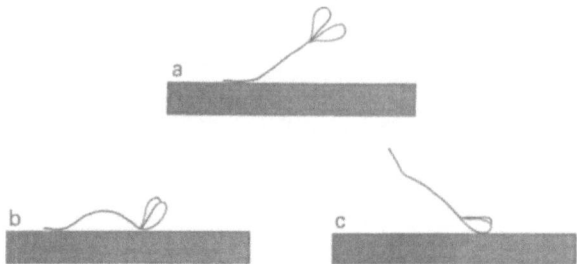

Fig. 7. Cartoons showing possible ways of binding of HMM to the nitrocellulose film. *a*, HMM is bound to the substrate by the C terminal end of long S2. *b*, same as *a* but neck portion of S1 is also tethered. *c*, HMM is bound to the substrate at some part of S1.

of long S2 in the released products means that some HMM may be tethered only by S1 portion (Fig. 7c).

This study showed that S1 and HMM are bound to the nitrocellulose film in several configurations and that only small fractions of them are responsible for the motility. The importance of the tail part of S1 and HMM-LMM junction for the motility was again demonstrated. Flexible nature of those regions indicated by the clustering of digestion sites and other studies[16][17], suggests that the flexibility is important critically in generating active movement.

ACKNOWLEDGEMENTS

I thank Dr. W.F. Harrington for his generous gift of anti-S2 antibody. I also thank Dr. C. Toyoshima for discussions.

REFERENCES

1. Toyoshima, Y.Y. et al. *Nature* **328**, 536-539 (1987).
2. Kron, S.J., Toyoshima, Y.Y., Uyeda, T.Q.P. & Spudich, J.A. *Methods Enzymol.* **196**, 399-416 (1990).
3. Manstein, D.J., Ruppel, K.M. & Spudich, J.A. *Science* **246**, 656-658 (1989).
4. Collins, K., Sellers, J.R. & Matsudaira, P. *J. Cell Biol.* **110**, 1137-1147 (1990).
5. Schwyter, D.H., Kron, S.J., Toyoshima, Y.Y., Spudich, J.A. & Reisler, E. *J. Cell Biol.* **111**, 465-470 (1990).
6. Sutoh, K., Ando, M. Sutoh, K. & Toyoshima, Y.Y. *Proc. Natl. Acad. Sci. USA* **88**, 7711-7714 (1991).
7. Huxley, H.E. *Science* **164**,1356-1366 (1969).
8. Toyoshima, Y.Y., Kron, S.J. & Spudich, J.A. *Proc. Natl. Acad.Sci. USA* **87**, 7130-7134 (1990).
9. Harada, Y., Sakurada, K., Aoki, T., Thomas, D.D. & Yanagida, T. *J. Mol. Biol.* **216**,49-68 (1990).

10. Uyeda, T.Q.P., Kron, S.J. & Spudich, J.A. *J. Mol. Biol.* **214**, 699-710 (1990).
11. Ishijima, A., Doi, T., Sakurada, K. & Yanagida, T. *Nature* **352**, 301-306 (1991).
12. Uyeda, T.Q.P., Warrick, H.M., Kron, S.J. & Spudich, J.A. *Nature* **352**, 307-311 (1991).
13. Toyoshima, Y.Y., Toyoshima, C. & Spudich, J.A. *Nature* **341**, 154-156 (1989).
14. Toyoshima, Y.Y. *J. Cell Sci. suppl* . **14**, 83-85 (1991).
15. Ueno, H. & Harrington, W.F. *J. Mol. Biol.* **180**, 667-701 (1984).
16. Kinosita, K., Ishiwata, S., Yoshimura, H., Asai, H. & Ikegami, A. *Biochemistry* **23**, 5963-5975 (1984).
17. Miyanishi, T., Toyoshima, C., Wakabayashi, T. & Matsuda, G. *J. Biochem. (Tokyo)* **103**, 458-462 (1988).

Discussion

Kawai: I was not sure about the second kind of binding, where the neck portion of the S-1 — S-2 is bound to the nitrocellulose. Did you say it is active in promoting motility?

Toyoshima: I think it is not responsible for motility.

IN VITRO STUDIES OF DETERMINANTS OF SMOOTH MUSCLE MECHANICS

James R. Sellers, Seiji Umemoto and Giovanni Cuda

Laboratory of Molecular Cardiology
National Heart, Lung, and Blood Institute
National Institutes of Health
Bethesda, MD 20892

ABSTRACT

Smooth muscle contraction is dependent upon phosphorylation of the 20,000 Da light chain subunits of myosin. Whereas the kinetics of the hydrolysis of MgATP by smooth muscle myosin suggest a simple phosphorylation-dependent "on-off" mechanism, the contractile response in smooth muscle tissue is complex. Experiments to unravel this complexity have been performed *in vitro* using a combination of motility assays and kinetic techniques. Some insight into this complexity is obtained, but the mechanism and the regulation of smooth muscle contraction is still not completely known.

Smooth muscle contains some of the same basic contractile elements that are so well documented in striated muscle. Actin and myosin are arranged in a more loosely packed interdigitating arrays than is found in striated muscle. Actin filaments are attached to α-actinin containing dense bodies[1] and the myosin filament may be side-polar as opposed to bi-polar[2]. Nevertheless, force is generated by cyclical actomyosin interactions in both muscle systems.

Smooth muscle contracts much more slowly than skeletal muscle which is undoubtedly due to the slower actomyosin MgATPase kinetic reactions in the former case[3][4]. Also the mode of the calcium dependent regulation is vastly different in the two muscles. Smooth muscle is thought to be primarily regulated by a calcium-calmodulin dependent phosphorylation of serine-19 on the 20 kDa light chain subunit of myosin[5][6]. In addition, two possible thin filament regulatory systems have been proposed. One of these is caldesmon, a calmodulin binding protein with a molecular weight of about 88,000 Da[7][8] and the other is a more recently discovered smooth muscle actin binding protein, calponin, which also binds calmodulin in a calcium-dependent manner[9][10]. Both of these proteins can inhibit the actin-activated MgATPase of smooth muscle myosin.

Mechanism of Myofilament Sliding in Muscle Contraction, Edited by
H. Sugi and G.H Pollack, Plenum Press, New York, 1993

267

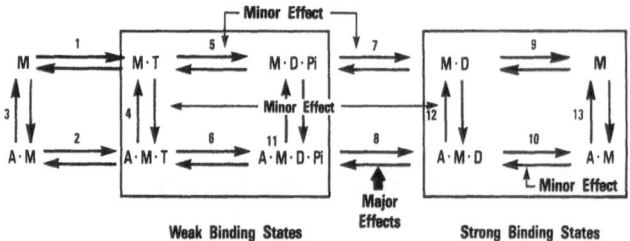

Fig. 1. Kinetic cycle for the hydrolysis of ATP by smooth muscle HMM. M represents myosin; A, actin; T, ATP; D, ADP; Pi, inorganic phosphate.

Kinetic studies of the regulation of smooth muscle heavy meromyosin (HMM) demonstrated that phosphorylation has a major effect on only a single step in the kinetic cycle, namely Pi release from an AM•ADP•Pi complex[4] (Fig. 1). Phosphorylation has a smaller effect at several other steps in the cycle such as the equilibrium between M•ATP and M•ADP•Pi with actin[11] and the equilibrium of M•ADP with actin[12]. The rate of product release from M•ADP•Pi in the absence of actin is also slightly affected by phosphorylation[4]. In the absence of phosphorylation there is no activation of the actin activated MgATPase activity of myosin when examined in single-turnover experiments.

Such a kinetic scheme suggests that phosphorylation should act as a simple switch in regulating smooth muscle contraction. However, the contractile response in smooth muscles is very complicated[13][14] (Fig. 2). Upon stimulation of a smooth muscle fiber, myosin becomes phosphorylated and generates tension. However, there is not a unique relationship between extent of phosphorylation and tension[13][14] (Fig. 2). Phosphorylation is high during the period of force generation

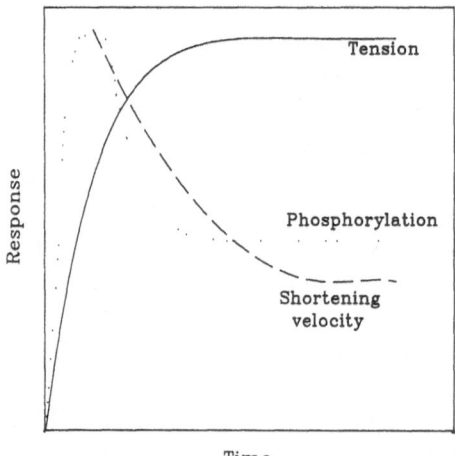

Fig. 2. Schematic representation of the time course of a smooth muscle contraction. This figure was adapted from Kamm and Stull[14].

(typically reaching a value of 0.5-0.8 mol Pi/mol LC) and declines to a steady-state value later in the contraction while force is maintained at maximal values. The steady state level of phosphorylation is shown schematically in Fig. 2, as about half of the maximal value, however experimentally, it can vary from about 0.6 mol/mol to values less that 0.2 mol/mol depending on the agonist and the tissue. Isotonic shortening velocity measured at small or zero loads also decreases from peak values during the period of force generation to levels around four to five fold less during the period of force maintainance. Therefore, during the period of force maintainance in smooth muscle fibers, tension is being maintained by slowly cycling myosin which is less than maximally phosphorylated. This condition was termed "latch" by Murphy and colleagues[13)15)]. It contributes to the economy of smooth muscle contraction.

To date, there have been many hypotheses to explain this behavior, but no definitive mechanism has been demonstrated. Below, is listed some of the mechanisms that have been considered to account for the modulated shortening velocity observed during a smooth muscle contraction. 1) Calcium binding to myosin may directly affect the cycling rate. There is a transient increase in the free calcium ion concentration at the onset of smooth muscle contraction which is followed by a submaximal, but still superbasal concentration during force maintainance. 2) Unphosphorylated myosin may be capable of cycling slowly by itself and that in relaxed muscle a thin filament regulatory system maintains the "off" state. The decline in shortening velocity observed during latch may thus be due to a larger percentage of cycling, unphosphorylated myosin. 3) Myosin may be phosphorylated during latch at sites other than serine-19 and these phosphorylations may affect the cycling rate. 4) Thin filament proteins such as tropomyosin, caldesmon or calponin may somehow play a role in latch. 5) Finally, the most widely disseminated model suggests that latch is a phenomena that can be explained solely on the basis of phosphorylation/dephosphorylation of myosin.

Attempts have been made to study various components of the latch behavior *in vitro* using biochemical approaches and *in vitro* motility assays[16-18)]. The motility assays are thought to be *in vitro* models of the unloaded shortening velocity of muscle fibers and are thus a possible model for the changes in unloaded shortening velocity that occur in latch[19)]. We will now address the biochemical evidence for or against each of the above hypotheses.

1) Direct Effect of Calcium on Myosin

Smooth muscle myosin is known to bind calcium ions[5)]. Using the sliding actin assay we have measured the effect of increasing calcium ions on the movement of fluorescently-labeled actin by turkey gizzard smooth muscle myosin. Over a range of free calcium ion concentration from 10^{-8} to 10^{-4} M there was no change in the rate of actin translocation[20)]. Thus, it is not likely that calcium has a direct effect on cycling rate of myosin during a smooth muscle contraction.

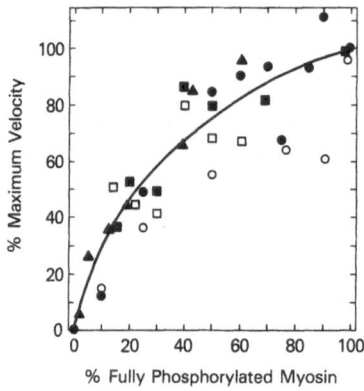

Fig. 3. Effect of unphosphorylated myosin on the velocity of phosphorylated myosin in the *Nitella* assay. Different symbols represent different preparations of myosin. Some of this data were taken from Sellers et al[24].

Active Unphosphorylated Myosin Combined with a Thin Filament Regulatory System

This hypothesis is dependent on unphosphorylated smooth muscle myosin having a measurable cycling rate that could be increased upon phosphorylation. Experiments using smooth muscle HMM demonstrate that the unphosphorylated state has no actin activated MgATPase activity above the basal MgATPase activity of the myosin when studied using single turnover experiments[4]. On the other hand, Wagner and colleagues have presented data that suggest that the unphosphorylated myosin has some activity and that phosphorylation is primarily affecting the affinity for actin[21][22]. More recent single turnover studies by Trybus[23] using stabilized thick filaments under more physiological conditions provided evidence against an active unphosphorylated myosin and provided rate constants that were in quantitative aggreement with the studies of Sellers[4]. Studies using both the *Nitella* assay and the sliding actin assay have typically shown that the movement induced by smooth muscle myosin is completely dependent on phosphorylation[20][24-27].

Experiments where unphosphorylated myosin is mixed with phosphorylated myosin show that unphosphorylated myosin may be able to modulate the velocity of active actin filament sliding by phosphorylated smooth muscle myosin presumably through weak binding interactions[24][26]. Therefore even though unphosphorylated myosin is itself inactive, its weak interaction with actin may modulate velocity.

Phosphorylation at Other Sites on Myosin

Myosin light chain kinase (MLCK) readily phosphorylates serine-19 on the 20,000 Da light chain which serves as the activating switch. Even though the overall levels of phosphorylation decline during latch, it is possible that other sites are being phosphorylated which may alter the kinetics of acto-myosin interaction. The 20,000

Da light chain of myosin can be phosphorylated at several other sites. MLCK can also phosphorylate threonine-18 at a much slower rate to give di-phosphorylated myosin[28] and protein kinase C (PKC) can phosphorylate serines-1, -2 and -9[29][30]. The additional phosphorylation by MLCK increases the actin-activated MgATPase activity through an effect on the affinity of actin for myosin[31]. Protein kinase C

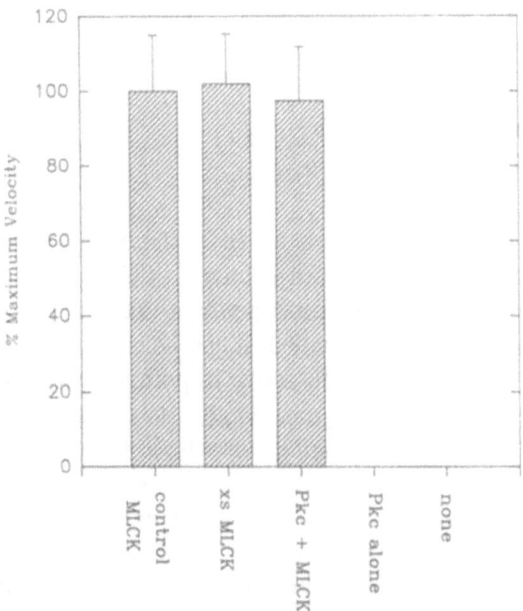

Fig. 4. Effect of various phosphorylations on the movement of myosin-coated beads in the *Nitella* motility assay. Control MLCK was myosin phosphorylated by myosin light chain kinase at serine-19. xs MLCK was myosin phosphorylated at serine-19 and threonine-18 by myosin light chain kinase. PKC + MLCK was myosin phosphorylated by myosin light chain kinase at serine-19 followed by phosphorylation by protein kinase C at serine-1 (or -2) and threonine-9. PKC myosin was phosphorylated at serine-1 (or-2) and threonine-9 by protein kinase C. None represents unphosphorylated smooth muscle myosin.

decreases the actin-activated MgATPase activity of myosin that is already phosphorylated at serine-19 also through an effect on the affinity of actin for myosin[32]. None of these phosphorylations affect the movement of serine-19 phosphorylated myosins in the *Nitella* assay (Fig. 4) and protein kinase C phosphorylation of myosin in the absence of serine-19 phosphorylation does not activate myosin[25]. These data suggest that phosphorylation of myosin light chain at sites other than serine-19 is not a likely mechanism to explain the slower shortening velocity observed in latch.

Table 1. Effect of Caldesmon on the Binding of HMM to Actin

HMM	Caldesmon	K_B
Phosphorylated	-	2.96×10^4 M^{-1}
Phosphorylated	+	1.06×10^6 M^{-1}
Unphosphorylated	-	5.8×10^4 M^{-1}
Unphosphorylated	+	2.38×10^5 M^{-1}

Data taken from Lash et al[35].

Effects of Thin Filament Regulatory Proteins

There is good *in vitro* biochemical evidence that thin filament proteins such as caldesmon and calponin can regulate the activity of phosphorylated smooth muscle myosin in a calcium- and calmodulin-dependent manner[7-10]. The mechanism of action of caldesmon is quite well studied. It can be subdivided into two domains; a carboxyl-terminal region that binds actin, tropomyosin and calmodulin and an amino-terminal myosin binding region. Fragments containing various portions of the carboxyl-terminal region can inhibit the actin-activated MgATPase activity of myosin in a calcium-calmodulin-dependent manner[7][8]. There is evidence that caldesmon competes with myosin for the same site on actin and thus may function as a steric blocker[33]. The role of the putative myosin binding region is less well characterized although the subfragment-2 region (S-2) of myosin appears to be the binding site[34].

Caldesmon greatly increases the binding of HMM to actin (Table 1) even though caldesmon is inhibitory to the actin-activated MgATPase activity[35]. It was speculated that this effect may be due in part to an inhibition of the rate of ADP release from AMADP (see Fig. 1). If this were occurring, several features of latch may be explained. First, the higher affinity binding of HMM to actin-caldesmon would be explained if inhibited ADP release rates caused the high affinity AMADP complex to comprise part of the steady-state intermediate complexes. Second, the rate of ADP release is a likely determinate of shortening velocity and inhibition of the ADP release rate would account in part for the slower shortening velocity observed in latch. However, direct measurements of the effect of caldesmon on the rate of ADP release failed to show any inhibition (Sellers and Hathaway, unpublished). It is more likely that the increased affinity is due to tethering of HMM to actin by bifunctional caldesmon binding between actin and myosin[34][36]. This mechanism raises the question of whether a caldesmon-induced tethering of actin and myosin could be responsible for either the decrease in shortening velocity or the phosphorylation-independent force maintainance during latch (Fig. 5). Okagaki et al.[27] showed that caldesmon can inhibit actin filament translocation in the sliding actin assay and we have preliminary data to suggest the same[37]. The inhibition can

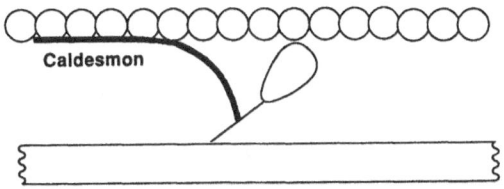

Fig. 5. Schematic representation of a possible tethering role of caldesmon between actin and myosin filaments.

be reversed by the presence of calcium and calmodulin[27]. Since the *in vitro* motility assays do not directly give a measure of force, we cannot determine whether the observed decline in sliding velocity is due to an internal load imposed by a caldesmon-mediated tethering of actin and myosin filaments. It should be possible to test this behavior using a carboxyl-terminal fragment of caldesmon that does not contain the myosin-binding domains. It would be predicted that this fragment should inhibit movement by competing with myosin binding sites on actin and, therefore, result in dissociation of actin from the surface.

Tropomyosin alone is able to modulate the velocity of actin sliding[20)27)] and actin-activated MgATPase activity[38] (Table 2). The ATPase results show that tropomyosin can be "pushed" into an even more activating position by the addition of rigor myosin heads in the form of N-ethyl-maleimide-modified subfragment-1 (NEM S-1). The effect of NEM S-1 addition to sliding actin filaments is not yet known, but if it gives rise to a further two-fold increase in the rate of sliding, then potentially a six-fold range of velocities could be obtained by positioning of the tropomyosin on the actin filament. This magnitude of change might be sufficient to account for the differences in shortening velocity between the two extremes in latch. Perhaps a role of caldesmon or calponin is to control the position of tropomyosin similar to the way in which troponin functions in skeletal muscle[39].

Table 2. Effect of Tropomyosin on the Actin-activated MgATPase Activity and Sliding Velocity of Smooth Muscle Myosin

Sample	MgATPase Activity (s^{-1})		Velocity of Sliding+ ($\mu m/s$)
	(P HMM)	(deP HMM)	(P myosin)
Actin	0.27	0.014	0.269 ± 0.033
Actin-TM	1.26	0.010	0.735 ± 0.081
Actin-TM+NEM S-1	2.00	0.011	ND

+Data taken from Umemoto and Sellers[20].

The actin concentration in the ATPase assays was 10 μM.

When NEM S-1 was added to a ratio of 1 NEM S-1 per 3 actin monomers, the concentration of free actin monomers was maintained at 10 μM.

274 J.R. Sellers et al.

High concentrations of calmodulin are required for reversal of caldesmon inhibition *in vitro* which call into question as to whether this is the physiological mechanism[27)35)]. There may exist an as yet unidentified calcium binding protein to interact with caldesmon. Alternatively, caldesmon has been shown to be phosphorylated by a number of kinases[7)8)]. Kinases that phosphorylate the carboxy-terminal region appear to abolish inhibition by reducing the affinity of caldesmon for actin[40)]. Kinases that phosphorylate the amino-terminal region similarly reduce the affinity of caldesmon for myosin[36)]. Perhaps these phosphorylations regulate the interaction of caldesmon with myosin and actin *in vivo*.

The Slowly Dissociating Dephosphorylated Crossbridge Model

This model is widely referred to as the "latch bridge model" and has received the most attention[13)]. It is based on a very specific kinetic model which postulates that if an active cross-bridge is dephosphorylated while attached to actin, its dissociation rate constant decreases by a factor of 10. Using this assumption and other estimates such as the kinase and phosphatase activity during the two phases of smooth muscle contraction, and cross-bridges association and dissociation rates, the Hai and Murphy[13)] kinetic model sucessfully explains many of the phenomena associated with smooth muscle contraction including the steep dependence of force on phosphorylation, the relationship between shortening velocity and phosphorylation and the reasons for the very high kinase and phosphatase activities that occur during contraction. The ability of dephosphorylated myosin to slow the movement of phosphorylated myosin[24)26)] has been taken as evidence for support of this model, however, in these experiments the phosphorylation state is static and myosin is not being dephosphorylated while attached to actin. Therefore, those experiments do not adequately test the model. We sought to directly test this model *in vitro*. To do this, we first performed experiments in which either phosphorylated or unphosphorylated smooth muscle myosin was mixed with skeletal muscle HMM (Fig. 6). We then set up an experimental condition where there was 70% skeletal HMM and 30% phosphorylated myosin (i.e where the velocity was about 30% of the maximum) and included myosin light chain kinase, calmodulin, calcium and the catalytic subunit of a myosin light chain phosphatase (gift of Dr. Paul Wagner, NCI, NIH) along with the normal constituents. If latch bridges where produced the velocity would be expected to decline due to the slower detachment rate constant proposed according to the Hai and Murphy model[13)] The ratios of kinase to phosphatase was varied over a wide range. In none of the cases did the velocity fall significantly. While this experiment is preliminary and needs more controls, the implication is that there was no evidence for slowly dissociating, dephosphorylated latch bridges in our *in vitro* assay.

In summary, the *in vitro* studies have as of yet to produce any conclusive evidence for the mechanism of the latch state. However, it is clear that sliding velocity of actin filaments can be modulated by several mechanisms including the presence or absence of tropomyosin or caldesmon on the thin filament and the ratio

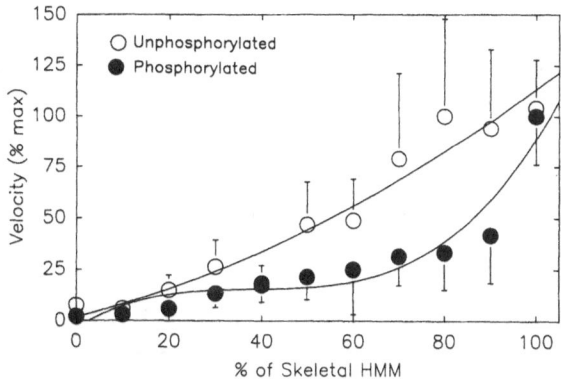

Fig. 6. Effect of phosphorylated and unphosphorylated smooth muscle myosin on the velocity of skeletal muscle HMM. Smooth muscle myosin and skeletal muscle HMM were mixed in varying molar ratios.

of phosphorylated to unphosphorylated myosin. Other hypothetical modulators such as direct calcium binding to myosin and phosphorylation of myosin at sites other than serine-19 appear to be ruled out.

REFERENCES

1. Kargacin, G. J., Cooke, P.H., Abramson, S.B. & Fay, F.S. *J. Cell Biol.* **108**, 1465-1475 (1989).
2. Cooke, P.H., Fay, F.S. & Craig, R. *J. Muscle Res. Cell Motility* **10**, 206-220 (1989).
3. Marston, S.B. & Taylor, E.W. *J. Mol. Biol.* **139**, 573-600 (1980).
4. Sellers, J.R. *J. Biol. Chem.* **260**, 15815-15819 (1985).
5. Hartshorne, D.J. in *Physiology of the Gastrointestinal Tract* (ed Johnson, L.R.) Vol. 2nd, 423-482 (Raven Press, New York, 1987).
6. Sellers, J.R. *Curr. Opin. Cell Biol.* **3**, 98-104 (1991).
7. Marston, S.B. & Redwood, C.S. *Biochem. J.* **279**, 1-16 (1991).
8. Sobue, K. & Sellers, J.R. *J. Biol. Chem.* **266**, 12115-12118 (1991).
9. Winder, S.J. & Walsh, M.P. *J. Biol. Chem.* **265**, 10148-10155 (1990).
10. Abe, M., Takahashi, K. & Hiwada, K. *J. Biochem. (Tokyo)* **108**, 835-838 (1990).
11. Sellers, J.R., Eisenberg, E. & Adelstein, R.S. *J. Biol. Chem.* **257**, 13880-13883 (1982).
12. Greene, L.E. & Sellers, J.R. *J. Biol. Chem.* **262**, 4177-4181 (1987).
13. Hai, C.-M. & Murphy, R.A. *Ann. Rev. Physiol.* **51**, 285-298 (1989).
14. Kamm, K.E. & Stull, J.T. *Ann. Rev. Physiol.* **51**, 299-313 (1989).
15. Dillon, P.F., Aksoy, M.O., Driska, S.P. & Murphy, R.A. *Science* **211**, 495-497 (1981).
16. Sheetz, M.P. & Spudich, J.A. *Nature* **303**, 31-35 (1983).
17. Kron, S.J. & Spudich, J.A. *Proc. Natl. Acad. Sci. USA* **83**, 6272-6276 (1986).
18. Harada, Y., Noguchi, A., Kishino, A. & Yanagida, T. *Nature* **326**, 8005-808 (1987).
19. Homsher, E., Wang, F. & Sellers, J.R. *Am. J. Physiol.* in press, (1991).
20. Umemoto, S. & Sellers, J.R. *J. Biol. Chem.* **265**, 14864-14869 (1990).
21. Wagner, P.D. & Vu, N.D. *J. Biol. Chem.* **262**, 15556-15562 (1987).
22. Wagner, P.D. & Vu, N D. *J. Biol. Chem.* **261**, 7778-7783 (1986).

23. Trybus, K.M. *J. Cell Biol.* **109**, 2887-2894 (1989).
24. Sellers, J.R., Spudich, J.A. & Sheetz, M.P. *J. Cell Biol.* **101**, 1897-1902 (1985).
25. Umemoto, S., Bengur, A.R. & Sellers, J.R. *J. Biol. Chem.* **264**, 1431-1436 (1989).
26. Warshaw, D.M., Desrosiers, J.M., Work, S.S. & Trybus, K.M. *J. Cell Biol.* **111**, 453-463 (1990).
27. Okagaki, T., Higashi-Fujime, S., Ishikawa, R., Takano-Ohmuro, H. & Kohama, K. *J. Biochem. (Tokyo)* **109**, 858-866 (1991).
28. Ikebe, M., Hartshorne, D.J. & Elzinga, M. *J. Biol. Chem.* **261**, 36-39 (1986).
29. Bengur, A.R., Robinson, E.A., Appella, E. & Sellers, J.R. *J. Biol. Chem.* **262**, 7613-7617 (1987).
30. Ikebe, M., Hartshorne, D.J. & Elzinga, M. *J. Biol. Chem.* **262**, 9569-9573 (1987).
31. Ikebe, M., Koretz, J. & Hartshorne, D.J. *J. Biol. Chem.* **263**, 6432-6437 (1988).
32. Nishikawa, M., Sellers, J.R., Adelstein, R.S. & Hidaka, H. *J. Biol. Chem.* **259**, 8808-8814 (1984).
33. Chalovich, J.M., Cornelius, P. & Benson, C.E. *J. Biol. Chem.* **262**, 5711-5716 (1987).
34. Hemric, M.E. & Chalovich, J.M. *J. Biol. Chem.* **265**, 19672-19678 (1990).
35. Lash, J.A., Sellers, J.R. & Hathaway, D.R. *J. Biol. Chem.* **261**, 16155-16160 (1986).
36. Sutherland, C. & Walsh, M.P. *J. Biol. Chem.* **264**, 578-583 (1989).
37. Shirinsky, V., Birukov, K.G., Hettasch, J.M. & Sellers, J.R. *J. Biol. Chem.* **267**, 15886-15892 (1992).
38. Chacko, S. & Eisenberg, E. *J. Biol. Chem.* **265**, 2105-2110 (1990).
39. Williams, D.L.,Jr. & Greene, L.E. *Biochemistry* **22**, 2770-2774 (1983).
40. Mak, A.S., Watson, M.H., Litwin, C.M.E. & Wang, J.H. *J. Biol. Chem.* **266**, 6678-6681 (1991).

Discussion

Pfitzer: Would you suggest that caldesmon can also support tension when it cross-links actin and myosin in the scheme you showed? This would be the other requirement for the latch state, and I guess you are suggesting that caldesmon regulates the latch state.

Sellers: I would think that if caldesmon is able to slow it down, it is most likely imposing a load. I don't have any means of calibrating how much of a load or how much tension is being supported per caldesmon, but I do think it is imposing some load. We can test this by trying to make the C-terminal fragment alone and repeating the assays. That is what we will be doing next.

Katayama: All the data you showed included the inhibition of movement by calponin or caldesmon, but without tropomyosin. What happens with tropomyosin?

Sellers: Actually, tropomyosin was present in both cases. It has no effect on the inhibition by calponin. Calponin will inhibit just as well with or without tropomyosin. Caldesmon will also inhibit without tropomyosin, but caldesmon concentrations about five times higher are required to obtain the same degree of inhibition.

Pfitzer: What was the free magnesium concentration when you looked at the calcium effect on motility?

Sellers: The free magnesium concentration was 4 mM in all experiments.

Pfitzer: The free magnesium concentration may be important. That calcium directly affects unloaded shortening velocity and depends on the free magnesium concentration was shown by Hartshorne and colleagues (Barsotti et al. *Am J. Physiol.* **252**, C543-C554, 1987).

Sellers: Hartshorne's study did show some effect. We should repeat our studies at lower free magnesium concentrations in order to be consistent with their work.

Reedy: Did the assay with *Lethocerus* myosin involve rabbit actin?

Sellers: Yes, and gizzard smooth muscle tropomyosin.

Sellers: We also found that *Limulus* muscle is phosphorylation-dependent and tropomyosin from *Limulus* and gizzard are equally effective in enhancing the movement in that assay.

Reedy: To what extent did you achieve the de-phosphorylation that made Lethocerus myosin totally inactive?

Sellers: We incubated the myosin with alkaline phosphatase overnight.

FACTORS AFFECTING FILAMENT VELOCITY IN *IN VITRO* MOTILITY ASSAYS AND THEIR RELATION TO UNLOADED SHORTENING VELOCITY IN MUSCLE FIBERS

Earl Homsher*, Fei Wang and James Sellers

Section of Cellular and Molecular Motility
Laboratory of Molecular Cardiology
Building 10, Rm. 8N202, NIH
Bethesda, Maryland 20892
*Department of Physiology
School of Medicine, UCLA
Los Angeles, California, 90024

ABSTRACT

The measurement of fluorescently labeled actin filament movement driven by mechanoenzymes (e.g. myosin) is an important methodology for the study of molecular motors. It is assumed that the filament velocity, V_f, is analogous to the unloaded shortening velocity, V_u, seen in muscle fibers. To evaluate this assumption we compared V_f to literature values for V_u with regard to the effects of [ATP], [ADP], [Pi], pH, ionic strength (10-150 mM) and temperature (15-30°C). V_f and V_u are quantitatively similar with respect to the effects of substrate and product concentrations and temperatures > 20°C. However, V_f is more sensitive to decreases in pH and temperatures < 20°C than is V_u. At ionic strengths of 50-150 mM, V_f and V_u exhibit similar ionic strength dependencies (decreasing with ionic strength). At ionic strengths < 50 mM, V_f is markedly reduced. Thus while V_f is a good analogue for V_u under certain conditions (elevated ionic strength and temperatures > 20°C), under others it is not. The results of motility assays must be cautiously interpreted.

INTRODUCTION

The reactions shown in Fig. 1 are generally accepted as those which produce force, shortening, and work in muscle[1]. In this model chemical states involving AM•ATP and AM•ADP•Pi are thought to be "weakly" attached; i.e., detaching and attaching rapidly (ca. > 1000 s^{-1}) and incapable of generating or sustaining significant force under normal physiological conditions. AM*•ΛDP, and A*M are "strongly" attached states which remain attached for a significant time and produce

Mechanism of Myofilament Sliding in Muscle Contraction, Edited by
H. Sugi and G.H Pollack, Plenum Press, New York, 1993

279

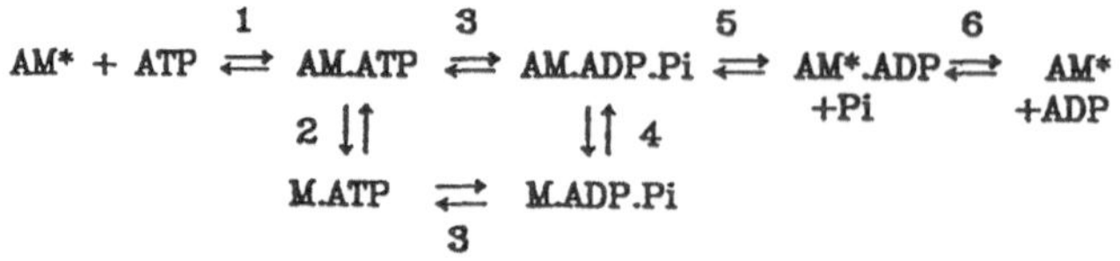

$$
\begin{array}{ccccccccc}
 & & \overset{1}{} & & \overset{3}{} & & \overset{5}{} & & \overset{6}{} \\
AM^* + ATP & \rightleftarrows & AM.ATP & \rightleftarrows & AM.ADP.Pi & \rightleftarrows & AM^*.ADP & \rightleftarrows & AM^* \\
 & & 2 \updownarrow & & \updownarrow 4 & & +Pi & & +ADP \\
 & & M.ATP & \rightleftarrows & M.ADP.Pi & & & & \\
 & & & 3 & & & & &
\end{array}
$$

Fig. 1. The mechanism of MgATP hydrolysis by actin (A) and myosin (M). The asterisk (*) indicates a strongly attached or force bearing state.

or bear force. The rates of transition between these states are thought to depend on the intracellular milieu and on the strain on the crossbridge, and it is assumed that maximal shortening velocity is primarily limited by the rate at which the attached crossbridge can dissociate from the sliding thin filament; i.e., reactions 6 and 1 in Fig. 1[1-5], and reaction 5 is thought to be related to force production[3)4)6]. Current muscle research seeks to characterize the factors that affect the rate constants in this reaction scheme and elucidate the relationship between these states and the structural and mechanical changes which produce contraction.

In vitro motility assays[7-9] allow greater control of the contractile protein composition and chemical environment than is possible in studies of whole or demembranated single fibers. One can genetically or chemically modify or replace contractile proteins to examine the sequelae of these changes on a contractile analog, the velocity of movement (V_f) of fluorescently labeled actin filaments. It is generally assumed that motility assays represent an accurate paradigm of unloaded shortening velocity. This assumption is more an article of faith than demonstrated fact.

The work described below aimed to test this assumption by comparing the effects of ionic strength, temperature, [MgATP], [MgADP], and [Pi] on the V_f produced by skeletal HMM in an effort to learn how closely V_f approximates the unloaded shortening velocity (V_u) observed in muscle fibers. We find that both ionic strength (10-150 mM) and pH have more profound effects on V_f (velocity increases with ionic strength and pH) than on V_u, that the activation energy for V_f is greater than that for skeletal muscle V_u, and that the effects of MgATP, MgADP, and Pi are quantitatively similar to those reported for skeletal muscle.

METHODS

Rabbit skeletal heavy meromyosin (HMM) and actin labelled with rhodamine-phalloidin were prepared as previously described[10]. The preparation of the assay chamber, the visualization of the labelled actin filaments, and the thermostating of the reaction chamber were also as described earlier[10]. In most experiments, the following basic conditions were maintained in the Assay Buffer: 1 mM MgATP, 1 mM Mg^{2+}, 10 mM MOPS, 0.1 mM K_2EGTA, pH, 7.35. Ionic strength, I, (computed after calculation of the concentrations and charge of the various ion species) was varied by changing the KCl concentration. When ATP, ADP or Pi concentrations were varied, $MgCl_2$ and KCl concentrations were altered to maintain the basic conditions above and the ionic strength. In the presence of ADP or low concentrations of ATP, P^1,P^5-di(adenosine-5')pentaphosphate (AP5A, Sigma Co.) (250 μM) was added to prevent ADP and ATP concentrations from changing. Solution compositions were computed using a QuickBasic computer program based on that of Fabiato and Fabiato[11] written by Neil Millar and E. Homsher.

Image Analysis

The goal of these studies was to measure the velocity of labeled actin filaments movement under different experimental conditions. Typically in our system the fraction of fluorescent actin filaments moving varied between 80-95%. However, as reported by others[8] much lower percentages were observed if rigor-like heads were not removed by a centrifugation against F-actin and ATP. A reduced percentage was also associated with other experimental variables (e.g. ionic strength, temperature, etc; described below). For filament velocity measurements, we used an Expert-Vision motion analysis system (Motion Analysis Corp, Santa Rosa, Ca)[10]. Using this system it is possible to monitor and quantitate the movement of every filament in the visual field; i.e., optical fields are acquired at pre-determined rates, stored in the computer's memory, and a centroid calculated for each filament in the field. By recording the position of the filament centroids in successive fields, the velocity for each filament is calculated. There are a variety of different ways in which the velocity for a given experimental condition could be reported. After examining a number of options we decided to calculate filament velocity by averaging the velocity for only those filaments that exhibited *uniform* movement (i.e., the S.D. for the mean of a given filament must be less than 0.3 of its mean velocity). This criterion was adopted because sarcomere shortening exhibits uniform shortening. Further, this method eliminates those filaments which do not move, those which start and stop, and those exhibiting pin-wheel or contorted

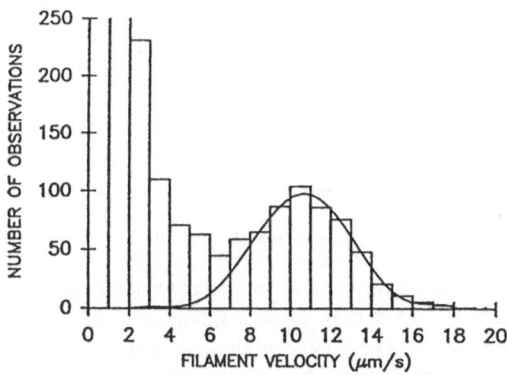

Fig. 2. A histogram of the number of filaments moving at an average velocity for conditions given in text. The solid line is a Gaussian fit to the data. Its mean and S.D. deviation is within 10% of those computed for uniformly moving filaments.

motion. It does not however bias for or against fast or slow velocities. Figure 2 shows the distribution of all filaments moving at 25°C and 75 mM ionic strength. There is a Gaussian distribution of the higher velocities, and the mean and S.D. of a Gaussian fit to this data yields values not different from those computed by selecting for uniformity of movement. Thus consistent data are obtained using this approach to compute the filament velocity.

RESULTS

The variables we examined were the [MgATP], [MgADP], [Pi], pH, ionic strength and temperature.

The Effect of MgATP Concentration on Velocity

Earlier workers showed that unloaded shortening velocity or filament velocity increases in proportion to the ATP concentration in both muscle fibers[4][12] and in *in vitro* motility assays respectively[8][13-15]. We repeated these experiments using HMM at an ionic strength of 50 mM and 25°C over a range of [MgATP] from 7.8 μM to 5 mM with the results shown in the inset to Figure 3. Here the solid line represents a least-squares fit to the equation:

$$V_f = V_{fmax}/(1 + K_{mapp} [MgATP])$$

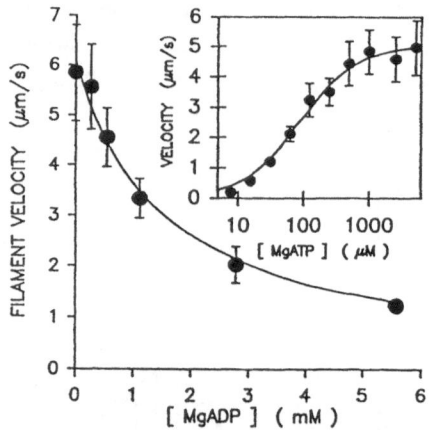

Fig. 3. The effect of MgADP on the mean actin filament velocity produced by c-HMM at 27°C and 50 mM ionic strength. The data is given as the mean ± S.D. and each data point is the average of 14 - 70 filaments. The MgADP concentration was varied while holding the MgATP concentration at 1 mM and the free Mg^{2+} at 1 mM. The inset is a semi-log plot of the mean filament velocity (± S.D.) produced by HMM versus the concentration of MgATP in the motility buffer at 25°C, an ionic strength of 50 mM. The lines in the plots are least-squares fit to the data.

where V_{fmax} is the filament velocity at saturating MgATP and K_{mapp} is the apparent K_m for filament velocity. The data is well fit by the equation and K_{mapp} (88 µM) is in reasonable agreement with data from earlier studies[7][8][13] and that from rabbit psoas muscle fibers at 10°C where the apparent K_m for V_u is 150 µM MgATP[4].

The Effect of [MgADP] Filament Velocity

Studies of unloaded shortening velocity in glycerinated rabbit psoas muscle fibers[4][6] at 10°C have shown that MgADP inhibits V_u with an apparent inhibition constant, K_{iapp}, of 200-300 µM. Figure 3 from experiments using HMM at 25°C and 50 mM ionic strength shows that the data is well fit using a K_{iapp} of 120 µM.

The Effect of Inorganic Phosphate on Filament Velocity

An increase in inorganic phosphate concentration, [Pi], to 10 mM is known to increase the unloaded shortening velocity of rabbit psoas fibers by about 10 %[17] In our experiments the total [Pi] was raised to 11.4 mM (free Pi = 10 mM). At 50 mM ionic strength, and in two different preparations of HMM, the filament velocity increased by an average of 18.5% (361 filaments). To be certain that the increased V_f was a consequence of the [Pi] increase and not the removal of Cl⁻ from the assay medium, KCl was replaced by K-acetate in the control solution. No significant change in V_f was observed.

The Effect of pH on Filament Velocity

A reduction in intracellular pH is thought to be a significant factor in muscle fatigue and earlier studies of the motility assay system showed that V_f is sensitive to

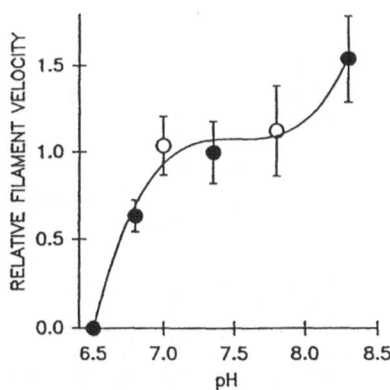

Fig. 4. The effect of pH on V_f. The data is from two different HMM preparations (given by the different symbols. In each case this V_f at pH 7.35 was assigned the value 1 and the V_f observed at different pH's were expressed relative to the appropriate value. The line is a third order best-fit polynomial.

pH[8)13)]. Figure 4 shows the results from two HMM preparations in which pH was varied. The data show that as the filament velocity slows as the pH is reduced below the control value of 7.35, falling to zero at 6.5 while filament velocity increases markedly and at 8.3 is 50% greater than that at 7.35. Although this type of behavior agrees with earlier reports[8)13)] and is in the same direction as that seen in the intact muscle fiber[18)], pH exerts a greater effect on V_f than V_u (see Discussion).

The Effect of Ionic Strength on Filament Velocity

While ionic strength exerts a profound effect on isometric tension in skeletal muscle, it has relatively little effect on unloaded shortening velocity in the range of 0.1 to 0.25 M[19)]. At ionic strengths < 100 mM unloaded shortening velocity declines[19)]. Earlier studies[7)8)20)] of the effects of ionic strength have shown that as the ionic strength is raised to 50 mM, actin filaments tend to dissociate from the mechanoenzyme coated surface and V_f falls. Previous studies at ionic strengths less than 40 mM yielded inconsistent results[7)8)14)15)20)21)] as ionic strength is lowered. To examine the role ionic strength plays in determining V_f, ionic strength and temperature were varied between 10-60 mM and 15-30°C respectively. The results of these experiments are shown in Figure 5.

V_f increases with increasing ionic strength at all temperatures studied (15-30°C). The quality of the filament movement becomes more erratic at higher ionic strengths and *fewer* filaments remain attached to the surface for a long duration of time. Thus at 25°C and 60 mM ionic strength, only 10.9% of the filaments moved a distance of more than 10 μm at a uniform velocity (S.D./mean < 0.3), but at 30 mM this value increased to 34.8%. At higher ionic strengths filaments do not attach to the surface making it impossible to compare the effects of ionic strength on filament velocity to unloaded shortening velocities in muscle fibers. Similar behavior is seen

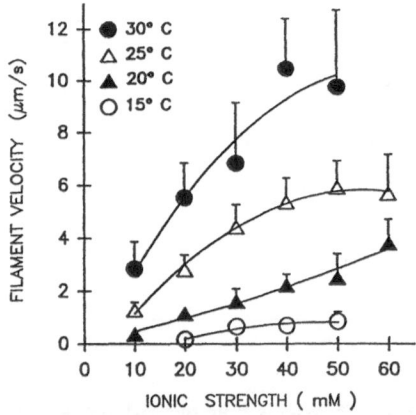

Fig. 5. The effect of ionic strength and temperature on actin filament velocity produced by HMM. Values given are the mean ± S.D. The lines through the data points are second order polynomial least-square fits to the data.

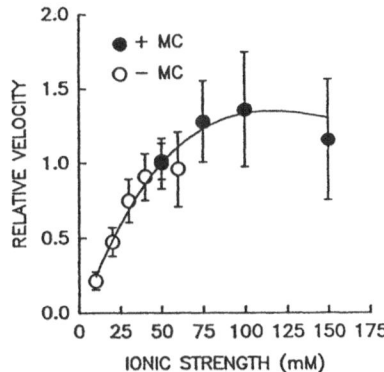

Fig. 6. The effect of ionic strength on relative filament velocity. The data is given as the mean ± S.D. The data is taken from two different HMM preparations. In the first, given by open circles, the data is that with no added methylcellulose and is from Fig.7 at 25°C. In that case V_f at 50 mM was 5.90 ± 1.02 μm/s (n = 29). In the second , which was performed at 25°C the V_f at 50 (without methylcellulose) mM was 4.22 ± 0.69 μm/s (n = 77). In both instances this value was assigned a value of 1.0 and the V_f at the other ionic strengths were referenced to this velocity. The filled circles indicate measurements in the presence of methylcellulose.

at each temperature examined. Thus the weak binding may become too weak to permit filaments to remain bound to the surface coated with myosin heads. Uyeda et al[22] used methylcellulose (0.7% w/w) addition to the assay buffer to increase viscosity and reduce the rate of lateral diffusion of filaments from the surface. We used this approach to examine the movement of filaments at higher and more physiological ionic strengths. Figure 6 shows that the addition of 0.7% methylcellulose had no effect on V_f at 50 mM, but V_f rose with ionic strength to 100 mM where 65% of the observed filaments moved at a uniform rate. However by 150 mM, V_f had begun to drop and the fraction of filaments moving at a uniform velocity had declined to 32%. Thus V_f approaches a maximal value at physiological ionic strengths.

When the ionic strength is lowered to less than 40 mM (in the absence of methylcellulose), V_f decreases markedly. The velocities obtained at a given temperature at ionic strengths less than 40 mM are significantly different from each other. At 10 mM, the velocities observed at 30, 25, 20, and 15°C were respectively 27.3%, 24.1%, 16.1%, and 0% of the velocities obtained at 40 mM. Therefore reduced ionic strength slows V_f.

The Effect of Temperature

The data contained in Figure 6 was used to construct Arrhenius plots to examine the effect of temperature on V_f. Because the data at the various ionic strengths exhibit similar behavior with respect to temperature, only data obtained at 20 mM is shown in Figure 7. The activation energy E_a from linear regression of log(velocity) against 1/K at 20, 30, 40 and 50 mM yielded values averaging 128 ± 18 kJ/mole

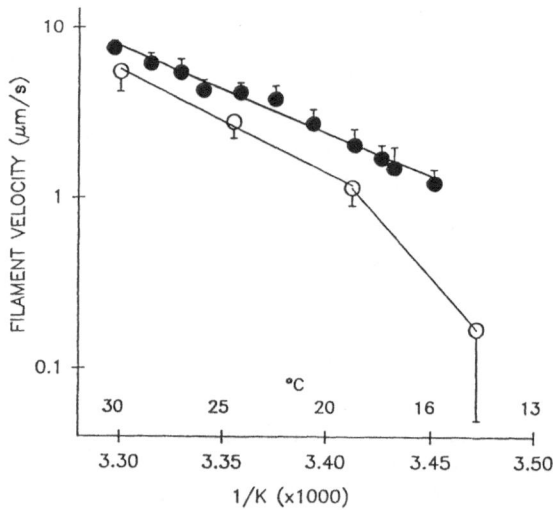

Fig. 7. The effect of temperature on filament velocity at ionic strengths of 20 and 50 mM. The data at 20 mM is from Fig.7. The filled circles are data taken at 50 mM ionic strength at about 1.5°C increments on a different HMM preparation to that used for the remaining data. The lines are linear regressions to fit the data.

which is close to the E_a (103 ± 21 kJ/mole) reported for the actin-activated myosin ATPase V_{max} recorded between 15-25°C[23]. These activation energies exceed the values observed for V_u[24][25]. Further, the plot at 20 mM ionic strength suggests the presence of a change in E_a at 20°C; over the range 20-30°C, the activation energy is 115 ± 8 kJ/mole and 269 kJ/mole from 15-20°C. To more carefully examine the behavior of V_f with temperature, measurements of V_f were made at ca. 1.5°C intervals in a temperature range of 16.7-30.3°C at 50 mM as seen in Fig. 7. A regression against all the data is well fit by a straight line with an E_a of 98 ± 5 kJ/mole ($r^2 = 0.996$).

DISCUSSION

Our basic conclusion is that in certain cases (at higher temperature, ionic strength, pH, and the effects of [MgATP], [MgADP], and [Pi]), the agreement between V_f and V_u is excellent. The similarity of V_f and V_u at higher ionic strength (permitted by the use of methylcellulose), higher temperature (> 20°C) and pH(> 7.3) suggests these conditions should be used when studying the mechanism of normal mechanoenzyme function. However, there are conditions (lower temperature, ionic strength, and pH) in which the correlation between V_f and V_u fails. These observations suggest that motility assays must interpreted cautiously. Our inability to account for these changes shows that much remains to be learned of the factors that control filament movement. Further since the effects of ionic strength, pH, and temperature are fully reversible, the structural arrangement of the mechanoenzymes on the assay surface, the interaction of thin filaments with the surface itself or the BSA used to coat the myosin free surfaces, the flexibility of proteins, and/or charge distributions may play a role in the behavior of the motility assay.

The Effect of Temperature

While a direct comparison of V_f to V_u in rabbit psoas fibers is not possible (a careful study of the temperature sensitivity of V_u has not been performed). Data available from living muscles shortening over a temperature range 4-35°C yield an E_a ranging from 44-69 kJ/mole[18)24)25)]. Where direct comparisons are available V_u and V_f are *not* the same; e.g., at 12°C, for glycerinated rabbit psoas muscle V_u is 2.07 muscle lengths/s[26)] which corresponds to 2.3 μm/s per half-sarcomere. In our experiments V_f was 0.59 ± 0.13 μm/s at 11.6°C (n = 14). This value is significantly less than the fiber value. However in whole muscles, V_u at 20-25°C is ca. 3-6 μm/s at 20-25°C[24)25)] which compares well to the values observed here and else where[27)]. These data suggest the need for more direct comparisons of V_u and V_f. Apparent differences could stem from differences in the source of myosin, the lack of regulatory proteins on the thin filaments, or most probably differences in ionic strength.

The Effect of pH

In muscle fibers, work has centered on pH values of 7 or less because the intracellular pH is 6.9-7.0 under normal conditions and during muscular fatigue pH declines[18)]. Both isometric force and V_u decline with a fall in pH at 10°C. Cooke et al[18)] obtained a roughly linear relationship between pH and V_u; i.e., at pH 7, V_u was 1.63 muscle lengths/s but fell to 1.23 muscle lengths/s at pH 6.0. Thus HMM is more sensitive to pH ($V_f = 0$ μm/s at pH 6.5) than are skeletal fibers.

The Effect of Ionic Strength

The data in Figs. 7 and 8 show that V_f is remarkably sensitive to ionic strength at values less than 100 mM. Comparable experiments have not been performed in glycerinated rabbit psoas fibers because it is difficult to maintain the ADP level in the μmolar range without the use of PCr which contributes significantly to ionic strength. However, in skinned frog muscle fibers, Gulati and Podolsky[19)] find that V_u is independent of ionic strength between 100-260 but falls to about 55% of control values when the ionic strength was reduced to 50 mM. This ionic strength dependence in muscle fibers similar to that observed in the motility assay. Possible mechanisms for this ionic strength dependence include: 1) reduction in ionic strength may increase the number and affinity of weakly attached crossbridge states so as to exert a retarding force and slow filament movement[15)25)]; 2) thin filaments may interact electrostatically with the slide surface or the BSA coating it. If so, elevated ionic strength will tend to screen, and thus reduce, this type of interaction; 3) the reduction of ionic strength may reduce the rate of ATP binding to the AM state and/or decrease the rate of ADP release from AM•ADP. In any case, the lack of an explanation for the data emphasizes our lack of understanding of the factors which determine the filament velocity in the *in vitro* motility assay.

ACKNOWLEDGEMENTS

We are grateful to Estelle Harvey, William Anderson, and Nick Richiutti for technical assistance. F.W. is a Fogarty post-doctoral scholar. This work was supported in part by a National Institutes of Health grant AR 30988 to E.H. Figures 2, 3, 4, 5, 6 and 7 are from ref. 10.

REFERENCES

1. Homsher, E. & Millar, N. *Ann. Rev. Physiol.* **52**, 875-896 (1990).
2. Brenner, B., M. Schoenberg, M., Chalovich, J.M., Greene, L.E. & Eisenberg, E. *Proc. Natl. Acad. Sci. USA* **79**,7288-7291 (1982).
3. Hibberd, M., Dantzig, J.A., Trentham, D.R. & Goldman, Y.E. *Science* **228**, 1317-1319 (1985).
4. Pate, E. & Cooke, R. *J. Muscle Res. Cell Motility* **10**,181-196 (1989).
5. Siemankowski, R. F, Wiseman, M.O. & White, H.D. *Proc. Natl. Acad. Sci. USA* **82**,658-662 (1985).
6. Cooke, R. & Pate, E. *Biophys. J.* **48**, 789-798 (1985).
7. Harada, Y., Noguchi, A., Kishino, A. & Yanagida, T. *Nature* **326**,805-808 (1987).
8. Kron, S.J. & Spudich, J.A. *Proc. Natl. Acad. Sci. USA* **83**,6272-6276 (1986).
9. Sheetz, M. P. & Spudich, J.A. *Nature* **303**, 31-35 (1983).
10. Homsher, E., Wang, F. & Sellers, J. *Am. J. Physiol. (Cell)* in press
11. Fabiato, A. & Fabiato, F. *J. Physiol. (Paris)* **75**, 463-505 (1979).
12. Ferenczi, M.A., Goldman, Y.E. & Simmons, R.M. *J. Physiol. (Lond.)* **350**, 519-543 (1984).
13. Sheetz, N.P., Chasan, R. & Spudich, J.A. *J. Cell Biol.* **99**,1867-1871 (1984).
14. Umemoto, S. & Sellers, J.R. *J. Biol. Chem.* **265**, 14864-14869 (1990).
15. Warshaw, D.M., Desrosiers, J.M., Work, S.S. & Trybus, K.M. *J. Cell Biol.* **111**, 453-463 (1990).
16. Eisenberg, E. & Keilley, W. *J. Biol. Chem.* **249**, 4742-4748. (1974)
17. Pate, E., Nakamaya, K., Franks-Skiba, K., Yount, R.G. & Cooke, R. *Biophys. J.* **59**, 598-605 (1991).
18. Cooke, R., Franks, K., Luciani, G. & Pate, E. *J. Physiol. (Lond.)* **395**, 77-97 (1988).
19. Gulati, J. & Podolsky, R.J. *J. Gen. Physiol.* **78**, 233-257 (1981).
20. Takiguchi, K., Hayashi, H., Kurimoto, E. & Higashi-Fujime, S. *J. Biochem.* **107**, 671-679 (1990).
21. Vale, R.D. & Oosawa, F. *Adv. Biophys.* **26**,97-134 (1990).

22. Uyeda, T.Q.P., Kron, S.J. & J.A. Spudich, J.A. *J. Mol. Biol.* **214**, 699-710 (1990).
23. Tesi, C., Kitagishi, K., Travers, F. & Barman, T. *Biochemistry* **30**, 4061-4067 (1991).
24. Ranatunga, K.W. *J. Physiol. (Lond.)* **329**, 465-483 (1982).
25. Stein, R.B., Gordon, T. & Shriver, J. *Biophys. J.* **40**, 97-107 (1982).
26. Sweeney, H.L., Kushmerick, M.J., Mabuchi, K., Sreter, F. & Gergely, J. *J. Biol. Chem.* **263**, 9034-9039 (1988).
27. Toyoshima, Y.Y., Kron, S.J. & Spudich, J.A. *Proc. Natl. Acad. Sci. USA* **87**, 7130-7134 (1990).

Discussion

Rall: How do the velocities you see compare with the absolute velocity one sees in fibers?

Homsher: If we are near 20-25°C, the argument is reasonable in that the velocity one sees in fibers is about 3-6 μm/sec [Ranatunga, *J. Physiol. (Lond.)* **329**, 465-483]. Where we have a bigger problem is at temperatures of 10-15°C and where the filament sliding velocity is only about 0.3-0.4 μm/sec. At such temperatures, the filament velocity in skinned fibers is on the order of 2 μm/sec (Sweeney et al. *J. Biol. Chem.* **263**, 9034-9039, 1988).

Katayama: What kind of sub-fragments or intact myosin did you use?

Homsher: We didn't use intact myosin. We used a-chymotrypsin-generated HMM.

Katayama: So, you have no idea about the case of intact myosin?

Homsher: Yes, I do have an idea, and that's why we used HMM. At least in our system, we have difficulty maintaining a good, brisk sliding velocity when we use myosin. If we get the myosin really fresh and use it right away, we find that the HMM and the myosin have pretty much the same velocity. However, if the myosin ages a day or two, filament velocity drops and movement is erratic. Even spinning the myosin against actin in the presence of ATP to pull out dead heads doesn't seem to remedy the situation. So, at least in our hands, using a nitrocellulose surface, our preps do not demonstrate a consistently good behavior. This is the case even if the EDTA-ATPase is constant. I know Toshio Yanagida's experience is totally different.

Reedy: As you mentioned the velocities growing slower as you cooled, I found myself thinking about frog myosin, where a lot of the experiments have been done in cold muscle. Is frog myosin ever used, or has it been used, in these assays?

Homsher: It hasn't, and there's a good reason. Frog myosin, when you isolate it, becomes inactive quickly. When you look at the actin-activated ATPase of myosin or S-1, it is constantly going downhill, and one is fighting a losing battle. Also, we know that frog muscle at 0° has a shortening velocity that is virtually the same as rabbit muscle at 20°. So, as a means of comparison, that might be a way to think about it.

Gillis: Could you manipulate the ATP, ADP, pH, and so on, in order to calculate what is the minimum free energy you need for movement?

Homsher: I'm sure you could do that. It would be just a matter of doing the calculations and then setting up the conditions and seeing what happens.

Tregear: I understand very clearly how you selected which ones to look at. I am just interested to know whether the irregularly moving ones gave the impression of moving much slower or moving much faster.

Homsher: If you do a histogram plot of the ones we keep for analysis and the ones that are moving fast, they overlap remarkably. The ones that are moving irregularly are invariably slow. That makes sense—if they are stopping for a while, their velocity has to decrease.

Godt: What salt did you use to vary ionic strength?

Homsher: We used two salts. First, we normally use potassium chloride. We also did experiments in which we replaced the potassium chloride with potassium acetate; we didn't see any change in velocity.

ter Keurs: It was a pleasure to see the Q_{10} for the velocity of shortening in this assay. We measured that in cardiac muscle and found a value between 4.6 and 4.9.

Homsher: Wow! Actually, the acto-S-1 ATPase shows rather high activation energies as well. It's just the living fiber that doesn't want to play along, at least skeletal fibers.

Rall: Earl, would you care to comment about this interesting difference with regard to low pH effects on the velocity of shortening, as one sees in a fiber versus in the motility assay. What do you think is going on there?

Homsher: I don't really know, but here are a number of things we need to try. One is to increase the ionic strength and see if we observe the same behavior. We have had a start at that already, and the preliminary indication is that the effect is much less than it appears at the lower ionic strength. The other is that these are thin filaments that don't have any regulatory proteins on them. Addition of these proteins may change the behavior.

DYNAMICAL ROLE OF "PROTEIN FRICTION" IN THE SLIDING MOVEMENT OF PROTEIN MOTORS *IN VITRO*

Katsuhisa Tawada

Department of Biology
Faculty of Science
Kyushu University
Fukuoka, Fukuoka 812, Japan

ABSTRACT

When protein motors interact with a sliding cytoplasmic-filament through a weak-binding interaction (thus, without ATP splitting), this interaction cycle results in friction opposing the sliding movement. The friction is owing to the flexible nature of the heads of these motors as globular proteins. Under a certain condition, the friction becomes proportional to the sliding velocity. This viscous-like friction by protein motor is called protein friction. Since the protein friction is more than 10 times larger than the hydrodynamic viscous drag, we propose that the sliding velocity in the *in vitro* motility system is limited when the active sliding force generated by protein motors is balanced by the protein friction. The model of the protein friction hypothesis is consistent with many experimental data of the *in vitro* motility systems such as those of mixture experiments with different myosins and the ATP-concentration dependence of the sliding velocity. By relating the coefficient of the protein friction to the diffusion coefficient, we show that the model is consistent with the data on the one-dimensional Brownian movement of a microtubule on a dynein-coated glass surface in the presence of vanadate and ATP. The model also shows that the Brownian movement is driven directly by the thermally-generated structural fluctuations of the dynein heads rather than the atomic collision of solvent molecules. Thus, the model implies that the thermal structural fluctuations of the protein motor heads underlie the ATP-induced sliding movement by protein motors and hence protein motors are a Brownian actuator.

Protein Friction

Myosin, dynein and kinesin are ATPase and protein motors, which generate a force for the sliding movement with the chemical energy derived from the ATP hydrolysis by them. The head portion of these motors are globular proteins. The Young's modulus of elasticity of globular proteins such as the myosin head is $10^8 \sim$

Mechanism of Myofilament Sliding in Muscle Contraction, Edited by
H. Sugi and G.H Pollack, Plenum Press, New York, 1993

291

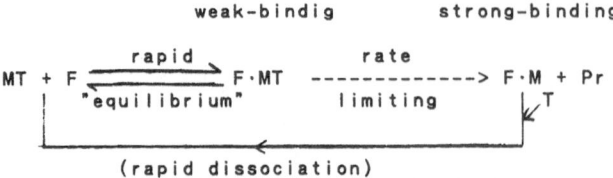

Fig. 1. Kinetic scheme of cytoplasmic filament-activated ATPase activities of protein motor enzymes. M: motor ATPase; F: cytoplasmic filament; T: ATP or ADP•Pi; Pr: ADP + Pi

10^9 (dyn/cm^2)[1],[2]), suggesting that they are relatively flexible. Their protein structure is significantly fluctuated even by thermal motion, and more significantly deformed by a force of pico Newton level[1], which force is produced by a protein motor.

Suppose a cycle of reaction in which a single protein motor attaches to its corresponding cytoplasmic filament (for example, actin in the case of myosin motor) and detaches from it without ATP splitting. Such interaction cycle takes place in the rapid "equilibrium" of the weak-binding interaction in the cytoplasmic filament-activated ATPase activity of motor enzymes (Figure 1). If there is a relative sliding movement between a motor protein and its corresponding cytoplasmic filament, such a flexible protein motor while being bound to the cytoplasmic filament is deformed by the sliding motion, and the elastic deformation energy is hence stored in the motor protein structure during this attachment (Figure 2). As soon as the motor detaches from the cytoplasmic filament, the stored deformation energy is dissipated into heat. Such attachment/detachment cycle thus

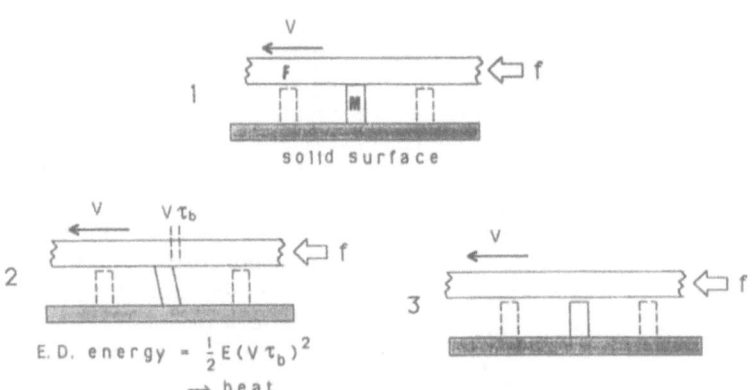

Fig. 2. Schematic model of "protein friction". (1) Protein motor (M) attachment to a cytoplasmic filament (F) that is subjected to an external force (f) and driven to side with velocity (V); (2) passive deformation of the motor head induced by the sliding movement of the filament, which results in the storage of elastic deformation (E.D.) energy in the head; (3) dissipation of the stored energy into heat after the motor dissociation from the filament. The coefficient of the protein friction, ζ_P, does not depend on the sliding velocity.

dissipates a fraction of the kinetic energy of the relative sliding motion into heat, resulting in friction. If the life time of the bound state of the motor in the cycle is constant and does not depend on the sliding velocity, which we assume here, we can show that the friction is proportional to the sliding velocity[3][4]. This viscous-like friction is called protein friction.

The protein friction exerted on a single cytoplasmic filament that is sliding with a constant velocity V is given by

$$-\zeta_P V, \tag{Eq.1}$$

where ζ_P is proportional to the length of a cytoplasmic filament, the life time of the bound state of a protein motor, and the elastic stiffness constant of the protein motor as well as the number of protein motor thus bound at a moment[3][4]. By energetic and another considerations of the protein friction, we estimate that the protein friction is more than 10 times greater than the hydrodynamic viscous drag[3][4]. Thus, we can ignore the hydrodynamic viscous drag in the following.

Velocity-Limiting Mechanism in the *in vitro* Motility

As shown by the kinetic scheme of the cytoplasmic filament-activated ATPase activity of protein motors (Figure 1), a motor enzyme repeats many cycles of attachment/detachment of the weak-binding interaction before going on along the rate-limiting step leading to the ATP hydrolysis and the sliding force generation. Thus, a fraction of the protein motor in the weakly attached state exerts the protein friction, which is a negative force, while the other fraction goes on along a rate-limiting step to generate an active sliding force. We propose that the sliding velocity of the *in vitro* movement by protein motors is limited when the active sliding force is balanced by the protein friction[3];

$$F_P = \zeta_P V,$$

and hence,

$$V = F_P / \zeta_P, \tag{Eq.2}$$

where F_P is the sliding force generated by protein motors.

Mixture Experiments with Different Myosins

Sellers et al. reported that (1) phosphorylated form of smooth muscle myosin supports the *in vitro* motility and the sliding velocity is about 10% of that by skeletal muscle myosin, (2) de-phosphorylated form of smooth muscle myosin does not support the *in vitro* motility, but (3) its addition to the phosphorylated form of smooth muscle myosin slows down the movement generated by the latter myosin[5].

Table 1. Characteristics of various types of myosin in MgATP.

Myosin	Weak-binding interaction	Protein friction	Active force	Sliding
skeletal	+	+	+	fast[5]
phosph.smooth	+	+	+	slow[5]
de-phos.smooth	+	+	−	no[5,8]
pPDM skeletal	+	+	−	no[8]

As also shown by Sellers[6], the ATPase activity of de-phosphorylated form of smooth muscle myosin is not activated, although it can be bound to actin by a weak-binding interaction[7]. Their third finding is hence consistent with our proposal of the protein friction, because de-phosphorylated form of smooth muscle myosin can be expected to exert a protein friction through the weak-binding interaction with actin. Our model can quantitatively explain the data of these mixture experiments by Seller et al.[5].

If analyzed by our model, experimental results of the three possible combinations with these three different types of myosin are not independent, and the results of one of the three combinations can be predicted by those of the other two combinations. Such three possible experiments with three different myosins were carried out by Warshaw et al.[8]. Their results are quantitatively consistent with our model[3]. Warshaw et al. also reported that pPDM-myosin of skeletal muscle does not support the *in vitro* motility but its addition to skeletal muscle myosin slows down the sliding velocity[8]. Although the ATPase activity of pPDM-myosin is not activated by actin, the myosin can bind to actin with a weak-binding interaction, so that it can exert a protein friction. This is our explanation for the suppressive effect on the motility by pPDM-myosin (Table 1).

One-Dimensional Brownian Motion of Microtubules

Vale et al. reported that microtubules that are placed on a dynein-coated solid surface undergo bi-directional Brownian motion in the presence of vanadate and ATP[9]. The diffusion coefficient experimentally determined in this system was about 10% of that of a microtubule freely suspended in a solution. Vanadate is a strong inhibitor for dynein ATPase, because an inactive ternary complex with vanadate and ADP is formed on the enzyme. The ternary complex interacts with a microtubule with a weak-binding interaction. Thus, the interaction can be expected to exert a protein friction on a moving microtubule. The coefficient of the protein friction can be related to a diffusion coefficient (D_P) by using the Einstein-Smolkofsky equation[4]:

$$D_P = k_B T / \zeta_P,$$ (Eq.3)

where k_B and T are the Bolzmann's constant and temperature. From the characteristics of ζ_P, we derive those of D_P as follows. (1) The magnitude of D_P is about 10% or less of that exerted for a microtubule suspended freely in a solution because ζ_P is more than 10 times larger than that of a hydrodynamic viscous drag; (2) D_P is inversely proportional to the length of a microtubule; (3) D_P is smaller when the interaction between a microtubule and dynein is stronger. ζ_P is proportional to the life time of the weak-binding interaction. The inverse of the life time is the dissociation rate of the bound dynein from a microtubule. Since the dissociation rate is proportional to the inverse of the solvent viscosity, ζ_P is proportional to the solvent viscosity. Hence, (4) D_P is inversely proportional to the solvent viscosity. All these four conclusions derived from our model[4] are consistent with the experimental results reported by Vale et al.[9].

The motion of a microtubule in Vale's system can be expressed by the following Langevan equation[4]:

$$m(dV/dt) = -(\zeta_S + \zeta_P)V + f_S(t) + f_P(t), \qquad (Eq.4)$$

where m is the mass of the microtubule, V is the sliding velocity, ζ_S and ζ_P are the coefficients of the solvent and protein frictions, $f_S(t)$ is the random force owing to the solvent particle, and $f_P(t)$ is the random force owing to the thermally-generated structural fluctuations of the dynein heads which are bound to the microtubule. By following the textbook physics, we can show that ζ_S is proportional to the square of the strength of f_S, and ζ_P is proportional to the square of the strength of f_P. Since $\zeta_P \gg \zeta_S$, $f_P \gg f_S$. The Brownian motion of the microtubule is hence driven directly by the thermally-generated structural fluctuations of the protein structure of the dynein heads bound to a microtubule rather than by the atomic collision of solvent particles[4].

Conclusion

Eqs. 2 and 3 are combined to give

$$V = F_P \cdot D_P/(k_B T), \qquad (Eq.5)$$

which describes the velocity of the *in vitro* sliding movement driven by protein motors. This is an equation of a diffusion process with a drift. Unlike in the usual case, however, both the force and the diffusion constant originate from protein motors in Eq.5. This is one of the characteristics of the *in vitro* sliding movement.

ACKNOWLEDGEMENTS

This work was supported by a Grant-in Aid for Scientific Research on Priority Area of "Molecular Mechanism of Biological Sliding Movements (No. 01657002)" for the Ministry of Education, Science, and Culture in Japan.

REFERENCES

1. Tawada, K. & Kimura, M. *J. Muscle Res. Cell Motility* **7**, 339-350 (1986).
2. Nishikawa, T. & Go, N. *Proteins; Struc. Func. and Gen.* **2**, 308-329 (1987).
3. Tawada, K. & Sekimoto, K. *Biophys. J.* **59**, 343-356 (1991).
4. Tawada, K. & Sekimoto, K. *J. theor. Biol.* **150**, 193-200 (1991).
5. Sellers J.R., Spudich, J.A., & Sheetz, M.P. *J. Cell Biol.* **101**, 1897-1902 (1985).
6. Sellers, J.R. *J. Biol. Chem.* **260**, 15815-15819 (1985).
7. Sellers J.R., Eisenberg, E. & Adelstein, R.S. *J. Biol. Chem.* **257**, 13880-13883 (1982).
8. Warshaw D.M., Desrosiers, J.M., Work, S.S. & Trybus, K.M. *J. Cell Biol.* **111**, 453-463 (1990).
9. Vale, R.D., Soll, D.R. & Gibbons, I.R. *Cell.* **59**, 915-925 (1989).

Discussion

Sellers: Dave Warshaw performed experiments where fast and slow myosin were mixed at higher ionic strengths, and found that the unphosphorylated myosin has even a more powerful effect, if you want to put it this way, on the velocity of the phosphorylated myosin under those conditions. Is that explained by your model?

Tawada: No, it is not.

Sellers: What I am talking specifically about is when Warshaw mixed phosphorylated smooth muscle and unphosphorylated smooth muscle at twice or three times the ionic strength, getting a stronger effect with the unphosphorylated myosin.

Tawada: I can write an equation fitted to their data, but I cannot explain why that effect is stronger.

Godt: You showed that the sliding velocity should be directly proportional to the inverse of temperature. Is that correct?

Tawada: Sure, because protein friction is related to the diffusion coefficient and absolute temperature.

Godt: Isn't that different from what Earl Homsher just showed, though?

Tawada: No. If you change temperature, you change the diffusion coefficient as well. So, it is not directly proportional to only the inverse of temperature.

Huxley: I have always supposed that most of the resistance to what prevents sliding at greater speeds is a negative force, when a firmly-attached cross-bridge has gone beyond its zero-force position. It appears from what you have been saying that you assume the resistance comes entirely from weakly-binding cross-bridges. Have you considered the possibility that much or most of the force comes from firmly-attached bridges beyond their zero-force position?

Tawada: Yes, this model does not exclude the possibility of the contribution of overstretched myosin cross-bridges. In my model, you have to subtract the negative force due to the overstretched heads to get the net force. My model is not inconsistent with your spring model.

Homsher: Doesn't your model then predict that as you decrease the number of heads that can interact with a filament, a discontinuity appears as you approach one cross-bridge trying to pull things?

Tawada: A discontinuity?

Homsher: Yes, because if you have just one cross-bridge pulling, you have no weakly-attached cross-bridges; therefore, you will only have the force of that single pulling cross-bridge. So velocity will be entirely quantized.

Tawada: My equation is averaged with the number of myosin heads. So, we need at least ten or twenty heads to get that steady-state equation. We can't talk about the interaction of a single head with a single actin filament.

V. PROPERTIES OF ACTIN-MYOSIN SLIDING AND FORCE GENERATION STUDIED BY *IN VITRO* FORCE-MOVEMENT ASSAY SYSTEMS

INTRODUCTION

The *in vitro* motility assay systems used by the authors in Chapter IV deals with the unloaded sliding velocity between actin and myosin, and therefore the experimental data obtained an only be compared with the maximum shortening velocity of muscle fibers under zero external load. Since the main function of muscle is to do work by lifting a load, it is essential to measure forces arising from actin-myosin sliding in the experiments with *in vitro* assay systems, so that the data obtained can be directly comparable with muscle mechanics data. This has become possible in many ways as described in the papers in this chapter.

Sugi, Oiwa and Chaen developed an assay system, in which a myosin-coated glass microneedle having an appropriate compliance was made to slide on actin cables of giant algal cells. With this force-movement assay system, they measured the amount of work done by actin-myosin sliding in response to iontophoretically applied ATP, and showed that the amount of work done with a constant amount of ATP application increased with increasing initial baseline force, thus the work versus baseline force was bell-shaped as with contracting skeletal muscle. Kinosita and others attempted to detect change in orientation of actin monomers in actin filaments sliding past myosin, in addition to measurement of fluorescence polarization of probe attached to actin, by microscopic observation of a bead attached to the tail of sliding actin filament, and observed no axial rotation of the bead. They could also estimate the force of actin-myosin sliding and its fluctuation by trapping the bead with the technique of "optical trap", which is also used by Simmons' group. Simmons and coworkers give a detailed description of the optical trap technique, in which a polystyrene bead attached to a single actin filament is trapped with a laser beam of varying intensity. After careful calibration of trap strength, they could measure the force exerted by a single actin filament sliding on HMM by trapping a bead attached to the tail of actin filament. They also discuss how the technique can be improved in future for studying the molecular mechanism of contraction.

Yanagida and others recorded force and its fluctuation by use of a microneedle attached to the tail of a single actin filament sliding on HMM or S-1. According to them, actomyosin ATPase cycle and force-generating attachment-detachment cycle between actin and myosin may not be tightly coupled, but may vary according to mechanical conditions. Chaen and others constructed a force-movement assay system by combining the myosin-coated bead versus actin cable system with the centrifuge microscope. Unlike all the other assay systems hitherto developed, this system can observe steady actin-myosin sliding under constant external load. Thus,

302

they obtained the steady-state force-velocity curve directly comparable with that in contracting muscle. With this system, it is also possible forces in the same direction as that of bead movement. Unexpectedly, such forces did not accelerate the bead movement, but slowed it down.

The last two papers are concerned with theories of actin-myosin sliding in muscle contraction. Ando, Kobayashi and Munekata propose that muscle contraction results from electrostatic force between actin and myosin with some experimental evidence. Mitsui and Oshima, on the other hand, assume that binding of ATP-activated myosin head to actin changes potential energy profile along the thin filament which in turn induces actin-myosin sliding. Their model predicts the slowing of actin-myosin sliding velocity by "negative" force observed by Chaen and others.

DEPENDENCE OF THE WORK DONE BY ATP-INDUCED ACTIN-MYOSIN SLIDING ON THE INITIAL BASELINE FORCE: ITS IMPLICATIONS FOR KINETIC PROPERTIES OF MYOSIN HEADS IN MUSCLE CONTRACTION

H. Sugi, K. Oiwa and S. Chaen

Department of Physiology
School of Medicine
Teikyo University
Tokyo 173, Japan

ABSTRACT

The properties of the ATP-dependent actin-myosin sliding responsible for muscle contraction was studied using an *in vitro* force-movement assay system, in which a myosin-coated glass microneedle was made to slide on actin filament arrays (actin cables) in the giant algal cell with iontophoretic application of ATP. With a constant amount of ATP application, the amount of work done by the actin-myosin sliding increased with increasing baseline force from zero to 0.4-0.6 P_0, and then decreased with further increasing baseline force, thus giving a bell-shaped work versus baseline force relation. The result that the maximum actin-myosin sliding velocity did not change appreciably with increasing baseline force up to 0.4-0.6 P_0 implies, together with the limited number of myosin heads involved, that (1) the rate of power output of actin-myosin sliding is determined primarily by the amount of external load rather than the velocity of actin-myosin sliding, and (2) the bell shaped work versus baseline force relation (and also the hyperbolic force-velocity relation) results from the kinetic properties of individual myosin head rather than the change in the number of myosin heads involved.

INTRODUCTION

The function of skeletal muscle in the animal body is primarily to do work, and the main objects of research work in the field of muscle mechanics have been the force-velocity (*P-V*) and the force-energy relations in contracting muscle[1]. Based on the above knowledge, A.F. Huxley constructed a contraction model which has been central in the field of muscle physiology. In this model, the *P-V* and the force-energy relations are explained by the change in the number of myosin heads

Mechanism of Myofilament Sliding in Muscle Contraction, Edited by
H. Sugi and G.H Pollack, Plenum Press, New York, 1993

303

involved depending on the velocity of myofilaments sliding[2]. From the above standpoint, it seems rather strange that so-called myosin step size, i.e. the distance of actin-myosin sliding per hydrolysis of one ATP molecule under unloaded condition[3][4], attracts attention of investigators, while the most important function of actin-myosin sliding to do work is totally ignored.

In the present study, we investigated kinetic properties of the ATP-dependent actin-myosin sliding using an *in vitro* force-movement assay system developed in our laboratory, in which a myosin-coated glass microneedle was made to slide on actin filament arrays (actin cables) in the giant algal cell in response to iontophoretically applied ATP.

MATERIALS AND METHODS

Fig. 1 shows the experimental arrangement. The internodal cell strip preparation (P), prepared from the internodal cell of a green alga *Nitellopsis obtusa*, was mounted with inner surface up in the experimental chamber (E) filled with ATP-free solution (5 mM $MgCl_2$, 4 mM EGTA, 1 mM ditiothreitol, 10 mM Pipes,

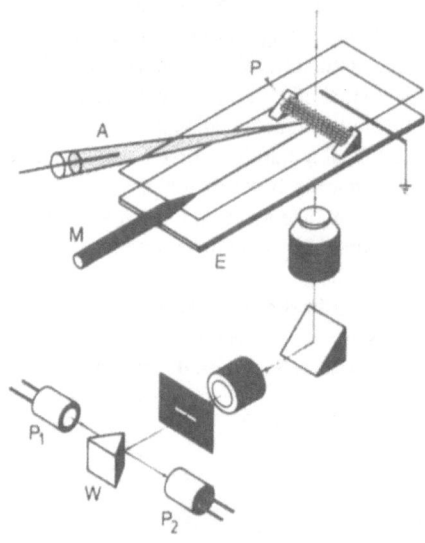

Fig. 1. Experimental arrangement. The myosin-coated tip of a glass microneedle (M) is put in contact with the inner surface of the internodal cell strip preparation (P) at right angles to the chloroplast rows on which the actin cables are located. The experimental chamber (E) filled with ATP-free solution was mounted on an inverted light microscope. The needle was made to slide along the actin cables by applying negative current pulses to the microelectrode (A) filled with 100 mM-ATP. The needle tip movement was recorded by splitting the needle tip image with the wedge-shaped mirror (W) into two parts, each of which was projected on a photodiode (P_1 and P_2). From Oiwa et al.[5].

0.225 mM P^1-P^5-di(adenosine-5')pentaphosphate, 200 mM D-sorbitol, pH 7.0). In some experiments, ATP-free solution further contained 50 unit/ml hexokinase and 2 mM D-glucose. A glass microneedle (M) (tip diameter $\approx$ 10 μm, elastic coefficient 150-170 pN/μm) was coated with rabbit skeletal muscle myosin at the tip, and was brought into contact with actin cables. The myosin heads on the needle tip, which formed rigor linkages with actin cables in ATP-free solution, were activated to interact with actin cables by iontophoretic application of ATP through a glass capillary electrode (ATP electrode, A) filled with 100 mM ATP. The distance between the myosin-coated needle and the ATP electrode was $\approx$ 50 μm. A linear relation was always obtained between the electrical charge passed through the ATP electrode and the amount of released ATP, measured with a luminometer (Analytical Luminescence Laboratory, Monolight 2010). The ATP-induced needle movement on actin cables was recorded by splitting the image of the needle tip with a wedge-shaped mirror (W) into two parts, each of which was projected on a photodiode (P_1 and P_2) (Hamamatsu Photonics, S2386-5K). Further details of the methods used have been described elsewhere[5]. All experiments were performed at room temperature (20-23°C).

RESULTS

Change in the Apparent Efficiency of ATP-induced Actin-Myosin Sliding in Producing Work Depending on the Initial Baseline Force

Reproducible movements of the myosin-coated needle on actin cables were induced by iontophoretic application of ATP. As shown in Fig. 2A, the needle started moving with a delay of 0.2-0.5 s after the onset of a negative current pulse (ATP current pulse) applied to the ATP electrode, and eventually stopped moving to stay in position, indicating that, when the applied ATP diffused away from the needle tip and became no longer available for actin-myosin interaction, the myosin heads on the needle tip again formed rigor linkages with actin cables to inhibit elastic recoil of the bent needle. A subsequent application of ATP induced the needle movement starting from the baseline force attained by the preceding needle movement. Thus, a stepwise needle movement was produced by repeated application of ATP (see Fig. 3).

At the early stage of the experiments, it was noticed that, with two successive application of the same amount of ATP, the magnitude of the second needle movement was nearly as large as that of the first one (Fig. 2A). The amount of work (W) done by an ATP-induced needle movement from position x_1 to position x_2 along actin cable is:

$$W = \int_{x_1}^{x_2} F(x)\,dx = \int_{x_1}^{x_2} Kx\,dx = K(x_2^2 - x_1^2)/2,$$

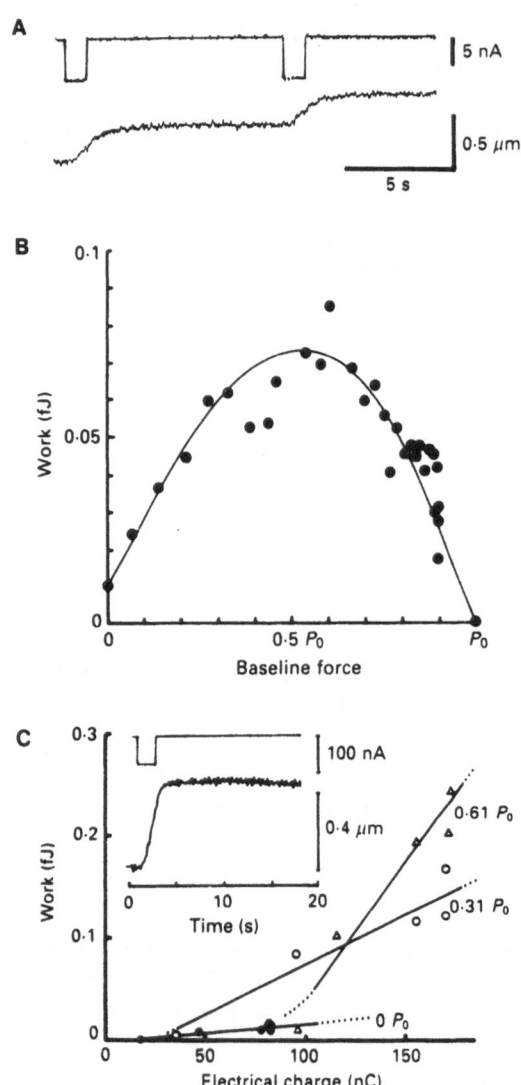

Fig. 2. Change in the apparent efficiency of ATP-induced actin-myosin sliding in producing work. (A) Stepwise needle movement in response to two successive 1 s ATP current pulses (15 nA). (B) Relation between the amount of work done by actin-myosin sliding induced by a constant amount of ATP application and the amount of initial baseline force. (C) Relation between the amount of work done by ATP-induced actin-myosin sliding and the amount of electrical charge passed through the ATP electrode. Inset shows a typical needle movement in the presence of hexokinase and D-glucose. From Oiwa et al.[5].

where $F(x)$ is the force exerted by the bent needle and K is the elastic coefficient of the needle. It follows from the above equation that the amount of work done by the second needle movement is about three times the amount of work done by the first one, though the same amount of ATP was applied in both cases, indicating that the apparent efficiency of ATP-induced actin-myosin sliding in producing work increases with increasing initial baseline force.

In Fig. 2B, the amount of work done by the needle movement induced by a 1 s ATP current pulse (15 nA) is plotted against the initial baseline force from which the needle movement started. In accordance with the result shown in Fig. 2A, the amount of work done by the needle movement increased with increasing baseline force from zero to 0.4-0.6 P_0, and then decreased with further increasing baseline force, reaching zero when the baseline force was equal to P_0. Thus, the work versus baseline force relation was bell-shaped.

The experiments were also made, in which myosin heads on the needle were activated repeatedly with various amounts of iontophoretically applied ATP starting from the same baseline force. After each needle movement, the initial needle position was restored by moving the mechanical stage carrying the internodal cell preparation. Hexokinase and D-glucose were added to ATP-free solution to avoid accumulation of ATP in the experimental chamber. As shown in Fig. 2C, the slope of the line relating the amount of work done by the needle movement with the amount of electrical charge passed through the ATP electrode (the amount of ATP applied) was steeper as the initial baseline force was increased, again indicating the increase in the apparent efficiency of ATP-induced actin-myosin sliding in producing work with increasing baseline force from zero to a certain value.

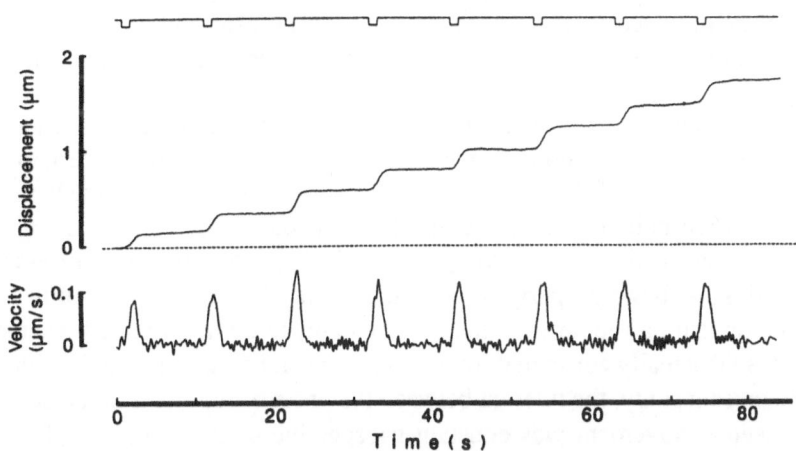

Fig. 3. Stepwise needle displacement in response to repeated application of 1 s ATP current pulses (34 nA) together with its velocity changes. Note that both the maximum velocity and the time course of ATP-induced actin-myosin sliding do not change markedly with increasing initial baseline force.

Time Course of the ATP-induced Actin-Myosin Sliding Starting from Various Baseline Forces

Fig. 3 shows an example of stepwise needle displacement in response to repeated application of 1 s ATP current pulses (34 nA) together with its velocity changes. The magnitude of baseline force attained after the last application of ATP was $\approx 0.15\ P_0$. Therefore, the stepwise needle movement constitute the ascending part of the work versus baseline force relation (Fig. 2B), where the apparent efficiency of actin-myosin sliding in producing work with applied ATP increased with increasing baseline force. As can be seen in the velocity trace, both the maximum velocity and the time course of ATP-induced actin-myosin sliding did not change markedly with increasing baseline force at which actin-myosin sliding started.

DISCUSSION

Possible Relation of the Present Results with the Fenn Effect

In the present experiments, myosin heads on the needle were activated to interact with actin cables for a limited period determined by the ATP current pulse. This situation may be comparable to contraction of stimulated muscle, except that the initiation of actin-myosin sliding in muscle is mediated by Ca^{2+} released from the sarcoplasmic reticulum. The value of P_0 in the present experiments was around 500 pN, implying that the number of myosin heads involved is also around 500. This figure is less that $1/10^6$ of the number of myosin heads within a half sarcomere of a single muscle fiber generating P_0 of 1.5 mN. The extremely small number of myosin heads involved in the ATP-induced actin-myosin sliding may provide a favorable condition to obtain information about the kinetic properties of individual myosin head.

In 1923, Fenn[6] showed that the energy (heat + work) output of a tetanized skeletal muscle increases with increasing external load from zero to a certain value, a phenomenon known as "Fenn effect"; both the work versus load and the heat versus load relations are bell-shaped. Thus, the bell-shaped work versus baseline force relation obtained in the present study (Fig. 2B) indicates that the present assay system retains the basic property of contracting muscle, despite myosin heads are randomly oriented on the needle, though it remains to be investigated whether the amount of ATP actually consumed for actin-myosin sliding also parallels with work.

In the present study, the interval between the onset of ATP current pulse and the onset of needle movement was constant irrespective of the initial baseline force (Figs. 2A and 3), indicating that myosin heads are almost simultaneously activated by iontophoretically applied ATP. The almost simultaneous activation of extremely limited number of myosin heads on the needle may be taken to imply that the increase in the apparent efficiency of actin-myosin sliding in producing work with applied ATP (Fig. 2) reflects the kinetic properties of individual myosin head rather than the change in the number of myosin heads involved; in other words, each myosin head may utilize ATP to produce work with various apparent efficiencies depending on the baseline force.

Factors Determining the Rate of Power and Energy Output in the ATP-dependent Actin-Myosin Sliding

A most basic characteristic of contracting muscle is the steady-state hyperbolic force-velocity $(P$-$V)$ relation[1]. Since the power is the product of shortening velocity and force (= load), the power versus load relation during steady shortening is directly obtained from the P-V curve. The power versus load relation obtained in this way is bell-shaped, as the power is zero when the load is zero or P_0. Recently, we have shown that the steady-state P-V relation of myosin-coated beads moving on actin cables is also hyperbolic in shape except for the high force region[7]. This means that the power versus load relation in the above assay system is also bell-shaped. Thus, the bell-shaped work versus baseline force relation obtained in the present study (Fig. 2B) indicates that, in both our assay systems and contracting muscle, the power versus load and the work versus load relations are bell-shaped.

In the contraction model of A.F. Huxley[2], the P-V relation (and also the power versus load and the work versus load relations) is associated with the change in the number of myosin heads interacting with the thin filament, which in turn is primarily determined by the velocity with which the thick and thin filaments slide past each other. In the present experiments, however, both the maximum velocity and the time course of ATP-induced actin-myosin sliding did not change markedly with increasing initial baseline force, while the apparent efficiency of actin-myosin sliding in producing work increased as the baseline force was increased (Fig. 3). This may be take to indicate that the power output of the ATP-dependent actin-myosin interaction in contracting muscle may be primarily determined by the amount of external load rather than the velocity of actin-myosin sliding.

REFERENCES

1. Hill, A.V. *Proc. R. Soc. Ser. B* **126**, 136-195 (1936).
2. Huxley, A.F. *Prog. Biophys. Biophys. Chem.* **7**, 255-318 (1957).
3. Toyoshima, Y.Y., Kron, S.J. & Spudich, J.A. *Proc. Natl. Acad. Sci. USA* **87**, 7130-7134 (1990).
4. Harada, Y., Sakurada, K., Aoki, T., Thomas, D.D. & Yanagida, T. *J. Mol. Biol.* **216**, 49-68 (1990).
5. Oiwa, K., Chaen, S. & Sugi, H. *J. Physiol.(Lond.)* **437**, 751-763 (1991).
6. Fenn, W.O. *J. Physiol. (Lond.)* **58**, 175-203 (1923).
7. Oiwa, K., Chaen, S., Kamitsubo, E., Shimmen, T. & Sugi, H. *Proc. Natl. Acad. Sci. USA* **87**, 7893-7897 (1990).

Discussion

Huxley: Can you estimate the actual ATP concentration around the tip of your needle? It is a complicated situation with diffusion and removal of the ATP by your enzyme system and by the myosin, but have you a rough figure?

Sugi: We calculated the ATP concentration changes around the needle tip by diffusion equations without taking the enzyme system and myosin into consideration [Oiwa et al. *J. Physiol. (Lond.)* **437**, 751-763, 1991]. What I am trying to do with Drs. Oiwa and Chaen now is to measure indirectly the amount of ATP actually split for producing work. Our plan is to confine this system in a small droplet of experimental solution. First, the system is kept in rigor, and then we inject a small amount of ATP into the droplet to induce actin-myosin sliding.

Homsher: Have you tried to vary the amount of myosin that is on the tip of your needle?

Sugi: Yes. This is what Dr. Oiwa has done recently. According to him, some quantal nature of actin-myosin sliding appears when the amount of myosin on the needle tip is reduced.

Tregear: One effect that was seen in the late 70s on myosin heads in rigor was that if you pulled them, you changed their affinity for ATP or for AMPPNP. I was wondering if part of what you are seeing, as you put the load on starting from rigor, might be your influence on the relative·tendency of the ATP to bind to myosin.

Sugi: So, you think the affinity of myosin to ATP may depend on its mechanical condition?

Tregear: Yes, there is evidence from Kuhn (*J. Muscle Res. Cell Motility* **2**, 7-44, 1981) and there is evidence from us (Marston et al. *J. Mol. Biol.* **128**, 111-126, 1979).

Sugi: Thank you. We should certainly consider this possibility.

ter Keurs: Would it be possible to measure the stiffness of the link between actin and myosin on your needle, for example, by using a piezo-element to modulate the position of the support for the actin filaments?

Sugi: It would be possible, though technically very difficult. In the case of rigor linkages, if you bend them, then actin cables will break before rigor linkages break, according to our experiment of stretching rigor fibers (Suzuki, S. and Sugi, H. *J. Gen. Physiol.* **81**, 531-546, 1983).

Kawai: I have a question for Dr. Sugi regarding the ATP-induced stepwise sliding. I wonder if this keeps on going forever if you keep on applying ATP?

Sugi: No. Efficiency eventually reaches an upper limit at the top of the bell-shaped work versus baseline force curve. Then, of course, it decreases as it approaches the isometric condition.

Morales: At the end of your talk, you made a very important statement, but you didn't have time to explain it. I wonder if you would say a few more words. You said that you felt that the bell-shaped work versus baseline force relation was not due to variations in the number of cross-bridges, but that it was an intrinsic property of each cross-bridge. Can you say again why you think this?

Sugi: Yes, one reason is that the number of myosin heads involved is very small. In our myosin-coated needle versus actin cable system, the force is around 500 pN. This probably represents about 500 cross-bridges involved. This is many millions of times smaller than the cross-bridge number within a half-sarcomere of an intact muscle fiber. And in our centrifuge microscope system, the maximum isometric force is only 10 pN or less, and still we get hyperbolic-shaped force-velocity curves. So, these results give us a strong impression that each cross-bridge would sense its mechanical condition to change its mode of performance. There appears to be a kind of load sensor within a cross-bridge, and it may affect the ATPase or chemo-mechanical conversion rate of a cross-bridge. Of course, I cannot definitely prove that at present.

Morales: Just now it's a feeling?

Sugi: Yes, of course.

UNITARY DISTANCE OF ATP-INDUCED ACTIN-MYOSIN SLIDING STUDIED WITH AN *IN VITRO* FORCE-MOVEMENT ASSAY SYSTEM

K. Oiwa, T. Kawakami and H. Sugi

Department of Physiology
School of Medicine
Teikyo University
Itabashi-ku, Tokyo 173, Japan

ABSTRACT

We studied the unitary distance of ATP-induced actin-myosin sliding using an *in vitro* force-movement assay system consisting of a myosin-coated glass microneedle and well organized actin filament arrays (actin cables) in the internodal cell of an alga *Nitellopsis obtusa*[1]. The number of myosin heads interacting with actin cables was reduced to about 100, as judged from the isometric force of about 100 pN attained in the presence of 2 mM ATP. When the amount of iontophoretically applied ATP was reduced by decreasing the amount of charge passed through the ATP electrode from 80 to 2 nC, the distance of the ATP-induced actin-myosin sliding decreased almost linearly from about 100 to about 10 nm, no detectable sliding being observed with further reduction of charge through the electrode. The sliding distances with small amounts of ATP (7-16 nC) distributed around integral multiples of 10 nm, suggesting the unitary distance of actin-myosin sliding of about 10 nm.

INTRODUCTION

A number of investigations have recently been made to estimate the distance of actin-myosin sliding per one ATP molecule (myosin step size) either from the relation between the velocity of actin-myosin sliding and the rate of ATP consumption in *in vitro* motility assays[2-4], or from the relation between distance of muscle fiber shortening and amount of photochemically released ATP[5)6]. In spite of these contributions, the myosin step size still remains to be a matter for debate and speculation.

In 1989, we developed an *in vitro* assay system, in which a myosin-coated glass microneedle was made to slide on well organized actin filament arrays (actin cables) in the internodal cell of an alga *Nitellopsis obtusa* in the presence of ATP, and succeeded in determining the force-velocity relation of the ATP-dependent actin-

Mechanism of Myofilament Sliding in Muscle Contraction, Edited by
H. Sugi and G.H Pollack, Plenum Press, New York, 1993

myosin sliding[1]. In the present experiment, we used this assay system to study the unitary distance of ATP-induced actin-myosin sliding by reducing the amount of iontophoretically applied ATP to a minimum. In this paper, we use the term, unitary distance of actin-myosin sliding, instead of the term, myosin step size, since the amount of ATP actually consumed for actin-myosin sliding is not measured.

MATERIALS AND METHODS

The experimental arrangement has been described in a previous paper[7]. In summary, a glass microneedle (tip diameter 5 μm, elastic coefficient 20-50 pN/μm, time constant of mechanical response 40-100 ms) coated with rabbit skeletal muscle myosin was brought into contact with actin cables in the strip of the *Nitellopsis* internodal cell, mounted in a chamber filled with ATP-free solution of the following composition (mM): $MgCl_2$, 5; EGTA, 4; dithiothreitol, 1; PIPES, 10; P^1-P^5-di(adenosine-5) pentaphosphate, 0.225; D-sorbitol, 200 (pH 7.0). The ATP-free solution also contained 50 unit/ml hexokinase and 2 mM D-glucose to facilitate exhaustion of iontophoretically applied ATP around the myosin-coated needle[7].

The myosin heads on the needle, forming rigor linkages with actin cables, were activated to interact with actin with iontophoretic application of ATP through the ATP electrode (resistance 100 MΩ, filled with 100 mM ATP) placed about 20 μm from the myosin-coated needle. Displacement of the needle was recorded with a photoelectric sensor with an accuracy of about 1 nm[7]. All experiments were made at room temperature (20-23°C).

RESULTS

Variation of Number of Myosin Heads Interacting with Actin at Various Ratios of Myosin-Myosin Rod Mixture

In the present study, we coated needles with myosin-myosin rod mixtures to reduce the number of myosin heads interacting with actin. Fig. 1 shows the relation between the isometric force generated by the myosin heads on the needle in the presence of 2 mM ATP (inset) and the myosin content of myosin-myosin rod mixtures. If the needle was coated only the myosin (myosin content 100%), it started moving on actin cables until come to a complete stop (inset) to generate the isometric force of 640 ± 440 pN (mean ± S.D., n = 18). When myosin content was reduced to 10%, the isometric force was 123 ± 76 pN (n = 9). With further reduction of myosin content to 1%, the isometric force was 43 ± 21 pN (n = 3), but the needle tended to detach from actin cables. We therefore used the myosin-myosin rod mixture containing 10% myosin in the present study. Assuming that each myosin head may generate an isometric force of 1 pN, the number of myosin heads involved in the needle movement is of the order of 100.

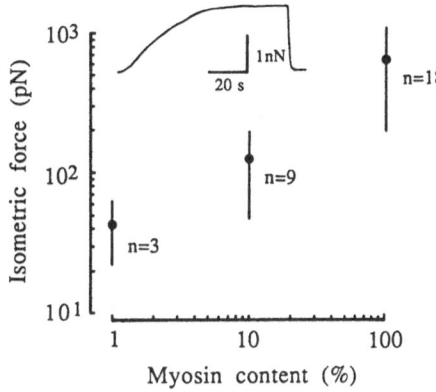

Fig. 1. Dependence of isometric force generated by myosin heads on myosin content of myosin-myosin rod mixture on the needle. Data points represent mean values with vertical bars indicating S.D. Number of experiments is given for each data point. Inset shows a typical record of movement of myosin-coated needle on actin cables in the presence of 2 mM ATP in the experimental solution. The needle was coated only with myosin (myosin content 100 %).

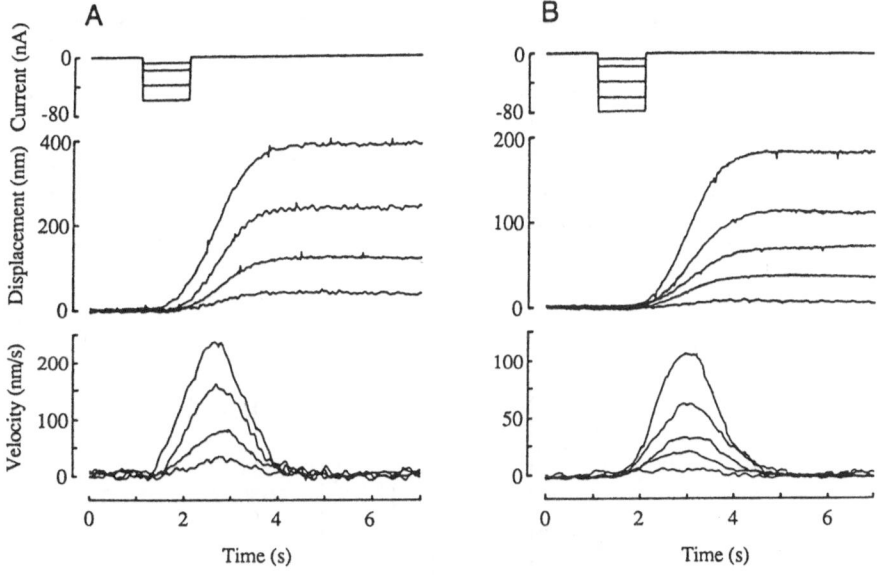

Fig. 2. Time course of ATP-induced movements of a needle coated only with myosin (myosin content 100 %) (A) and of a microneedle coated with myosin-myosin rod mixture (myosin content 10 %) (B) in response to different amounts of iontophoretically applied ATP. In both A and B, upper, middle and lower traces show the negative current pulses applied to the ATP electrode, the displacement of the needle and the velocity of needle movement, respectively. The amounts of charge passed through the ATP electrode were ; 7, 16, 35 and 53 nC in A; 7, 16, 35, 53 and 70 nC in B. Small spike-like fluctuations on the displacement traces are artifacts arising from data processing.

Dependence of ATP-induced Needle Movement on the Number of Myosin Heads Interacting with Actin

Typical records of needle movement in response to various amounts of iontophoretically applied ATP are shown in Fig. 2. When the needle was coated only with myosin (myosin content 100%), the interval between the onset of negative current pulse passed through the ATP electrode and the onset of needle movement increased with decreasing current intensity, and the time required for the velocity of actin-myosin sliding to reach a maximum also increased with decreasing current intensity (A). When the needle was coated with myosin-myosin rod mixture (myosin content 10%), on the other hand, both the time of onset of needle movement and the time at which the velocity reached a maximum remained constant irrespective of the current intensity (B). This may be taken to indicate that, if the number of myosin heads interacting with actin is appropriately reduced, the heads can be activated almost simultaneously with iontophoretically applied ATP.

Relation between Distance of Needle Movement and Amount of Applied ATP

Fig. 3 shows the relation between the distance of ATP-induced needle movement and the amount of iontophoretically applied ATP, which is proportional to the amount of charge passed through the ATP electrode[7]. The sliding distance of needles coated with myosin-myosin rod mixture (myosin content 10%) decreased linearly with decreasing amount of applied ATP from 80 to 7 nC, while the sliding distance of needles coated only with myosin decreased sharply with decreasing amount of applied ATP below 15 nC. In both cases, no needle movement was detected in response to applied ATP below 7 nC.

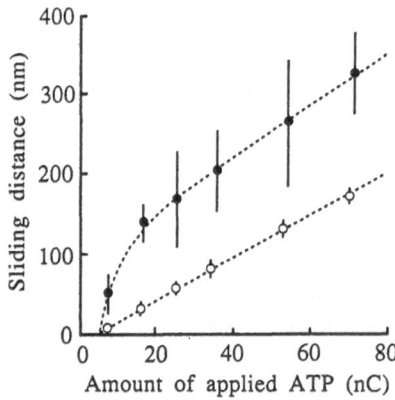

Fig. 3. Relation between the distance of ATP-induced needle movement and the amount of iontophoretically applied ATP. Filled and open circles show data points (mean values with S.D.) obtained with a needle coated only with myosin (myosin content 100 %) and with a needle coated with myosin-myosin rod mixture (myosin content 10 %) respectively.

Quantal Distribution of Needle Movement Distances Induced by Small Amounts of Applied ATP

If a myosin head slides past actin filament for a unitary distance per hydrolysis of one ATP molecule, the distribution of actin-myosin sliding distance in response to small amounts of applied ATP may exhibit quantal nature. To ascertain the above possibility, we induced movements of a needle coated with myosin-myosin rod mixture (myosin content 10%) repeatedly (70-120 times) by applying ATP with charges of 7 or 16 nC passed through the ATP electrode.

Fig. 4 shows histograms of distribution of actin-myosin sliding distance. On the assumption that the sliding distances are built up statistically of unitary distance of actin-myosin sliding, a few Gaussian curves were fitted to the histograms by the simplex method described by Nelder and Mead[8]. The statistical analysis indicates that the distribution of sliding distance with 7 nC ATP consists of two Gaussian curves with mean values of 5 and 10 nm respectively (A). Similarly, the distance distributions with 16 nC ATP consist of Gaussian curves with mean values equal to integral multiples of 10 nm (B and C). The mean peak of 5 nm with 7 nC ATP (A) is explained as being due to mechanical vibration of the needle, since the needle vibration with amplitude around 5 nm is continuously recorded without ATP application; in case that 7 nC ATP happens to fail inducing actin-myosin sliding, only this vibration is recorded. As a matter of fact, the mean peak of 5 nm was never observed in the distance distributions with 16 nC ATP.

DISCUSSION

The present experiments have shown that, when a small number of myosin heads are activated almost simultaneously with small amounts of applied ATP, the distribution of actin-myosin sliding distance is built up statistically of unitary actin-myosin sliding distance of about 10 nm (Fig. 4). We shall first consider the experimental conditions under which the above result has been obtained.

Calculation of ATP concentration changes around the tip of a myosin-coated needle by solving three-dimensional differential equation for diffusion of ATP from the ATP electrode[7] indicates that the ATP concentration reaches a maximum after termination of an ATP current pulse, and then decays with time. The presence of hexokinase-D-glucose system not only increases the critical amount of ATP to induce actin-myosin sliding, but also facilitates removal of ATP around the needle after termination of an ATP current pulse. As a matter of fact, the duration of needle movement in response to a given amount of applied ATP is markedly shortened in the presence of hexokinase-D-glucose system[7]. On this basis, when the amount of iontophoretically applied ATP is reduced to or near the minimum value, the myosin heads on the needle may be able to utilize ATP for a very limited period, so that each myosin head may hydrolyze ATP only once or at most a few times. The above consideration may validate the statistical analysis of actin-myosin sliding distance distribution (Fig. 4).

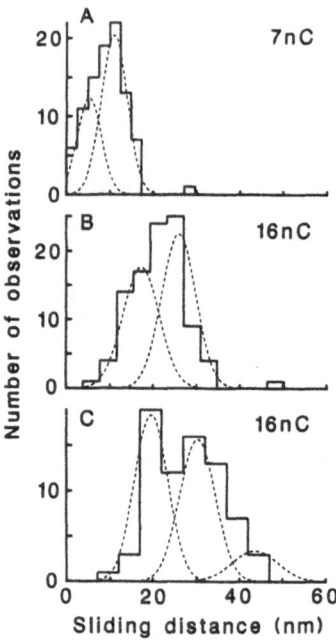

Fig. 4. Histograms showing distribution of actin-myosin sliding distance induced by small amounts of iontophoretically applied ATP. The amount of charge passed through the ATP electrode is given in each histogram. Broken lines indicate Gaussian distribution curves chosen to fit the histograms obtained. Mean values ± S.D. of the Gaussian curves are 5.1 ± 2.7 nm and 10.8 ± 3.9 nm in A, 18.0 ± 4.2 nm and 26.5 ± 3.9 nm in B, and 19.5 ± 3.9 nm, 30.2 ± 4.1 nm and 43.5 ± 5.4 nm in C.

In this connection, it is of interest that the minimum shortening of a glycerinated rabbit psoas muscle fiber induced by photolysis of caged ATP has been found to be about 10 nm/half sarcomere[9]. In this study, the minimum fiber shortening is induced with 75 μM of photochemically released ATP; as the total myosin head concentration in the fiber is about 150 μM, each myosin molecule can utilize only one ATP molecule when 75 μM ATP is released. Thus, both the present result and the result of the above caged ATP experiment suggest that the unitary distance of actin-myosin sliding is of order of 10 nm, though much more experimental work is needed to settle the problem.

REFERENCES

1. Chaen, S., Oiwa, K., Shimmen T., Iwamoto, H. & Sugi, H. *Proc. Natl. Acad. Sci. USA* **86**, 1510-1514 (1989).

2. Harada, Y., Sakurada, K., Aoki, T., Thomas, D.D. & Yanagida, T. *J. Mol. Biol.* **216**, 49-68 (1990).
3. Toyoshima, Y. Y., Kron, S.J. & Spudich, J.A. *Proc. Natl. Acad. Sci. USA* **87**, 7130-7134 (1990).
4. Uyeda, T.Q.P., Warrick, H.M., Kron, S.J. & Spudich, J.A. *Nature* **352**, 307-311 (1991).
5. Higuchi, H. & Goldman, Y.E. *Nature (London)*, **352** 352-354 (1991).
6. Yamada, T., Abe, O. & Sugi, H. *Biophys. J.* **59**, 417a (1991).
7. Oiwa, K., Chaen, S. & Sugi, H. *J. Physiol. (London)*, **437**, 751-63 (1991).
8. Nelder, J. A. & Mead, R. *Computer J.* **7**, 308-313 (1965).
9. Yamada, T., Abe, O., Kobayashi, T. & Sugi, H. *J. Physiol. (London)* in press (1993).

Discussion

Kinosita: You said at the end of your talk that you activated 100 myosin molecules synchronously, but could it be the case that most of the myosin molecules were in rigor and you activated one molecule of myosin after another? If so, the rigor myosin head would be your load.

Oiwa: The value we estimated for the number of myosin molecules participating in motility corresponds to the upper limit. A computer simulation showed that the ATP concentration around the needle tip rose rapidly at the onset of an ATP pulse. So we think that is not the case.

Kinosita: You said that the ATP concentration was about one μM. Is that enough to activate all the myosin heads simultaneously?

Homsher: They would bind at a rate of about one per second at that concentration.

Sugi: We started with a simplified assumption. If you consider that the number of myosin molecules on the needle tip was very limited—100 or fewer myosin heads—and they were surrounded by a large amount of ATP, it is permissible to assume that the activation occurred simultaneously because of the rapid rate of detachment of the rigor linkages by ATP.

Pollack: Was the movement always continuous, or did you ever see discrete stepping?

Oiwa: My needle system does not have enough time resolution to detect such fluctuation.

ORIENTATION OF ACTIN MONOMERS IN MOVING ACTIN FILAMENTS

Kazuhiko Kinosita, Jr., Naoya Suzuki, Shin'ichi Ishiwata*, Takayuki Nishizaka*, Hiroyasu Itoh**, Hiroyuki Hakozaki, Gerard Marriott and Hidetake Miyata

Department of Physics
Faculty of Science and Technology
Keio University
Hiyoshi 3-14-1, Kohoku-ku
Yokohama 223, Japan
*Department of Physics
School of Science and Engineering
Waseda University
Okubo 3-4-1, Shinjuku-ku
Tokyo 169, Japan
**Tsukuba Research Laboratory
Hamamatsu Photonics K. K.
Tokodai 5-9-2
Tsukuba-shi, Ibaraki 300-26, Japan

ABSTRACT

We have visualized, under an optical microscope, the orientations of actin monomers in individual actin filaments undergoing Brownian motion in solution, actively sliding past myosin molecules, or immobile on a surface. For the visualization, two strategies have been adopted. One is to exploit the fluorescence polarization of a fluorescent probe firmly attached to actin. Using the probe phalloidin-tetramethylrhodamine, the fluorescence was clearly polarized along the filament axis, showing alignment of the probe molecules along the filament axis. Within our temporal resolution of 33 ms and spatial resolution of better than 1 μm (average over ~ 10^2 actin monomers), the orientation of the probe (hence of actin monomers) did not change upon interaction of the filament with heavy meromyosin; myosin-induced reorientation was estimated to be a few degrees at most. This first method, while highly sensitive to small reorientations of monomers off or toward the filament axis, does not report on reorientations around the axis. To detect rotation around the filament axis, we adopted the second strategy in which we attached small plastic beads to the actin filaments. Axial turns would be immediately apparent from

Mechanism of Myofilament Sliding in Muscle Contraction, Edited by
H. Sugi and G.H Pollack, Plenum Press, New York, 1993

321

the movement of the beads. Preliminary observations indicate that actin filaments can slide over a heavy meromyosin-coated surface without axial rotations. Since rotations have been implicated in different experiments, we are currently investigating the source of the apparent discrepancy. The attached bead also serves as a handle through which we can apply force, via optical tweezers, on the filament. By letting the sliding actin filament pull the bead against the optical force, we were able to estimate the sliding force and its fluctuation.

INTRODUCTION

The establishment of the sliding theory[1][2] was a break-through in the study of muscle contraction. However, unraveling the molecular mechanism of sliding, or the molecular events at the myosin-actin interface, is yet a challenging task. A recent progress toward this goal has been the introduction of the *in vitro* motility assay systems[3-5], which allow direct and continuous observation of molecular events under an optical microscope. Here we use this system in an attempt at elucidating several aspects of molecular motions at the myosin-actin interface during sliding.

ORIENTATION OF ACTIN PROTOMERS IN SLIDING ACTIN FILAMENTS[6]

Orientation Measurement by Dual-Polarization Microscopy

To reveal the orientation of actin protomers, we used the technique of fluorescence polarization: when a fluorophore emits fluorescence, the emitted light is fully polarized along the emission transition moment of the fluorophore. By measuring the polarization of the fluorescence, therefore, one can determine the orientation of the fluorophore at the moment of emission. If the fluorophore is firmly attached to a protein molecule, the fluorescence polarization serves as a marker of the protein orientation. In our case, we labeled actin filaments with the fluorescent probe phalloidin-tetramethylrhodamine at the ratio of one fluorophore per one actin protomer. The binding appeared to be rigid.

To quantify the polarization of fluorescence under a microscope, we invented what we call "W(double view video) microscopy", in which two different images of an object are projected side by side on a video camera. The two images in the present case were two mutually orthogonal components of the polarized fluorescence, as shown in Figs. 1A and 1B. The figures clearly indicate that the fluorescence, excited with unpolarized light, was polarized along the axis of actin filaments. The transition moment of the rhodamine dye was aligned nearly parallel to the filament axis.

The polarization of fluorescence, calculated from Figs. 1A and 1B, is shown in Fig. 1C. Positive polarization shown in white color indicates that the dye molecules lay along the vertical axis in the figure, and negative polarization in black along the horizontal axis. The parallelism between the dye axis and filament axis is immediately clear in this representation. Quantitative analysis of the polarization

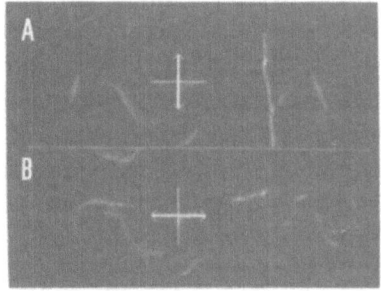
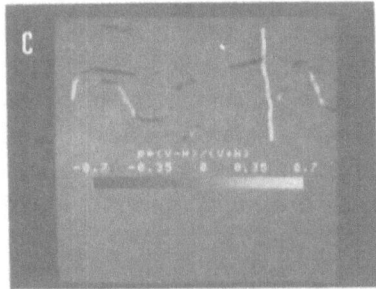

Fig. 1. (A and B), a pair of polarized fluorescence images of actin filaments stained with phalloidin-tetramethylrhodamine. Thick arrows indicate the direction of polarization. The filaments were bound, in the absence of ATP, by heavy meromyosin lying on a nitrocellulose-coated coverglass. (C), polarization, p, of fluorescence calculated from A and B. V, vertically polarized fluorescence; H, horizontal. For details, see ref. 6.

values suggests that the angle between the emission transition moment of the dye and the filament axis was about 30°.

DOES ACTIN PROTOMER ROTATE IN THE FILAMENT?

Our interest is whether actin protomers rotate in the actin filament when the filament interacts with myosin. If the sliding motion is caused by the rotation of bound myosin heads relative to the filament axis, we expect some rotation also in the actin part, which would be detected in the fluorescence polarization of the bound rhodamine. The rotation of rhodamine toward the filament axis would result in an increase in the absolute value of the polarization (in the polarization image as in Fig. 1C, vertical parts of the filaments would become more white and horizontal parts more black), and the rotation off the filament axis would reduce the polarization.

The actin filaments shown in Fig. 1 lay on a layer of heavy meromyosin which was adsorbed on a glass surface coated with a nitrocellulose film[7]. In the absence of ATP, the filaments rested still forming the rigor complex. When ATP was added, the filaments started to slide as shown in Fig. 2A. The fluorescence polarization of the sliding filaments, however, was not significantly different from that of the rigor complex in Fig. 1C. Actin filaments alone, undergoing Brownian motion in solution, also showed the same polarization (Fig. 2B). Apparently, interaction with myosin did not induce detectable reorientation of actin protomers either off or toward the filament axis.

On the nitrocellulose film, as discussed below, the number of myosin molecules actively interacting with an actin filament might be quite small. The fluorescence polarization shown in Figs. 1 and 2, on the other hand, is the average over $\sim 10^2$ actin protomers, since 1 μm of the filament contains 363 protomers and the resolution in the images is of the order of the wavelength of light. Reorientation of a small number of actin protomers may thus have taken place without affecting the polarization. To test this possibility, we made a preparation in which excess heavy meromyosin was added to actin filaments in the absence of ATP. In this fully decorated rigor complex, too, the fluorescence polarization was indistinguishable from that in other samples (Fig. 2C). Myosin-induced reorientation of actin protomers appears to be at most a few degrees. A study on muscle fibers has indicated a reorientation by a few degrees[8], which is not inconsistent with our present results.

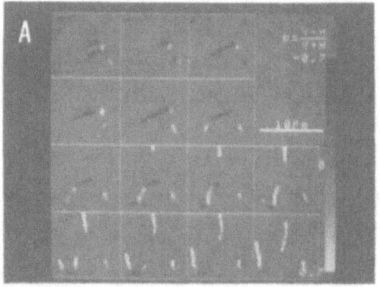

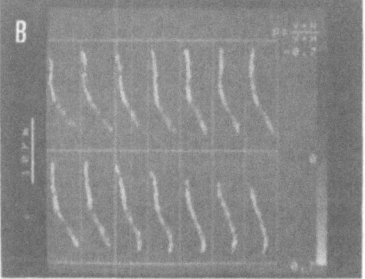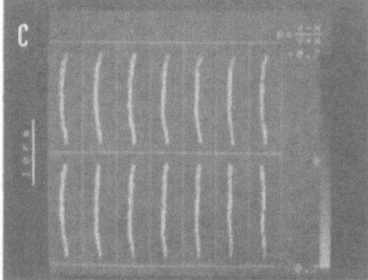

Fig. 2. (A), fluorescence polarization of actin filaments sliding over heavy meromyosin. ATP was added to the sample in Fig. 1C. Snap shots at intervals of 0.2 s, from left to right and then from top to bottom. (B), an actin filament undergoing Brownian motion in solution. Part of the filament went out of focus in some images. 33 ms intervals. (C), an actin filament undergoing Brownian motion in the presence of excess heavy meromyosin and in the absence of ATP. 33 ms intervals.

Among several models of the sliding mechanism proposed by Huxley and Simmons[9], one postulates that the force arises from the rotation of actin protomers in the filament. Now this particular model seems rather unlikely. If, on the other hand, it is part of myosin that rotates and produces force, associated reorientation of actin protomers may be quite small. This possibility cannot be excluded at the current precision.

DOES SLIDING ACTIN FILAMENT ROTATE AXIALLY?

The fluorescence polarization measurements above are not sensitive to rotations of actin protomers (or of the whole filament) around the filament axis. Since an actin filament has a helical structure, one may expect that the filament rotates axially when it slides past myosin. The axial rotation has in fact been indicated by Tanaka et al.[10] for filaments in which the sliding motion of the front part was slower than that of the rear part. To see the axial rotation in a smoothly sliding actin filament, we attached a visible marker on the filament.

The marker we chose was a polystyrene bead with a diameter of 1 μm. By coating the bead surface with gelsolin, we were able to attach the bead selectively at the barbed end of actin filaments. Since the barbed end is the rear end when the filament slides over myosin, we call this preparation "bead-tailed F-actin". The bead was visualized by lightly staining it with tetramethylrhodamine, while the actin filament was labeled with the phalloidin-tetramethylrhodamine. On a layer of heavy meromyosin on a nitrocellulose-coated glass surface, the bead-tailed F-actin was seen to slide smoothly, trailing the bead at its end. The sliding velocity was indistinguishable from that of the unbeaded filaments.

The single bead at the tail end did not serve as the marker for axial rotation, but occasionally we found another bead or two firmly attached to the tail bead (Fig. 3). Then the orientation of the bead aggregate could be followed continuously on a video monitor while the filament was sliding over heavy meromyosin. So far, we have not seen any evidence for axial rotation apart from random fluctuation: actin filaments slid over distances of many tens of micrometers without showing a complete turn of the bead aggregate.

A possible reason for the failure in observing a turn of the bead aggregate is the

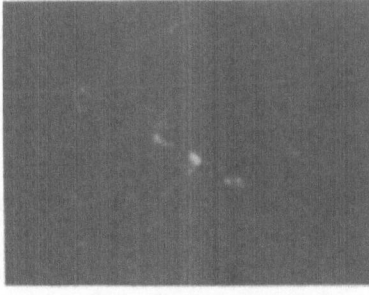

Fig. 3. An actin filament with two polystyrene beads attached to its tail. All filaments are sliding over heavy meromyosin.

hydrodynamic friction. To rotate double beads in water at one turn per second around the center of one of the double beads, one needs a torque of 100 pN•nm. At the surface of an actin filament with a radius about 5 nm, the required force is 20 pN. This force, which should be produced by the actin-myosin interaction and should operate in a direction perpendicular to the filament axis, is higher than the sliding force along the filament axis which, as shown below, is about 1 pN per μm of actin filament. Thus the bead aggregate appears to be too heavy a load for the force generators. (Note that the double beads produce a friction against the translational motion of only 0.1 pN at the sliding velocity of 5 μm/s. This friction is negligible compared to the sliding force. The friction against rotation is significant because the torque has to be generated at the surface of the thin actin filament.)

The friction against rotation is proportional to the cube of the bead radius. We therefore tested beads with a diameter of 0.5 μm, half the previous size. The small beads, too, did not show any sign of unidirectional rotation.

Thus the actin-myosin interaction does not produce a large torque, at least in the present system of heavy meromyosin lying on a nitrocellulose film. Or, the torque produced by individual power strokes does not accumulate to a level sufficient to rotate the beads, presumably because of some kind of slipping at the actin-myosin interface.

MEASUREMENT OF SLIDING FORCE

The bead at the tail served as a handle, with which we were able to manipulate a single actin filament. We used the technique of optical tweezers[11], in which a tightly focused laser beam produces an attractive force toward the light spot. The bead could be trapped in the light spot, and by moving the spot we were able to pull the bead-tailed F-actin in a desired direction, either along the myosin-coated surface or off the surface. That is, the trapping force was stronger than the sliding force.

When we turned on the laser while bead-tailed actin filaments were freely sliding over heavy meromyosin, a bead which happened to come close to the light spot was drawn to the spot and trapped. If the laser beam was sufficiently weak, a long actin filament could pull its bead out of the trap once the filament became straight (Fig. 4). From the result of this tug of war, we estimated that the sliding force was about 1 pN per μm of actin filament.

A bead attached to a short actin filament (or to a long filament but in a stronger laser beam) remained trapped in the light spot even after the actin filament became straight. In this case, the bead exhibited noticeable fluctuation in the direction of the filament. Precise analysis of the bead location, from the phase-contrast image of the bead, indicated that the amplitude of the fluctuation was several tens of nm. The average bead position was ~100 nm away from the center of the light trap and toward the actin filament, implying that the filament was always pulling the bead but the pulling force fluctuated. The average pulling force was again estimated to be about 1 pN per μm of actin filament. The pattern of the fluctuation was similar to those reported by Ishijima et al[12] in their isometric conditions.

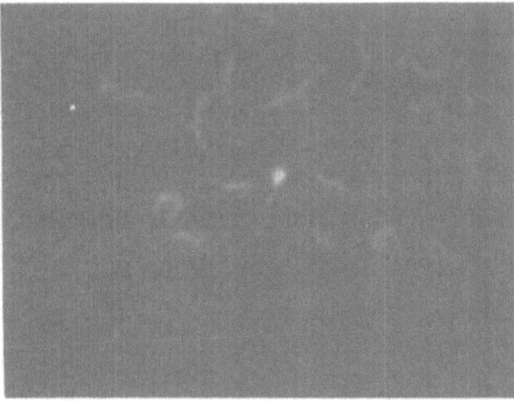

Fig. 4. A bead-tailed F-actin trapped by optical tweezers. Other actin filaments are sliding over heavy meromyosin.

In muscles, one myosin head is considered to produce an average force of the order of 1 pN[13]. The above results may then imply that, in our system, only several myosin heads were actively interacting with the actin filament which was, on the average, several μm long. With such a small number of heads, the fluctuation in generated force is an expected consequence.

Measurement of the EDTA-ATPase of the heavy meromyosin, however, suggested that the myosin density was of the order of 10^3 molecules/μm^2, as has already been reported[14]. This number suggests that tens of myosin heads must be available for 1 μm of actin filament. To reconcile this with the force data above, two extreme views are to be considered. One is that the EDTA-ATPase correctly estimated the number of active myosin heads and therefore the observed force was produced by tens of heads per μm of actin, all working more or less uniformly but probably in a more or less sequential manner. In this view, the average force produced by a single head on the nitrocellulose film is far less than the value in native muscles. The other extreme view is to suppose that, somehow, the number of really active heads was far fewer than the number suggested by the EDTA-ATPase assay. Misoriented heads, for example, should be discounted. If the active heads were as small as one or a few per μm of actin filament, the force per head was normal. In this case, each head should have spent a considerable portion of time pulling actin, since otherwise the bead would have been drawn to the trap center. The length of the stroke by each head should also have been large to account for the observed fluctuation over tens of nm.

The first of the two views above is obviously based on the view of Spudich group[14]. The second view is on the side of Yanagida group[15]. These two groups debate on the "step size", the length of the force generating movement of actin past myosin upon the hydrolysis of one ATP molecule. Spudich group maintains, on the basis of experiments using the nitrocellulose system, that the step size is well within 40 nm, whereas Yanagida group, using a glass surface covered with Sigma coat, claims that the step size exceeds 100 nm under unloaded conditions. Our present result might shed some light on this discrepancy. If myosin heads on the nitrocellulose film cannot produce the normal force, as in the first view above, it may imply that the head motion on this substrate is severely limited (e.g., by tight binding at the neck part) because, at least in the Huxley model[13], the magnitude of generated force is proportional to the size of the head motion. Or, if the number of active heads on the nitrocellulose film is much less than the estimation from the ATPase assay, as in the second view, the conclusion of Spudich group needs be revised (toward a larger step size), because their estimate critically depends on the myosin density.

At present, we cannot draw any definite conclusion. Our present experimental system, however, should prove useful in the analysis of molecular details, since it allows simultaneous measurements of displacement at nm resolution and force at subpiconewton resolution, together with real-time imaging of the filament motion.

ACKNOWLEDGEMENTS

This work was supported in part by Grants-in-Aid from the Ministry of Education, Science and Culture of Japan, and in part by Special Coordination Funds for Promoting Science and Technology given by the Agency of Science and Technology of Japan.

REFERENCES

1. Huxley, A.F. & Niedergerke, R. *Nature* **173**, 971-973 (1954).
2. Huxley, H.E. & Hanson, J. *Nature* **173**, 973-976 (1954).
3. Kron, S.J. & Spudich, J.A. *Proc. Natl. Acad. Sci. USA* **83**, 6272-6276 (1986).
4. Honda, H., Nagashima, H. & Asakura, S. *J. Mol. Biol.* **191**, 131-133 (1986).
5. Harada, Y. Noguchi, A., Kishino, A. & Yanagida, T. *Nature* **326**, 605-608 (1987).
6. Kinosita, K., Jr., Itoh, H., Ishiwata, S., Hirano, K., Nishizaka, T. & Hayakawa, T. *J. Cell Biol.* **115**, 67-73 (1991).
7. Kron, S.J., Toyoshima, Y.Y., Uyeda, T.Q.P. & Spudich, J.A. *Methods Enzymol.* **196**, 399-416 (1991).
8. Prochniewicz-Nakayama, E., Yanagida, T. & Oosawa, F. *J. Cell Biol.* **97**, 1663-1667 (1983).
9. Huxley, A.F. & Simmons, R.M. *Cold Spring Harbor Symp. Quant. Biol.* **37**, 669-680 (1972).
10. Tanaka, Y., Ishijima, A. & Ishiwata, S. *Biochim. Biophys. Acta* **115**, 94-98 (1992).
11. Ashkin, A., Dziedzic, J.M., Bjorkholm, J.E. & Chu, S. *Optics Lett.* **11**, 288-290 (1986).
12. Ishijima, A., Doi, T., Sakurada, K. & Yanagida, T. *Nature* **352**, 301-306 (1991).
13. Huxley, A.F. *Prog. Biophys. Biophys. Chem.* **7**, 255-318 (1957).
14. Uyeda, T.Q.P., Kron, S.J. & Spudich, J.A. *J. Mol. Biol.* **214**, 699-710 (1990).
15. Harada, Y., Sakurada, K., Aoki, T., Thomas, D.D. & Yanagida, T. *J. Mol. Biol.* **216**, 49-68 (1990).

Discussion

Edman: Do you think that the rotation you observed would take place *in situ* in a muscle fiber? Wouldn't it be difficult, since one end is fixed?

Kinosita: Yes, but in theory, the other can rotate.

Edman: Wouldn't it put a constraint on the sliding movement of actin?

Kinosita: The actin filament has a helical structure, so to me it's natural that the actin rotate to some extent when the myosin pulls the actin or the actin goes over the myosin.

Tregear: What is the limit of resolution of that rhodamine-polarization measurement? Supposing they were all turning around, how small an angle could you detect?

Kinosita: I would say one or two degrees. The current uncertainty is several degrees. The reason for this is that we do not know whether the filament is really straight or not. So, now we pull the filament with optical tweezers and repeat the measurement, making the precision at least two degrees.

Tregear: So, then even with the very low duty cycle that Spudich suggested in his recent paper, you would still be able to see it?

Kinosita: Yes, if the rotation is very large, something like 40° or 50°.

Homsher: Dr. Kinosita, did you say that the step size at 3 pN was 30 nm distance per ATP?

Kinosita: I used wrong words. In a sense it is a step size, but it is not a step size per one ATP hydrolysis.

Gillis: When you show that the filament with the bead is in a laser trap, it fluctuates in some cases. Does the frequency of fluctuation depend on the density of the myosin molecules?

Kinosita: That is the kind of experiment we are now in the process of conducting, so I cannot give you the answer yet. I can tell you one difference, however: when the trap force is strong, the speed of forward motion slows down, and the amplitude of the motion decreases. But we haven't changed the myosin density yet.

FORCE ON SINGLE ACTIN FILAMENTS IN A MOTILITY ASSAY MEASURED WITH AN OPTICAL TRAP

R.M. Simmons[†], J.T. Finer[*], H.M. Warrick[*], B. Kralik[**], S. Chu [**] and J.A. Spudich[*].

Department of *Biochemistry and **Physics
Stanford University
Stanford, CA 94305 USA
†MRC Muscle and Motility Unit
Kings's College London
London WC2 5RL UK

ABSTRACT

We have used an optical trap to measure or exert a force on single actin filaments via the attachment of polystyrene beads which were coated with NEM-modified HMM. In the simplest experiment, beads were attached to rhodamine phalloidin labelled actin filaments and observed to move on an HMM coated surface in the presence of ATP. Moving beads were steered into the vicinity of the trap using a PZT operated microscope stage. The minimum force needed to stop a moving bead was measured by lowering the trap strength until the bead resumed movement. By aligning the optical trap with the centre of a quadrant detector placed in an image plane of the microscope, it was possible to measure the force exerted on a filament by measuring the displacement of the bead position from the centre of the trap. In each of these experiments, the trap was calibrated by applying a Stokes force to a bead in free solution. The characteristics of the trap were studied, and the displacement of the bead from the centre of the trap was shown to be directly proportional to the applied force over a large part of the total range of the trap. The compliance of the trap could be substantially reduced by the use of feedback control to deflect the laser beam via an acousto-optic modulator. The advantages and limitations of this technique will be discussed.

INTRODUCTION

In the optical trap (or optical tweezers) technique, a beam of light from a laser is brought to a focus by a microscope objective to a diffraction-limited spot. A dielectric particle in solution that comes into the proximity of the focus is drawn into

Mechanism of Myofilament Sliding in Muscle Contraction, Edited by
H. Sugi and G.H Pollack, Plenum Press, New York, 1993

331

the laser beam and may become trapped[1]. The force exerted by the trap on the particle depends on, among other things, the particle size and the power of the laser beam, but for example a 10 mW Nd-YAG laser beam (with the power measured before the objective) produces a force from 1-10 pN on a polystyrene bead 1μm in diameter. Forces of at least 100 pN can be attained, though the maximum useful force is not known, and it is probably limited by softening of the cement at the front optical element of the objective. The range of forces, 1-100 pN, is what would be predicted for force on a single actin filament in a muscle or in a motility assay[2]. The technique has in fact been used recently to measure comparable forces exerted by a few kinesin molecules on a bead interacting with a microtubule[3]. We have now made analogous measurements of the force exerted on single actin filaments on a myosin-coated surface, and we have developed the technique to make it more quantitative.

Development of the Technique

The optical arrangement is shown in outline in Fig 1. Briefly, the attenuated beam from a 10 W Nd-YAG laser (1.06 μm) is passed through an acousto-optic modulator (AOM), and then through a series of lenses and mirrors, finally to enter a microscope (Zeiss Axioplan) to be reflected into the back aperture of the objective (Zeiss 1.25 NA Plan-Neofluar) via a dichroic mirror. Two of the mirrors are motorised so that the position of the trap can be moved, relatively slowly, over distances of about 50 μm. The AOM can be used to move the position of the trap

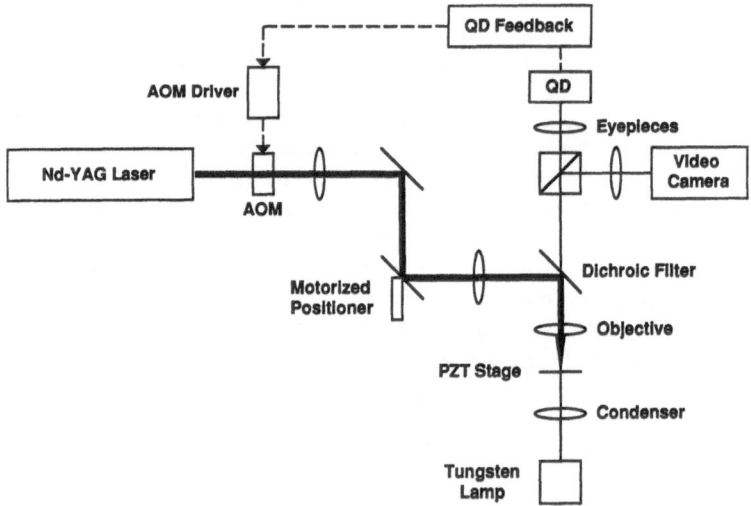

Fig. 1. Outline diagram of the optical path and associated electronics. QD, quadrant detector. AOM, acousto-optic modulator. PZT, piezo-electric transducer. Only one axis shown and fluorescence light path omitted for clarity.

over a distance of up to 5 μm with a delay of only a few microseconds. The position of an object in the trap is detected by a photodiode quadrant detector using brightfield tungsten illumination. The total magnification used is either 500x or 1000x. The microscope is also equipped for epifluorescence. The specimen is placed on the microscope stage and its position controlled over a range of about 30 μm using piezo-electric transducers (PZTs) controlled by a joystick.

Force in Motility Assay

We attached beads of diameter 0.5-2.0 μm to actin filaments labelled with rhodamine phalloidin, using NEM-modified HMM as the linking agent. Conditions were adjusted so that on average roughly 1 bead was attached to each actin filament. Coverslips were coated with nitrocellulose and rabbit skeletal muscle HMM was added[4]. At a high density of HMM (eg 800 HMM/μm^2), most filaments with beads attached to them were observed to move when they diffused on to the HMM-coated surface, in the presence of ATP. Most filaments with beads were 2-5 μm in length and their velocity was approximately 6 μm/s. This was about the same as for filaments moving without beads.

Using the joystick controlled stage, it was possible to steer a moving filament so that its bead moved into the close vicinity of the trap. When the trap strength was sufficiently low, the filament moved through the position of the trap, though with reduced velocity. When the trap strength was high, in the majority of cases the bead-filament complex was detached from the surface, and if the bead remained in the trap when the laser power was reduced to zero via the AOM or with a shutter, the filament no longer moved on the surface. In about 20-30% of the trials, the filaments remained in contact with the surface and usually these resumed movement immediately when the laser power was lowered. In such cases we measured the "escape force" by lowering the laser power until the bead escaped. The escape force varied from 1-5 pN. The trap strength was calibrated by trapping a bead in free solution and flowing water past it using a triangular waveform input to one of the PZTs.

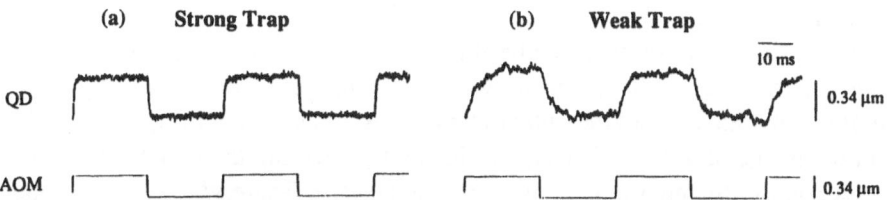

Fig. 2. Trapped bead (2 μm diameter) in free solution: output from quadrant detector (QD) and position (input signal) of acousto-optic modulator (AOM). Trap position displaced with square wave into AOM driver amplifier. (a) Laser beam power 60 mW. (b) Laser beam power 15 mW.

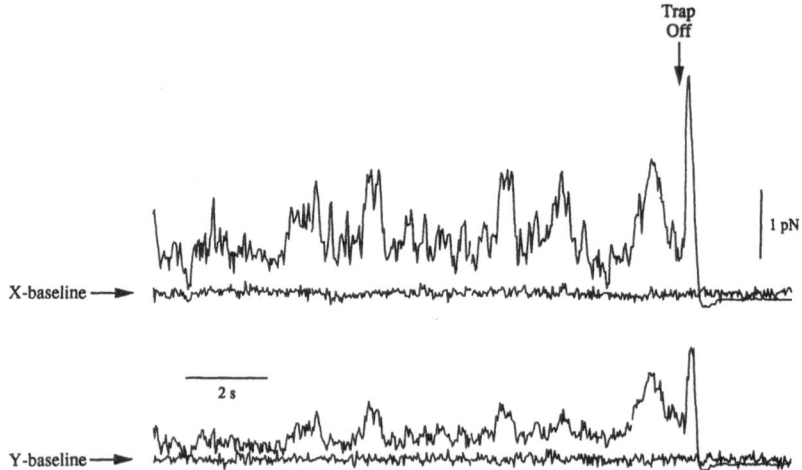

Fig. 3. Position of bead (1 µm diameter) attached to actin filament in motility assay. Records of the X and Y positions from the quadrant detector. Towards the end of the record the laser beam strength was reduced to zero. Original data smoothed by integration over successive 32 ms intervals.

Real Time Force Measurements

The technique so far described gives reliable force values but no real time dependence. We explored whether it would be possible to use the trap as a force transducer, based on the observation that in a weak trap a bead was displaced from the centre of the trap by a considerable distance (ca 0.1 µm) when a Stokes force was applied. In fact, over a range of movements the bead behaved as if it were in a potential well with a parabolic profile. Its behaviour was analogous to that of a highly damped mass on a spring, and this can be seen from the records shown in Fig. 2, in which the position of the trap was suddenly moved by a small amount by applying a square wave to the AOM driver amplifier. The movement of the bead, recorded from the quadrant detector, shows a first order delay with a time constant of the order of a millisecond. The displacement of the bead from the centre of the trap was calibrated in terms of force again by flowing solution past a trapped bead, as shown in Fig. 4a.

Now using the position of the bead in the trap to measure force, we repeated the same experiments on beads on moving filaments, but this time using tungsten illumination. Fig. 3 shows the output from the two channels of the quadrant detector from the image of a trapped bead, terminated by lowering the trap strength to zero, whereupon the filament moved away on the surface. There was a severe potential artefact in this type of experiment: the bead position became less stable in the trap when the focus of the laser beam was too close to the surface, but if this was avoided there was a sustained offset of the bead position from the centre of the trap. The corresponding forces were similar to those measured by the "escape force" method.

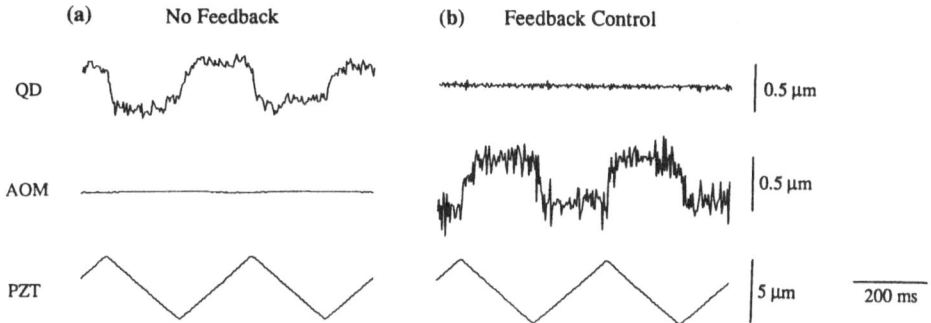

Fig. 4. Trapped bead (2 μm diameter) with and without feedback. (a) Triangular wave input to microscope stage piezo-electric transducer (PZT). No feedback. QD shows square wave response corresponding to viscous (Stokes) force. (b) As in (a) but with feedback. AOM now shows square wave response.

Limitations and Improvements

A comparatively weak trap had to be used in these experiments in order to obtain a satisfactory signal to noise ratio (see below), and the stiffness was only about 5pN/μm . The stiffness can be greatly increased by the use of feedback from the quadrant detector output to the AOM driver amplifier. With sufficiently high gain, the image of the bead remains centred on the quadrant detector when an external force is applied, as the position of the trap moves instead of the bead. Figs. 4a & 4b show records of the movement of a 2 μm diameter bead and the position of the trap with and without feedback respectively. Over a range of forces up to about half the "escape force", the movement of the trap was proportional to the external force and it can therefore be used as a force transducer.

We have also made some preliminary measurements of the characteristics of the optical trap technique such as noise on the position signals, frequency response and strength of the trap. Much of the noise at low trap strength for beads of 1 μm or greater derives from the Brownian motion of a trapped bead. Although increasing the trap strength reduces the amount of such motion, the signal to noise ratio gets worse. This is in part because the mean displacement of the bead depends on $1/k^{1/2}$, where k is the stiffness of the trap, while the displacement due to a given force depends on $1/k$. So the signal to noise ratio depends on $1/k^{1/2}$ and thus decreases as the laser power is increased. The measurements are affected by other factors such as bandwidth and instrumental noise, but at present the optimum trap strength appears to be the minimum necessary to contain the forces exerted on the bead. Size of bead is another factor: the smaller the bead, the greater is the Brownian motion, and also the greater is the effect of shot noise in the detection system. So in principle the larger the bead the better the signal to noise ratio, though there is an optimum around 1 μm for the stiffness of the trap and for the maximum trap strength.

The feedback system can separate out displacements due to external forces on the bead from changes of position *per se*. With a weak trap in the absence of feedback, the noise on the displacement signal is very large (typically for a 2 μm bead the noise has a standard deviation of about 10 nm). However, applying feedback also has the effect of transferring the Brownian motion on the bead to the force signal (ie movement of the trap by the AOM), so that in principle the noise on the displacement signal can be reduced to the level of the shot noise on the quadrant detector. The noise on the force signal with feedback has a typical standard deviation of the order of 0.05 pN. These noise figures are larger than is desirable, but some signal averaging may be possible, and it is therefore likely that this system or modifications of it will be useful in measuring individual forces and displacements between molecules involved in motile processes.

ACKNOWLEDGEMENTS

This work was supported in part by grants from the NSF and AFOSR to S. Chu, from the NIH to J.A. Spudich, and from the Fulbright Commission to R. M. Simmons.

REFERENCES

1. Ashkin, A., Dziedzic, J.M., Bjorkholm, J.E. & Chu, S. *Optics Lett.* **11**, 288-290.
2. Ishijima A., Doi, T., Sakurada, K. & Yanagida, T. *Nature* **352**, 301-306 (1991).
3. Block, S.M., Goldstein, L.S.B. & Schnapp, B.J. *Nature* **348**, 348-352 (1990).
4. Kron, S.J. & Spudich, J.A. *Proc. Natl. Acad. Sci. USA* **83**, 6272-6276 (1981).
5. Uyeda, T.Q.P., Kron, S.J. & Spudich, J.A. *J. Mol. Biol.* **214**, 669-710 (1990).

Discussion

Yanagida: Have you done the noise analysis of force fluctuation, as we did previously?

Simmons: We just got the software working for that, so not yet.

Oiwa: I have a question about the calibration of force. Near the surface, the drag force is quite different from that far from the surface. How far was the bead trapped at the calibration?

Simmons: When we do the calibration, we usually go down to between five and ten diameters of the bead.

Oiwa: Is that enough?

Simmons: It doesn't seem to change very much beyond that, and because of aberrations, the trap stops working after about 20 diameters.

Lombardi: Can you figure out what the difference is between the force you measured and the one you would expect under stiffer conditions?

Simmons: No, I don't have a model for that. I have thought about it, but I really have no idea. I suspect what we are going to find is what Dr. Yanagida has already found using needles.

Kinosita: When you measured a force of about one pN, what was the length of the actin filament pulling the bead?

Simmons: The range was between about one micron and five microns, and the beads ranged from one end to the other in position.

Kinosita: So it basically agrees with our result of about one pN/μm. I think the difference between ours and Yanagida's is the difference in the density of myosin. Do you agree with that?

Simmons: I asked him earlier and it seemed to work out about right. We haven't tried with a very high density, because it chops up the filaments rather quickly.

Kinosita: You are working with a nitrocellulose film. That makes a difference. We are also working on a nitrocellulose film, and on Sigma-coat you can put many myosin molecules.

Simmons: Right. It still chops up the filaments rather quickly and in our present stage of inexpertise, we need a lot of time to do the manipulations.

Huxley: What ATP concentration did you use?

Simmons: I think they are all about a millimolar.

COUPLING BETWEEN ATPASE AND FORCE-GENERATING ATTACHMENT-DETACHMENT CYCLES OF ACTOMYOSIN *IN VITRO*

Toshio Yanagida, Akihiko Ishijima*, Kiwamu Saito and Yoshie Harada

Department of Biophysical Engineering
Osaka University
Toyonaka, Osaka
*Honda R & D
Wako, Saitama, Japan

ABSTRACT

We have developed a high resolution force measurement system *in vitro* by manipulating a single actin filament attached to a microneedle. The system could resolve forces less than a piconewton, and has time resolution in the submillisecond range. We have used this system to detect force fluctuations produced by individual molecular interactions. We observed large force fluctuations during isometric force generations. Noise analysis of the force fluctuations showed that the force was produced by stochastic and independent attachment-detachment cycles between actin and myosin heads, and one force-generating attachment-detachment cycle corresponded to each ATPase cycle. But, the force fluctuations almost completely disappeared during sliding at the velocities of 20 to 70% of the maximum one at zero load. The analysis indicated that myosin heads produced an almost constant force for most (probably > 70%) of the ATPase cycle time, i. e., the duty ratio > 0.7. Since the myosin step size was given as (velocity) x (the duty ratio) x (the ATPase cycle time, 30 ms), it was calculated to be 40 to 110 nm, corresponding to velocities of 20 to 70 % of the maximum one (9μm/s), respectively. These values are much greater than the displacement by a single attachment-detachment cycle of actomyosin (10 -20 nm), indicating that multiple force-generating attachment-detachment cycles correspond to each ATPase cycle during sliding at velocities of > 20 % of the maximum one. In conclusion, the coupling between the ATPase and the force-generating attachment-detachment cycles of actomyosin is not rigidly determined in a one-to-one fashion but is variable depending on the load.

Mechanism of Myofilament Sliding in Muscle Contraction, Edited by
H. Sugi and G.H Pollack, Plenum Press, New York, 1993

339

INTRODUCTION

Muscle contraction is propelled by the relative sliding of actin and myosin filaments driven by the chemical energy liberated by hydrolysis of ATP[1)2)]. In the current crossbridge theory, it has been believed that the sliding force is produced by the tilting motion of myosin heads being attached to the actin filament and one attachment-tilting-detachment cycle of actomyosin corresponds to each ATPase cycle, independent of the load[3-5)]. In 1985, however, we reported that the myosin step size in myofibrils is > 60 nm near zero load, suggesting that many force-generating attachment-detachment cycles can occur during one ATPase cycle[6)]. Since then, the fundamental question of whether the coupling between the ATPase and force-generating attachment-detachment cycles of actomyosin is rigidly determined in a one-to-one fashion independent of the load has attracted much attention. This problem has been extensively studied using an *in vitro* motility assay that approaches the motion directly at the molecular level.

We have demonstrated that single actin filaments, labeled with fluorescent phalloidin, can be clearly and continuously seen under fluorescence microscopy[7)]. This has allowed us to directly observe the motions of actin when interacting with myosin in the presence of ATP. We showed that bending motion of actin filaments was modulated in solution by interacting with myosin fragments, HMM in the presence of ATP[7)]. Spudich and his colleagues found that unidirectional movement of actin filaments was observed when interacting with myosin or its subfragments (HMM and S-1) fixed on a glass surface[8)]. Using this myosin-coated surface assay, we measured the velocities of actin when its length was changed over a wide range of several μm to 40 nm. The analysis of the data indicated that the myosin step size was > 100 nm during sliding at zero load[9)]. On the other hand, using almost the same assay, the Spudich group estimated the step size to be 5 to 30 nm, which was in the range of the displacement produced by a single attachment-detachment cycle of actomyosin[10)11)].

It is unknown why the results were inconsistent with each other, although almost the same assays were used. In order to clarify this discrepancy, we have estimated the step size not from the movement at zero load, but from the force fluctuations during sliding at a loaded condition. The force fluctuations agreed with our previous conclusion. The discrepancy between Osaka and Stanford was discussed in terms of the number of myosin heads that can produce the normal sliding force in the randomly arranged system.

RESULTS

A High-Resolution Force Measurement System

Figure 1 shows a high resolution force measurement system. Single actin filaments were labeled with rhodamine phalloidin, attached to glass microneedles

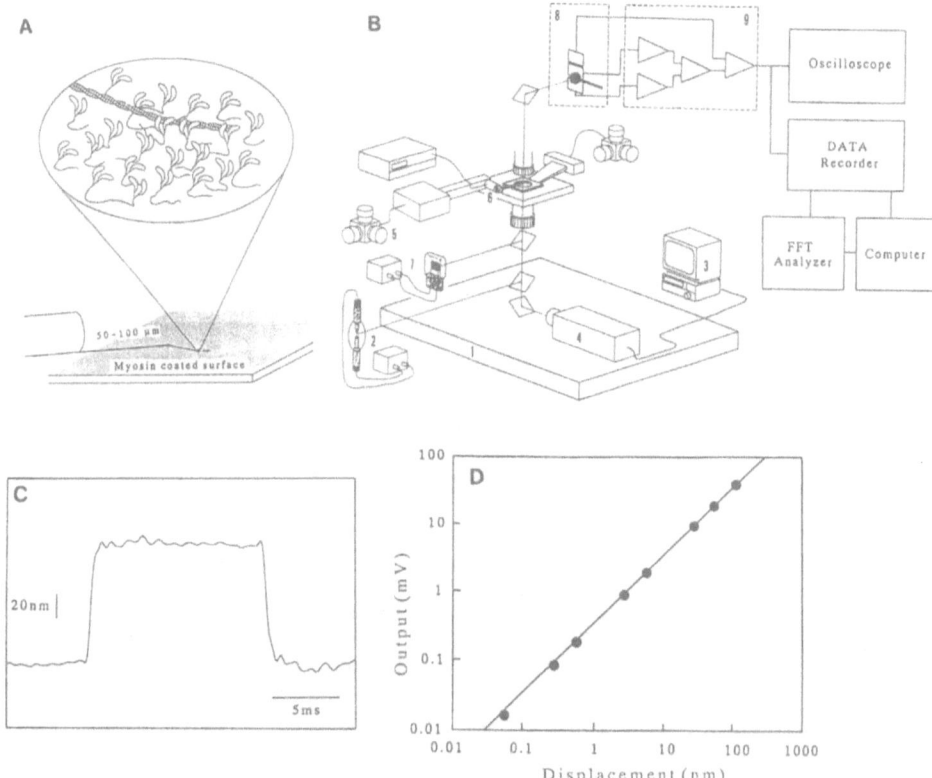

Fig. 1. A, Micromanipulation of a single actin filament by a glass microneedle. B, Schematic diagram of the ultra-high resolution force measurement system. C, Changes in the differential output when sinusoidal displacements were applied to the needle with amplitudes of 0.06 to 100 nm and a frequency of 100 Hz. D, Time course of the movement of the needle when sudden length changes of 100 nm were applied to the base of the needle by the piezo actuator.

and were manipulated in an inverted fluorescence microscope essentially as described by Kishino & Yanagida[12] (Fig. 1A). In the presence of ATP, actomyosin interactions caused sliding of the filament and bending of the needle. The force is determined by measuring the bending of the needle.

The bending of the needle was measured by the system shown in Fig. 1B. An image of the needle was projected on to a pair of photodiodes and the displacement of the needle was obtained from differential output of the photodetector. The differential output was very sensitive to movement of the particle as already shown[13-15] and was linear for motions from 0.1 nm to 100 nm (Fig. 1C). The lower limit of movement detection was measured to be less than 0.1 nm.

To minimize deterioration of the time resolution due to the needle's mass and viscous drag in the solution, we used very fine microneedles 50 to 100 μm long and about 0.3 μm in diameter, considerably shorter and stiffer than in our previous experiments[12]. The response time of the needle's motion driven by actomyosin interactions is estimated to be less than 1 ms (Fig. 1D).

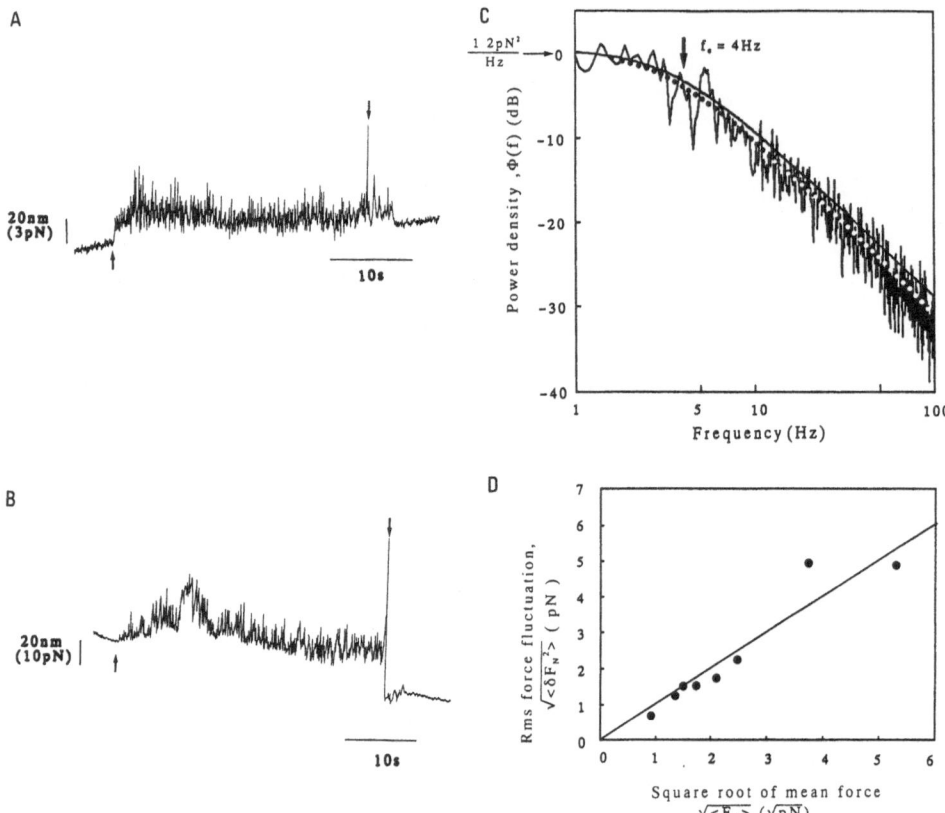

Fig. 2 . A, Force fluctuation near isometric conditions. B, The power density spectrum ($\Phi(f)$) of the force fluctuations. D, The relationship between the rms amplitude of the force fluctuations ($\sqrt{<\delta F_N^2>}$) and the square root of the mean force ($\sqrt{<F_N>}$)

Force Fluctuations near Isometric Conditions

Minimal sliding was obtained by measuring the displacements of needles due to a small number of myosin heads interacting with an actin filament. In this situation, large force fluctuations were observed immediately when the actin filament began to interact with the myosin-coated surface. Figures 2A and B show the force f fluctuations measured using needles with 0.15 and 0.5 pN/nm stiffness respectively.

Figure 2C shows the power density spectrum of the data from panel B. The power density spectrum ($\Phi(f)$) nearly fits a Lorentzian curve (solid line, $\Phi(f) \propto 1/(1 + (f/f_c)^2)$, in which the power density decreases with $1/f^2$ at high frequencies. The corner frequency (f_c), where $\Phi(f)$ is one-half the low-frequency asymptote, was 5.1 ± 0.4 Hz (S.E.M., $n = 8$) when the pH of the medium was 7.8. f_c decreased to < 2 Hz at pH 6.5. Since the actomyosin ATPase activity decreases several-fold when pH is reduced from 7.8 to 6.5[16], the force fluctuations seem to be closely coupled to the ATPase cycle.

A Lorentzian power spectrum would be expected if the actomyosin interactions occur randomly and independently of each other[17][18]. In that case, the root mean square (rms) deviation from the mean force should be proportional to the square root of the mean force[17][18]. Therefore in Fig. 2D we plotted the rms deviation $(\sqrt{<\delta F_N^2>})$ vs. the square root of the mean force $(\sqrt{<F_N>})$ for experiments with varying numbers of interacting heads. The linear relationship indicates that the force fluctuations are derived from independent, asynchronous force generators.

The force fluctuations are therefore analyzed in two kinds of random two states models. In the simple ON-OFF two state model, the power density spectrum is Lorentzian and the rms force fluctuation is expressed by

$$\sqrt{<\delta F_N^2>} = \sqrt{(1-p_0) \cdot F_0} \cdot \sqrt{<F_N>} \qquad (1)$$

where p_0 is the proportion of the time heads are in the ON state and F_0 is the force of a head in the ON state[17][18]. Therefore, the slope of the line shown in Figure 2D gives the value of $\sqrt{(1-p_0) \cdot F_0}$ From the corner frequency f_c of the power spectrum and the ATPase rate (V_{ATP}), the two solutions for K_+ and K_- are obtained to be 8.4 / s and 23 / s or 23 / s and 8.4/ s, respectively. The corresponding values of p_0, F_0 and $<F>$ are 0.28, 1.5 pN and 0.42 pN or 0.72, 3.6 pN and 2.6 pN, respectively (see ref. 19 for detail).

In several experiments, the mean force $<F_N>$ was less than 2.6 pN (Fig. 2D). The former case is probably the correct solution. If so, the mean force per myosin head $<F>$ is 0.4 pN, which is several times smaller than the value expected for a muscle fiber, 1 to 2 pN per head[3]. It is likely that the force vectors of myosin heads randomly oriented on the surface are not all parallel to the axis of actin filament, decreasing the axial force exerted on the filament. If this is correct, the true force would be larger than 0.4 pN, probably 1 pN.

Since the ON-OFF model is too simple to explain the actual events in force generation, we also analyzed the data by a computer simulation based on A. F. Huxley's (1957) model of the cross-bridge cycle[3]. The force fluctuations and the power density spectrum under isometric conditions could be simulated well by this model (Fig. 2C), as well. To fit the results, the parameters of the model were set to the following values : f, (ON rate at 0 < x < 10 nm) = 1 x s^{-1}. g_1(OFF rate at 0 < x < 10nm) = 2 x s^{-1}. g_2 (OFF rate at x < 0) = 390 s^{-1}.

With both models in order to fit the experimental data, it was necessary to relate each ON-OFF cycle to splitting to one ATP molecule. This result suggests that during isometric force generation, chemo-mechanical coupling is one-to-one.

Force Fluctuations during Sliding

When the length of actin filaments in contact with the myosin-coated surface was relatively long (0.5 to 3 μm) and the generated forces were high (20 to 100 pN), the

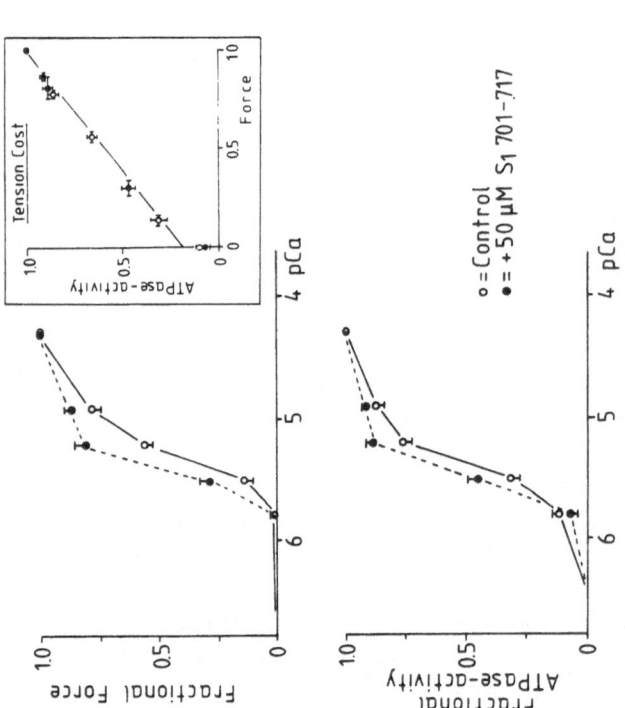

Fig. 3. Force fluctuations during the sliding at low (A), middle (B) and high velocities (C). Top row, the time course of force generation after an actin filament contacts the myosin-coated surface. Needle stiffness was 0.32 (A), 0.09 (B) and 0.15 pN/nm (C). The broken lines are averaged curves ($<X>$) obtained by fitting a fifth order polynomial to the displacement trace. Second row, the deviations ($\delta X = X - <X>$) of the needle position from the averaged curves. Third row, thermal vibrations of the needle alone ($\delta X_n = X_n - <X_n>$). The ordinates in second and third rows are scaled up 1.5-fold compared with the top ones. Fourth row, The time courses of the sliding velocities Bottom row, the values of the attenuation factor α at each time during sliding, calculated using the stiffness of actomyosin attachments ($K_{mN}(t)$) as described in the text. The sampling time was set to 2.5 ms for A and 0.25 ms for B and C.

actin filaments were observed to slide 150 to 600 nm at average velocities of 0.15 to 4 μm/s (Max; 6 μm /s). These velocities correspond to 1.6 to 45 % (Max; 70 %) of the velocity at zero load which is about 9 μm/s (27°C).

At the average velocities less than 0.2 μm/s the force fluctuated as much as under isometric conditions (Fig. 3A). But the force developed quite smoothly at velocities > 1 μm/s (Figs. 3B & C). The data during sliding were corrected for intrinsic thermal fluctuations of the needle and attenuation by the actomyosin attachments as described earlier. The ratios of rms force fluctuations to mean force averaged over the sliding period were $\sqrt{<\delta F_N{}^2>}/<F_N> = 0.036$ for Fig. 3B and 0.023 for Fig. 3C. These values are 10 and 6-fold smaller than those obtained under isometric conditions at the same mean forces (0.35 at 10 pN and 0.14 at 50 pN Fig. 2D). These results indicates that the myosin heads produce force much more smoothly during sliding than under isometric conditions. Therefore, the proportion (p_0) of the time the heads are in the force generating state seems much greater than that under isometric conditions, 0.28.

In the ON-OFF model, for example, rearrangement of Eq. 1, noting that $F_0 = <F> / p_0$ gives: $\sqrt{<\delta F_N{}^2>}/<F_N> = \sqrt{(1-p_0)/p_0} / \sqrt{N}$. Taking $<F_N>$ to be the overall average of force during the force development and N as (maximum mean force at the end of each trace) / 0.4 pN, p_0 during the sliding was found to be 0.94 for Fig. 3B and 0.88 for Fig. 3C. In these estimations, N was assumed to be maximal during the sliding period. If N was smaller, p_0 would be greater. Thus, the results imply that myosin heads produce an almost constant force for most (> 70 %) of the ATPase cycle time during moderate to high velocity sliding.

DISCUSSION

Large fluctuations were observed under nearly isometric conditions, similar to membrane current fluctuations due to channel gating in electrophysiological systems with small numbers of channels[20]. The noise analysis of the force fluctuations showed that the fluctuations seen in Figure 2a were produced by only about 5 heads with one force-generating attachment-detachment cycle corresponding to each ATPase cycle. Therefore, some of the force impulses seen in Figure 2a should be produced by single myosin heads. Thus, individual mechanical interactions of protein molecules coupled with the ATPase cycle could be directly observed for the first time. In this sense, the force impulses correspond to motions of individual ions moving across the membrane produced by ion-pump proteins coupled with the ATP hydrolysis rather than membrane current flowing through ion channels.

The force fluctuations under isometric conditions were consistent with the widely accepted idea that muscle contraction is produced by stochastic and independent interactions between actin and myosin heads, and one interaction corresponds to each ATPase cycle[3]. Our data thus provided direct experimental substantiation of these basic principles.

However, this did not apply when the filaments were sliding. At velocities greater than 1 μm/s, the force fluctuations almost completely disappeared (Fig. 3B & C). It is unlikely that smooth force development is due to a precisely regular sequence of mechanical impulses from the interacting molecules. The analysis implies that myosin produces an almost constant force for ~70 % of the ATPase cycle time during sliding, in disagreement with the usually accepted hypotheses[21-23].

In these hypotheses, it is assumed that one force-generating attachment-detachment cycle occurs during hydrolysis of one molecule of ATP, independent of the load[3)5)24-27]. Therefore, the force-generating time, obtained as (myosin step size, 10 - 15 nm) / (sliding velocity, v), decreases as the velocity increases. The duty ratio, p_0 (the force-generating time / the ATPase cycle time) also decreases. For example, at v = 4 μm/s, the duration of the working stroke would be 2.5 - 4 ms and p_0 = 0.08 - 0.13. These values are about ten-fold smaller than the values obtained here. If the myosin step size is much greater than 10-15 nm, then the proportion of cycle time spent in the force generating state would be increased. But, it is unlikely that the myosin step size is much greater than the size of a myosin head (~20 nm)[28]. Therefore, the large value of p_0 suggests that multiple force-generating attachment-detachment cycles between actin and myosin correspond to each ATPase cycle during sliding at velocities of > 20 % of the maximum one. Thus, we infer that the coupling ratio (mechanical-to-chemical) is one-to-one in nearly isometric conditions and increases up to several-to-one as sliding velocity increases.

Next, we discuss the discrepancy between us and Stanford group. It is probably due to ambiguity in estimation of myosin heads that can produce normal fast motion of actin in the myosin-coated surface, in which myosin heads are randomly arranged. When the myosin step size was estimated from the analysis of sliding movement, all myosin heads in the vicinity of the actin were counted, assuming that they could produce the normal fast motion of actin. But it has been shown when using giant thick filaments of molluscan smooth muscle[29)30)] and synthetic filaments of rabbit skeletal myosin[31] that only myosin heads correctly oriented with respect to actin filaments can produce normal motion. Therefore, it is expected that the number of myosin heads that can produce normal fast motion of actin is considerably smaller than that in the vicinity of the actin because the myosin heads were randomly arranged. Preliminary force measurements indicate that the force in the HMM-coated nitrocellulose film assay is unexpectedly small although that in the myosin-coated sillicone surface assay is not so much small, suggesting that the number of effective heads is much smaller than that which had been estimated from the density of HMM on the surface. If this situation is considered, the myosin step size could be much larger than their estimations. Stanford's group has recently indicated from slow motion of actin on a very sparse HMM-coated surface that the myosin step size is 5 to 28 nm[32]. But it is likely that the slow motion on the random system was due to myosin heads oriented to misfit the polarity of actin, so that the step size of correctly oriented myosin would be much larger. In order to confirm these points in detail, we are estimating the myosin step size using synthetic myosin filaments where myosin heads are regularly oriented.

ACKNOWLEDGEMENTS

We thank Y.E. Goldman for helpful comments about this paper and stimulating discussions.

REFERENCES

1. Huxley, A.F. & Niedergerke, R. *Nature* **173**, 971-973 (1954).
2. Huxley, H.E. & Hanson, J. *Nature* **173**, 973-976 (1954).
3. Huxley, A.F. *Prog. Biophys. Biophys. Chem.* **7**, 255-318 (1957)
4. Huxley, H.E. *Science* **164**, 1356-1366 (1969)
5. Huxley, A.F. & Simmons, R.M. *Nature* **233**, 533-538 (1971).
6. Yanagida, T., Arata, T. & Oosawa, F. *Nature* **316**, 366-369 (1985).
7. Yanagida, T., Nakase, M., Nishiyama, K. & Oosawa, F. *Nature* **307**, 58-60 (1984).
8. Kron, S.J. & Spudich, J.A. *Proc. Natl. Acad. Sci. USA* **83**, 6272-6276 (1986).
9. Harada, Y., Sakurada, K., Aoki, T. Thomas, D.D. & Yanagida, T. *J. Mol. Biol.* **216**, 49-68 (1990).
10. Uyeda, T.Q.P., Kron, S.J. & Spudich J.A. *J. Mol. Biol.* **214**, 699-710 (1990).
11. Yano-Toyoshima, Y., Kron, S.J. & Spudich, J.A. *Proc. Natl. Acad. Sci. USA* **87**, 7130-7134 (1990)
12. Kishino, A. & Yanagida, T. *Nature* **334**, 74-76 (1988).
13. Borejdo, J. & Morales, M.F. *Biophys. J.* **20**, 315-334 (1977).
14. Iwazumi, T. *Am. J. Physiol.* **252**, 253-262 (1987).
15. Kamimura, S. & Kamiya, R. *Nature* **340**, 476-478 (1989).
16. Ohno, T. & Kodama, T. *J. Physiol. (Lond.)* in the press.
17. Lee, Y.W. *Statistical Theory of Communication* **Chap. 8** (John Wiley & Sons, Inc. USA) (1969).
18. Stevens, C.F. *Biophys. J.* **12**, 1028-1047 (1972).
19. Ishijima, A., Doi, T. Sakurada, K. & Yanagida, T. *Nature*, **352**, 301-306 (1991)
20. Katz, B. & Miledi, R. *Nature New Biol.* **232**, 124-126 (1971).
21. Bagshaw, C.R. *Muscle Contraction* (Chapmann and Hall, London,1982).
22. Huxley, H.E. *J. Biol. Chem.* **265**, 8347-8350 (1990).
23. Spudich, J.A. *Nature* **348**, 284-285 (1990).
24. Taylor, E.W. *CRC Critical Rev. Biochem.* **6**, 103-164 (1979).
25. Eisenberg, E. & Greene, L.E. *Ann. Rev. Physiol.* **42**, 293-309 (1980).
26. Tonomura, Y. *Energy-transduing ATPases - Structure and Kinetics* (Cambridge Univ. Press) (1986).
27. Hibberd, M.G. & Trentham, D.R. *Ann Rev. Biophys. Biophys Chem.* **15**, 119-161(1986).
28. Cooke, R. *CRC Critical Rev. Biochem.* **21**, 53-118 (1986).
29. Sellers, J.R. & Kachar B. *Science* **249**, 406-408 (1990).
30. Yamada, A., Ishii, N & Takahashi, K. *J. Biochem.* **108**, 341-343 (1990).
31. Ishijima, A. & Yanagida, T. *Proc. of the 29th Biophys. Soc. Japan* S195 (1991).
32. Uyeda, T.Q.P., Warrick, H.M., Kron, S.J. & Spudich, J.A. *Nature* **352**, 307-311(1991).

Discussion

Morales: Is an alternative possibility that the system could slide and during a lot of the time it is going by thermal motion simply kept close to the track, and that occasionally it performs an ATPase and that gives it direction? In other words, it is somewhat like the DNA repressor system. The thermal business drives it back and forth. But in this case, it is as though you had a few valves. Is that a possibility?

Yanagida: Yes. I think each step in the unidirectional motion must be produced by some energy. In any case, the free energy must be fractionated for use in each step. It is not easy to explain this mechanism.

Simmons: I was worried when I read your paper about this. You seem to make the measurements under isometric conditions and during shortening under different conditions: do you have different needles or have different densities? I wondered why you didn't just use the stiffest needle you could get results from and then do something like an isotonic release with that needle?

Yanagida: Yes, Yale Goldman visited my lab and we tried that experiment, but unfortunately, we did not succeed. The situation of a myosin head interacting with the surface is very complicated. Therefore, we will do that experiment by using the thick filament, where the myosin heads are oriented. But, I showed that we observed the force fluctuation both during the sliding and after reaching the plateau. I am sure that the force fluctuation—the small fluctuation—is not due to, for example, the compliance of the needle or the compliance of the thick filament or the large number of myosin heads.

Huxley: Your theory assumes that each myosin acts at random time intervals. Could the small scale of fluctuations be due to myosin operating at regular intervals, perhaps once per half turn of the actin or something of that kind?

Yanagida: If, for example, 50 myosin heads produce a force in turn regularly, then the force fluctuation is very small even if the step size is only 10 nm or the duty ratio is only 0.05. I think that myosin heads, at least under isometric conditions, produce the force independently and randomly. So, I think it is unlikely that such a strong cooperative activity takes place during sliding.

Sugi: If you put your system in rigor, then what kind of external noise do you pick up? In other words, to what extent might the fluctuation be external in origin?

Yanagida: Of course, the force fluctuation in rigor is much smaller than the force fluctuation of the micro-needle alone.

Sugi: What does the frequency spectrum look like?

Yanagida: The noise is very small, so it is not easy to take a perspective on.

Homsher: When I read your paper in *Nature*, describing your analysis based on the on/off cross-bridge model, I found that each cross-bridge in the isometric state spends only about 30% of its time strongly attached (Ishijima et al. *Nature* **352**, 301-306, 1991). How does that square with the stiffness measurements that Huxley and Goldman have made in which they reckon something like 70% of the cross-bridges are attached at any one time?

Yanagida: Of course, in this analysis we used an on/off two-state model. One possibility is that the model is too simple. We measured stiffness during force generation in the *in vitro* motility assay, and we found that the stiffness is unexpectedly small.

Goldman: Just a point that relates to what you are asking, Earl (Homsher). Toshio's on/off model distinguishes between force-generating and non-force-generating, whereas our stiffness measurements, if they measure attachment, show the interaction between actin and myosin, whether they are producing force or not. So you get more stiffness at a lower on/off ratio, and are still consistent that way.

KINETIC PROPERTIES OF THE ATP-DEPENDENT ACTIN-MYOSIN SLIDING AS REVEALED BY THE FORCE-MOVEMENT ASSAY SYSTEM WITH A CENTRIFUGE MICROSCOPE

S. Chaen, K. Oiwa, T. Kobayashi, T. Gross, E. Kamitsubo*, T. Shimmen**, and H. Sugi

Department of Physiology
School of Medicine
Teikyo University
Tokyo 173, Japan
*Biological Laboratory
Hitotsubashi University
Tokyo 186, Japan
**Department of Life Science
Faculty of Science
Himeji Institute of Technology
Himeji 671-22, Japan

ABSTRACT

To study the kinetic properties of the ATP-dependent actin-myosin sliding responsible for muscle contraction, we developed an *in vitro* force-movement assay system, in which centrifugal forces were applied to myosin-coated polystyrene beads sliding along actin cables of giant algal cells in the presence of ATP. Under constant centrifugal forces directed opposite to the bead movement ("positive" loads), the beads moved with constant velocities. The steady-state force-velocity (P-V) curve thus obtained was double-hyperbolic in shape, being analogous to the P-V curve of single muscle fibers. Under constant centrifugal forces in the direction of the bead movement ("negative" loads), on the other hand, the beads also moved with constant velocities. Unexpectedly, the velocity of bead movement did not increase with increasing negative loads, but decreased markedly (by 20-60%). We also studied the effect of centrifugal forces at right angles with actin cables on the bead movement.

Mechanism of Myofilament Sliding in Muscle Contraction, Edited by
H. Sugi and G.H Pollack, Plenum Press, New York, 1993

351

INTRODUCTION

It has become fashionable to use an *in vitro* motility assay system, consisting of fluorescently labeled F-actin and myosin heads (subfragment-1, S-1) fixed on a glass surface[1], for studying the mechanism of muscle contraction. From the standpoint of muscle physiology, however, the system is totally inadequate for studying the mechanism of muscle contraction for the following reasons: (1) F-actin should continuously "search" and "find" new myosin heads by changing its configulation, a condition that differs too far from the myofilament sliding; (2) F-actin movement is claimed to be without external load, a condition that never happens in muscle in the animal body; (3) it has been argued whether F-actin movement is actually free-loaded or not, implying that the mechanical condition of F-actin movement is ill defined; (4) the system is unsuitable for studying basic kinetic properties of actin-myosin sliding (such as *P-V* relation); and (5) even though F-actin versus S-1 sliding is shown to generate force[2], it does not necessarily mean that the events taking place on a glass surface is the same in quality as those taking place in muscle[3]. As a matter of fact, it has been shown that myosin subfragment-2 plays an essential role in muscle contraction.

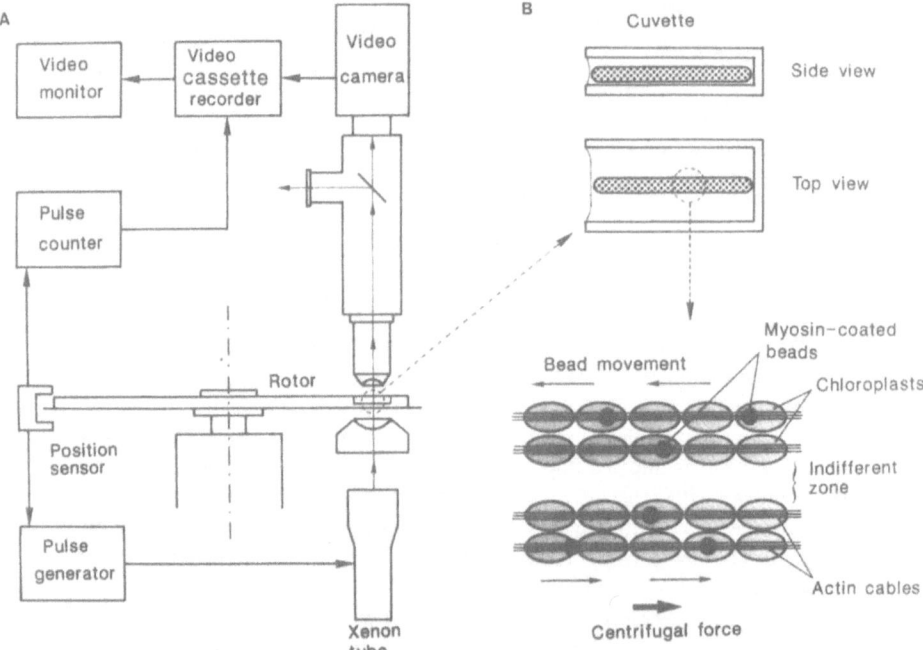

Fig. 1. Diagram of the experimental arrangement. (*A*) Composition of the centrifuge microscope and the video system. (*B*) Principle of application of centrifugal forces serving as positive or negative loads to myosin-coated beads moving on actin cables. Note that the direction of bead movement is reversed across the indifferent zone. From Oiwa et al.[6].

A most basic characteristic of the ATP-dependent actin-myosin sliding in muscle is the *P-V* relation, which shows how the rate of work (and energy) production is influenced by external loads[4]. To eliminate the gap between the biochemistry of actomyosin in solution and the physiology of contracting muscle, we developed a new *in vitro* assay system, which enabled us to study the steady-state *P-V* relation of myosin-coated beads sliding on the well-organized actin filament bundles (actin cables) in the giant algal cell by use of a centrifuge microscope.

MATERIALS AND METHODS

Diagram of the experimental arrangement is shown in Fig. 1. The centrifuge microscope consisted of a light microscope, a rotor (diameter 16 cm) that was rotated at 250-5000 rpm (4 - 1900 x *g*), and a stroboscopic light source[5] (Fig. 1A). An acrylite centrifuge cuvette containing an internodal cell preparation was mounted on the rotor (Fig. 1B). The internodal cell preparation (diameter ~ 0.5 mm, length ≈ 1 cm) was prepared first by cutting open an internodal cell at both ends, then introducing polystyrene beads coated with rabbit skeletal muscle myosin (diameter 2.8 μm, specific gravity 1.3) together with MgATP solution (5 mM EGTA, 6 mM $MgCl_2$, 1 mM ATP, 200 mM sorbitol, 50 mM KOH, and 20 mM Pipes, pH 7.0), and finally ligating both ends with polyester thread. Further details of the methods used have been described elsewhere[6]. Since the polarity of actin cables determining the direction of bead movement is reversed across the indifferent zone, it was possible to observe beads moving on actin cables either in the direction opposite to, or the same as, the direction of applied centrifugal forces (Fig. 1B).

The amount of centrifugal force (= load) *F* on the bead was calculated as,

$$F = \Delta\rho V r \omega^2,$$

where $\Delta\rho$ is the difference in density between the bead and the surrounding medium, *V* is the bead volume, *r* is the effective radius of centrifugation, and ω is the angular velocity of the rotor[6]. All experiments were made at room temperature (20-23°C).

RESULTS

Steady-state *P-V* Relation of Myosin-coated Bead versus Actin Cable Sliding in Response to Positive Loads

Fig. 2 shows selected video frames of myosin-coated beads sliding on actin cables in one direction determined by the polarity of actin filaments under various centrifugal forces directed opposite to the bead movement. When the amount of centrifugal forces serving as positive loads were gradually increased, the response of the beads was not uniform, but differed from bead to bead; under a given centrifugal

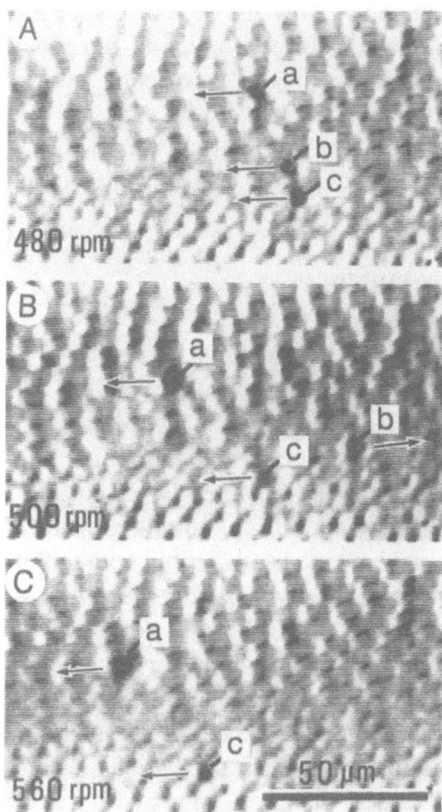

Fig. 2. Selected video frames showing myosin-coated beads moving on actin cables under three different rotation rates of the rotor indicated in each frame. Arrows show the direction of bead movement. (A) Beads a, b, and c move forward with nearly the same velocities. (B) Beads a and c continue to move forward, while bead b just starts to go backward because the amount of positive load by the applied centrifugal force exceeds P_0 of bead b. (C) Beads a and c still move forward. From Oiwa et al.[6].

force, some beads were detached from actin cables to flow away in the direction of centrifugal force, while the other beads still continued moving (Fig. 1B). This indicates that, though the unloaded velocity of movement is nearly the same[7], a considerable variation exists in the load-bearing ability among the beads moving on actin cables.

As shown in Fig. 3, the bead moved with a constant velocity under a constant centrifugal force, indicating the presence of a definite steady-state relation between the force (= load) generated by actin-myosin interaction and the velocity of actin-myosin sliding. When the load on the bead eventually reached a value equal to the maximum isometric force (P_0) generated by myosin molecules interacting with actin cables, the bead stopped moving to stay at the same position for 5-10 s until it was

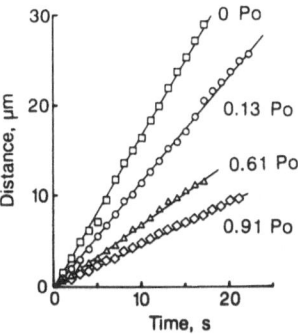

Fig. 3. Constant velocity movements of a myosin-coated bead on actin cables under four different amounts of positive load. The amount of load on the bead is expressed relative to P_0. Regression lines were drawn by the least-squares method. The amount of P_0 was 15 pN. From Oiwa et al.[6].

detached from actin cables to flow away in the direction of centrifugal force. The values of P_0 showed a considerable variation from 1.9 to 39 pN (mean 19 pN, n = 18). If the force exerted by each myosin head is assumed to be 1 pN, the number of myosin heads involved would be only 2 to 40. The maximum velocity of bead movement (V_{max}) showed a much smaller variation of 1.6 - 3.6 μm/s (mean 2.0 μm/s, n = 18).

Fig. 4 is a typical *P-V* relation obtained from the beads with high values of P_0 (10-39 pN). The curve was fitted to a hyperbola in the low force range from zero to 0.3 - 0.5 P_0, while in the higher force range the velocity decreased steeply with increasing forces. The above shape of *P-V* curve resembles that of the double-hyperbolic *P-V* curve of single skeletal muscle fibers[8], indicating that this assay system retains basic characteristics of contracting muscle. Fig. 5 is an example of *P-*

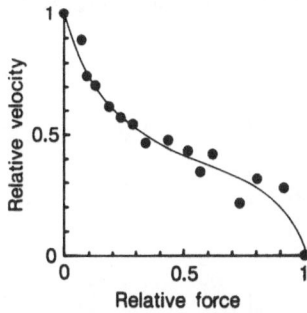

Fig. 4. Typical example of the steady-state *P-V* relations of bead movement in the positive load region, obtained from the beads with high P_0 values (10-39 pN). Data points were obtained in a random order. The velocity and the force (= load) values are expressed relatively to V_{max} and P_0 respectively. The curve in the low-force range (from 0 to 0.4 P_0 in this case) was fitted to a hyperbola ($a/P_0 = 8$) by the nonlinear least-squares method. The results shown in Fig. 3 and 4 were obtained from one and the same bead. From Oiwa et al.[6].

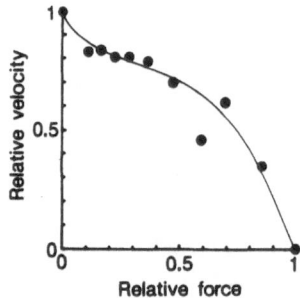

Fig. 5. Example of the steady-state *P-V* relations of bead movement obtained from the beads with low P_0 values (3.5-5.9 pN). The data points were obtained in a random order from a bead with P_0 of 3.8 pN. Note the difference in shape between the curves shown in Figs. 4 and 5. From Oiwa et al.[6].

V relation obtained from the beads with low values of P_0 (3.5-5.9 pN). The hyperbolic part of the curve in the low force range was mush less pronounced, being confined to the velocities above 0.6-0.8 V_{max}.

Steady-state *P-V* Relation of Myosin-coated Bead versus Actin Cable Sliding in Response to Negative Loads

Next, we applied centrifugal forces in the same direction as the bead movement serving as "negative" loads, so that the beads moving on actin cables were "pushed" forward instead of being "pulled" backwards with ordinary positive loads. To our surprise, the velocity of bead movement was not increased but was decreased with increasing negative loads, as shown in Fig. 6. As with positive loads, the bead moved with a constant velocity under a constant negative load. An example of the *P-V* relation of the bead subjected to negative loads is presented in Fig. 7. The velocity of bead movement decreased by 20-60% from V_{max} if the amount of negative load was

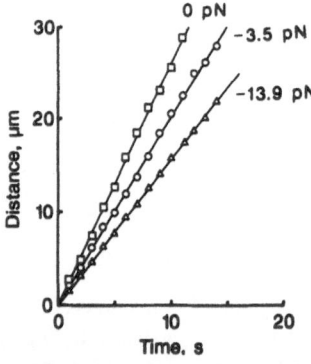

Fig. 6. Constant velocity movements of a myosin-coated bead on actin cables under three different amounts of negative load. The amount of load on the bead is expressed in pN. Regression lines were dawn by the least-squares method. Note that the velocity of bead movement decreases with decreasing amount of negative load. From Oiwa et al.[6].

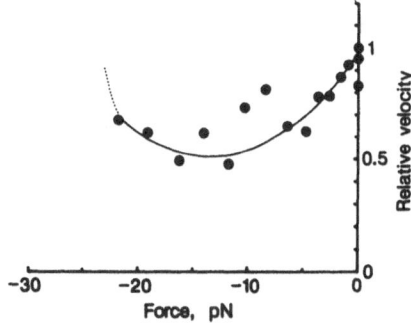

Fig. 7. Steady-state *P-V* relation of bead movement in the negative-load region. Data points were obtained in a random order. The velocity values are expressed relative to V_{max}, while the force (= load) values are expressed in pN. The results shown in Figs. 6 and 7 were obtained from the same bead. The bead was detached from the actin cables with a negative load of ≈25 pN. From Oiwa et al.[6].

gradually increased. The bead was eventually detached from actin cables to flow away in the direction of centrifugal force.

Steady-state *P-V* Relation of Myosin-coated Bead versus Actin Cable Sliding in Response to Laterally Applied Forces

The experiments were also made, in which the beads moving on actin cables were subjected to centrifugal forces in the direction at right angles with actin cables (and parallel to the rotor plane), by rotating the position of the internodal cell preparation in the centrifuge cuvette by 90°. The results obtained are summarized in Fig. 8. The velocity of bead movement decreased with increasing laterally applied

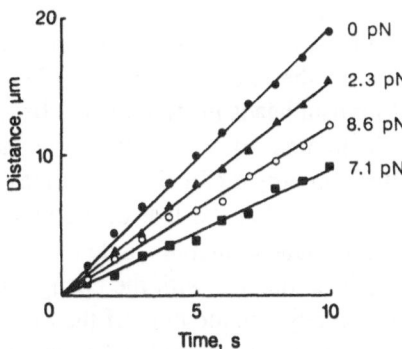

Fig. 8. Constant velocity movements of a myosin-coated bead on actin cables under four different centrifugal forces at right angles to actin cables. The lateral forces were directed towards the left side of the bead. Regression lines were drawn by the least-squares method. The velocity of bead movement decreases under laterally applied forces.

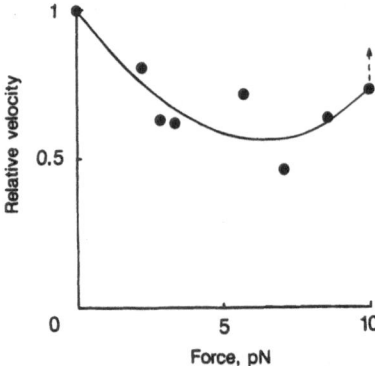

Fig. 9. Steady-state *P-V* relation of bead movement under laterally applied centrifugal forces. Data points were obtained in a random order. The results shown in Figs. 8 and 9 were obtained from the same bead.

forces, though the bead moved with a constant velocity under a constant lateral force (Fig. 8). The velocity decreased by 20-55% from V_{max} with increasing lateral forces until the bead was eventually detached from actin cables to flow away in the direction of centrifugal force (Fig. 9). *P-V* curves were obtained irrespective of whether centrifugal forces were applied from the right side or from the left side of the bead moving on actin cables.

DISCUSSION

In the present study, we have succeeded in obtaining the steady-state *P-V* relation of the ATP-dependent actin-myosin sliding under a condition much simpler than that of contracting muscle. Under a constant positive load, the bead continued to move on actin cables over many micrometers (Fig. 3), whereas in contracting muscle the distance of actin-myosin sliding in each half sarcomere is limited to ≈ 1 μm and is associated with the change in the amount of overlap between the thick and thin filaments. The number of myosin heads involved in the bead movement is estimated to be only 2 to 40, while the number of actin-myosin linkages involved in isometric force generation in each half sarcomere of a singe fiber may be ≈10^9.

On this basis, it is of interest that the *P-V* curve of bead movement on actin cables (Fig. 4) is analogous in shape to that of single fibers. In the contraction model of Huxley[9], the *P-V* curve is associated with the change in the number of myosin heads interacting with actin at any one moment. If the extremely limited number of myosin heads involved in the bead movement is taken into consideration, the present results suggest the possibility that the *P-V* relation reflects kinetic properties of individual myosin head interacting with actin rather than the change in the number of myosin heads involved in force generation and myofilament sliding.

Our force-movement assay system not only provides experimental results comparable with muscle mechanics data, but also enable us to perform experiments that are totally impossible by any other means. Such experiments would yield unexpected results, thus creating new "mysteries" to be solved in future. As a matter of fact, the most remarkable result brought about in the present study is that the bead movement is slowed down by the application of negative loads, though the bead is being pushed forwards by centrifugal forces (Figs. 6 and 7). It is impossible to explain the above results on the basis of the Huxley contraction model, in which V_{max} is determined by a balance between positive and negative forces generated by actin-myosin linkages in the unloaded condition. As far as we know, the only model which can predict the decrease of bead movement by negative loads is the self-induced translation model of Mitsui and Oshima[10], in which attachment of myosin head induces local conformational changes in the thin filament to produce an axial gradient of electrostatic force to cause actin-myosin sliding. Though it does not necessarily mean that their model is correct, such a mechanism should seriously be considered in future.

Using the slack test method, Edman[11] showed that the shortening velocity of a single muscle fiber can be increased above V_{max} if resting force is present. It seems possible that, in his experiments, the actin-myosin interaction is inactivated by a quick release, so that the slackened fiber can be made taut quickly by elastic restoring force other than the actin-myosin interaction.

We also examine the effect of lateral forces on the bead movement, in the hope that this type of experiments would reveal "helical" nature of actin-myosin sliding reflecting the helical structure of actin filaments. The velocity of bead movement, however, decreased by laterally applied centrifugal forces irrespective of their direction (from right or from left) (Figs. 8 and 9). It may be that, at least in our assay system, myosin heads would slide straight along actin cables, despite the helical arrangement of actin monomers in actin filaments.

REFERENCES

 1. Kron, S.J. & Spudich, J.A. *Proc. Natl. Acad. Sci. USA* **83**, 6272-6276 (1986).
 2. Kishino, A. & Yanagida, T. *Nature* **334**, 74-76 (1988).
 3. Sugi, H., Kobayashi, T., Gross, T., Noguchi, K., Karr, T. & Harrington, W.F. *Proc. Natl. Acad. Sci. USA* **89**, 6134-6137 (1992).
 4. Hill, A.V. *Proc. R. Soc. London Ser. B* **126**, 136-195 (1938).
 5. Kamitsubo, E., Ohashi, Y. & Kikuyama, M. Protoplasma 152, 148-155 (1989).
 6. Oiwa, K., Chaen, S., Kamitsubo, E., Shimmen, T. & Sugi, H. *Proc. Natl. Acad. Sci. USA* **87**, 7893-7897 (1990).
 7. Sheetz, M.P. & Spudich, J.A. *Nature* **303**, 31-35 (1983).
 8. Edman, K.A.P. *J. Physiol. (London)* **404**, 301-321 (1988).
 9. Huxley, A.F. *Prog. Biophys. Byophys. Chem.* **7**, 257-318 (1957).
10. Mitsui, T. & Oshima, H. *J. Muscle Res. Cell Motility* **9**, 248-260 (1988).
11. Edman, K.A.P. *J. Physiol (London)* **291**, 143-159 (1979).

Discussion

Brenner: Did you calculate the magnitude of Coriolis forces involved in these measurements? At the very high velocities, a significant Coriolis force might affect the observed velocity. This might change the shape of the force-velocity relationship.

Sugi: Please explain the nature of such a force.

Brenner: It is the type of force you experience when something moves radially on a spinning platform. Like on a roundabout where you experience lateral force when you walk from a higher radius to a smaller radius. Significant Coriolis forces would act as lateral forces on your moving system, which is shown to be decreasing the movement.

Sugi: In our experiment, the diameter of the rotor is 8 cm, while the range of excursion of the bead is very small, being within the microscopic field of a 40x objective. I think the force is very small, if I understand you correctly.

Brenner: But the velocity, I think, is very high. . .

Sugi: But still the measurement is made within a microscopic field, with a 40x objective. (Note added after discussion: We calculated the Coriolis force in our experiments, and it actually proved to be negligibly small.)

Edman: The P-V relation in your system is quite different from what we see in the living system, particularly concerning as to the velocities at negative loads. I'm surprised that you do not obtain velocities exceeding V_{max} with negative loads.

Sugi: It is an experimental fact: if you push the bead, it always slows down.

Gillis: I have a comment on Dr. Chaen's nomenclature. I am a little concerned about the use of the word "negative load," because I suppose that the work made by the beads being pushed by the centrifugation force—"negative work"—is getting confused with classical muscle mechanics, where the expression of "negative work" reflects the fact that the muscle is stretched forcibly. But the direction of the movement of the actin filament is just the opposite of this one. So I would suggest not using this expression, because it's going to introduce a lot of difficulties. What you do, in fact, is to add some force to the normal force produced by the actomyosin interaction and perhaps you should call it "additive load" or something like that. I think the effort is to make a bridge between classical muscle mechanics and what you can get with a very simplified assay, but the use of the word "negative load" can be extremely misleading.

ELECTROSTATIC POTENTIAL AROUND ACTIN

Toshio Ando, Naohiro Kobayashi* and Eisuke Munekata*

Department of Physics
Faculty of Science
Kanazawa University
Kanazawa, 920, Japan
**Department of Applied Biochemistry*
Tsukuba University
Tsukuba 305, Japan

ABSTRACT

We presume that tension of contracting muscle originates from electrostatic force experienced by actin and myosin. We suppose that a high-energy state of myosin-ADP-Pi interacts with actin, transferring the stored energy to actin, and that the actin excited in this way develops around itself electric field which exerts sliding force against charged myosin heads. To explore the idea, first we conjectured how electric charges on actin produce electric field in the axial direction, and second we experimentally examined electrostatic circumstances around actin in solution and in muscle fibers by optimizing diffusion-enhanced fluorescence energy transfer. In the experiments, Tb ion, which has a long excited-state lifetime, was used as donor. To introduce Tb to actin, Tb-DTPA-phalloin and Tb-DTPA-maleimide were synthesized. As acceptors with electric charges ($Za = -3$ to $+2$), rhodamine B that was conjugated with various amino acids or their derivatives was used. The fluorescence energy transfer efficiency (ET) was estimated from the shortening in the lifetime of Tb. The electrostatic circumstances around actin were inferred from the ET-Za relation. When Tb was introduced at Cys-374 of actin, the Tb-site was found in negative electric potential. S-1 binding to the labeled actin neutralized the electric potential almost completely. Tb-DTPA-phalloin bound to actin seemed to reside in the vicinity of tryptophan residue(s). Electric potential around the phalloin site was negative. S-1 binding to the actin slightly reduced the negativity. In glycerinated fibers in the rigor state, the phalloin site was again found in negative potential. When fibers were transferred from an ADP-rigor solution to an active solution, the negative electric potential was neutralized to some extent. The direction of this change could not be explained by detachment of crossbridges from actin, since the detachment should have given an opposite direction of changes in the electric potential. Thus, this observation may indicate that electric potential characteristic of the active state occurs at actin surfaces.

Mechanism of Myofilament Sliding in Muscle Contraction, Edited by
H. Sugi and G.H Pollack, Plenum Press, New York, 1993

INTRODUCTION

Recent studies on microtubule-based protein motors have shown that on the microtubule protein motors move towards the plus end or the minus end of the microtubule, depending upon the property of the each motor. Kinesin is a plus-end-directed motor[1]. The protein encoded by the Drosophila melanogaster claret nondisjunctional gene (termed ncd) contains a microtubule-binding domain whose amino acid sequence is quite similar to that of kinesin[2]. However, the ncd protein moves towards the minus end of the microtubule[3][4] Organelles from Reticulomyxa which show features characteristic of dynein-powered movements can move bidirectionally along microtubules[5]. Probably the direction may be determined by modification of the dynein, for instance, by phosphorylation. There has not been found the structural basis for directionality. On the supposition that sliding force is generated by electrostatic force exerted between protein motors and microtubules, we can naturally interpret the above observations. Suppose a motor interacts with a microtubule and thereon generates local electric field along the microtubule, the charged protein motor experiences force from the electric field. In this context the direction of sliding is determined by the sign of electric charges on a key position of the motor. So, protein motors may be able to switch the direction of sliding by reversing the sign of the electric charge. As Malik and Vale suggested[6], the conventional explanation of force-generation by means of a large-scale transformation in the motor seems difficult to reconcile with the above observations. Other types of force produced by hydrophobic interaction or hydrogen bonding interaction cannot readily account for the observations, either. We believe that actin-myosin system works on the same principle as that of adopted by microtubule-based protein motor systems. From the foregoing considerations, we think that sliding force in actin-myosin system is generated by electrostatic interaction between the two proteins.

Electric charges locate in clusters at acto-S-1 interfaces[7-9]. S-1 affinity for actin in the presence of nucleotides is reduced with increasing ionic strength of the solution[10][11]. So, people believe that electrostatic interaction plays an important role in association of actin and S-1. But, except for a small number of people such as Elliot et al.[12] and Iwazumi[13], most of the people have never seriously considered that electrostatic force may drive sliding between actin and myosin, since it is hard for one to imagine how electrostatic force is developed in the axial direction of an actin filament. In this paper, first we present a model in which a uni-directional electric field is generated along the each single strand of the double-stranded helix of an actin filament (i.e., roughly in the axial direction of an actin filament). Second, we demonstrate how we can experimentally estimate electrostatic circumstances around a site on protein surface. Third, we present data which reveal that electrostatic potential characteristic of the active state of glycerinated rabbit psoas muscle fibers is developed on actin surfaces and its development is sensed around the phalloidin-binding.

MATERIALS AND METHODS

Synthesis of Acceptors

To introduce various electric charges to rhodamine B, the following amino acids or their derivatives were reacted with rhodamine B isothiocyanate; Glu-Glu, Glu, Gly, Gly-methyl ester, Arg, Lys, Lys-Lys, Arg-methyl ester, Arg-Lys-methyl ester, Lys-Lys-methyl ester. The carboxy-methylation of amino acids or dipeptides were carried out as follows. Briefly, 0.25 ml of thionyl chloride was slowly added to 1 ml of anhydrous methyl alcohol cooled at -10°C. 10 min later, 1 mmoles of amino acids or dipeptides were added in solid to this solution. The mixture were stirred for two days at room temperature. They were then concentrated under reduced pressure, washed several times with methanol, and freed of solvent in vacuo. 2 mM amino acids or their derivatives were added to 0.2 mM rhodamine B isothiocyanate dissolved in 30-50 % methanol and they were incubated in dark for 10 hours at room temperature. The reacted compounds were purified either by silica gel chromatography or DEAE chromatography.

Preparation of Tb-DTPA-phalloin

δ-aminophalloin was synthesized according to Wieland et al.[14]. First, tosylphalloin was prepared. Tosyl chloride (12 mg, 10 equiv.) was added to the solution of phalloin (50 mg, 0.063 mmol) in pyridine (1 ml) at 0°C, and the solution was stirred for 5 hours at room temperature. The solution was concentrated in vacuo and tosylphalloin was purified by reverse phase HPLC. Yield, 25 mg (41.9 %). Tosylphalloin was treated with 2.5 N NH_3 in methanol at room temperature for 6 hours. The solution was diluted with excess water and lyophilized. Finally, δ-aminophalloin was purified by reverse phase HPLC. DTPA anhydride (3.6 mg, 0.01 mmol) was added to the solution of δ-aminophalloin (1 mg, 1.27 μmol) in dry DMSO (0.2 ml), and the solution was stirred for 2 hours at room temperature. The solution was diluted with methanol and DTPA-phalloin was purified by reverse phase HPLC. Yield was about 20 %. The purified DTPA-phalloin was stored as solid at -35°C. Before use, DTPA-phalloin was dissolved in dry DMSO and an aliquot of the solution to be used was mixed with equimolar solution of $TbCl_3$. The rest of the dissolved DTPA-phalloin was stored at -35°C.

Fibers Preparation

Fibers from rabbit psoas muscle were glycerinated for periods ranging from a week to several months in a solution containing 80 mM KCl, 2 mM $MgCl_2$, 2 mM EGTA, 5 mM K_2HPO_4 (pH 7.0), 50 % (v/v) glycerol (GL-solution) at -20°C. Bundles of 10 to 20 fibers were dissected and washed with an ice-cold ADP-free rigor solution (80 mM KCl, 20 mM TES (pH 7.0), 5 mM $MgCl_2$, 0.5 mM leupeptin). The fibers were then stained with 0.2 mM Tb-DTPA-phalloin in the ADP-free rigor solution for 1 hour at 4°C. The free Tb-chelate was removed by washing the fibers with the ADP-free rigor solution for 1 hour at 0°C with changing the solution several times. Part of the stained fibers was stored in GL-solution at -20°C. The stained fibers were fixed horizontally at both ends by a fast-drying glue to stainless steel rectangle that fits inside a fused-silica quartz fluorescence cuvette. The horizontally fixed fibers ran diagonally in the cuvette. The cuvette holder was maintained at 26°C by circulation of water from a constant temperature bath. Ten to twenty fibers were mounted in the instrument at one time. Only the middle section of the fiber bundle was illuminated with a focused excitation beam. When fluorescence photon counts were measured in a solution containing an acceptor dye, the fiber bundle was preincubated for 30 min at 0°C in a ADP-rigor solution containing the dye whose concentration was the same as those used in photon counting measurements.

Time-Resolved Fluorescence Measurements

For time-resolved fluorescence measurements, the excitation source used was a bulb type xenon flashtube (EG&G, FX-801U) triggered by TTL signals from a multichannel scaler (NAIG, E564 Process Memory). When Tb^{3+} chelated to dipicholinate (DPA) was excited, the pulsed light was passed through an interference filter (Vacuum Optical Corp. of Japan, MIF-W, λmax = 267 nm, $\Delta\lambda_{1/2}$ = 14.5 nm, Tmax = 17.5 %). When Tb-DTPA-phalloin was excited, the pulsed light was first passed through a UV-transmitting black glass filter (Sigma Koki, UTVAF-36U, λmax = 360 nm, $\Delta\lambda_{1/2}$ = 40 nm, Tmax = 60 %). The filtered light was focused with two symmetric-convex fused-silica lenses onto the sample in a 1 cm x 1 cm fused-silica quartz cuvette. The resulting excitation pulse at the sample was 5 μsec wide, illuminating a 1 mm x 3 mm rectangle. The repetition rate used varied between 39.063 Hz and 97.656 Hz, depending on the lifetime of the sample. The fluorescence emission at right angles was passed through an all-dielectric band-pass filter (Vacuum Optical Corp. of Japan, DIF-BPF-4, λmax = 547.2 nm, $\Delta\lambda_{1/2}$ = 23 nm, Tmax = 81 %), passed through a tri-acetyl cellulose sharp-cut filter (Fuji Film, SC-52), and focused with a symmetric-convex fused-silica lens onto a H1161 photomultiplier (Hamamatsu Photonics). The anode signal was preamplified with a model 2005 preamplifier (Camberra), amplified with a E511A linear amp (NAIG), and discriminated with a E512A single-channel analyzer (NAIG). The resulting pulses were then counted synchronously with trigger by using a multichannel scaler which was connected to a personal computer (Epson, 286VG). The photon counting rate ranged between 1 kHz and 4 kHz. The stored data were transferred to a UNIX work station (Sony, NEWS NA-564 system) and the fluorescence lifetime was then determined with a least-square fit of the data (from which the background data had been subtracted) to a single-exponential decay, where the first 20 to 30 points of the total 512 points of the photon counting data were omitted from the analysis.

Precaution

Surfaces of glass tubes, fluorescence cuvettes and pipette tips usually carry electric charges. So, to keep the charged acceptors from being adsorbed onto these surfaces, surfaces of all glassware and plasticware, with which acceptor solutions were handled, were freshly coated with silicon using Sigma-coat, and commercially-siliconized pipette tips were used.

RESULTS AND DISCUSSION

<A Model for Generation of Uni-directional Electric Field>

Evidently there are electric charges on the surfaces of an actin molecule and a myosin head. Because an actin filament has an ordered helical structure, the electric potential which is formed by electric charges on actin subunits must be distributed regularly around an actin filament. It is, therefore, impossible for the electric potential to have a uni-directional gradient which spans from one actin to the next one. That is, there always exist both plus- and minus-signed gradients between

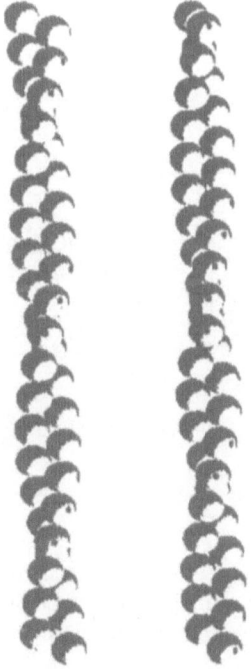

Fig. 1. The configurations of actin filaments with non-distorted and distorted twists. The left illustration shows a non-distorted actin filament, and the right shows a distorted one. The distortion was given according to eq.(1) with $\Delta \times \exp(-1/N) = 14°$ and $N = 7$.

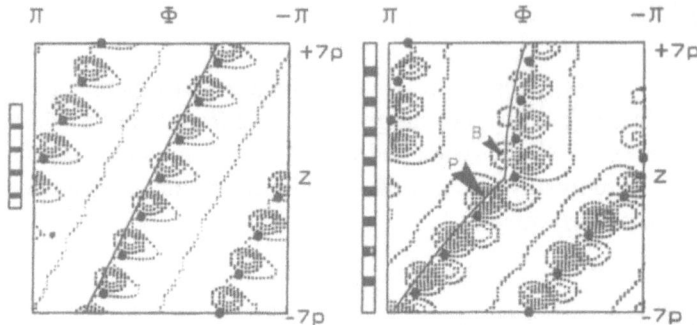

Fig. 2. The contour line maps of the distribution of electric potential. The left is a illustration for a non-distorted filament, and the right is for a distorted filament. The closed circles indicate the centers of actin subunits. Since this black-and-white illustration was obtained from a copy of colared one, the potential height became ambiguous. The arrow heads P and B point the peak and the bottom of electric potential, respectively. The center of the central actin was put on the X-axis. Thus, the position vector of the center of the central actin, C, is given by $C = (x,0,0)$, where $x = 24.03$ Å. The location, relative to the center of an actin subunit, of a positive charge and a negative charge is expressed using the polar coordinates, $\theta\pm$, $\Phi\pm$. The position vectors, $r_0^\pm$, of these charges on the central actin are given by

$$r_0^\pm = \begin{pmatrix} x + r_0\sin\theta^\pm \cos\Phi^\pm \\ r_0\sin\theta^\pm \sin\Phi^\pm \\ r_0\cos\theta^\pm \end{pmatrix} \qquad \text{(i)}$$

The position vectors, $r_n^\pm$, of the charges on the n-th actin are given by

$$r_n^\pm = A \times r_0^\pm + t \qquad \text{(ii)},$$

where $t = \begin{pmatrix} 0 \\ 0 \\ np \end{pmatrix}$, $A = \begin{pmatrix} \cos\xi_n & -\sin\xi_n & 0 \\ \sin\xi_n & \cos\xi_n & 0 \\ 0 & 0 & 1 \end{pmatrix}$, $\xi_n = \begin{cases} \sum_{i=1}^{n}\Psi_n (n>0) \\ -1 \\ \sum_{i=n}\Psi_n (n < 0) \end{cases}$

The electric potential at $r = (\rho\cos\Phi, \rho\sin\Phi, Z)$ is given by

$$U(r) = \sum_{n=-\infty}^{\infty}U_n^+(r) + \sum_{n=-\infty}^{\infty}U_n^-(r) \qquad \text{(iii)}$$

where, $U_n^+(r)$ and $U_n^-(r)$ are the electric potential made by a positive charge and a negative charge on the n-th actin, respectively. These are given by

$$U_n^\pm(r) = \frac{Z^\pm e_0 \exp[-(|r - r_n^\pm| -a)/\lambda]}{D|r - r_n^\pm| (1 + a/\lambda)} \qquad \text{(iv)}, \rightarrow$$

neighboring actin protomers. So, it is quite hard to anticipate that the electric field around an actin filament can drive a charged myosin head from one actin to the next one. But, if the structure of an actin filament changes asymmetrically, it must distort the ordered distribution of electric potential around an actin filament, which may result in generation of uni-directional electric field. (We should note that about three decades ago Oosawa et al.[15] have first described a possibility of an asymmetric change in the F-actin conformation.) We examined this possibility by calculating the electric potential around an idealized actin filament, assuming certain locations of the electric charges. There are several possibilities as to the conformational change in an actin filament. For instance, (a) change in twist of the helical structure, (b) change in the actin-actin bond length, (c) rotation of actin subunits within the fixed helical structure (as suggested by Yanagida[16]), (d) hinging within individual actin monomers (as suggested by Erickson[17]), (e) protonation or deprotonation of some amino acid residues of actin subunits. Here, we employed the type (a). Actually, it has been recognized that an actin filament is flexible and has various twists, from the studies of quasielastic light scattering[18], phosphorescence depolarization[19] and electron microscopic observations[20]. However, it should be emphasized here that other types of the conformational changes also might give more or less the same result as the one presented below.

When an actin filament is looked upon as a linear left-handed helix, the structure is characterized by two parameters:

p; the intersubunit axial translation (translational advance)

Ψ; the intersubunit angle (angular advance).

Let's give a number, $n = 0$, to a central actin (where a high energy state of myosin-ADP-Pi binds), and successively $n = \pm 1, \pm 2, \text{---}$ to the left- and right-sided actin subunits. The positive values of n are given to the side to which the left-handed helix proceeds. An asymmetric change in Ψ is specified as follows. When $n > 0$ $|\Psi|$ increases, and when $n < 0$ $|\Psi|$ decreases. The extent of change in $|\Psi|$ decreases exponentially with increasing n. That is,

$$\Psi = \text{sign}(n) \times \overline{\Psi} + \Delta x \exp(-n/N) \tag{1},$$

where $\overline{\Psi} = 166.7°$, $\Delta x \exp(-1/N)$ is the maximum change in $|\Psi|$, and N is the persistence length of change in Ψ as measured by the number of actin subunits. An actin subunit was approximated by a sphere with the radius of $r_0 = 27.5$ Å. The

---

→where $Z^{\pm}$ the valence of charges, e_0 unit charge, D the dielectric constant of the medium, a the nearest distance between ions, λ the Debye radius. The distribution of electric potential shown here was obtained under conditions; $D = 80$, $\lambda = 12$ Å, $a = 2$ Å, $\theta^- = 20°$, $\Phi^- = 20°$, $\theta^+ = 150°$, $\Phi^+ = 320°$. The solid line (called "track") in the right figure passes the minimum point of the potential in the region $Z>0$ and runs along the single strand of the double-stranded helix of the actin filament. The track in the left figure passes a point identical to the minimum point in the right figure.

configuration of actin filaments with non-distorted and distorted twists is depicted in Fig. 1. Here the distortion was given according to eq.1 in which N and Δx exp(-1/N) were 7 and 14°, respectively. At first glance the lower half of the distorted actin filament may seem to be twisted tightly. Virtually larger |Ψ| was given to the upper half. This is because we looked at the filament as a right-handed double-stranded helix.

We assume here that an actin subunit possesses on its surface a plus electric charge and a negative electric charge, and also that the positive charge is located near a negative charge of a neighboring actin subunit. (We should note that this is true, because actin-dimer, -trimer, etc. are produced when F-actin is treated with a zero-length cross-linker, EDC[7].) On the latter assumption, electric potential around actin will be altered in a relatively large extent by twisting the actin filament. Since an atomic structure of actin appeared in 1990[21], we should have considered this when putting electric charges on the surface of an idealized actin. But, for the reason that we will mention at the end of this section, we did not. Electric potential, U(Φ,Z), on the surface of a coaxial cylinder that wrapped around actin filament was calculated. The contour line maps of electric potential are shown in Fig. 2. The left figure was obtained with no-distortion, and the right figure was obtained with the same distortion as employed in Fig. 1. U(Φ,Z) with no distortion distributed regularly along with the helical arrangements of actin subunits. U(Φ,Z) with distortion, on the other hand, did not necessarily accord with the distorted helical arrangements of actin subunits. While in the region of Z > 0 there are peaks to the right of the central positions of actins, in the region of Z < 0 they are to the left. It is unknown where the myosin•ADP•Pi has been bound to the central actin. On the right hand map, we now draw a line which passes a minimum point of the electric potential in the region Z > 0 and runs along the single strand of the double-stranded helix of F-actin. This is shown by a continuous line in the right hand map. We call this line "track". Then, we assume that the myosin head had been bound to the central actin at a position on the track, and that it will migrate on the track. It is apparent that a uni-directional field is developed along the track, ranging from the foot to the head of the central actin. This uni-directional field spans the distance between the neighboring actin subunits. Thus, if the myosin head travels along this gradient, it can shift from the central actin to the neighboring actin subunit. The force exerted against a unit charge by the electric field is about 0.2 pN. Since myosin heads have a lysine-rich sequence in the actin-binding site, a myosin head experiences sliding force of about 1 pN from the electric field.

People may criticize the model proposed here for the reason that in reality an actin subunit possesses charges more than two, and moreover for the reason that these charges reside not only the actin-actin-contact areas but also on other places, especially on the S-1-binding site. But, electric potential formed by charges located on places other than the actin-actin-contact areas will change little by alteration of the filament twists, and therefore, will not contribute to development of electric field that drives myosin heads in the axial direction. This additive electric potential may play a role in keeping myosin heads in contact with actin surfaces while they are sliding, and/or in activating myosin ATPase reaction.

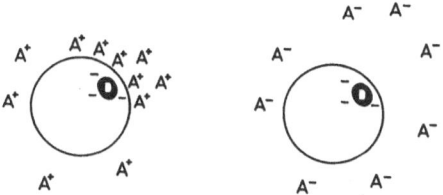

Fig. 3. Distribution of charged acceptor molecules around the donor-site on protein. The large circles denote protein. D and $A^{\pm}$ denote donor and charged acceptors, respectively. The donor (D) attached to protein resides in negative electric potential.

\<Dependence of Fluorescence Energy Transfer on Electrostatic Interaction\>

In order to estimate electrostatic circumstances around a site on protein surface, we optimized diffusion-enhanced fluorescence energy transfer. The basic idea was originally proposed by Wensel and Meares[22]. Principle of the method is illustrated in Fig. 3. A site on protein to be studied is labeled with a fluorescence donor (D). Let's suppose that this site is in negative electric potential. Charged acceptor molecules ($A^{\pm}$) in the medium distribute around the protein, depending upon the sign of the electric charge. Positively charged acceptor gets closer to the donor site, resulting in higher efficiency of energy transfer. Negatively charged acceptor, on the other hand, gets away from the donor site, resulting in lower efficiency of energy transfer. So, by measuring energy transfer efficiency (ET) as a function of charges (Za) that acceptor carries with it, we can estimate the electrostatic circumstances in which the donor site on protein locates. When donor has a long excited-state lifetime, a large number of acceptor molecules are available for the energy transfer, because acceptor molecules that are too far from donor for efficient transfer at the moment of excitation may be brought into close proximity to the donor by diffusion while the donor is still in the excited state. When the lifetime is short (nsec order), we need a very high concentration of acceptor to attain significant efficiency of energy transfer, which is virtually unpractical. So, we used a lanthanide ion, terbium, as donor, since it has a lifetime longer than 1 msec.

To observe how sensitively fluorescence energy transfer probes electrostatic interaction between acceptor and donor, we studied the following simple system; charged donor: Tb^{3+} chelated to DPA, $Tb(DPA)_3$, which has the net charge, $Zd = -3$, acceptors: variously charged rhodamine B derivatives. The spectroscopic properties of the rhodamine B compounds did not depend on what amino acid derivatives were conjugated, which is very advantageous since under this condition energy transfer efficiency solely depends upon the electric charges of the acceptors. The optical density of the acceptors at 556 nm was kept constant at 0.072 for all the acceptors used (the concentration of acceptors was about 1.4 μM). As shown in Fig.

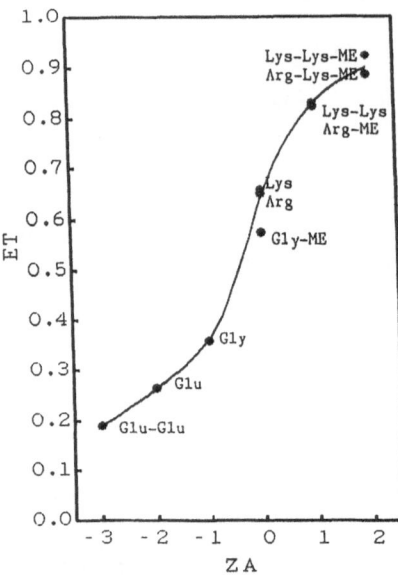

Fig. 4. The dependence of energy transfer efficiency on the electrostatic interaction between donor and acceptor. Donor: $[Tb(DPA)_3]^{3-}$; Acceptor:Rhodamine B conjugated with amino acids or dipeptides. The conjugated amino acids or dipeptides are indicated beside each of the data points. To ascertain complete chelation, 10 μM Tb^{3+} was mixed with excess amount of DPA^{2-} (100 μM) in 20 mM TES (pH 7.0). The optical density at 556 nm of all of the acceptors were kept at 0.072. The energy transfer efficiency was determined by $1-\tau_0/\tau$, where τ_0 and τ are the lifetimes of the donor in the absence and presence of acceptor, respectively. τ_0 was about 2.09 msec.

4, the energy transfer efficiency (ET) varied from 0.2 to 0.9 according to Za (from -3 to +2). This result demonstrated that ET-Za relation allowed us to deduce an unknown Zd or electric circumstances around a Tb-site on protein.

<Electric Potential on Actin in Solution>

In order to attach Tb specifically and selectively to actin, Tb-DTPA-phalloin was used. When Tb-DTPA-phalloin was attached to F-actin, the height of the near UV excitation bands was enhanced by 30-50 %, and moreover, a very large excitation band around 280 nm appeared (Fig. 5). The increase in the fluorescence at 547 nm with excitation at 280 nm was about 13-fold. This large enhancement cannot be explained simply by an effect of binding of the Tb chelate to F-actin, since the increment at other excitation bands was not so extensive. It is very likely that the large enhancement around 280 nm is due to energy transfer from excited Trp residue(s) of actin. Since optical density of Tb-DTPA-phalloin at wavelength where tryptophan emits its fluorescence is quite small, the tryptophan residue(s) must be in immediate proximity to the Tb site. (It has also been known that tryptophan fluorescence of actin is diminished when phalloin is attached to F-

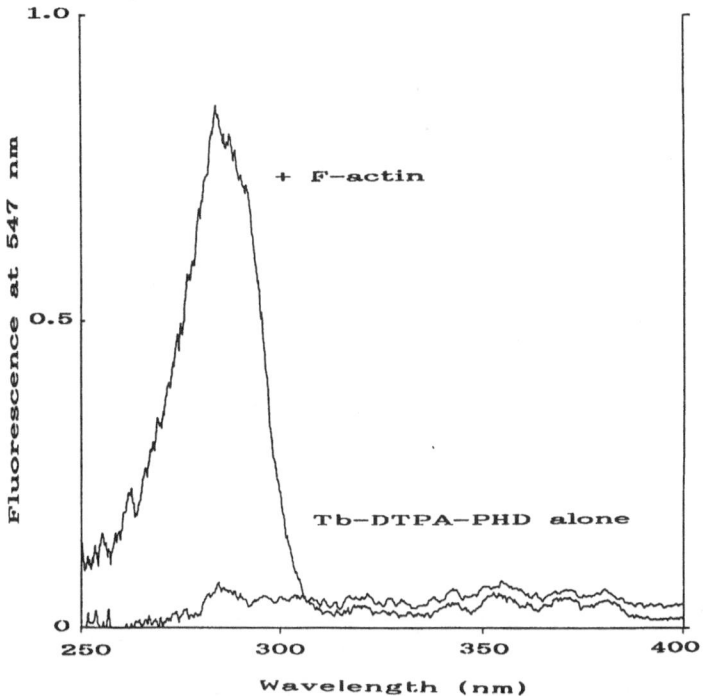

Fig. 5. Excitation spectra of free Tb-DTPA-phalloin and F-actin-bound Tb-DTPA-phalloidin. The fluorescence was observed at 547 nm. The lower spectrum was obtained with 5 μM free Tb-DTPA-phalloin in a buffer solution containing 50 mM KCl, 2 mM MgCl$_2$, 5 mM TES (pH 7.0), 0.2 mM CaCl$_2$, 0.2 mM ATP and 0.1 mM NaN$_3$. The higher spectrum was obtained with 5 μM Tb-DTPA-phalloin plus 5 μM F-actin in the same buffer solution.

actin.) The phalloin binding site on actin has been identified by means of chemical cross-linking[23]. The cross-linked amino acid residues are Glu-117, Met-119 and Met-355. Glu-117 and Met-119 are located on a large alpha-helix starting from Glu-107 and ending with Glu-125. According to the atomic structure of actin presented by Kabsh et al.[21], this alpha-helix sits in the vicinity of a massive alpha-helix starting from Trp-79 and ending with Asn-92 where two tryptophan residues (Trp-79 and Trp-86) are located. Met-356 resides next to Trp-356. The atomic structure of actin indicates that Met-356 is rather far from Glu-117 and Met-119. Even in the atomic model of the filament[24] it is far from the two residues within the neighboring actin subunits. So, it is difficult to assign which tryptophan residue contributes to the energy transfer to Tb.

Tb-DTPA-phalloin itself carries a unit negative charge. So, as Za is varied, the ET must change in a manner similar to that observed with Tb(DPA)$_3$. As shown in Fig. 6, the ET actually behaved in such an expected way. When Tb-DTPA-phalloin was introduced to F-actin, the ET augmented much more sharply with increasing Za. With Za > 0 the ET was larger than that of free Tb-DTPA-phalloidin, while with Za < 0 it was smaller (Fig. 6). We can, therefore, conclude that the phalloidin-site

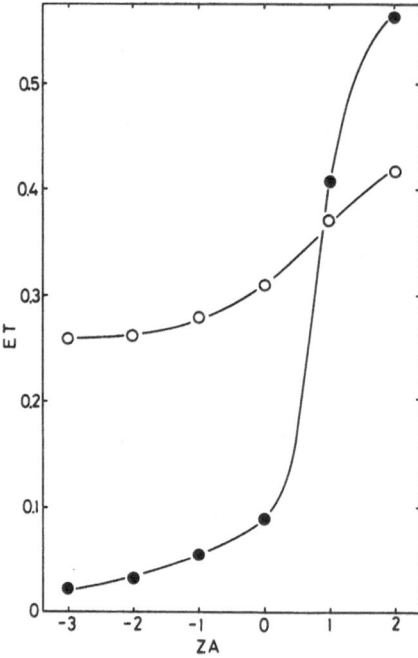

Fig. 6. Energy transfer efficiency of free Tb-DTPA-phalloin and F-actin-bound Tb-DTPA-phalloin as a function of electric charges of acceptor. The optical density at 556 nm of all of the acceptors used was kept at 0.035. The sample solutions were 5 μM Tb-DTPA-phalloin alone (open circles) or 5 μM F-actin plus 5 μM Tb-DTPA-phalloin (closed circles) dissolved in 5 mM TES (pH 7.0), 0.2 mM $CaCl_2$ and 1 mM $MgCl_2$.

within F-actin is situated in an environment of highly negative electric potential. When Tb-EDTA-maleimide was attached to Cys-374 of actin, we observed a similar response of ET to Za (data not shown).

It has been reported that negative charges reside in clusters at the N- and C-terminal regions of actin. Based on several studies these regions are believed to be close to each other in the tertiary structure. So, this place is in a highly-negative electrostatic environment (there are -5 charges as net). This place is also known as a binding site of myosin S-1. Positive charges (+5 as total) are situated in clusters at an actin-binding site on S-1. Chemical cross-linking experiments with a water-soluble carbodiimide have indicated that in the rigor state of acto-S-1, actin and S-1 contact with each other via electrostatic interaction between these charged groups. It is interesting to see how the phalloidin-binding site senses such electric neutralization when S-1 is bound to F-actin. As shown in Fig. 7, stoichiometric binding of S-1 to actin slightly neutralized the negative electric potential observed with F-actin alone. When Tb-EDTA-maleimide was attached to Cys-374 of F-actin, the negative potential at this C-terminal region was almost completely neutralized by S-1 binding to F-actin (by T. Yamamoto, data not shown). So, these results indicate that negative charges at the N- and C-terminal regions of actin contribute only a little to the

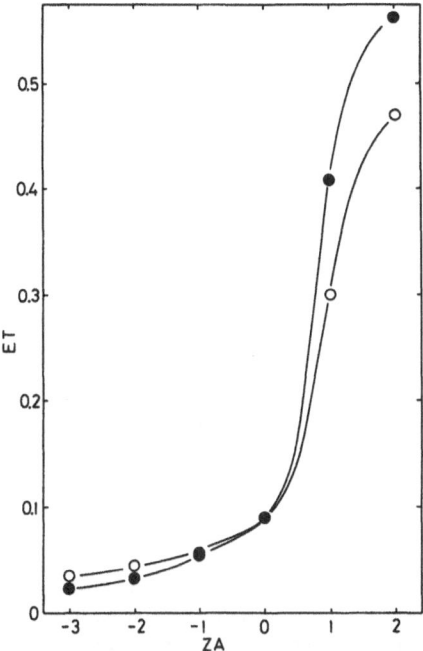

Fig. 7. Effect of S-1 binding to F-actin on the energy transfer efficiency. The optical density at 556 nm of all of the acceptors used was kept at 0.05. Closed circles: 3 μM F-actin-bound Tb-DTPA-phalloin in 80 mM KCl, 20 mM TES (pH 7.0) and 2 mM MgCl$_2$. Open circles: 3 μM S-1 was added to the above sample solutions.

negative electric potential at the phalloin-binding site. This also implies that the phalloin-binding site is not in the vicinity of the terminal regions of actin. Prof. Holmes suggested in this symposium that a massive alpha-helix starting from Trp-79 and ending with Asn-92 might be an additional binding site for S-1 (that is, S-1 may be able to bind to two actin subunits), and that this alpha-helix had negative charges which might form salt-bridges with positive charges of S-1. The negative charges may contribute to the negative electric potential observed at the phalloin-binding site, since the massive alpha-helix is close to the phalloin-binding site as mentioned above. However, S-1 binding to F-actin neutralized only a little the negative electric potential at the phalloin site, which may suggest that S-1 does not bind to the massive alpha-helix or that electrostatic interaction is not involved in the rigor binding of S-1 to the alpha-helix.

<Electric Potential on Actin in Fibers>

In experiments with glycerinated rabbit psoas fibers, precaution has to be taken concerning the diffusion of acceptor molecules into the lattice space within fibers. Firstly, we estimated a time in which equilibrium is attained after Tb-DTPA-phalloin-stained fibers were transferred from an acceptor-free solution to an

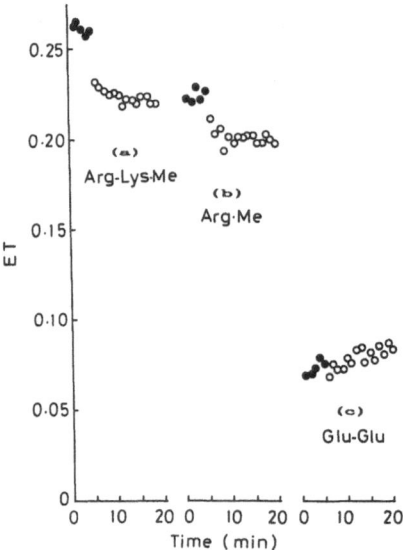

Fig. 8. Effect of fibers activation on the energy transfer efficiency of actin-bound Tb-DTPA-phalloin in glycerinated rabbit psoas muscle fibers. The acceptors used are indicated in each of the data. The optical density at 556 nm of the acceptors was kept at 0.07. Closed circles: in an ADP-rigor solution containing 80 mM KCl, 20 mM TES (pH 7.0), 4 mM MgCl$_2$, 0.2 mM CaCl$_2$, 4 mM ADP, 10 mM Glucose, 30 µg/ml hexokinase, 0.5 mM leupeptin. Open circles: in an active solution containing 80 mM KCl, 20 mM TES (pH 7.0), 4 mM MgCl$_2$, 0.2 mM CaCl$_2$, 4 mM ATP, 5 mM phosphoenol pyruvate, 30 µg/ml pyruvate kinase. When the fibers were transferred from the ADP-rigor solution to the active solution, they were quickly washed twice with the active solution and then photon counting was re-started.

acceptor-containing solution, by measuring a time course of decrease in the lifetime of Tb. Equilibrium was reached in 30 min at longest. So, whenever we measure ET with fibers in an acceptor-containing solution, the fibers were preincubated for 30 min at 0°C in an ADP-rigor solution containing the same acceptor whose concentration was the same as the one used for ET measurements. In the ADP-rigor state of fibers the phalloin site on actin was found in negative-electrostatic environment. When fibers were transferred from an ADP-rigor solution to an active solution, ET began to be diminished and then within a few minutes it reached a steady level (Fig. 8a). In this measurement rhodamine B-Arg-Lys-methyl ester (Za = +2) was used as the acceptor. When rhodamine B-Arg-methyl ester (Za = +1) was used, ET again decreased, but with a smaller extent (Fig. 8b). In case where rhodamine B-Glu-Glu (Za = -3) was used, ET, on the other hand, increased (Fig. 8c). The size of the increment was small, but still sufficient for its affirmation. These changes in ET were reversible, i.e., when fibers were returned to a previous ADP-rigor solution, the ET value recovered to its previous one. From these observations we can conclude that on activation of rigor muscle fibers, the negative electric potential around the phalloin site on actin is neutralized to some extent. Electric charges on myosin heads must contribute to electric potential at actin surfaces. Their contribution is to add positive electric potential, based on the foregoing results with F-actin and acto-S-1 solutions. On activation of fibers a part of myosin heads dissociates from actin. So, we would expect that electrostatic environment at the phalloin site on actin would become more negative when fibers were activated. But, the fact was contrary to this presumption. So, this unforeseen result indicates that tension generation in fibers is associated with alteration in actin conformation, and thereby is also associated with transformation of electric potential around actin filaments. We still do not know whether electric potential around actin characteristic of the active state of muscle fibers drives myosin movement. By genetic engineering technique Sutoh et al.[25] have recently evidenced that disruption of electric charges at the N-terminus of actin inhibits myosin-powered sliding movement of the actin filaments. This excellent study may suggest that the mutations distort inherent electric potential around actin and the distorted electric field can no longer exert sliding force against charged myosin heads.

REFERENCES

1. Vale, R.D., Schnapp, B.J., Mitchinson, T., Steuer, E., Reese, T.S. & Sheetz, M.P. *Cell* **43**, 623-632 (1985).
2. Endow, S.A., Henikoff, S. & Soler-Niedziela, L. *Nature* **345**, 81-83 (1990).
3. Walker, R.A., Salmon, E.D. & Endow, S.A. *Nature* **347**, 780-782 (1990).
4. McDonald, H.B., Stewart, R.J. & Goldstein, L.S.B. *Cell* **63**, 1159-1165 (1990).
5. Schliwa, M., Shimizu, T., Vale, R.D. & Euteneuer, U. *J. Cell Biol.* **112**, 1199-1203 (1991).
6. Malik, F. & Vale, R.D. *Nature* **347**, 713-714 (1990).
7. Mornet, D., Bertrand, R., Pantel, P., Audemard, E. & Kassab, R. *Nature* **292**, 301-306 (1981).
8. Sutoh, K. *Biochemistry* **22**, 1579-1585 (1983).
9. Yamamoto, K. *Biochemistry* **28**, 5573-5577 (1989).
10. Katoh, T. & Morita, F. *J. Biochem.* **96**, 1223-1230 (1984).
11. Highsmith, S. *Biochemistry* **29**, 10690-10694 (1990).
12. Elliot, G.F., Rome, E. & Spencer, M. *Nature* **226**, 417-420 (1970).
13. Iwazumi, T. in *Cross-Bridge Mechanism in Muscle Contraction* (eds. Sugi & Pollack, Univ. of Tokyo Press,Tokyo, 1979).
14. Wieland, T., Hollosi, M., & Nassal, M. *Liebigs Ann. Chem.* **1983**, 1533-1540 (1983).
15. Oosawa, F., Asakura, S. & Ooi, T. *Suppl. Prog. Theor. Phys.* **17**, 14-34 (1961).
16. Yanagida, T. *J. Muscle Res. Cell Motility* **6**,43-52 (1985).
17. Erickson, H.P. *J. Mol. Biol.* **206**, 465-474 (1989).
18. Fujime, S., Ishiwata, S. & Maeda, T. *Biochim. Biophys. Acta* **283**, 351-363 (1972).
19. Mihashi, K., Yoshimura, H., Nishio, T., Ikegami, A. & Kinoshita, K.Jr. *J. Biochem.* **93**, 1705-1707 (1983).
20. Stokes, D.L. & DeRosier, D.J. *J. Cell Biol.* **104**, 1005-1017 (1987).
21. Kabsch, W., Mannherz, H.G., Suck, D., Pai, E.F. & Holmes, K.C. *Nature* **347**, 37-44 (1990).
22. Wensel, T.G. & Meares, C.F. *Biochemistry* **22**, 6247-6254 (1983).
23. Vandekerckhove, J., Deboben, A., Nassal, M. & Wieland, T. *EMBO J.* **4**, 2815-2818 (1985).
24. Holmes, K.C., Popp, D., Gebhard, W. & Kabsch, W. *Nature,* **347**, 44-49 (1990).

25. Sutoh, K., Ando, M., Sutoh, K. & Yano-Toyoshima, Y. *Proc. Natl. Acad. Sci. USA* **88**, 7711-7714 (1991).

Discussion

Godt: It has been known for a long time: that elevation of ionic strength decreases the force of skinned fibers, which would be consistent with electro-static interactions. What I will be talking about later is something that I can't understand on that basis: the introduction of certain zwitterions into the system counteracts the effect of ionic strength. So I hope you will be listening when I talk; perhaps you can explain this in terms of your model.

Pollack: I'm sorry, but it's not obvious to me how, from the electro-static potentials surrounding the actin filament, you get axial force generation. Can you explain that?

Ando: Of course, an electro-static field may work just for binding between the actin and myosin, since actin filaments have a regular periodic structure. Electric potential distributes along the actin helices in a periodic way. So, you cannot get axial force generation. In my simple model—I actually don't stress that this is true— you can get it if you twist the actin filament asymmetrically about an actin subunit (centered actin). You tighten one side of the actin filament, and loosen the other side; then you have a new electric field around the distorted actin filament (see Figs. 1 and 2). A uni-directional electric field is formed around the centered actin, roughly in the axial direction. The electric field spans the centered actin and the neighboring actin. If we suppose myosin has been bound to the centered actin, then the electric field propels charged myosin along the actin filament.

FURTHER STUDIES OF THE SELF-INDUCED TRANSLATION MODEL OF MYOSIN HEAD MOTION ALONG THE ACTIN FILAMENT

Toshio Mitsui and Hiroyuki Ohshima*

Department physics
School of Science and Technology
Meiji Univerity
Higashimita, Tama-ku, Kawasaki 214, Japan
*Faculty of Pharmaceutical Sciences
Science University of Tokyo
Ichigaya, Shinjuku-ku 162, Tokyo, Japan

ABSTRACT

We have extended and refined the model of myosin head motion along the actin filament which we proposed in 1988 (*J. Mucscle Res. Cell Motility* 9, 248-260), and obtained the following results. (1) We assumed that the height of the induced potential depends upon tension with a maximum around the isometric tension, and got a force-velocity relation similar to the observation by Oiwa et al. (1990) that the velocity of the myosin-coated beads along the actin cables decreases with increasing centrifugal force applied in the direction of bead movement and then the velocity tends to increase when the force increases in the same direction beyond a certain value. (2) We introduced a correction factor in the relation between the measured tension and the microscopic tension produced by myosin head, and got a feature in force-velocity relation similar to the observation by Edman (1988) that the velocity drops sharply as the tension approches to about 80% of the isometric tension. (3) We assumed that binding of an ATP-activated myosin head to an actin filament causes a local structural change extended roughly 20nm long along the filament, which provides a potential well spread over about 16nm for the myosin head, with three shallow potential wells in it. We studied kinetics of the myosin head in the potential well of about 16nm in order to explain the early tension recovery after the sudden change of muscle length observed by Ford, Huxley and Simmons (1977), with results in good agreement with their experimental data.

Mechanism of Myofilament Sliding in Muscle Contraction, Edited by
H. Sugi and G.H Pollack, Plenum Press, New York, 1993

INTRODUCTION

In 1957 Huxley[1] proposed the theory of muscle contraction and pointed out importance of thermal motion of molecules in producing tension. In 1971 Huxley and Simmons[2] studied response of frog muscle fibre to stepwise length change, and proposed a model to explain obsorved results assuming that there is internal freedom in motion of myosin head attached to actin filament. In 1985 Yanagida et al.[3] measured sliding velocity of free thin filament on thick filament and the rate of ATP hydrolysis, and estimated the sliding distance of the thin filament during one ATP hydrolysis cycle as more than 60 nm. In 1988 we[4] proposed a model to explain the result by Yanagida et al. Basic assumptions in our model were as follows. Potential of force U_i with height of H_i is induced on actin flament by binding of an ATP-activated myosin head. U_i remains for a while after detachment of the head, and thus H_i acts as a potential barrier for backward movement of the head. We have made further studies of the model. Results obtained are reported below.

MODIFICATION OF THE MODEL AND RESULTS OF CALCULATION

Tension Dependence of Height H_i of the Induced Potential U_i

In the previous paper[4] we assumed that H_i depends upon p linearly, where p is a force produced by a myosin head or exerted on a myosin head by the principle of action and reaction. But, if muscle would be designed as an energetically efficient molecular machine, H_i might increase with increasing p and reach a maximum arounnd the isometric tension p_0. We assumed such a feature of H_i and got results shown in Figs. 1, 2 and 3.

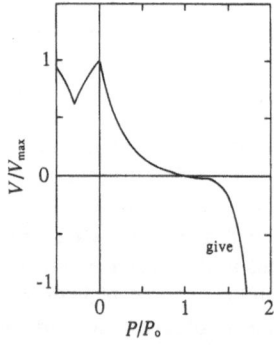

Fig. 1. Calculated V/V_{max} vs. P/Po relation, where V/V_{max} is relative sliding velocity and P/P_0 is relative tension. Calculation was made in the same way as in our previous paper[4], by using the formula v = (jf-jb) L, where v is filament sliding velocity, j_f and j_b are forward and backward jumping frequencies, respectively, and L is the period of actin strand projected onto the filament axis, 5.46 nm. The feature for $P/P_0 < 0$ is similar to the experimental observation by Oiwa et al.[5]

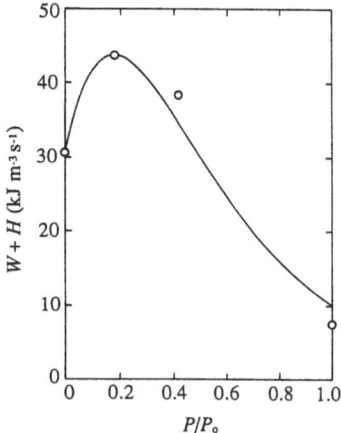

Fig. 2. Calculated rate of energy liberation W + H as a function of P/P_0 (solid line) in comparison with experimental data reported by Yamada and Homsher[6]. Caluculation was made in the same way as in our previous paper4, by using the formula W + H = (($A + B_p)j_f$ + Aj_b) {2(0.4N_{hs} / s)} with A = 0.4 kT and B = 13 nm, where N_{hs} (1.76 x 10^{17} m^{-2}) is number of myosin heads in the volume of which the base is $1m^2$ and the thickness is the half sarcomere length, and s stands for sarcomere length, 2.05 µm.

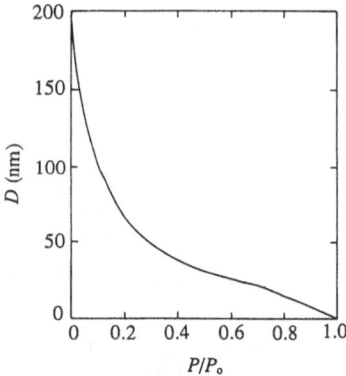

Fig. 3. Calculated distance D over which a myosin molecule moves along the thin filament utilizing energy liberated by hydrolysis of one ATP molecule, as a function of P/P_0 Calculation was made in the same way as in our previous paper4 by using the formula D = (j_f - j_b)Lt_ε where t_ε = ε/{(A + Bp)j_f + Aj_b}. Here ε is energy liberated by hydrolysis of one ATP molecule, 12 kcal mol^{-1}. This result indicated D for free shortening can be as large as about 200 nm.

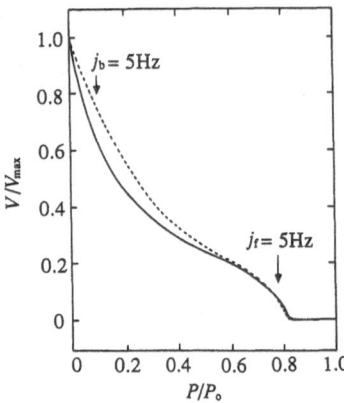

Fig. 4. Furce-velocity relation when possible rest time of a myosin head just after backward jump is taken into account. Calculation was made by using Eq. (2) with $a/b_f = 1.85$ and $b_b/b_f = 30$. Solid line shows results of our calculation. Dashed line is due to the data points in Fig. 6A of the paper by Edman[7]. In this calculation j_f and j_b became 5 Hz at the values of P/P_0 indicated by arrows, respectively, suggesting that tension fluctuation of about 5 Hz could exist around the P/P_0 region, as actually observed by Ishijima et. al.[8].

Correction Conceming Possible Rest Time of a Myosin Head Just after Backward Jump

In our previous paper[4] we assumed that the macroscopically observed tension P is related with p by

$$P = (0.4\, H_{hs})\, p \qquad (1)$$

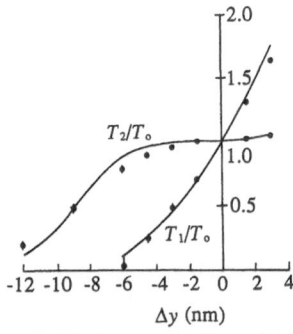

Fig. 5. T_1/T_0 and T_2/T_0 as functions of Δy in early tension recovery in phase 2 after sudden length change in muscle, where To is the isometric tension, T_1 is extreme tension, T_2 is tension approached during the early recovery phase and Δy is the step amplitude per half sarcomere. Solid lines represent results of our calculation. Data point are due to the experimental data given in Fig. 13 of the paper by Ford et. al.[9].

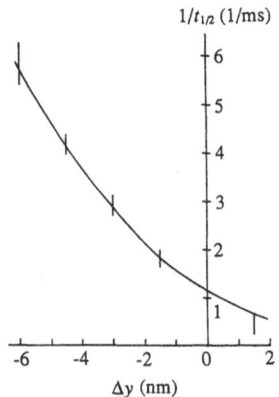

Fig. 6. Reciprocal of half-time $t_{1/2}$ of the early tension recovery in phase 2 as a function of Δy. Solid line represents results of our calculation. The bars show the range of distribution of data points in Fig. 24 of the paper by Ford et al.[9].

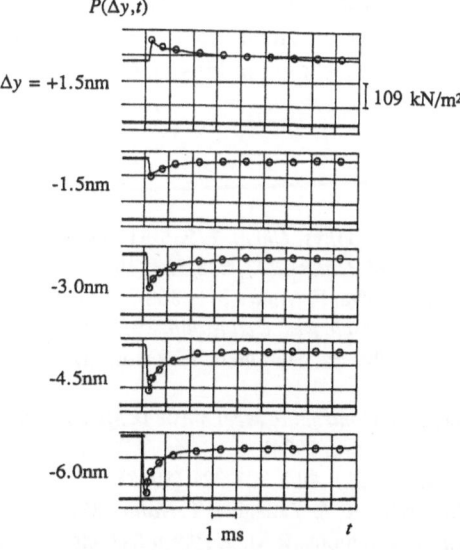

Fig. 7. Tension $P(\Delta y, t)$ in early recovery phase (phase 2) for various step amplitudes Δy, as functions of time t. Circles show results of our calculation. Solid lines are due to the experimental data given in Fig. 23 of the paper by Ford et al.[9].

Recent experimental results[7][8] appear to suggest the possibility that some myosin heads do not contribute to tension production just after backward jump, so that we replaced Eq. (l) by

$$P = \frac{a + b_f j_f}{a + b_f J_f + b_b J_b} (0 \ 4 \ N_{hs}) \ p, \qquad (2)$$

where a, b_f and b_j are constants. Result of the calculation based upon Eq. (2) is shown in Fig. 4.

A MODEL TO EXPLAIN TENSION RESPONSES TO SUDDEN LENGTH CHANGE IN MUSCLE

We tried to explain the experimental data on early tension recovery after sudded change of muscle length reported in the paper by Ford, Huxley and Simmons[9]. After many trial calculations we were forced to assume the followings: Binding of an ATP-activated myosin head to an actin filament causes a local structural change extended roughly 20 nm long along the actin filament. A part of the local structural change is assumed to be responsible to produce the induced potential U_i. The other provides a potential well spread over about 16nm for the myosin head, with three shallow potential wells in it. We treated kinetics of a myosin head in the potential well of about 16nm, with results shown in Figs. 5, 6 and 7.

REFERENCES

1. Huxley, A.F. *Prog. Biophys. Biophys. Chem.* **7**, 255-318 (1957).
2. Huxley, A.F. & Simmons, R.M. *Nature* **233**, 533-538 (1971).
3. Yanagida, T., Arata, T. & Oosawa, F. *Nature* **316**, 366-369 (1985).
4. Mitsui, T. & Ohshima, H. *J. Muscle Res. Cell Motility* **9**, 248-260 (1988).
5. Oiwa, K., Chaen, S., Kamitsubo, E., Shimmen, T. & Sugi, H. *Proc. Natl. Acad. Sci. USA* **87**, 7893-7897 (1990).
6. Yamada, T. & Homsher, E. in *Contractile Mechanism in Muscle* (eds. Pollack, G.H. & Sugi, H.) 883-885 (Plenum, New York, 1984).
7. Edman, K.A.P. *J. Physiol. (Lond.)* **404**, 301-321 (1988).
8. Ishijima. A., Doi, T., Sakurada, K. & Yanagida, T. *Nature* **352**, 301-306 (1991).
9. Ford, L.E., Huxley, A.F. & Simmons, R.M. *J. Physiol. (Lond.)* **269**, 441-515 (1977).

Discussion

Morales: I have a comment about Dr. Mitsui's and also Dr. Ando's presentations. I think it is certainly possible to invent a model that leads to a potential field that would cause the force and would maybe match certain experiments. But, it seems to me that now—when we know that the free energy for the force is coming ultimately from ATP hydrolysis and that the thing is an enzyme, and we know something about proteins—you cannot just invent the potential field. You have to explain how it is related to ATP hydrolysis in a mechanical way.

Mitsui: Professor Sugi said this is not a conference for biochemistry.

Morales: Touché!

Simmons: I'm not sure I understood the last thing you explained, but you seem to replace the three different angles of the head that Andrew Huxley and I had in our model (Huxley, A.F. and Simmons, R.M. *Nature* **233**, 533-537, 1971) with three different successive actin monomers. I think that Terrell Hill had a model of that kind (*Prog. Biophys. Mol. Biol.* **29**, 105-159, 1975).

Mitsui: I read, for instance, the articles by Huxley and Simmons, and by Hill. Our model is a modification of their models. Without the modification, we found it very difficult to get such good agreement with experimental data as shown in Figs. 6 and 7 of our paper in this book. We will discuss the modification in detail in our forthcoming paper.

Discussion on the Myosin Step Size

Homsher: Dr. Toshio Yanagida, would you consider for us the possible reasons for the large differences in step size between your and Spudich's groups?

Yanagida: Perhaps you wonder why the myosin-step size is so greatly dependent upon the investigators although similar *in vitro* motility assays are used. In addition to the difference in the sliding velocity of actin on myosin between our group and Spudich's group, Spudich's group analyzed the data based on the assumption that velocity of the actin filament is independent of orientation of the myosin head with respect to the actin filament axis (Uyeda et al. *Nature* **352**, 307-311, 1991). But as I have shown in my talk, the velocity is greatly dependent on the orientation of the myosin head. Therefore, it is likely that the number of myosin heads actually involved in the maximum sliding movement of actin filament may be several times smaller than the number they used for their estimation. As the myosin-step size is roughly proportional to the reciprocal number of myosin heads involved, it is possible that the actual myosin step size is several times larger than they estimated. So, if they analyze the velocity of actin filament produced only by correctly oriented myosin heads, the step size must be more than several times larger than what they observed, perhaps 500 nm or more, a value close to the values we obtained.

Sellers: When we were studying the movement of actin filaments by purified, monomeric clam myosin and by clam native thick filaments, the monomers oriented randomly on the surface were moving the actin filaments at least 75% as fast as the oriented thick filaments were. So, I don't know if you can use that same argument in those cases.

Yanagida: Yes, but to obtain the myosin-step size we have to estimate the number of myosin heads involved in the maximum speed of movement.

Tregear: If I understand what you have just said, you're saying that the 0.31 μm/second figure that appears in the *Nature* paper of Spudich's group is too low, because of the disorientation of the myosin heads.

Yanagida: Yes, it is possible.

Tregear: Let's hold onto that argument a moment, because that the peak in velocity, at least in their *Nature* paper, is fairly definite. If it is a definite maximum velocity, would you expect a maximum to occur for a misoriented head? I would have thought that must indicate, however obscurely, the oriented ones.

Yanagida: But if myosin heads can produce movement of the actin filament at any angle, it is unlikely that such a definite peak appears.

Tregear: But there is a peak. You have to accept that there is a peak there.

Toyoshima: The measurement of the myosin step size I did with Dr. Spudich (Toyoshima et al. *Proc. Natl. Acad. Sci. USA* **87**, 7130-7134, 1990) was under a condition different from that of Yanagida's group. As Dr. Yanagida explained, this condition may not be good for estimating maximum step size. However, I don't agree with some parts of Dr. Yanagida's explanation, because from my experiment using HMM tracks or native thick filaments from crustacean muscle, I got the same velocity of actin filament sliding with regular and opposite polarities of the thick filament. The force is different, but with regard to velocity, I didn't get a difference. So, if HMM molecules are scattered on the surface, I'm not sure that velocity is affected by disorientation.

Sugi: I have one question about the measurement of ATPase activity in the experiment to estimate the myosin step size. As far as I understand, ATPase activity is not so much different—only by a factor of two or so between the systems with and without actin filaments, because of the very limited degree of actin activation in the latter. I think this may cause considerable variation in the results.

Yanagida: There are two kinds of methods used to measure ATPase activity. When we apply a large amount of actin filaments—a saturated concentration of actin filaments—to the myosin-coated surface, the actin-activated ATPase activity is much higher than that of the myosin-coated surface alone.

Sugi: Do you make ATPase measurements in such conditions?

Yanagida: Yes, we did in both conditions.

Homsher: Can I just add one thing about your own measurements? Yoko Toyoshima and you calculate the step size in a rather different way. If I take your values at 30° C and use Yoko's formula for calculating the step size—at least when I've tried taking the data from your paper—I get a step size of 25 nm. But if I use your data at 22° C and your ATPase, I get something like 100 nm, using Yoko's equation.

Yanagida: Yes, we also estimate the myosin-step size with values at 30° C. The myosin-step size is much larger than her estimation.

Homsher: That's using your formula. Using her formula, it is much lower.

Brenner: If you want to explain the discrepancy between your data and the Spudich data by saying that they did not take into account myosin head misorientation, then I assume you have taken that into account. Have you introduced some correction factor in your calculations to consider the misorientation?

Yanagida: Just recently, we found that our velocity is dependent upon the orientation of myosin head. My group is now doing that experiment.

Sellers: One other big difference between your data (Ishijima et al. *Nature* **352**, 301-303, 1991) and Uyeda's data (Uyeda et al. *Nature* **352**, 307-311, 1991) is that when filaments are sliding rapidly, your data indicates that 90% of the cross-bridges are attached, whereas his whole argument supposes that the cross-bridge is only attached for 5% of its cycle. Could you comment on that?

Yanagida: The duty ratio is obtained by calculation. If we calculate the duty ratio by using their data in this way the duty ratio becomes more than 0.5.

Sellers: One observation that Uyeda made was that he had to use methylcellulose, because in the absence of methylcellulose the actin filaments would not remain bound to the surface. This would also seem to indicate, at least under his conditions, that he is having very infrequent occupancy.

Yanagida: In our experiment, we did not use methylcellulose. But, the 40 nm long actin filaments can move. So, if we use methylcellulose, still shorter actin filaments might move.

Gillis: You were using synthetic filaments. I suppose that the local density of myosin heads is enormous compared to simply sticking of myosin on the nitrocellulose plate—would that make the difference?

Yanagida: We have to do the experiments by using myosin filaments where the myosin heads are oriented.

VI. STRUCTURAL CHANGES DURING MUSCLE CONTRACTION STUDIED BY X-RAY DIFFRACTION

INTRODUCTION

Changes in the X-ray diffraction pattern from striated muscle give information about the structural changes during contraction. The use of intense X-rays in the synchrotron radiation has made it possible to dynamic structural changes in contracting muscle with a high time resolution, though there are a number of problems in interpreting the data obtained.

In an attempt to demonstrate the ability of myosin head to rotate in muscle fibers, Iwamoto and Podolsky examined changes in the intensity of equatorial reflections from glycerinated rabbit psoas fibers after crosslinking myosin heads to actin. The results obtained in the presence of various ligands, such as ATP, ATP-γ-S etc., were interpreted in terms of transition between weak- and strong-binding states of the crosslinked heads, Griffiths and others performed time-resolved studies on changes in equatorial reflections and stiffness during isometric contraction and during unloaded shortening of intact single frog muscle fibers with interesting results. They also observed a compression of myofilament lattice during force development, suggesting radial component of cross-bridge force. Yagi, Takemori and Watanabe succeeded in recording a two-dimensional diffraction pattern from a frog muscle shortening at the maximum velocity using a Fuji imaging plate. The results obtained suggest that myosin head arrangement during shortening is different from that during isometric contraction.

In the paper of Squire and Harford, the advantages of bony fish skeletal muscle for X-ray diffraction studies are experimental results to date strongly support swinging of myosin heads on actin as part of the contractile cycle. Wakabayashi and coworkers reported their detailed studies on the intensity changes of actin-based layer lines from frog muscles with full or none myofilament overlap, and discussed possible changes of myosin head configuration and thin filament structure during contraction. To obtain evidence for the rotation of myosin heads during contraction, Brenner and Yu made carefully recorded high resolution equatorial X-ray diffraction patterns from single chemically skinned rabbit psoas fibers under various conditions. The results suggest the force-generating cross-bridges are not a mixture of the weakly-bound and rigor cross-bridges.

CROSSBRIDGE ROTATION IN EDC-CROSSLINKED STRIATED MUSCLE FIBERS

H. Iwamoto* and R.J. Podolsky

National Institute of Arthritis and Musculoskeletal and Skin Diseases
National Institutes of Health
Bethesda, Maryland 20892 USA

ABSTRACT

The rotatability of the strong- and weak-binding myosin heads was tested by stretching glycerinated rabbit psoas fibers after crosslinking the heads to actin by using a carbodiimide EDC. The equatorial 1,1 reflection intensity ($I_{1,1}$) decreased by ~10% upon 1% stretch in the presence of various ligands (ATP, ATP-γ-S, pyrophosphate and AMPPNP). As the action of ligands to dissociate actomyosin increased, the relaxation of tension response to stretch and the $I_{1,1}$ decrease were accelerated. This result is best explained if the ligand converts the crosslinked head to a weak-binding state, in which the head is rotatable because of its acquired elasticity.

Conversely, the weak-to-strong transition was induced in the crosslinked system by removing a ligand (ATP-γ-S) from myosin. Force was produced upon weak-to-strong transition and was accounted for by the increased stiffness of each crosslinked myosin head. However, the comparison of stress-strain curves for the weak- and strong-binding myosin showed that the equilibrium angle of myosin attachment was unchanged, making it unlikely that the weak-to-strong transition is the sole mechanism for active contraction. The calcium-activated force of the same crosslinked fibers showed several features in marked contrast to the force produced by the weak-to-strong transition. This leads to a possibility that the active force is supported by a third class of intermediate which is distinct not only from the weak-binding but also from the strong-binding intermediates in a classical sense.

---

* present address: Department of Physiology, School of Medicine, Teikyo University, Itabashi-ku, Tokyo 173 Japan

Mechanism of Myofilament Sliding in Muscle Contraction, Edited by
H. Sugi and G.H Pollack, Plenum Press, New York, 1993

INTRODUCTION

In insect flight muscles, the rigor myosin heads have been shown to be at a 45° angle with respect to the filament axis and relaxed myosins at a 90° angle[1]. This observation has been incorporated into the rotating crossbridge theories of muscle contraction (e.g., ref. 2), in which a myosin head rotation from 90° to 45° is postulated to produce contractile force.

In a reverse type of experiment, stretching a rigor muscle was expected to rotate the heads back from 45° to 90° configurations. Calculations have shown that this type of rotation should accompany a large change in the X-ray equatorial intensity ratio, $I_{1,0}/I_{1,1}$[3], but this effect has not been confirmed experimentally[4][5]. Spectroscopic measurements[6-10] and electron microscopy[11] have also failed to show this type of rotation.

The fast dissociation rate of actomyosin at saturating concentrations of ATP[12] suggests that rigor complexes do not exist significantly during contraction. Rather, the rotation problem could be better conceived in the light of weak- and strong-binding intermediates[13] of the actomyosin ATP hydrolysis cycle. Besides the association constant for myosin and actin, the strong- and weak-binding myosins are also contrasted in biochemical characteristics (e.g., the ability to activate regulated actin in a cooperative manner[14], the susceptibility to proteolysis and sulfhydryl modification[15], and the fluorescence of pyrene attached to actin[16]), suggesting that the higher order structure of myosin is different in the strong- and weak-binding forms. Thus, attempts to test rotatability should also be directed towards the weak-binding heads. At the same time, it should be possible to compare the equilibrium angles of attachment of weak- and strong-binding heads by stretching the fibers by different amounts and comparing the fiber length at which the stress starts to rise.

However, the small association constant for actin and myosin makes it difficult to test the rotatability of weak-binding myosin heads, because the heads may detach from actin before any rotation could be observed. To overcome this difficulty, we crosslinked the heads to actin in muscle fibers with a zero-length crosslinker, EDC (1-ethyl-3-[3-dimethylaminopropyl] carbodiimide), which is known to crosslink specific sites on myosin to actin while leaving actomyosin's basic properties intact (e.g., ATPase activity)[17].

MATERIALS AND METHODS

Preparation

Rabbit psoas muscles were chemically skinned in a relaxing solution and stored for up to 1 month in a freezer in a 50% mixture of the relaxing solution and glycerol. Thin bundles (~20 fibers, 10 mm) were dissected from the muscle and transferred to an experimental chamber placed in the X-ray beam path. The ends were connected to the force transducer (Akers, AE801) and the extension of a servo motor using aluminum T-clips[18] and cyanoacrylate glue. Sarcomere length was adjusted to 2.6 μm by a He-Ne laser beam.

Solutions

The compositions of solutions in mM were as follows. Relaxing solution: 5 $MgCl_2$, 4 ATP, 5 EGTA, 50 I.U. creatine kinase (Sigma) and 15 phosphocreatine (Sigma); rigor solution: 2.5 EGTA, 2.5 EDTA; Mg-rigor solution: 5 EGTA, 2 $MgCl_2$; ATP-γ-S solution: 5 $MgCl_2$, 4 ATP-γ-S (Boehringer Mannheim), 5 EGTA, 10 I.U. apyrase (Sigma), which has ATPase and ADPase activities, 10 I.U./ml hexokinase (Sigma), 2 glucose; PPi (pyrophosphate) solution: 4 $MgCl_2$, 4 PPi, 10 I.U./ml hexokinase, 2 glucose; low-EGTA relaxing solution: relaxing solution containing 0.05 EGTA; contracting solution: relaxing solution containing 0.1 free Ca^{2+}. The relaxing solution used for crosslinked fibers contained 20 I.U./ml creatine kinase and 15 BDM (butanedione monoxime, Sigma), which inhibits actomyosin ATPase by blocking reattachment of myosin to actin[19][20]. All the solutions contained 120 K-acetate, 10 PIPES (pH 7.2). The temperature was kept at 5°C.

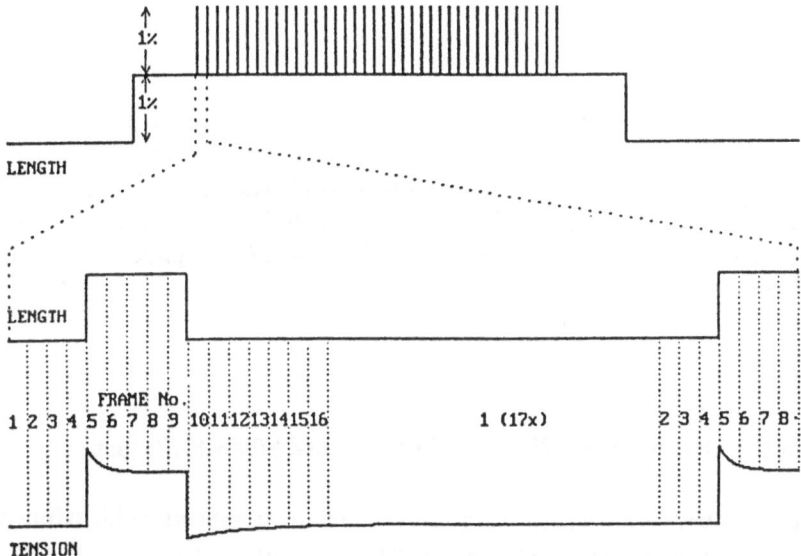

Fig. 1. Procedure of time-resolved X-ray experiments. Top, length trace for the entire experiment; Middle, enlargement of a part of the top trace, including a single cycle of stretch and release; Bottom, tension response corresponding to the middle trace. The dotted line connecting the middle and bottom lines indicate the margins of time-slicing frames. The time frames had an equal duration except for the first one, which was 17 times longer than the others. The duration of a frame (2nd - 16th) ranged from 10 ms to 20 s. Note that fiber length was increased by 1% before cycles of stretch and release were applied.

Mechanical Measurement

Fiber length was altered by using a servomotor (300B, Cambridge Technology). To observe the stress relaxation during a stretch period, the fibers were stretched by 1% for 20 s after 1% prestretch. The instantaneous stiffness was measured by imposing quick stretches on fibers (0.2% fiber length, rise time 1 ms, duration 1 s, at an interval of 120 s). The maximum amplitude of the tension response was recorded by using a Nicolet 3091 digital oscilloscope.

Equatorial X-ray Diffraction Pattern

The fibers were exposed to the X-ray beam produced by an Elliott GX-13 generator operated at 35 kV x 65 mA. The diffraction patterns were collected by a 1-dimensional position-sensitive proportional counter (1050X, Ordela) and sent to a Tracor-Northern multi-channel analyzer. The data were further sent to an IBM PS/2 computer for analysis. The time-resolved measurements were performed by switching 16 time-frames in the analyzer by using a digital timing circuit which was also used to trigger stretches. For the time-resolved X-ray measurements, the fibers were prestretched by 1% in advance. The procedure for time-resolved measurements is shown in Figure 1.

Crosslinking

The crosslinking procedure basically followed that of Tawada & Kimura[21]. After putting fibers in rigor, the crosslinking was carried out in a rigor solution containing 8 mM EDC (Pierce) for 20 min at 20°C. The reaction was quenched by replacing the crosslinking solution by a rigor solution containing 0.2% (v/v) β-mercaptoethanol (Pierce). The fibers were crosslinked under these standard conditions unless otherwise stated.

RESULTS

Tension Response of Uncrosslinked and Crosslinked Fibers to Stretch.

Figure 2a shows the tension responses of the uncrosslinked rabbit skinned fibers to 20 s, 1% stretch. The ligands accelerate the relaxation in the order of 1 mM AMPPNP, 4 mM PPi and 4 mM ATP. This order generally coincides with the ability of ligands to dissociate actomyosin complex and is consistent with the previous results[22].

The responses of crosslinked fibers (Fig. 2b) have an offset of steady component, but the ability of ligands to accelerate relaxation is unaffected. Since most of the myosins are crosslinked (shown later), the crosslinked myosin molecules should be responsible for accelerated relaxation.

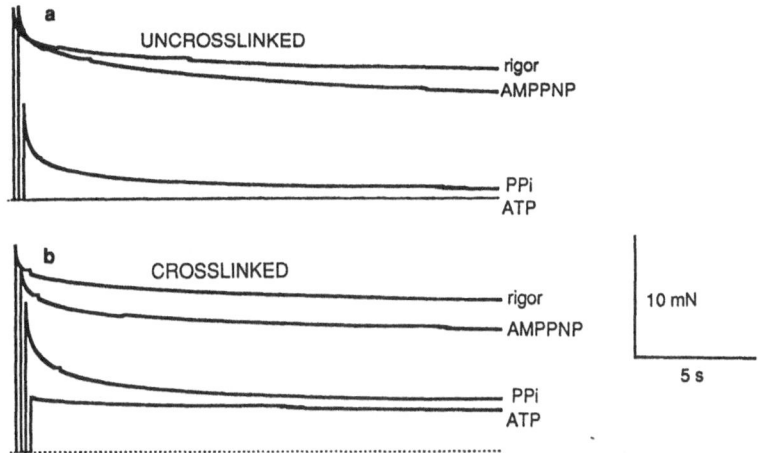

Fig. 2. The tension responses of a fiber bundle to 1%, 20 sec stretch before and after crosslinking. (a), before crosslinking; (b), after crosslinking. In each, traces are for rigor, 1 mM AMPPNP, 4 mM PPi and 4 mM ATP from above. Output of the digital oscilloscope. The traces are horizontally offset for clarity. The broken line shows the tension level before stretch. The little steps seen on the traces are due to the conversion error of the oscilloscope.

This behavior of the crosslinked fibers could be best explained in the light of two-step dissociation[23-25] i.e., the ligand-specific strong-to-weak transition and the breakage of weak interaction. Since the latter is impossible in the crosslinked system, it would be reasonable to consider that the ligand-induced strong-to-weak transition makes the myosin molecules more elastic and leads to muscle relaxation.

Time-resolved Measurement of X-ray Equatorial Change upon Stretch

Figure 3a, b shows the changes in the equatorial reflection intensities of uncrosslinked fibers upon 1% stretch. There is no clear equatorial intensity changes upon stretch either in rigor (Fig. 3a) or in the presence of PPi (Fig. 3b). This is consistent with the previous results[4].

Figure 3c-h shows the changes in the equatorial reflection intensities upon stretch after crosslinking. Now the $I_{1,1}$ decreases by about 10% in a ligand-dependent manner. In the presence of ATP, the $I_{1,1}$ decrease is complete within the first 20 ms of stretch (Fig. 3c) and persists for at least 100 s (Fig. 3d). In the presence of PPi, the $I_{1,1}$ decrease is slower (Fig. 3e). In rigor, little $I_{1,1}$ decrease is observed for 5 s stretch (Fig. 3g), but the prolongation of stretch makes the decrease clearer (Fig. 3h).

The dependence of the rate of $I_{1,1}$ decrease on ligand is more clearly seen by plotting the averaged amplitude of $I_{1,1}$ decrease against the duration of stretch (Fig. 4). It shows that the ligands accelerate the $I_{1,1}$ decrease in the order of AMPPNP, PPi, and ATP, in agreement with the order for mechanical relaxation (Fig. 2). Since calculations show that the preferential decrease of $I_{1,1}$ is explained if the heads rotate

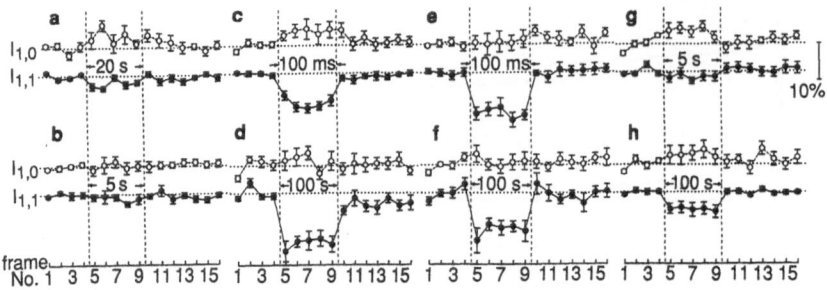

Fig. 3. The time course of $I_{1,0}$ and $I_{1,1}$ changes during stretch-release cycles. (a) and (b), before crosslinking. (c)-(h), after crosslinking under standard conditions. (a), PPi, 5 s stretch; (b), rigor, 20 s stretch; (c) and (d), ATP, 100 ms and 100 s stretch; (e) and (f), PPi, 100 ms and 100 s stretch; (g) and (h), rigor, 5 s and 100 s stretch respectively. The values are the mean from 8 to 14 specimens after normalization to the average values of the first four frames. The stretch is given between the two vertical broken lines (frames 5-9). Open circle, $I_{1,0}$; filled circle, $I_{1,1}$. The vertical bars associated with circles indicate SEM.

in a three-dimensional manner (not shown), the results are in agreement with the possibility that the increased elasticity of the myosin molecules upon strong-to-weak transition is due to the acquisition of rotatability by the heads.

Force Produced by Weak-to-Strong Transition

The results described above predict that a weak-binding myosin molecule will become stiffer if it is deprived of its ligand. We tested whether this weak-to-strong transition leads to force production. In the present study ATP-γ-S was used as the ligand to exclude possible contributions of nucleotide hydrolysis. Experiments were

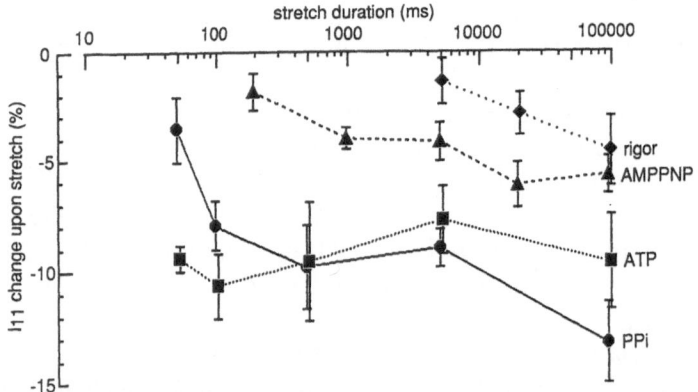

Fig. 4. Dependence of $I_{1,1}$ decrease on stretch duration in fiber bundles crosslinked under standard conditions. Square, ATP; circle, PPi; triangle, AMPPNP; diamond, rigor. The average of the $I_{1,1}$ decrease for frames 5-9 (in Fig. 3) is plotted against the logarithm of the stretch duration. Vertical bars indicate SEM (n = 5-14).

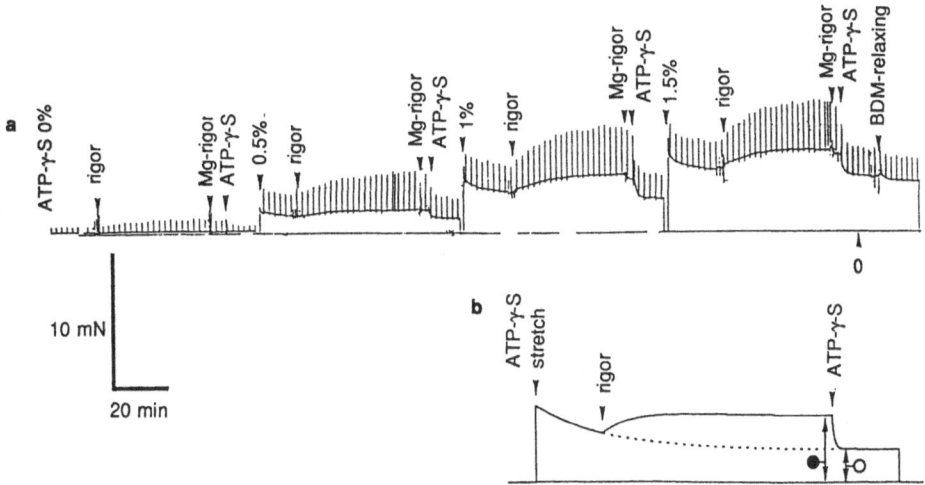

Fig. 5. Force produced by withdrawal of ATP-γ-S (weak-to-strong transition) in a crosslinked fiber bundle. (a), the whole time course of tension during a single experiment using a fiber bundle crosslinked under standard conditions (8 mM EDC, 20°C, 20 min reaction). The bundle is prestretched from 0% to 1.5% fiber length by 0.5% steps as indicated. The bottom line indicates zero tension level. The short vertical lines are due to the quick stretches (0.2%) to measure instantaneous stiffness. (b), schematic illustration of the time course of force production by the weak-to-strong transition at a prestretch level. The tension response to the prestretch is considered to decay slowly as indicated by a broken line. The amplitudes of the tension before and after the readdition of ATP-γ-S (two vertical arrows) are regarded to represent the steady tension supported by each myosin molecule in strong- (filled circle) and weak-binding forms (open circle) respectively.

carried out with increasing levels of prestretch from 0 to 1.5% of fiber length in a procedure similar to that of Tawada & Emoto[26].

Upon removal of ATP-γ-S, tension starts to rise, and its amplitude of increase is apparently proportional to the amount of prestretch (Fig. 5a). No tension rise is observed at 0% prestretch. This behavior is thus very similar to that of rigor tension observed by Tawada and Emoto[26]. Even the replacement of ATP-γ-S by ADP, or the removal of PPi to induce rigor, resulted in force production (not shown).

The Stress-Strain Curves of Weak- and Strong-binding Myosins

The rise of tension upon weak-to-strong transition is believed to originate only from crosslinked myosins, because the time scale of experiment is long enough to allow all uncrosslinked myosins to go through cycles of detachment and reattachment, during which they would loose their strain. Thus, the tension levels in the presence and absence of ATP-γ-S (arrows in Fig. 5b) are believed to reflect directly the force supported by each crosslinked myosin molecule in weak- and strong-binding form respectively.

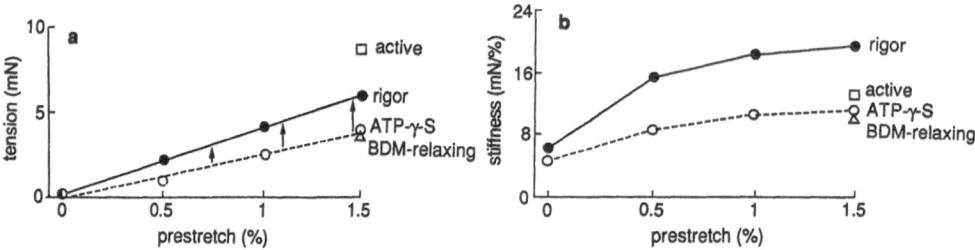

Fig. 6. Dependence of steady tension level (a) and instantaneous stiffness (b) on prestretch level. Obtained from a single bundle. Lines in a, obtained by least-squares regression, represent the stress-strain curves of the myosin elasticity. Filled and open circles represent the values before and after the readdition of ATP-γ-S (see Fig. 5b) respectively. The triangles represent the values for ATP. The squares represent the values for calcium activation. From the same fiber bundle as in Fig. 1a and b.

On this ground of argument, the plot of the steady tension level against the prestretch level in the presence and absence of ATP-γ-S (Fig. 6a) should give the stress-strain curves of the weak- and strong-binding myosin molecules respectively.

The curves are linear, indicating that the elasticity obeys Hooke's law, with a slope 2.11 ± 0.18 times steeper for strong-binding molecules than for the weak-binding molecules (mean ± S.D., n = 6). This means that a strong-binding molecule is about twice stiffer than a weak-binding molecule. This increased stiffness accounts for the force production upon weak-to-strong transition (arrows in Fig. 6a).

An important observation is that both lines pass through the origin, indicating that the equilibrium angle of attachment is the same for the strong- and weak-binding heads. The difference in the x-intercept was 0.01 ± 0.06% of fiber length (n = 6), or 0.13 ± 0.78 nm/half sarcomere, corresponding to 0.4 ± 2.2 degrees of rotation if the head is assumed to be 20 nm long. This amount is too small to account for sliding.

Extent of Crosslinking

The instantaneous stiffness (Fig. 6b), measured by applying 0.2% quick stretches, is regarded proportional to the number of attached myosin molecules, whether they are crosslinked or not. In rigor, all the myosin molecules are bound to actin[27)28)], whereas only the crosslinked myosin molecules would be the bound population in the presence of ATP-γ-S. According to this argument one can estimate the extent of crosslinking (fraction of crosslinked myosin molecules, see DISCUSSION) by using the following equation:

$$\text{extent} = (S_W / S_S)/(T_W / T_S)$$

where S_W and S_S are the values of instantaneous stiffness in the presence and absence of ATP-γ-S, and T_W and T_S are the steady tension levels (Fig. 6a) in the presence and absence of ATP-γ-S respectively. In case of the standard 8 mM EDC, 20°C 20 min

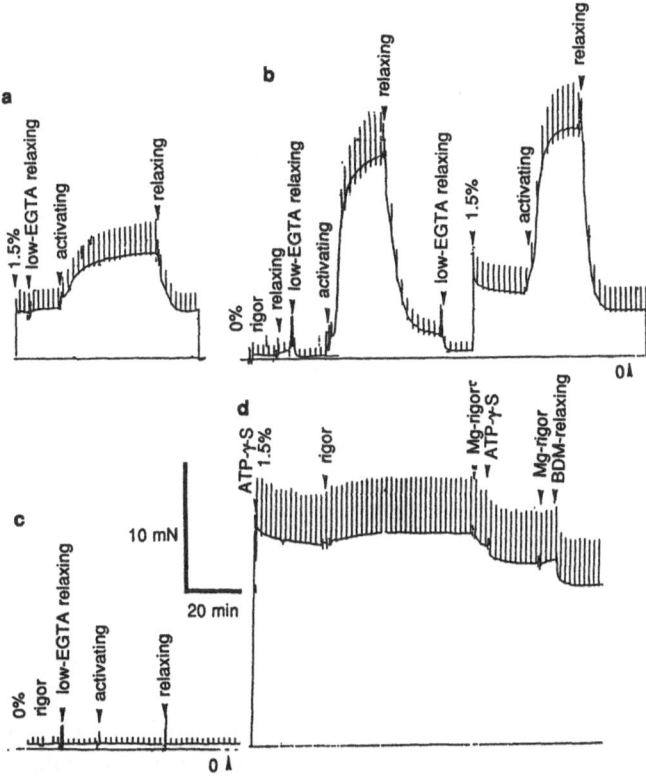

Fig. 7. Calcium-activated force of crosslinked fibers. (a), calcium activated force produced at 1.5% prestretch level. Continuation of the trace in Fig. 5a. (b), comparison of the calcium-activated forces at different prestretch levels. (c), effect of calcium-activating solution on a fiber bundle crosslinked for 2 hr. (d), weak-to-strong transition force produced by the fiber bundle crosslinked for 2 hr. Each of (b)-(d) is from separate fiber bundles.

reaction, $91 \pm 8\%$ of molecules are crosslinked (mean $\pm$ S.D., n = 6). In case of 2 mM EDC, the $56 \pm 9\%$ of the myosin molecules are crosslinked (n = 3).

Calcium-activated Force of Crosslinked Fibers

The possibility was tested whether the force produced by weak-to-strong transition accounts for active contraction. At the end of experiments, the fibers are calcium activated at 1.5 % prestretch (Fig. 7a). The fibers produce force much larger than that produced by the weak-to-strong transition in spite of almost complete crosslinking (square in Fig. 6a). On the other hand, the stiffness is not as large as in rigor (square in Fig. 6b).

The active force is produced even at 0% prestretch and its magnitude is almost the same as that at 1.5% prestretch (Fig. 7b). The active forces produced at 0% and 1.5% prestretch level are 49 ± 9 and $45 \pm 11\%$ of the active force before crosslinking respectively. Thus, the calcium activated force is independent of the prestretch level.

After excessive crosslinking for 2 hrs, the active force disappears (Fig. 7c), whereas the force produced by weak-to-strong transition is still present (Fig. 7d). Therefore, contrary to the force produced by weak-to-strong transition, the calcium-activated force is inversely related to the extent of crosslinking.

DISCUSSION

The use of EDC crosslinking made it possible to compare the rotatability of myosin heads in weak-binding form with that in strong-binding form. To summarize the results, the weak-binding heads do rotate upon straining but the strong-binding heads do not rotate appreciably. The results also showed that the rotation of the weak-binding heads is due to the increased elasticity without a change in the equilibrium angle of attachment.

Rationale for Using EDC

The main conclusions of the present study rest on the assumption that the EDC-crosslinked myosin mimics the intact weak-binding myosin, which is typically found in the presence of ATP at low ionic strengths[29], whereas it takes the strong-binding form in the absence of ligands. The validity of this assumption is verified by the following evidence from the literature:

1) All the characteristics of weak- and strong-binding actomyosin cited in INTRODUCTION are preserved in EDC-crosslinked actomyosin[14][30] as well as the ability to hydrolyze ATP[17].

2) The crosslinking is highly specific, and it is reported that the crosslinking site on actin is in the mainly acidic first 11 amino acids from the N-terminus[31]. The antibody raised against this segment of actin is known to inhibit actomyosin ATPase activity[32]. This has led to a possibility that this part of actin plays an important role in the weak interaction of actomyosin. Further evidence that this part of actin is implicated in the weak interaction is that the actin-binding fragment of caldesmon, which competes for the binding to actin with the same antibody[33], also inhibits stiffness generation at low ionic strength in the presence of ATP[34].

The ability of crosslinked actomyosin to produce active force in a calcium-dependent manner in the fiber system (Fig. 7) also shows that the EDC crosslinking does not impair the basic properties of myosin or the regulatory system on actin filaments.

The considerations above make it likely that the EDC crosslinking replaces the electrostatic interactions at the first 11 amino acids by covalent linkages while keeping the ability of making or breaking the strong interaction, which leads to the changes in the higher order structure of myosin head. A model has recently been proposed in which weak and strong actin-myosin interactions occur at several distinct sites[35].

Increased Elasticity of Weak-binding Heads

The increased elasticity of the weak-binding heads as suggested in the present study is consistent with many lines of evidence in the literature. EPR studies show that the weak-binding heads take a wide range of angles while attached (or crosslinked) to actin[36][37]. Electron micrographs of negatively stained crosslinked actomyosin or quick-frozen uncrosslinked actomyosin also showed a variety of angles of attachment[38][39]. Two-dimensional X-ray diffraction pattern of fibers at low ionic strength shows myosin layer lines[40][41], indicating that the weak-binding heads are flexibly attached to actin.

Myosin Molecule as a Structural Unit to Support Force

It is a matter of open debate as to whether a myosin head (subfragment-1, or S-1) or a myosin molecule as a whole acts as an independent force-supporting unit. The presence of elasticity in the subfragment-2 (S-2)[42], which is common to both heads, and the paucity of evidence for head rotatability make the latter possibility more likely in the strong-binding form. This is in agreement with the observation that a myosin molecule does not release its strain until both heads detach from actin[43]. When the elasticity doubles upon strong-to-weak transition, the elasticity will be equally distributed in the S-1 and S-2, but as long as the S-2 stays elastic the head cannot be an independent force-supporting unit.

In spite of the nearly complete crosslinking, the equatorial ratio $I_{1,0}/I_{1,1}$ is still sensitive to ligands (data not shown). The simplest explanation for this is that only one of the heads is crosslinked to actin due to some negative cooperativity between the heads. If this is the case, the removal of ATP-γ-S will convert a weak-binding myosin molecule attached with one head to a strong-binding molecule attached with both heads, resulting in the reduction of elasticity to a half. Then the elasticity of a weak-binding head will be comparable to that of S-2.

Tawada & Kimura[21] have reported an extent of crosslinking much lower than ours under comparable conditions. However, the present data are not in disagreement with theirs if there is some cooperativity between the two heads.

The Calcium-activated Force

The present study suggests that the weak-to-strong transition leads to force production as envisaged in some theories[25][26][44][45]. However, the invariability of equilibrium angle of attachment makes it unlikely that this is the sole mechanism for active force production. A recent EPR study has also shown that the averaged angle of attached weak-binding heads is the same as that of rigor heads[37].

The calcium-activated force qualitatively differs from the force produced by weak-to-strong transition in many respects. The inverse relationship between the extent of crosslinking and the active force production raises a possibility that association-dissociation cycles at the weak-binding site are needed for the production of active force while they are not necessary for the force produced by weak-to-strong transition. Thus, it is reasonable to consider that calcium-activated force is supported by a third class of intermediate which is distinct not only from weak-binding but also from strong-binding species in a classical sense. This idea is also consistent with the results by Brenner & Yu[46] in this volume.

The active heads rather resemble weak-binding heads in that they do not accompany high stiffness (Fig. 6b). In support of this, X-ray[47], EPR[48], biochemical[49] and ultrasonic[50] evidence shows that there is a relatively small amount of rigor-like heads (< 20%) during contraction. The active heads may also be rotatable because 14.3 nm meridional reflection reversibly decreases upon stretch or release[51]. The way of rotation however may be different because this rotation does not affect the equatorial reflections.

Although the weak-to-strong transition may not be directly coupled to the initiation of force production, it may play a role in augmenting the force produced by another mechanism. Variable stiffness of a single myosin will also have profound implications in understanding the mechanical behavior of muscle.

ACKNOWLEDGEMENTS

We would like to express our sincere thanks to Dr. Katsuhisa Tawada for his kind instructions for the use of EDC and Drs. Mark Schoenberg and Manuel Morales for their valuable discussions.

REFERENCES

1. Reedy, M.K., Holmes, K.A. & Tregear, R.T. *Nature* **207**, 1276-1280 (1967).
2. Huxley, H.E. *Science* **164**, 1133-1136 (1969).
3. Lymn, R.W. *Biophys. J.* **21**, 93-98 (1978).
4. Naylor, G.R. & Podolsky, R.J. *Proc. Natl. Acad. Sci. USA* **78**, 5559-5563 (1981).
5. Tanaka, H., Hashizume, H. & Sugi, H. in *Contractile Mechanism in Muscle* (eds Pollack, G.H. & Sugi, H.) 203-205 (Plenum Press, New York, 1984).
6. dos Remedios, C.G., Millikan, R.G.C. & Morales, M.F. *J. Gen. Physiol.* **59**, 103-120 (1972).
7. Cooke, R. *Nature* **294**, 570-571 (1981).
8. Crowder, M.S. & Cooke, R. *Biophys. J.*, **51**, 323-333 (1987).
9. Yanagida, T. in *Contractile Mechanism in Muscle* (eds Pollack, G.H. & Sugi, H.) 397-412 (Plenum Press, New York, 1984).

10. Hambly, B. Franks. K. & Cooke, R. *Biophys. J.* **59**, 127-138 (1991).

11. Suzuki, S., Oshimi, Y. & Sugi, H. This volume, 57-70.

12. Marston, S.B. *Biochem. J.* **203**, 453-460 (1982).

13. Stein, L.A., Schwarz, R.P., Chock, P.B. & Eisenberg, E. *Biochemistry* **18**, 3895-3909 (1979).

14. Greene, L.E. & Eisenberg, E. *Proc. Natl. Acad. Sci. USA* **77**, 2616-2620 (1980).

15. Duong, A.M. & Reisler, E. *Biochemistry* **28**, 3502-3509 (1989).

16. Coates, J.H., Criddle, A.H. & Geeves, M.A. *Biochem. J.* **232**, 351-356 (1985).

17. Mornet, D., Bertrand, R., Pantel, P. Andemard, E. & Kassab, R. *Nature* **292**, 301-306 (1981).

18. Ford. L.E., Huxley, A.F. & Simmons, R.M. *J. Physiol. (Lond.)* **269**, 441-515 (1977).

19. Horiuti, K., Higuchi, H., Umazume, Y., Konishi, M., Okazaki, O. & Kurihara, S. *J. Muscle Res. Cell Motility* **9**, 156-164 (1988).

20. Lenart, T.D., Tanner, J.W. & Goldman, Y.E. *Biophys. J.* **55**, 260a (1989).

21. Tawada, K. & Kimura, M. *J. Muscle Res. Cell Motility* **7**, 339-350 (1986).

22. Schoenberg, M. & Eisenberg. E. *Biophys. J.* **48**, 863-871 (1985).

23. Trybus, K.M. &Taylor, E.W. *Proc. Natl. Acad. Sci. USA* **77**, 7209-7213 (1980).

24. Geeves, M.A., Goody, R.S. & Gutfreund, H. *J. Muscle Res. Cell Motility* **5**, 351-361 (1984).

25. Geeves, M.A. *Biochem. J.* **274**, 1-14 (1991).

26. Tawada, K. & Emoto. Y. in *Molecular Mechanism of Muscle Contraction* (eds Sugi, H. & Pollack, G.H.) 219-226 (Plenum Press, New York, 1988).

27. Cooke, R. & Franks, K.E. *Biochem. J.* **19**, 2265-2269 (1980).

28. Lovell, S. J. & Harrington, W.F. *J. Mol. Biol.* **149**, 659-674 (1981).

29. Brenner, B., Schoenberg, M., Chalovich, J.M., Greene, L.E. & Eisenberg, E. *Proc. Natl. Acad. Sci. USA* **79**, 7288-7291 (1982).

30. King, R.T. & Greene, L.E. *J. Biol. Chem.* **262**, 6128-6134 (1987).

31. Sutoh, K. *Biochemistry* **21**, 3654-3661 (1982).

32. DasGupta, G. & Reisler, E. *J. Mol. Biol.* **207**, 833-836 (1989).

33. Adams, S., DasGupta, G., Chalovich, J.M. & Reisler, E. *J. Biol. Chem.* **265**, 2231-2237 (1990).

34. Brenner, B., Yu, L.C. & Chalovich, J.M. *Proc. Natl. Acad. Sci. USA* **88**, 5739-5743 (1991).

35. Labbé, J.-P., Méjean, C., Benyamin, Y. & Roustan, C. *Biochem. J.* **271**, 407-413 (1990).

36. Svensson, E.C. & Thomas, D.D. *Biophys. J.* **50**, 999-1002 (1986).

37. Fajer,P.G., Fajer, E.A., Schoenberg, M. & Thomas, D.D. *Biophys. J.* **60**, 642-649 (1991).

38. Craig, R., Greene, L.E. & Eisenberg, E. *Proc. Natl. Acad. Sci. USA* **82**, 3247-3251 (1985).

39. Applegate, D. & Flicker, P. *J. Biol. Chem.* **262**, 6856-6863 (1987).

40. Matsuda, T. & Podolsky, R.J. *Proc. Natl. Acad. Sci. USA* **81**, 2364-2368 (1984).

41. Collett, B. & Podolsky, R.J. personal communication.

42. Highsmith, S., Kretzschmar, KM., O'Konski, C.T. & Morales, M. *Proc. Natl. Acad. Sci. USA* **74**, 4986-4990 (1977).

43. Anderson, M., & Schoenberg, M. *Biophys. J.* **52**, 1077-1082 (1987).

44. Eisenberg, E. & Hill, T.L. *Science* **227**, 999-1006 (1985).

45. Podolsky, R.J. & Arata, T. in *Molecular Mechanism of Muscle Contraction* (eds Sugi, H. & Pollack, G.H.) 319-330 (Plenum Press, New York, 1988).

46. Brenner, B. & Yu, L.C. This volume, 461-469.

47. Huxley, H.E. & Kress, M. *J. Muscle Res. Cell Motility* **6**, 153-161 (1985).

48. Cooke, R., Crowder, M.S. & Thomas, D.D. *Nature* **300**, 776-778 (1982).

49. Duong, A.M. & Reisler, E. *Biochemistry* **28**, 1307-1313 (1989).

50. Hatta, I., Sugi, H. & Tamura, Y. *J. Physiol. (Lond.)* **403**, 193-209 (1988).

51. Huxley, H.E., Faruqi, A.R., Kress, M., Bordas, J. & Koch, M.H.J. *J. Mol. Biol.* **158**, 637-684 (1982).

Discussion

Pollack: I was wondering whether it's possible that the interpretation of the I(1,1) intensity decrease could be somewhat different than the one you gave. If you have only perhaps 90% of the fiber cross-linked, is it possible that, by stretching this fiber, you introduce some non-uniformity, and the reason for the diminished intensity is because the stretch is not uniform?

Iwamoto: The reason that I believe the equatorial change comes from the cross-bridge is that the behavior of the decreased I(1,1) is ligand-dependent, and the type of change is what is expected from the ability of ligands to dissociate the actomyosin complex.

Holmes: I wonder to what extent your arguments are affected by the statement that the cross-linker forces the weak-binding state, because the N-terminus seems very likely to be in the middle of the rigor-binding site and, therefore, is part of a strong-binding site. The results you quote, based on peptide studies, have to be treated with extreme caution, and it is very likely that, in fact, you are generating some approximation to the rigor state by your cross-linking.

Huxley: I wondered whether the low tension that you obtained in rigor, compared with activated, might be due to series compliance at the damaged ends of the fiber where an activated fiber can detach and reattach further along, but in a rigor fiber it can't. It might be an artifact due to series compliance.

Iwamoto: In the case of cross-linked fiber there is an extensive cross-linking all along the length of the fiber, and the influence of the end compliance is less likely than in the uncross-linked fiber.

Tregear: Dr. Iwamoto, if I understood you right, you had a state that was weaker—it wasn't as stiff as ours—and it was a weak cross-bridge state. How does that relate to the weak cross-bridge states that we have been hearing about from Dr. Brenner and Dr. Schoenberg? I thought that they were actually as stiff—provided you shook them fast enough—as any other state. Can you tell me what the status of this new state is?

Iwamoto: I think the weakly binding state I observed is basically the same species as Dr. Brenner observed by measuring relaxed stiffness (Brenner et al. *Proc. Nat. Acad. Sci. USA* **79**, 7288-91, 1982) because I cross-linked the fiber, and the cross-linking site on actin is suggested to be at the same position as the site for weak interaction. Part of the reason for that is Dr. Brenner's work (Brenner et al. *Proc. Nat. Acad. Sci. USA* **88**, 5739-43, 1991) showing that caldesmon fragments that compete with an antibody raised against the N(1-7) sequence of actin for binding to actin also inhibit relaxed stiffness.

K. Yamada: Dr. Iwamoto, you cross-linked at perhaps the 90% level, and according to Dr. Tawada's paper, with that much cross-linking, the rigor force is very small. What do you think about that?

Iwamoto: I think Dr. Tawada's basic assumption is that the stiffness of the rigor bridge and the stiffness of the weakly bound bridge are the same, but according to my study, the weakly binding cross-bridge is more compliant. That means if you estimate the extent of cross-linking by measuring the stiffness, you will underestimate the extent of cross-linking.

TIME-RESOLVED EQUATORIAL X-RAY DIFFRACTION MEASUREMENTS IN SINGLE INTACT MUSCLE FIBRES

P.J. Griffiths, C.C. Ashley, M.A. Bagni*, G. Cecchi* and Y. Maèda**

University Laboratory of Physiology
Parks Road
Oxford, OX1 3PT, UK.
*Dipartemento di Scienze Fisiologiche
Università degli Studi di Firenze
Viale Morgagni 63
Florence I-50432, Italy
**EMBL Outstation
Deutsches Elektronen Synchrotron
Notkestraße 85
52 Hamburg D-W2000, Germany.

ABSTRACT

Equatorial X-ray diffraction techniques have been successfully applied to the intact single muscle fibre preparation under length clamp and "fixed end" conditions. 10 and 11 intensity changes and stiffness have been measured in the same preparation. Under isometric conditions, equatorial signals and stiffness led force by 14-20ms during the rise of tetanic tension. During relaxation, stiffness and equatorial signals lagged force. The time course of the intensity changes suggests a low force crossbridge state is present to a greater extent during the rise of tetanic tension and during relaxation than at the tetanus plateau. During isotonic shortening at V_{max}, stiffness fell to 30% of its isometric level, while equatorial signals fell to 60%. Since stiffness and equatorial signals are thought to detect attached crossbridges, either the average stiffness per attached bridge measured at 4kHz during shortening is less than at the plateau, or the relation between equatorial intensities and the proportion of attached crossbridges during isotonic shortening differs from that measured under isometric conditions.

Active tension also affects the lattice spacing. The myosin lattice was compressed during the development of longitudinal force. This implies a radial component of crossbridge tension. The lattice compression was smaller in a compressed lattice and larger in an expanded lattice.

Mechanism of Myofilament Sliding in Muscle Contraction, Edited by
H. Sugi and G.H Pollack, Plenum Press, New York, 1993

INTRODUCTION

The molecular conformational changes which are the fundamental event in muscle contraction cannot be observed directly in a living cell. However X-ray diffraction patterns from relaxed and activated muscle do provide information at the molecular level about the changes in molecular structure which accompany activation. In early work on intact muscle cells, the laboratory X-ray sources used were suitable only for static patterns[1]. More recently the use of synchrotron radiation sources has permitted steady progress in both the time resolution of structural events detected by the X-ray diffraction pattern, and in the size of the preparation needed to produce reflections intense enough for study.

According to the model of Huxley and Simmons[2], instantaneous stiffness of a muscle fibre is proportional to the fraction of crossbridges attached to actin. Stiffness and force do not change together during activation and relaxation[3]. One explanation of this observation is that attachment and force development are separate steps in the crossbridge cycle. Crossbridge attachment is a change in muscle structure which may be detected in the X-ray diffraction pattern of the muscle. It was therefore of interest to compare the time course of changes in fibre stiffness with X-ray diffraction pattern intensities. Stiffness measurements on which the Huxley-Simmons model is based have been performed on frog single fibre preparations[4], while X-ray diffraction studies in the past have used frog whole muscle[5], so direct comparison of X-ray diffraction studies and stiffness were not completely satisfying. We therefore attempted to apply X-ray diffraction techniques and stiffness measurements to the same intact single fibre preparation using synchrotron radiation. Because the equatorial 10 and 11 reflections are the strongest in the muscle pattern and their interpretation is well established[6], we have concentrated our studies on these reflections. We discovered that changes in 10 and 11 intensities and spacings from an intact single fibre are detectable with a time resolution (5 ms) approaching that of mechanical measurements.

During the rise of tetanic tension and during relaxation, stiffness and equatorial reflections are of similar time course. During isotonic shortening, however, stiffness and equatorial intensities report different fractions of attached bridges. The myosin lattice behaviour during tension development suggests a radial component to crossbridge force.

METHODS

Single muscle fibres were isolated from the tibialis anterior muscles of *Rana temporaria* with aluminium clips attached to their tendons to reduce series compliance. A fibre was mounted in the X-ray chamber using hooks on a capacitance force transducer and motor which were inserted into holes in the aluminium clips. The electronics of the force transducer[7] and the characteristics of the motor are described elsewhere[8]. Synchrotron radiation (wavelength 0.15 nm, beam dimensions 3 x 0.4 mm) entered the chamber through a Kapton window positioned within 200 μm of the fibre. The resulting equatorial X-ray diffraction pattern was recorded on a one dimensional detector at a distance of 3.8 - 4 m from the preparation. A He/Ne laser beam entered the chamber through a glass window and struck the fibre in the centre of the region exposed to the X-ray beam. The position of first order laser diffraction band from the fibre was monitored by a position-sensitive photodiode, and converted to sarcomere length by a simple analogue computer circuit[9]. Stiffness was calculated from the force response to sinusoidal motion of the stretcher at 4 kHz (0.05% of fibre length peak to peak amplitude).

Data collection occurred at 5 or 10ms sampling rates during tetani of 400 ms duration. Typically 20 tetani were averaged to obtain sufficient counts on the one dimensional detector for analysis of the equatorial signals. 5 minute intervals were allowed between tetani to prevent fibre fatigue and to avoid X-ray-induced deterioration in fibre excitability. More detailed descriptions of the experimental methods have been published elsewhere[10].

Equatorial intensities were calculated either by integration of the whole area under a peak (including background) or by subtraction of an exponential or polynomial fitting of the background before integration. The time course of equatorial signals measured by both procedures were the same, but generally the whole area integration gave smoother traces, and has been used for the figures presented here. Lattice spacing measurements were performed after background subtraction. Either the integated intensity under a peak was computed, and the point at which 50% of the integrated intensity was reached was taken as the centre of the reflection, or the reflections were fitted to a Gaussian distribution, and the centre of the Gaussian computed by a Marquardt-Levenberg method. The time resolved lattice spacing changes were the same in both methods. Because the smaller 20 and Z reflections lie close to the 11 reflection, any broadening of the 11 may result in overlap with these smaller reflections and potential errors in the computed lattice spacing. However, when the spacing was computed using only intensities above 20% of the reflection peak, which avoids contributions from 20 and the Z reflection, results were not significantly altered. Unless otherwise stated, ± values are the standard deviation from the mean.

RESULTS

A. Intensity Measurements.

i) **Under Isometric Conditions.** The time course of stiffness changes during the rise and relaxation of tetanic tension has been previously described. Stiffness leads tension during the tetanus rise and lags during relaxation. If equatorial signals (ie: the fall in 10 intensity and the rise in 11 intensity that accompany activation) arise from crossbridge displacement from the myosin filament to the actin filament as a result of attachment, then stiffness and equatorial signals time courses should be similar, assuming that the initial attached state would be detectable by stiffness

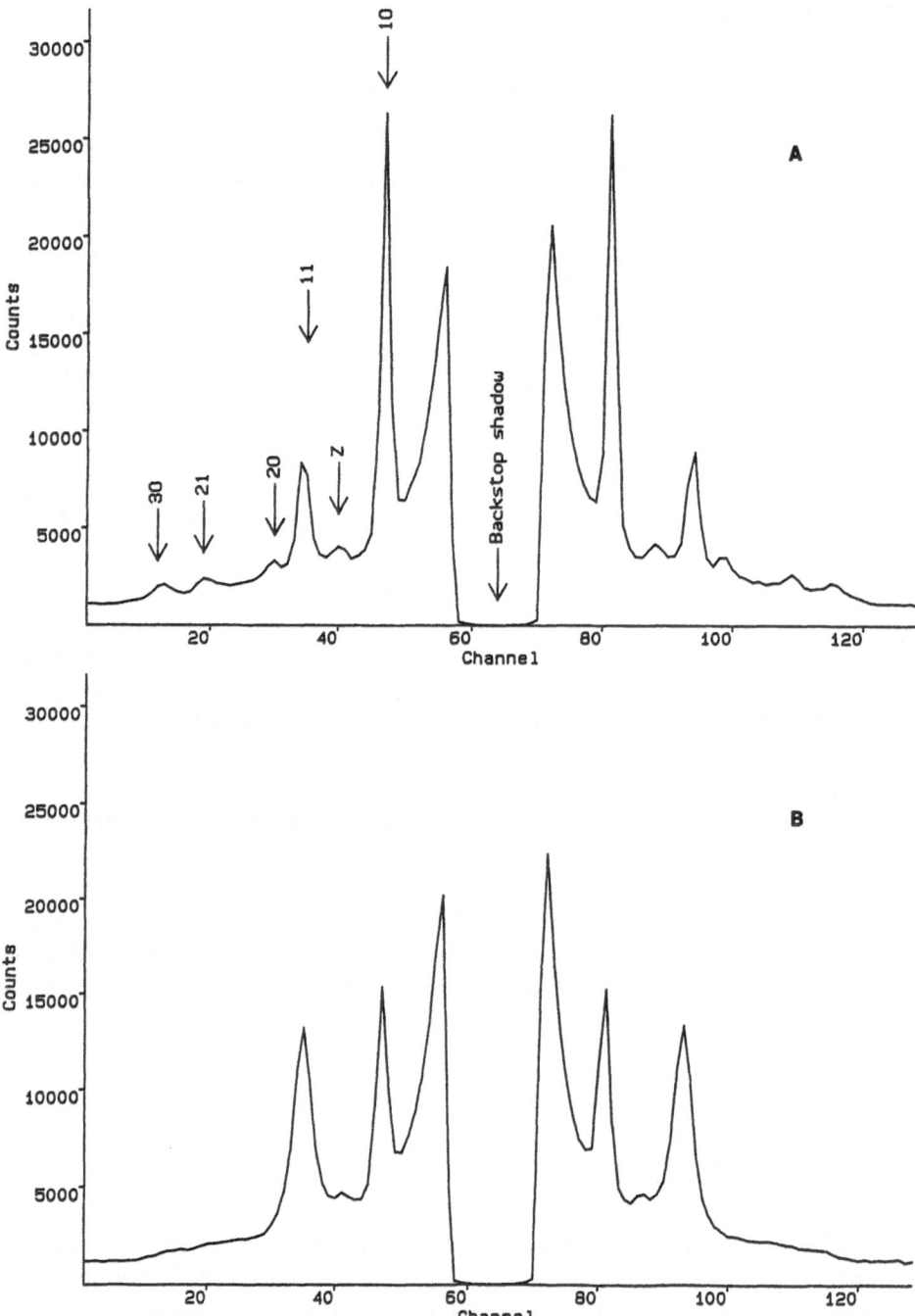

Fig. 1. Equatorial diffraction pattern from a single fibre at rest (A) and during an isometric tetanus (B). Total exposure time 3s for both spectra. Temperature 4°C in all experiments.

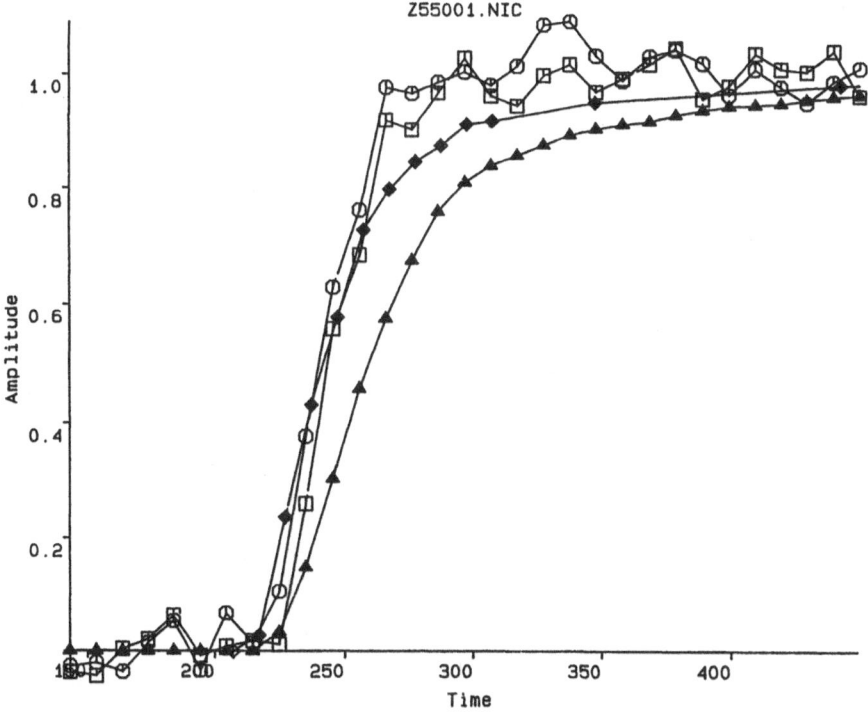

Fig. 2. Time course of changes in force (▲), equatorial signals (10,□;11,○) and stiffness (◆) during the rise of tetanic tension. Sampling rate 10 ms. Time zero indicates the point at which the X-ray shutter was opened. Average of 20 tetani. Stiffness meaasured at 4 kHz. All traces normalised with respect to their resting and activated levels. Sarcomere length 2.15 μm. Length clamped using the sarcomere lentgth signal from the position sensitive photodiode to control motor position.

measurements. Figure 1 shows a typical equatorial pattern from a relaxed and activated intact single fibre. Our early experiments did not involve length clamp control of the sarcomere length, and suggested that there might be a difference between the time course of 10 and 11 signals during the rise of tetanic tension. Subsequent use of length clamping via our laser diffractometer system indicated that the time courses of 10 and 11 signals were very similar to one another, and close to that of stiffness. The half time for the rise of stiffness was 13ms ahead of that for force at 4°C. The half time for the reflections was 20ms ahead of force (figure 2). Relaxation from a tetanus occurs in two phases, the first nearly isometric with a time constant of about 1s, the other much quicker (time constant 0.1s) during which sarcomere length in different regions of the fibre becomes unstable and changes in a complex way. During the isometric phase, stiffness and equatorial signals declined once again with a similar time course, and more slowly than tension.

During activation, the higher order equatorials become weaker due to a 'smearing out' which makes them more difficult to measure accurately. The 11 reflection also undergoes an increase in width as well as an increase in amplitude,

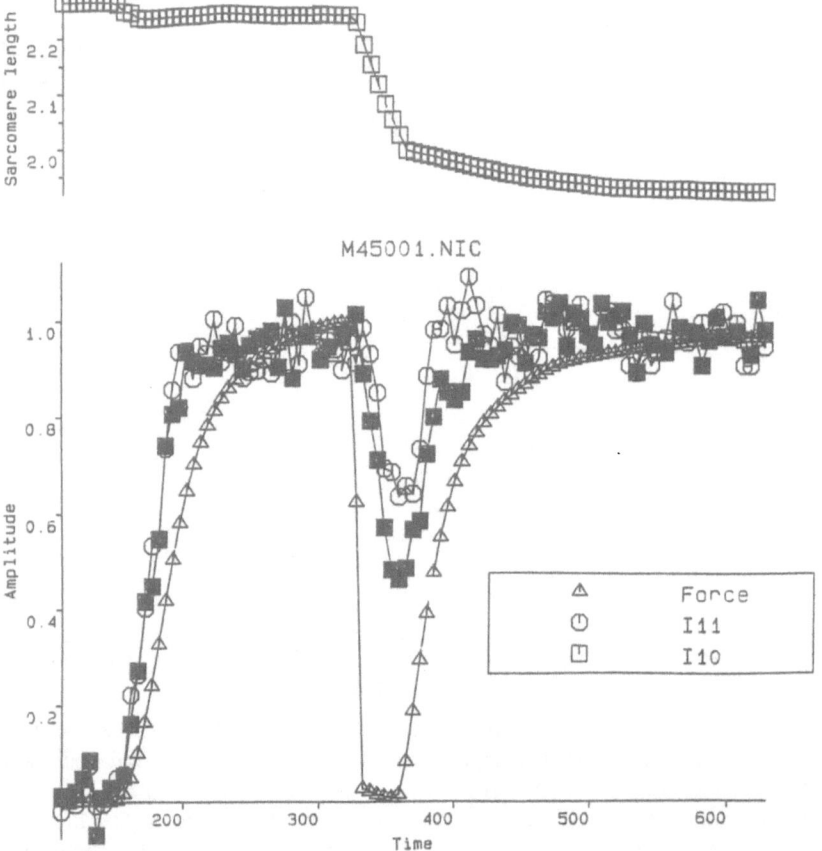

Fig. 3. Upper panel, sarcomere length (μm); lower panel, equatorial signals (10,■;11,○) and force (Δ) during shortening at a velocity close to V_{max} (2.96 nm.(half sarcomere)$^{-1}$.s^{-1}). Time in milliseconds.

which causes it to partially overlap the 20 and the Z reflection which lie on either side of it. The origin of this broadening may be a disordering of the myosin lattice as a result of force development[11].

ii) **Isotonic Conditions.** In contrast to the situation in the isometric case, during isotonic shortening the changes in equatorial signals and stiffness did not agree. Figure 3 shows the equatorial signals during shortening at close to V_{max} (the velocity of unloaded shortening). Stiffness fell gradually as the velocity of shortening increased, while equatorial signals remained at their isometric level until velocities of $0.4V_{max}$ were reached. In the range $0.4V_{max}$ to V_{max} the equatorial signals declined to reach 60% of their isometric value at V_{max}. Stiffness at V_{max} was 30% of its isometric value. Although both stiffness and equatorial intensities are putative indicators of the fraction of attached crossbridges, under isotonic conditions we appear to have a contradication between these two measures.

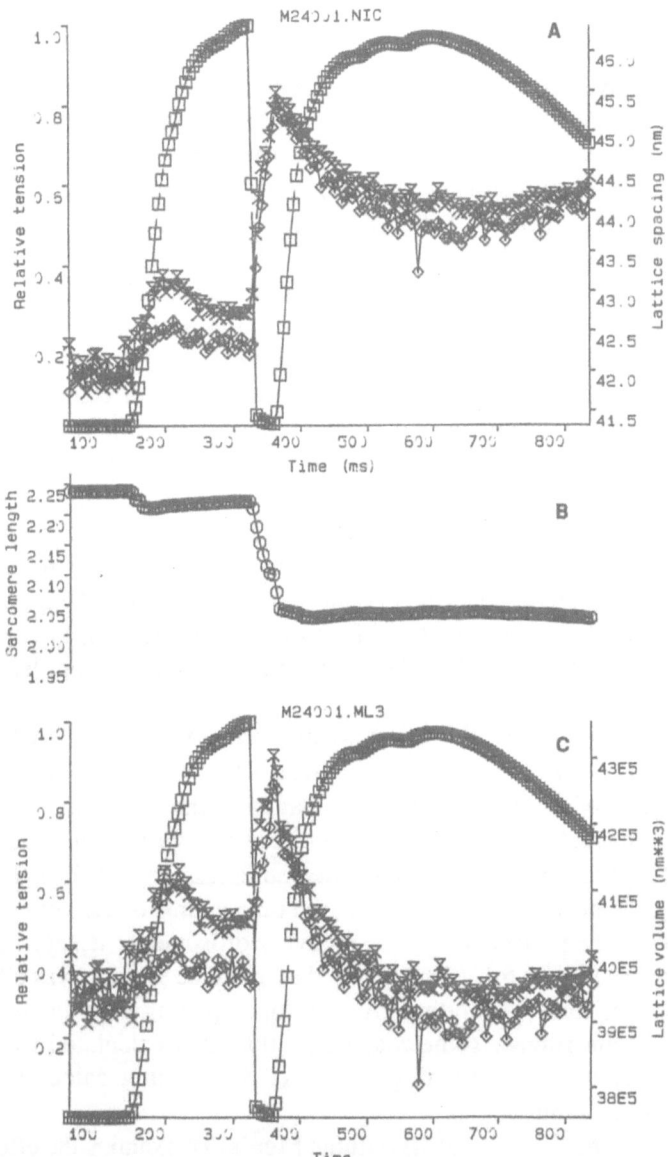

Fig. 4. A) lattice spacing changes during recovery from a period of isotonic shortening in unmodified Ringer's solution. B) simultaneously recorded sarcomere length changes, C) lattice volume calculated from the square of the spacing and the sarcomere length. In this fibre, lattice volume still shows an expansion on activation, but less than predicted from the lattice spacing change. Spacing and volume calculated from 10 (◆) and 11 (✗) reflections. Spacing obtained from least squares fit of points greater than 20% of peak to avoid contributions from minor reflections. 28 tetani averaged.

B. Lattice Spacing

i) **During Fixed End and Isometric Contractions.** The spacing between the origin and the other equatorials provides information about dimensions of the myosin lattice. Because of their much greater intensity, we have only used the 10 and 11 reflections for computation of lattice spacing. We express the lattice spacing as the centre to centre distance between adjacent myosin filaments rather than the spacing between lattice planes. At rest we observe a spacing of 41.9nm at 2.15 μm sarcomere length. Upon activation under fixed end conditions, we find that this spacing increases during the rise of tetanic tension, but before plateau tension is reached, the expansion stops, and a slower compression of the lattice is observed. A shortening of sarcomere length is also detected. Since shortening *per se* may give rise to a lattice expansion, we repeated these experiments under length clamp control, where sarcomere length changes are minimised. We found that the expansion of the lattice was greatly reduced, but the slow compression of the lattice was still evident. As in fixed end conditions, the lattice compression was of slower time course than the rise of tension. A similar compression of the lattice with a reduced expansion phase could be obtained from the fixed end experiments, if the lattice spacing changes were corrected for sarcomere shortening by expressing them as lattice volume. This is good evidence of the effectiveness of our length clamping technique.

ii) **Isotonic Conditions.** When fibres shortened at velocities close to V_{max}, force fell abruptly to below 5% of isometric tension. This was accompanied by a similar abrupt expansion of the lattice. Upon termination of the shortening, force recovered rapidly to its isometric level. During the force recovery, which occurred under isometric conditions, the lattice underwent a compression of 1.02 ± 0.39nm (figure 4). As during the onset of tetanic tension, the time course of the lattice compression was slower than the time course of the recovery of isometric tension. The half time of force recovery was 21.6 ± 2.7ms, while that of lattice spacing was 67.5 ± 14.3ms. In the activated, force-generating fibre, lattice spacing calculated from the 11 reflection was consistently somewhat larger than that calculated from 10. In the relaxed fibre or during shortening at V_{max}, the spacing calculated from both reflections was the same.

iii) **Exposure to Hyper- and Hypotonic Media.** To examine the effects of hyper- and hypotonicity on the lattice spacing changes which accompany activation or isotonic shortening, we prepared Ringer's solutions which were rendered either hyper- (1.4x) or hypo- (0.8x) tonic by addition or omission of NaCl from the basic recipe. Under hypotonic conditions, the myosin lattice increased its spacing to 44nm. During tetani, the lattice spacing was compressed to a greater degree than in unmodified Ringer's (figure 5). Force increased by 5%. When the fibre shortened at V_{max}, the expansion of the lattice was greater. During the isometric recovery from shortening, the recompression of the lattice was also greater (2.70 ± 1.17nm on average). Under hypertonic conditions, lattice compression was virtually eliminated

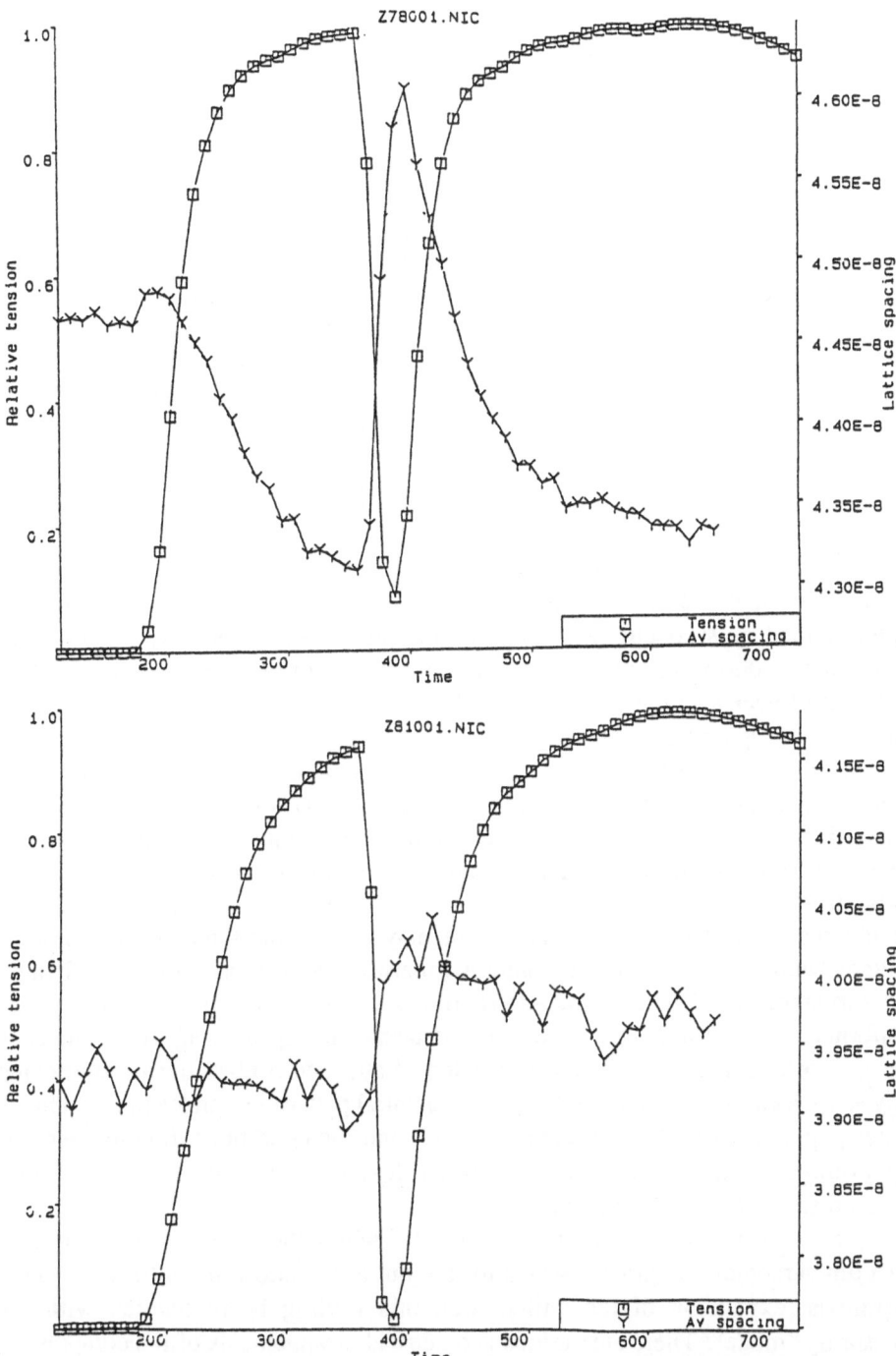

Fig. 5. Lattice spacing changes (Y) and force (□) during recovery from isotonic shortening under hypotonic (upper panel, 0.8x normal tonicity, average of 15 tetani) and hypertonic (lower panel, 1.4x normal tonicity, average of 18 tetani) conditions. Same fibre used in both cases.

$(0.34 \pm 0.24$nm), and the expansion of the lattice during shortening at V_{max} was also reduced (figure 5). Isometric tension was decreased by 25%.

DISCUSSION

Over the last 20 years, refinements in the technique of X-ray diffraction studies on muscle has permitted a reduction of the size of preparation required in order to obtain a strong diffraction pattern, and a marked improvement in time resolution. Since our original report of the successful application of time resolved X-ray diffraction techniques to the study of single intact muscle fibres[12], two other groups have reported success in using this preparation for x-ray diffraction studies[13][14][19]. The viability of this preparation for X-ray diffraction studies using synchrotron radiation is therefore confirmed.

During the rise of tetanic tension, stiffness (4 kHz) and equatorial signals had a similar time course, which suggests that both are detecting the same event, i.e.: crossbridge formation (equatorial signals do have a small lead over stiffness, and reach their peak before maximum stiffness is reached). If so, then tension development must occur at a step in the cycle subsequent to attachment. Furthermore, during relaxation it would appear that once again attached bridges exist with lower average force than at the tetanus plateau, since equatorials and stiffness now lag tension.

During isotonic shortening, the agreement between stiffness and equatorial signals was not observed. Stiffness reports that only 30% of bridges attached in the isometric state are present during shortening, while equatorials report closer to 60%. The 1957 model of A.F. Huxley predicts values closer to the stiffness data, taking the parameters of the model from a best fit to the force velocity curve which we obtained from the fibres we used. This may result from the existence of some bridges whose attachment detachment kinetics are too fast for our stiffness measurements to detect them. Alternatively detached heads may have some conformation that causes them, even when detached, to appear as attached bridges in the equatorial pattern, or a small population of attached bridges during shortening may adopt a conformation such that their equatorial pattern resembles that of a much larger population of attached bridges under isometric conditions. It could even be that average stiffness of the attached bridges is less during shortening. At present we cannot distinguish between these possibilities.

Our observation that axial force is associated with lattice compression (at least at 2.15 μm sarcomere length)[20] seems to rule out any models of contraction which require an expansion of the lattice. Can this finding be reconciled with the crossbridge model? There are various geometrical arrangements of the components of the crossbridge which would give rise to a lattice compression associated with axial force[15], where the radially compressive force would arise as a component of the force vector in the S2 element. Similar compression of the lattice associated with crossbridge formation has also been reported in skinned fibres[16][17], but in skinned

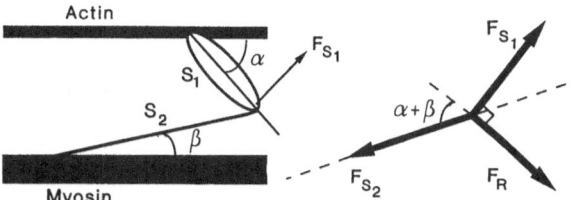

Fig. 6. Schematic representation of the origin of radial force. Using F_R to represent the reaction in the head to forces F_{S1} and F_{S2}, summing force vectors parallel and orthogonal to F_{S2}, at equilibrium we may obtain the relations:

$$F_x = \frac{F_S \cos(\beta)}{\cos(\alpha)\sin(\beta) + \cos(\beta)\sin(\alpha)} \qquad F_r = \frac{F_S \sin(\beta)}{\cos(\alpha)\sin(\beta) + \cos(\beta)\sin(\alpha)}$$

from which a variation in radial force (F_r) would be expected, but also a variation in axial force (F_x).

mouse fibres it would appear that the radial force accompanying the formation of rigor bridges is independent of axial force, and results simply from crossbridge attachment[18]. In our experiments, crossbridge attachment occurred much more quickly than lattice compression during recovery from a period of isotonic shortening or during the tetanus rise. The time course of the compression is more similar to that of axial force development than stiffness, though somewhat slower. Furthermore, the radial force is detectable during recovery from shortening velocities at which equatorial signals do not differ significantly from the isometric case, implying that crossbridge attachment alone is not determining the magnitude of the radial force. This leads us to suppose that the radial component of a force vector directed along S2 could be the origin of the radial force, being greater when lattice spacing is larger, and reduced when spacing is already compressed. But if this is true, it must follow that axial force should also show a dependence on lattice spacing. Representing the system by the very simple model in figure 6, we see that axial force falls with increasing β, except in the case $\alpha = \pi/2$. Providing β is small, then the effect of lattice spacing on axial force will be smaller than on radial. Nevertheless, effects of lattice spacing on axial force of this kind have been reported[9]. Shortening at V_{max} reduces the axial force to zero, and should simultaneously discharge the radial component. However, the time course of the lattice compression during recovery from isotonic shortening is slower than the rise of force. In the simple model of figure 6, radial and axial forces would develop simultaneously, so the slow lattice compression observed on recovery of axial force implies a damping influence on the movement of the filaments. It is possible that the viscosity of the medium acts provides this damping, but this seems unlikely since the lattice expansion which accompanied the initial fall of force is very rapid. It seems more likely that the radial force itself is transmitted through a viscoelastic element.

The existance of a considerable radial force must be accounted for by any molecular model of muscle contraction. Its possible influence on rapid tension transients or on the behaviour of single actin filament motility studies, where the effective lattice spacing is unknown, has yet to be determined. The authors express their gratitude to the MRC, EMBL and CNR for their support of this project.

REFERENCES

1. Huxley, H.E. & Brown, W. *J. Mol. Biol.* **30**, 383-434 (1967).
2. Huxley, A.F. & Simmons, R.M. *Nature* **233**, 533-538 (1971).
3. Cecchi, G., Griffiths, P.J. & Taylor, S.R. in *Contractile Mechanisms in Muscle* (Eds. Pollack, G.H. and Sugi, H.) 641-655 (Plenum Publishing Corporation, 1984).
4. Ford, L.E., Huxley, A.F. & Simmons, R.M. *J. Physiol. (Lond.)* **269**, 441-515 (1977)
5. Haselgrove, J.C. & Huxley, H.E. *J. Mol. Biol.* **92**, 113-143 (1973).
6. Yu, L.C., Hartt, J.E. & Podolsky, R.J. *J. Mol. Biol.* **132**, 53-67 (1979).
7. Cecchi, G. *Arch. Ital. de Biol.* **121**, 215-217 (1983).
8. Cecchi, G., Colomo, F. & Lombardi, V. *Boll. Soc. Ital. Biol. Sper.*, **52**, 733-736 (1976).
9. Bagni, M.A., Cecchi, G. & Colomo, F. *J. Physiol. (Lond.)* **430**, 61-75 (1990).
10. Cecchi, G., Griffiths, P.J., Bagni, M.A., Ashley, C.C. & Maéda, Y. *Biophys. J.* **59**, 1273-1283 (1991).
11. de Graaf, H. *Acta Cryst.* **A45**, 861-870 (1989).
12. Cecchi, G., Griffiths, P.J., Bagni, M.A., Ashley, C.C. & Maéda, Y. *J. Physiol. (Lond.)* **418**, 58P (1989).
13. Konishi, M., Wakabayashi, K., Kurihara, S., Higuchi, H., Onodera, N., Umazume, Y., Tanaka, H., Hamnaka, T. & Amemiya, Y. *Biophys. Chem.* **39**, 287-297 (1991).
14. Irving, M., Mannson, A., Simmons, R.M., Piazessi, G., Lombardi, V., Ferenczi, M.A. & Harries, J. *J. Physiol. (Lond.)* **438**, 147P (1991).
15. Schoenberg, M. *Biophys. J.* **30**, 69-78 (1980).
16. Matsubara, I., Goldman, Y.E. & Simmons, R.M. *J. Mol. Biol.* **173**, 15-33 (1984).
17. Brenner, B. & Yu, L.C. *Biophys. J.* **48**, 829-834 (1985).
18. Matsubara, I., Umazume, Y. & Yagi, N. *J. Physiol. (Lond.)* **360**, 135-148 (1985).
19. Irving, M., Lombardi, V., Piazzesi, G. & Ferenczi, M. *Nature* **357**, 156-158 (1992).
20. Cecchi, G., Bagni, M.A., Griffiths, P.J., Ashley, C.C. & Maéda, Y. *Science* **250**, 1409-1411 (1990).

Discussion

Edman: Have you done these experiments at different sarcomere lengths?

Griffiths: No, we haven't had a chance yet.

Simmons: I think I saw in one of your records that the time course of the change of the equatorial intensities at the onset of shortening is delayed, which is something that Hugh Huxley, Wasi Faruqi, and I observed. I wonder if you had tried to compare them with the time course of stiffness change, and whether they're the same or different.

Griffiths: We haven't tried to compare them with the time course of stiffness changes, but we have tried to compare them with the model solution of Zahalak which gives the transient solution to the Huxley '57 model, allowing you to predict the time course to attain steady state, in proportion of attached cross-bridges, after you start a period of ramp shortening (Zahalak, G. I. *Mathematical Biosciences* 55, 89-114, 1981). The agreement with that sort of modelling is reasonably good. We predict about 15 milliseconds to reach steady state at high shortening velocity and we tend to see our reflections equilibrating within 15 to 20 milliseconds. So the agreement isn't too bad.

Pollack: If I understood your data correctly, early during the rise of tension, you see an increase of interfilament spacing, followed by a decrease. You dismiss the increase by saying you expect it because a little sarcomere shortening under constant volume ought to give you this increase. But there is some force that's responsible for this increase. How do you envision this kind of force that is causing the early increase?

Griffiths: I don't dismiss it. In fact, I don't say that it is entirely accounted for by sarcomere length changes. What I do say is that when we use a length clamp control, where we prevent sarcomere length changes, this phase is much smaller. So some of it is due to shortening, which suggests that there is some sort of constant volume behavior of the fiber. But I don't say that is entirely the reason for it. It's possible there are other things we don't know about yet that could also play a role.

Edman: Would you expect from your findings that there is a decrease in lattice volume if you have a loaded shortening along the descending limb?

Griffiths: You mean if we have decreased the lattice-spacing by stretching the fiber?

Edman: No, I'm referring to the radial force you postulated. During a loaded shortening along the descending limb of the length-tension curve, the radial force produced by the cross-bridges would steadily increase because of greater overlap. This would tend to decrease the lattice-spacing.

Griffiths: You mean increasing the radial force by increasing the amount of overlap? We plan to do some experiments of that kind, looking at variations of overlap on these responses.

Edman: Didn't Elliott, Lowy and Worthington (*J. Mol. Biol.* 6, 295-305, 1963) show a long time ago that there was a constant volume behavior during shortening?

Squire: More or less, yes.

Griffiths: I think if you're looking at the steady-state situation maybe it is, but I think the transient situation need not necessarily be constant volume, so perhaps the volume returns to the same level when the shortening is complete. But during the shortening I think there is a deviation.

Iwamoto: I have a comment about the discrepancy between the stiffness measurement and the equatorial reflection during shortening. I showed evidence that the stiffness of the weakly bound bridges is different from that of the strongly binding bridges, and there will be a lot of weakly binding bridges during shortening. This would explain the discrepancy between the stiffness and the equatorial reflection.

Kawai: You indicated that there is this strongly attached, low force-producing state while tension is developing. What is the status of it? Is it a chemical state different from the strongly attached, force-generating state?

Griffiths: It is a strongly bound state inasmuch as there is a lot of it relative to a weakly bound state, but I can't tell you in terms of a biochemical model what sort of state it is.

Kawai: Could it be the same as a strongly bound tension-generating state, but somewhere in the structure it is slack, so that you don't see tension while you change from no tension to full tension?

Griffiths: That's certainly possible.

Kawai: In other words, the chemical state is the same, but you don't see it?

Griffiths: Yes, the same chemical state—not necessarily a weakly bound state that has become more strongly bound.

Kawai: So, if you run the experiment during steady-state tension, you might not see this state. Is that correct?

Griffiths: I would assume not. I would assume the number of bridges in this state is much smaller anyway as a proportion of the total under steady-state conditions than during the rise of tetanic tension.

Squire: I think it is very dangerous at the moment to try and relate biochemical states and structural states, because there is no reason why a head, for example, in one structural state shouldn't change its biochemical state and stay there. It may want to do that, but it may not be able to do it. So, when we see a particular structural state, it could be a mixed biochemical state.

Kawai: Well, you need evidence to assert such things. If there is a state change, you need to have some evidence for it.

Squire: I don't follow that. What X-ray diffraction can tell you is whether mass is shifting from one position to another. It doesn't tell you anything about the biochemical state or the kinetics. One has to be a bit careful.

CURRENT X-RAY DIFFRACTION EXPERIMENTS USING A SYNCHROTRON RADIATION SOURCE

N. Yagi, S. Takemori* and M. Watanabe*

Department of Pharmacology
Tohoku University School of Medicine
*Department of Physiology
The Jikei University School of Medicine

ABSTRACT

A Fuji imaging plate and synchrotron radiation are the most distinct innovations of the last twenty years in the X-ray diffraction experiments on biological materials. Here we present results of recent experiments on skeletal muscles made at Photon Factory, Tsukuba.

It is now possible to record a two-dimensional X-ray diffraction pattern from a rabbit or frog single skinned fiber with a 30-sec exposure. Although weaker compared with those from whole muscles, it shows layer-lines up to 5.1 nm. When the fiber is activated by Ca^{2+}, the pattern changes in a way similar to that observed when a live muscle is electrically stimulated.

Use of single fibers makes various types of structural experiments much easier than using whole muscles or fiber bundles. Not only suitable for physiological experiments, better diffusion makes it also suitable for biochemical experiments using various kinds of labels.

Time-resolved experiments with imaging plates are possible by using an imaging-plate exchanger devised by Dr. Y. Amemiya. By combining this and a fast-acting mechanical shutter, it is possible to record a two-dimensional diffraction pattern from a frog whole muscle shortening at the maximum speed.

The pattern thus obtained shows weakening of the 5.1 and 5.9-nm actin layer-lines and the third (14.3 nm) and the sixth (7.2 nm) myosin meridional reflections, compared with the pattern from isometrically contracting muscles. On the other hand, the second meridional reflection from the thick filament is intensified. These results suggest very different arrangement of myosin heads during active shortening from that during isometric contraction.

Synchrotron radiation and a Fuji imaging plate are most important technologies introduced to the field of X-ray diffraction in the last twenty years. By combining these two innovations, important results have been reported in structural and physiological studies on muscle contraction[1)2)].

Mechanism of Myofilament Sliding in Muscle Contraction, Edited by
H. Sugi and G.H Pollack, Plenum Press, New York, 1993

We have extended the use of these innovations in X-ray diffraction experiments on single skinned muscle fibers and shortening muscles. The following experiments were made at the beam line 15A in Photon Factory, Tsukuba using a small-angle camera described[3]. Imaging plates were read by a scanner[4]. The images were analyzed on an engineering workstation (NEWS-3460, Sony, Tokyo) using an interactive graphic program written by N.Yagi.

1. X-ray Diffraction Studies on Single Skinned Muscle Fibers

Single skinned fibers[5] and bundles made of a few skinned fibers have been used in many muscle experiments both with physiological and biochemical purposes. The advantage in the use of skinned fibers over intact muscles is that it is possible to change the cytoplasmic environment. For this purpose, it is desirable to use a single fiber rather than whole muscle because the diffusion of ions or molecules may not be

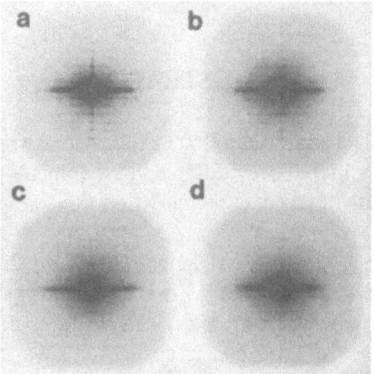

Fig. 1. X-ray diffraction patterns recorded from single skinned fibers of frog sartorius muscle. (a) In relaxing solution which contains 26.1 mM K-methanesulphonate, 5.66 mM Mg(methanesulphonate)$_2$, 4.36 mM Na$_2$ATP, 10 mM EGTA, 10 mM Na$_2$phosphocreatine, 24 units/ml creatine phosphokinase, 20 mM PIPES (pH 7.0). (b) In activating solution with a pCa of 5.5. (c) In activating solution with a pCa of 4.4. (d) In relaxing solution after contraction at pCa 4.4. Sarcomere length of the muscle fiber was 2.4 μm. The temperature was 4°C. Fibers were treated by 0.5 % Triton X-100 in the relaxing solution for 30 min to make the sarcolemma permeable. The patterns were recorded on Fuji imaging plates at the beam line 15A in Photon Factory, Tsukuba. The ring current was 250 - 350 mA. Exposure on each fiber was 30 sec. For each state, patterns from several fibers were summed before intensity measurements. The number of patterns summed for (a) was 4, 3 for (b), 5 for (c) and 4 for (d). For the analysis, each pattern was displayed on a computer screen. The 14.3-nm meridional reflection from the thick filament was located, and the pattern was rotated so that the position of the direct beam became the center of the pattern and the meridian parallel to the vertical axis. Then the four quadrants were averaged. The specimen-to-plate distance was calibrated by the position of the 14.3-nm reflection in the resting state (1/14.34 nm^{-1}, Haselgrove, 1975). The area of integration was 0 - 0.010 nm^{-1} in the meridional direction for the equatorial reflections, 0 - 0.018 nm^{-1} in the equatorial direction for the meridional reflections, 0.042 - 0.090 nm^{-1} in the equatorial direction for the 5.1- and 5.9-nm layer-lines, 0.17 - 0.23 nm^{-1} in the equatorial direction for the second actin layer-line.

complete in thick specimens. Especially when the molecule is consumed in the muscle (for example, ATP), it is difficult to keep its concentration at a fixed value throughout the specimen. Thus single skinned fibers have many important applications.

To investigate the structural changes that accompanies physiological and biochemical experiments, an X-ray diffraction study on the same sample is necessary. However, the size of a single muscle fiber, especially its diameter (40 - 200 μm) makes X-ray studies difficult. On the small angle camera at BL15A of Photon Factory[3], the size of the X-ray beam at the sample position is about 8 mm in width and 1 mm in height with a camera length of 2400 mm. This implies a single muscle fiber occupies less than 10 % of the cross section of the beam. Nevertheless, using imaging plates, it is possible to record two-dimensional diffraction patterns from single muscle fibers in various conditions.

Figure 1 shows X-ray diffraction patterns from single fibers from frog sartorius muscle in various states. These were obtained by summing 3 to 6 patterns from different fibers. An exposure on each fiber in one state was 30 sec.

Fig. 1a is a pattern from fibers in a relaxing solution. The myosin layer-lines, meridional reflections and the 5.9-nm actin layer-line are visible. Fig. 1b is a pattern from fibers activated at a pCa of 5.5 ; the fiber developed about 35 % of its maximum tension. The myosin layer-lines are markedly weakened. Fig. 1c is from fibers developing maximum tension at a pCa of 4.4. The myosin layer-lines almost disappeared but the third (at $1/14.5$ nm^{-1}) and sixth (at $1/73$ nm^{-1}) are strong and broader in the lateral direction. The intensity changes seen in these diffraction patterns from activated skinned muscle fibers are similar to those observed in intact whole muscles which are electrically stimulated. Indications of rigor patterns, namely enhancement of the first actin layer-line at $1/36$ nm^{-1} and the shift of intensity distribution along the 5.9-nm actin layer-line, were not evident when the ATP-regenerating system was included in the activating solution. Without the regenerating system, a layer-line was observed at $1/36$ nm^{-1}.

One disappointing finding in X-ray experiments using single skinned fibers was that the diffraction pattern never recovers to the original relaxed image after activation. Fig. 1d is a pattern in relaxing solution after activation at pCa 4.4. The myosin layer-lines and meridional reflections are weaker. This is probably the result of mechanical damage due to the contraction : skinned muscle fibers are more fragile than intact fibers because of the loss of sarcolemma. Also, the activation may not be uniform throughout the fiber at the early stage of activation, although the fiber was incubated in a relaxing solution containing low concentration of calcium buffer before activation.

Considering the contribution of skinned single muscle fibers in understanding the regulation and mechanism of muscle contraction, the necessity for X-ray diffraction experiments using single fibers is obvious. It is now possible to make X-ray diffraction experiments which have been previously done on skinned whole muscles or bundles of skinned fibers using single muscle fibers. Examples of such experiments are : (1) antibody labeling[6][7] (2) changing fiber volume using PVP or

Dextran[8)9)] (3) changing ionic strength[10)11)] (4) ATP-analogues[12-15)] (5) chemical modifications using sulfhydryl-reagents (Yagi, unpublished; K.Wakabayashi, unpublished) (6) labeling of actin with myosin subfragment-1[16)]. Since most of these experiments does not cause the muscle fiber to contract, the difficulties accompanied with the activation experiment will not be serious.

2. X-ray Diffraction from Shortening Muscles

Structural studies on muscles which are actively shortening at high velocity have been of increasing importance because of the current interest in the sliding mechanism[17-19)]. However, most of the X-ray diffraction experiments on muscles have been made on three states : resting, contracting and rigor. These states are relatively stable and last long enough to record X-ray diffraction patterns of high quality. Shortening muscles are more difficult for X-ray diffraction experiments because shortening with low load lasts only less than 100 msec. Most of the X-ray diffraction experiments on shortening muscles to date have studied only equatorial reflections and the results showed no change with high load[20)] and a small intensity change towards the relaxed state with low load[21)].

We have been trying to record a two-dimensional X-ray diffraction pattern from frog skeletal muscle during shortening with small load. When a muscle is allowed to shorten with a fixed load, the shortening velocity becomes constant after a short transient phase[22)]. Thus a muscle shortening at a constant velocity may be regarded as in a steady state. Supporting this assumption, studies on X-ray equatorial intensities have showed that, after a short transient phase, the equatorial intensity ratio stays constant during the rest of shortening[23)].

2.1. Experimental Techniques

In order to record diffraction patterns during resting, isometrically contracting, and shortening states from the same muscle, an imaging plate exchanger[1)] was used. This mechanism can replace an imaging plate at the position for X-ray exposure within a time of 200 msec.

The specimen was frog sartorius muscle which was stimulated electrically for 1 sec (20 Hz stimulation) at 4°C. The sarcomere length was adjusted to 2.6 μm in a resting state. The tendon end of the specimen was connected to a solenoid with a stainless-steel thread on which a small LED was fixed. The LED was used to monitor length of the muscle. The solenoid was activated to make the thread slack at 280 msec after the first stimulus. Then the muscle shortened about 16 % of its length until the thread became taut again. The load, which included weight of the thread and LED, was small (about 1 gram) compared with the tension the specimen developed (300 - 500 grams). Thus the specimen shortened with virtually no load, with a velocity of about 6 μm per sarcomere per sec. The shortening lasted for about 70 msec.

We made exposures at 5 different phases with 300 msec separations : (1) in a resting state before contraction, (2) during isometric contraction before shortening,

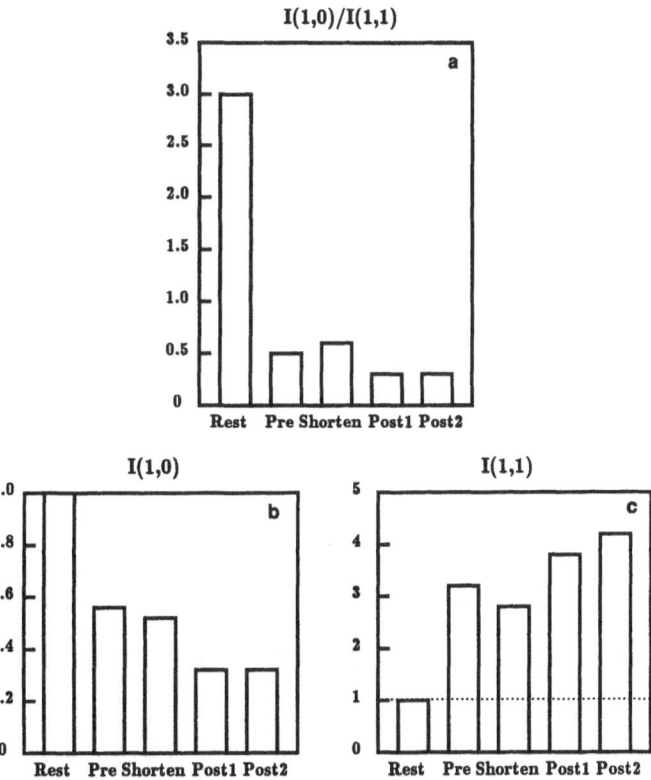

Fig. 2. Intensities and intensity ratio of equatorial reflections (a) Intensity ratio of the (1,0) and (1,1) equatorial reflections. "Rest" is the value in a resting state, "Pre" is that during isometric contraction before the muscle was allowed to shorten, "Shorten" is that during shortening with virtually no load, "Post1" is that during isometric contraction 330 msec after the beginning of shortening, "Post2" is that during isometric contraction 630 msec after the beginning of shortening. (b) Intensity of the (1,0) reflection. The values are normalized by that in the resting state. (c) Intensity of the (1,1) reflection. The values are normalized by that in the resting state.

(3) during shortening (4) during isometric contraction 330 msec after shortening, (5) during isometric contraction 630 msec after shortening. For each exposure, a fast-acting shutter was opened for 30 msec. In (3), to avoid the transient phase immediately after the release, the shutter was opened at 30 msec after the muscle was released.

Each specimen was subjected to 20 contractions and the diffraction patterns were accumulated on the plates. Thus the total exposure on each imaging plate was 600 msec. Further, patterns from four muscles were summed before intensity of each reflection was measured.

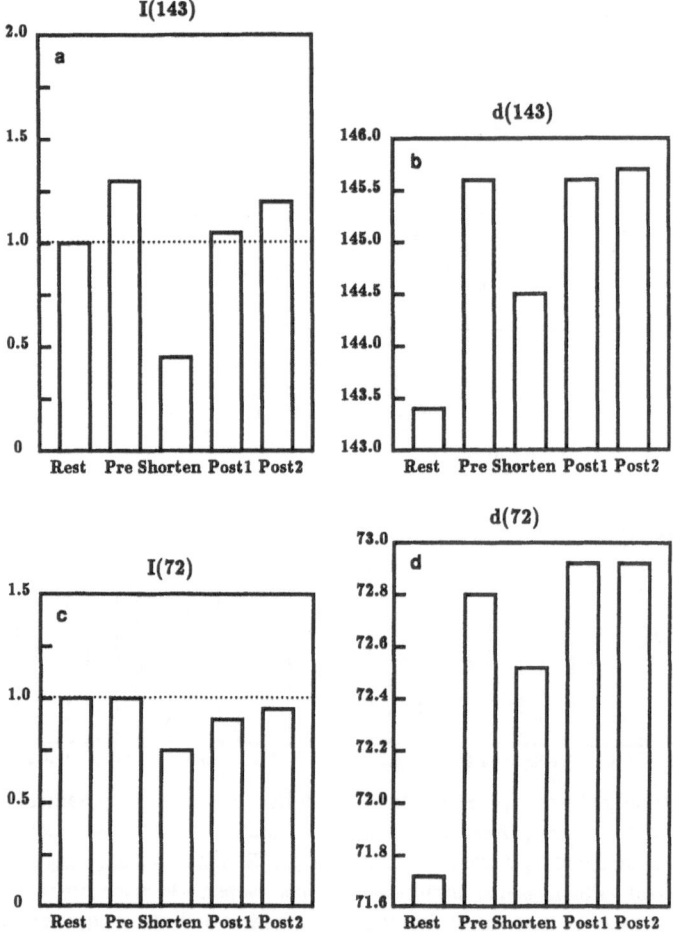

Fig. 3. Intensities and spacing of meridional reflections from the thick filament (a) Intensity of the third meridional reflection at $1/14.3$ nm^{-1}. The values are normalized by that in the resting state. (b) Bragg spacing of the third meridional reflection (in A). (c) Intensity of the sixth meridional reflection. The values are normalized by that in the resting state. (d) Bragg spacing of the sixth meridional reflection (in A).

3. Isometric Contraction

The intensity changes found in the diffraction pattern when the muscle contracted isometrically were similar to those reported previously[2,4]. Namely, the intensities of the myosin layer-lines and meridional reflections of the 42.9-nm repeat decrease[24] (Fig. 4c) except the meridional reflections at the orders of 14.3 nm (Fig. 3a), which were intensified[25] and shifted towards the origin by about 1.5 %[26] (Fig. 3b). The intensities of the actin layer-lines at $1/5.9$ and $1/5.1$ nm^{-1} increased by approximately 30 and 120 %[27] (Fig. 5a, 5b). The second actin layer-line at $1/18$ nm^{-1}, which is too weak to observe in a relaxed state, was also intensified[28] (Fig.

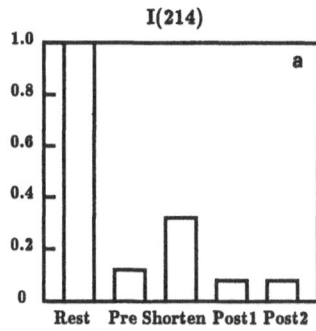 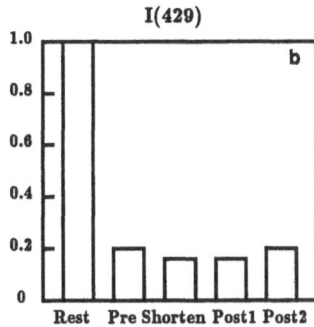

Fig. 4. Intensities of reflections from the thick filament (a) Intensity of the second meridional reflection at 1/21.5 nm^{-1}. The values are normalized by that in the resting state. (b) Intensity of the off-meridional part of the first layer-line. The values are normalized by that in the resting state.

5c). The intensities of the equatorial (1,0) and (1,1) reflections were reversed (Fig. 2b, 2c), changing the intensity ratio (I(1,0)/I(1,1)) from 3.0 in the resting state to 0.5 during isometric contraction[29] (Fig. 2a).

3.1. Shortening

Equatorial Reflections: The intensity of the (1,1) equatorial reflection decreased during shortening (Fig. 2c), and that of the (1,0) reflection also showed slight decrease (Fig. 2b) making the intensity ratio slightly increase to 0.6 (Fig. 2c). Since the intensities of the equatorial reflections depend on the sarcomere length, these values need correction. After the shortening is completed, the intensity of the (1,0) reflection is lower than that before shortening and that of the (1,1) reflection is higher. By interpolating these two values, it is found that the (1,0) intensity is actually larger than that expected for the sarcomere length during shortening, and the intensity of the (1,1) reflection and the intensity ratio of the (1,0) and (1,1) reflections are underestimated.

Layer-Lines from the Thick Filament: During shortening, the 14.3-nm and 7.2-nm meridional reflections decreased in intensity; the intensity of the 14.3-nm reflection was less than half of that during isometric contraction (Fig. 3a). A similar observation has been reported previously after a quick release of muscles[30] and during shortening and lengthening at lower velocities[31]. These reflections moved away from the origin by about 0.5 % during shortening (Fig. 3b). The off-meridional part of the myosin layer-lines did not change in intensity (Fig. 4b), while the meridional reflection at 1/21.4 nm^{-1}, which is the meridional part of the second order myosin layer-line, increased in intensity during shortening (Fig. 4a). Increase in the intensity of this reflection has been reported to occur after a quick release[30].

Layer-Lines from the Thin Filament: The intensities of the actin layer-lines at 1/5.9 and 1/5.1 nm^{-1} decreased during shortening (Fig. 5a, 5b). Especially, the 5.1-nm layer-line, whose intensity was 2.2 times of the resting value during isometric

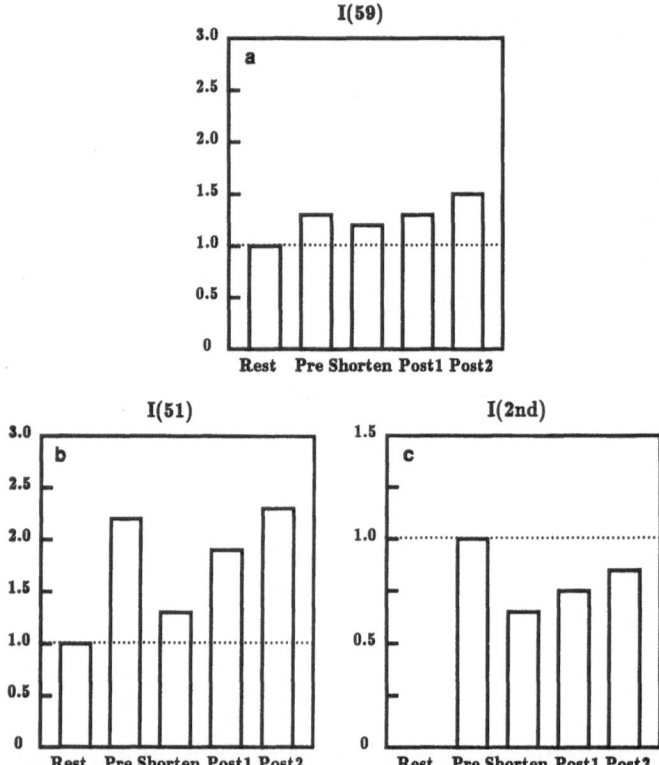

Fig. 5. Intensities of reflections from the thin filament (a) Intensity of the 5.9-nm layer-line. The values are normalized by that in the resting state. (b) Intensity of the 5.1-nm layer-line. The values are normalized by that in the resting state. (c) Intensity of the second layer-line. The values are normalized by that during isometric contraction before shortening.

contraction, decreased to 1.2 time of the resting value. Intensity decrease of the 5.9-nm layer-line during shortening has been reported[32]. The intensity of the second actin layer-line was, compared with that during isometric contraction, found to decrease by about 30 % during shortening (Fig. 5c). However, no intensity change of this layer-line was found using a one-dimentional detector[32]. Since this layer-line is broader during shortening, its intensity is difficult to measure with accuracy and the discrepancy may result from such a technical difficulty.

4. Structural Changes during Shortening

The intensity changes observed during shortening indicate extensive structural changes of the myofilaments. Concerning the binding of myosin heads to actin, two groups of reflections are most interesting : the equatorial reflections and the 5.9 and 5.1-nm actin layer-lines.

The intensity changes of the equatorial reflections are small considering that the tension is almost zero during shortening. This observation may be interpreted to indicate most of the myosin heads are bound to the thin filament even during shortening near the maximum velocity.

On the other hand, intensities of the 5.1- and 5.9-nm layer-lines showed a large drop during shortening. Enhancement of these layer-lines during isometric contraction, especially that of the 5.1-nm layer-line, has been regarded as indication of binding of myosin heads to actin[27)28)33)]. Thus the large drop suggests that the number of myosin heads bound to the thin filament during shortening is small. If it is assumed that the enhancement of the 5.1-nm layer-line is proportional to the number of myosin heads bound to the thin filament, the number of attached heads during shortening is as small as 20 % of that during isometric contraction.

These apparently conflicting results may be explained if it is assumed that the myosin heads not bound to the thin filament are still in the vicinity of the thin filament. The intensities of the actin layer-lines are influenced by the myosin heads bound at the same moment. If many heads are undergoing a rapid attach-detach cycle and spending longer time in a detached state, the enhancement of the actin layer-line is small. If this cycle is rapid enough, myosin heads may stay close to the thin filament even in a detached state. For instance, if a myosin head can be bound to one actin molecule over an axial distance of 10 nm, at the shortening velocity used in this study (3 μm per half sarcomere per sec), it takes 3 msec to pass this distance. If 20 % of the myosin heads are bound at a same moment, the heads spend only 12 msec in a detached state. This may be too short for them to return to the vicinity of the thick filament. Moreover, if the heads bind to actin in a transient manner as suggested by A.F. Huxley[34)], chance for the heads to return to the thick filament would be even smaller.

It is interesting to note that in the cross-bridge model of A.F. Huxley[35)], the number of cross-bridges decreases as the shortening velocity increases and at the maximum velocity, it is about 25 % that during isometric contraction. The present results are consistent with this prediction.

An alternative interpretation of the present results is to presume that, in a shortening muscle, most of the myosin heads are attached to the thin filament in a conformation that produces no tension and contributes little to the intensity of the actin layer-lines. This interpretation is interesting in the view of current hypothesis of the sliding mechanism[18)] in which myosin heads can change actin molecules before one ATP hydrolysis cycle is completed.

The axial repeat of the thick filament (14.34 nm in a relaxed state[36)]) increases about 1.5 % during isometric contraction and this increase is halved during shortening. This axial repeat is determined by the axial stagger between myosin molecules in the shaft of the thick filament. The axial stagger may increase if the a part of the light-meromyosin portion of the thick filament is stretched by the tension. The decrease in the axial repeat during shortening may be taken as evidence for such a mechanical extension of a myosin molecule. However, the mechanical extension does not account for the whole of the increase in the axial repeat since it does not change in proportion to the tension.

The intensity of the second layer-line from the thin filament is influenced by the regulation of the thin filament by calcium[28)33)37-39]. It has been suggested that calcium is released from troponin when the number of cross-bridges is decreased by the release of muscle[40]. The decrease in the intensity of the second layer-line may support this. However, it is necessary to measure the intensity of this layer-line with higher accuracy to establish it.

ACKNOWLEDGEMENTS

These studies were supported by a Grant-in-Aid for Scientific Research on Priority Areas of "Molecular Mechanism of Biological Sliding Movements" from the Ministry of Education, Science and Culture of Japan. The X-ray experiments were performed with the approval of the Photon Factory Advisory Committee (proposals 89-059, 91-064). We thank Dr. Y. Amemiya for help in the use of the X-ray camera and imaging plates.

REFERENCES

1. Amemiya, Y. & Wakabayashi, K. *Adv. Biophys.* **27**, 115-128 (1991).
2. Wakabayashi, K., Tanaka, H., Saito, H., Moriwaki, N., Ueno, Y., & Amemiya, Y. *Adv. Biophys.* **27**, 3-13 (1991).
3. Amemiya, Y., Wakabayashi, K., Hamanaka, T., Wakabayashi, T., Matsushita, T. & Hashizume, H. *Nucl. Instr. Meth.* **208**, 471-477 (1983).
4. Amemiya, Y., Wakabayashi, K., Tanaka, H., Ueno, Y. & Miyahara, J. *Science* **237**, 164-168 (1987).
5. Natori, R. *Jikeikai Med. J.* **33 (suppl. 1)**, 1-74 (1986).
6. Rome, E., Offer, G. & Pepe, F.A. *Nature New Biol.* **244**, 152-154 (1973).
7. Rome, E., Hirabayashi, H. & Perry, S.V. *Nature New Biol.* **244**, 154-155 (1973).
8. Magid, A. & Reedy, M.K. *Biophys. J.* **30**, 27-40 (1980).
9. Maeda, Y. *Nature* **277**, 670-672 (1979).
10. Matsuda, T. & Podolsky, R.J. *Proc. Natl. Acad. Sci. U.S.A.* **81**, 2364-2368 (1984).
11. Xu, S., Kress, M., Huxley, H.E. *J. Muscle Res. Cell Motility* **8**, 39-54 (1987).
12. Lymn R.W. & Huxley, H.E. *Cold Spring Harbor Symp. Quant. Biol.* **37**, 449-453 (1973).
13. Rodger, C.D. & Tregear, R.T. *J. Mol. Biol.* **86**, 495-497 (1974).
14. Goody, R.S., Holmes, K.C., Mannherz, H.G., Barrington-Leigh, J. & Rosenbaum, G. *Biophys. J.* **15**, 687-705 (1975).
15. Padron, R. & Huxley, H.E. *J. Muscle Res. Cell Motility* **5**, 613-655 (1984).
16. Goody, R.S., Reedy, M.C., Hofmann, W., Holmes, K.C. & Reedy, M.K. *Biophys. J.* **47**, 151-169 (1985).
17. Yanagida, T., Nakase, M., Nishiyama, K. & Oosawa, F. *Nature* **307**, 58-60 (1984).
18. Harada, Y., Sakurada, K., Aoki, T., Thomas, D.D. & Yanagida, T. *J. Mol. Biol.* **216**, 49-68 (1990).
19. Ueda, Q.P., Warrick, H.M., Kron, S.J. & Spudich, J.A. *Nature* **352**, 307-311 (1991).

20. Podolsky, R.J., St.Onge, R., Yu, L. & Lymn, R.W. *Proc. natl. Acad. Sci. U.S.A.* **73**, 813-817 (1976).
21. Huxley, H.E. in *Cross-bridge Mechanism in Muscle Contraction* (eds. Sugi, H. & Pollack, G.H.) 391-405 (University of Tokyo Press, Tokyo, 1979).
22. Podolsky, R.J. *Nature* **188**, 666-668 (1960).
23. Yagi, N. *Photon Factory Activity Report* **8**, 302 (1990).
24. Huxley, H.E., Faruqi, A.R., Kress, M., Bordas, J. & Koch, M.H.J. *J. Mol. Biol.* **158**, 637-684 (1982).
25. Yagi, N., O'Brien, E.J. & Matsubara, I. *Biophys. J.* **33**, 121-138 (1981).
26. Huxley, H.E. & Brown, W. *J. Mol. Biol.* **30**, 383-434 (1967).
27. Matsubara, I., Yagi, N., Miura, H., Ozeki, M. & Izumi, T. *Nature* **312**, 471-47 (1984).
28. Kress, M., Huxley, H.E., Faruqi, A.R. & Hendrix, J. *J. Mol. Biol.* **188**, 325-342 (1986).
29. Matsubara, I., Yagi, N. & Hashizume, H. *Nature* **255**, 728-729 (1975).
30. Huxley, H.E., Simmons, R.M., Faruqi, A.R., Kress, M., Bordas, J. & Koch, M.H.J. *J. Mol. Biol.* **169**, 469-506 (1983).
31. Matsubara, I. & Yagi, N. *J. Physiol. (Lond.)* **360**, 135-148 (1985).
32. Huxley, H.E., Kress, M., Faruqi, A.F. & Simmons, R.M. in *Molecular Mechanism of Muscle Contraction* (eds. Sugi, H. & Pollack, G.H.) 347-352 (Plenum Press, New York and London, 1988).
33. Parry, D.A.D. & Squire, J.M. *J. Mol. Biol.* **75**, 33-55 (1973).
34. Huxley, A.F. *Proc. R. Soc. Lond. B* **183**, 83-86 (1973).
35. Huxley, A.F. *Prog. Biophys. Biophys. Chem.* **7**, 255-318 (1957).
36. Haselgrove, J.C. *J. Mol. Biol.* **92**, 113-143 (1975).
37. Haselgrove, J.C. *Cold Spring Harbor Symp. Quant. Biol.* **37**, 341-352 (1973).
38. Huxley, H.E. *Cold Spring Harbor Symp. Quant. Biol.* **37**, 361-376 (1973).
39. Yagi, N. & Matsubara, I. *J. Mol. Biol.* **208**, 359-363 (1989).
40. Gordon, A.M. & Ridgeway, E.B. *J. Gen. Physiol.* **90**, 321-340 (1987).

Discussion

Maéda: Just one technical question. When you compare the intensity during shortening and during isometric contraction with the resting state, how do you normalize?

Yagi: What we are currently doing is to normalize intensity by the intensity of the whole diffraction pattern.

Maéda: Assuming that the intensity over the entire area of the pattern is proportional to the mass in the beam?

Yagi: Yes. That gives very consistent results, so far.

TIME-RESOLVED STUDIES OF CROSSBRIDGE MOVEMENT: WHY USE X-RAYS? WHY USE FISH MUSCLE?

John Squire and Jeff Harford

Biophysics Section
Blackett Laboratory
Imperial College
London SW7 2BZ, UK

ABSTRACT

The advantages of using time-resolved X-ray diffraction as a means of probing myosin cross-bridge behaviour in active muscle are outlined, together with the reasons that bony fish muscle has advantages in such studies. We show that the observed X-ray diffraction patterns from fish muscle can be analysed in a way that is rigorous enough to allow reliable information about crossbridge activity to be defined. Among the advantages of this muscle are that diffraction patterns from resting, active and rigor muscles are all well-sampled at least out to the 30 row-line, that the resting myosin layer-line pattern can be 'solved' crystallographically to define the starting position of the crossbridges in resting muscle, and that the equatorial intensity distribution, which in all patterns from vertebrate skeletal muscles comprises overlapping peaks from the A-band and the Z-band, can be analysed sufficiently rigorously to allow separation of the the two patterns, both of which change when the muscle is active. Finally, we present results both on a new set of myosin-based layer-lines in patterns from active muscle (consistent with the presence of low-force bridges as also indicated by the time-courses of the intensity changes on the equator and the changing mass distribution in the A-band unit cell) and also on changes of the actin-based layer-lines (consistent with stereospecific labelling of the actin filaments by force-producing crossbridges). Our results to date, which demonstrate the enormous power of time-resolved X-ray diffraction studies, strongly support the swinging of myosin heads on actin as part of the contractile cycle.

INTRODUCTION

Since the seminal papers by A.F. Huxley[1] and H.E. Huxley[2] on the crossbridge theory, most studies of the mechanism of force production in muscle have been

Mechanism of Myofilament Sliding in Muscle Contraction, Edited by
H. Sugi and G.H Pollack, Plenum Press, New York, 1993

based on the postulate that, once bound to actin, myosin crossbridges change their interaction with actin so as to generate axial sliding, or a tendency to slide, of actin and myosin filaments. However, despite an enormous battery of experimental tests of this basic idea about force production, current views about what crossbridges actually do are still rather vague. In this paper we assess what we believe is needed in structural terms to define the crossbridge mechanism at low resolution (say 50Å), and we discuss methods of achieving what is needed. In particular, we discuss the unique advantages of time-resolved X-ray diffraction studies of muscle and why we prefer to use the muscles of bony fish in such studies. The results described here are very much part of "work in progress", but they show the direction in which we are going, and the end goal of our studies.

STRUCTURAL STUDIES OF CROSSBRIDGE STATES

As far as structural studies of muscle are concerned, it is useful to distinguish studies of static or equilibrium states and studies of transitions between dynamic states, because techniques appropriate for studying one are not necessarily ideal for studying the other. The main techniques providing information about the arrangements of myosin heads under different conditions are electron microscopy, probe studies, birefringence measurements, X-ray diffraction and neutron scattering[3-6].

Electron Microscopy

Classical electron microscopy using plastic embedded tissue has provided and is providing a great deal of very useful information about the organisation of protein density in different muscles. However this classical technique is highly invasive, and requires rather harsh treatment (i.e. fixation, dehydration, embedding, sectioning and staining) of the tissue before it can be viewed. Even when viewed, the sections suffer a very great deal of radiation damage because electron beam interactions with matter are considerable. The technique, clearly is also restricted to studying "static" states and, even then, preserving the "*in vivo*" structure by chemical fixation is by no means 100% successful.

The use in recent times of frozen hydrated material viewed directly in the electron microscope alleviates some of the difficulties with plastic embedded material. Freeze-fixation of defined static or transient states by rapid freezing is possible[7][8] and the harsh chemical processing mentioned above is not required. Furthermore radiation damage of such preparations held at, for example, liquid N_2 temperatures is much reduced compared with room temperature. Another advantage of such frozen preparations is that they seem to be much better at preserving the inherent order in the tissue; we have found with a test specimen (Z-crystals or nemaline rods in nemaline myopathy) that about twice as good resolution is obtained with negatively stained cryosections rather than with conventional plastic sections[7)9)]. Even though, in this case, the cryosections were not still frozen when being viewed, the freeze-fixation process itself seems to be much better than chemical fixation.

A kind of compromise preparation is that of freeze-fixation followed by freeze substitution and plastic embedding. This has great advantages over chemical fixation[8)], but the problem remains that very high resolution (e.g. better than, say, 20Å) is difficult to achieve with plastic sections.

In summary, good progress is being made with e.m. techniques and the final images can be interpreted directly (unlike X-ray diffraction patterns: see below) and are providing very valuable information. But, whatever preparation method is used, only static or trapped states can be studied, the technique is always invasive and radiation damage is inevitable.

Probe Studies

Studies of crossbridges labelled with extrinsic probes can also be very fruitful[10)], but, if they are to provide information which is of direct relevance to states of muscle found *in vivo*, then it seems to be essential that the skinned fibre preparations that have to be used should also be monitored by other techniques, such as X-ray diffraction, to make sure that the same "relaxed", "rigor" or active states are being studied by the different methods. Some probe studies are non-invasive and make use of the intrinsic properties of the myofibril components[4)5)]. Once again such studies are providing very useful information, but they suffer from the problem that interpretation of the observations is model-dependent. Unlike electron microscopy, these various probes have the great advantage that they can be used to study dynamic states directly, but they suffer from providing only limited data (e.g. an angle; an angular spread; a speed of motion).

Diffraction Studies

X-ray and neutron diffraction can also be entirely non-invasive techniques, since intact living muscles can be studied using either technique[11)12)13-23)]. However, they each have their own advantages and disadvantages. X-ray diffraction using powerful synchrotron sources provides a wealth of structural information in exposures now down to less than 1 ms. Carefully controlled dynamic experiments are routinely carried out and large structural changes, shown by changes in the X-ray diffraction patterns, are evident[13-17)]. The main problem with such diffraction techniques is that diffraction patterns need to be "solved" to provide some kind of "image" of the diffracting object. This is no trivial matter; solving the first protein structure by X-ray crystallography occurred in 1959 about 25 years after the first protein crystal diffraction pattern had been recorded! This is by far the major problem with diffraction methods—we discuss below our approach to it.

Neutron beams could, in principle, be used, like X-rays, to provide information about muscle structure *in vivo*. However, the ability to carry out time-resolved experiments is rather limited—it takes a long time (at least minutes) to record a usable diffraction pattern[12]. Where neutrons score is that their unique interaction with matter (with the nuclei rather than the electron clouds which scatter X-rays), and the fact that neutron scattering by D_2O and H_2O are totally different, allows the technique of contrast matching to be carried out[6]. In this way appropriately deuterated proteins in a protonated environment (or vice versa) can be made invisible by adjusting the mixture of D_2O and H_2O in the bathing medium to have just the right scattering power to "match out" that particular protein. Unlike X-ray diffraction, where such a technique is much more difficult, neutrons can be used to probe selected molecules in multicomponent systems[6].

Time-Resolved X-ray Diffraction

Because of the ability of X-ray diffraction to probe dynamic states of muscle non-invasively (provided the exposure of the muscle to the X-ray beam is not so long that significant radiation damage occurs) it has great advantages in structural studies. However, what about the problem of "solving" the observed diffraction pattern? It is this problem that led us to use fish muscle for time-resolved X-ray diffraction studies[16].

THE ADVANTAGES OF BONY FISH MUSCLE

Background

Because of the enormous amount of excellent research into the mechanical and physiological properties of frog muscle[1)5], and the seminal structural studies of frog muscle by Hugh Huxley and his colleagues[11)19-22], the early work in this laboratory on vertebrate skeletal muscle was all done on frog sartorius and semitendinosus muscles[3)23)24]. This included both electron microscopy and X-ray diffraction studies. It was the combination of both of these approaches that led eventually to the realisation that frog muscles, along with most other vertebrate skeletal muscles, are relatively disordered structures[24]; there is not a totally predictable arrangement of myosin filament orientations across the hexagonal lattice of the A-band (Figure 1(a)). In fact there is a statistical superlattice unit cell. It is a basic tennet of X-ray crystallography that the more order there is in the specimen, the better the chances of solving the structure uniquely. The superlattice in frog, rabbit, chicken and other muscles was therefore very bad news for those interested in structural studies of muscle. The aim is to solve muscle structure as rigorously as is done, for example, with protein crystals. Disordered protein crystals are of limited use; only if the order in them is good enough are they used for solving protein structures. Why should muscle be any different? That is not to say that these superlattice muscles should not be studied (they must be studied—mammalian cardiac muscle is of this type!) or that they will not provide useful structural information. What it does say is that to come to rigorously derived unique solutions (i.e. images) of the observed X-ray diffraction patterns is going to be extremely difficult.

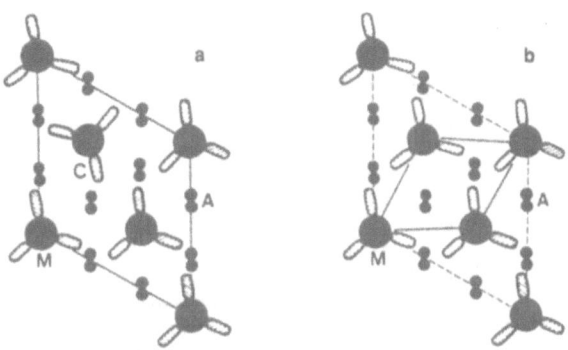

Fig. 1. Filament lattices in the A-bands of (a) frog skeletal muscle (and other muscles from higher vertebrates) and (b) bony fish muscle. Both are shown in cross-section. M are the 3-stranded myosin filaments (c represents pairs of myosin heads) and A are actin filaments. Frog muscle (and all amphibian, mammalian, avian, reptilian skeletal muscles studied to date) has a disordered statistical superlattice structure (a), compared with the regular simple lattice in bony fish muscle A-bands (b).

When we happened across the highly ordered myosin filament arrangement in bony fish muscle[3)25)], where in a given A-band lattice all of the myosin filaments have the same orientation (Figure 1(b)), it became clear to us that fish muscle would be an easier structure to solve than frog muscle. (Not easy, but easier!) Some of the reasons for this are identified individually in the next sections.

Fish Muscle Diffraction Patterns

Figure 2 shows a schematic diagram of the kind of low-angle diffraction pattern that is obtained from bony fish muscle. Characteristic features are (i) the equator with the well-known 100, 110 etc reflections; (ii) the meridian with the characteristic 143Å reflection based on the myosin crossbridge axial repeat on the myosin filaments, and with other characteristic meridional reflections (e.g. 442Å and 418Å from C-protein; 392Å and 385Å from troponin; 215Å region from the crossbridge array on the myosin filament[3)]); (iii) the myosin-based layer-line pattern (spacings 429Å, 215Å, 143Å etc) and (iv) the actin layer-line pattern (spacings of about 360 to 370Å, 180 to 185Å, 120 to 123Å......59Å, 51Å etc).

The "conventional" diffraction story[3)11)19)] *now requiring modification*, was (a) that the 100 and 110 equatorial reflections change intensity in a *reciprocal* way as actin attachment by crossbridge occurs; the more attachment there is, the larger is the 110 reflection and the weaker the 100 reflection. Thus when a muscle is activated the 110 intensity increases and the 100 reflection decreases. At the same time (b) the myosin-based layer-lines (429Å and orders) get weaker and disappear[11)], consistent with myosin heads moving off the resting helix on the myosin filaments. Finally (c), if actin attachment by myosin heads occurs, then the actin layer-line pattern *ought* to

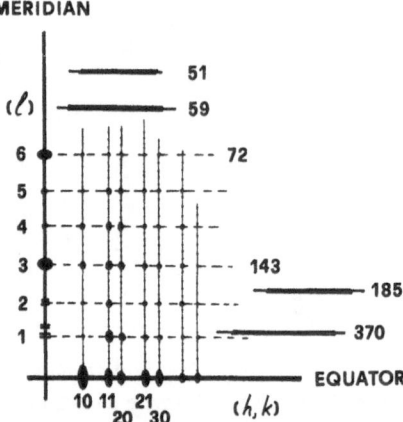

Fig. 2. Summary sketch of the low-angle X-ray diffraction pattern from turbot fin muscle[16)26)27)37)] (fibre axis vertical) showing the myosin-based layer-lines (labelled l = 1, 2, 3 ..) sampled by vertical row-lines (labelled h k = 10, 11, 20, 21 ..) together with some of the actin-based layer-lines at 370, 185, 59 and 51 Å. Only the top righthand quadrant of the pattern is shown; the pattern is not drawn exactly to scale.

get stronger, just as it does in patterns from rigor muscle. In fact, this latter event has proved difficult to demonstrate, although now several studies have proved the point; the actin layer lines from active muscle *are* different from relaxed muscle.

One of the distinguishing features of diffraction patterns from fish muscle is that all the low-angle myosin based layer-lines are sampled in the same way as the equator because of the simple lattice of myosin filaments (Fig. 1(b)). This means that reflections of the type 10l, 11l, 20l, etc. are seen on all of these layer-lines; 100, 110 etc. on the equator; 101, 111 etc. on the 429 Å layer-line; 103, 113 etc. on the 143 Å layer-line and so on). It is this simple sampling that makes the fish muscle diffraction pattern easier to handle than patterns from other vertebrate muscles.

EQUATORIAL DIFFRACTION FROM FISH MUSCLE

Analysis of the Equator

The conventional view about the reciprocal changes in the intensities of the 100 and 110 equatorial reflections is based on the work of Elliott et al.[18)] and Huxley[19)]. Using a simple hexagonal filament lattice as in Figure 1(b), Elliott et al.[18)] showed that the structure factor F(hk0) for the equatorial reflections could be calculated knowing the positions of the myosin and actin filaments in the unit cell. F(hk0) can be determined using the expression:

$$F(hk0) = \sum_{j=1}^{N} f_j \exp 2\Pi i(hx_j + ky_j).$$

(N objects in unit cell, jth object at x_j, y_j has scattering factor f_j, and the fractional unit cell coordinates (x_j, y_j) of the myosin filament are 0, 0 and of the two actin filaments are 1/3, 2/3 and 2/3, 1/3). Hence $F(100) = f_M - f_A$ and $F(110) = f_M + 2.f_A$.

This was an extremely useful starting point in the analysis of muscle equatorial diffraction, and it is expressions like this which, in the first place, were taken to indicate that if mass moves from myosin to actin the changes in $F(100)$ and $F(110)$ might be linearly related to the amount of mass moved (i.e. the number of attached heads). Unfortunately, it is only part of the story. At the very least it is now necessary to consider in a specific way what the crossbridges are doing. In the resting muscle the crossbridges can be assigned 'average' (mass weighted) myosin-centred scattering factors $f_C(\text{Rel};100)$ and $f_C(\text{Rel};110)$ for the 100 and 110 reflections respectively, allowing for a directional component of f_C. Similarly, making the simplest possible assumption that there is only one structurally distinguishable actin-attached crossbridge state in active muscle, then the attached heads can be assigned 'average' (mass weighted) actin-centred scattering factors $f_C(\text{Cont};100)$ and $f_C(\text{Cont};110)$ for the 100 and 110 reflections.

Thus if fraction x of the crossbridges are in their resting state and fraction 1-x are in the active configuration on actin then the structure factors are:

$$F(100) = f_M + x \bullet f_C(\text{Rel};100) - f_A - 1/2(1-x) \bullet f_C(\text{Cont};100)$$

$$F(110) = f_M + x \bullet f_C(\text{Rel};110) + 2 \bullet f_A + (1-x) \bullet f_C(\text{Cont};110)$$

In this case it is by no means clear that $F(100)$ and $F(110)$ or the corresponding intensities $I(100) = |F(100)|^2$ and $I(110) = |F(110)|^2$) should be directly reciprocally related; the intensities $I(100)$ and $I(110)$ and their ratio depend not just on f_A and f_M but also on x, on $f_C(\text{Rel};100)$, $f_C(\text{Rel};110)$, $f_C(\text{Cont};100)$ and $f_C(\text{Cont};110)$.

In summary: *we should not normally expect changes in the 100 and 110 reflections to necessarily be linearly related to the number of attached heads or to have the same time sequence on activation.* Note, however, that in some special cases it may be that such simple relationships are approximately true. Even so, the assumption cannot be made that because it holds sometimes it will always be true. This is even more important if there is more than one structurally distinguishable attached state of crossbridges on actin. These and other calculations are presented in detail elsewhere (Squire & Harford, in preparation). We have already shown[16] that the time courses of the changes of the 100 and 110 reflections from fish muscle during the rising phase of a tetanus are quite distinct; the times $(t_{1/2})$ for half maximum change in these reflections to occur[26] are about 15 ms different $[t_{1/2}(100) = 35 \pm 8$ ms, $t_{1/2}(110) = 21 \pm 4$ ms], and the 110 change precedes the tension change by about 20 ms $[t_{1/2}(\text{Tension}) = 41 \pm 3$ ms]. Note here that analysis of changes in sarcomere length during the tetanic contractions of turbot muscle has indicated very little contribution to the observed intensity time-courses[26]. Note finally that the conclusion that the 100 and 110 intensities are not necessarily reciprocally related, or a linear function of the number of attached heads (i.e. 1 - x) is *not* a *negative* one; the most important conclusion from the expressions above for F(100) and F(110) is that they carry information *both* about attachment number *and* about the configuration of the heads. After all, it is information about both of these things that structuralists are searching for. It is only by solving the X-ray patterns properly that the full information can be obtained.

Solving the Equatorial Pattern

The main point of X-ray diffraction studies of muscle is to obtain 'images' of muscle under different conditions. What protein crystallographers do with their data is to combine observed intensities I (actually amplitudes $|F| = \sqrt{I}$ are used) with phases determined by some indirect method such as Multiple Isomorphous Replacement (MIR) or Multiple wavelength Anomalous Diffraction (MAD). Both amplitudes and phases are needed to carry out Fourier synthesis. The process is analogous to what goes on in the image-forming part (e.g. eyepiece) of a light microscope, but here the process is carried out by computer. Since it is not possible to interpret muscle equatorial intensities directly (i.e. changes in the intensities of the 100 and 110 reflections are not related linearly to attachment number of crossbridges on actin[27], as used to be assumed), one approach to analysing the equatorial data, as with protein crystallography, is to produce Fourier synthesis 'images'. This is not some arbitrary procedure; it is well tried and tested and is essential for progress on muscle to be obtained.

We have attempted Fourier synthesis with the equatorial diffraction patterns from fish muscle in various states. However there are four problems, some more serious than others:

[1] The A-band is not the only part of the sarcomere that contributes to the equator. The Z-band and I-band close to the Z-band make a major contribution as well.

[2] The A-band equatorial peaks, as with all vertebrate muscle diffraction patterns, partially overlap each other. Careful peak stripping is needed to extract reliable intensity values.

[3] As yet neither MIR nor MAD seem appropriate as phasing tools, but the equatorial phases are needed.

[4] The available equatorial information is of limited resolution (say 50 Å at best), but only some of the reflections out to this resolution have been phased.

The Z-band/I-band Contribution

An equatorial reflection of spacing about 290 Å occurs between the 100 and 110 reflections from the muscle A-band. This has been ascribed to Z-band and I-band diffraction[28]. However, the Z-band and the actin filament array in the I-band close to the Z-band has a quasi-tetragonal lattice (Figure 3) which need not give just a 100 relection at 290 Å, but also could contribute 110, 200 reflections and other higher orders. Unfortunately, the Z-band 100 reflection (Z_{110}) happens to coincide with the A-band 200 reflection (A_{200}; Figure 3) and the Z-band 200 reflection (Z_{200}) is just between the closely spaced A-band 210 and 300 reflections (A_{210} and A_{300}). Therefore, in order to do anything useful with the equatorial diffraction data from *any* vertebrate skeletal muscle, these two overlapping diffraction patterns need to be separated. Fish muscle has an advantage here because it is highly ordered and it remains highly ordered even when the muscle is fully active. Thus the equatorial

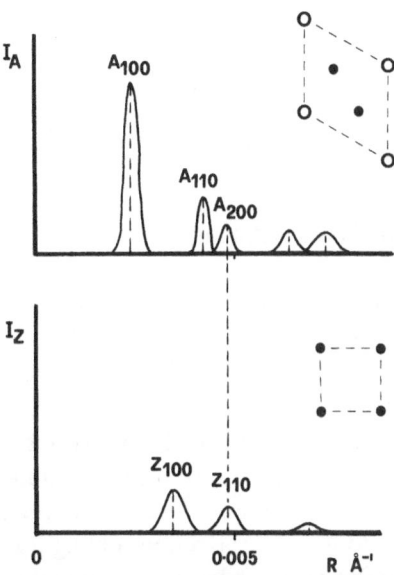

Fig. 3. The hexagonal filament lattice in the A-band (inset top), the square actin filament array near to the vertebrate Z-band (inset bottom) and sketches of their expected diffraction patterns along the equator (as in Figure 2). Intensities I_A and I_Z are plotted upwards (not drawn to scale) against radial distance R Å^{-1} in the expected diffraction patterns. The diagram shows the coincidence of the A_{200} and Z_{110} reflections. These need to be separated[29] (see text) before the equator can be analysed properly.

diffraction pattern from active fish muscle is still very sharp and shows equatorial peaks out to the A-band A_{300} and well beyond. It so happens that when the observed overlapping equatorial reflections are rigorously fitted by an exponential plus polynomial background together with overlapping Gaussian peak profiles, the A-band and Z-band contributions to the fish muscle equator *can* be separated[16)26]. The peak widths of the Z-band peaks are slightly broader than the A-band reflections and their contributions to the total equatorial pattern can be determined. When this is done, and this emphasises the importance of careful analysis, it is found that over 50% of the traditional A_{200} reflection from resting fish muscle is actually due to the Z-band, not to the A-band. However, when fish muscle is activated, the Z-band contribution to this peak changes and becomes almost negligeable. The Z_{200} reflection appears to be rather weak in both cases, since an additional peak close to the A_{210} and A_{300} peaks is not needed to explain the observed intensity profiles.

The take home message from this is that the Z-band contributes significantly to the observed muscle equatorial diffraction pattern and that its contribution changes according to the state of the muscle. Even had the Z-band not contributed to the observed 200 reflection, this peak and the Z_{100} peak both overlap the A_{110} peak and attempts to estimate the intensity of the A_{110} reflection without proper curve fitting are to be deplored. The changing contribution of the Z-band makes it essential to analyse the equatorial peaks properly. The great benefit of so doing is not just that changes in the A-band reflections can be accurately monitored and used reliably for Fourier synthesis to show mass movements in the A-band unit cell, but also that exactly the same can be done for the Z-band to show dynamic changes there too.

Equatorial Fourier Synthesis of the A-band

In order to compute electron density maps of the A-band in a view down the filament axis, it is necessary to know the phases of the equatorial A-band peaks. No 'direct' determination of these phases has yet been reported, but we have used indirect methods[16] which have selected the phase set 0, 0, Π, 0, 0 for the 100, 110, 200, 210 and 300 A-band reflections. (An alternative set which is just possible is 0, 0, Π, Π, 0). With such phases and the observed amplitudes, electron density maps have been computed for the rising phase of tetanic contractions in turbot muscle. So far we have used only up to the 300 reflection, giving a resolution of about 130 Å, but we know that many reflections can be seen beyond this. These have not yet been phased and so cannot, as yet, be reliably included, but their omission from a Fourier synthesis computation could lead to spurious peaks in the map due to Series Termination Errors. In order to avoid the possibility of seeing such peaks, we have used the well-tried crystallographic procedure of applying an artificial temperature factor to the observed equatorial peaks[30]. It is dangerous to attempt to interpret maps where this has not been done. As it is, we can rely on what we see, even though the resolution is limited. In particular we can start to measure the amount of mass within a given radius (8 nm) of the actin filament axis during the rising phase of the tetanus in fish muscle. The $t_{1/2}$ for the mass change at actin is about 25 ms; it is about 16 ms before $t_{1/2}$(Tension)[26], consistent with myosin heads arriving at the actin filament well before they produce tension, as previously concluded from stiffness measurements[31][32]. They must be heads in some form of non force-producing attached state[33][34].

We are now in the process of phasing reflections beyond the 300 A-band peak to give higher resolution electron density maps.

THE ACTIN LAYER-LINE PATTERN

Evidence for Stereospecific Actin Labelling

The search for evidence for actin-attached crossbridges in active muscle has been largely based on results from rigor muscles where large intensity changes on the actin-based layer-lines, especially the first layer-line at about 360 to 370 Å (Figure 2), are known to be due to myosin heads labelling actin filaments with a well-defined interaction geometry at the actin/ myosin interface. We term this stereospecific labelling; the myosin head configuration is dictated by the symmetry of the actin filaments, the heads are actin-centred, and the presence of a strong interaction between the heads and actin is indicated. For a long time changes in the actin pattern due to myosin head labelling were hard to substantiate. However with improved X-ray sources (synchrotrons) and with better detectors, changes in the actin layer-line pattern have now been reported[14)17)35)]. But why are the changes so relatively small if the number of heads labelling actin stereospecifically is relatively large? One of the problems, only recently acknowledged, is that the "classic" tropomyosin shift[20)21)36)], associated with regulation, causes both an increase in the second actin layer-line *and a decrease in the first actin layer-line*[37)38)]. Thus, any intensification of this first layer-line due to crossbridge labelling of actin has to overcome the *decrease* due to thin filament activation before any NET increase will be apparent. In rigor, where more heads are stereospecifically bound to actin, the drop due to regulation will be swamped by the large increase in mass due to crossbridge labelling.

In the case of fish muscle diffraction patterns, both an increase in the 2nd layer-line associated with tropomyosin movement, and increases on the 1st and 6th layer-lines, probably mostly associated with stereospecific myosin head attachment, have been observed[16)26)]. These patterns are now being analysed in detail.

THE MYOSIN-BASED LAYER-LINES

The Resting Pattern

As discussed earlier, the myosin-based layer-lines which are orders of the 429 Å myosin repeat in patterns from resting muscle (Figure 2) have been ascribed to the 3-stranded quasi-helix of myosin heads on the 'resting' myosin filaments[3)39)]. We have already used simple myosin head models to determine the crossbridge arrangements on the myosin filament surfaces in resting frog and fish muscles[23)39)40)]. However, once again, fish muscle has a distinct advantage over other muscles; the myosin layer-line pattern is sampled (Figure 2) by peaks related to the simple lattice A-band unit cell (Figure 1(b)). Because the sampling is regular (the same on all layer-lines) and there is little unsampled layer-line, the observed intensities can be rigorously determined and corrected. In this way, not only can the head geometry on the myosin filaments be defined, but also the absolute orientation of the whole myosin filament around its long axis can be determined relative to the lattice vectors of the A-band unit cell. Unsampled layer-lines do not allow this, but this information is needed in order to determine the geometry of interaction of myosin crossbridges with actin target areas[40].

Changes on Activation

The second interesting feature of these layer-lines from fish muscle is that they do not only come from myosin heads in the resting helix on myosin. We have found that, in patterns from fully active muscle, the myosin layer-line pattern, although much weaker, is still present and still well sampled. That this is not due to a part of the muscle that has not been activated properly is indicated by the fact that the relative intensities of the peaks along the 429 Å layer-line (e.g. the 101, 111 and 201 reflections) are different in patterns from resting and active muscles[26]. We interpret this new myosin-based layer-line pattern as being due to the non-force-producing bridges evident from analysis of the equatorial diffraction data. If these heads are in some sort of weak-binding state, we presume that they are not stereospecifically bound to actin, but that their neck ends point back to the head origins on the myosin filament. These heads, although probably actin-attached at least transiently, are therefore still myosin-centred and give a myosin-like layer-line pattern[40]. The intensity changes along the layer-lines are consistent with this view.

CONCLUSION

In summary, the results to date from time-resolved X-ray diffraction studies of bony fish muscle (mainly turbot fin muscle) have shown the existence of two states (or groups of states) of myosin heads in active muscle that are *structurally* different from each other and from the resting state. Further analysis will help to define the details of these states and, indeed, whether further structural states can be defined. However, in order to test the swinging crossbridge model it is necessary to show whether the observed states have the myosin heads attached to actin with different axial tilts. Our answer to this is that they probably do. Our analysis so far of the new 'active' myosin-based layer-line pattern is consistent with the heads extending between the myosin and actin filaments with their neck ends on the myosin filament surface and their outer ends at least transiently attached to actin (our data provide no information about the kinetics of this process). The actin layer-line pattern, showing heads that are stereospecifically bound to actin, is more consistent with the presence of rigor-like (i.e. tilted) myosin heads. Evidence supporting this view comes from the equatorial Fourier synthesis electron density maps for two different times during the early part of the tetanus, when the amount of mass near to actin (within the 8nm radius) is virtually constant, but the tension levels are sub maximal (80%) or at the maximum (tetanus plateau) value (100%). The difference maps obtained by subtracting these two observed electron density distributions show mass redistributing around actin in a manner consistent with axial swinging of attached heads[26].

The bones of the story are already there. We believe that further work using fish muscle X-ray diffraction, supplemented by electron microscopy where appropriate, promises to provide the flesh to complete the structural picture at least to a resolution of about 50 Å.

REFERENCES

1. Huxley, A.F. *Prog. Biophys.* **7**, 255-313 (1957).
2. Huxley, H.E. *Science* **164**, 1356-1366 (1969).
3. Squire, J.M. in *The Structural Basis of Muscular Contraction.* (Plenum Press, New York, 1981).
4 Squire, J.M. in *Molecular Mechanisms in Muscular Contraction* (Macmillan, 1990).
5. Irving, M. in *Fibrous Protein Structure* (eds Squire, J.M. & Vibert, P.J. Academic Press, 1987).
6. Curmi, P M.G., Stone, D.B., Schneider, D.K., Spudich, J.A. & Mendelson, R.A. *J. Mol. Biol.* **203**, 781-798 (1988).
7. Sjostrom, M., Squire, J.M., Luther, P.K., Morris, E.P. & Edman, A.-C. *J. Microsc.* **163**, 29-42 (1991).
8. Padron, R., Alamo, R., Craig, R. & Caputo, C. *J. Microsc.* **151**, 81-102 (1988).
9. Morris, E.P., Nneji, G. & Squire, J.M. *J. Cell Biol.* **111**, 2961-2978 (1990).
10. Cooke, R., Crowder, M. S. & Thomas, D.D. *Nature* **300**, 776-778 (1982).
11. Huxley, H.E. & Brown, W. *J. Mol. Biol.* **30**, 383-434 (1967).
12. Worcester, D.L., Gillis, J.M., O'Brien, E.J. & Ibel, K. *Brookhaven Symp. Biol.* **27**, 101-114 (1975).
13 Huxley, H.E., Faruqi, A.R., Bordas, J., Koch, M.H.J. & Milch, J.R. *Nature* **284**, 140-143 (1980).
14. Huxley, H.E., Faruqi, A.R., Kress, M., Bordas, J. & Koch, M.H.J. *J. Mol. Biol.* **158**, 637-684 (1982).
15. Bordas, J., Diakun, G.P., Harries, J.E., Lewis, R.A., Mant, G.R., Martinez-Fernandez, M.L. & Towns-Andrews, E. *Adv.Biophys.,* **27**, 15-33 (1991).
16. Harford, J.J. & Squire, J.M. in *Molecular Mechanisms in Muscular Contraction* (ed Squire, J.M.) 287-320 (Macmillan Press, 1990).
17. Wakabayashi, K., Ueno, Y., Amemiya, Y. & Tanaka, H. in *Molecular Mechanisms of Muscle Contraction* (eds Sugi, H. & Pollack, G.H.) 353-367 (Plenum. London, 1988).
Wakabayashi, K., Tanaka, H., Amemiya, Y., Fujishima, A., Kobayashi, T., Sugi, H. & Mitsui, T. *Biophys. J.* **47**, 847 (1985).
18. Elliott, G.F., Lowy, J. & Worthington, C.R. *J. Mol. Biol.* **6**, 295-305 (1963).
19. Huxley, H.E. *J. Mol. Biol.* **37**, 507-520 (1968).
20. Haselgrove, J.C. *Cold Spring Harbor Symp. Quant. Biol.* **37**, 341-352 (1973).
21. Huxley, H.E. *Cold Spring Harbor Symp. Quant. Biol.* **37**, 361-376 (1973).
22. Huxley, H.E. & Kress, M. *J. Muscle Res. Cell Motility* **6**, 153-161 (1985).
23. Squire, J.M. *Ann. Rev Biophys. Bioeng.* **4**, 137-163 (1975).
24. Luther, P.K. & Squire, J.M. *J. Mol. Biol.* **141**, 409-439 (1980).
25. Luther, P.K., Munro, P.M.G. & Squire, J.M. *J. Mol. Biol.* **151**, 703-730 (1981).
26. Harford, J.J. & Squire, J.M. *Biophys. J.,* **63** (in press)

27. Lymn, R.W. *Biophys. J.* **21**, 93-98 (1978).

28. Yu, L.C., Lymn, R.W. & Podolsky, R.J. *J. Mol. Biol.* **115**, 455-464 (1977).

29. Squire, J.M., Harford, J.J., Chew, M.W.K. & Towns-Andrews, E. in *Synchrotron Radiation Appendix to 1991 Daresbury Annual Report.* 171 (1991)

30. Franks, N.P., Melchior, V., Kirschner, D.A. & Caspar, D.L.D. *J. Mol. Biol.* **155**, 133-153 (1982).

31. Ford, L.E., Huxley, H.E. & Simmons, R.M. *J. Physiol. (Lond.)* **372**, 595-609 (1986).

32. Bagni, M. A., Cecchi, G. & Schoenberg, M. *Biophys. J.* **54**, 1105-1114 (1988).

33. Brenner, B., Schoenberg, M., Chalovich, J.M., Greene, L. & Eisenberg, E. *Proc. Natl. Acad. Sci. U.S.A.* **79**, 7288-7291 (1982).

34. Brenner, B., Yu, L.C. & Podolsky, R.J. *Biophys. J.* **46**, 299-306 (1984).

35. Matsubara, I., Yagi, N., Miura, H., Ozeki, M. & Izumi, T. *Nature, Lond.* **312**, 471-473 (1984).

36. Parry, D. A.D. & Squire, J.M. *J. Mol. Biol.* **75**, 33-55 (1973)

37. Yagi, N. *Biophysics* **27**, 35-43 (1991).

38. Yagi, N. & Matsubara, I. *J. Mol. Biol.* **208**, 359-363 (1989).

39. Harford, J.J. & Squire, J.M. *Biophys. J.* **50**, 145-15 (1986).

40. Squire, J.M. & Harford, J.J. *J. Muscle Res. Cell Motility* **9**, 344-358 (1988).

Discussion

Pollack: John, you showed very nicely that something as small in mass as the Z-line can make an important contribution to the X-ray pattern. There is another element that I wonder what you think about. For example, titin, which may be about 8% or 10% of total muscle mass, sits along the same axis as myosin. Your analysis implicitly assumes that there is not much contribution. Do you think that's reasonable?

Squire: That is a hard one to answer. My approach is to try to interpret what we see, making as few assumptions as possible. In other words, we explain what we can in terms of myosin-filament structure as we know it, and the actin filament structure near the Z-line. The fact that we have had to invoke diffraction from the Z-line is an example of what you have to do if you can't explain what you see in terms of your original assumptions. We couldn't explain some reflections, nor the fact that the (2,0) intensity was much broader than it should have been. We had to invoke another structure. But until we are forced into that situation, I think it would be hard to make any comment about what titin is doing. As yet there is no obvious reflection from titin. However, out-phasing of the equatorial pattern suggests that titin contributes as part of the myosin filament backbone (Harford, J.J. and Squire, J.M. *Molecular Mechanisms in Muscular Contraction,* pp. 1-48, ed. Squire J.M., MacMillan, 1990).

Maéda: You deduced that the sarcomere length stayed unchanged, based on the constant mass in the unit cell. Did you measure the sarcomere length of the specimen at site?

Squire: No.

Maéda: I have a question regarding the differentiated time course of I(1,0) and I(1,1). I agree that there is no theoretical reason that the time courses of these two reflections should be associated with each other. Experimentally, however, it is not difficult to observe differentiated time courses. Actually, using single fibres from frog, we often obtained such results, unless the sarcomere length was clamped. We think the internal movements are the cause of the anomalous behavior. So we have decided to be very cautious about it. What is your comment?

Squire: My comment is that I think you're right. We are actually designing experiments to control sarcomere length, because it is something we want to test. But I think the X-ray data themselves do actually contain a lot of information about that same problem, and as I said, the total mass in the unit cell does not appear to change. We haven't assumed that it is constant. This is just something that comes out when we do the analysis. I think it is something that would be hard to explain away if there were a gross change in sarcomere length.

Maéda: Another question. You said that the time course of mass distribution, deduced from the ratio $I(1,1)/I(1,0)$ must be close to the time course of stiffness. Do you have a record of stiffness?

Squire: No, what I said to you is that I think the mass should bear a better correlation to stiffness than to the $(1,0)$ or the $(1,1)$ intensities. We haven't done the stiffness experiment on fish, but it has been done on frog muscle by Sir Andrew and other people. The results are very similar to what we get with our mass time course.

Iwamoto: Did you do some experiments using skinned fibers, lowering the ionic strength and looking at the pattern? Is there any resemblance between the pattern with weakly bound bridges produced by a low salt condition and the pattern during the rise of contraction?

Squire: Yes, I did some experiments at NIH with Podolsky, Yu, and Brenner on skinned fish muscle fibers (Squire, J.M. et al. *J. Struct. Biol.* **107**, 221-226, 1991). Lowering ionic strength had very little or no effect, and there is no evidence for much of a weak-binding state. I don't know why that is.

One thing that seems to have cropped up is the apparent question about whether there is actin labelling, or actin-enhanced layer lines, in "honest" frog and "dishonest" fish, because we seem to have a contradiction of results. It just occurred to me that there is a possible reason for that. As you know, the lattices in the two muscles are different. In the fish muscle, if you look at a particular actin filament, you have three sets of myosin heads coming to the actin at the same axial position and then a similar group of heads 370 Å further up (cf. Figure 1 of paper). In the super-lattice muscles, there is a more even distribution of heads along the actin (Squire, J.M. and Harford, J.J. *J. Muscle Res. Cell. Motility* **9**, 344-350, 1988). This means that in active muscle, you could have more of a marked ladder arrangement of labelled actin every 370 Å or so in fish, and a much more helical arrangement of heads in frog. Therefore, you would get a stronger first layer line from fish than you would from frog. Maybe those results are not incompatible, though these possibilities need to be modelled in detail.

Simmons: On the same point on the frog muscle pattern and the question of whether the actin contribution to the first layer line is intensified during contraction, there are several relevant papers in *Advances in Biophysics*: one by Bordas, J et al. (**27**, 15-33, 1991), as well as one by Wakabayashi and others in the same journal (Wakabayashi, K. et al. *Adv. Biophys.* **27**, 3-13, 1991), who obtained a negative result. Joan Bordas and his colleagues did find an increase and so did Naoto Yagi (*Adv. Biophys* **27**,35-43, 1991). Hugh Huxley, Wasi Faruqi, and I collected a data set, which has been partly analyzed by Alex Stewart. In our data the actin contribution to the layer line is very weak in a resting muscle, and does not increase much, if at all, during contraction. The proportion of the layer line that is ascribed to actin differs in the four data sets, but it is not clear whether the differences are real or result from the procedures used for background subtraction.

THE FIRST THIN FILAMENT LAYER LINE DECREASES IN INTENSITY DURING AN ISOMETRIC CONTRACTION OF FROG SKELETAL MUSCLE

Katsuzo Wakabayashi*, Hideaki Saito*, Noriyoshi Moriwaki*, Takakazu Kobayashi** and Hidehiro Tanaka**†

*Department Biophysical Engineering
Faculty of Engineering Science
Osaka University,
Toyonaka, Osaka 560, Japan
** Department of Physiology
School of Medicine
Teikyo University
Itabashi-ku, Tokyo 113, Japan

ABSTRACT

During isometric contraction/activation of full-overlap and non-overlap live frog skeletal muscles, the intensity of the first thin filament layer line at the axial spacing of ~1/37 nm^{-1}, when separated from the partially overlapping first thick filament layer-line at ~1/43 nm^{-1}, remained unchanged in the inner radial region (0.02-0.08 nm^{-1}) where a large intensity increase is observed in the rigor state. The intensity decreased in the outer radial region (0.08-0.18 nm^{-1}) where this layer line is expected to peak in the resting state. The intensity decrease in the outer region became larger with increasing filament overlap; on activation of the non-overlap muscle, it was about half that of the full-overlap muscle. Thus the first thin filament layer line decreases in intensity and any indication of the rigor-like intensification is not observed at all during contraction. This intensity decrease can be attributed to the same structural changes giving rise to the intensity increase of the second thin filament layer line.

The results indicate that the configuration of the myosin heads interacting with actin during contraction differs significantly from that of the rigor state. Four-fold rotational symmetry of the thin filament structure as a whole becomes more pronounced during isometric contraction of the overlap muscle.

---

† Present address: School of Nursing, Teikyo Heisei College, Ichihara, Chiba 290-01, Japan

Mechanism of Myofilament Sliding in Muscle Contraction, Edited by
H. Sugi and G.H Pollack, Plenum Press, New York, 1993

INTRODUCTION

During isometric contraction of vertebrate skeletal muscles, the intensities of the thin filament-based outer layer lines increase without meridional shift of their centroids, being dissimilar to the intensification taking place in the rigor state[1-4]. However, it has been generally interpreted that such intensification is caused by the specific formation of the actomyosin complex similar to the rigor complex. If a substantial fraction of the myosin heads interacts with the thin filaments in a contracting muscle stereospecifically as in a rigor muscle, the first thin filament layer line is expected to be enhanced, because the very big intensification of this layer line takes place when the muscle is put into the rigor state[4,5]. Unfortunately, the first thin filament layer line which should appear at $\sim 1/37$ nm^{-1} axial spacing is inherently diffuse and partially overlapped with the strong first thick filament layer line at $\sim 1/42.9$ nm^{-1} in the resting state. During isometric contraction the first thick filament layer line decreases remarkably in intensity. However, these two layer lines remain generally unresolved and appear as a single layer line at an axial spacing of $1/40 \sim 1/41$ nm^{-1} near the meridian. Huxley and Brown[5] discussed the possibility that such an appearance of the layer line arises from the first thin filament layer line, superimposed on the remnant of the first thick filament layer line. This seems to be the case, since the axial profile of the apparent first layer line is asymmetric with an outward shift of its centroid as the radial coordinate increases.

We have separated both layer-line components from this overlapping layer line in the resting and contracting states and examined whether the first thin filament layer line becomes more intense during contraction, as observed in the other layer lines. We found that during isometric contraction the first thin filament layer line is weakened in the outer region and no rigor-like intensification is observed in the inner region.

EXPERIMENTAL

Semitendinosus and sartorius muscles from the bullfrogs (*Rana catesbeiana*) were used for studies. X-ray experiments were done using the synchrotron radiation at the Photon Factory, Tsukuba. Synchrotron X-rays were collimated with the double focusing small-angle optics[3,6]. The two-dimensional diffraction patterns were recorded on the storage phosphor detector (Fuji imaging plates)[7] at a camera length of 2.5 m. The muscles were stimulated with 33 Hz electrical pulses (1 ms duration) for 1.3 s at 10°C. The diffraction patterns were recorded for 1 s respectively during the resting phase and at the plateau of isometric tension and/or on activation. This stimulation was repeated 10-20 times, during which the diffraction patterns were summed on the same imaging plate. We used these minimal contraction cycles with a rather short rest time (~ 30 s) in between them and the diffraction measurements were started after applying a few pre-tetanic stimulations on muscles to reduce the strong lattice sampling effects. The studies were made on muscles at full-overlap of the thin and thick filaments (~ 2.3 μm sarcomere length) and non-overlap (≥ 4 μm) of the thin and thick filaments which were monitored by

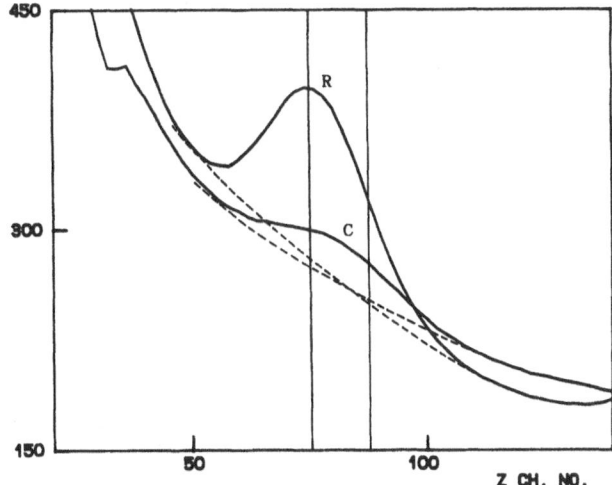

Fig. 1. Axial intensity profiles around the overlapping first layer line in the patterns from the full-overlap semitendinosus muscle. The intensities in the radial spacing of 0.02-0.08 nm^{-1} (in the inner region of the layer line) are integrated. C, in the contracting state; R, in the resting state. The ordinate is the intensity on an arbitrary unit; the abscissa, the channel number in the axial direction (one channel corresponds to 100 μm of a pixel size). The dashed curves denote the background. The inner vertical line indicates the peak position at an axial spacing of 1/43 nm^{-1} and the outer one, the peak position at an axial spacing of 1/37 nm^{-1}.

the isometric force level and the optical diffraction pattern. Image data were analyzed using an NEC ACOS 850 computer at the Institute for Protein Research of Osaka University and high-resolution personal computers (NEC PC98-XA and RL). Two or three sets of high quality diffraction patterns from different muscles were collected and analyzed. The intensities were scaled by normalizing the total intensities recorded in the imaging plates between the resting and contracting/activating patterns from the same muscles and between different muscles.

RESULTS AND DISCUSSIONS

The intensity distributions across the overlapping first layer line were measured in a direction parallel to the meridian in the resting and contracting states after integrating the intensities radially in the appropriate radial ranges (Fig. 1). The background intensity was subtracted from each measured intensity distribution. In the resting pattern, a strong intensity peak was observed at the axial position of 1/43 nm^{-1}. In the contracting pattern from the overlapped muscles, the intensities were markedly weakened due to the large intensity decrease of the first thick filament

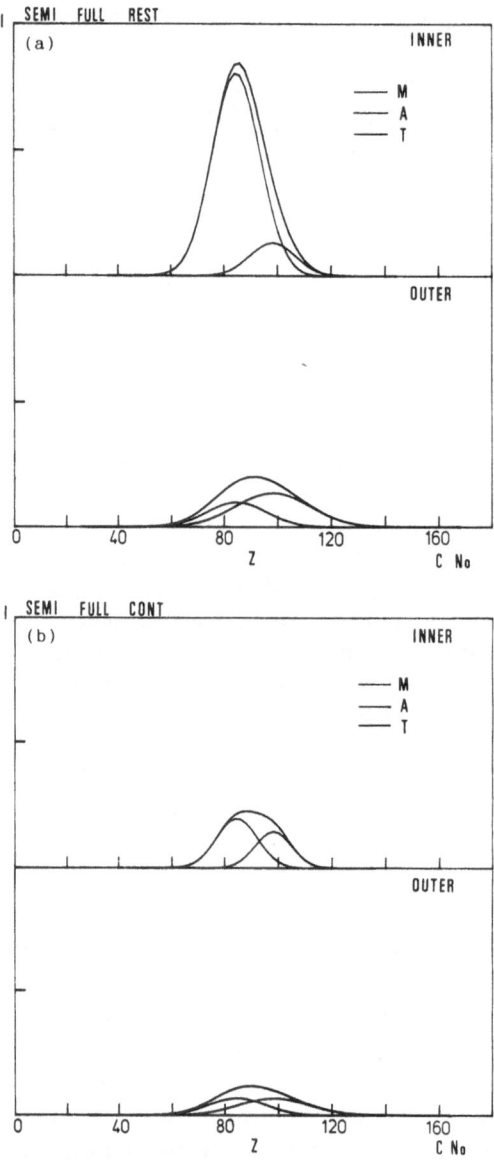

Fig. 2. The Gaussian separation of the axial cross-sectional intensity profiles of the observed first layer line from the full-overlap muscle. The highest curve is the observed (total) intensity profile and the lower two curves are the calculated best-fit intensity profiles: the curve on the left, the first thick filament layer-line component; the one on the right, the first thin filament layer-line component. (a) in the resting state and (b) during contraction. Inner, the inner radial region at the spacing of 0.02-0.08 nm^{-1} and outer, the outer radial region at the spacing of 0.08-0.18 nm^{-1}. The ordinate is the intensity on an arbitrary scale; the abscissa, the channel number in the axial direction.

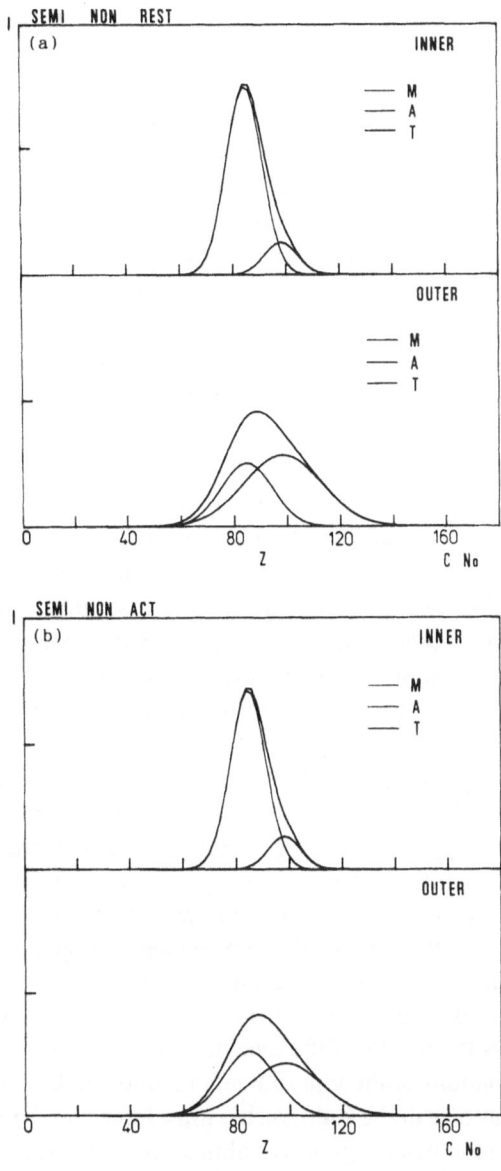

Fig. 3. The Gaussian separation of the axial cross-sectional intensity profiles of the observed first layer line from the overstretched muscle. The curves are expressed in the same way as in Fig. 2. (a) in the resting state and (b) on activation.

layer line and a peak was observed at an axial position of $1/40 \sim 1/41$ nm^{-1}. The axial intensity profile was asymmetric with a more gentle slope at the high-angle side of the peak and became broader with an outward shift of its centroid as the radial coordinate increased (see Figs. 2 and 3). This feature indicates that the apparent first low-angle layer line consists mainly of two layer lines, the first thick and first thin filament layer lines and their relative contributions are different in the different radial ranges and in the resting and contracting states.

In order to examine the intensity changes of the first thin filament layer line during contraction/activation, the axial intensity profiles of this overlapping layer line were measured by radial integration, on stripes parallel to the meridian, at two different radial regions of 0.02-0.08 nm^{-1} and 0.08-0.18 nm^{-1} in the resting and contracting/activating states, respectively. The inner region is the range where the first thick filament layer line is the most intense in the resting state and also the first thin filament layer line has the most enhanced peak in the rigor state[3)5)]. The outer region is the range where the first thin filament layer line is expected to peak in the resting state. The intensities of these two regions were analyzed using two Gaussian functions by a least-squares fitting procedure. The axial widths, heights of the two Gaussians were used as variable parameters and axial positions of these were first constrained to predicted positions and then allowed to vary to achieve better fits.

Fig. 2 shows the examples of the two layer-line components deconvoluted from the overlapping first layer lines in the resting and contracting patterns from the full-overlap muscles. In the inner region of the resting state (a), the thin filament component was very small, about 10% of the total intensity. In the outer region, it was as large as 50% of the total intensity. In the contracting state (b), the total intensity of the inner region decreased to less than 25% of that in the resting state. This large decrease is due to a large intensity reduction of the thick filament layer-line component. In the outer region, the total intensity decreased to lesser extent than in the inner region. The intensity of the thin filament component was as large as that of the thick filament component. Thus the intensity of the thin filament layer line appreciably decreased in the outer region. This result is consistent with that of Yagi[8)]. Fig. 3 shows the results of the overstretched muscles. In the resting state (a), the intensity contributions of the thin filament component both in the inner and outer regions were the same as in the full-overlap muscle, respectively. On activation (b), the total intensity in the inner region was almost the same as in the resting state. The change was observed in the outer region where the total intensity decreased to about 75% of that in the resting state. This is due to the intensity reduction of the thin filament component[9)].

The Gaussian deconvolution procedure revealed that the axial spacing of the first thin fialment layer line was 37.0 ± 0.5 nm both in the resting and contracting states and that of the thick filament layer line was 43.0 ± 0.2 nm in the resting state and 44.0 ± 0.3 in the contracting state, thus shifting in a direction similar to the other thick filament layer lines.

The results were summarized as the histograms in Fig. 4, where each intensity is expressed as a fraction of the total resting intensity in the respective regions. During contraction of the full-overlap muscle, the intensity of the thick filament layer line decreases remarkably in the inner region and does not disappear completely[10)]. In

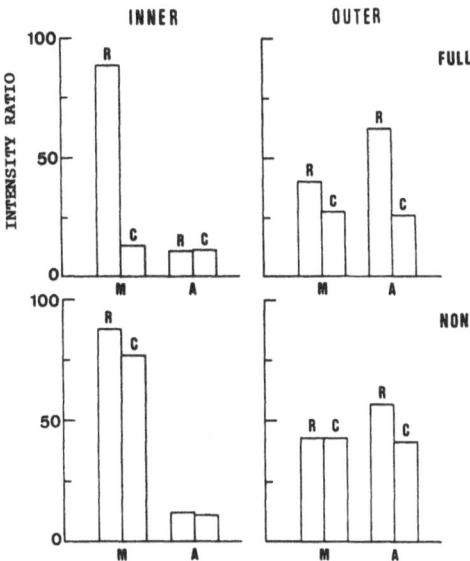

Fig. 4. Histograms of the intensity changes of the first thin and first thick filament layer-line components in the transition from rest to contraction/activation. The upper histogram, the full-overlap muscle; the lower one, the overstretched muscle. Inner, in the inner radial region; outer, the outer radial region. Full, the full-overlap muscle; non, the overstretched muscle. The intensity histograms are based on the data of Figs. 2 and 3. A, the thin (actin) layer-line component; M, the thick (myosin) layer-line compnent; R, at rest; C, contraction/activation. The heights of the histograms are expressed as a fraction of the total intensity of the first overlapping layer line in the resting state.

the outer region, the decrease is smaller than that in the inner region. The resting intensity ratio (~10%) of the thin filament layer line in the inner region is unchanged during contraction. In the outer region the intensity ratio decreases to a value less than half the resting value. The results from the semitendinosus muscles were quite similar to those from the sartorius muscles. Thus, in the overlap muscles the thin filament layer line did not show an increase in the inner region (though a small increase has been suggested previously[8][11]) and a fairly large decrease occurred in the outer region when the muscles contract. In the overstretched muscles, the intensity of the thick filament layer line was almost unchanged both in the inner and outer regions, indicating that the myosin projections do not move on activation, as observed previously[12]. The intensity of the thin filament layer line remained unchanged in the inner region but decreased by about 20% of the resting value in the outer region. The amount of the intensity decrease on activation of the overstretched muscles was significantly smaller (about half) than that taking place in the full-overlap muscle during force generation.

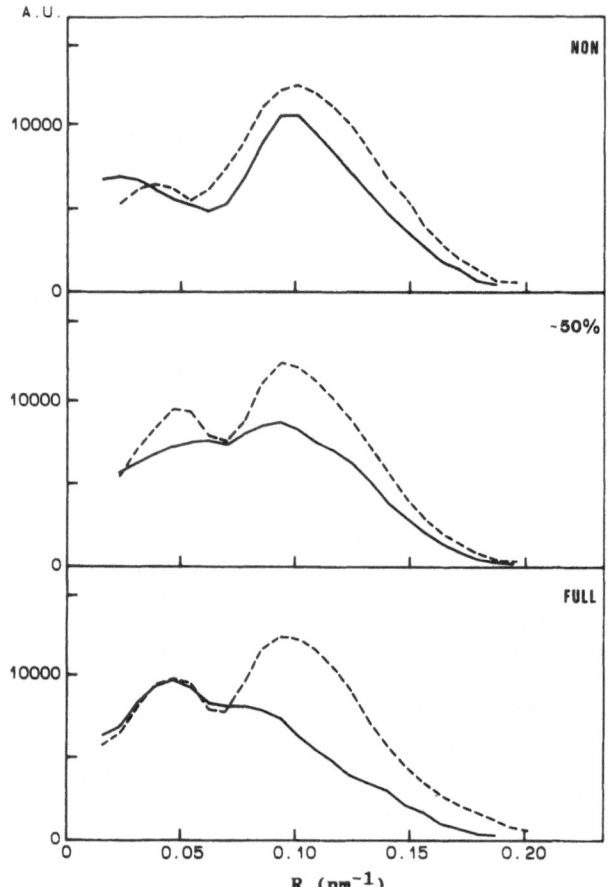

Fig. 5. Intensity distributions of the first thin filament layer lines. They were derived by the Gaussian separation in every 0.01 nm^{-1} step. The chain line, the resting distribution; the solid line, the contraction/activation ones. Non, the overstretched muscle; ~50%, the half-overlap muscle; full, the full-overlap muscle.

Finally, the changes of the intensity distribution of the first thin filament layer line obtained in a similar way by the Gaussian deconvolution procedure are shown in Fig. 5. The layer line appears to have a hump in the inner region. As mentioned above, in the activated state the marked intensity reduction in the outer region is seen, resulting an apparent inner shift of the outer peak. Integrated intensities of three data in Fig. 5 showed roughly a linear dependence on the filament overlap. For a reference, Fig. 6 shows the intensity changes of the second thin filament layer line during contraction/activation. The intensity of this layer line is too weak and diffuse to measure in the resting state but increases markedly in the high radial range of around 0.2 nm^{-1} during contraction[13]: the integrated intensity in the full-overlap muscle was about 40% of the resting intensity of the 5.9 nm layer-line reflection and on activation of the overstretched muscle the intensity increase was about half that of

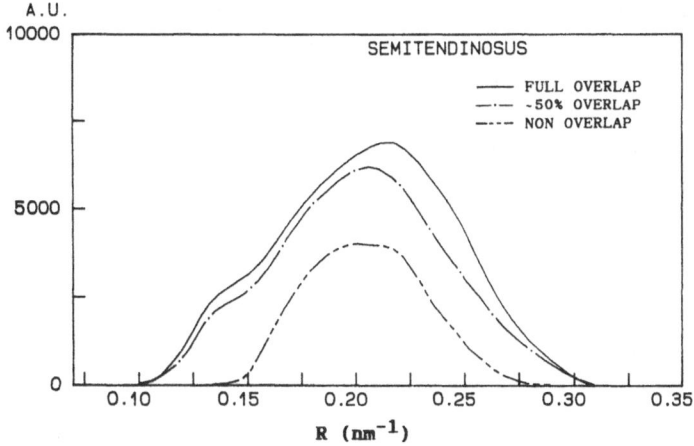

Fig. 6. Intensity distributions of the the second thin filament layer line at the axial spacing of ~1/18.5 nm⁻¹ during contraction of the full- and ~50%-overlap muscles and on activation of the overstretched muscle.

the full-overlap muscle, as suggested previously[3)4)]. Although the intensity increase in the overstetched muscle was slightly smaller than the value extrapolated to the zero filament overlap length of the muscle, the linear dependence of the intensity increase on the filament overlap agreed well with the result of Maeda et al.[14)] Thus the behavior of the intensity drop of the first thin filament layer line during contraction/activation is strongly correlated with that of the intensity rise of the second layer line. This means that Ca-activation alone cannot induce the full changes of the first and second thin filament layer-line intensities but they are caused by actomyosin interaction. The intensity change of the second layer line at a high radial spacing has been attributed with a tropomyosin movement[13)] and/or a conformational change in actin[3)4)9)]. The intensity change of the first layer line may have the same origin. The reciprocal intensity change of the first and second layer lines indicates that four-fold rotational symmetry in the thin filament is strengthened on activation and becomes more pronounced during force generation, since the first layer line is contributed by the Bessel function J_2 and the second layer line by J_4. Although the expected intensity reduction of the inner part of the first layer line due to an enhancement in the four-fold rotational symmetry was not always observed, it never showed an increase. Since this behavior was the same in both overlapped and oversretched muscles, the lack of an increase in the first layer line in this region shows that there is no increase in intensity caused by attachment of myosin heads. However, the data from the inner region of the first layer line should be treated with caution. Firstly, the first troponin layer line is present near the meridian at an axial spacing of 1/38 nm⁻¹. This layer line component was not removed from the first layer line in the present analysis. The intensity change in the inner region of the overlap muscle might be masked by complicated changes of the troponin layer line[15)]. Secondly, the residual sampling effect might affect the intensity change in this region. Main part of the inner hump of the first layer line in Fig. 5 may be due to the troponin component or caused by the residual sampling effect. Whatever the cause, the lack of any appreciable intensity increase in the inner region of the first layer line during contraction indicates that there is very little sign of the rigor-like specific binding of the myosin heads to the actin filaments in native muscles. This finding suggests strongly that interacting myosin heads assume a wide range of configuration to actin during contraction. The intensity change of the first layer line can be attributed with the same structural changes giving rise to the intensity change of the second layer line, revealing that structural changes within the thin filament also occur in the force generating process. Our present results contrast with the observations by rapid-freezing experiments of contracting muscles which showed the development of the first layer line[16-18]. Argument of this point should be referred to the paper of Huxley et al.[18]

REFERENCES

1. Amemiya, Y., Wakabayashi, K., Tanaka, H., Ueno, Y. & Miyahara, J. *Science* **237**, 164-168 (1987).
2. Wakabayashi, K., Ueno, Y., Amemiya, Y. & Tanaka, H. *Adv. Exp. Med. Biol.* **226**, 353-367 (1988).
3. Wakabayashi, K. & Amemiya, Y. in *Handbook on Synchrotron Radiation Vol. 4* (eds. Ebashi, S., Koch, M. & Rubenstein, E.) 597-678 (North-Holland, Amsterdam, 1991).
4. Wakabayashi, K., Tanaka, H., Saito, H., Moriwaki, N., Ueno, Y. & Amemiya, Y. *Adv. Biophys.* **27**, 3-13 (1991).
5. Huxley, H.E. & Brown, W. *J. Mol. Biol.* **30**, 383-434 (1967).
6. Amemiya, Y., Wakabayashi, K., Hamanaka, T., Wakabayashi, T., Matsushita, T. & Hashizume, H. *Nucl. Instrum. Methods* **208**, 471-477 (1983).
7. Amemiya, Y. & Wakabayashi, K. *Adv. Biophys.* **27**, 115-128 (1991).
8. Yagi, N. *Adv. Biophys.* **27**, 35-43 (1991).
9. Yagi, N. & Matsubara, I. *J. Mol. Biol.* **208**, 359-363 (1989).
10. Huxley, H.E., Faruqi, A.R., Kress, M., Bordas, J. & Koch, M.H.J. *J. Mol. Biol.* **158**, 637-684 (1982).
11. Bordas, J., Diakun, G.P., Harries, J.E., Lewis, R.A., Mant, G.R., Martin-Fernandez, M.L. & Towns-Andrews, E. *Adv. Biophys.* **27**, 15-33 (1991).
12. Yagi, N. & Matsubara, I. *Science* **207**, 307-308 (1980).
13. Kress. M., Huxley, H.E., Faruqi, A.R. & Hendrix, J. *J. Mol. Biol.* **188**, 325-342 (1986).
14. Maeda, Y., Popp, D. & McLaughlin, S.M. *Adv. Exp. Med. Biol.* **226**, 381-390 (1988).
15. Popp, D., Maeda, Y., Stewart, A.E.A. & Holmes, K.C. *Adv. Biophys.* **27**, 89-103 (1991).
16. Tsukita, S. & Yano, M. *Nature* **317**, 182-184 (1985).
17. Lenart, T.D., Allen, T.S., Barsotti, R.J., Ellis-Davies, G.C.R., Kaplan, J.H., Franzini-Armstrong, C. & Goldman, Y.E. This volume, 475-487.
18. Huxley, H.E., Kress, M., Faruqi, A.R. & Simmons, R.M. *Adv. Exp. Med. Biol.* **226**, 347-352 (1988).

EVIDENCE FOR STRUCTURAL CHANGES IN CROSSBRIDGES DURING FORCE GENERATION

B. Brenner and L.C. Yu*

Department of General Physiology
University of Ulm, Germany
*National Institutes of Health
Bethesda, MD USA

ABSTRACT

During muscle contraction, it is generally thought that myosin heads undergo large scale conformational changes, such as an oar-like rotation between 90° and 45° while attached to actin. However, evidence for conformational changes of the *attached* crossbridges associated with force generation has been ambiguous. In this study, we compared the conformations of attached crossbridges in (i) the pre-force generating state, (ii) force generating state, (ii) rigor state.

High resolution equatorial X-ray diffraction patterns have been obtained from single chemically skinned rabbit psoas fibers under relaxed, fully Ca^{2+}-activated and rigor conditions. The experimental condition was chosen (ionic strength = 50 mM and temperature = 5°C) such that there are large fractions (80 -100%) of crossbridges attached in all the three states, and the attached crossbridges in the relaxed muscle represent the pre-force generating state.

Upon activation, changes in the two innermost intensities I_{10} and I_{11} did not follow the familiar reciprocal changes. Instead, there was almost no change in I_{11} while I_{10} decreased by 30%. Similarly, greater changes were found in I_{10} as the fiber goes into rigor from the activated state. Changes were also found in the higher order reflections suggesting that the structure of the force generating crossbridges is not a mixture of those found in the weakly bound and rigor crossbridges. Therefore, our data provides evidence that the average conformation of the force generating crossbridges is different from the weakly attached and from rigor crossbridges.

INTRODUCTION

It is generally accepted that force generation in muscle is a result of the cyclic interaction between myosin (crossbridge) and actin while ATP is hydrolyzed. The structural model of crossbridge action was proposed over twenty years ago[1-3]). To generate force the crossbridge, after it attaches to actin, goes through a large scale conformational change such as an oar-like rotation of the myosin head from an orientation perpendicular to the filament axis (90°) to the one tilted approximately 45° to the filament axis. This proposal was based on an earlier electron microscopic observation from insect flight muscle by Reedy, Holmes and Tregear[4]. In that study, it was shown that the orientation of the *detached* crossbridge in a relaxed muscle was perpendicular to the fiber axis and the attached crossbridge in rigor was orientated at 45°. Thus far, however, direct evidence for conformational change in the *attached* myosin heads associated with force generation has not been conclusive. It should be emphasized that it is the *attached* myosin heads that are of interest. Ideally, one would like to compare the conformations of the attached crossbridge before it generates force, during force production and at the end of the crossbridge cycle.

In this report, evidence is presented suggesting that there are multiple conformations of crossbridges attached to actin during the process of force generation.

METHODS AND MATERIAL

Equatorial X-ray Diffraction

Patterns were obtained from single, chemically skinned rabbit psoas muscle fibers. X-ray source was the synchrotron radiation at DESY (Deutsches Elektronen Synchrotron), Hamburg, FRG. Beamline X33 of European Molecular Biology Laboratory was used.

Fiber Preparation

Skinned fibers were prepared according to methods described earlier[5][6]. In the past, ionic strengths of the skinning solution and subsequent bathing media were adjusted by adding K propionate or KCl[5][6]. However, we have since found that lattice order of the skinned fibers is better preserved by using phosphocreatine, naturally occurring in muscle in 20 - 40 mM concentration, as the anion instead. For example, Fig. 1 shows two diffraction patterns obtained from the same fiber at the same ionic strength of 120 mM, except that the anions differed. All the reflections (recorded up to [3,0] in this study) in phosphocreatine containing solution were stronger and slightly sharper than those obtained in K propionate containing solution. Therefore, throughout this study, phosphocreatine was used to adjust the ionic strength in solutions containing ATP. Skinning solution contained (mM): 3 ATP, 3 Mg Acetate, 5 KH_2PO_4, 5 EGTA, 33 PCr, pH = 7.0, μ (ionic strength) = 145 mM; relaxing solution: 10 Imidazole, 1 EGTA, 1 MgATP, 2 $MgCl_2$, 10 PCr, pH = 7.0, μ = 50 mM; rigor solution: 10 Imidazole, 2.5 EGTA, 2.5 EDTA, 30 KProp, pH = 7.0, μ = 50 mM.

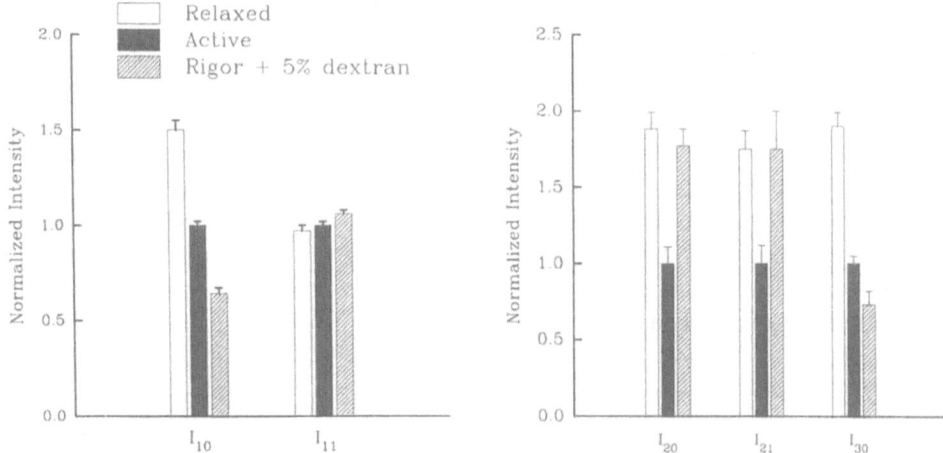

Fig. 1. Summary of equatorial intensities from single skinned rabbit psoas fibers under (i) relaxed ($\square$), (ii) fully Ca^{2+}-activated ($\blacksquare$), (iii) rigor + 5% dextran T_{500} ($\boxtimes$). Dextran T_{500} (Pharmacia Fine Chemicals Inc., Uppsala, Sweden) was added to rigor solution such that the lattice spacing d_{10} remained constant to within 5 Å at 375 Å under all three conditions. Diffraction patterns from each fiber were recorded first under relaxing condition, followed by Ca^{2+}- activation. Another relaxed pattern was recorded afterwards. Diffraction patterns under rigor condition were taken last. Error bars are standard errors of the mean. Intensities of each reflection under relaxing and rigor conditions are normalized to those under Ca^{++}-activation condition.

Experimental Conditions

Equatorial diffraction patterns from single skinned fibers under (i) relaxing (ii) fully Ca^{2+}-activated and (iii) rigor conditions were recorded. The key element of the experimental conditions was that the experiments were performed at ionic strength of 50 mM and 5°C.

Rationale of the Experimental Design

The goal of this study is to compare attached crossbridges in the (a) preforce generating states; (b) force generating state; (c) rigor state. For studying (b), the muscle fiber was fully activated by Ca^{2+}; for (c), ATP was absent in the bathing solution. Attached crossbridges in the preforce generating state (a) were represented by those crossbridges attached to actin in the relaxed fiber. It was shown that crossbridges in muscle can attach to actin with low affinity (weak binding states) under relaxing conditions[7)8)]. Recently, we showed that for crossbridges to generate force, weak attachment to actin must precede force generation. If we first inhibited the weak crossbridge attachment to actin under relaxing conditions, subsequent force production by Ca^{2+} activation was also inhibited[9)10)]. Therefore the weakly bound states are essential intermediates to force generation in the crossbridge cycle.

Another crucial advantage of using low ionic strength and low temperature is that the fractions of crossbridges attached to actin are large and kept approximately constant in the three states. In a relaxed fiber under such conditions, 80% - 90% of the crossbridges are estimated to be attached to actin[6)7)11-15]. During Ca^{2+}-activated contraction, the affinity of crossbridge for actin is even stronger, and hence nearly all of the crossbridges are attached. In rigor, it has been shown that all of the crossbridges are attached to actin[16)17]. Holding the fraction of attached crossbridges constant is critical since intensities of the equatorial reflections are known to be affected not only by the conformation of attached crossbridges but also by the number of attached crossbridges[18)19]. By limiting the changes in the fraction of attached crossbridges to within 10-20%, ambiguity in interpreting the results is greatly reduced.

RESULTS

Under relaxing, fully activated and rigor conditions, the single skinned fibers show well defined reflection peaks [1,0], [1,1], [2,0], [2,1], and [3,0] (number of recorded peaks were limited by the length of the detector). The data is summarized in Fig. 2. Upon activation from relaxing conditions, I_{10} decreases by 33% and it decreases a further 35% in going into rigor. During Ca^{2+}-activation I_{20} and I_{21} are weaker than during either relaxation or rigor while I_{30} has a value intermediate of those two conditions. However, the most remarkable feature of the result is that I_{11} is almost constant under all three conditions.

Some qualitative conclusions may be made based on the reflection intensities alone: (A) The conformation of the attached force-generating crossbridges is different from that of the weakly bound crossbridges. If the fraction of crossbridges attached to actin remains unchanged between relaxing and activating conditions, the differences in the intensities (Fig. 2) clearly originate from differences in the attachment conformation. Even if a small increase in the fraction occurs upon activation, the conclusion is the same, since it was shown earlier that an increase in the fraction of weakly attached crossbridge caused an increase in I_{11} and proportionately less decrease in I_{10}. However, in this study we find that the magnitude of decrease in I_{10} is greater than the increase in I_{11}. In addition, changes in the higher order reflections beyond [1,1] induced by force generating crossbridges are different from those induced by attachment of weak binding crossbridges. Therefore, we conclude that there is a structural change in the attached crossbridge in transition from relaxing to Ca^{2+}-activating conditions. (B) A further conformational change occurs in the transition from the Ca^{2+}-activation to rigor condition. The reasoning for this conclusion is similar to that used in (A): the fraction of crossbridge bound is approximately the same and yet the diffraction patterns are distinct from each other.

X-ray diffraction intensities reflect structures averaged over the entire population of sarcomeres exposed to radiation. It is possible that the force generating crossbridges do not have a unique conformation, but the intensities

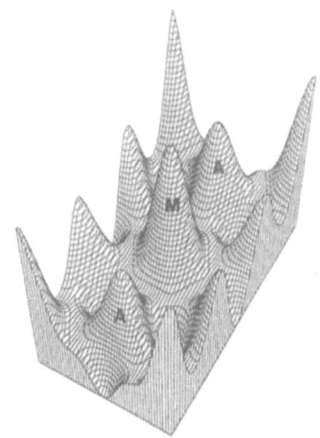

Fig. 2. Electron density maps of muscle unit cells in axial projection based on Fourier synthesis of the reflection intensities ([1,0], [1,1], [2,0], [2,1], [3,0]) obtained under fully Ca^{2+}-activated conditions. The phase set (0°, 0°,π, 0°, 0°) was used for the reconstruction. "M" = thick filament, "A" = thin filament. The six-fold symmetry shown in the density map could be due to effects caused by series termination with limited (5) terms[19].

observed during Ca^{2+}-activation is the result of a mixture of structures of weakly attahced and rigor crossbridges. This is unlikely. Recently it was shown that the attached crossbridges exhibited distinct elasticities in the radial direction during full Ca^{2+}-activation and in rigor, while radial elasticity in the relaxed fiber was not detectable[20]. The diverse radial elasticities most likely result from different molecular structures of the crossbridges or different attachment mode, as suggested by reports at this Symposium (Katayama; Wakabayashi).

The apparent widths of each individual reflection is somewhat broader during contraction than those found in the relaxed state, indicating that lattice disorder is increased. It is possible that the observed changes in transition from the relaxed condition to the Ca^{2+}-activated condition are not caused by conformational changes but by an increase in lattice disorder. However, if we make corrections for the lattice disorder by normalizing I_{10} of the Ca^{2+}-activated state to be the same as the relaxed state, I_{11} would be higher than that for the rigor state and the other reflections would be inconsistent with those observed for the weakly attached crossbridges in the relaxed muscle[6]. Therefore, even if lattice disorder is taken into account, the results are not consistent with the possibility that crossbridge structure remains the same in transition from weak attachment to force generation.

Although we may conclude that the attached crossbridges have several distinct structures depending on the state of the muscle, the nature of the difference cannot yet be characterized unambiguously due to the lack of phase information. However, the phase set (0°,0°,π,0°,0°) appears to be the most likely one for the relaxed muscle[6][21][22]. With the assumption that there is no phase change upon activation

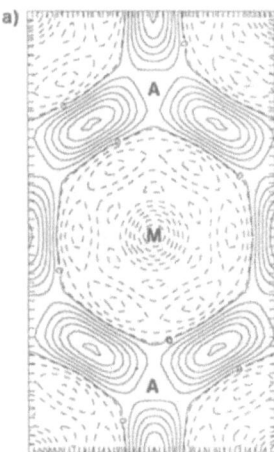 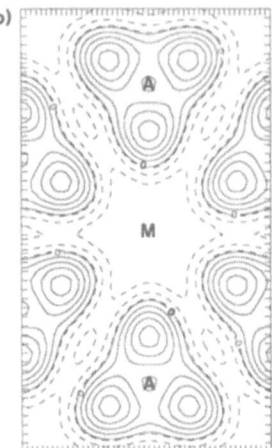

Fig. 3. Difference density maps between (a) activated and relaxing conditions, and (b) rigor and activated conditions. The phase set used is kept as the same as those used in Fig. 2. Solid lines indicate gain in mass whereas dashed lines indicate loss of mass. The contour line where no change occurs is marked by "0". The six-fold symmetry in the difference maps could also be due to series termination effects.

and going into rigor, one may construct density maps based on the experimental intensities for those three states. Fig. 3 shows the two dimensional density maps for the Ca^{2+}-activated muscle. Features of the density maps for the relaxed and rigor conditions are rather similar. Differences in mass distribution are detected more conveniently by studying the difference maps. Fig. 4a is a difference map of the active and the relaxed conditions; Fig. 4b is the difference between the rigor and active conditions. The transition from the weakly-attached states to the force-generating states appears to involve an outward radial shift of mass away from the thick filament backbone and there is little change in the immediate vicinity of the thin filament. It should be pointed out that such characteristics are not sensitive to which phase combination is used for the density map of the activated muscle (data not shown). This is consistent with recent modelling work which shows that I_{10} is very sensitive to crossbridge mass moving away from the surface of the thick filament, while I_{11} is much less sensitive[19]. The transition from active to rigor state, on the other hand, involves increase in mass mostly around the thin filament.

SUMMARY

In the process of producing force, there are at least three distinct conformations of the attached crossbridges: (i) the weakly bound, (ii) the force generating, strongly bound and (iii) the rigor conformations. In transition from the weakly bound to the force generating conformation, there is probably a slight shift of mass radially away from the surface of the thick filament backbone but little change occurs at the surface of the thin filament. The mass redistribution from active to rigor state appears to mostly involve mass increase onto the thin filament.

REFERENCES

1. Pringle, J.W.S. *Prog. Biophys. Biophys. Chem.* **17**, 1-60 (1967).
2. Huxley, H.E. *Science* **164**, 1356-1366 (1969).
3. Huxley, A.F. & Simmons, R.M. *Nature* **233**, 533-538 (1971).
4. Reedy, M.K., Holmes, K.C. & Tregear, R.T. *Nature* **207**, 1276-1280 (1965).
5. Brenner, B. *Biophys. J.* **41**, 99-102 (1983).
6. Yu, L.C. & Brenner, B. *Biophys. J.* **55**, 441-453 (1989).
7. Brenner, B., Schoenberg, M., Chalovich, J.M., Greene, L.E. & Eisenberg, E. *Proc. Natl. Acad. Sci. USA* **79**, 7288-7291 (1982).
8. Brenner, B., Yu, L.C. & Podolsky, R.J. *Biophys. J.* **46**, 299-306 (1984).
9. Brenner, B., Yu, L.C. & Chalovich, J.M. *Proc. Natl. Acad. Sci. USA* **88**, 5739-5743 (1991).
10. Kraft, Th., Chalovich, J.M., Yu, L.C. & Brenner, B. *Biophys. J.* **59**, 375a (1991).
11. Chalovich, J.M., Chock, P.B. & Eisenberg, E. *J. Biol. Chem.* **256**, 575-578 (1981).
12. Chalovich, J.M. & Eisenberg, E. *J. Biol. Chem* **257**, 2431-2437 (1982).
13. Brenner, B., Chalovich, J.M., Greene, L.E., Eisenberg, E. & Schoenberg, M. *Biophys. J.* **50**, 685-691 (1986).
14. Brenner, B., Yu, L.C., Greene, L.E., Eisenberg, E. & Schoenberg, M. *Biophys. J.* **50**, 1101-1108 (1986).
15. Schoenberg, M. *Biophys. J.* **54**, 135 (1988).
16. Lovell, S.J. & Harrington, W.F. *J. Mol. Biol.* **149**, 659-674 (1981).
17. Cooke, R. & Franks, K. *Biochemistry* **19**, 2265-2269 (1980).
18. Lymn, R.W. *Biophys. J.* **21**, 93-98 (1978).
19. Yu, L C. *Biophys. J.* **55**, 433-440 (1989).
20. Brenner, B. & Yu, L.C. *J. Physiol. (Lond.)* **441**, 703-718 (1991).
21. Yu, L C., Steven, A.C., Naylor, G.R.S., Gamble, R.C. & Podolsky, R.J. *Biophys. J.* **47**, 311-321 (1985).
22. Harford, J.J. & Squire, J.M. in *Molecular mechanisms in musclular contraction* (ed Squire, J.M.) 287-320 (Macmillan Press, London, 1989).

Discussion

Kawai: Do you equate the pre-force state with the weakly bound state? Is that the conclusion?

Yu: This is not a conclusion from this data. We used the findings from the recent publication by Brenner et al. (*Proc. Natl. Acad. Sci USA* **88**, 5739-5743, 1991). Pre-force generation is the attached state in the weakly binding configuration.

Kawai: So they are the same?

Yu: Yes.

Gillis: So you do confirm Dr. Wakabayashi's earlier statement that no rigor-like cross-bridges exist during tetanus?

Yu: I can't exclude that, but they are different from rigor, yes. The average conformation of the tetanus state is different from the rigor state.

Gillis: So the messages from the two papers are coherent.

Yu: I have to really digest his talk. But that is my conclusion, yes.

Katayama: What appears if you make the difference map of rigor minus active?

Yu: From active to rigor the mass moves closer to actin.

Tregear: How far can you tell that what you are seeing is a different form of the individual cross-bridge? How far could it be a distribution onto different actin monomers of the same form of cross-bridge?

Yu: We always look at average conformations. There certainly could be a distribution of conformation in that state. But the average of the pre-force-generating state is different from the active state.

Tregear: So, just to make it clear on that point, you are saying that you could have the same sort of cross-bridges, but attached to different actin monomers, and that could give you the change that you see?

Yu: No. On the equator, it is a projected view of mass distribution. You move along the actin, you cannot tell the difference between the top one and the bottom one, but the conformation on these actins should be different.

Gergely: Can you say from your data that in the active state, the myosin cross-bridges remain indexed on the myosin thick filament rather than targeting the actin?

Yu: I cannot say, because this is on the equator again. Active is different from the weak-binding states we have seen in the low ionic strength relaxed muscle. As you know, Richard Podolsky and Matsuda (*Proc. Natl. Acad. Sci. USA* **81**, 2364-2368, 1984) made a study on the low ionic strength, low salt, relaxed muscle where the myosin layer line is still present even though we know there is attachment. For the active state I do not know, but I think in the relaxed state the center of mass is still centered around the myosin. In the active state, what I can say is it has moved out a little bit. But whether it still remains indexed on the myosin, I cannot say.

Holmes: Leepo, Terrell Hill says—I think this must be right—to get force generation you must have a force-generating state. You can't say going from weak to strong generates force; you've got to have a force-generating state, but could that force-generating state not be bent, distorted rigor?

Yu: Clearly, it has to be distorted to gain some strength, but what I'm saying is this strain is different from the rigor state.

Holmes: Yes, but it's got to be different, it's got to be strained.

Brenner: Ken, if your statement were correct, we should be able to obtain equatorial information when stretching rigor fibers that match the intensities observed during isometric contraction. We have tried to impose strain on rigor fibers to get as much tension as during isometric contraction. Applying this much strain, we see significant change on the equator compared to the unstrained rigor. So it seems to be a bit unlikely—it doesn't rule it out—but it seems unlikely that the structure we see in activated fibers is just a strained rigor structure.

Maéda: I'm curious, according to your data, the fully activated states didn't give rise to a (1,1) reflection that was very intensified compared to resting, although you said the muscle fibers produced a lot of tension. On the other hand, in frog—at least living frog muscle—there is an enormous change of intensity, not only (1,0) but also (1,1). Do you think this is due to a species difference?

Yu: The evidence shows that there are very few cross-bridges attached in the relaxed frog muscle. So, on activation you have a five or tenfold increase of the number of attached bridges. That fact alone could change the $I(1,0)/I(1,1)$ by a lot. I have evidence that shows skinned rabbit psoas muscle fibers activated at 170 mM ionic strength at the same termperature, 5°C. You see the familiar reciprocal change in (1,0/1,1). Here there is simultaneous change of the number and maybe conformation. In this study we have largely avoided the ambiguity arising from changing number and configuration simultaneously.

Squire: Perhaps I could just comment that the fish results show the same kind of thing—the (1,1) changes extremely fast, and its change is complete when the tension is only about 50%; then the (1,0) changes afterwards.

Yagi: Maybe the difference is due to temperature. In rabbit fibers, when the temperature is lowered from 20°C to 5°C, we lose all myosin layer lines, and the (1,1) reflection is very intensified. So I would presume that most of the heads are spread out from the thick filament at normal ionic strength in rabbit, but not in frog. So the starting situation is different. Is that correct?

Yu: That could be, but we have looked at the rabbit psoas fiber at low temperature, 5°C, but different ionic strength, 170 mM. You see the familiar reciprocal change (1,0/1,1), and we know, in that case, the number of cross-bridges changes by five or tenfold. So there you have a case where the number and conformation changes are occurring at the same time.

Kushmerick: Does this mean that if you go to higher ionic strength there is a second conformation of the weakly attached states, or is there yet a fourth conformation that is completely detached?

Yu: No, I'm just saying that the reciprocal change we see at high ionic strength is mainly due to a change in the number attached. That's all.

VII. KINETIC PROPERTIES OF ACTIN-MYOSIN SLIDING IN MUSCLE STUDIED BY FLASH PHOTOLYSIS OF CAGED SUBSTANCES AND TEMPERATURE JUMP

INTRODUCTION

Recent development of the technique of rapid flash photolysis of caged substances, i.e. biologically inert and photolabile precursors of substances involved in functionally important reactions in cells and tissues, has provided a powerful tool to obtain information about the relation between the steps of biochemical reaction to the cell and tissue function, The use of the caged substances, such as caged-Ca^{2+}, caged ATP, caged Pi, is, especially useful for studying the direct correlation between the steps of actomyosin ATPase reaction and the characteristics of contracting muscle.

Lenart and others activated skinned frog muscle fibers by photolysis of caged Ca^{2+}, and the fibers were rapidly frozen at various times after the flash activation for electron microscopic observation. Optical diffraction patterns of electron micrographs of activated fibers were consistent with the results of time-resolved X-ray diffraction of contracting muscle, while the shape of cross-bridges was highly variable. Yamada and others investigated the force decay in the EDC-crosslinked rabbit psoas fibers in rigor state following photolysis of caged ATP, and observed that the rate of force decay was dependent on ionic strength and the amount of ATP released.

It is generally held that force generation in muscle is coupled with the release of Pi from AM•ADP•Pi cross-bridge. On this basis, Homsher and Millar studied the effect of rapid photolysis of caged Pi on the force in rabbit slow twitch and fast twitch muscle fibers. The release of Pi reduced steady isometric force, but was less effective in slow fibers than in fast fibers. The four phases of the force response to released Pi was analyzed. Yamada and others almost free-loaded shortening of glycerinated rabbit psoas fibers following photolysis of caged ATP, attention being focused on the distance of shortening when the amount of release ATP was around the total myosin head concentration within the fiber (150 µM). Uniform fiber shortening reached a minimum (about 10 nm/half sarcomere) when the amount of released ATP was reduced to 75 µM, i.e. one ATP molecule per one myosin molecule, suggesting that the distance of a myosin head powerstroke per hydrolysis of one ATP molecule is about 10 nm.

Finally, Davis and Harrington examined the temperature dependence of the isometric force transients following quick releases (L-jump) in rabbit psoas fibers, the sub divided the phase 2 of the transients into fast and slow phases, arising from perturbation of the elastic element of a cross-bridge and transition of the endothermic order-disorder transition responsible for force generation. They also used the technique of laser-temperature jump to study force generation and the isometric force transients in rigor fibers following the temperature-jump.

MECHANICS AND STRUCTURE OF CROSS-BRIDGES DURING CONTRACTIONS INITIATED BY PHOTOLYSIS OF CAGED Ca2+

Thomas D. Lenart, Taylor StClaire Allen, Robert J. Barsotti, Graham C. R. Ellis-Davies, Jack H. Kaplan, Clara Franzini-Armstrong, and Yale E. Goldman

University of Pennsylvania
Philadelphia, PA. 19104, USA

ABSTRACT

Cross-bridge structure and mechanics were studied during development of skinned frog muscle fiber contractions initiated by photolysis of DM-nitrophen (a caged Ca^{2+}). Stiffness rises earlier than tension following photo-release of Ca^{2+}. A similar lead of stiffness in electrically stimulated fibers and the early rise of the I_{11}/I_{10} ratio of equatorial X-ray reflections are thought to signal attachment of cross-bridges into states with lower force than in steady-state contraction. We investigated the structure of the early attachments by electron microscopy of fibers activated by photolysis of DM-nitrophen and then ultra-rapidly frozen and freeze substituted with tannic acid and OsO_4. Sections from relaxed fibers show helical tracks of myosin heads on the thick filaments surface. Optical diffraction patterns show strong meridional intensities and layer lines up to the 6th order of 1/43 nm, indicating preservation and resolution of periodic structures smaller than 10 nm. Following photo-release of Ca^{2+}, the 1/43 nm myosin layer line becomes less intense, and higher orders disappear. A ~1/36 nm layer line appears early (12-15 ms) and becomes stronger at later times. The 1/14.3 nm meridional spot weakens initially and recovers at a later time, while it broadens laterally. The 1/43 nm meridional spot is present during contraction, but the 2nd order meridional spot (1/21.5 nm) is weak or absent. These results are consistent with time resolved X-ray diffraction data on the periodic structures within the fiber. In sections along the 1,1 plane of activated fibers, the individual cross-bridges have a wide range of shapes and angles, perpendicular to the fiber axis or pointing toward or away from the Z-line. Fibers frozen at 13 ms, 33 ms, and 220 ms after photolysis all show surprisingly similar cross-bridges. Thus, a highly variable distribution of cross-bridge shapes and angles is established early in contraction.

Mechanism of Myofilament Sliding in Muscle Contraction, Edited by
H. Sugi and G.H. Pollack, Plenum Press, New York, 1993

475

INTRODUCTION

In order to solve the contraction mechanism, we must correlate the biochemical, mechanical and structural details of the cross-bridge cycle on the millisecond time scale. The transitions between cross-bridge attachment and generation of the sliding force are crucial events. Most of the cross-bridges are detached from actin in relaxed muscle and attach during activation, so investigation of the transient state immediately following activation might elucidate the mechano-chemical and structural changes during the force generating transition. Mechanical and structural studies using intact muscle have shown that the cross-bridges formed just after the onset of electrical stimulation are different from those during steady contraction. The stiffness of fibers rises substantially faster than tension[1-5], suggesting that the cross-bridges attach in low-force or zero-force states and then generate force a few milliseconds later. Both the ratio I_{11}/I_{10} of equatorial X-ray reflections[5-8] and the birefringence[9] increase towards their steady-state contraction values faster than tension, supporting the idea that the cross-bridges initially generate low or zero force on attachment.

With skinned fibers, the closest analog of the physiological switch between relaxation and contraction is the sudden liberation of Ca^{2+} by photolysis of caged Ca^{2+}. Here again, the I_{11}/I_{10} equatorial intensity ratio increases earlier than tension[10]. We measured the time course of tension and stiffness following photolysis of caged Ca^{2+} in skinned fibers of frog muscle and also found that the stiffness increase leads tension development.

A recently developed method to obtain structural data on the cross-bridges is to freeze muscles rapidly against a cryogenically cooled metal block and then to examine sections or fracture planes by electron microscopy. Several studies have indicated that the cross-bridges present during active contractions have a highly variable shape and a wide angular distribution[11][12]. We combined the ultra-rapid freezing technique with activation of fibers by caged Ca^{2+} photolysis to investigate the structural basis for the shift from the initial low-force attachment to the high-force steady-state configuration of cross-bridges. The samples were freeze-substituted, sectioned, and examined by electron microscopy. Optical diffraction patterns of the electron micrographs qualitatively showed many characteristics of the X-ray diffraction patterns recorded from intact, electrically stimulated frog muscle. The shapes and angles of individual cross-bridges are highly variable during contraction, as shown previously[11][12]. The present time-resolved study shows that this characteristic disorder is established very early during force development. Some of the data have been reported in preliminary form[13].

MECHANICAL EXPERIMENTS

The kinetics of the rise in tension and stiffness following photolysis of caged Ca^{2+} were measured for correlation with the electron microscopic observations. Frog sartorius or semitendinosus muscle fibers were skinned chemically[14] or

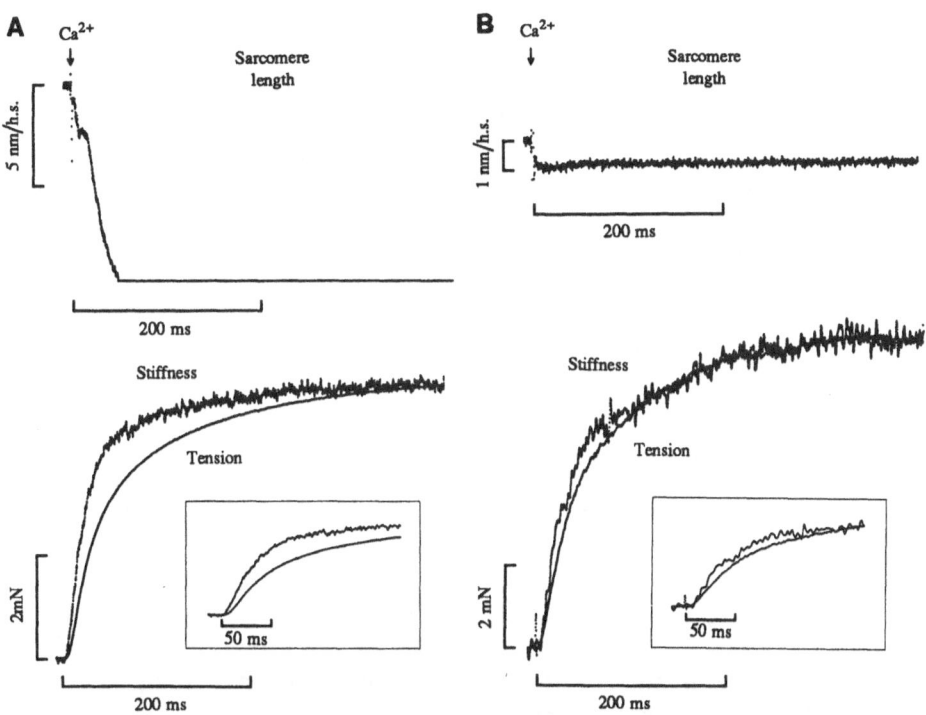

Fig. 1. Sarcomere length (upper traces), force and stiffness recordings (lower traces) from skinned single fibers from frog semitendinosus muscle during activation by laser photolysis of DM-nitrophen. The striation spacing was clamped isometric for the fiber in panel B, but not in panel A. The fibers were initially relaxed in a solution containing ~2 mM DM-nitrophen and 1.6 mM $CaCl_2$ total, giving a free Ca^{2+} concentration of 0.25 mM. The arrow indicates the time of the laser pulse which produced a free Ca^{2+} concentration of approximately 16 mM. The half-times for the development of stiffness and tension were 41 and 60 ms in A and 26 and 38 ms in B. The inset in each panel shows the stiffness and tension recordings on a more rapid time base.

mechanically[15]), and the ends were treated briefly with 5% glutaraldehyde to stiffen the attachment points[16]). Single fibers were attached between a tension transducer[15]) and a moving coil motor[17]), which applied 2 kHz sinusoidal length oscillations, 0.4% of fiber length, for measurement of stiffness. The amplitude of the resulting 2 kHz force oscillation was determined by a computer program that emulates a lock-in amplifier[18]).

A fiber was initially relaxed in the presence of EGTA and ATP, and then the EGTA was washed out and replaced with 1-(2-nitro-4,5-dimethoxyphenyl)-N,N,N',N'-tetrakis-[(oxycarbonyl)methyl]-1,2-ethanediamine[19]) (DM-nitrophen). The photolysis solution contained (total concentrations in mM) ~2 DM-nitrophen, 3 Na_2ATP, 1.2 $MgCl_2$, 1.6 $CaCl_2$, 100 TES buffer, 26 HDTA, 20 Na_2CP (creatine phosphate), 10 reduced glutathione and 20 g/l polyvinylpyrrolidone, pH 7.1, 10 °C.

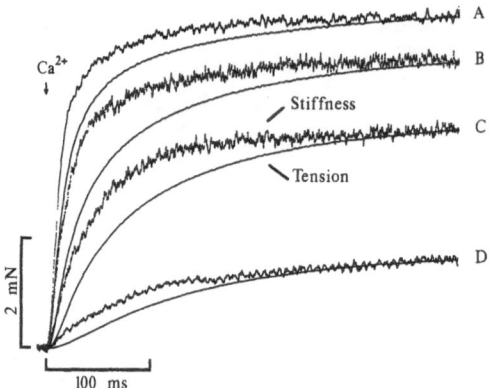

Fig. 2. Tension and stiffness recording from single skinned frog fiber which was activated at various free Ca^{2+} concentrations by altering the intensity of the laser pulse. The final pCa values were 4.5, 4.8, 5.3 and 5.6 for the recordings labeled A to D respectively. Note that both the rates of development and the final levels of stiffness and tension increased with free Ca^{2+} concentration.

The free Ca^{2+} concentration before photolysis in this solution is estimated to be 0.25 μM (pCa 6.6) from the binding constants of the constituents and fluorescence of the Ca^{2+} indicator fluo-3. The fibers was fully relaxed. Upon photolysis of ~30 % of the DM-nitrophen by a 50 ns, 400 mJ.cm^{-2} pulse from a frequency doubled ruby laser[18], the Ca^{2+} concentration increased to ~16 μM (pCa 4.8), fully activating the fiber. Stiffness and tension increased rapidly on photolysis (Fig. 1A) with stiffness leading tension by about 15 ms at the midpoint of the rise. The initial lag before tension begins to rise is much smaller, less than 2 ms. The mean half-times for stiffness and tension rise were 22.8 ± 1.6 ms and 38.2 ± 2.9 ms (S.E.M., n = 7).

Observations of the striation spacing by white light diffraction[20] showed that sarcomeres shortened considerably following activation by photoreleased Ca^{2+} (Fig. 1A). To eliminate the delay in tension development caused by internal shortening, we clamped the striation spacing by feedback to the moving coil motor (Fig. 1B). The tension rises faster in this situation but still later than stiffness. The stiffness record is more noisy in feedback control because noise on the striation spacing signal perturbs the feedback loop.

Striation clamping is not readily adaptable to the ultra-rapid freezing apparatus, so the electron microscopy experiments were not conducted in the length-clamped condition. For comparison, most of the fibers used for the mechanical measurements were therefore not length clamped. Since stiffness rises faster than tension, the ratio of tension to stiffness is lower during the rise of the contraction than at the full steady-state activation. At 15 and 35 ms, the tension to stiffness ratio is about 75% of that during the steady contraction.

At lower laser energies, activation is slower (Fig. 2), suggesting that the rates of the initial steps of the attachment and/or force generating mechanisms are $[Ca^{2+}]$-

dependent. The rate of Ca^{2+} release from DM-nitrophen after photolysis (half-time < 200 µs) does not limit the kinetics[21]). Stiffness rises faster than tension at all [Ca^{2+}] tested (pCa 5.6 - 4.5). The half-times of tension and stiffness decreased gradually as [Ca^{2+}] was increased, and we did not obtain evidence of saturation of the rate before saturation of steady tension. This result contrasts with that of Ashley and coworkers[22]), who reported that the rate of tension development saturates at Ca^{2+} concentrations producing only about half-maximal steady tension. The reason for this discrepancy is not clear. The the fiber type and conditions of the experiments are similar, although Ashley et al. used a different caged Ca^{2+}, nitr-5.

The mechanical experiments suggest that upon release of Ca^{2+} by photolysis of DM-nitrophen, cross-bridges rapidly attach to the thin filaments in a configuration with less force per attachment than in the steady state. The structural change in the attached crosss-bridges responsible for the subsequent increase in force presumably corresponds to the protein motion that leads to the generation of force.

ULTRA-RAPID FREEZING FOR ELECTRON MICROSCOPY

Skinned frog sartorius muscle fibers were activated by photolysis of DM-nitrophen and ultra-rapidly frozen by contact with a gold-plated copper block, cooled by liquid helium. The freezing apparatus was a modified version of the commercially available "cryo-press" of Heuser et al[23]). In this setup, the specimen holder falls under gravity and brings the sample into contact with a stationary metal block cooled by liquid helium. We mounted a tension transducer on the specimen holder to record muscle fiber tension up to the moment of freezing[24]). The muscle fiber was backed by a thin strip of agar. A frequency-doubled ruby laser was arranged to photolyze DM-nitrophen at pre-selected times from 12 to 220 ms before the fiber contacted the cold metal surface.

Frozen fibers were freeze-substituted at -80°C for two days with 0.1 - 0.5 % tannic acid in acetone, washed extensively with acetone, warmed to room temperature in the presence of 4% OsO$_4$ in acetone, stained en bloc with uranyl acetate and then embedded in araldite. Ultra-thin sections were cut and examined in a Phillips transmission electron microscope.

Fig. 3A shows an image of a frog sartorius fiber ultra-rapidly frozen while relaxed at 11°C. The prominent ~43 nm periodicity apparent in the A-band represents mainly C-protein and other accessory proteins superimposed on the helical structure of the thick filament. The optical diffraction pattern (Fig. 3C) of the image in panel A is dominated by 1/43 nm myosin layer lines due to the helical ordering of the myosin heads on the thick filaments and by sharply localized meridional intensities indexing on 1/43 nm spacing. Diffraction peaks are apparent up to the 5th or 6th order of the 1/43 nm diffraction peak indicating preservation of the periodic structures in the fiber to 7-8 nm resolution. The prominent diffraction peaks due to the thick filament backbone and myosin heads are also observed in X-ray diffraction patterns from intact, relaxed frog muscle. This similarity indicates

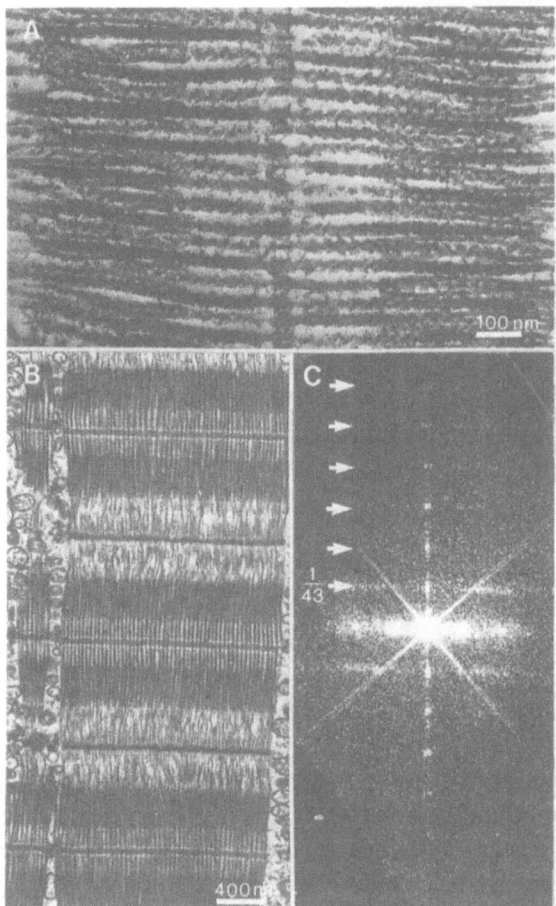

Fig. 3. Electron micrographs and an optical diffraction pattern from a frog sartorius fiber ultra-rapidly frozen in relaxing solution. The section in A was stained with 0.5% tannic acid during freeze-substitution. Oblique lines over the outlines of the thick filaments represent the myosin heads helically arranged with a 43 nm repeat. B, thicker section of the same sample. C, optical diffraction pattern of the image (the A-band of the central sarcomere) in panel A Note that the myosin meridional reflections extend out to six or seven orders of 43 nm.

that the images in the micrographs preserve the disposition of the cross-bridges present in live relaxed muscle. The helical tracks of the myosin heads on the surface of the thick filament are shown directly in a micrograph of a thinner section from a relaxed fiber (Fig. 3B) both in the overlap and non-overlap (H-zone) region of the thick filaments. Profiles at the edges of the thin filaments are smooth indicating that few cross-bridges are attached to the thin filaments. Images from fibers frozen 12 - 15 ms after photolysis (Fig. 4B) appear markedly different from micrographs of relaxed fibers (c.f. Fig. 3). Following release of Ca^{2+} the helical order of the thick filaments is reduced in both the overlap zone and H-zone. A large number of cross-

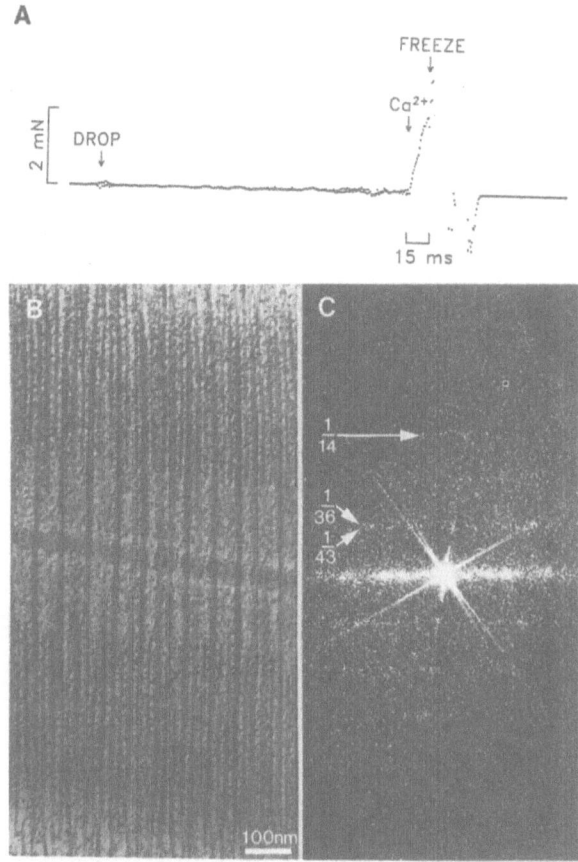

Fig. 4. Force recording (A), electron micrograph (B) and optical diffraction pattern (C) from an experiment in which a frog sartorius fiber was frozen early in contraction. The fiber was initially bathed in a solution containing 1.4 mM CaCl$_2$ and ~2 mM DM-nitrophen giving a free Ca^{2+} concentration of 0.12 μM. The plunger rod was released at the time indicated by "DROP". Approximately 200 ms later the laser pulse was triggered (indicated "Ca^{2+}") liberating 1.8 μM free Ca^{2+} by photolysis of DM-nitrophen. During the rising phase of the tension response, at 13 ms after the laser pulse, the fiber was ultra-rapidly frozen ("FREEZE"), followed by processing for electron microscopy.

bridges appear to be attached to the thin filaments with a wide range of shapes and angles. There is little indication of discrete classes of shapes.

The optical diffraction pattern from the image for Fig. 4B is shown in panel C. The diffraction pattern extends out only to the third order of 1/43 nm. The intensities of the meridional reflections relative to that of the 1/43 nm layer line are lower at 12 - 15 ms than in the pattern from relaxed muscle. The third order meridional intensity (1/14.3 nm) is spread laterally from the meridian. A layer line at 1/36 nm spacing is visible in the diffraction pattern and is separated from the 1/43

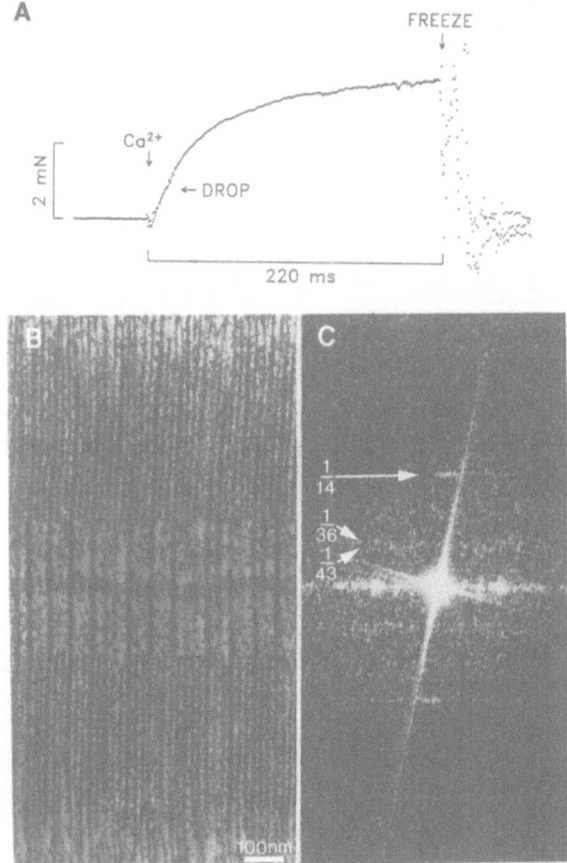

Fig. 5. Force recording (A), electron micrograph (B) and optical diffraction pattern (C) from a fiber frozen during steady isometric contraction . The fiber was initially relaxed at 1.2 mM $CaCl_2$ and ~2 mM DM-nitrophen, free Ca^{2+}, 0.05 μM. The laser was triggered at the time point indicated by "Ca^{2+}", releasing 0.6 μM free Ca^{2+} to activate the fiber. At "DROP" the plunger rod was released and at "FREEZE", the fiber contacted the cold metal block. The time between the laser pulse and the freezing was 220 ms. In C, note the distinct layer lines at spacings of 1/14.3 nm and 1/36 nm and the lateral spread of the 1/14.3 nm meridional spot.

nm layer line by a dark region. The 1/36 nm layer line is assigned to the cross-over repeat of the long-pitch actin helix.

Tension recordings made in the freezing experiments (Figs. 4A and 5A) were similar to those in the mechanical experiments without sarcomere length clamping. The half-time for tension development following photolysis of DM-nitrophen was 37.0 ± 3.2 ms at a temperature of 11°C measured at the agar surface. Tension reached the steady level approximately 200 ms after the laser pulse.

Images from fibers frozen 220 ms after photolysis have the most apparent cross-bridge attachments (Fig. 5). The shapes and angles of the cross-bridges at 220 ms are similar to those at 12 - 15 ms. The optical diffraction pattern shows the first myosin (1/43 nm) and first actin (1/36 nm) layer lines clearly separated by a dark region. Both the 1/21.5 nm and the 1/43 nm meridional intensities are weak or absent. Compared to the diffraction pattern at 12 - 15 ms, the pattern at 220 ms shows a stronger 1/14.3 nm meridional intensity and this spot is spread laterally ~1/169 nm, approximately 4-fold wider than the 1/14.3 nm spot at 12-15 ms. The 1/36 nm layer line is more intense at 220 ms and the gap between the 1/36 nm and 1/43 nm layer lines is more distinct than at 12 - 15 ms. These changes in the optical diffraction pattern occur even though the individual cross-bridges look alike at the early and late time points.

DISCUSSION

The early increase of 2 kHz stiffness following release of Ca^{2+} into the filament lattice (Figs. 1 & 2) confirms the results of several studies on intact muscles and muscle fibers[1-5]. The ratio I_{11}/I_{10} of the intensities of equatorial X-ray reflections also rises ahead of tension in intact[5-8] and skinned[9] fibers. Stiffness and the I_{11}/I_{10} ratio probably both provide qualitative indications of cross-bridge attachment to the thin filaments, but neither signal provides a quantitative estimate of attachment. Despite this reservation, the lag of tension behind stiffness and I_{11}/I_{10} at the onset of Ca^{2+} activation suggests that the tension per attached cross-bridge is lower during the rising phase of the contraction than during the steady state.

What structural features in the cross-bridges correlate with the increase in force of attached cross-bridges? The present electron micrographs help address this question, but first it is important to consider whether the images represent the native structure with enough fidelity to obtain the correct answer. Distortions of the structure on freezing, during freeze substitution and preparation of the sections, and in the electron microscope are important problems. An approach toward testing the image fidelity is to compare the optical diffraction (OD) patterns of the electron micrographs with X-ray diffraction patterns of intact muscles. Since the X-ray diffraction patterns are obtained directly from intact, physiologically contracting muscles, the reflections represent the periodic aspects of the native structures. Even with ideal preservation and staining, OD patterns of electron micrographs are not expected to correspond exactly to X-ray diffraction patterns for several reasons. Only one azimuthal orientation of the myofibril is contained in the micrograph, whereas the X-ray pattern samples many myofibrils with virtually random azimuthal placements in the muscle fiber. The diffraction intensities in the OD pattern are very sensitive to the axial orientation of the electron micrograph, because the OD represents a narrow slice in reciprocal space through the three-dimensional Fourier transform of the structure. There is also the difficulty of selecting appropriate regions in the micrographs for analysis. In the present work we have taken the initial approach of selecting a particular azimuthal orientation (parallel to the 1,1 planes of the filament lattice) and have chosen as the closest representation of the native structure typical regions of the micrographs that have sufficient order to produce clear intensities on the OD patterns. Quantitative analysis would require more views and objective statistics. Spacings of the diffraction spots are calibrated in reference to the third order (1/14.3 - 1/14.5 nm) meridional intensity to correct for shrinkage and compression of the structures during preparation and sectioning.

Many qualitative features of the OD patterns of the micrographs correspond to the results of X-ray diffraction studies from relaxed and physiologically contracting frog muscle. The OD patterns from relaxed muscle are dominated by meridional peaks and layer lines from the thick filaments. The strong 1/43 nm layer line represents the helical ordering of myosin heads on the surface of the thick filaments. In the OD patterns from muscles frozen 12 - 15 ms following activation by photolysis of DM-nitrophen, the meridional spots and layer lines are more diffuse and the first three meridional intensities are less intense relative to the 1/43 nm layer line. The 1/14.3 nm spot is broadened across the meridian. Between the 12 - 15 ms and 220 ms time points the intensity of the 1/14.3 nm meridional spot recovers and spreads laterally. In contrast, the first two meridional peaks (1/43 nm and 1/21.5 nm) become less intense or disappear. The layer line at 1/43 nm remains intense at 220 ms.

The 1/14.3 nm meridional X-ray reflection becomes weak shortly after electrical activation, but then recovers and broadens laterally[26)27)] as in our OD patterns. The 1/43 nm and 1/21.5 nm X-ray meridional reflections become weak and remain so during contraction. The 1/43 nm layer line is strong during contraction in both X-ray and OD patterns. This layer line is absent in X-ray patterns of muscles in rigor[25)] and also in OD patterns we have obtained from fibers frozen in rigor (data not shown). Thus the most prominent changes of the low order meridional and layer line intensities observed by X-ray diffraction are preserved in the OD patterns of our micrographs. The maintenance of the 1/14.3 nm meridional and 1/43 nm layer line intensities during contraction indicates that the cross-bridges can attach to the thin filaments while retaining the periodic spacing determined by the thick filament structure. Whether this periodicity of the cross-bridges extends all the way across the interfilament space to the surface of the thin filaments is not known.

An additional prominent feature of the present OD patterns obtained during contraction is a layer line at 1/36 nm. This spacing corresponds to the cross-over repeat of the long pitch helix of actin. The 1/36 nm layer line is much more intense in OD and X-ray patterns from rigor muscles and is attributed to cross-bridges adopting the helical periodicity of the actin filaments when they attach[25)28)]. Whether this layer line is more intense in X-ray diffraction patterns of active muscle relative to the relaxed pattern is difficult to determine because it overlaps with the 1/43 nm layer line[26)29)], but Yagi[30)] reported that it does become stronger near the meridian during contraction. In the OD patterns from our fibers activated using DM-nitrophen, a dark region separates the 1/36 nm and 1/43 nm layer lines, so the increased intensity in the 1/36 nm region is clear. Tsukita and Yano[11)] also reported that this layer line is present in OD patterns obtained from electron micrographs of rabbit psoas fibers frozen during contraction. In their case, the 1/36 nm layer line was more intense than the 1/43 nm layer line, but in the present OD patterns the 1/43 nm layer line is more intense.

These comparisons between the OD patterns of the electron micrographs and X-ray diffraction data help to validate the preservation and resolution of the native structures in the electron micrographs. Although time resolution is more difficult, and potential distortion or disruption of the structure is a problem with electron microscopy, if the structural preservation is sufficient, electron microscopy has several major advantages. The phases of the diffraction peaks can be determined, so the periodic parts of the structure can be reconstructed unambiguously from selected regions of the OD. Which regions of the image produce various aspects of the OD pattern can be determined by computing the OD from selected regions of the image. The individual elements can be viewed, so that non-periodic aspects of the structure can be examined without averaging.

The variability of cross-bridge shapes and angles observed in micrographs of fibers frozen 220 ms following photolysis of DM-nitrophen is similar to that reported by Hirose and Wakabayashi[12] in rabbit psoas fibers rapidly frozen during active contraction. In their electron micrographs the cross-bridges showed a wide range of angles and shapes. The distribution of cross-bridge mass across the space between the thick and thin filaments was rather uniform compared to the rigor situation where more of the mass of the attached cross-bridge was shifted toward the thin filament. The thick cross-bridge density near the thin filament in rigor images is probably due to firm attachment of both heads of the myosin molecules to actin which has been observed both *in vitro*[31] and in the intact filament lattice[32-35]. In agreement with the data of Hirose & Wakabayashi[12], the cross-bridge mass in our micrographs of actively contracting muscle is more uniformly distributed across the interfilament space than in rigor.

The individual cross-bridges in micrographs obtained from muscles frozen 12 - 15 ms after photolysis are remarkably similar to those at 220 ms. At 12 - 15 ms the cross-bridges are attached with many shapes and angles and the mass distribution appears fairly uniform across the interfilament space. Although the diffraction patterns at the two time points differ markedly, there is no conspicuous change in the shape or angle distribution of the individual cross-bridges. The variability of attachment configuration present during the steady contraction is thus established very early after Ca^{2+} release. Tension is only about 1/4 and stiffness 1/3 of their maximum values at the 12 - 15 ms time point. The structural change in the cross-bridges that leads to the subsequent increase in force is evidently more subtle than a major shift in the angular distribution. We are applying image processing methods to determine what specific aspects of the structure are associated with the changes in the OD patterns and whether the individual cross-bridges change shape when tension increases after attachment.

ACKNOWLEDGEMENTS

This work was supported by the Muscular Dystrophy Associations of America and by NIH grant HL15835 to the Pennsylvania Muscle Institute. We thank Nosta

Glaser and Xinhui Sun for technical help, Marcus Bell for electronic and software support and Joseph Pili for skillful construction of the mechanical apparatus.

REFERENCES

1. Hill, A.V. *Proc. R. Soc. B* **137**, 320-329 (1950).
2. Cecchi, G., Griffiths, P.J. & Taylor, S.R. *Science* **217**, 70-72 (1982).
3. Ford, L.E., Huxley, A.F. & Simmons, R.M. *J. Physiol. (Lond.)* **372**, 595-609 (1986).
4. Cecchi, G., Colomo, F., Lombardi, V. & Piazzesi, G. *Pflügers Arch.* **409**, 39-46 (1987).
5. Cecchi, G., Griffiths, P.J., Bagni, M.A., Ashley, C.C. & Maéda, Y. *Biophys. J.* **59**, 1273-1283 (1991).
6. Huxley, H.E. *Acta Anat. Nippon.* **50**, 310-325 (1975).
7. Huxley, H.E. in *Cross-Bridge Mechanism in Muscle Contraction* (eds. Sugi, H. & Pollack, G.H.) 391-405 (1979).
8. Matsubara, I. & Yagi, N. *J. Physiol. (Lond.)* **278**, 297-307 (1978).
9. Irving, M. *J. Physiol. (Lond.)* **353**, 64P (1984).
10. Poole, K.J.V., Maéda, Y., Rapp, G. & Goody, R.S. *Adv. Biophys.* **27**, 63-75 (1991).
11. Tsukita, S. & Yano, M. *Nature* **317**, 182-184 (1985).
12. Hirose, K. & Wakabayashi, T. *Adv. Biophys.* **27**, 197-203 (1991).
13. Lenart, T.D., Franzini-Armstrong C. & Goldman, Y.E. *Biophys. J.* **61**, A286 (1992).
14. Padrón, R. & Huxley, H.E. *J. Muscle Res. Cell Motility* **5**, 613-655 (1984).
15. Goldman, Y.E. & Simmons, R.M. *J. Physiol. (Lond.)* **350**, 497-518 (1984).
16. Chase, P.B. & Kushmerick, M.J. *Biophys. J.* **53**, 935-946 (1988).
17. Cecchi, G., Colomo, F. & Lombardi, V. *Bol. Soc. Ital. Biol. Sper.* **52**, 733-736 (1976).
18. Goldman, Y.E., Hibberd, M.G. & Trentham, D.R. *J. Physiol. (Lond.)* **354**, 577-604 (1984).
19. Kaplan, J.H. & Ellis-Davies, G.C.R. *Proc. Natl. Acad. Sci. USA* **85**, 6571-6575 (1988).
20. Goldman, Y.E. *Biophys. J.* **52**, 57-68 (1987).
21. Fidler, N., Ellis-Davies, G., Kaplan, J.H. & McCray, J.A. *Biophys. J.* **53**, 599a (1988).
22. Ashley, C.C., Mulligan, I.P. & Lea, T.J. *Quart. Rev. Biophys.* **24**, 1-73 (1991).
23. Heuser, J.E., Reese, T.S., Dennis, M.J., Jan, Y., Jan, L. & Evans, L. *J. Cell Biol.* **81**, 275-300 (1979).
24. Padrón, R., Alamo, L., Craig, R. & Caputo, C. *J. Microsc.* **151**, 81-102 (1988).
25. Huxley, H.E. & Brown, W. *J. Mol. Biol.* **30**, 383-434 (1967).
26. Huxley, H.E., Faruqi, A.R., Kress, M., Bordas, J. & Koch, M.H.J. *J. Mol. Biol.* **158**, 637-684 (1982).
27. Bordas, J., Diakun, G.P., Harries, J.E., Lewis, R.A., Mant, G.R., Martin-Fernandez, M.L. & Towns-Andrews, E. *Adv. Biophys.* **27**, 15-33 (1991).
28. Reedy, M.K. *Am. Zool.* **7**, 465-481 (1967).
29. Wakabayashi, K., Tanaka, H., Saito, H., Moriwaki, N., Ueno, Y. & Amemiya, Y. *Adv. Biophys.* **27**, 3-13 (1991).
30. Yagi, N. *Adv. Biophys.* **27**, 35-43 (1991).
31. Craig, R., Szent-Györgyi, A.G., Beese, L., Flicker, P., Vibert, P. & Cohen, C. *J. Mol. Biol.* **140**, 35-55 (1980).
32. Varriano-Marston, E., Franzini-Armstrong, C. & Haselgrove, J.C. *J. Muscle Res. Cell Motility* **5**, 363-386 (1984).
33. Taylor, K.A., Reedy, M.C., Córdova, L. & Reedy, M.K. *Nature* **310**, 285-291 (1984).
34. Reedy, M.K. & Reedy, M.C. *J. Mol. Biol.* **185**, 145-176 (1985).
35. Bard, F., Franzini-Armstrong, C. & Ip, W. *J. Cell Biol.* **105**, 2225-2234 (1987).

Discussion

Pollack: When you look at the 43-nm layer line and you see after some time that apparently there are two components—one at 43 nm, the other at about 36—can you rule out the possibility that the 36 is due to some regions along thick filaments that have shortened? The thick filament itself would then be giving you a dual component, one at 43 nm and one at 36, the 36 corresponding to some local regions that may have shortened.

Goldman: We haven't actually measured the length of the filaments. That would be difficult using our micrographs because a filament isn't always confined to the plane of the section. We see both 1/36 and 1/43 nm reflections in the overlap zone, so if the shortening of the thick filament occurs in a well-defined region as you have suggested, we would be able to detect the altered structure in different regions of the filament. Otherwise, we cannot completely rule out some length change. Other experiments would be better for that.

Huxley: Would it not show up in the 143?

Goldman: That's a good point. Thank you.

Morales: Since you are concerned with variability—although you impose all these things as well as you can and as rapidly as possible—you do have to consider that there is nothing that coordinates the movement of one bridge with that of another, so that they are going to respond stochastically in your time frame. Molecularly, these are very long times, so one could always expect some variability, I would think. One would not anticipate lock-step behavior of the cross-bridges.

Goldman: We don't expect them all to look the same at any time point. In fact, the range of shapes we observe might be just the sort of variability expected on the basis of the Huxley-Simmons (*Nature* **233**, 533, 1971) idea that cross-bridges oscillate among various states. What surprised us is that this variability is established so early in the contraction. The shapes and angles present 12 - 15 ms after Ca^{2+} release are similar to those present later even though the force per attachment is much lower. We expected to see a big change in the distribution of shapes as the force per attachment increased. There must be a difference in that distribution, and the difference is contained in our images, because the optical diffraction patterns do change. But, we haven't yet done the considerable image analysis that will be required to detect the alterations of the cross-bridge shapes accounting for the progression observed in those optical diffraction patterns.

Brenner: In the traces with the clamped sarcomere length, the stiffness and tension were very close to each other. That means that nowhere in your time course you have a large, pure population of these pre-force, or force-generating states. So in principle, wouldn't it be very difficult to identify in a mixture of cross-bridges which ones are the force-generating ones and which the pre-force-generating ones?

Goldman: The tension and stiffness traces are closer under length clamp than under fixed-end conditions, but they are not the same. The stiffness trace definitely rises earlier than the tension. There is a difference of 10 or 15 milliseconds between the two traces.

EFFECTS OF IONIC STRENGTH ON FORCE TRANSIENTS INDUCED BY FLASH PHOTOLYSIS OF CAGED ATP IN COVALENTLY CROSSLINKED RABBIT PSOAS MUSCLE FIBERS

K. Yamada, Y. Emoto, K. Horiuti, and K. Tawada*

Department of Physiology
Medical College of Oita
Oita 879-55 Japan
*Department of Biology
Faculty of Science
Kyushu University
Fukuoka 812, Japan

ABSTRACT

Single fibers from glycerinated rabbit psoas muscle were treated with 1-ethyl-3[3-(dimethylamino) propyl] carbodiimide (EDC), after rigor was induced, to crosslink myosin heads to actin. The optimally pre-stretched (~1.8%), partially crosslinked fibers produce a large force when MgATP is depleted, and this force is abolished when MgATP is reintroduced, even in high ionic strength solution of 0.5 M (Tawada et al. 1989). We investigated the rate of force decay in the crosslinked, force-producing fibers using pulse photolysis of caged ATP (Goldman et al. 1984). The decay of force was fast, the rate of which depending both on the ionic strength and on the amount of ATP released (0.2-2.2 mM) with the second-order rate constant of $0.5 - 1 \times 10^5$ $M^{-1}s^{-1}$ at the ionic strength of 0.5 M. At high ionic strength (1-2M) force decayed at lower rate. At low ionic strength (0.1-0.2 M), however, force decayed more rapidly, but force redeveloped subsequently, which is probably caused by uncrosslinked myosin heads.

INTRODUCTION

Mornet et al.[1] developed a method for covalently crosslinking myosin subfragment 1 (S1) to actin using the zero-length crosslinker 1-ethyl-3-[3-(dimethylamino) propyl] carbodiimide (EDC). Tawada & Kimura[2] have shown that EDC crosslinks myosin filaments at the inter-rod portion and. therefore, EDC-

Mechanism of Myofilament Sliding in Muscle Contraction, Edited by
H. Sugi and G.H. Pollack, Plenum Press, New York, 1993

490 K. Yamada et al.

treated skinned fibers retain their sarcomere structure even if they are placed in high
ionic strength solutions.

Tawada & Emoto[3] showed that, after suitably pre-stretched, such EDC-treated
fibers produced force when transferred from a relaxing solution to a rigor solution
even in 0.5 M KCl. The force produced in EDC-treated fibers in high ionic strength
rigor solution may be attributed to a structural change of the crosslinked myosin
heads upon transition of these heads into the rigor configuration[4].

On rebinding of ATP, the strain of the crosslinked crossbridges and thus force is
abolished. In order to investigate the nature of such force production in EDC-treated
fibers, we observed the time course of the force decay induced by ATP in EDC-
crosslinked, force-producing fibers using pulse photolysis of caged ATP[5].

MATERIALS AND METHODS

Preparation of single chemically skinned rabbit psoas muscle fibers, the
procedure for crosslinking with EDC, and the determination of the degree of
crosslink were the same as those described previously[2]. The compositions of
solutions (in mM) at pH 7.0 were as follows: 'relaxing' solution; 20 PIPES, 5 EGTA,
2-5 $MgCl_2$, 5-10 ATP: EDTA rigor solution; 20 PIPES, 5 EDTA: caged ATP
solution; 50 PIPES, 5 EGTA, 5 $MgCl_2$, 10-20 DTT, 10-20 creatine phosphate, 0.5-1

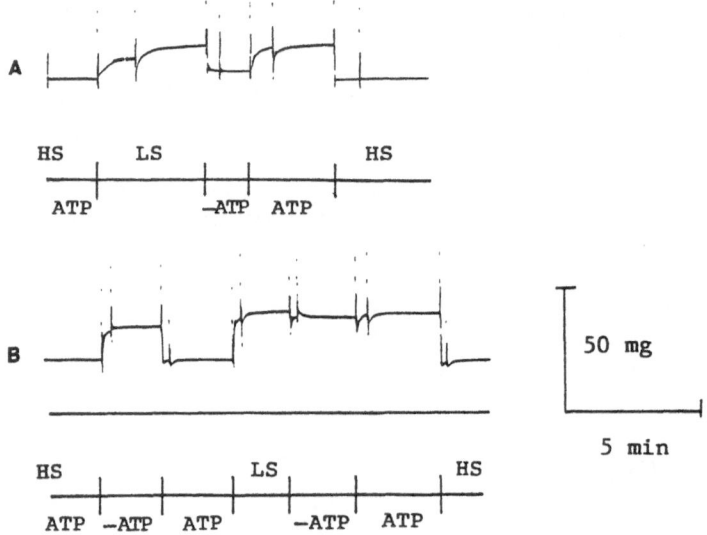

Fig. 1. Steady-state force produced by a demembranated and EDC-crosslinked rabbit psoas
muscle fiber. The extent of crosslink, 38%. The fiber was held at its slack length in A, and it
was stretched beforehand by about 1.8% in B. Bottom horizontal line in B shows the level of
zero force. The ionic strength was either 0.5 M or 0.1 M as indicated. The fiber was in the
relaxing solution except during short periods (R) where the EDTA rigor solution was applied.

mg/ml creatine kinase, 4-10 caged ATP. Ionic strength (I) was adjusted by adding KCl. The experimental setup including the system for flash photolysis and the determination of the amount of liberated ATP will be described elsewhere[6]. Temperature was 16°C.

RESULTS

Figure 1 shows chart records of the steady-state isometric force produced by an EDC-treated, partially crosslinked fiber (38% of S1 crosslinked to actin) and shows that the fiber produces force in low ionic strength ($I = 0.1$ M) relaxing solution even in the absence of Ca^{2+}. These fibers relax, however, when ionic strength is high ($I >$ 0.5 M). When fibers are held at their slack length, force produced in rigor solution is very small (Fig. 1A), indicating that both crosslinked and uncrosslinked heads produce only a small extra force when rigor is induced in EDC-treated, partially crosslinked fibers.

When optimally pre-stretched[4] (~2%), however, a rigor solution produced almost independently of its ionic strength a large extra force above the force in the relaxing solution at high ionic strength, comparable to the extra force production in low ionic strength relaxing solution (Fig. 1B). The fact that the rigor force in such a fiber is large when optimally pre-stretched indicated that this large force is mainly

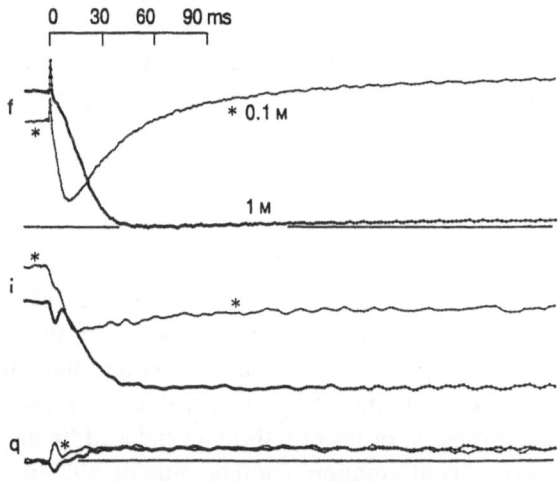

Fig. 2. Time courses of transient force and stiffness (1 kHz; traces i for in-phase and q for 90° out-of-phase or quadrature stiffnesses) changes on pulse photolysis of caged ATP of a partially crosslinked (40%) and force producing muscle fiber. The fiber was optimally pre-stretched after EDC-crosslinked. After the force was induced in rigor solution at high ionic strength (0.5 M), the ionic strength was changed as indicated in the figure. Time zero corresponds to the time of photoflash to release ATP (about 1 mM) from caged ATP. The temperature was 16°C.

produced by crosslinked heads, because uncrosslinked heads should behave in the same way irrespective of the extent of stretch.

Figure 2 shows the force transients induced by ATP released on pulse photolysis of caged ATP. After fibers were treated with EDC they were optimally pre-stretched in the relaxing solution at the ionic strength of 0.5 M. Force was then produced by changing the solution to the rigor solution at the same ionic strength (0.5 M). Ionic strength was then changed to carry out the photolysis experiment at different ionic strengths. It is noticeable in the figure that lowering the ionic strength decreases the force but increases the stiffness (trace i).

When ATP was released by pulse photolysis of caged ATP, the force decayed rapidly. The rate of force decay at very high ionic strengths ($I = 1$-2 M) depended on the amount of ATP released. At low ionic strengths (0.1-0.5 M), however, force decayed more rapidly but force was redeveloped subsequently. The lower the ionic strengths the greater the force redeveloped. This force redevelopment can be attributed to uncrosslinked myosin heads.

The rate of force decay depended on the amount of ATP released (0.2-2.2 mM). The second-order rate constant can be obtained from the relationship between the reciplocal half-time of force decay and the ATP concentration[5]. The second-order rate constant was estimated to be 0.5-1×10^5 M^{-1} S^{-1} at $I = 0.5$ M at $16°C$[7].

In-phase stiffness (1 kHz) increased in rigor solution, reflecting the formation of rigor link between uncrosslinked myosin heads and actin sites. In-phase stiffness changed also rapidly on pulse photolysis of caged ATP. It decreased almost in parallel with force at high ionic strengths. On the other hand, it decreased to an intermediate level at lower ionic strengths.

The quadrature stiffness decreased in rigor solution and increased again when ATP was released from caged ATP. The time course of the increase of the quadrature stiffness (trace q) on flash was almost in parallel with that of the force and in-phase stiffness decay. The extent of the changes appeared to be the same at both high and low ionic strengths.

DISCUSSION

Tawada & Kawai[8] showed that the crosslinked fibers exhibited oscillatory work utilizing ATP in high ionic strength solutions. Therefore, the crosslinked myosin heads should be distributed between the weak- and strong-binding states when ATP is present. The extra force produced by crossbridges crosslinked to actin should reflect the elastic deformation of the crossbridges induced by the transition of these crossbridges to the rigor configuration. On rebinding of ATP, the extra strain of the crossbridges and thus force is abolished.

The present study is the first demonstration of the rate of these transitions not limited by diffusion of ATP. However, the fact that the rate depended on the amount of ATP released both at high and low ionic strengths indicated that the rate may be limited by ATP binding to myosin heads. The fact that the rate of the force decay depended on ionic strengths indicates that the rate of binding of ATP to actomyosin depends on ionic strengths in the range between 0.1 to 2 M.

ACKNOWLEDGEMENTS

This work was supported in part by grants from the Ministry of Education, Science and Culture, Japan, and the Uehara Memorial Foundation.

REFERENCES

1. Mornet, D., Bertrand, R., Pantel, P., Audemard, E. & Kassab R. *Nature* **292**, 301-306 (1981).
2. Tawada, K. & Kimura, M. *J. Muscle Res. Cell Motility* **7**, 339-350 (1986).
3. Tawada, K. & Emoto, Y. in *Molecular Mechanisms of Muscle Contraction* (eds. Sugi, H. & Pollack, H.) 219-224 (Plenum Press, New York, 1988).
4. Tawada, K., Huang, Y.-P. & Emoto, Y. *Muscle Energetics* (eds. Paul, R. J., Elzinga, G. & Yamada, K.) 37-43, (Alan R. Liss, New York, 1989).
5. Goldman, Y.E., Hibberd, M G. & Trentham, D.R. *J. Physiol. (Lond.)* **354**, 577-604 (1984).
6. Horiuti, K., Sakoda, T., Takei, M. & Yamada, K. *J. Muscle Res. Cell Motility*, **13**, 199-205 (1992).
7. Yamada, K. Emoto, Y., Horiuti, K. & Tawada, K. *J. Physiol. (Lond.)* **446**, 264P (1992).
8. Tawada, K. & Kawai, M. *Biophys. J.* **57**, 643-647 (1990).

Discussion

Morales: What electrolyte provided the ionic strength that went up to two molar?

K. Yamada: Potassium chloride.

Morales: How did you determine the amount of cross-linking?

K. Yamada: By comparing the stiffness in the absence and presence of ATP.

Morales: It was not chemically determined by the amount of the gel?

K. Yamada: That's right.

Morales: It is hard to see how, when actin and myosin are presumably locked together in one of the states, it can nevertheless go on and do work. That is remarkable to me.

Reedy: I did not quite understand the basis for force development in the low ionic strength. I was trying to guess whether the force development was being triggered by calcium in the solution or whether you supposed it was being triggered by rigor attachments that were already in place in activating the thin filament. Which of these accounts for the force development after the release of caged ATP in low ionic strength?

K. Yamada: No calcium was present.

Reedy: And that was not the case at high ionic strength because that suppresses these structural changes?

K. Yamada: At high ionic strength, uncross-linked cross-bridges do not produce force. When cross-linked, the fiber is active without calcium, so that at low ionic strength uncross-linked bridges produce steady force, but not at high ionic strength.

Kawai: When you measured K_d, you had at least two kinds of bridges—the freely-mobile bridges and the cross-linked ones. Did you differentiate these? Does your signal come from mobile cross-bridges or both?

K. Yamada: From cross-linked, but also from uncross-linked at low ionic strength.

Kawai: Is K_d ionic strength-dependent?

K. Yamada: K_d is greater at low ionic strength.

Iwamoto: I put a cross-linked fiber in ATP and the fiber produced large active force without calcium. There is some discrepancy between our results and yours. The question is, what is the extent of the cross-linking at pre-stretch?

K. Yamada: We always observe force in low ionic strength relaxing solution, but as I suppose in your preparation and ours the extent of cross-linking is quite different. We have about 40% cross-linking.

Iwamoto: But you compare stiffness in rigor and in the presence of ATP at high ionic strength. That is how you calculate the extent of cross-linking.

K. Yamada: Yes, that's right.

Goldman: Do you know whether the difference in extent of cross-linking is an actual difference between the fibers, or is it in the manner of analysis of the cross-linking?

Iwamoto: The manner of analyzing the extent of cross-linking is quite different for ours and for theirs.

Goldman: Is the condition of cross-linking the same, though?

Iwamoto: I think so.

KINETICS OF FORCE GENERATION AND Pi RELEASE IN RABBIT SOLEUS MUSCLE FIBERS

Earl Homsher and Neil Millar

Department of Physiology
School of Medicine, UCLA
Los Angeles, California, 90024

ABSTRACT

Recent studies have shown that the release of Pi from the AM.ADP.Pi cross-bridge occurs in two steps: an initial force-generating isomerization which is followed by the release *per se* of Pi. These two steps have been shown to be little affected by the presence or absence of Ca^{2+} but both steps are significantly altered by temperature. In this study, the kinetics of the force generating and Pi release steps of the actomyosin ATPase cycle have been compared in Ca^{2+}-activated skinned fibers of rabbit soleus (slow twitch) and psoas (fast twitch) muscle. Pi was rapidly photogenerated within the fiber lattice by laser flash photolysis of caged Pi (1-(2-nitro)phenylethyl phosphate). Pi reduces isometric tension in the steady state, but is less effective in slow muscle than in fast (eg. 14 mM Pi reduces tension by $29 \pm 4.6\%$ in slow muscle and by $47 \pm 5.3\%$ in fast). As in fast muscle, the tension response to a sudden increase in [Pi] in slow muscle has four phases. As in fast muscle, only phase II (an exponential decline in force) appears to be caused by Pi binding to cross-bridges, while the other three phases are probably indirect effects caused by caged Pi photolysis. The amplitude of phase II is consistent with the steady state reduction in force by Pi. The rate of phase II (k_{pi}) is 3.9 ± 0.33 s^{-1} at 20°C and 0.28 ± 0.02 s^{-1} at 10°C (1 mM Pi). k_{pi} is ca. 33 times slower in slow muscle than in fast at 20°C and 84 times slower at 10°C. In contrast to fast muscle, k_{pi} in slow muscle is so slow that it may partially limit the ATPase turnover rate.

INTRODUCTION

The reactions shown in Fig. 1 are generally accepted as those which produce force, shortening, and work in muscle[1]. In this model chemical states involving AM•ATP and AM•ADP•Pi are thought to be "weakly" attached; i.e., detaching and attaching rapidly (ca.> 1000 s^{-1}) and incapable of generating or sustaining significant force under normal physiological conditions. AM*•ADP, and A*M are "strongly" attached states which remain attached for a significant time and produce

Mechanism of Myofilament Sliding in Muscle Contraction, Edited by
H. Sugi and G.H Pollack, Plenum Press, New York, 1993

495

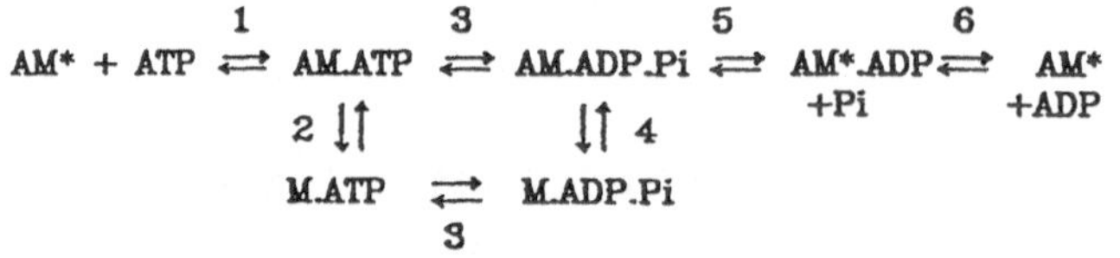

$$\begin{array}{ccccccc}
& 1 & & 3 & & 5 & & 6 \\
AM^* + ATP & \rightleftarrows & AM.ATP & \rightleftarrows & AM.ADP.Pi & \rightleftarrows & AM^*.ADP & \rightleftarrows & AM^* \\
& & & & & & +Pi & & +ADP \\
& 2 \updownarrow & & \updownarrow 4 & & & & & \\
& & M.ATP & \rightleftarrows & M.ADP.Pi & & & & \\
& & & 3 & & & & &
\end{array}$$

Fig. 1. Mechanism of ATP hydrolysis by actomyosin.

or bear force. The rates of transition between these states are thought to depend on the intracellular milieu and on the strain on the crossbridge, and it is assumed that maximal shortening velocity is primarily limited by the rate at which the attached crossbridge can dissociate from the sliding thin filament; i.e., reactions 6 and 1 in Fig. 1[1-5]. Reaction 5 is thought to be related to force production[3)4)6]. It is interesting to note that the rates governing reactions 1-4 appear to be similar in both solution studies and in the glycerinated contracting muscle fiber[1]. The phosphate release step (5) is, however, different. The addition of inorganic phosphate to fibers is known to reduce isometric force and stiffness, and the dependence of this effect on Pi concentration yields a dissociation constant (K_D) for Pi of ca. 10 mM[6]. this is 10^4 times tighter than the K_D estimated in solution (100 M) and suggests that in the fiber in physiological [Pi] this step is close to equilibrium. Other steady state studies have shown that Pi slightly increases unloaded shortening velocity and only mildly depresses the steady state rate of ATP hydrolysis[7-9]. The large free-energy change associated with Pi release naturally lead to the suggestion that this step is directly associated with energy production and perhaps the force producing step itself[10].

Evidence in support of this idea was obtained by Hibberd et al.[3] who examined the rate of force decline in fibers (initially in rigor) following the photogeneration of ATP in the fibers. They observed that the initial rate of force decline (crossbridge detachment) increased from 20 s[-1] at [Pi] ~ 1 mM to 60 s[-1] at 10 mM Pi. Similarly in the presence of Ca^{+2} 10 mM Pi accelerated the rate of force production from 80 s[-1] to 140 s[-1]. To explain these results, Hibberd et al proposed that force production accompanies Pi release. This means that the observed rate of force production is given by the sum of the forward and reverse rate constants for Pi release ($k_{+5} + k_{-5}$ [Pi]), and the extent of force production will be given by the equilibrium constant for Pi release. This hypothesis is consistent with: the observed rates of force decline following ATP photogeneration in the absence of Ca^{2+}; 2.) the force rise following ATP photogeneration in the presence of Ca^{2+}; and 3.) the reduction of isometric force as [Pi] rises[11)12].

To directly examine the kinetics of the Pi release step and the factors that affect it, one can use caged phosphate. In caged phosphate experiments a muscle fiber is activated in a solution containing caged Pi and allowed to reach a steady-state isometric force, whereupon Pi is rapidly photogenerated within the fiber lattice. This sudden increase in Pi produces an exponential decline in force which can be used to characterize the force generating step(s) in the fiber. Below we summarize the results of our work on glycerinated fast (psoas)[13] and slow (soleus)[14] muscle fibers of the rabbit.

METHODS

The preparations of muscle fibers, recording of force, synthesis and purification of caged Pi, photolysis apparatus, and analytic procedures were as previously described[13][14]

RESULTS

Figure 2 shows the time course of the tension change accompanying the photogeneration of Pi from caged Pi in an isometrically contracting psoas muscle fiber when the phosphate concentration is stepped from 0.7 mM to 2.7 mN. The mechanical response is composed of 4 phases: I,a short lag (ca. 1 ms) during which force does not change: II, a period during which force declines exponentially and whose amplitude, AII, is dependent on the amount of Pi released and which can be as great as 20% of the initial force; III, a slow rise in force (amplitude AIII); and IV, a slow decline to a steady state force. Control experiments[13] in which the initial Pi concentration, final Pi concentration, caged Pi concentration, and extent of caged Pi photolysis were varied have yielded the following conclusions. First the steady state isometric force is inhibited by Pi to an extent quantitatively similar to that described earlier[6][8]. second the caged Pi itself depresses force slightly but to a lesser extent than Pi. The effects of both Pi and caged Pi are fully reversible. Third, the rate of phase II force decline is a function only of the final [Pi] after the photoliberation of Pi; e.g., the rate is independent of whether the [Pi] increases from o.7 to 2.0 mM or from 1.5 to 2.0 mM, or whether the caged Pi decreases from 5mM to 3.7 mM or from 2mM to 1.5 mM in the process of reaching a final [Pi] of 2.0 mM. Fourth, the amplitude of phase II is equivalent to the change in steady state force occasioned by the simple addition of Pi. Fifth the amplitude of phase III increases as the extent of caged Pi photolysis is increased but its amplitude and time course are independent of the [Pi]. Its time course is independent of the extent of caged Pi photolysis. These results have lead to the hypothesis that phase III is associated with the loss of caged Pi inhibition of the force when its concentration is reduced by photolysis[13] Its independence of the [Pi] suggest that it is not directly associated with the reactions involved in Pi release from the crossbridge. Finally the neither the amplitude or rate of phase IV is associated with either [Pi] caged Pi. Its amplitude is often of equal and opposite magnitude to that of phase III and it is often essentially absent. These results have lead to the conclusion that only phase II which is affected by the timecourse of Pi binding and release from the crossbridge in step 5. Additional control experiments have shown that the rate of phase II is independent of thick and thin filament overlap but, as expected, its amplitude is proportional to the amount of

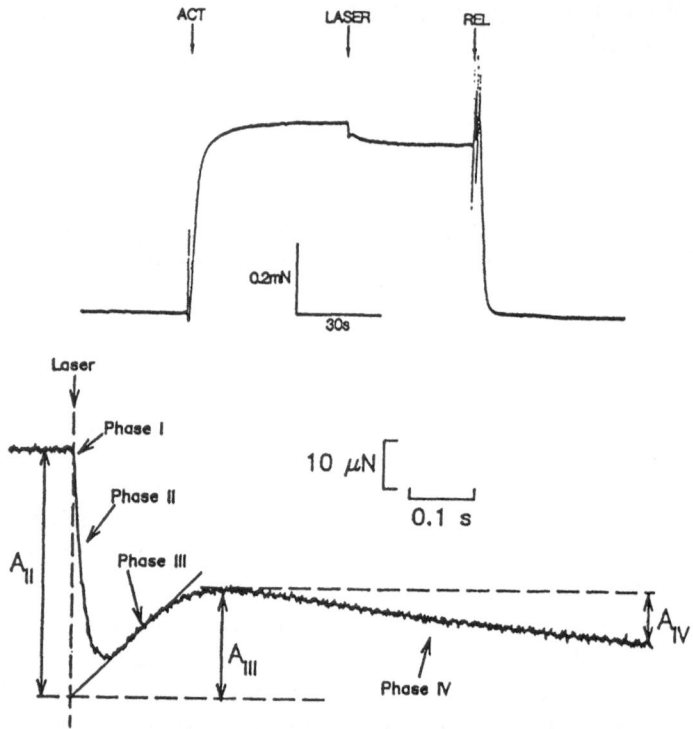

Fig. 2. Slow and fast tine base recordings of tension during a caged Pi photolysis experiment, 10°C. *Left panel*, strip chart recording. The fiber was initially equilibrated in a pre-activating solution (2 mM EGTA, pCa 9.0) containing 5 mM caged Pi. At *ACT* the fiber was moved into an activating solution (20 mM Ca^{2+}-EGTA, pCa 4.5) containing 5 mM caged Pi. At *LASER*, the laser pulse released 1.8 mM Pi. At Rel, the fiber was moved into a relaxing solution (20 mM EGTA, pCa 9.0). *Right panel*, enlargement of the photolysis tension transient. Phase I (the lag phase) cannot be seen on this time scale.

thick and thin filament overlap. Finally the photolysis of caged ATP in an isometrically contracting muscle fiber does not produce transients that are in any way similar to those seen during the photolysis of caged Pi. Thus other photolytic by-products are not responsible for the phosphate transient.

The reaction scheme in Fig. 1 predicts that a plot of the rate of phase II, k_{pi}, versus the phosphate concentration should yield a straight line whose intercept is equal to k_{-5} and whose slope is k_5. Figure 3 shows the type of plots obtained when k_{piu} F is measured as a function of [Pi]. This plot shows a saturation of k_{pi} at high concentrations of Pi. Such behavior is a signature of a two-step process. The data is most easily explained by a reaction mechanism (Fig. 4) which includes a force generating isomerization (step 5) which is followed by the release of phosphate (step 6)[13]. In this mechanism, the "weak" (non-force bearing) states have been grouped with their detached counterparts in brackets to indicate a fast equilibrium. Using this model the data in Fig. 3 is nicely fit by rate constants which at 10°C are $k_a = 20$ s^{-1},

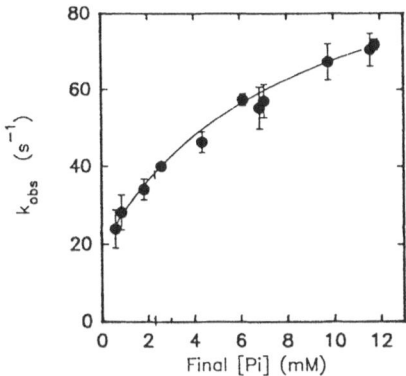

Fig. 3. A plot of the rate of phase II (k_{obs}) versus final [Pi] at 10°C. Bars represent sem.

$k_b = 100$ s^{-1}, and $K_c = 12$ mM[13]). At 20°C similar experiments yield values of k, = 80 s^{-1}, $k_b = 115$ s^{-1}, and $K_c = 4$ mM. The fitted rate and equilibrium constants reveal that the two steps are closely coupled, so that at low [Pi] the effect of the Pi release step is to drive the reaction toward the force-generating states. It is of interest to note that the observed k_{pi} is about two or 3 times faster than the so-called k_{tr}[15)16) the rate of rise of force in an active muscle following the detachment of crossbridges by a quick release followed by a quick restretching of the muscle. This result shows that k_{tr} is does not simply monitor a force-generating step, but is actually determined by several rate constants in the reaction scheme.

Slow Muscle Fibers

Like fast muscle, slow muscle isometric force production is inhibited by both Pi[14)17) and caged Pi[14). However, the inhibitory effect of Pi is only about 60% as great as in fast muscle. On the other hand, the inhibitory effect of caged Pi in slow muscle is about 50% greater for a given caged Pi concentration than in fast muscle. The consequence of these effects are that the decline in force (phase II) following the photogeneration of Pi from caged Pi is less than is seen in fast muscles, and phase III has a relatively larger amplitude. This behavior is seen in Figure 5 Here one sees that in slow muscle the same phases seen in fast muscle (phase I is too small to be seen on the time base in the figure) are present. However the amplitude of phase II is only

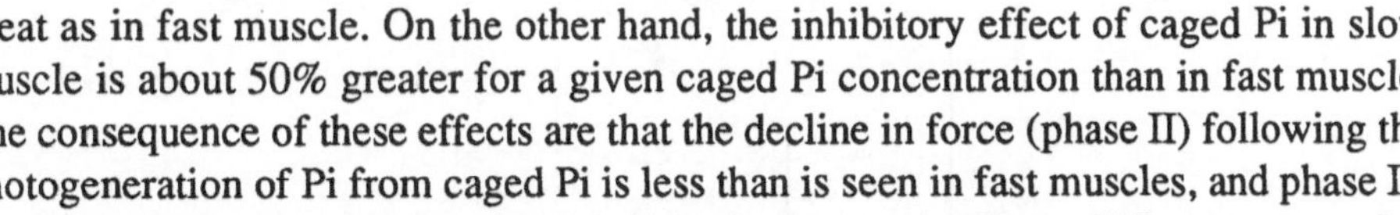

Fig. 4. Reaction mechanism for actomyosin ATPase in the fiber which includes a force generating isomerization (reaction 5) and the phosphate release step (reaction 6). The states in brackets ({}) are weakly attached states in a rapid equilibrium.

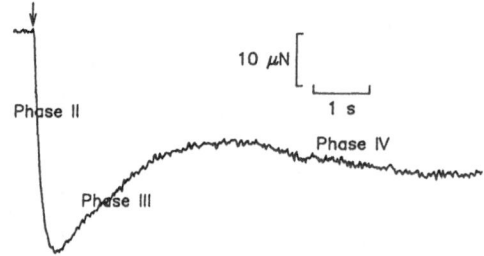

Fig. 5. Phosphate transient in a slow muscle fiber. Caged Pi photolysis occurred at the vertical arrow. Force developed in this fiber was 300 μN. Temperature is 20°C and the change in phosphate is from 0.8 mM to 1.8 mM.

about half as great as might be expected in fast muscles and the amplitude of phase III is relatively large. The consequence of this behavior is that it is very difficult to examine transients at [Pi] much greater than 3 mM Pi because phase III essentially obliterates phase II and useful data can not be obtained[17]. Fig. 5 plots k_{pi} at different [Pi] and temperature, and shows three interesting features. First, k_{pi} is more than 30 times slower in slow muscle than in fast. For example at 20°C, 1 mM Pi, k_{pi} is 100 s^{-1} in fast muscles and 3.9 s^{-1} in slow. This indicates that the rates of the force generating and Pi release steps of the cross-bridge cycle are much slower in the slow muscles. Second, in slow muscle, k_{pi} is extremely temperature dependent, increasing 14 fold between 10 and 20°C. Third, k_{pi} increases approximately linearly with

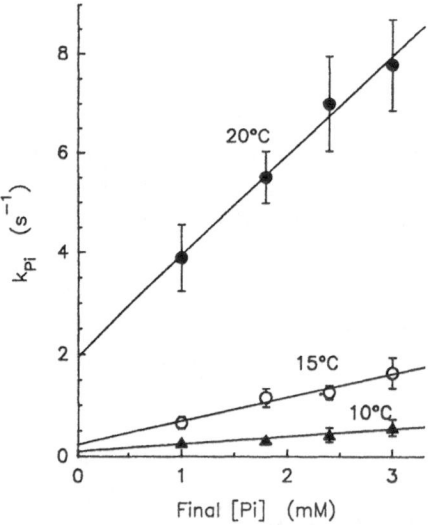

Fig. 6. Effect of [Pi] and temperature on the rate of phase II (k_{pi}). The data points are mean ± S.E.M. for 5 - 15 observations. The best fit straight lines are superimposed. 20°C: intercept = 1.96 s^{-1}, slope = 1.99 $mM^{-1}s^{-1}$; 15°C: intercept = 0.24 s^{-1}, slope = 0.46 $mM^{-1}s^{-1}$; 10°C: intercept = 0.11 s^{-1}, slope = 0.14 $mM^{-1}s^{-1}$.

increasing [Pi] over the range 1 - 3 mM. In fast muscle k_{pi} increased in a hyperbolic fashion with increasing [Pi] up to 12 mM[13]. This indicated that Pi binding was a two-step process: a fast Pi binding equilibrium followed by a slower isomerization (steps 6 and 5 in reverse). In slow muscle we cannot measure k_{pi} at [Pi] > 3 mM, so we have no direct evidence for two-step Pi binding. However in other respects the tension transient behaves similarly in the two muscle types, and it is likely that the mechanism of Pi binding is the same. In that case the straight lines in Fig. 6 represent the initial part of the hyperbola, so the intercept corresponds to the forward rate of the force-generating step (k_a), and the slope to the second order rate constant for Pi binding (k_b / K_c; note that K_c refers to Fig. 4 and is thus a dissociation constant). These values are listed in Table 1, together with the corresponding values for fast muscle.

Rate of Tension Redevelopment

Since k_{pi} in slow muscle is so slow, it is of interest to compare this rate with another measurement of the rate of force development in these muscles. k_{tr}, the rate of tension redevelopment following a rapid release and restretch, is 8 times slower in slow than fast muscles of the rat[18], so we made k_{tr} measurements in rabbit soleus fibers under the same conditions as the photolysis experiments. The values of k_{tr} obtained (at 1 mM Pi) were 2.32 ± 0.24 s^{-1} at 20° and 0.57 ± 0.17 s^{-1} at 20°C (n = 6). We did not control the sarcomere length during the k_{tr} measurement so the true rates could be up to 30% faster than the values obtained here[19]. These rates are 10 - 20 times slower than k_{tr} in fast muscle, and are very similar to k_{pi} in the same muscle (see Table 1). This is in sharp contrast to fast muscle, where k_{tr} is significantly slower than k_{pi}.

DISCUSSION

The Pi-transient is qualitatively similar in the two muscle types, and the results are consistent with the hypothesis that only phase II is caused by Pi binding to cross-bridges. The limited [Pi] range in slow muscle precludes a rigorous kinetic analysis,

Table 1. Rate constants for the Pi release steps in fast and slow muscle. The rate constants refer to Scheme I. Psoas rate constants from ref. 13, soleus rate constants from ref.14

	k_a (s^{-1})	k_b/K_c (M^{-1}s^{-1})
Psoas, 20°C	79.2	31,000
Psoas, 10°C	21.1	8,317
Soleus, 20°C	1.96	1,994
Soleus, 15°C	0.24	459
	0.11	144

but the mechanism is assumed to be the same in the two muscles. The kinetics of phase II therefore reflect the kinetics of the steps involved in force generation and Pi release. The results show that in soleus muscle these steps are 30 - 80 times slower than in psoas muscle. Estimates for the forward rate of the force-generating step (k_a) and for the second order rate constant of Pi binding (k_b/K_c) have been obtained (Table 1). These show that much of the difference between the fast and slow muscle kinetics lies in k_a, which is 40 times slower at 20°C and 190 times slower at 10°C. The second order rate constant is also 15 - 58 times slower in the slow muscles, but we do not know whether the difference lies in k_b, K_c, or both.

ACKNOWLEDGEMENTS

The authors gratefully acknowledge Mr. Nicholas V. Ricchiuti for expert technical assistance, and Drs. Yale Goldman and Jody Dantzig for helpful discussions. This work was supported by NIH grant AR30988 to EH.

REFERENCES

1. Homsher, E. & Millar, N. *Ann. Rev. Physiol.* **52**, 875-896 (1990).
2. Brenner, B., Schoenberg, M., Chalovich, J.M., Greene, L.E. & Eisenberg, E. *Proc. Natl. Acad. Sci. USA* **79**, 7288-7291 (1982).
3. Hibberd, M., Dantzig, J.A., Trentham, D.R. & Goldman, Y.E. *Science* **228**, 1317-1319 (1985).
4. Pate, E. & Cooke, R. *J. Muscle Res. Cell Motility* **10**, 181-196 (1989)
5. Siemankowski, R.F., Wiseman, M.O. & White, H.D. *Proc. Natl. Acad. Sci. USA* **82**, 658-662 (1985).
6. Cooke, R. & Pate, E. *Biophys. J.* **48**, 789-798 (1985).
7. Pate, E., Nakamaya, K., Franks-Skiba, K., Yount, R.G. & Cooke, R. *Biophys. J.* **59**, 598-605 (1991).
8. Kawai, M., Güth, K., Winnikes, K., Haist, C. & Rüegg, C. *Pflügers Arch.* **408**, 1-9 (1987).
9. Webb, M., Hibberd, M., Goldman, Y. & Trentham, D. *J. Biol. Chem.* **261**, 15557-15564 (1986).
10. White, H. & Taylor, E. *Biochemistry* **15**, 5818-5826 (1976).
11. Goldman, Y, Hibberd, M. & Trentham, D. *J. Physiol. (Lond.)* **354**, 577-604 (1984).
12. Goldman, Y., Hibberd, M. & Trentham, D. *J. Physiol. (Lond.)* **354**, 605-624 (1984).
13. Dantzig, J., Goldman, Y., Millar, N., Lacktis, J. & Homsher, E. *J. Physiol. (Lond.)* in the press.
14. Millar, N. & Homsher, E. *Am. J. Physiol.* submitted.
15. Brenner, B. *Proc. Natl. Acad. Sci. USA* **85**, 3265-3269 (1988).
16. Metzger, J., Greaser, M. & Moss, R. *J. Gen. Physiol.* **93**, 855-883 (1989).
17. Chase, P. & Kushmerick, M. *Biophys. J.* **49**, 10a (1986).
18. Metzger, J. & Moss, R. *Science* **247**, 1088-1090 (1990).
19. Brenner, B. & Eisenberg, E. *Proc. Natl. Acad. Sci. USA.* **83**, 3542-3546 (1986).

Discussion

Godt: Earl, when you talk about isomerization in terms of your model, do you mean the same kind of thing that Mike Geeves means in his model of the cross-bridge cycle (Geeves, M.A. et al. *J. Muscle Res. Cell. Motility* **5**, 351-361, 1984; Geeves, M.A. *Biochem J.* **274**, 1-14, 1991)? You use the same words, but do they mean the same thing?

Homsher: The way he represents it, there is a detached state, an attached state, and a strongly-attached state. The values we put after those AM•ADP•P$_i$ are the same; one is a non-force-bearing state and another is a force-bearing state. So to that extent, they are. Whether Mike (Geeves) thinks that way or not, I can't say. It fits naturally with his model, but I am not aware that there is a way I could distinguish them for you. There is great similarity, if not identity. For example our AM•ADP•P$_i$ state is a weakly attached cross-bridge (non-force generating) and is thus similar to Mike's A—M•D•P state. Both of us then have an isomerization, which Mike designates as a rigor-like (R) state, A•M•D•P and which I call AM'•ADP•P$_i$ which is strongly attached and exerting force.

DISTANCE OF MYOFILAMENT SLIDING PER ATP MOLECULE IN SKELETAL MUSCLE FIBERS STUDIED USING LASER FLASH PHOTOLYSIS OF CAGED ATP

T. Yamada, O. Abe*, T. Kobayashi and H. Sugi

Department of Physiology
School of Medicine
Teikyo University
Tokyo 173, Japan

ABSTRACT

We studied the distance of myofilament sliding per hydrolysis of one ATP molecule by recording shortening of single glycerinated muscle fibers induced by laser flash photolysis of caged ATP, diffusion of photochemically released ATP out of the fiber being prevented by surrounding the fiber with silicone oil. With 75 μM ATP released (one half of the total myosin head concentration within the fiber), the fiber showed the minimum shortening (10 ± 2 nm/half sarcomere, n = 10) taking place uniformly in each sarcomere in the fiber. Comparison of the initial flash-induced shortening velocity with the force-velocity relation of maximally Ca^{2+}-activated fibers indicated that the above minimum fiber shortening took place under an internal load nearly equal to P_0. These results may be taken to indicate that, under a nearly isometric condition, the distance of myofilament sliding per hydrolysis of one ATP molecule is of the order of 10 nm.

INTRODUCTION

In the cross-bridge model of muscle contraction[1][2], the myosin heads (cross-bridges) extending from the thick filaments attach to actin in the thin filament, and change their angle of attachment to actin (power stroke) leading to myofilament sliding and force generation in muscle, and then detach from actin; each attachment-detachment cycle between a myosin head and actin is coupled with hydrolysis of one ATP molecule.

---

*Present address: Seikai National Fisheries Research Institute, Shimonoseki Branch, Fisheries Agency of Japan, Shimonoseki-shi, Yamaguchi 750, Japan

Mechanism of Myofilament Sliding in Muscle Contraction, Edited by
H. Sugi and G.H Pollack, Plenum Press, New York, 1993

Recent development of *in vitro* motility assay systems has enabled us to estimate the distance of a myosin head power stroke (or myosin step size) directly by measuring the total distance of actin-myosin sliding and the amount of ATP utilized. The results obtained are, however, extremely divergent: Spudich's group claim that the step size is about 10 nm[3-5], which Yanagida's group report that the step size is more than 100 nm[6]. This discrepancy may result from a number of uncertainties about the mode of actin-myosin sliding in the assay systems.

On the other hand, the technique of laser flash photolysis of caged ATP (P^3-1-(2-nitro) phenylethyladenosine-5'-triphosphate), a biologically inert and photolabile precursor of ATP[7], has made it possible to induce actin-myosin sliding simultaneously within the whole muscle fiber[8-10]. With this technique, Higuchi and Goldman[11] have studied the ATP-induced isotonic shortening of muscle fibers, and estimated the myosin step size to be at least 40 nm under a load of 0.07 P_0. We also used the same technique to study the distance of myofilament sliding per hydrolysis of one ATP molecule, by inducing fiber shortening with photochemically released ATP in concentrations around or equal to the total myosin head concentration within the fiber.

MATERIALS AND METHODS

Single glycerinated muscle fibers, prepared from rabbit psoas, were mounted horizontally in a glass experimental chamber filled with a 50% glycerol solution containing 50 mM KCl, 4 mM $MgCl_2$, 4 mM EGTA and 20 mM Tris-maleate (pH 7.0); one end was glued to a very flexible glass needle (elastic coefficient 400 pN/μm), while the other end was glued to a rigid stainless-steel rod. Then glycerol solution was replaced by a Ca^{2+} rigor solution containing 122 mM KCl, 2.1 mM $CaCl_2$, 5 mM $MgCl_2$, 2 mM EGTA, and 100 mM TES (N-Tris (hydroxymethyl) methyl-2-aminoethanesulfonic acid) (pH 7.0) and further by a photolysis solution containing 35 mM KCl, 2.1 mM $CaCl_2$, 5 mM $MgCl_2$, 2 mM EGTA, 2-4 mM caged ATP and 100 mM TES (pH 7.0). The fiber (sarcomere length, 2.4 μm) was kept in caged ATP solution for 30 s to diffuse uniformly into the myofilament lattice.

Prior to laser flash irradiation, caged ATP solution was replaced by silicone oil to prevent diffusion of photochemically released ATP out of fiber. The fiber was then irradiated with a light flash (duration 10 ns, wavelength 355 nm, intensity up to 200 mJ) from a Nd:YAG laser system (Spectra Physics, DCR-3). Details of the method have been described elsewhere[12] together with estimation of the amount of released ATP with a reverse-phase HPLC system (Hitachi, L-6200). The fiber shortening following photolysis of caged ATP was recorded with a high-speed video system (Nac, HSV-200) at 200 frames/s. All experiments were made at 20-22°C.

RESULTS

Fig. 1 shows selected frames from a video record of a single muscle fiber before and during the shortening in response to a laser flash releasing 200 μM ATP. All the fiber segments, divided by carbon surface markers, shortened uniformly for a distance, and eventually stopped shortening when the released ATP was used up by the myofilament sliding within the fiber. Fig. 2 shows typical time courses of shortening of fiber segments; each segmental shortening first proceeded linearly with time, and then slowed down to come to a complete stop. Electron micrographs of the fibers fixed either before or after the flash-induced shortening indicated that

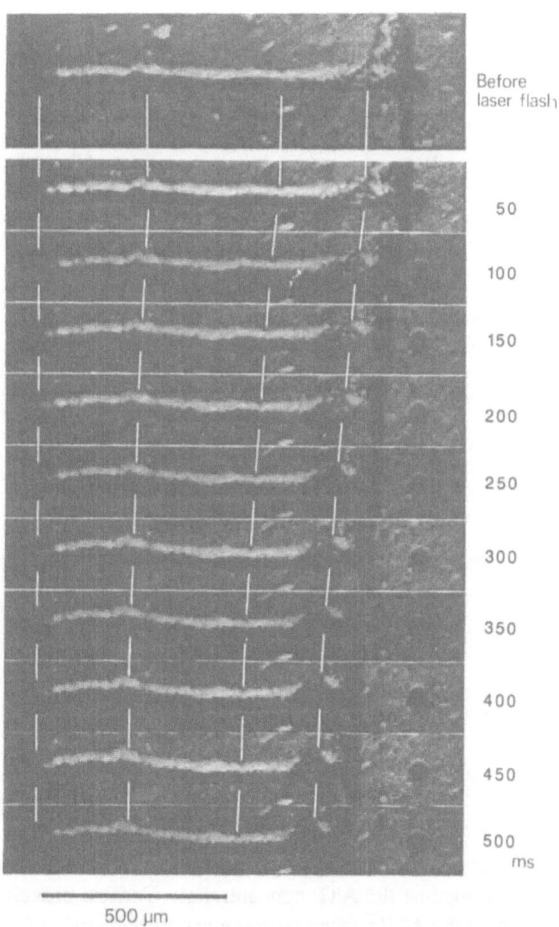

Before
laser flash

50

100

150

200

250

300

350

400

450

500
ms

500 μm

Fig. 1. Selected frames from a video record of laser-flash-induced muscle fiber shortening. Time after a laser flash irradiation is indicated alongside each frame. White lines across the adjacent frames indicate the movement of carbon markers attached to the fiber surface. Note uniform shortening of the fiber segments divided by the carbon markers.

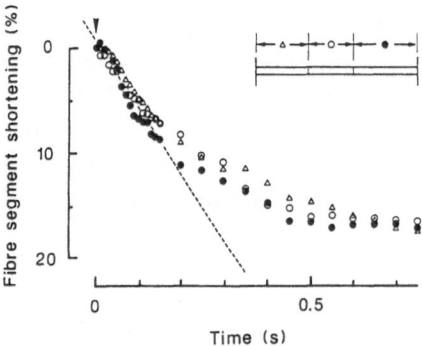

Fig. 2. Time course of fiber segment shortening following a laser flash releasing 200 μM ATP. The fiber was divided into three elementary segments with carbon surface markers (inset). The broken line indicates that the segments initially shorten with a constant velocity.

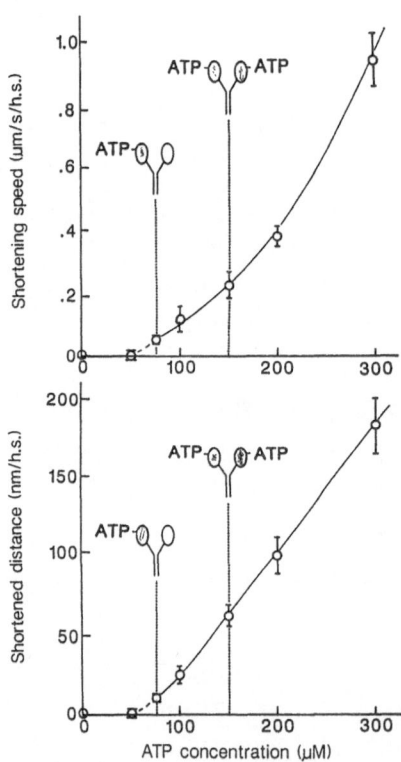

Fig. 3. Relations between the concentration of released ATP and the velocity and distance shortened are plotted against the ATP concentration. Vertical broken lines and diagrams of myosin molecule show the ATP concentrations equal to, and one half of, the total myosin head concentration within the fiber.

the shortening took place uniformly in every sarcomere within the fiber[12].

Fig. 3 summarizes the results of experiments, in which the fibers (initial sarcomere length, 2.4 µm) were made to shorten with laser flashes releasing 50-300 µM ATP. The fibers did not shorten appreciably with 50 µM ATP, except for slight changes in their shape. With higher concentrations of ATP, all the fibers examined showed distinct uniform shortening along the fiber length. As shown in Fig. 3, the initial linear shortening velocity increased steeply with increasing ATP concentrations from 75 to 300 µM. The values (mean ± S.E.M., n = 10) were 0.95 ± 0.08, 0.38 ± 0.03, 0.23 ± 0.04, 0.12 ± 0.04 and 0.05 ± 0.01 µm/s/half sarcomere with 300, 200, 150, 100 and 75 µM ATP, respectively. As also shown in Fig. 3, on the other hand, the distance of fiber shortening increased nearly in proportion to increasing ATP concentrations from 75 to 300 µM. The values (mean ± S.E.M., n = 10) were 180 ± 20, 100 ± 10, 65 ± 7, 25 ± 5 and 10 ± 2 nm/half sarcomere with 300, 200, 150, 100 and 75 µM ATP, respectively.

In order to estimate the amount of internal resistance against the flash-induced fiber shortening, we obtained the force-velocity relation of maximally Ca^{2+}-activated fibers, by applying ramp decreases in force from P_0 to zero to isometrically contracting fibers, and recording the resulting fiber length changes (Fig. 4). The maximum shortening velocity was about 5.5 µm/s/half sarcomere, being many times larger than the initial shortening velocity of about 0.95 µm/s/half sarcomere with 300 µM released ATP (Fig. 3). The latter value was equal to the shortening velocity of maximally Ca^{2+}-activated fibers under an external load of about 0.55 P_0. Similarly, the initial shortening velocity with 150 and 75 µM ATP was equal to the shortening velocity under a load of 0.93 and 0.98 P_0, respectively. The above comparison indicates that, although the external load exerted by the bent needle did not exceed 0.0005 P_0, considerable internal resistance against the flash-induced fiber shortening exists within the fiber.

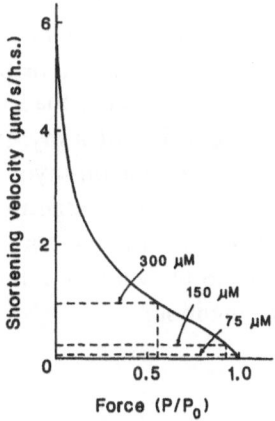

Fig. 4. Force-velocity relation of maximally Ca^{2+}-activated muscle fibers. Horizontal and vertical broken lines illustrate the way of estimating the internal resistance against the laser flash-induced fiber shortening from the measured values of initial shortening velocity.

DISCUSSION

Based on the force-velocity curve of maximally Ca^{2+}-activated fibers (Fig. 4), the internal resistance against fiber shortening with 300 μM ATP was estimated to be about 0.55 P_0. The large internal resistance may largely arise from rigor actin-myosin linkages, as photochemical release of ATP is not instantaneous but with a rate constant of about 100/s. In agreement with the above result, Higuchi and Goldman[11] could not saturate the initial shortening velocity of the fibers in response to laser flashes releasing up to 1 mM ATP. They also ascribe the internal resistance to rigor linkages.

In the present study, it was found that 75 μM ATP produced the minimum uniform fiber shortening along the fiber length (Fig. 3). As this critical ATP concentration corresponds to one half of the total myosin head concentration within the fiber, we will hereafter focus attention on the distance of fiber shortening with 75 μM ATP. As shown in Fig. 4, the amount of internal resistance against fiber shortening with 75 μM ATP was about 0.98 P_0, indicating that, at the beginning of fiber shortening, the total force generated by myosin heads producing myofilament sliding is only slightly larger than the internal resistance exerted by rigor myosin heads. If it is assumed that one rigor linkage can be made to slip along the thin filament with one or two myosin heads per myosin molecule generating myofilament sliding force, the distance of fiber shortening with 75 μM ATP (about 10 nm/half sarcomere) may be taken to represent the distance of myofilament sliding per hydrolysis of one ATP molecule or the distance of a myosin head power stroke under a nearly isometric condition, since the majority of myosin heads producing myofilament sliding may utilize ATP only once following release of 75 μM ATP. This value is similar to that supposed in the cross-bridge model of muscle contraction[1][2] though it is not possible to exclude the possibility that the distance of a myosin head power stroke is not constant but increases with decreasing internal resistance or external load.

Higuchi and Goldman[13] recently estimated the distance of isotonic sliding per hydrolysis of one ATP molecule during the interaction between actin and myosin (interaction distance) under a large external load of about 0.8 P_0 to be 11 nm. Though the interaction distance consists of a myosin head power stroke and the distance of myofilament sliding when a actin-myosin linkage bears negative drag force[1], the latter distance is small under large external loads. Therefore, their figure of 11 nm under a load of 0.8 P_0 may be close to the distance of a myosin head power stroke, a value very close to the value of a myosin head power stroke of 10 nm under the nearly isometric condition.

REFERENCES

1. Huxley, A.F. *Prog. Biophys. Biophys. Chem.* **7**, 255-318 (1957).
2. Huxley, A.F. & Simmons, R.M. *Nature* **233**, 533-538 (1971).

3. Toyoshima, Y., Kron. S.J. & Spudich, J.A. *Proc. Natl. Acad. Sci. USA* **87**, 7130-7134 (1990).

4. Uyeda, T.Q.P., Kron S.J. & Spudich, J.A. *J. Mol. Biol.* **214**, 699-710 (1990).

5. Uyeda, T.Q.P., Warrick, H.M., Kron, S.J. & Spudich, J.A. *Nature* **352**, 307-311 (1991).

6. Harada, Y., Sakurada, K., Aoki, T., Thomas, D.D. & Yanagida, T. *J. Mol. Biol.* **216**, 49-68 (1990).

7. Kaplan, J.H., Forbush, B. & Hoffman, J.F. *Biochemistry* **17**, 1929-1935 (1978).

8. Goldman, Y.E., Hibberd, M.G., McCray, J.A. & Trentham, D.R. *Nature* **300**, 701-705 (1982).

9. Goldman, Y.E., Hibberd, M.G. & Trentham, D.R. *J. Physiol. (Lond.)* **354**, 577-604 (1984).

10. Goldman, Y.E., Hibberd, M.G. & Trentham, D.R. *J. Physiol. (Lond.)* **354**, 605-624 (1984).

11. Higuchi, H. & Goldman, Y.E. *Nature* **352**, 352-354 (1991).

12. Yamada, T., Abe, O., Kobayashi, T. & Sugi, H. *J. Physiol. (Lond.)* in press (1993).

13. Higuchi, H. & Goldman, Y.E. *Biophys. J.* **61**, A140 (1992).

Discussion

Edman: Have you considered that you have an unstirred layer of solution around the fiber that would contain caged ATP that you may release?

T. Yamada: We carefully removed the solution by blotting with filter paper.

Goldman: When the fiber is shortening upon release of, say, 200 μM ATP and then it stops, during the last part of the shortening there will be bridges with no ATP and other bridges pulling them along into the negatively-strained region. So, it seems a good opportunity to observe negatively-strained cross-bridges in your electron micrographs. Have you looked at that magnification and seen what the cross-bridges actually look like?

T. Yamada: During the very end of the shortening?

Goldman: Well, when you fix these fibers and section them as you showed, what happens in much higher-magnification images?

T. Yamada: We could not get good cross-bridge structure.

Sugi: We made many good electron micrographs, but they were mainly at lower magnification to observe uniform sarcomere shortening. We will make observation with higher magnifications in future.

KINETIC AND PHYSICAL CHARACTERIZATION OF FORCE GENERATION IN MUSCLE: A LASER TEMPERATURE-JUMP AND LENGTH-JUMP STUDY ON ACTIVATED AND CONTRACTING RIGOR FIBERS

Julien S. Davis and William F. Harrington

Department of Biology
The Johns Hopkins University
34th and Charles Streets
Baltimore, Maryland 21218

ABSTRACT

Experiments are presented that probe the mechanism of contraction in normal activated muscle fibers and in heated rigor fibers. In activated fibers we subdivide the partial recovery of isometric tension during the Huxley-Simmons phase 2 into temperature-independent and temperature-dependent steps termed, respectively, phase 2_{fast} and phase 2_{slow}. Evidence is presented to show that phase 2_{fast} arises from the perturbation of a damped elastic element in the cross-bridge and that phase 2_{slow} is the manifestation of an endothermic, order-disorder transition responsible for *de novo* tension generation. These responses are common to both frog and rabbit fibers. The only difference between animals is that the kinetics of phase 2_{slow} appears to scale with the working temperature of the muscle and not absolute temperature. Rigor fibers heated above the working temperature of the muscle contract. Tension generation is, as with activated fibers, endothermic. Tension transients following a laser temperature-jump of activated and heated rigor fibers are virtually indistinguishable on the basis of either the form or magnitude of the response. In length-jump experiments, tension recovery by heated rigor fibers consists of three exponentials with a tension-dependent rate for the medium speed step. Preliminary data indicate that the rigor cross-bridge operates over a distance of between 13.5 and 18 nm. Collectively, these data imply that tension generation in muscle arises from accessible conformational states in the proteins of the cross-bridge alone. ATP hydrolysis in active fibers and the heating of rigor fibers simply serve to shift these intrinsic conformational equilibria towards tension generation.

Mechanism of Myofilament Sliding in Muscle Contraction, Edited by
H. Sugi and G.H Pollack, Plenum Press, New York, 1993

INTRODUCTION

Our approach to the study of the mechanism of muscle contraction is to focus on the effect of temperature on the contractile process—the rationale being that many protein conformational changes, and possibly those associated with contraction, might respond in a unique and mechanistically revealing way to temperature. The interest was prompted by earlier experiments in which we showed that rigor fibers depleted of ATP will contract when heated a few degrees above the working temperature of the muscle[1]. At the time, laser T-jump (temperature-jump) experiments revealed that the kinetics of tension generation by activated fibers and by heated rigor fibers are remarkably similar in form. The heat-induced contraction of rigor fibers indicates that an endothermic, order-disorder transition is responsible for tension generation in the system. The central question that arose at this juncture was whether tension generation in activated and heated rigor fibers derives from the same entropy-driven reaction. In this contribution, published and unpublished data are presented that provide support for the notion that contraction has a common origin in both systems. If this ultimately proves to be the case, rigor fiber contraction could provide a useful and much simplified experimental system for detailed studies on the conformational changes that underlie muscle contraction. An equilibrium-based contractile system that can be switched to generate tension by a laser T-jump in less than 1 μs clearly has potential in this context.

Our first task was to investigate the primary events of tension generation (embodied in the kinetics of the Huxley-Simmons phase 2[2)3)]) in activated fibers and to characterize each individual step in as great a detail as possible. We were particularly interested in determining which steps correlated with endothermic reactions. A combination of L-jump (length jump/step) and laser T-jump perturbations were applied to activated fibers in these experiments[4)5)] A protocol, based on the temperature-dependence of the kinetics of tension generation, that would allow kinetic step/s arising from *de novo* tension generation to be identified and separated from those arising from elastic and damped elastic elements and other reactions in the cross-bridge was developed[5)]. In these experiments, the well researched kinetic and mechanical responses that occur during the Huxley-Simmons phases 1 & 2 in frog fibers served as a reference system[2)3)]. The technique was chosen for the purpose because changes in length perturb *all* states that generate tension or bear tension (either transiently or continuously) in a muscle fiber. A maximum number of kinetic steps therefore respond. Response to a T-jump is more limited in that the additional property of temperature sensitivity is required[4)5)]. The relationship of the two techniques to each other is evident when the responses of fibers to a L-jump and to a T-jump are compared. A four exponential response in the millisecond and longer time domain results from a length-step, while the T-jump of a fiber results in a simpler biexponential response[5)]

A complication we encountered in experiments on rabbit fibers was that reports in the literature suggested that tension recovery during phase 2 in the frog differed from that in rabbit psoas fibers. Phase 2 kinetics have been variously described as virtually temperature-independent in rabbit fibers[6)7)] and temperature-dependent in frog fibers[3)8)]. Since it was our stated intention to use the extensive studies on the mechanical and kinetic properties of phase 2 in the frog in the interpretation of our data, this issue had to be dealt with before progress could be made. The apparent paradox was resolved by showing that the H-S phase 2 is comprised of two physically different reactions in both frog and rabbit fibers[4)5)]. The one step, termed phase 2_{fast}, has a rate that is virtually temperature-insensitive while the other, phase 2_{slow}, has a rate that increases markedly with temperature above $5°C$[4)5)]. The key realization was that the rate of phase 2_{slow} in both rabbit and frog fibers scales not with absolute temperature but with the working temperature of the muscle. Contradictory reports in the literature[6)7)] had arisen because phase 2_{slow} had been excluded from the H-S phase 2 in rabbit fibers resulting in Q_{10} values close to 1.0—a value typical of the included temperature-insensitive step, phase 2_{fast}. The final conclusion from the experiments, which will be described in greater detail below, was that *de novo* tension generation in both rabbit and frog fibers results from a single endothermic order-disorder transition in the time domain of the H-S phase 2[4)5)]. Tension generation in both contracting rigor and activated fibers appears to be entropy driven.

Studies on the kinetics and mechanics of tension generation by contracting rigor fibers have been extended in recent unpublished T-jump and L-jump experiments. Considerable effort was expended to test whether contaminating ATP, or for that matter ADP[9)], could have given rise to the observed tension transients that bear an otherwise remarkable resemblance to those seen when activated fibers are similarly perturbed. The application of various protocols to remove all nucleotide from the fibers left the kinetics of tension generation unchanged from responses observed in earlier experiments[1)]. In recent T-jump experiments single fibers were used in place of four fiber bundles and the pre-load tension was doubled. Comparison of these improved data with T-jump data on activated single fibers[5)] revealed that the tension transients of both are remarkably similar in form and magnitude. It is next to impossible to distinguish T-jump tension transients obtained from perturbing heated rigor fibers from similarly perturbed activated fibers without prior knowledge of the source. These results are of interest because they imply that the T-jump transients in normal activated fibers and contracting rigor fibers arise from accessible conformational equilibria in the proteins of the cross-bridge alone. ATP hydrolysis or heating rigor fibers to above the working temperature of the muscle simply serves to shift these extant conformational equilibria towards tension generation.

Subjecting contracting rigor fibers to step changes in length provides useful insights into the mechanics and kinetics of tension generation. Experiments are presented to show that the L-jump response of a heated rigor fiber is quite different to that seen at lower temperatures where tension recovery is minimal. In many respects, the response observed with contracting rigor fibers resembles that of similarly perturbed activated fibers. Features that serve to distinguish the two systems from one another include a slower overall rate of tension recovery and a kinetic response consisting of three, rather than the four, exponentials in the rigor system. Data is also presented on the distance over which the cross-bridge operates in heated rigor fibers. The results reported are preliminary insofar as they were performed on four fiber bundles subjected to low pre-load tensions. The rigor fiber equivalent of the Huxley-Simmons T_2 appears to decline in a linear manner as the size of the length step is increased. A correction for the contribution from passive tension yields a cross-bridge throw of 13.5 to 18 nm. Experiments on single fibers, subjected to higher pre-load tensions and the inclusion of a procedure to account for the contribution from passive tension at elevated temperatures, will allow this estimate to be improved upon.

The order-disorder conformational change that generates tension can belong to one of two classes. The entropy increase can arise either from ordered water molecules displaced from inter- or intra-protein bonding domains during an assembly type reaction or from a denaturation-like reaction in which there is an increase in the conformational entropy of the polypeptide chain on going from an ordered state to random-coil[10]. It remains to be seen which of these two mechanisms operates during contraction—the main distinguishing feature is that tension generation by a denaturation type mechanism should result in a decrease in fiber stiffness as a result of random-coil formation[5].

TENSION GENERATION IN ACTIVATED FIBERS

Response of an Elastic Element and a Tension Generating Transition in the Cross-bridge to Temperature

The experiments described here were carried out in order to establish a reference for the temperature dependence of an elastic element and a tension generating transition in the cross-bridge. Detailed experiments of others have attributed phase 1 to the response of an elastic element in the cross-bridge[11] and phase 2 to the primary events of tension generation[2,3]. We carried out L-jump experiments at various temperatures and found that the constituent reactions phases 1 & 2 responded in monotonic, but quite different ways, to temperature. Single skinned fibers of rabbit psoas muscle were used in the experiments, details of which can be found elsewhere[5]. The response of the elastic element to temperature is considered first.

The response of the Huxley-Simmons T_1 to temperature (see Fig. 1) serves as the benchmark for the temperature-dependence of an elastic element in the cross-bridge. These data were obtained from L-jump experiments[5] in which Huxley-Simmons phases 2, 3 and 4 were fitted to the sum of four exponentials using a non-linear least-squares routine[12]. This method of analysis allows the contribution of each kinetic step to the overall reaction to be rigorously determined. T_1 values were obtained by the back extrapolation of all kinetic steps to zero time set at the midpoint of phase 1. It is evident from the data in Fig. 1 that the value of T_1 normalized to isometric tension increases with temperature. Our interest is in the related quantity, $1 - T_1/P_0$, which is a measure of the instantaneous stiffness of the cross-bridge. Its temperature dependence provides a 'signature' for the behavior of an elastic element in the cross-bridge. An equivalent response was first observed in frog fibers[3]. In a later section, we correlate these stiffness data with the similar form of the response of phase 2_{fast} (the fast kinetic component of phase 2) to increasing temperature.

The T_2 response illustrated in Fig. 1 provides the benchmark for the temperature dependence of *de novo* tension generation in the cross-bridge. The T_2 values shown are the computed end-points of the two kinetic steps that constitute phase 2 under our experimental conditions[5]. It is generally accepted that T_2 arises

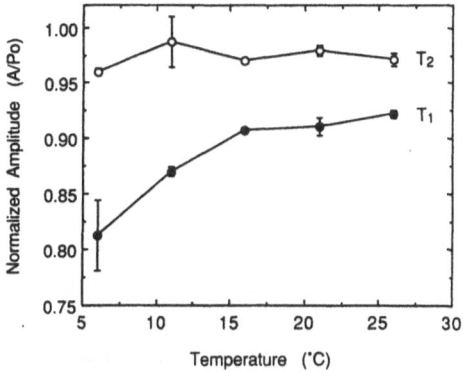

Fig. 1. Temperature dependence of T_1, the tension minimum attained during phase 1, and T_2, the tension maximum reached after phase 2 recovery, in rabbit psoas fibers. T_1 values were determined by the back extrapolation of the kinetics of phases 2, 3 & 4 to zero time; T_2 values from the computed end-point of the reactions of phase 2. Tensions are normalized to isometric tension (P_0) at the temperature of the experiment; a -1.5 nm per half sarcomere L-jump was applied in 160 μs. These data are from reference 5.

from the functioning of the primary events of tension generation[2][3]. Fig. 1 shows that T_2, unlike T_1, appears to remain in fixed ratio to isometric tension regardless of fiber temperature. The mechanistic implication is that the cross-bridge generates tension over a fixed distance regardless of temperature (assuming cross-bridge stiffness remains linear). It therefore appears that a conformational change of definite size generates tension under isometric conditions. These results establish the use of the temperature dependence of T_2 values normalized to isometric tension as a 'signature' of a reaction in the cross-bridge that results in de *novo* tension generation—much the same way as the T_1 response is used as an indicator of an elastic or damped elastic element in the cross-bridge. We later correlate the T_2 response of Fig. 1 with that of phase 2_{slow}.

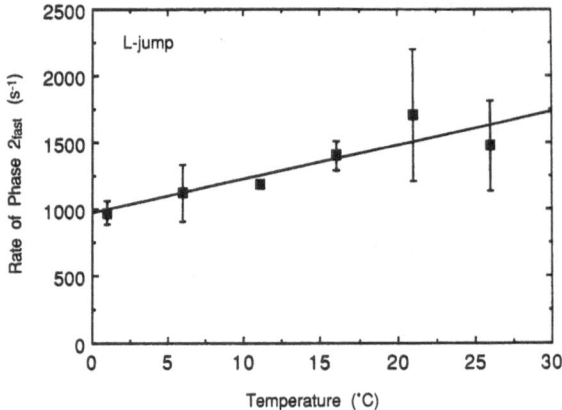

Fig. 2. Temperature dependence of the L-jump kinetics of phase 2_{fast}. The temperature dependence of the rate is non-Arrhenius in form with an approximate Q_{10} of 1.2. Reaction amplitudes were normalized to isometric tension (P_0) at the temperature of the experiment; a step release of -1.5 nm per half sarcomere was applied in 160 μs. These data are from reference 5.

Phase 2_{fast} and Phase 2_{slow} are Physically Different Reactions

The heterogeneity inherent in the kinetics of phase 2 is apparent when T-jump and L-jump kinetic data obtained from isometrically contracting fibers are compared[4)5)]. In L-jump experiments[5)], phase 2 was resolved into two exponential components designated k'_1 and k'_2. We refer to the kinetic steps that give rise to these rate constants as phase 2_{fast} and phase 2_{slow}. In T-jump experiments phase 2_{slow} manifests itself as the faster component ($1/\tau_2$ of reference 5) of a predominantly biexponential process of tension generation. Occasionally a third, small-amplitude fast relaxation ($1/\tau_1$) can be detected at lower temperatures. It differs from the dominant biexponential tension transient because it is thought to arise from the equivalent of a mini L-jump imposed by fiber expansion on heating, and not from a direct effect of temperature on the contractile cycle. In other words, $1/\tau_1$ is thought to arise from phase 2_{fast}. Selected experiments that provided evidence for these various assignments are presented below. Reference 5 should be consulted if it is felt necessary to examine the T- & L-jump transients on which these secondary plots are based. After the evidence for the assignments has been presented, the mechanistic implications of the temperature-dependence of the reaction amplitudes are considered in detail.

The temperature dependence of the apparent rate constant of phase 2_{fast} obtained from L-jump experiments on single skinned rabbit psoas fibers is illustrated in Fig. 2. The apparent rate constant (k'_1, reference 5) of phase 2_{fast} is virtually temperature-independent with a Q_{10} of 1.2. As mentioned earlier, the only reason it appears as a minor component in the T-jump is that the step is perturbed by the equivalent of a L-jump imposed by fiber expansion and not from the direct effect of

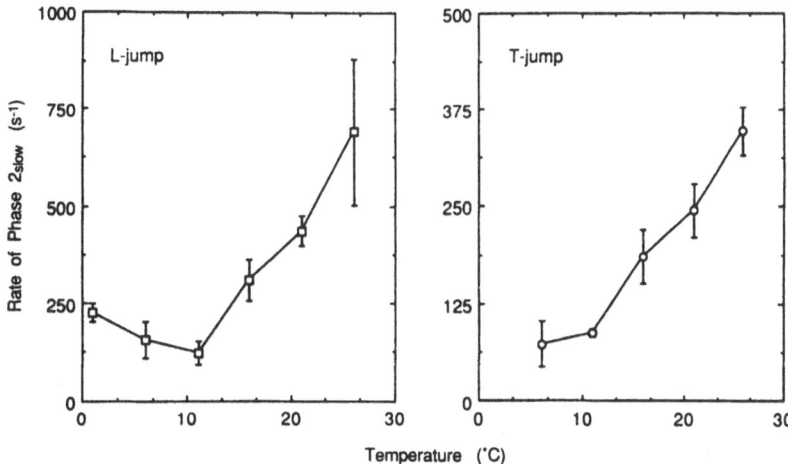

Fig. 3. Temperature dependence of the L-jump and T-jump kinetics of phase 2_{slow}. Note the similar form of the temperature dependence of the apparent rate constants (see k'_2 for the L-jump and $1/\tau_2$ for the T-jump in reference 5). A L-jump of -1.5 nm per half sarcomere was applied in 160 μs. A T-jump of 5°C with a rise time of < 1 μs was applied to the fiber; these data are plotted at the post-jump temperature.

temperature. We could only detect it at low temperatures, conditions where, as we shall later show, its amplitude relative to isometric tension is large.

Inspection of the L-jump data of Fig. 3 shows that the temperature dependence of the kinetics of phase 2_{slow} is quite different from that of phase 2_{fast} (Fig. 2). The apparent rate constant of phase 2_{slow} declines in value between 1 and 11°C; the trend is reversed above 11°C where it increases in a non-Arrhenius manner with an approximate Q_{10} of 3.0. A similar response was observed for the medium speed relaxation ($1/\tau_2$ of reference 5) in T-jump experiments (Fig. 3, right panel) The break in the kinetics at 11°C is less marked in the T-jump data but.the linear dependence of the apparent rate constants on temperature at and above 11°C is evident in both data sets. The increase in rate below 11°C is indicative of a coupling of phase 2_{slow} to an adjacent reaction. We conclude[5] that the apparent rate constant of phase 2_{slow} is temperature dependent. Evidence for the assignment the of the T- and L-jump rate data to individual reactions summarized in the opening paragraph is now complete and we can turn our attention to a discussion of the amplitudes of phases 2_{fast} and 2_{slow}.

Amplitude data are used to establish whether elastic or tension generating elements in the cross-bridge give rise to phase 2_{fast} and phase 2_{slow}. The assignment was made by comparing data presented in Figs. 4 & 5 with the benchmark and similarly normalized T_1 (elastic) and T_2 (*de novo* tension generation) data of Fig. 1.The temperature dependence of the amplitude of phase 2_{fast} (Fig. 4.) obtained from L-jump experiments is considered first. It is evident that the reaction amplitude normalized to isometric tension declines as temperature rises. Comparison of this curve (Fig. 4) and the temperature dependence of T_1 illustrated

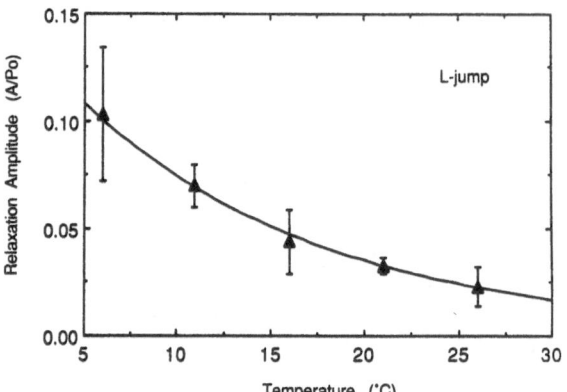

Fig. 4. Temperature dependence of the L-jump amplitude of phase 2_{fast}. Note that the decline in the magnitude of the normalized relaxation amplitude is similar in form to the increase in T_1 illustrated in Fig. 1. Reaction amplitudes are normalized to isometric tension (P_0) at the temperature of the experiment. These data are from reference 5.

in Fig. 1 show that both share a similar form. This correlation, and the virtual absence of a response of either phase 1 or phase 2_{fast} to a T-jump, points to phase 2_{fast} having the physical properties of a damped elastic element and not tension generation.

The temperature dependence of the amplitude of phase 2_{slow} is illustrated in Fig. 5. Above 6°C (the point at which the reaction amplitudes become uncoupled from, and independent of, slower processes) the reaction amplitude remains in fixed ratio to isometric tension[5]. This behavior is similar to the temperature dependence of T_2

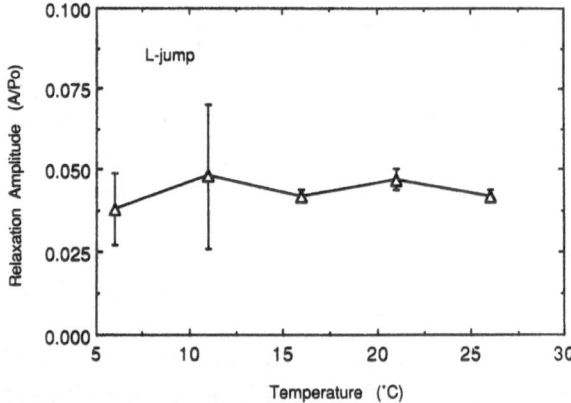

Fig. 5. Temperature dependence of the L-jump amplitude of phase 2_{slow}. Note that the temperature independence of the amplitude of the step is similar to that observed for T_2 in Fig. 1. Reaction amplitudes were normalized to isometric tension (P_0) at the temperature of the experiment. These data are from reference[5].

(see Fig. 1), the earlier established benchmark for *de novo* tension generation in the cross-bridge. Isometric tension rises with increasing temperature and the amplitude of phase 2_{slow} rises in concert. This provides support for the contention that phase 2_{slow} is the one primary tension generating process in muscle fibers—faster tension transients with similar characteristics have not been detected in either L-jump or T-jump experiments on active isometric fibers.

TENSION GENERATION IN RIGOR FIBERS

Tension generation by the rigor state is observed when a fiber is heated above the working temperature of the muscle[1]. A low ionic strength medium consisting of 0.1 mM $MgCl_2$; 20 mM $Na_2B_4O_7$ and 10 mM NaCl at pH 8.3 at 20°C is routinely used in these experiments. If the low ionic strength medium is replaced by one of physiological ionic strength, the temperature at which the onset of tension generation occurs is elevated. It is not clear at this point whether the effect of ionic strength is the result of changes to actomyosin interaction or a modulation of the release of myosin subfragment-2 from the surface of the thick filament as we originally suggested[1]. Possible contamination of the fibers by residual ATP, or even ADP, has been a major concern. Various protocols including the addition of a myokinase inhibitor, ATP and ADP hydrolyzing enzymes, detergents and numerous other strategies have all failed to alter the nature of the contractile response first documented when we described the phenomenon[1]. We feel that, for all intents and purposes, the fibers used in the rigor contraction experiments were free of ATP.

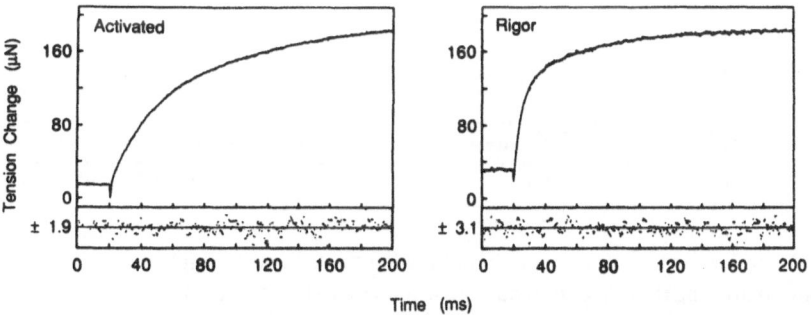

Fig. 6. Kinetics of tension generation following the T-jump of a single activated fiber, and a single contracting rigor fiber of rabbit psoas muscle. The tension transient of the activated fiber at a post-jump temperature of 6°C was fitted with $1/\tau_2$ of 46.5 s^{-1} and $1/\tau_3$ of 9.6 s^{-1}; the sarcomere spacing was 2.6 μm (see reference 5 for details). Two transients were averaged. The tension transient of the contracting rigor fiber at a post-jump temperature of 51°C was fitted with relaxation times of 233 s^{-1} and 23 s^{-1}; the sarcomere spacing was 2.6 μm. A single tension transient was recorded. Both fibers were subjected to a T-jump of 5°C. Residuals obtained by subtracting the fitted line from the data are plotted below each transient

Laser T-Jump of Rigor Fibers

Tension records similar to that of Fig. 6 prompted the research described here. They show that both activated fibers at 1°C, and rigor fibers at 46°C, generate tension in a kinetically well controlled manner when subjected to a T-jump of 5°C. The quality of the rigor T-jump tension transients has been enhanced by the use of single fibers and an improved method for rapidly heating (1 to 50°C in 40 s) the fiber cuvette to the pre-jump temperature[5]. The T-jump of the activated fiber illustrated in Fig. 6 was performed at low temperatures; conditions under which the isometric tension of rabbit fibers is particularly sensitive to temperature. The amplitude of the observed tension transient is therefore the maximum attainable at temperatures above 1°C. Comparably large amounts of tension are generated by the rigor fiber when it is subjected to a T-jump. In activated fibers ATP is the energy source; in rigor fibers thermal energy drives contraction. Biexponential tension transients were recorded for both activated and rigor fibers. The high quality of the fit can be judged from the residuals plotted below each tension transient. Rigor fibers can clearly generate similar amounts of tension to fibers contracting normally. The rigor fiber would have generated twice the tension of the activated fiber had it been matched to the activated fiber on the basis of the similar kinetic constants. The observation that tension generation in activated and rigor fibers is so remarkably similar in form and amplitude points to a common source of tension generation in both contractile systems. That contraction is not a spontaneous process but is induced by heating shows the reaction to be entropy-driven and an order-disorder transition.

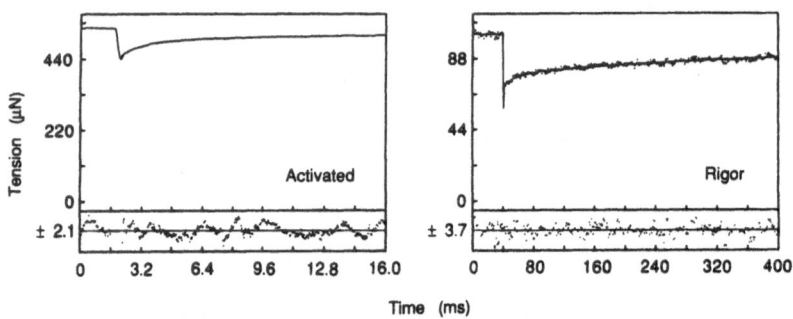

Fig. 7. Kinetics of tension generation following the L-jump of single activated fiber, and a single contracting rigor fiber of rabbit psoas muscle. Activated fiber tension recovered 96% of isometric tension at 6°C. Sarcomere spacing was 2.6 μm and a L-jump of -1.5 nm per half sarcomere was applied. Tension recovered with apparent rate constants of 1076 s^{-1} for phase 2_{fast} and 163 s^{-1} for phase 2_{slow}. Eight tension transients were averaged. Rigor fiber tension recovered 91.5% of the pre-jump tension at 51°C. Sarcomere spacing was 2.4 μm and a L-jump of -1.5 nm per half sarcomere was applied. Tension recovery is triexponential with k'_1 = 2298 s^{-1}; k'_2 = 62 s^{-1}; k'_3 = 3.0 s^{-1}. A single tension transient was recorded. Residuals obtained by subtracting the fitted line from the data are plotted below each transient.

L-Jump Experiments on Contracting Rigor Fibers

The L-jump experiments on contracting rigor fibers were carried out in order to determine whether the response resembled that of an activated fiber. Fig. 7 illustrates the response of an activated and a contracting rigor fiber subjected to a similar -1.5 nm step change in length. It is apparent that the recovery of tension by a single rigor fiber at 51°C resembles, in both form and extent, that seen for activated fibers. Tension recovery by rigor fibers consists of a slower three, rather than four, exponential process. The fall in tension that occurs during phase 1 (the instantaneous stiffness, dF/dL) in activated fibers, and in the rigor fiber equivalent of phase 1, are of comparable magnitude. Tension recovery by a rigor fiber at 51°C is quite distinct from that seen at lower temperatures where tension recovery following the length-step is minimal.

The response of contracting rigor fibers to step changes in length ranging from -1.5 nm to -13.5 nm are illustrated in Fig. 8. These experiments were carried out in order to determine the distance over which the heated rigor cross-bridge operates. The extent of tension recovery without cross-bridge detachment (the rigor fiber equivalent of T_2 in activated fibers) declines linearly over the range studied and can

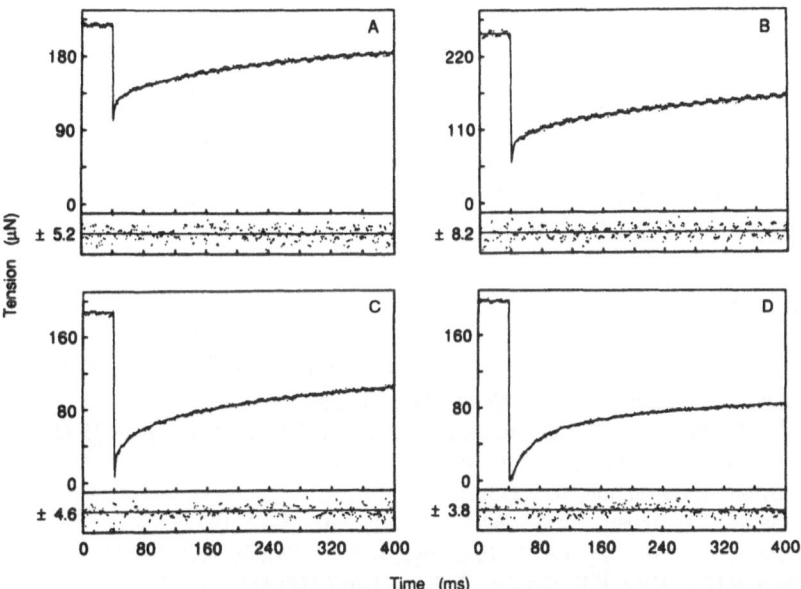

Fig. 8. Kinetics of tension generation by contracting rigor fibers after the application of L-jumps of different sizes. Four fiber bundles with a sarcomere spacing of 2.6 µm were used in the experiment as and a tension pre-load of 60 µN per fiber was applied at a temperature of 51°C. All data are fitted to a three exponential tension transient. Residuals obtained by subtracting the fitted line from the data are plotted below each transient. The extent of tension recovery and the size of the L-jump (applied in 180 µs) per half sarcomere were A: 92%, -1.5 nm; B: 80%, -3.0 nm; C: 66%, -9.0 nm; D: 47%, -13.5 nm.

be extrapolated to a zero tension intercept at -27 nm. This is almost certainly an over estimate of the distance because the pre-load tension includes a proportionately large contribution from passive tension. In these experiments four fiber bundles were still being used. Each individual fiber in the bundle carried a pre-loaded tension of only 60 μN with an estimated passive tension of 20 to 30 μN. The passive tension component raises the zero intercept thereby reducing the throw of the cross-bridge to the 13.5 to 18 nm range. Low tensions were used in the experiments because it was thought that higher tensions might inhibit tension generation. The L-jump data of Fig. 7 on single fibers with a higher pre-load tension show this to be an exaggerated concern. Passive elements appear to respond instantaneously to a L-jump and are not perturbed by a 5°C T-jump[5]. Preliminary analysis of the kinetics indicates that the second exponential process is the primary tension generating step in heated rigor fibers since both its apparent rate constant and amplitude change in proportion to the size of the length-step. We are in the process of using single fibers at slack length with a higher pre-load tension to better define these parameters. Whether this will be sufficient to reduce the throw of the cross-bridge close to the accepted value of 12 nm for fibers contracting isometrically remains to be seen. It is evident that fibers in full rigor (no nucleotide) have some negatively strained cross-bridges[9]; this might contribute to the somewhat larger throw were all rigor cross-bridges to contribute to tension generation and recovery.

ACKNOWLEDGEMENTS

We thank Dr. Michael E. Rodgers for comments on the manuscript and Ms. Connie Clark for assistance with the preparation of the figures. The research was supported by NIH grant AR-04349 to W. F. H.

REFERENCES

1. Davis, J.S. & Harrington, W.F. *Proc. Natl. Acad. Sci. USA* **84**, 975-979 (1987).
2. Huxley, A.F. & Simmons, R.M. *Nature* **233**, 533-538 (1971).
3. Ford, L.E. Huxley, A.F. & Simmons, R M. *J. Physiol. (Lond.)* 269, 441-515 (1977).
4. Davis, J.S. & Harrington, W.F. *Biophys. J.* **59**, 35a. (1991).
5. Davis, J.S. & Harrington, W F. (submitted).
6. Abbott, R.T. & Steiger, G.J. *J. Physiol. (Lond.)* **266**, 13-42 (1977).
7. Goldman, Y.E., McCray, J.A. & Ranatunga, K.W. *J. Physiol. (Lond.)* **392**, 71-95 (1987).
8. Gilbert, S.H. & Ford, L.E. *Biophys. J.* **54**, 611-617 (1988).
9. Dantzig, J.A., Hibberd, M.G., Trentham, D.R. & Goldman, Y.E. *J. Physiol. (Lond.)* **432**, 639-680 (1991).
10. Harrington, W.F. Rodgers, M.E. & Davis, J.S. in *Molecular Mechanisms in Muscular Contraction. Topics in Molecular and Structural Biology* (J. W. Squire, editor) **14**, 241-257 (Macmillan Press, London, 1990).
11. Ford, L.E. Huxley, A.F. & Simmons, R.M. *J. Physiol. (Lond.)* **311**, 219-249 (1981).
12. Johnson, M.L. & Frazier, S.G. *Methods Enzymol.* **117**, 301-342 (1985).

Discussion

Kawai: I believe that your argument correlating phase 2_{fast} to a damped elastic element and phase 2_{slow} to a tension-generating step is based on the temperature sensitivity of these rate constants. Is that right?

Davis: Yes, the temperature sensitivity of rate constants, and particularly the amplitudes.

Kawai: I think it is dangerous to make a correlation based on temperature sensitivity alone. Have you looked at the temperature sensitivity of phase 3?

Davis: Yes, unpublished.

Kawai: Is it higher than phase 2?

Davis: Yes.

Kawai: So, in that sense, phase 3 is a better candidate for the force-generation step.

Davis: You have to be very careful that your reactions are kinetically uncoupled from each other. In this particular instance here, there are no closely-coupled reactions within an order of magnitude of rate of phase 2, whereas phase 3 and phase 4 are at certain temperatures close to each other in rate, and could easily interfere.

Kawai: Actually, I point this out because our interpretation of phase 2_{fast} is the isomerization of the ATP-bound collagen complex. Phase 2_{slow} is a cross-bridge detachment. (See Kawai and Zhao, this volume).

Davis: I agree with you that phase 2_{fast} is an isomerization of some type.

Kawai: It is an isomerization of the AM•ATP complex, and there is a piece of supporting evidence in biochemical literature (Trybus, K.M. and Taylor, E.W. *Biochemistry* **21**, 1284-1294, 1982). Phase 2_{fast} has a low Q_{10} such as 1.4; your 1.3 is very close to it. So, I think a bit more consideration is needed when you try to map observed phenomena to underlying mechanisms.

Davis: I would disagree with you, and I think we should discuss it in detail, because obviously there are many more arguments— dottings of 'i's and crossings of 't's—than I have presented here. There are direct measurements of heat uptake by Gilbert and Ford in the time domain of the Huxley and Simmons phase 2 in frog fibers (Gilbert, S.H. and Ford, L.E. *Biophys. J.* **54**, 611-617, 1988). So, we do have direct evidence for an endothermic reaction. I didn't mention it, but I have done similar experiments to those that I have described on frog fibers, and the kinetics correlate to this extent: the rate of the temperature-sensitive component of phase 2_{slow} in the frog and in the rabbit relates not to absolute temperature, but to the working temperature of the muscle. If you take this into account, the phase 2 kinetics of the two systems correlate very nicely. We also have direct evidence for an endothermic reaction in that time domain. Of course, a process with a temperature-sensitive rate constant and amplitude that generates tension in fixed ratio and directly proportional to fiber tension would support strongly the assignment I have made of phase 2_{slow} to tension generation.

Griffiths: Could you just clarify something for me? If you are making temperature jumps that affect endothermic reactions, are there any changes occurring in the solution in which the fiber is being bathed? For example, does the free calcium change or does the pH change?

Davis: The binding of calcium by EGTA is temperature-sensitive. Sufficient calcium is added to maintain full activation of the fiber. A temperature-insensitive buffer is used to prevent changes in pH.

Lombardi: What is the value of the total change in length you need to go to zero tension?

Davis: In the rigor system a length-step of between 13.5 and 18.0 nm per half sarcomere is required to reduce the T2 recovery to zero. These distances have been corrected for the contribution from passive tension.

Huxley: What was the duration of your step? What I have in mind is something we discussed in our paper—that the extent of the T1 tension change is limited by truncation of the effect. When things get faster at higher temperature, this is more severe. I suspect that much of the reduction of T1 is loss of the early part in the duration of the step.

Davis: The duration of the length-step is of the order of 160 microseconds. I don't think that's the case, because the rate of phase 2_{fast} changes little ($1000/s^{-1}$ to $1700/s^{-1}$) over the temperature range. This means that the amplitude of the step and therefore the zero-time intercept should be relatively well determined. The amplitude of T1, computed by the back extrapolation of the kinetics to zero time at the mid-point of the phase 1 drop is in fact extrapolated. I therefore don't think we have truncation.

VIII. KINETIC PROPERTIES OF ACTIN-MYOSIN SLIDING STUDIED WITH DEMEMBRANATED SYSTEMS

INTRODUCTION

When we use intact muscle or muscle fibers as experimental material, it is difficult to change the environment of myofilaments at will because of the presence of cell membrane. This difficulty is eliminated by use of "demembranated systems", i.e. muscle fibers in which cell membrane is mechanically removed by skinning the fibers or is made freely permeable to small molecules by treating the fibers with glycerol or detergent.

Brenner made careful and extensive studies on force responses of Ca^{2+}-activated permealized rabbit psoas fibers to stretches and releases not only during steady isometric contraction but also during steady isotonic shortening. He suggests that the cross-bridges dissociate from, and reassociate to, actin with a time scale much faster than that of cross-bridge cycling. Ishiwata and coworkers investigated spontaneous oscillatory contractions of glycerinated skeletal and cardiac muscle fibers at various concentrations of MgATP, MgADP and Pi, suggesting that the oscillatory contractions result from transitions between strong- and weak-force-generating cross-bridge states. Tregear and others described their unpublished evidence for cross-bridge movement in insect flight muscles, and discussed possible modes of operation of cross-bridges in muscle contraction.

The next two papers of Kawai and coworkers are concerned with the influence of factors such as MgATP, MgADP and filament lattice spacing on the rate constants of actomyosin ATPase reaction, based on their reaction scheme including six states of attached cross-bridges. Although they claim that their data can best be explained by this scheme, its validity should be verified by techniques other than the sinusoidal analysis that they always use. Gulati studied the role of TnC in the length-dependence of Ca^{2+}-sensitivity in hamster skinned trabeculae using the technique of genetic engineering with interesting results.

Despite the demonstration that myosin head alone can generate force and slide actin filaments with *in vitro* motility assay systems, it is not clear whether actin-myosin sliding observed in *in vitro* assay systems is the same as that actually taking place in muscle. Harrington, Karr and Busa explored the role of myosin subfragment 2 (S-2) in muscle contraction by studying the effect of crosslinking rabbit psoas muscle myofibrils on their shortening induced by rapid photolysis of caged ATP, and suggested that the inhibitory effect of crosslinking the myofibrils results in part from immobilization of myosin S-2 as it is linked to the thick filament. Kobayashi and others also explored the role of myosin S-2 in muscle contraction by studying the effect of anti-S-2 antibody, prepared in the laboratory of Harrington, on the contraction characteristics and ATPase of glycerinated rabbit psoas fibers. The antibody produced parallel decrease of Ca^{2+}-activated isometric force and fiber stiffness in a dose- and time-dependent manner, while the unloaded fiber shortening velocity remained unchanged. In addition, they found that ATPase activity of the fibers remained unchanged even when Ca^{2+}-activated isometric force was reduced to zero in the presence of antibody. These results strongly suggest the essential role of myosin S-2 in muscle contraction.

DYNAMIC ACTIN INTERACTION OF CROSS-BRIDGES DURING FORCE GENERATION: IMPLICATIONS FOR CROSS-BRIDGE ACTION IN MUSCLE

Bernhard Brenner

Department of General Physiology
University of Ulm
Albert-Einstein Allee 11, D-7900 Ulm, FRG.

ABSTRACT

The force response of Ca^{2+}-activated, permeabilized segments of rabbit psoas muscle fibers to stretches and releases was studied. These length changes were imposed (i) during isometric steady state contraction, (ii) as a restretch at the end of a ramp-shaped prerelease, and (iii) during isotonic steady state shortening. The speed of the stretches/releases was varied between about 10 and 10^5 (nm/half-sarcomere)/s. At physiological ionic strength and at low temperature (5°C), the force response to stretches apparently is neither affected by cross-bridges that occupy weak-binding states nor by redistribution among various attached force-generating states. Plots of force vs. imposed length change ("T-plots") and plots of apparent fiber stiffness vs. speed of the imposed length change ("stiffness-speed relations") recorded under all these conditions suggest that cross-bridges, even during force-generation, dissociate and reassociate from and to actin on a time scale that is fast compared to active cross-bridge cycling ($> 50\text{-}1000s^{-1}$ vs. $1\text{-}10s^{-1}$). This rapid dissociation/reassociation of force-generating cross-bridges may provide a mechanism to account for the unexpectedly low ATPase activity during high-speed shortening and for filament sliding exceeding 10-20nm while a cross-bridge passes through the force-generating states.

We previously demonstrated that in permeabilized segments of rabbit psoas muscle fibers under relaxing conditions and in the presence of nucleotide analogues cross-bridges dynamically interact with actin, i.e., dissociate and reassociate from and to actin rapidly on the time scale of active cross-bridge cycling under the same experimental conditions[1-3).

From preliminary stiffness measurements on Ca^{2+}-activated fibers we subsequently proposed that even during force generation cross-bridges dynamically

Mechanism of Myofilament Sliding in Muscle Contraction, Edited by
H. Sugi and G.H Pollack, Plenum Press, New York, 1993

531

interact with actin, again on a time scale that is fast compared to active turnover[4][5]. In these earlier experiments, however, possible redistribution among attached force-generating states[6] prevented us from providing unambiguous evidence for dynamic actin interaction of force-generating cross-bridges. We now established experimental conditions where redistribution among different force-generating states is suppressed to a minimum. This allows us to study and characterize possible dynamic actin interaction of force-generating cross-bridges, again by (i) recording plots of instantaneous force vs. imposed length change (T-plots) during stretches (and releases) of different speed, and (ii) from stiffness-speed relations derived from such T-plots[7][8].

In applying this approach we can provide experimental evidence in support of our proposal that force-generating cross-bridges can dynamically interact with actin, i.e., detach and reattach not just in response to strain imposed on the cross-bridge by stretching or releasing the fiber but even during steady state isometric contraction without imposed additional cross-bridge strain[7-9].

In this study, in addition, we present experiments that allow for testing whether dynamic actin interaction is a general feature of force-generating cross-bridge states or may be limited to only the first of the force-generating states which is the state mainly populated during isometric steady state contraction under our experimental conditions.

MATERIALS AND METHODS

Fiber Preparation and Solutions

Experiments were performed with permeabilized fiber segments of the M. psoas major of rabbits. Fibers were permeabilized by incubation in skinning solution containing Triton X-100. Details of the skinning procedure and isolation of fiber segments were previously described[10][11]. Fiber segments were attached to force transducer and lever system with surgical glue (Histoacryl, Braun Melsungen, FRG) after fixing their ends, while in rigor, with glutaraldehyde dissolved in a rigor solution (glutaraldehyde concentration 5%) containing toluidine blue as a marker (modified from ref. 12).

Preactivating and activating solutions contained (in mM) 10 imidazole, 2 $MgCl_2$, 1MgATP, 1EGTA or CaEGTA, 10 creatine phosphate, and creatine kinase (200 U/ml). Rigor solution contained (in mM) 10 imidazole, 2.5 EGTA, 2.5 EDTA. Ionic strength was set to 170 mM by adding potassium propionate; pH was adjusted to 7.0 at the experimental temperature (5°C).

Experimental Setup

The mechanical apparatus with the system for laser light diffraction was previously described[13][14].

Fiber Stiffness

Apparent fiber stiffness was determined from plots of instantaneous force vs. imposed length change ("T-plots"; ref. 15). Fiber stiffness was defined as the change in force when imposed filament sliding has reached 1 (or 2) nm/half-sarcomere (chord stiffness).

Stiffness-Speed Relations

Plots of apparent fiber stiffness vs. speed of imposed length change ("Stiffness-speed relations") were derived from T-plots recorded during ramp-shaped length changes (stretches or releases) with the speed of length change ranging from 10^{-1} up to 10^5 (nm/half-sarcomere)/s.

Experimental Protocol

The permeabilized fibers were incubated in preactivating solution for about 10-15 min prior to activation. During Ca^{2+}-activation fibers were cycled between isometric conditions (for about 3-5 s) and a 200 ms period of lightly loaded (or unloaded) shortening, followed by restretch to the isometric sarcomere length. This cycling procedure ensures structural stability even during prolonged full Ca^{2+}-activation, as judged from the stability of the striation pattern (10) and the X-ray diffraction pattern (16) of the activated fibers. During one such cycle only one stretch (or release) was imposed on the fiber such that all measurements started from identical steady state conditions.

Experiments were performed at an ionic strength of 170 mM. Our previous studies showed that at this ionic strength only little if any contributions are expected from cross-bridges that dynamically interact with actin in weak-binding states whether at low Ca^{2+}-concentrations[1,2] or at saturating Ca^{2+}-concentrations[17].

RESULTS AND DISCUSSION

(I) Distinction between Redistribution among Force-generating States and Dynamic Actin Interaction during Force Generation

To separate effects of redistribution among force-generating states during and after imposed length changes from effects due to dynamic dissociation and reassociation of force-generating cross-bridges from and to actin we attempted to establish conditions where redistribution among force-generating states is suppressed to a minimum. To establish such conditions, we first examined what effects should be expected upon T-plots if redistribution among force-generating cross-bridge states occur during imposed length changes[15].

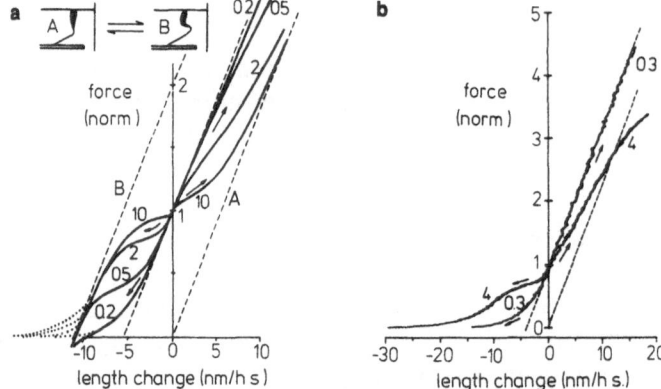

Fig. 1. Plots of force vs. imposed length change ("T-plots"; ref. 15) for Ca^{2+}-activated fibers. (a) Calculated T-plots with redistribution among attached force-generating states (6) assumed to take place. Two attached force-generating states are assumed ("A" and "B" in the schematic diagram at the top). Calculations according to ref. 15. Forward rate constant (A→B)is assumed to be strain dependent. For average strain in isometric steady state for both rate constants 500 s^{-1} is assumed resulting in 1:1 occupancy of the two states. For all cross-bridges in state A, no redistribution, T-plots would correspond to dashed line labeled "A". All cross-bridges in state B, no redistribution, T-plots would correspond to dashed line labeled "B". With 1:1 occupancy, no redistribution, the response during stretch and release would follow dashed line drawn through the isometric steady state force level. With redistribution, during releases dashed line B is approached, during stretches dashed line A. Dotted lines illustrate the behaviour expected in real experiments with fibers buckling rather than excerting negative forces upon the force- transducer. (b) Experimentally observed T-plots. Note the asymmetric appearance with no significant characteristics of redistribution during the stretches. T-plots during stretches do not approach the dashed line drawn through the origin (or any parallel line that were expected if state A already makes contribution to active force). Numbers next to each trace gives time taken for a length change of 5 nm/half-sarcomere (h.s.). Arrows indicate direction of imposed length change. All force values normalized to isometric steady state level. Reproduced from ref. 8.

Expected Effects of Redistribution upon T-Plots. Effects upon T-plots, expected from redistribution among force-generating cross-bridge states, were derived by model calculation based on the proposal of Huxley and Simmons[6)15)]. Cross-bridges which occupy a series of (at least two) force-generating states are assumed to redistribute between these states with rate constant of about 500 s^{-1}. Such redistribution results in characteristic nonlinearity of the T-plots during both stretches and releases (Fig. 1a; ref. 15).

T-Plots Recorded from Ca^{2+}-activated Fibers. Different from the calculated T-plots (Fig. 1a), T-plots recorded from Ca^{2+}-activated fibers under our experimental conditions are highly asymmetric when comparing stretch with release. The nonlinearity characteristic for redistribution among attached force-generating states (e.g. A ⇌ B) is only significant during releases, while during stretches T-plots are

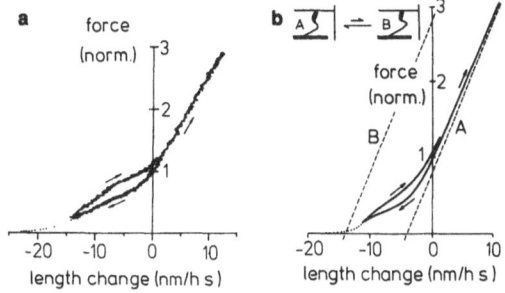

Fig. 2. T-plots during restretch; (a) experimental (b) expected from model calculations. Arrows indicate direction of imposed length change. Dotted line, T-plot expected for larger release amplitude (without restretch). The assumed system for model calculation is schematically shown at the top. Note, occupancy of state A vs. state B is assumed to be 10:1 (under isometric steady state, forward rate constant, A→B, is 100 s^{-1}, reverse rate constant 1000 s^{-1}). Forward rate constant strain dependent. To account for the experimentally observed isometric force, already state A has to make significant contribution to force generation. Thus, an average strain under isometric conditions of 4nm is assumed for state A. All cross-bridges in state A, no redistribution, response during stretch/release would follow dashed line A, all cross- bridges in state B, no redistribution, response would follow dashed line B. Reproduced from ref. 8.

remarkably similar to T-plots recorded under relaxing conditions or in the presence of nucleotide analogues[1][2].

To test whether this indicates that redistribution among force-generating states only occurs (to a detectible extent) during releases, possibly due to insignificant occupancy of a second force-generating state (e.g. "B" in Fig. 1a), we studied the response during a restretch after prereleasing the fiber. Speed and amplitude of the prerelease were chosen such that redistribution favoring occupancy of a subsequent force-generating state ("B" in Fig. 1a) is possible (Fig. 2a). Occurrence of such redistribution is evidenced by the characteristic nonlinearity in the T-plot recorded during the prerelease. Restretching the fiber with the speed of the prerelease now results in a T-plot that clearly shows the characteristic nonlinearity arising from redistribution among attached force-generating cross-bridge states. This demonstrates that once redistribution of cross-bridges among force-generating states is possible during a stretch, the T-plots recorded during such stretch do show the characteristics expected from the redistribution reaction.

In conclusion, the highly asymmetric response when starting both stretch and release from isometric steady state contraction suggests that during isometric steady state contraction only the first of those force-generating states is significantly occupied among which such redistribution reactions can occur. Redistribution during stretch can only occur once a subsequent force-generating state becomes significantly occupied during a prerelease.

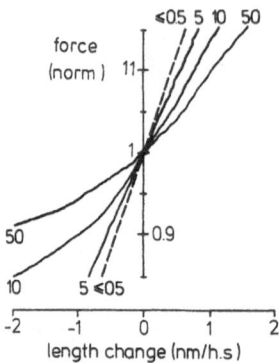

Fig. 3. Near-isometric part of T-plots recorded during stretches and releases imposed during isometric steady state contraction. For each plot 10 recordings were averaged. Noise was further reduced by a 5-point weighted averaging routine. Reproduced from ref. 8.

(II) Evidence for Dynamic Actin Interaction during Force Generation

Having established conditions at which redistributions among various attached force-generating states are insignificant, we again can make use of T-plots recorded during stretches of different speed and stiffness-speed relations derived from such T-plots to probe for possible dynamic actin interaction during force generation.

(1) T-Plots and Stiffness-Speed Relations Obtained from Stretches Imposed during Isometric Steady State Contraction. The T-plots recorded from Ca^{2+}-activated fibers during stretches (redistribution reactions suppressed) show features that were unexpected without rapid dissociation and reassociation of force-generating cross-bridges from and to actin. (i) T-plots recorded during stretches are nonlinear, but the nonlinearity is quite different from that expected from redistribution among force-generating cross-bridges states (cf. Fig. 1a, b). Instead, the nonlinearity is quite similar to that observed in relaxed fibers or in the presence of nucleotide analogues. (ii) The slope of the T-plots depends on the speed of the imposed length change even for $\Delta SL \rightarrow 0$ (Fig. 3). This again is very similar to conditions where cross-bridges dissociate and reassociate from and to actin, e.g. in relaxed fibers or in the presence of nucleotide analogues.

The shape of stiffness-speed relations derived from such T-plots of Ca^{2+}-activated fibers is also quite similar to stiffness-speed relations obtained in relaxed fibers or in the presence of nucleotide analogues (Fig. 4), and is consistent with rapid dissociation/reassociation of force generating cross-bridges.

From the position of the stiffness-speed relation along the abscissa we can estimate the magnitude of the rate constant for cross-bridge dissociation (k^-) during force generation[18)19)]. During force generation we find k^- in a range from about 50 s^{-1} to $\geq 1000 \ s^{-1}$). This is more than an order of magnitude larger than the rate constants for active cross-bridge cycling under the same experimental conditions (1-10 s^{-1}; ref. 20). The wide range of rate constants presumably results from different

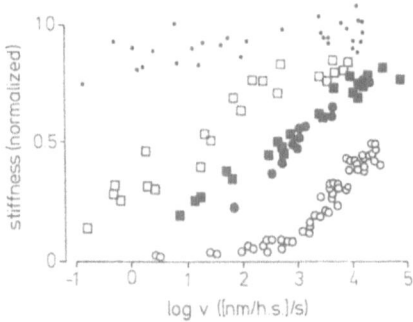

Fig. 4. Plots of apparent fiber stiffness vs. speed of imposed length changes ("stiffness-speed-relations"). ■T-plots recorded from Ca^{2+}-activated fiber during stretch, ●T-plots from same fiber during release. For comparison stiffness-speed relations for fiber in rigor (•), fiber relaxed, ionic strength 20 mM (O), in the presence of 4 mM MgPPi (□). Reproduced from ref. 8.

strain experienced by the force-generating cross-bridges during isometric contraction[21]. From the high actin affinity (strong-binding cross-bridge states; $k^+/k^- >> 1$) the rate constant for reattachment has to be at least an order of magnitude larger than k^-.

(2) T-Plots and Stiffness-Speed Relations Obtained from Releases Imposed during Isometric Steady State Contraction. T-plots recorded during releases (Fig. 1b) show the distinct nonlinearity characteristic for redistribution between attached force-generating states. However, the very initial part of the T-plots is again dependent on the speed of the imposed release which disagrees with the model calculation if *only* redistribution between attached force generating states is assumed (ref. 15; Fig. 1a). More importantly, however, for all speeds of release/stretch, the slope of the very initial part of the T-plots recorded during releases is essentially identical with the slope of the initial part of T-plots recorded during stretches (Fig. 3); i.e., for $\Delta SL \rightarrow 0$ there is no distinct discontinuity in the slope of T-plots for stretch vs. release of comparable speed. This becomes evident from the almost identical stiffness-speed-relation derived from stretches compared with that derived from releases. Only for slow releases apparent fiber stiffness decreases more than for slow stretches. This is expected from increasing contribution to chord stiffness of redistribution among force-generating states that are also present during releases.

To fully account for all these observations, for T-plots recorded during releases both type of reactions have to be assumed, i.e., redistribution among different attached force-generating states[6] and rapid dissociation and reassociation from and to actin of cross-bridges during force-generation.

(3) T-Plots and Stiffness-Speed Relations Recorded during Restretch. The apparently insignificant incidence of redistribution reactions between different attached force-generating states during stretches of any speed that were imposed during isometric steady state contraction (Fig. 1b) suggested that only the first of those force-generating cross-bridge states that can become involved in such

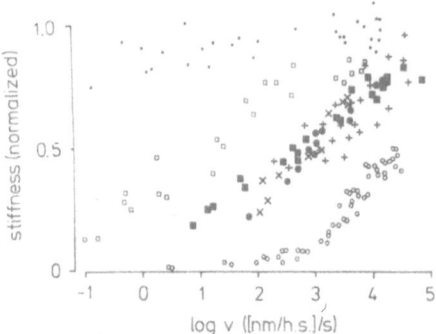

Fig. 5. Stiffness-speed relations for restretch from prerelease (x) and from isotonic steady state shortening (+). For comparison stiffness speed relations for stretch/release of isometrically contracting fiber, fiber in rigor, relaxed at 20mM ionic strength and in presence of 4 mM MgPPi are also shown. Symbols as in Fig. 4. Data for restretch from isotonic steady state shortening (+) scaled to match data for isometric condition (■) at speed of 10^4-10^5 nm/half-sarcomere.

redistribution is significantly occupied. This is further supported by the quite prominent signs of redistribution reactions during a *restretch* subsequent to a release of appropriate speed such that subsequent force-generating states (e.g. state "B" in Fig. 2) could become significantly occupied during this prerelease (Fig. 2).

Thus, evidence for dynamic cross-bridge interaction during force-generation presented so far only applies for this first force-generating state. To test for dynamic actin interaction in subsequent force-generating states of the cross-bridge cycle, we studied T-plots and stiffness-speed relations recorded during restretch following a prerelease or following isotonic steady state contraction. Both interventions are thought to shift cross-bridge distribution toward subsequent force-generating states (cf. Fig. 2; 22).

(a) Restretch from a Prerelease Imposed on Isometrically Contracting Fibers. T-plots during restretch were recorded as illustrated in Fig. 2a. Speed of restretch was varied over a wide range while the prerelease was identical for all measurements. Velocity and amplitude of the prerelease was set such that significant redistribution could occur (pronounced nonlinearity characteristic for the redistribution reactions). Characteristically for dynamic actin interaction, again the initial slope (ΔSL→0) of the T-plots recorded during the restretch is affected by the speed of the restretch, and stiffness-speed relations derived from these T-plots (x in Fig. 5) were essentially indistinguishable from those obtained during stretch (or release) of fibers during isometric steady state contraction. At slower restretch velocities the data points of the stiffness-speed relation are essentially identical with those obtained during release of isometrically contracting fibers. This, again, presumably is due to beginning (and increasing) contributions from the redistribution reactions (as evidenced from the increasing non-linearity of the T-plots characteristic for redistribution reactions; see Fig. 2a) that tend to decrease the chord stiffness (cf. Fig.

4). This suggests that during restretch from a prerelease the same type of reactions occur with similar kinetics as during the release.

(b) Restretch from Isotonic Steady State Contraction. T-plots for restretch from isotonic steady state contraction were recorded after fibers had been allowed to shorten at low load (about 5% of isometric steady state force) for 200 ms. Again velocity of restretch was varied over a wide range while the period of isotonic shortening prior to restretch was always identical. Stiffness-speed relations derived from these measurements (+ in Fig. 5) are essentially indistinguishable from those recorded during stretches (or releases) of isometrically contracting fibers. This again is unexpected if there were no dynamic actin interaction taking place in the states occupied during isotonic steady state shortening with dissociation kinetics quite similar to those observed under isometric steady state conditions.

In conclusion, (1) the results obtained from stretches of isometrically contracting fibers (redistribution among force-generating cross-bridge states suppressed) suggest that cross-bridges, even during force-generation, dynamically interact with actin, i.e., dissociate and reassociate from and to actin on a time scale that is fast compared to active cross-bridge turnover.

(2) Dependence of the very initial part of the T-plots on speed of stretch or release and the lack of detectible discontinuity between T-plots recorded during stretch and release when $\Delta SL \rightarrow 0$ (Fig. 3), resulting in essentially identical stiffness-speed relations for both stretches and releases, suggests that during both stretches and releases force-generating cross-bridges detach and reattach from and to actin with comparable dissociation and reassociation kinetics. The speed-dependence of the very initial part of the T-plots ($\Delta SL \rightarrow 0$) is suggestive that the rate constant for dissociation is non-zero even during steady state isometric contraction. Thus, dissociation with subsequent reassociation are not the result of forcing cross-bridges to detach by imposing strain by stretch or release of the fiber, and therefore the dissociation/reassociation reaction described here is fundametally different from cross-bridge "slippage" that was proposed to occur only after actomyosin bonds are broken by stretching (or releasing) attached cross-bridges beyond a critical level, the "yield" point[23-25].

(3) During releases apparently both type of reactions occur, dissociation and reassociation from and to actin and redistribution among force-generating cross-bridge states.

(4) Stiffness-speed relations for restretch from prerelease or from isotonic steady state shortening, are again essentially identical with those obtained from T-plots recorded during stretch or release from isometric steady state contraction. This is suggestive that detachment/reattachment also occurs in force-generating states subsequent to the state that is mainly populated during isometric contraction* .

---

*One may raise the question as to how force is maintained while cross-bridges in the force-generating states detach and reattach rapidly from and to actin. Thermodynamically, dissociation/reassociation is a reversible reaction and no energy is dissipated. A cross-bridge in the strong-binding states, therefore, may well be able to return to its force-generating configuration several times. That force generating cross-bridges can resume contribution to active force almost

As discussed elsewhere, dynamic dissociation/reassociation of force generating cross-bridges from and to actin may have several implications for cross-bridge action in muscle:

(i) No fixed coupling exists between attachment to actin and transition into the force-generating states, and, more importantly, detachment is not coupled to return from strong-binding to weak-binding states, as they were in the previous quantitative cross-bridge models[22][29][30]. In all these previous models a strong-binding ("force-generating") cross-bridge, in order to detach, was forced to return to the weak-binding states. Thus, in these models a cross-bridge can only remain in the force-generating states as long as filament sliding does not exceed the radius of action of an attached cross-bridge which, from the size of the myosin head, is thought to be at most some 10 nm.

As a consequence of the present results the terms "attachment" and "detachment" can no longer be correlated with cross-bridge cycling between weak- and strong-binding states.

(ii) As an essential difference to the concept of A.F. Huxley[29] and related concepts for cross-bridge action in muscle[22][30] the rapid dissociation/reassociation reaction provides a possible mechanism for the rapid detachment of force-generating cross-bridges during high-speed shortening. With this mechanism for rapid detachment before filament sliding induces large opposing forces in the attached strong-binding cross-bridge, the force-generating cross-bridges, during high-speed shortening, are not forced to return rapidly to the weak-binding states for the rapid detachment. Thus, high-speed shortening does not necessarily entail several-fold larger ATPase activity than under isometric conditions. As a consequence, rapid dissociation and reassociation in force-generating cross-bridge states may account for the apparent discrepancy of high occupancy of force generating states (from stiffness measurements; refs. 31-34) at still low ATPase activity during high-speed shortening[34][35]. Furthermore, with this mechanism filament sliding while a cross-bridge passes through the force-generating states needs not to be limited to some 10 nm, the distance over which a cross-bridge can remain attached to actin.

This concept is also different from that of Podolsky and Nolan[36] who proposed a return to non-force-generating (weak-binding-type) states during high-speed shortening without completion of the ATPase cycle. The concept of Podolsky and Nolan was able to account for the unexpectedly low ATPase during high-speed shortening, however, the distance of filament sliding while a cross-bridge passes through the strong-binding states is still limited to some 10 nm.

---

immediately after a detachment/reattachment reaction was previously demonstrated by studying force redevelopment after periods of isotonic shortening under nonzero loads[7][9]. To resume force-generation after reattachment, we propose that for the strong-binding cross-bridge states the detachment/reattachment reaction is a two-step process, e.g. via a low actin affinity intermediate, similar to the two-step reaction for actin binding that was previously proposed on the basis of biochemical studies[26-28].

(iii) Upon releases, apparently both rapid dissociation with reassociation to a subsequent site on the actin filaments and redistribution among force-generating states can occur. Thus, the rapid early recovery of force in response to a stepwise release, different from the proposal of Huxley and Simmons[6], is expected to result from both type of reactions. Early tension decay after stepwise stretches from isometric steady state contraction, at least in rabbit fibers under our experimental conditions, apparently is mainly the result of dissociation of force-generating cross-bridges with reassociation to a subsequent site on actin. Redistribution reactions, in contrast, are suppressed to a minimum. Effects of substrate concentration upon early recovery[37] suggests that even turnover reactions between strong- and weak-binding states may make some contribution to early tension recovery.

REFERENCES

1. Brenner, B., Schoenberg, M., Chalovich, J.M., Greene, L.E. & Eisenberg, E. *Proc. Natl. Acad. Sci. USA* **79**, 7288-7291 (1982).
2. Brenner, B., Chalovich, J.M., Greene, L.E., Eisenberg, E. & Schoenberg, M. *Biophys. J.* **50**, 685-691 (1986).
3. Schoenberg, M. & Eisenberg, E. *Biophys. J.* **48**, 863-871. (1985)
4. Brenner, B. *Basic Res. Cardiol.* **81**,1, 1-15 (1986).
5. Brenner, B. *Ann. Rev. Physiol.* **49**, 655-672 (1987).
6. Huxley, A.F. & Simmons, R.M. *Nature* **233**, 533-538 (1971).
7. Brenner, B. *Pflügers. Arch.* **412**, R79 (1988).
8. Brenner, B. *Proc. Natl. Acad. Sci. USA*, **88**, 10490-10494 (1991).
9. Brenner, B. *J. Physiol. (Lond.)* **426**, 40P (1990).
10. Brenner, B. *Biophys. J.* **41**, 99-103 (1983).
11. Yu, L.C. & Brenner, B. *Biophys. J.* **55**, 441-453 (1989).
12. Chase, B. & Kushmerick *Biophys. J*. **53**, 935-946 (1988).
13. Brenner, B. *J. Muscle Res. Cell Motility* **1**, 409-428 (1980).
14. Brenner, B. & Eisenberg, E. *Proc. Natl. Acad. Sci. USA* **83**, 3542-3546 (1986).
15. Kuhn, H.J., Güth, K., Drexler, B., Berberich, W., & Rüegg, J.C. *Biophys. Struct. Mech.* **6**, 1-29 (1979).
16. Brenner, B., Yu, L.C. *Biophys. J.* **48**, 829-834 (1985).
17. Kraft, Th., Yu, L.C., Brenner, B. *Biophys. J.* **57**, 410a (1990).
18. Schoenberg, M. *Biophys. J.* **48**, 467-475 (1985).
19. Brenner, B. In: *Molecular Mechanisms of Muscular Contraction.* (ed. J.M. Squire. Macmillan Press Ltd, London, 1990).
20. Brenner, B. *Proc. Natl. Acad. Sci. USA* **85**, 3265-3269 (1988).
21. Hill, T.L. *Prog. Biophys. Mol. Biol.* **28**, 267-340 (1974).
22. Eisenberg E., Hill, T.L. & Chen, Y. *Biophys. J.* **29**, 195- (1980).
23. Güth, K. & Kuhn, H.J. *Biophys. Struct. Mech.* **4**, 223-236 (1978).
24. Lombardi, V. & Piazzesi, G. *J. Physiol. (Lond.)* **431**, 141-171 (1990).
25. Cooke, R. & Bialek, W. *Biophys. J.* **28**, 214-258 (1979).
26. Marston, S.B. & Taylor, E.W. *J. Mol. Biol.* **139** 573-600 (1980).
27. Geeves, M.A., Goody, R.S., Gutfreund H. *J. Muscle Res. Cell Motility* **5**, 351-361 (1984).
28. Taylor, E.W. *J. Biol. Chem.* **266**, 294-302 (1991).
29. Huxley, A.F. *Prog. Biophys.* **7**, 255-318 (1957).

30. Eisenberg, E. & Hill, T.L. *Science* **227**, 999-1006 (1985).
31. Julian, F.J. & Sollins, M.R. *J. Gen Physiol.* **66**, 287-302 (1975).
32. Ford, L.E., Huxley, A.F. & Simmons, R.M. *J. Physiol. (Lond.)* **361**, 131-150 (1985).
33. Brenner, B. *Biophys. J.* **41**, 33a (1983).
34. Brenner, B. *Pflügers Arch.* **411**, R186 (1988).
35. Kushmerick, M.J., & Davies, R.E. *Proc. R. Soc. B* **174**, 315-353 (1969).
36. Podolsky, R.J. & Nolan, A.C. *Cold Spring Harb. Symp. Quant. Biol.* **37**, 661-668 (1973).
37. Kawai, M. *Biophys. J.* **22**, 97-103 (1978).

Discussion

Kawai: What fraction of cross-bridges was in a weakly-attached state?

Brenner: The fraction of cross-bridges in the weak-binding state was somewhere around 20%. The ionic strength was chosen to be 170 mM, because, as we previously showed only about 1/20 of the cross-bridges in the weak-binding states are attached to actin at any given moment under those conditions (Brenner, B. et al. *Biophys. J.* **50**, 685-691, 1986).

Huxley: Lombardi and others have shown very clearly that rapid stretch applied to an isometric fiber causes breakage, if you like, of links, followed by very rapid re-attachment—orders of magnitude more rapid than attachment of a cross-bridge that has gone through the whole cycle [Lombardi, V. and Piazzesi, G. *J. Physiol. (Lond).* **431**,141-171, 1990]. I wonder whether you are just seeing the same phenomenon, or whether you really have evidence of a different kind of detachment.

Brenner: Yes. What Lombardi and his co-workers talked about is the slippage that Hans Kuhn discussed many years ago (Güth, K. and Kuhn, H. *J. Biophys Struct. Mech.* **4**, 223-236, 1978). They proposed that slippage occurs only when a cross-bridge is strained beyond a critical level. The minimum stretch required to strain cross-bridges up to this yield point was proposed to be about 5 nm/half-sarcomere. Thus, the slippage reaction should only occur when isometrically contracting fibers are stretched beyond some 5 nm/half-sarcomere. The expanded part of the T-plots around the isometric range (Fig. 3) shows that under our conditions even the very initial part of these T-plots ($\Delta SL \rightarrow O$) is sensitive to speed of stretch. According to Kuhn, with cross-bridge slippage, the initial part of the T-plots should not be sensitive to speed of stretch. This suggests, to me at least, that the dissociation-reassociation reaction that I have proposed over the years and which does not require stretching the cross-bridge beyond a critical level but occurs during isometric and obviously also during isotonic contraction, is something quite different from the slippage Kuhn and Lombardi have been discussing.

Goldman: Bernhard, you have arranged the conditions so that what you call state B is not very highly populated. I think I remember from hearing you talk about this before, that this occurs at a very low temperature. First, I want to know what the temperature is in your experiment and how much force the fiber produces per cross-sectional area. Then, if the force is very low because of the low temperature, what is the situation when you raise the temperature? You have more force in the fiber and potentially more population of state B.

Brenner: The temperature is 5°; the amount of tension is 1-1.2 kg/cm^2. That's at least about a third of what a frog would produce under such conditions. So the tension is not very low. That reminds me of a point I wanted to make. To account for significant isometric force while only state A (Figs. 1, 2) is occupied means that this state is already of the strong-binding (force-generating) type. At higher temperatures the higher force cannot simply result from higher occupancy of the force-generating states (Brenner, B. *Basic. Res. Cardiol.* **81**, 1-15, 1986). Instead, I think redistribution between states A and B might be different, with state B more favored. Thus, at higher temperatures I would expect that also during stretch a redistribution reaction between states A and B would occur. We no longer have a simplified situation as at low temperatures, where we can study dissociation/reassociation of force-generating cross-bridges during stretch, due to the essential absence of redistribution between states A and B on the stretch side at low temperature.

SPONTANEOUS TENSION OSCILLATION (SPOC) OF MUSCLE FIBERS AND MYOFIBRILS MINIMUM REQUIREMENTS FOR SPOC

Shin'ichi Ishiwata, Takashi Anazawa, Takashi Fujita, Norio Fukuda, Hideharu Shimizu and Kenji Yasuda

Department of Physics
School of Science and Engineering
Waseda University
3-4-1 Okubo, Shinjuku-ku, Tokyo 169, Japan

ABSTRACT

Several years ago, we found a new chemical condition for the spontaneous oscillatory contraction of glycerinated skeletal muscle and named it "SPOC". The condition was such that MgATP coexists with its hydrolytic products, MgADP and inorganic phosphate (Pi). Micromolar concentrations of free Ca^{2+} were not necessarily required for this oscillation. Here, we summarize our recent work on the mechano-chemical properties of SPOC not only in glycerinated single fibers and myofibrils of skeletal muscle (fast type) but also in glycerinated small bundles of cardiac muscle; the isometric tension and its oscillation were examined at various concentrations of MgATP, MgADP and Pi while controlling the concentration of free Ca^{2+}; we constructed a three-dimensional "state diagram" taken against the concentrations of MgADP, Pi and free Ca^{2+}. The 3-D state diagram clearly showed the existence of three regions corresponding to three muscular states; the SPOC region was located in between the regions for contraction (without oscillation) and relaxation. Based on these results, we discuss the mechanism of SPOC, especially the minimum requirements for its occurrence. Finally, we suggest that slow shortening and quick lengthening repeatedly occur every half-sarcomere through the transition between the two states, where weak-force-generating complexes or strong-force-generating complexes are dominant; the transition may be induced by a coupling with the mechanical states of cross-bridges and/or thin filaments.

INTRODUCTION

The contractile system of muscle usually takes two states, i.e., contraction and

Mechanism of Myofilament Sliding in Muscle Contraction, Edited by
H. Sugi and G.H. Pollack, Plenum Press, New York, 1993

545

relaxation. It has been reported, however, that auto-oscillation of tension and sarcomere lengths of muscle fibers occurs with either intermediate pCa values[1-5], high pH in the absence of Ca^{2+},[6] or the coexistence of MgATP, MgADP and Pi[7-11]. We named the oscillation induced by the last condition "SPOC"[7)8]. The common characteristics to these oscillations are that the oscillatory state is intermediate between contraction and relaxation. The developed tension is not full but medium, irrespective of whether the oscillation is induced by partial Ca^{2+} activation or the coexistence of MgADP and Pi with MgATP; to distinguish these two conditions for oscillation, we here call the former "Ca-SPOC" and the latter "ADP-SPOC", respectively. To induce SPOC, it seems to be essential to realize a stable intermediate state. In the present work, we made an effort to find out the minimum requirements for SPOC, which may be common to all kinds of SPOCs, and to get deep insights into the molecular mechanism of SPOC.

MATERIALS AND METHODS

Preparation of Glycerinated Muscle and Myofibrils

Skeletal Muscle[12]: a bundle of muscle fibers (3-5 mm thick, 5-7 cm long) dissected from rabbit psoas muscle was tied to a thin glass rod (3-4 mm diam.) with a cotton thread which was boiled in distilled water and dried beforehand. It was then immediately immersed in cold glycerol solution (pH, about 7.5) containing 50 % (v/v) glycerol, 0.5 mM $NaHCO_3$ and 5 mM EGTA (usually, 1 mM leupeptin was also added to protect the proteolysis of elastic filaments[12]) and stored over night (longer than 12 hrs) at 0°C. The next day, the glycerol solution was exchanged for a fresh one and about 6 hrs later stored in a deep freeze (-20°C). Three weeks later we started to use it. To prepare myofibrils, several pieces of a small bundle of glycerinated fibers (about 1 mm thick, 5 mm long) were immersed in a few ml of solution A containing 60 mM KCl, 5 mM $MgCl_2$, 10 mM Tris-maleate buffer (pH 6.8) and 1 mM EGTA at 0°C. Then the solution was gently homogenized at a low speed for about 10 s with a homogenizer (Ultra turrax; IKA-WERK STAUFEN, Staufen, FRG; 8N type of edge) on ice. To remove glycerol and other materials released from fibers, the solution was centrifuged at 2000 rpm for 10 min and the pellet was dispersed in solution A. This procedure was repeated two to three times to prepare a homogeneous suspension of myofibrils. This washing procedure could be replaced by washing under an optical microscope. The Triton treatment was done by washing with solution A containing 1 % (v/v) Triton X-100 either under a microscope or with centrifugation.

Cardiac Muscle: A small segment of muscle bundle was dissected from a papillary muscle of a bovine heart and treated with a glycerol solution through the same procedure as described above. More details will be described elsewhere[14)15].

Tension Measurement

The isometric tension of fibers (single fibers for skeletal muscle and small bundles for cardiac muscle; both, 40-120 μm thick and a few mm long) was measured with a tension transducer using a standard method[10)15)]. To measure the tension of myofibrils, we devised a simple method: both ends of isolated myofibrils were held by a pair of glass micro-needles under an inverted phase-contrast microscope and the deflection of the flexible needle was measured with the accuracy of 50 nm by using a computer(image processor)-aided system of optical microscopy. Thus developed tension of an order of 10-100 μg (1 μg = 10 nN) could be measured with an accuracy of 50 ng (1 ng = 10 pN). The average sarcomere length of myofibrils was estimated from the total length divided by the number of sarcomeres. Details will be published elsewhere[11)].

RESULTS AND DISCUSSION

Figure 1 shows an example of a time course of length oscillation of sarcomeres in a myofibril obtained under a standard SPOC condition. The oscillation of sarcomeres consisted of a slow shortening phase and a quick lengthening phase with a period of 2 to 3 s. The sliding of thick and thin filaments could be recognized from the change in the width of the overlapping region and the accompanying change in the width of the H-zone[8)9)]. This occurred reversibly and repeatedly. Sometimes this oscillation propagated from one end to the other of the myofibril (we call such a state of SPOC an "organized" SPOC[9)11)]; cf. Fig. 5). Figure 1 shows an example of an "unorganized" SPOC, where there is no organization of oscillations among adjacent sarcomeres, although the oscillation of each sarcomere appears to be regular with time.

Next, we examined the effects of MgADP and Pi on the isometric tension of rabbit psoas glycerinated single muscle fibers and compared the results with those of cardiac muscle. The results are summarized schematically in Fig. 2. Figure 2a shows that in the presence of Ca^{2+}, the tension increased slightly with the addition of MgADP and then decreased. On the other hand, in the absence of Ca^{2+} (Fig. 2b), muscle, which is relaxed without MgADP, developed tension along a sigmoidal curve with the addition of MgADP (ADP-induced tension generation), reaching 80 to 90 % of the tension in the presence of Ca^{2+}. Although the final level of tension was high, the shortening velocity was slow, about one tenth of that at a normal contraction condition. The concentration of MgADP at which the increase of tension shows a half-maximum (critical MgADP concentration) increased in proportion to the concentration of MgATP, suggesting the competitive binding of MgADP and MgATP to the binding sites on the actomyosin complexes. Such features were common to both types of muscle. The only difference was that the critical MgADP concentration was lower in cardiac muscle irrespective of the presence or the

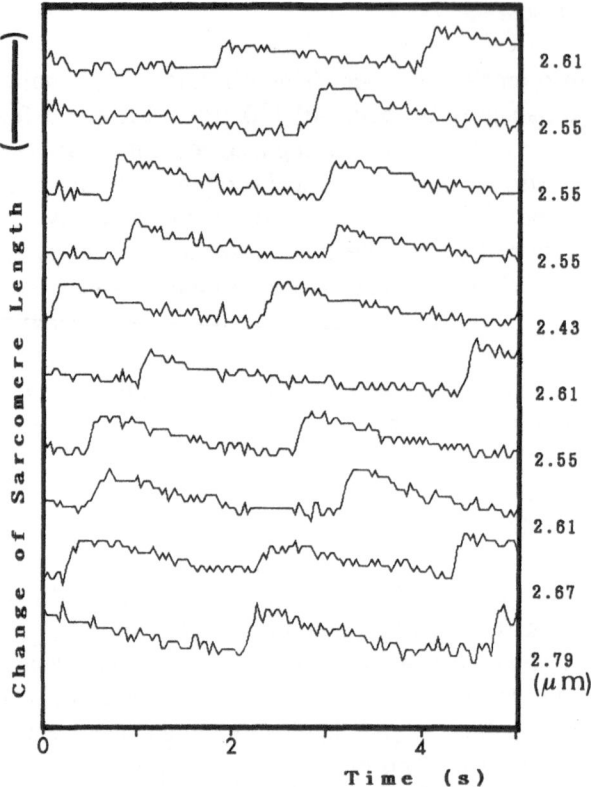

Fig. 1. Time course of length change during SPOC of 10 adjacent sarcomeres in a central part of a myofibril (rabbit psoas) of which both ends were attached to a glass surface; here, the myofibril is placed vertically. The myofibril was labeled with rhodamine-phalloidin to clearly identify the position of Z-lines under a fluorescence microscope 13) . The distances between the Z-lines of the sarcomeres (sarcomere lengths) were measured from a single frame of video tape by using an image processor every 1/30 s, and only the portion of the change is shown on the ordinate. The average length of each sarcomere is shown on the right (S.D. = 0.12 μm); the average position of each sarcomere is arbitrary. Conditions (a standard SPOC condition): 0.12 M KCl, 0.2 mM ATP, 4 mM ADP, 4 mM MgCl$_2$, 4 mM Pi, 4 mM EGTA, and 10 mM MOPS (pH 7.0) at room temperature. Vertical scale bar, 0.5 μm.

absence of Ca^{2+}. This suggests that the actomyosin system of cardiac muscle has relatively higher binding affinity for MgADP.

In living muscle, it has been established that contraction and relaxation are regulated by micromolar concentrations of free Ca^{2+},[16)17)], where it is essential to regulate the state of thin filaments, either on-state or off-state. The ADP-induced activation suggests that the state of thin filaments can be regulated through not only regulatory proteins but also cross-bridges.

Next, we examined the effect of Pi on ADP-induced tension. In the presence of Ca^{2+}, Pi decreased tension monotonically irrespective of the presence or absence of MgADP (Fig. 2a), although MgADP reduced the effect of Pi. In any case, in the

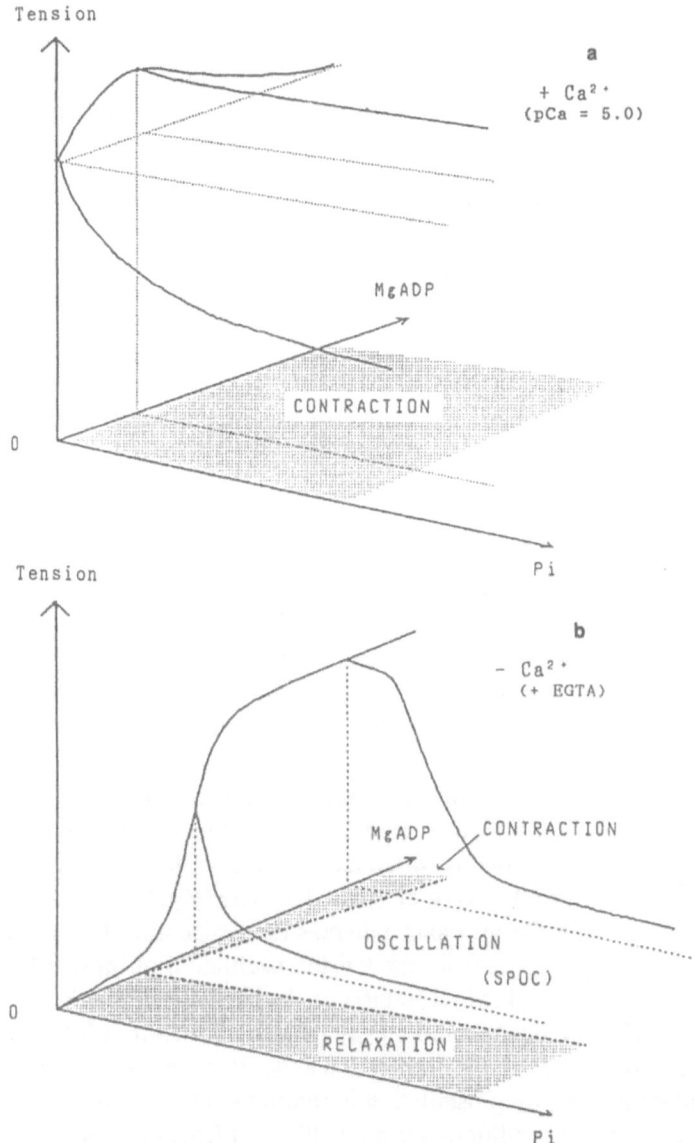

Fig. 2. Schematic diagram showing the isometric tension of fibers developed under various concentrations of MgADP and Pi in the presence of MgATP and the presence (a) or absence (b) of Ca^{2+}. The state of muscle is indicated on the MgADP-Pi plane. This diagram was constructed from data taken under the following conditions by using fibers: 1-3 mM MgATP, 0-20 mM MgADP, 2 mM Mg^{2+}, 0-20 mM Pi, 10 mM MOPS (pH 7.0), 0.1 mM AP_5A and (a) pCa = 5.0 or (b) pCa = 8 (+ 2 mM EGTA), and the ionic strength was maintained at 0.15 M with KCl at 25°C.

presence of Ca^{2+}, the muscle developed tension and contracted. In the absence of Ca^{2+}, on the other hand, the tension decreased sigmoidally and reached a very low level through the addition of Pi (Fig. 2b). Accompanying the decrease of tension, the spontaneous length oscillation of sarcomeres and also the tension oscillation of muscle fibers and myofibrils appeared. So, near the MgADP axis, contraction occurred without oscillation, but with the addition of Pi, the SPOC state appeared. Thus, in the absence of Ca^{2+}, the muscle took three states; the SPOC state was located in between contraction and relaxation.

The state of muscle under the above conditions is summarized schematically in the two-dimensional state diagram on the MgADP-Pi plane (Fig. 2). In the presence of Ca^{2+}, the state of the muscle was always contraction. In the absence of Ca^{2+}, the SPOC region occupied a large region between contraction and relaxation regions. The fact that the SPOC is not a transient state but a stable one suggests that it is a third state for the contractile system of muscle.

In summary, *MgADP* and *Pi* function respectively as an *activator* and an *inhibitor* of tension development. In the ADP-induced activation, *actomyosin(AM)-ADP complex* probably functions as an allosteric desuppressor for the regulatory system, similar to the so-called rigor complex[17].

Next, we examined the effect of Ca^{2+} concentration, in addition to MgADP and Pi, on the isometric tension. Through increasing the concentration of free Ca^{2+}, the tension gradually increased from that without Ca^{2+} to that with Ca^{2+}, that is, the tension curve moved between the two extremes schematically shown in Figs. 2a and b.

To summarize the above results, a three-dimensional state diagram of muscle was constructed. In Fig. 3, three regions corresponding to three states of muscle are schematically illustrated against three axes, pCa and the concentrations of Pi and MgADP. In the case of a fast type of skeletal muscle (Fig. 3a), the SPOC region was difficult to identify on the pCa axis, whereas in cardiac muscle (Fig. 3b) the SPOC region was present (judging from the tension oscillation of muscle fibers)[1)2)5)15]. In cardiac muscle, the ADP-SPOC region on the MgADP-Pi plane and the Ca-SPOC region on the pCa axis were continuously connected by a thin SPOC region, which was sandwiched in between the contraction and relaxation regions; this suggests that the molecular mechanism of oscillation is common to both types of SPOC, although the apparent chemical conditions are quite different from each other.

According to this diagram and the properties of SPOC described above, we present the following hypothesis concerning the minimum requirements for SPOC; these seem to be common to all types of SPOC reported so far. Because the SPOC state is located between contraction and relaxation, i.e., in an intermediate state, the state of cross-bridges will be, on the average, in between a non(or weak)-force-generating state and a (strong)-force-generating state. With the increase of either the concentration of free Ca^{2+} or the MgADP concentration, the proportion of such non(or weak)-force-generating species as AM•ADP•Pi (and/or AM•ATP) complex, which is dominant under relaxation, will decrease as shown in Fig. 4, whereas that of such (strong)-force-generating species as AM•ADP complex, which is dominant

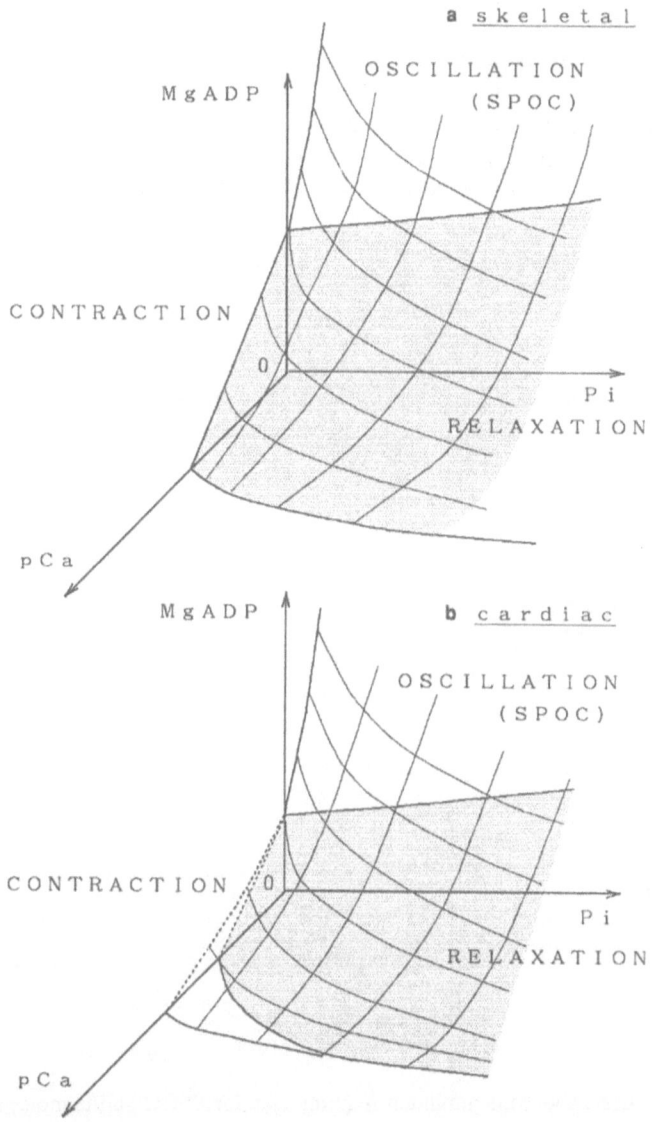

Fig. 3. Three-dimensional state diagram schematically showing the regions of three states of muscle, i.e., spontaneous oscillation (SPOC), contraction (without oscillation) and relaxation in the presence of MgATP. Three axes are taken for pCa (8-5) and the concentrations of Pi (0-20 mM) and MgADP (0-20 mM). Other conditions for the construction of this diagram were the same as those in Fig. 2. The contraction region is located in front of the meshed surface, the relaxation region is behind the dotted surface and the SPOC region is sandwiched between the meshed and dotted surfaces. Diagrams (a) and (b) are for skeletal and cardiac muscle, respectively. The dotted lines in (b) are drawn without experimental data.

under contraction, will increase. Under SPOC conditions, both kinds of complexes will coexist in some proportion (for details about the chemical species of actomyosin complex during ATPase cycle, see refs. 18 & 19).

Now, we infer that for SPOC to occur there is a minimum requirement for the proportions of such complexes: that is, we postulate that the SPOC state appears only when the proportions of both of the types of complexes indicated above are *beyond*

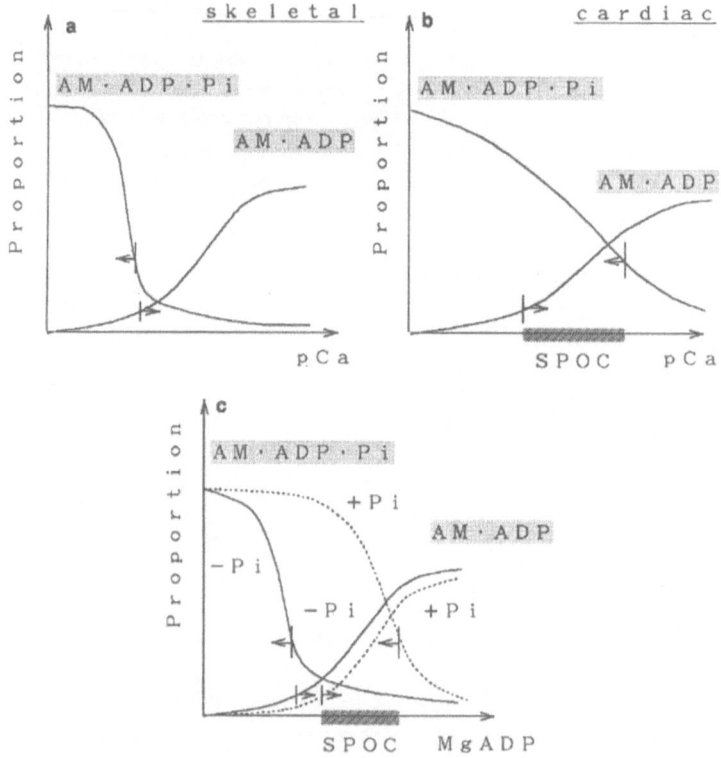

Fig. 4. Hypothesis about the minimum requirements for SPOC on the molecular level. We assume that a stable SPOC state appears only when the proportions of particular chemical species of cross-bridges, such as AM-ADP-Pi (and/or AM-ATP) and AM-ADP complexes, are beyond certain threshold values (indicated by arrows). For more details, see the text.

certain threshold values. This requirement may be fulfilled at submicromolar concentrations of free Ca^{2+} in cardiac muscle (Fig. 4b) because of low cooperativity for tension development, whereas in skeletal muscle the requirement may be difficult to be fulfilled because of high cooperativity (Fig. 4a). When Pi coexists with MgADP, however, the requirement can be fulfilled even in skeletal muscle, so that the SPOC region appears (dotted curves in Fig. 4c).

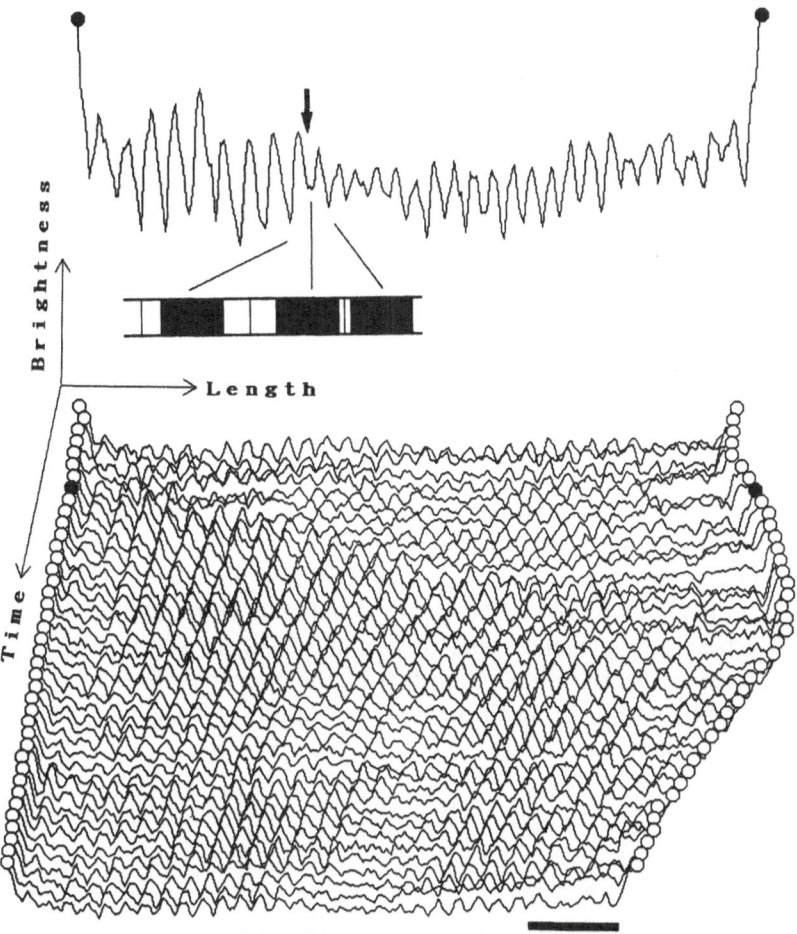

Fig. 5. Time course of the phase-contrast image profile of myofibrils showing the propagation of an "organized" SPOC wave along a myofibril of which both ends were attached to a pair of glass micro-needles. Serial images were taken every 1/15 s from top to bottom. The lengthening phase of sarcomeres propagated from the edge of a stiff micro-needle (the circle on the left end) to the edge of a flexible micro-needle (the circle on the right end). Some parts, such as the upper right and the lower middle, were out of focus, so that the image is not clear. A typical example of an image profile of the myofibril is shown at the top, where the lengthening phase (a few sarcomeres to the left part from the arrow) and the shortening phase (sarcomeres in the right half and the left end) of sarcomeres coexist; the arrow indicates the boundary between the two phases as schematically illustrated below. The developed tension can be estimated from the deflection of the flexible needle. In this example, a small bundle of myofibrils (about 6 μm thick, 32 sarcomeres long) composed of several single myofibrils was used in order to obtain an "organized" SPOC. The other conditions are the same as in Fig. 1. Scale bar, 10 μm.

Finally, we briefly describe the results of a microscopic analysis of "organized" SPOC obtained under an inverted phase-contrast microscope with image processing (Fig. 5). Under standard SPOC condition, a small bundle of myofibrils developed oscillatory tension steadily; correspondingly, a lengthening phase of sarcomeres (to be exact, a half-sarcomere is a unit of oscillation[9]), which appeared from the left end of a myofibril, propagated to the right end along the myofibril (SPOC wave).

Although it is difficult to determine the molecular mechanism of SPOC from such simple experimental analyses, and even though there are still many unresolved problems, we point out one thing deduced from the present study. First, we infer that every sarcomere bears the same load during the SPOC, because the velocity of a mechanical impulse traveling along a myofibril should be very fast. This means that every sarcomere develops the same tension at the same time, irrespective of whether it is in a stretching phase, shortening phase or nearly stopped phase (cf. Fig. 5). This would be possible only through the spontaneous arrangement of the proportions of different chemical states of cross-bridges, such as a non-force-generating state, a weak-force-generating state (e.g., AM•ADP•Pi and AM•ATP complexes) and a strong-force-generating state (e.g., AM•ADP complex), and also a state only sustaining a load (e.g., AM•ADP complex which may be formed by the exogenous ADP). Such an arrangement would be possible from the self-control mechanism through mechano-chemical coupling; that is, the chemical states of cross-bridges would be regulated by the mechanical states (strain) of cross-bridges and/or thin filaments, and *vice versa*.

REFERENCES

1. Fabiato, A. & Fabiato, F. *J. Gen. Physiol.* **72**, 667-699 (1978).
2. Brenner,B. *Basic Res. Cardiol.* **74**, 177-202 (1979).
3. Iwazumi, T. & Pollack, G.H. *J. Cell. Physiol.* **106**, 321-337 (1981).
4. Stephenson, D.G. & Williams, D.A. *J. Physiol. (Lond.)* **333**, 637-653 (1982).
5. Sweitzer, N.K. & Moss, R.L. *J. Gen. Physiol.* **96**, 1221-1245 (1990).
6. Onodera, S. *Jikeikai Med. J.* **37**, 447-455 (1990).
7. Ishiwata, S., Okamura, N. & Shimizu, H. *J.Muscle Res. Cell Motility* **8**, 275 (1987) (Abstr.).
8. Okamura, N. & Ishiwata, S. *J. Muscle Res. Cell Motility* **9**, 111-119 (1988) (Erratum, *ibid.*, **10**, 93 (1989)).
9. Ishiwata, S., Okamura, N., Shimizu, H., Anazawa, T.. & Yasuda, K. *Adv. Biophys.* **27**, 227-235 (1991).
10. Shimizu, H., Fujita, T. & Ishiwata, S. *Biophys. J.* **61**, 1087-1098 (1992).
11. Anazawa, T., Yasuda, K. & Ishiwata, S. *Biophys. J.* **61**, 1099-1108 (1992).
12. Ishiwata, S. & Funatsu, T. *J. Cell Biol.* **100**, 282-291 (1985).
13. Funatsu, T., Higuchi, H. & Ishiwata, S. *J. Cell Biol.* **110**, 53-62 (1990).
14. Fukuda, N., Fujita, T. & Ishiwata, S. *J. Muscle Res. Cell Motility* 12, **304** (1991) (Abstr.).
15. Fukuda, N., Fujita, T. & Ishiwata, S. (manuscript in preparation).
16. Ebashi, S. & Endo, M. *Prog. Biophys. Mol. Biol.* **18**, 123-183 (1968).
17. Weber, A. & Murray, J.M. *Physiol. Rev.* **53**, 612-673 (1973).
18. Goldman, Y. *Ann. Rev. Physiol.* **49**, 637-654 (1987).
19. Homsher, E. & Millar, N.C. *Ann. Rev. Physiol.* **52**, 875-895 (1990).

Discussion

Godt: Your work with SPOC is quite an enterprise. You can get these tension oscillations in skinned fibers, as shown by Richard Nichols at Emory University, in the absence of ATP, ADP, or P_i. By rapidly stretching slow muscle fibers, he produced an oscillation. It never happens in his hands in the fast fibers. I wonder if you think that is the same sort of situation you are looking at with SPOC.

Ishiwata: I don't know the work of Dr. Nichols, but the situation of oscillation seems to be similar to our SPOC. I would like to stress that under our SPOC condition you get a stable auto-oscillation even in fast muscle fibers. It is certain that the SPOC occurs more easily in slow muscle fibers. In other words, the SPOC region in the state diagram of muscle presented here is probably larger in the slow fibers than in the fast fibers.

Pollack: One of the intriguing features of your result is that the signal appears to be propagated very regularly along the myofibril, in spite of the fact that the force is apparently relatively constant. So, there seems to be some signal transmitted from one half-sarcomere or one sarcomere to the very next one, but not necessarily to ones far away. I wonder if you have some speculations. Do you think the work by Professor Ogata, in which he demonstrates that the structure of water is different in the activated versus the non-activated muscle is relevant? (*A collapse of the water structure during the muscle contraction*, poster, this symposium). I wonder if that is a possibility for signalling from one sarcomere to the very next, or if you have some other ideas.

Ishiwata: I think that is one possibility, but I don't have a definite idea about the molecular mechanism. I may have forgotten to point this out, but the propagation of the SPOC wave from one end to the other of the myofibril (organized SPOC) easily occurs when you apply a stretching force. If you release the muscle, tension develops slowly—probably because of the slow shortening velocity—and the organized SPOC occurs with difficulty. Only local propagation of oscillation occurs. Also, when the width of the myofibril bundle is larger, the organized SPOC occurs easily. This may be important information about the SPOC mechanism; it may suggest that mechanical properties such as tension and stiffness are important factors.

Morales: First, let me say I think you are studying a very clever and interesting system. I have one simple-minded idea that may fit very well, forgetting the calcium, in the simpler case of magnesium ADP. It seems to me that there are general rules stating that in a linear system whenever the thing can be approximated by a linear system, it has no periodic solution. But, when you put in ADP and orthophosphate at the same time, then you introduce a non-linearity that probably gives this non-linear auto-oscillation. That seems to fit very well with the way you are getting at this. I would think that to work out the differential equations for the calcium would be valuable. It might relate to pace-making.

Ishiwata: The non-linearity should exist in such an antooscillation mechanism, but I don't know how it plays a part. The oscillation of calcium concentration is not coupled with our SPOC; however, a similar differential equation may be worth considering in our SPOC system.

INFERENCES CONCERNING CROSSBRIDGES FROM WORK ON INSECT MUSCLE

R.T. Tregear, E. Townes*, J. Gabriel** and C. Ellington**

AFRC Institute of Animal Physiology
Babraham
Cambridge CB2 4AT, UK.
*SERC Laboratory
Daresbury, UK.
**Department of Zoology
Cambridge University, UK.

ABSTRACT

This paper presents a number of separate results concerning crossbridge attachment: [1] X-ray diffraction from live bumble bee flight muscle shows a set of layer lines distinct from that of relaxed *Lethocerus*, in which the apparent myosin helix is shorter than that of the actin. [2] Rigor crossbridges of *Lethocerus* are not rotatable by stretch. [3] Rabbit and *Lethocerus* fibres in rigor relaxed by ATP at -35°C show evidence of non-rigor crossbridge attachment.

INTRODUCTION

It is widely assumed that force and motion in muscle is generated by the rotation of a crossbridge on an actin monomer consequent on ATP hydrolysis, although this assumption has been repeatedly challenged and is currently once again under fire[1]. Much of the work that my colleagues and I have performed on *Lethocerus* flight muscle in the last 30 years was concerned with crossbridge movement. The present paper describes some hitherto unpublished evidence on this subject, and the inferences that may be drawn from it.

Mechanism of Myofilament Sliding in Muscle Contraction, Edited by
H. Sugi and G.H Pollack, Plenum Press, New York, 1993

METHODS

1] Material

Live *Bombus* dorsal longitudinal flight muscle (DLM) was prepared by hemidissection of the thorax, which was irrigated with insect saline, and irradiated with X-rays in air at room temperature. *Lethocerus* DLM and rabbit psoas muscle fibres were glycerol-extracted, mounted and observed as previously described[2]).

2] Diffraction Technique

(a) *Bombus* patterns were obtained from beam line 2 of SRS, Daresbury, at 4.8 m specimen-detector spacing. Diffraction was recorded either on a 2-dimensional proportional counter or on film; spacings were estimated from scans across a digital readout of the strongest film. (b) *Lethocerus* equatorials were obtained at EMBL, Hamburg, on a 1-dimensional counter during sinusoidal oscillation of the specimen. The major equatorial peaks (1,0 and 2,0) were collected at each of 8 phase points around the oscillatory cycle.

3] Low-Temperature Mechanical Measurements

Bundles of 1-4 rabbit psoas or *Lethocerus* DLM muscle fibres were mounted on a thermally-insulated mechanical apparatus (Fig. 1), immersed in a low ionic strength 50% ethylene glycol/ 50% water solution (50 mM KCl, 6 mM $MgCl_2$, 5 mM PIPES at pH 7.0, 5 mM EGTA, 1 mM NaN_3) at room temperature and extended by 0.5%

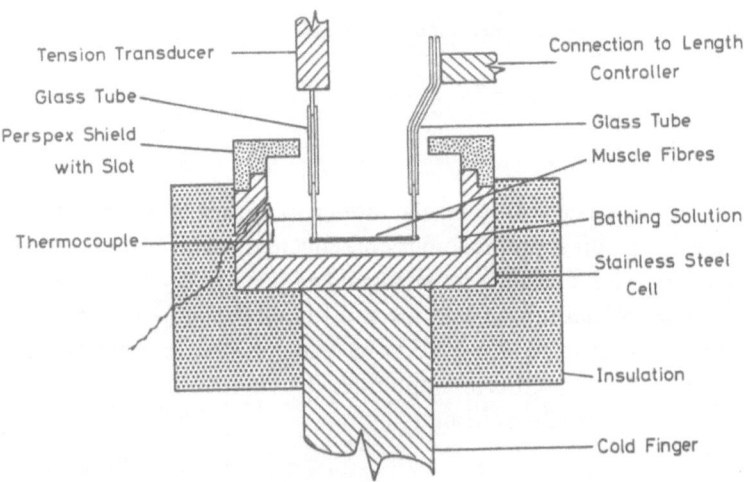

Fig. 1. Apparatus for observation of fibre mechanics at temperatures below 0°C. A bundle of 1-4 fibres was used; the temperature could be lowered as far as the freezing point of the solution allowed.

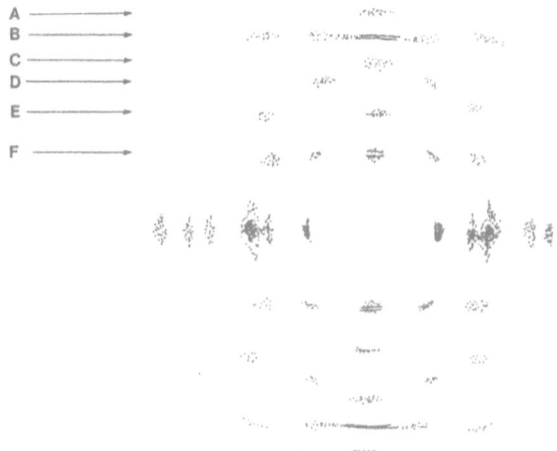

Fig. 2. Sketch of the X-ray diffraction pattern obtained on film from live, relaxed *Bombus* flight muscle at 10°C. The original was too weak to obtain a satisfactory half-tone reproduction.

over slack length. The apparatus was then cooled to -35°C, which is just above the freezing point of the solution. After thermal equilibration a small volume of 50 mM MgATP was added to the bath and mixed by bubbling in air. Isometric tension was monitored. Stiffness was measured in terms of the response to a small amplitude oscillation at 0.1 to 40 Hz. For fixation the fibre bundle was transferred to a second bath within the cooled apparatus, containing 2.5% glutaraldehyde and 0.2% tannic acid in addition to the other components of the solution described above. After 20 min fixation in the cold the fibres were rinsed in 50% ethylene glycol-water and warmed to room temperature.

RESULTS

1] X-ray Diffraction from Live, Relaxed Bumble-Bee Muscle

A characteristic pattern was obtained from 3 specimens, using the 2-dimensional proportional counter. Longer exposure on film gave a weak pattern, many features of which could only just be discerned by eye (Fig. 2); longer exposures were not possible because of radiation damage. The pattern consisted of layer lines sampled on the meridian and on the 1,0 and 2,0 row lines of the lattice spacing. There were sharp interference cuts in the meridian at approx 36.5 nm and 14.5 nm; their width was consistent with interference between adjacent half-sarcomeres. Meridional scans across a digitised film revealed the weakest visible features of the pattern and allowed all but one of the layer lines to be provisionally assigned to the actin or myosin-based order (Table 1). If the 2,0 sampling of layer-line E arises from the myosin origins on the thick filament, as has been supposed for the similar layer-line in relaxed *Lethocerus* muscle, its spacing suggests a helix of n times

Table 1. Spacings (nm) of reflections from live, relaxed *Bombus* flight muscle. The layer lines are identified by their position in Fig 2 and the row lines are given relative to the lattice spacing.

Layer line	Row line		
	0,0	1,0	2,0
A	12.81	-	-
B	14.51*	14.64	14.69
C	16.90	-	-
D	-	19.00	-
E	24.00	-	24.28
F	-	38.71	38.50

Layer lines B and E have been interpreted as related to the helix of crossbridge origins on the thick filament and layer lines A, D and E as related to the actin helix. The meridional C was not assigned. All spacings are cited relative to the '145' meridional (*), whose value is assumed from other work.

$(1/14.7-1/24.3)^{-1} = 37.2$ nm. This appears incommensurate with the likely value of 2 times 38.5 nm for the long helix of actin monomers, as judged either from layer line F of this picture, or the observed value in rigor in *Lethocerus* (or other muscles).

2] Equatorial X-ray Diffraction from Oscillated *Lethocerus* Fibres in Rigor

Sinusoidal oscillation of a small bundle of glycerol-extracted *Lethocerus* DLM fibres in rigor by 0.7% at 5Hz produced no greater variation in the intensities of the 1,0 and 2,0 equatorials than is expected statistically, and there was no obvious phasic relation in the resultant ratio change (Fig. 3). A similar result (not shown) was obtained in the presence of 0.5 mM MgAMPPNP. These data appear to exclude

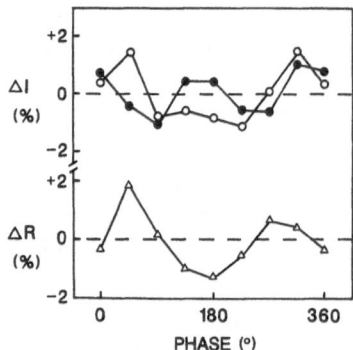

Fig. 3. Phasic changes in the normalised intensities of the equatorial 1,0 (O) and 2,0 (●) X-ray diffraction peaks during a 0.7% oscillation at 5 Hz of a bundle of *Lethocerus* DLM fibres in rigor at room temperature. The phasic change in the ratio (Δ) of the two normalised intensities is also plotted. Each point contained 20-25,000 counts, so the expected SD is 0.6-0.7%.

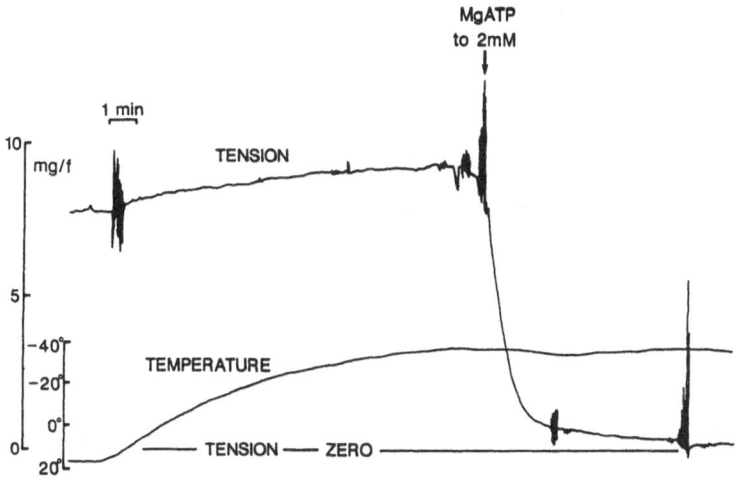

Fig. 4. Tension in a bundle of 2 rabbit psoas fibres in rigor-glycol (see methods for details of solution and technique). After mounting the temperature was lowered over 10 min; MgATP was added and mixed in at the arrow.

alteration of the equatorial ratio by more than 1% during high-stress producing extension.

3] Relaxation of Rabbit and *Lethocerus* Fibres at Sub-Zero Temperatures

On cooling glycerol-extracted rabbit psoas fibres in rigor immersed in a 50% ethylene glycol solution to -35°C, their tension remained nearly constant. When MgATP was added it fell to near zero (Fig. 4). A similar result was obtained with *Lethocerus* fibres. Extension caused only a transient rise in tension. Thus, in mechanical terms, relaxation occurs at this low temperature.

Cooling caused a large increase in rigor stiffness measured at the highest feasible frequency with this apparatus, 40 Hz (Table 2). On adding MgATP the stiffness fell only slightly. A simple interpretation of these data would be that the cold relaxed crossbridges still make contact with actin, but detach at a rate < 40 Hz. However,

Table 2. The effect of cooling rabbit psoas fibres in rigor from room temperature to -35°C, and then adding ATP, on their stiffness measured at 40 Hz.

Muscle condition	Stiffness (mN/fibre)
Rigor at 20°C	39 ± 6
Rigor at -35°C	86 ± 10
Relaxed at -35°C	56 ± 7

Mean ± S.E.M.; 15 experiments.

stiffness rises with cooling in rat tail tendon as well, so that this interpretation may be wrong.

Electron microscopy of *Lethocerus* fibres fixed at low temperature and examined by M.C. Reedy showed regular features, although the 50% glycol necessary in the fixative solution reduced the overall quality of fixation. The rigor fibres appeared to retain angled chevrons, while fibres fixed in relaxation did not; these results are consistent with those obtained previously at room temperature in the absence of glycol.

DISCUSSION

1] Crossbridge Activation

The principal specialisation of 'asynchronous' insect flight muscle is that it requires strain to activate, even in the presence of high calcium[3]. In *Lethocerus*, the helix of crossbridge origins is numerically close to the pitch of the actin helix in rigor[4][5] suggested that match of origin to insertion of the crossbridge caused the strain activation. This is an attractive idea, for which there is experimental support[6]. However it has been attacked recently on the grounds that the helices do not actually fit together in the *Lethocerus* lattice[7]. The present observation from *Bombus* diffraction is also difficult to reconcile with the idea since the two helices appear sufficiently different that match could not be retained over the length of filament overlap; however, the weakness of the film data renders the mismatch uncertain. The current availability of fluorescence-sensitised ('Fuji') plates should allow observations on insect flight muscle in a manner that does not destroy the specimen during irradiation before a strong pattern is obtained, and hence provide confirmation or denial of helical match in asynchronous flight muscle of different insect species.

2] Crossbridge Synchrony

Whatever the cause of strain activation, it occurs particularly dramatically in *Lethocerus* flight muscle[8]. Active tension in excess of 200 kN/m^2 can be obtained from strain applied to a single fibre, provided that the experimental conditions are adjusted to give a metastable but inactive state before stretch. Thus a large fraction of the crossbridges are probably activatable by strain, allowing a similar degree of synchrony to that occurring during electrical activation of intact vertebrate skeletal muscle. To date structural observation of activated flight muscle has not achieved nearly such a high degree of synchrony[9][10] due to the difficulty in observing X-ray diffraction quickly in a single fibre (the only condition under which such large activation can be obtained). However the advent of quick-freezing of mechanically-controlled single fibres followed by freeze-substitution should make electron microscopy of synchronously-activated *Lethocerus* fibres possible. This may show up better images of activated crossbridges than can be obtained from vertebrate muscle, because of the peculiar clarity of crossbridge observation in the insect lattice.

If the strain is released at a suitable rate (e.g. in a sinusoidal oscillation), then a great deal of work can be obtained from the activated crossbridges, and at a high efficiency[3]. Fixation during such shortening should show whether work-producing crossbridges can be distinguished from those producing isometric tension.

3] Crossbridge Deformation

Nucleotide-free rigor is the obvious model for the end state of the crossbridge cycle, prior to ATP binding and detachment. There have therefore been several efforts, over the years, to mechanically influence the rigor structure of *Lethocerus* flight muscle. My own (and, I believe, others) have all been unsuccessful; the structure proved immoveable by forces up to the maximum generated actively. In the experiments shown in Fig. 2 the index of structural change, the 1,0/2,0 ratio R, is a measure of redistribution of mass in the transverse plane of the lattice. Clearly R changed by much less than 1%, if at all. This is remarkable because of two correlated sets of observations. First, binding AMPPNP to the crossbridge changes R by 10-20%, at the same time causing a reversible loss of muscle tension[11]. Second, stretching *Lethocerus* fibres in rigor to a high tension increases their affinity for AMPPNP; thus there appears to be a reversible chemomechanical balance between nucleotide binding energy and mechanical energy[12]. Yet the present experiments indicate that the observable structural index is not altered by stretch. The only obvious way out of this impasse is to suppose that the change in R is unconnected with either tension or nucleotide binding. This is such a curious claim that it would be of great interest to check it, which should now be possible by single fibre equatorial X-ray diffraction.

4] Crossbridge Transition

Muscle in rigor is stiff and holds tension for a long time. These properties can be modified, in a form of graded relaxation, by binding unhydrolysable ATP analogues to the crossbridges. ADP has no obvious effect, AMPPNP speeds tension decay without lowering isotonic stiffness, and AMPPNP binding in 50% ethylene glycol completely relaxes the muscle[2][13]. In *Lethocerus* muscle these mechanical states show characteristic crossbridge conformations. AMPPNP gives a different form of the rigor chevron, in which the tail end of the crossbridge appears to take up a different orientation[14]. AMPPNP in a low glycol concentration, calculated so as to retain stiffness but to hold no tension, gives a completely different image, with thin crossbridges attached at right angles to the filaments[15] and X-ray diffraction confirms this impression[16].

The present results show that addition of ATP to rigor muscle in 50% ethylene glycol at low temperature also gives a state in which stiffness is high but tension is not held. Mechanically the low-temperature ATP-glycol state resembles the AMPPNP-intermediate glycol state at room temperature. The crossbridges bind predominantly unhydrolysed ATP at low temperature[17] so they are an obvious model for the first stage of the crossbridge cycle, as rigor is for the last. At present the structural data from *Lethocerus* says no more than that the crossbridges are different from those in rigor; the angled chevrons have disappeared. Freeze-substitution should provide better images of these ATP-binding crossbridges.

5] Prognosis for Work on Insect Flight Muscle

The first electron micrographs of *Lethocerus* muscle had an immediate impact on crossbridge concepts, because they provided a visual image of rotation[18]. This simple idea is now under considerable pressure, and indeed much of the later insect evidence does not itself readily fit into the concept of crossbridge rotation. With the advent of quick-freezing and freeze-substitution *Lethocerus* may again be of value, because it has shown so far the clearest crossbridge images of any muscle. I hope that activated crossbridge arrays may be analysable in terms of the static arrays seen during partial relaxation; in other words, local areas of similar order to the static arrays may appear in the active muscle.

REFERENCES

1. Vale, R.D. & Oosawa, F. *Adv. Biophys.* **26**, 97-134 (1990)
2. Somasundaram, B., Newport, A. & Tregear, R.T. *J. Muscle Res. Cell Motility* **10**, 360-368 (1989).
3. Tregear, R.T. in *Insect Muscle* (ed. by P.N.R. Usherwood) 357-403 (Academic Press, New York, 1975).
4. Miller, A. & Tregear, R.T *J. Mol. Biol.* **70**, 85-104 (1972).
5. Wray, J.S. *Nature* **280**, 325-326 (1979).
6. Abbott, R.H. & Cage, P.E. *J. Physiol. (Lond.)* **289**, 32-33P (1979).
7. Squire, J.M. *J. Muscle Res. Cell Motility* (in press) (1992).
8. Schaedler, M., Steiger, G.J. & Rüegg, J.C. *Pflügers Arch.* **330**, 217-229 (1971).
9. Armitage, P.M., Tregear, R.T. & Miller, A. *J.Mol. Biol.* **92**, 39-53 (1975).
10. Rapp, G., Guth, K., Maeda, A.Y., Poole, K.J.V. & Goody, R.S. *J. Muscle Res. Cell Motility* **12**, 208-215 (1991).
11. Marston, S.B., Rodger, C.D. & Tregear, R.T. *J. Mol. Biol.* **104**, 263-276 (1976).
12. Kuhn, H.J. *J. Muscle Res. Cell Motility* **2**, 7-44 (1981)
13. Tregear, R.T., Terry, C.S. & Sayers, A.J. *J. Muscle Res. Cell Motility* **5**, 687-696 (1984).
14. Reedy, M.C., Reedy, M.K. & Goody, R.S. *J. Muscle Res. Cell Motility* **8**, 473-503 (1987)
15. Reedy, M.C., Reedy, M.K. & Tregear, R.T. *J. Mol. Biol.* **204**, 357-383 (1988).
16. Tregear, R.T., Wakabayashi, K., Tanaka, H., Iwamoto, H., Reedy, M.C., Reedy, M.K., Sugi, H. & Amemiya, Y. *J. Mol. Biol.* **214**, 129-141 (1990).
17. Tregear, R.T. & Kellam, S. *Eur. J. Biochem.* **155**, 95-98 (1976).
18. Reedy, M.K., Holmes, K.C. & Tregear, R.T *Nature* **207**, 1276-1280 (1965).

Discussion

Pollack: If you look at what you refer to as the myosin reflections, or layer lines, it looks as though there are doublets. What is their significance?

Tregear: You mean the vertical separation? That is an interference crack. They are at the right dimension to be half-sarcomere interference.

Huxley: As regards your second theme about the absence of change in the equatorials, you claim that the rigor cross-bridge is not rotated. In the ordinary way, one interprets equatorials as indicating mass moving from the myosins to the actins. I didn't think of rotation of the kind that might be produced by stretch as showing up in the equatorials. Am I wrong?

Tregear: There are two calculations here, actually. The first is that if there were a change like the change when you bound AMPPNP, then very little of it could have occurred during the pulling. The second, applies to the rotation: you would calculate a fairly large change if you simply took an item which was allegedly at 45° to the thin filament and rotated it as a body, because then you would be moving much of its mass further away from the thin filament. But I agree, I ought to calculate that exactly. I didn't do so because I didn't really believe in that model.

Holmes: I would like to provoke Richard into mentioning his previous set of experiments, which concerned the same thing but looking at the whole of the diffraction pattern using film. In those experiments, I think you saw very few changes anywhere in the diffraction pattern. That set of experiments was never reported, yet they seem perfectly valid to me.

Tregear: I am sure they were valid. I confess that I can hardly remember them myself now after 15 years. Yes, that's right, you couldn't see any changes in the meridionals either. As far as I can remember, we certainly couldn't see any changes in the 143, either in spacing—which we were hoping for—or intensity.

Holmes: The corollary of this would seem to be that the molecules of the cross-bridges that are reporting in our diffraction patterns in rigor cannot be the cross-bridges on which you are pulling. I think that is the only way we can explain this conundrum.

Tregear: No, I don't think that is quite true. Another possibility is that the changes occurring within the molecules that are pulled on do not have an X-ray sign. That is to say, the internal motions are sufficiently small that they do not show up.

ELEMENTARY STEPS OF CONTRACTION PROBED BY SINUSOIDAL ANALYSIS TECHNIQUE IN RABBIT PSOAS FIBERS

Masataka Kawai, Yan Zhao, and Herbert R. Halvorson*

Department of Anatomy
The University of Iowa
Iowa City, IA 52242, U.S.A.
*Division of Physical Biochemistry
Henry Ford Hospital
Detroit, MI 48202, U.S.A.

ABSTRACT

Elementary steps of contraction were probed by sinusoidal analysis technique in skinned fibers from the rabbit psoas muscle during maximal Ca^{2+} activation (pCa 4.55-4.82) at 20°C and 200 mM ionic strength. Our study included the effects of MgATP, MgADP, and Pi concentrations, and an ATP hydrolysis rate measurement. We increased the frequency range up to 350 Hz, and resolved an extra process (D), in addition to well defined processes (A), (B), and (C). Based on these studies, we established a cross-bridge scheme consisting of six attached states, one detached state, and transitions between these states. We deduced all kinetic constants to specify the scheme. The scheme uniquely explains our data, and no other scheme with an equal degree of simplicity could explain our data. We correlated process (D) to ATP isomerization, process (C) to cross-bridge detachment, and process (B) to cross-bridge attachment. We deduced the tension per cross-bridge state, which indicates that force is generated on cross-bridge attachment and before Pi-release. We also found that the rate constants of elementary steps become progressively slower starting from ATP binding to the myosin head and ending by ADP isomerization, and this stepwise slowing may be the essential and integral part of the energy transduction mechanism by muscle.

Mechanism of Myofilament Sliding in Muscle Contraction, Edited by
H. Sugi and G.H Pollack, Plenum Press, New York, 1993

INTRODUCTION

It has been over 30 years since tension transients were initially described in response to sinusoidal length changes[1], and in response to step length changes[2]. At about the same time, length transients were also described in response to step force change[3]. These transients in general contain information on chemomechanical reactions that underlie cross-bridge cycle. Thus, it is possible to characterize elementary steps[4] of contraction in the cycle by studying the various aspects of transients. However, initial studies were based on intact muscle fibers, hence information these experiments offered was limited.

We chose to study tension transients, because of our recognition that force of cross-bridges from different states is additive, hence resolution of various states are easier with tension transients than length transients. We chose to use a sinusoidal waveform of varying frequencies, because temporal resolving power is greater[5] in this waveform than in step waveforms. We chose linear analysis rather than nonlinear analysis, because of our belief that the correct mapping between observed exponential processes and the underlying elementary steps is fundamental to our understanding of the contractile process. We chose to use skinned fibers[6] so that we could apply chemical perturbations as well as mechanical perturbations to the contractile system. The results from chemical perturbations would enhance our knowledge of contractility, because the detailed characterization of cross-bridge kinetics is possible, and comparison of our data to those from biochemical studies is readily available. The results obtained from skinned fiber experiments are essential, because steric constraints on contractile proteins are present so that a load limitation on reaction steps (Fenn effect[7]) is kept intact in this preparation.

As our knowledge on contractile proteins and interactions between actin, myosin, and nucleotides have increased, it has become apparent that cross-bridge cycle consists of multiple states, hence their analysis is complex. An additional complexity is that the rate constants of chemomechanical reactions are thought to be length sensitive[2,7], hence there are multiple combinations of parameters that specify a model. This produces a model which cannot be uniquely determined by experiments. A method which could simplify the cross-bridge cycle is sought, so that one can justify the presence of individual cross-bridge states. A simplification in mathematical formulation is also desired, so that one can comprehend the significance of individual parameters involved in the cross-bridge cycle.

We achieved these simplifications by applying principles developed by Hammes[8]. These are: (1) A reaction much faster than the speed of observation can be approximated by the mass action law, and (2) a reaction much slower than the speed of observation can be assumed not to occur. Simplification 1 is convenient, because this reduces the number of simultaneous differential equations, hence the number of exponential processes to solve. It also reduces the number of independent variables by using equilibrium constants instead of rate constants. Simplification 2 is convenient, because the cross-bridge cycle can be effectively opened at the reaction step slower (<<) than the speed of observation. Based on these principles, we deduced the following cross-bridge scheme 1[9,10].

Step 0   Step 1a   Step 1b   Step 2   Step 3   Step 4   Step 5   Step 6

$$
\text{AMD} \overset{D}{\underset{K_0}{\rightleftarrows}} \text{AM} \overset{K_{1a}}{\underset{S}{\rightleftarrows}} \text{AM}^+\text{S} \overset{k_{1b}}{\underset{k_{-1b}}{\rightleftarrows}} \text{AM}^*\text{S} \overset{k_2}{\underset{k_{-2}}{\rightleftarrows}} \left[\begin{array}{ccc} \text{AMS} & \leftrightarrow & \text{AMDP} \\ \updownarrow & & \updownarrow \\ \text{MS} & \leftrightarrow & \text{MDP} \end{array} \right] \overset{k_4}{\underset{k_{-4}}{\rightleftarrows}} \text{AM}^*\text{DP} \overset{P}{\underset{K_5}{\rightleftarrows}} \text{AM}^*\text{D} \overset{k_6}{\rightarrow}
$$

X_0 X_{1a} X_{1b} X_2 X_3 X_4 X_5 X_6

where A = actin, M = myosin head, S = MgATP, D = MgADP, and P = phosphate (Pi).This scheme consists of six attached states, one detached state, and transitions between these states. It is important to note that this scheme was developed based on our studies from skinned fibers with sinusoidal analysis technique, and it was not taken from biochemical studies. The difference of our scheme from that of biochemistry is that we resolved an extra force-generating state AM*DP, which was not recognized in biochemical studies. However, we do not resolve among cross-bridge states which are detached, hence all detached and weakly-attached states are lumped together in the Det state. Otherwise, scheme I is identical to those derived from biochemical studies[11-13]. With our sinusoidal analysis method in skinned fibers combined with the ATP hydrolysis rate measurement, we were able to determine seven rate constants and three association constants to specify scheme I. It is the purpose of this report to describe how we arrived at scheme I, why it is unique to describe our data, and to demonstrate additional properties of cross-bridges that we learned in the course of our investigation.

METHODS

Chemicals and solutions, experimental procedures, and the methods of calculating complex modulus are the same as in our published work[5)9)10], hence not repeated here. Tension and magnitude data were normalized by the control tension (T_c), which is the tension generated by the control activating solution (mM): 6 CaEGTA (pCa 4.82), 5.3 MgATP, 4.7 free ATP, 15 CP, 8 Pi, 35 KProp, 28 NaProp, 10 MOPS, and 160 units/ml CK. Averaged T_c was 207 ± 8 kN/m^2 (N = 112, ± S.E.M.). All other solutions were variations of the control activating solution. Ionic strength was maintained at 200 mM, pH was adjusted to 7.00, and all experiments were performed at 20°C on rabbit psoas muscle fibers (1-3 fibers/preparation).

The complex modulus Y(f) was fitted to Eq. 1. The extra process (D) was necessary because we expanded our frequency range up to 350 Hz, and we noticed the presence of an additional process.

$$
\begin{array}{ccccc}
 & \text{Process A} & \text{Process B} & \text{Process C} & \text{Process D} \\
Y(f) = H + & A/(1 + a/fi) & - B/(1 + b/fi) & + C/(1 + c/fi) & + D/(1 + d/fi) \\
 & \text{Phase 4} & \text{Phase 3} & \text{Phase 2s} & \text{Phase 2f}
\end{array} \qquad (1)
$$

$$Y_\infty = H + A - B + C + D$$
Phase 1

$\underline{f}$ is the frequency of length oscillation. Peak-to-peak amplitude was fixed to 0.25% L_0. We denote characteristic frequencies of respective processes by a, b, c, and d, and their magnitudes by A, B, C, and D. H is a constant, and i = $\sqrt{-1}$. 2π times characteristic frequencies represent apparent (observed) rate constants. Y_∞ is the complex modulus extrapolated to the infinite frequency. This quantity is referred to as stiffness in the following paper (Zao et al.). Correlation with the phases of step analysis[2][14] is indicated below Eq. 1. Phases 2s and 2f correspond to the slower and faster components, respectively, of phase 2.

RESULTS

Effect of MgATP on the Apparent Rate Constants

To characterize the elementary steps surrounding nucleotide binding, the MgATP concentration was varied, and the apparent rate constants of processes (B), (C), and (D) were studied. The averaged results with S.E.M. error bars (N = 12) are shown in Fig. 1. The plots in Fig. 1 reveal that their MgATP dependence is hyperbolic: all apparent rate constants increased at low millimolar concentrations, and they approached saturation by 5-10 mM. Such MgATP dependence can be explained by a partial cross-bridge scheme of scheme 1, that correlate step 1b to process (D), step 2 to process (C), and step 4 to process (B). Step 1a must be faster

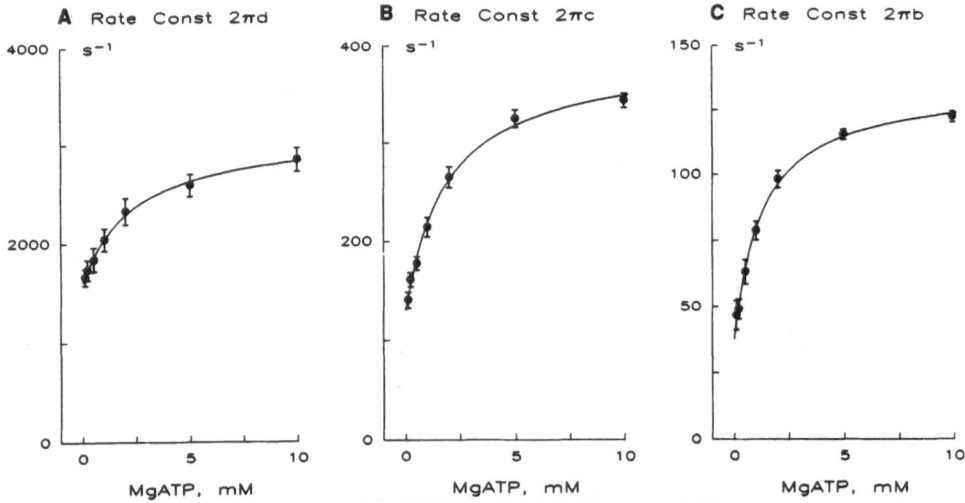

Fig. 1. Dependence of the apparent rate constants, $2\pi d$ (A),$2\pi c$(B) and $2\pi B$ (C) on MgATP concentration.

than our speed of observation, else the apparent rate constant would be linear to the MgATP concentration. The slowest reaction comes last, otherwise the last reaction becomes insensitive to the MgATP concentration[8)9)]. Thus, the above correlations are unique, and they cannot be transposed. The analytical form of equations that describe the apparent rate constants are[10)]:

$$2\pi d = \frac{K_{1a}S}{1 + K_0D + K_{1a}S} k_{1b} + k_{-1b} \tag{2}$$

$$2\pi c = \frac{K_{1b}K_{1a}S}{1 + K_0D + (1 + K_{1b})K_{1a}S} k_2 + k_{-2} \tag{3}$$

$$2\pi b = \sigma k_4 + \frac{K_5P}{1 + K_5P} k_{-4} \tag{4}$$

$$\text{where } \sigma = \frac{K_{1b}K_2K_{1a}S}{1 + K_0D + (1 + K_{1b} + K_{1b}K_2)K_{1a}S} \tag{5}$$

The data in Fig. 1A, B were fitted to Eqs.2-3 by assuming D = 0, and the following kinetic constants were deduced: $K_{1a} = 0.23 \pm 0.04$ mM^{-1} (N = 10, $\pm$ S.E.M.), $k_{1b} = 1880 \pm 220$ s^{-1}, $k_{-1b} = 1510 \pm 110$ s^{-1}, $K_{1b} = 1.29 \pm 0.15$, $k_2 = 510 \pm 30$ s^{-1}, $k_{-2} = 132 \pm 7$ s^{-1}, and $K_2 = 3.9 \pm 0.3$. The equilibrium constant is defined in the usual way (K = k_+/k_-). The results of the kinetic constants are summarized in Table 1 of the following paper. Theoretical projections based on Eq. 2-4 are entered by curves in Fig. 1.

Effect of MgADP on the Apparent Rate Constants

To characterize the step involved in the MgADP binding, the MgADP concentration was changed (0-8 mM) at a fixed MgATP concentration (5 mM), and the apparent rate constants were obtained. In Fig. 2A, $2\pi c$ is plotted against MgADP concentration. As seen in this figure, $2\pi c$ decreased hyperbolically with an increase in MgADP concentration. The same general ADP dependence was observed on $2\pi b$ and $2\pi d$. Such ADP dependence can be explained in the cross-bridge scheme which adds step 0 as shown in scheme 1. The analytical form of the rate constants that include MgADP concentration (D) is already shown (Eqs.2-4). The results were fitted to Eq. 3 with K_{1a}, k_{1b}, and k_{-1b} obtained from the MgATP study, and the theoretical projection is entered in Fig. 2A by a curve. K_0 was deduced to be 0.58 ± 0.09 (N = 7).

Effect of Pi on the Apparent Rate Constant $2\pi b$

To characterize reaction steps involved in the Pi binding, the Pi concentration was varied in the range from 0 to 16 mM at a fixed MgATP concentration (5 mM), and the apparent rate constants were studied. Averaged $2\pi b$ is plotted against Pi

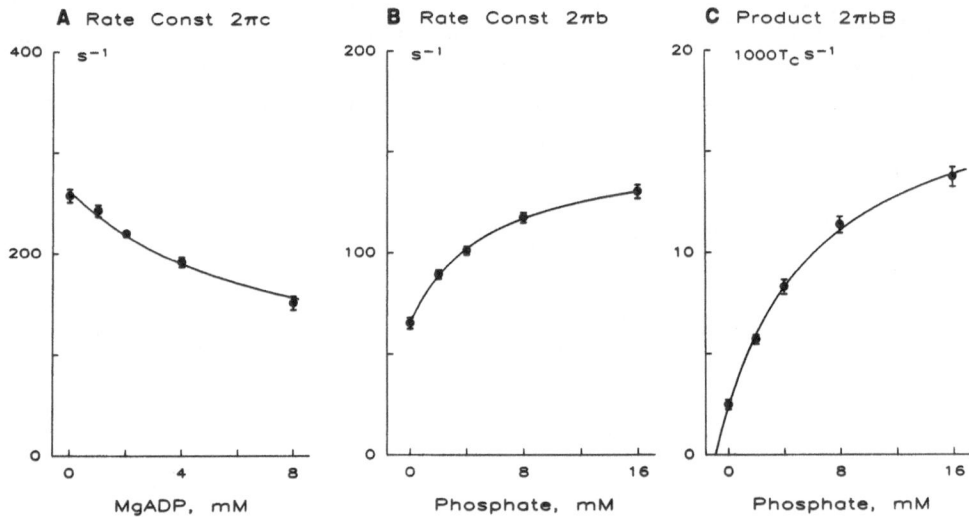

Fig. 2. Effect of MgADP and phosphate (Pi) on the apparent rate constants.

concentration in Fig. 2B ($N = 11$); $2\pi c$ and $2\pi d$ were much less sensitive to Pi (data not shown). Such Pi effect can be explained by adding the Pi-release step 5 (scheme 1), and correlating step 4 with observed exponential process (B). The analytical form for $2\pi b$ is shown in Eq. 4. The results were fitted to this equation ($D = 0$), and the following kinetic constants were deduced: $k_4 = 106 \pm 4\ s^{-1}$ ($N = 11$, $\pm$ S.E.M.), $k_{-4} = 90 \pm 5\ s^{-1}$, $K_4 = 1.20 \pm 0.07$, and $K_5 = 0.19 \pm 0.02\ mM^{-1}$ (Table 1 of the following paper).

The product $2\pi bB$ is proportional to oscillatory work, and it is important because extrapolation to 0 product indicates the average Pi concentration across the fiber when no exogenous Pi is added to the activating solution[9]. This is because the magnitude parameter B is proportional to the number of cross-bridges in the AM*DP state, hence the product diminishes to 0 in the absence of Pi. Fitting the product ($2\pi bB$) is easier than fitting the magnitude parameter (B), because the products Pi dependence follows simple saturation kinetics[9][10]. The product data are plotted in Fig. 2C, and the theoretical projection is the continous curve. From this fitting, we determined $P_0 = 1.0 \pm 0.2\ mM$ ($N = 11$) (observe the negative intercept to the abscissa in Fig. 2C). The negative intercept is the excess Pi concentration in the fibers compared to the solution. Apparently, higher Pi concentration in the fibers resulted, because of the continuous hydrolysis of ATP and the diffusion of Pi out of the fibers, creating a concentration gradient.

Distribution (Probability) of Cross-bridges among Various States.

Assuming that the cross-bridge scheme is open at step 6, and approximating the steady-state with equilibria at each step, we calculated the distribution of cross-bridges among various cross-bridge states (Fig. 3A). In these calculations, we

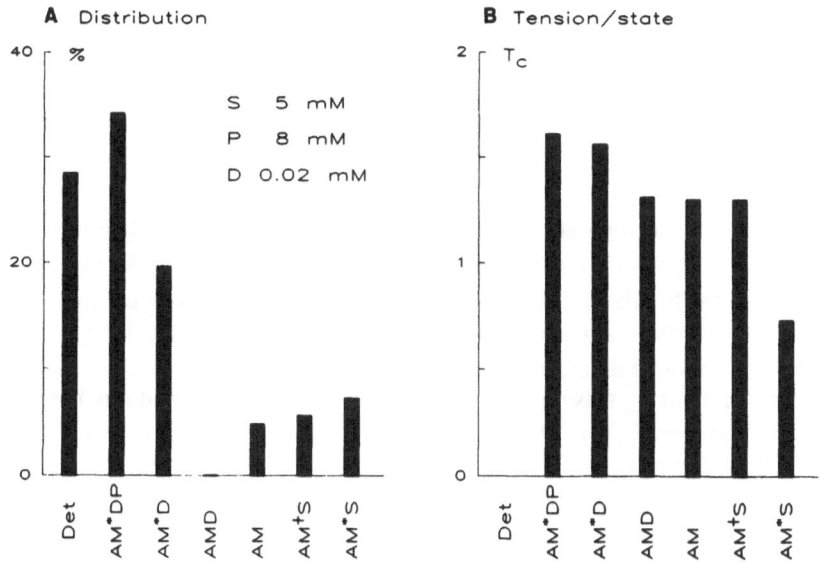

Fig. 3. Distribution of cross-bridges among various states (A) and tension per cross-bridge state (B).

assumed that S = 5 mM, D = 0.02 mM, and Pi = 8 mM (our standard activating condition), and Po = 1.0 as determined in Fig. 2C. Fig. 5A demonstrates that the probability of cross-bridges in states AM, AM+S, and AM*S is small, and it ranges from 5 to 7%. The cross-bridges are distributed primarily in the Det (28%), AM*DP (34%), and AM*D (20%) states. The probability for the AMD state is less than 0.1% because the estimated MgADP concentration is about 20 μM in the presence of the ATP regenerating system.

ATP Hydrolysis Rate and the Rate-limiting Step 6

Step 6 is the slowest of all forward reactions in the cross-bridge cycle[9], and it limits the ATP hydrolysis rate. Else, the above effects of MgATP, MgADP, and Pi could not be explained. The analytical form of the hydrolysis rate is $J = [AM*D]k_6$. Thus, to characterize the step 6, we studied the ATP hydrolysis rate with the method that used NADH fluorescence coupled with ATP consumption in the presence of phospho(enol)pyruvate, pyruvate kinase, and lactic dehydrogenase[15]. The hydrolysis rate was determined to be 0.64 ± 0.05 mMs^{-1} (N = 18), and the corresponding turn over number was 3.2 ± 0.3 s^{-1} assuming that the myosin head concentration is 0.2 mM[16]. Since 20% of cross-bridges are in the AM*D state (Fig. 3A), the rate constant k_6 is 16 s^{-1}.

Isometric Tension and Tension per Cross-bridge State

One way of describing isometric tension is as a linear combination of the probabilities of cross-bridge states:

$$\text{Tension} = T_0 X_0 + T_{1a}X_{1a} + T_{1b}X_{1b} + T_2X_2 + T_5X_5 + T_6X_6$$

$$= T_0 X_0 + T_{1a} X_{1a} + T_x X_{1b} + T_6 X_6 \qquad (6)$$

$$\text{where} \qquad T_x = T_{1b} + K_{1b}T_2 + K_{1b}K_2K_4T_5 \qquad (7)$$

X_i represents the steady-state probability of the cross-bridges in state i as defined in scheme 1, and T_i is the linear coefficient: T_i indicates tension per cross-bridge at state i. Eq. 6 was rewritten as shown based on the mass action law for steps 1b, 2, and 4. This is because X_{1b}, X_2, and X_5 have the same S, P, and D dependence, hence their coefficients cannot be determined independently.

The coefficient T_6 was determined from the Pi dependence of tension (Fig. 4A), T_{1a} was determined from the MgATP dependence (Fig. 4B), and T_0 from the MgADP dependence (Fig. 4C). We obtained (in T_c units): $T_0 = 1.31$, $T_{1a} = 1.30$, and $T_6 = 1.56$ (Fig. 3B). The above studies also determined T_x. When we carried out experiments that compressed muscle fibers with dextran T500, we found that K_{1b} and K_2 were not significantly affected by compression, whereas K_4 changed three fold[10] (following paper). This result was used to determine T_5 based on Eq. 7 (Fig. 4D). From the slope, we obtained $T_5 = 1.61$; from the intercept, we obtained $T_{1b} + K_{1b}T_2 = 2.24$. Although there is no convenient way to determine T_{1b} and T_2 independently at this time, it is useful to make an estimate. Since X_{1b} (AM+S) is a collision complex, the force in this cross-bridge may not be different from the force in the cross-bridge before the collision, X_{1a} (AM). Thus, we approximate by $T_{1b} = T_{1a} = 1.30$. From this, we determined $T_2 = 0.73$. These values are entered in Fig.3B.

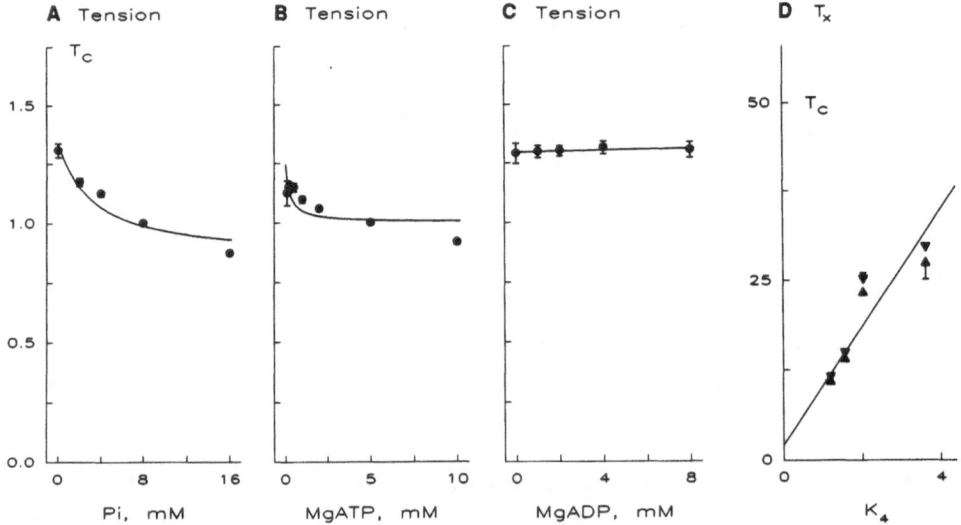

Fig. 4. Factors influencing isometric tension.

DISCUSSION

The most important finding in this report is that scheme 1 is not only consistent with our results, but also uniquely explains our data. Our results show that exponential process (D) correlates with step 1b, process (C) correlates with step 2, and process (B) correlates with step 4. If one transposes these correlations, there emerge inconsistencies in the MgATP, MgADP, and Pi effects on the apparent rate constants. The second important finding is that we can determine independently all the kinetic constants needed to specify scheme 1 with our methods. There is no guess work involved in specifying the scheme, and all the steps can be characterized with our method. The third important finding is that the nucleotide binding steps 0 and 1a, and the Pi binding step 5 are faster than our speed of observation. Because of this, all the binding steps can be approximated by the mass action law.

We found an extra process (D) (Eq. 1) that corresponds to the isomerization of the ATP-bound myosin head (step 1b). The existence of such isomerization step was demonstrated in biochemical studies of myosin subfragment-1 (SF-1) and acto-SF-1 using ATP analogs MgAMPPNP and MgADP[17]. Our temperature study indicates that Q_{10} of $2\pi d$ is 1.4, and much less than Q_{10} (2.9) of $2\pi c$[10)18]. This result is consistent with the report[17] that the isomerization step is much less sensitive to a change in temperature than the subsequent steps in solution studies.

From the studies of isometric tension as functions of MgATP, MgADP, and Pi concentrations, and by assuming a linear combination of the cross-bridge states (Eq. 6), we were able to determine tension per cross-bridge state for states AM*D, AMD, and AM (Fig. 3B). Tension for state AM*DP was determined by experiments that compressed fibers by dextran T500 (Fig. 4D). Our results unambiguously demonstrate that tension is generated by step 4 (Det → AM*DP) and not by Pi-release (step 5) as previously thought[12)19]. Thus, we identify the power stroke step to be the step 4.

In Fig. 3B, it is interesting to point out that tension does not change much before and after collision complex formation between a macromolecule (myosin) and small ions (MgADP, Pi) (steps 0 and 5). This is reasonable, because tension is the result of the macromolecular architecture of the contractile apparatus, and an addition of small ions would not influence the architecture. If we apply the same reasoning to step 1a, then tension in AM and AM+S would be the same as we assumed. This resulted in about half tension for the AM*S state (Fig. 3B). Thus, we can generalize that tension changes on isomerization after a collision complex is formed, and this occurs in step 1b (AM+S → AM*S), in step 4 (AM*DP → Det), and in step 6 (AMD → AM*D).

In the upper portion of Fig. 5A (including broken line), the energy levels demonstrate the free energy change based on the equilibrium constants listed in Table 1 in the following paper. Since we do not have the equilibrium constant for step 6, this was estimated ($K_6 = 100$) based on the data from biochemical studies[20]. The free energy levels of the lower portion of Fig. 5A (continuous lines) represent chemical energy. These energy levels are based on the fact that step 4 is the

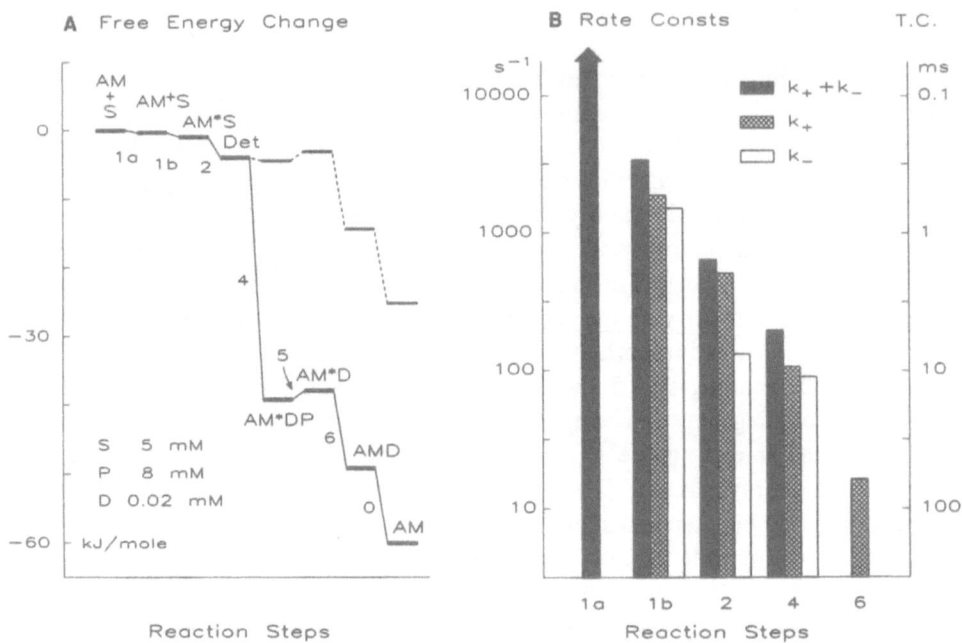

Fig. 5. Free energy change (A) and rate constants (B) depending on reaction steps.

transduction step, and that 60 kJ is liberated when 1 mole of MgATP is hydrolyzed[21]). The mechanical energy as the result of step 4 is retained as potential energy in the cross-bridge, and this is measured as tension. Unlike in solution biochemistry[12]), step 4 in skinned fibers is reversible, because the potential energy in the cross-bridge can be transduced back to chemical energy in the reversal of step 4.

Fig. 5B plots the rate constants of elementary steps starting from the binding of MgATP to the myosin head (step 1a). This figure demonstrates that the speed of reaction becomes progressively slower in each step. The only exception is the Pi-release step 5, which is faster than our speed of observation, and not included in Fig. 5B. It appears from this figure, that one purpose of the chemomechanical transduction is to achieve a slow speed. Usual chemical reactions occur in 1 msec or less, and this would be too fast to generate any motion. A slower step of force generation would be convenient, because it involves a macromolecular rearrangements in the myosin head. The presence of the slowest step 6 after the force generation would be convenient, because there is adequate time for the cross-bridge to perform work in this step.

In conclusion, scheme 1 and our analysis methods have yielded considerable information on cross-bridge kinetics, and there exists several other potential studies to examine additional properties of the contractile apparatus in skinned fibers. These future studies would include the effects of the lattice spacing (following paper), temperature, and ionic strength. These experiments are expected to yield additional insights into the molecular mechanisms of contraction.

ACKNOWLEDGEMENTS

The present work was supported by grants from NIH (AR21530), NSF (DCB90-18096), and Iowa Affiliate of AHA (IA-91-G-13).

REFERENCES

1. Machin, K.E. & Pringle, J.W.S. *Proc. Roy. Soc. (B)* **152**, 311-330 (1960).
2. Huxley, A.F. & Simmons, R.M. *Nature* **233**, 533-538 (1971).
3. Podolsky, R.J. *Nature* **188**, 666-668 (1960).
4. Goldman, Y. *Ann. Rev. Physiol.* **49**, 637-654 (1987).
5. Kawai, M. & Brandt, P.W. *J. Muscle Res. Cell Motility* **1**, 279-303 (1980).
6. Natori, R. *Jikeikai Med. J.* **1**, 119-126 (1954).
7. Fenn, F.O. *J. Physiol. (Lond.)* **58**, 175-203 (1923).
8. Hammes, G.G. *Adv. Prot. Chem.* **23**, 1-57 (1968).
9. Kawai, M. & Halvorson, H.R. *Biophys. J.* **59**, 329-342 (1991).
10. Zhao, Y. & Kawai, M. *Biophys. J.* (submitted).
11. Tonomura, Y., Nakamura, H., Kinoshita, N., Onishi, H & Shigekawa, M. *J. Biochem. (Tokyo)*, **66**, 599-618 (1969).
12. Taylor, E.W. *CRC Crit. Rev. Biochem.* **6**, 103-164 (1979).
13. Eisenberg, E. & Greene, L.E. *Ann. Rev. Physiol.* **42**, 293-309 (1980).
14. Heinl, P., Kuhn, H.J. & Rüegg, J.C. *J. Physiol. (Lond.)* **237**, 243-258 (1974).
15. Takashi, R. & Putnam, S. *Analyt. Biochem.* **92**, 375-382 (1979).
16. Tregear, R.T. & Squire, J.M. *J. Mol. Biol.* **77**, 279-290 (1973).
17. Trybus, K.M. & Taylor, E.W. *Biochemistry* **21**, 1284-1294 (1982).
18. Zhao, Y. & Kawai, M. *Biophys. J.* **61**, A267 (1992).
19. Huxley, A.F. *Reflections on Muscle.* 93-95 (Princeton University Press, Princeton, N.J., USA, 1980).
20. Sleep, J.A. & Hutton, R.L. *Biochemistry*, **19**, 1276-83 (1980).
21. Kushmerick, M.J. & Davies, R.E. *Proc. R. Soc. B* **174**, 315-353 (1969).

Discussion

Godt: How does your scheme compare with what Earl Homsher was talking about in his presentation in this symposium?

Kawai: I think the steps surrounding force generation and P_i-release are virtually identical. One thing I have to add is that the force-generating AM*DP complex was not recognized in biochemical studies. Biochemists' Pi-release step appears to include both step 4 (force generation) and step 5 (actual P_i-release).

Homsher: Have you set the muscle shortening into a steady state, and then tried the frequency analysis to get at the strain dependency?

Kawai: Yes. To find out which step is strain dependent is a consideration we are working on. We think that a strain sensitivity resides somewhere in the substrate-binding step 1a as Kuhn (1977) and Marston et al. (1979) have shown in insect muscles. If you assume that substrate binding increases as you stretch the muscle fiber, then all the tension-transient phases—2, 3, and 4—can be explained. You don't have to assume strain sensitivity on the force-generating step 4.

Simmons: I was wondering how unique you think your ordinary type scheme is, because I was remembering the work of David White and John Thorson (1984) to explain similar things—I can't remember, insect or rabbit—the effect of ATP on the different components that you see. My memory of the results was that, making different assumptions, they could match the experimental data fairly well. I think they were trying to distinguish whether your kind of scheme was right, or whether the force generation took place in the way that Andrew Huxley and I thought it did. An alarming aspect of what came out of their analysis was that they got these matches in both cases, but the rate constants that you would assume from looking at the different components bore no exact relationship to individual rate constants in the kinetic scheme.

Kawai: In the schemes other than mine, you will find inconsistency in the data of apparent rate constants versus MgATP concentration.

Brenner: Masataka, whenever you talk about detachment and attachment, you really mean transition from strong binding to weak binding for the detachment, and transition from weak binding to strong binding for attachment.

Kawai: It is a matter of definition of terms. We do not see weakly attached states directly under our experimental conditions (pCa 4.8, 200 mM ionic strength, 20° C). If you suppose that attachment/detachment takes place *via* the weakly-attached state, then what you and I are saying is the same.

Brenner: I just want to avoid confusion, because when I talk about detachment-reattachment, I consider that all these states in general can also be detached. So, attachment is a reaction like from your MDP_i to $AM*DP_i$, which is in general possible, while you talk about a different reaction, which I am afraid might cause a lot of confusion in the audience.

Kawai: What I presented is the majority of the states we can measure. Perhaps weakly attached states are present, but we don't see them and our estimate of the weakly attached state is about 1%; hence, we ignored them in our analysis.

Godt: Could you please clarify this business of attached and detached states that Dr. Brenner brought up?

Kawai: Our cross-bridge scheme is based on experimental observation and on our interpretation of the observation. We simplified the scheme as much as possible. Therefore, we do not recognize cross-bridges that are present in low concentrations. Cross-bridge states AMS and AMDP may exist, but because we don't recognize them in our experiments, we don't place them in the cross-bridge scheme. .

Brenner: I have nothing against simplifying kinetic schemes, but I think where we have to be very clear is with the terminology. In the cross-bridge scheme of Huxley (*Prog. Biophys.* **7**, 255-318, 1957) two states were defined: transition from the non-force-generating state to the force-generating state, identical to cross-bridge attachment; and return to the non-force-generating state, identical with detachment. So there the terms "attachment" and "detachment" could be used as synonyms for the turnover reactions.

By now, we have shown that cross-bridges in all states of the ATPase cycle dynamically interact with actin, i.e., detach and re-attach from and to actin with rate constants that are fast compared to the active cycling. In these recent, more complex cross-bridge schemes the terms "attachment" and "detachment" are no longer correlated with cross-bridge cycling between the non-force and force-generating states. Consequently, we can no longer use the terms "attachment" and "detachment" as synonyms for the transitions between the non-force and force-generating states. So we have to be careful to distinguish "turnover" reactions from "attachment"/"detachment" reactions.

Iwamoto: I measured the high frequency stiffness; I oscillated a living single frog fiber during contraction with a 1 kHz sine wave and found a phase advance that was initially high but decreased during high tension and then decreased again. I don't know if this high frequency component at 1 KHz corresponds to Dr. Kawai's process *D*, but it is high at the beginning of contraction. Therefore, I believe that this component reflects the reattachment process of the actomyosin interaction, rather than the detachment of the AM•ATP state.

Kawai: You can't measure the phase advance when force is changing significantly. You can carry out this perturbation analysis only when force is not changing. Therefore, I don't know how to interpret your results.

Taylor: Dr. Kawai, you say that only one step is involved in force generation. Is it step 5, where the phosphate ion comes off?

Kawai: No. Step 4, before the phosphate comes off, is involved in force generation. As Dr. Brenner pointed out, cross-bridges probably go through this weakly attached AMDP state. If so, step 4 is a conformational change; it is the step that generates force. Phosphate release or phosphate binding is the association of a small molecule with a macromolecule and seems not to make any difference in the tension.

Taylor: Would you be able to say that several force generation steps occur for splitting one ATP molecule, or does this limit you to one force step per ATP-splitting in your kinetic scheme?

Kawai: Our experiments were carried out under near-isometric conditions. It appears that there is one step of force generation per ATP split. But we don't have any idea about the isotonic condition.

Brenner: Masataka, you assume that there is only one force-generating step. But the tension, if I recall correctly, did not change very much in your different states. How then do you account for the isometric transients?

Kawai: Isometric transients can be explained by assuming a length-sensitivity at the ATP-binding step. If stretch increases the ATP-binding step 1a then AM$^+$S isomerizes to AM*S (step 1b, phase 2f), resulting in lower tension. Cross-bridges then detach (step 2, phase 2s) to increase the pool of the detached state. They then attach slowly in step 4 resulting in delayed force generation (phase 3).

Huxley: I think it is very difficult to explain the earlier part, the phase-2 early recovery, on the basis of detachment and attachment because the stiffness hardly changes. So I think bridges remain attached—apologies to Dr. Brenner—throughout the early part of the transient.

Kawai: We carried out the high frequency stiffness measurement and confirmed that stiffness did change in the slower portion of phase 2. We do not have a stiffness measurement for the faster component of phase 2. However, since we interpret phase 2f as an isomerization of the collision complex AM$^+$S to form AM*S, we don't expect stiffness to change in the faster component of phase 2.

Goldman: I'd like to pursue the point made by Bernhard Brenner and Prof. Huxley. The states you are calling "detached" include some in rapid equilibrium between myosin associated with actin and dissociated. For instance, you labeled one AM'•ADP•Pi. I agree with them that it is a confusing nomenclature to term such states "detached" if they include some actomyosin association. In your model, you assume that all of the strain dependence in the rate constants occurs at substrate binding. But, if the initial attached state (AM•ADP•P) has a different force than AM'•ADP•Pi, according to Terrell Hill's formalism, the different free energy contributions of mechanical strain would require there to be changes in either forward or reverse rate constants at different strains. Therefore I expect there is strain-dependence of the rate constants in that step. How can your model be compatible with the Hill formalism?

Kawai: I'm not saying that there aren't other strain sensitivities besides the ATP-binding step. There could be other strain sensitivities as well. But suppose the force generation step 4 has a strain sensitivity, stretch will then induce the reversal of force-generating step 4, or a tension decrease. That sort of mechanism does not explain phase 3, or the delayed tension rise. Therefore, the strain sensitivity of this force-generating step is not essential to explain the tension transient. Rather, we believe that the strain sensitivity during or after the ATP binding is essential to explain the transients.

Goldman: So then there are other strain-sensitive steps, and your model is compatible with that idea, but you're saying that those aren't exciting the transients that we observe.

Kawai: That's correct.

THE EFFECT OF LATTICE SPACING CHANGE ON CROSS-BRIDGE KINETICS IN RABBIT PSOAS FIBERS

Yan Zhao, Masataka Kawai and John Wray*

Department of Anatomy
The University of Iowa
Iowa City, IA 52242, U.S.A.
*Abteilung Biophysik
Max-Planck Institut für Medizinische Forschung
D-6900 Heidelberg 1, F.R.G.

ABSTRACT

The effect of compression on the elementary steps of the cross-bridge cycle was investigated with the sinusoidal analysis technique and ATP hydrolysis rate measurement. The lattice spacing of rabbit psoas muscle fibers was osmotically compressed with a macromolecule, dextran T-500 (0-16%). The effects of MgATP, MgADP, Pi on exponential processes (B), (C), (D), and isometric tension were studied at different dextran concentrations. The experiments were performed at the saturating Ca concentration (pCa 4.5-4.8), 200 mM ionic strength, pH 7.0, and 20°C. Our results show that the fiber width decreased linearly with an increase in the dextran concentration, and the width measurement was perfectly correlated with the lattice spacing measurement using equatorial x-ray diffraction studies. We find that the nucleotide binding steps, the ATP-isomerization step, and the cross-bridge detachment step were minimally affected by the compression. Our results indicate that the rate constant of the reverse power stroke step (k_{-4}) decreases with mild compression (0-6.3% dextran), presumably because of the stabilization of the attached cross-bridges in the AM*DP state. We also found that the rate constant of the power stroke step (k_4) decreases with higher compression (> 6.3% dextran), presumably because of increased difficulty in performing the power stroke reaction. Our results further show that the association constant (K_5) of phosphate to cross-bridges is not changed with compression. The ATP hydrolysis rate declined almost linearly with an increase in the dextran concentration. This observation indicates that the rate limiting step is also affected by the lattice spacing change so that the associated rate constant (k_6) becomes progressively less with compression.

Mechanism of Myofilament Sliding in Muscle Contraction, Edited by
H. Sugi and G.H Pollack, Plenum Press, New York, 1993

581

INTRODUCTION

During the past 30 years, it has been generally assumed that there is a compensatory mechanism in cross-bridges when the actin-myosin lattice spacing is altered by various physiological changes[1]. Early evidence came from a length-tension diagram that showed a linear tension decline at sarcomere lengths between 2.2 and 3.5 μm in electrically stimulated frog semitendinosus muscle fibers[2]. Because the actin-myosin lattice spacing of intact fibers changes in an isovolumetric manner when the sarcromere length changes, it has been assumed that the force per cross-bridge remains the same even though the distance between thick and thin filaments changes significantly[1]. More recently, length-tension diagrams that did not follow the linear relationship have been published[3][4]. At the same time it was reported that isometric tension changes as the lattice spacing of skinned fibers was compressed osmotically[5-7]. These results raise a possibility that the force generating mechanisms may be influenced by the lattice spacing change. Thus, we found it necessary to test directly the effect of the lattice spacing change on the elementary steps of the cross-bridge cycle to establish how the spacing change modifies the mechanisms of force generation.

As initially reported by Maughan and Godt[8][9] using skinned fibers, the lattice spacing can be compressed osmotically by macromolecules such as polyvinyl pyroridine and dextran T-500. These macromolecules increase the osmotic pressure of the bulk solution, but they are effectively excluded from the actomyosin lattice, and the consequent osmotic gradient causes water and other solutes to move out from the lattice space resulting in a compression of fibers. We previously reported that low compression increased isometric tension and high compression decreased isometric tension in skinned rabbit psoas fibers[7]. Because the observed rate constants decreased with compression, we hypothesized that low compression levels decreased the cross-bridge detachment rate and high compression levels decreased the cross-bridge attachment rate.

In our previous works with sinusoidal analysis in skinned muscle fibers, we established a cross-bridge scheme which is consistent with the MgATP, MgADP, and Pi effects on exponential processes (B), (C), and (D)[10][11]. These are summarized in scheme 1.

Step 0	Step 1a	Step 1b	Step 2	Step 3	Step 4	Step 5	Step 6

$$
\text{AMD} \underset{K_0}{\overset{D}{\rightleftharpoons}} \text{AM} \underset{S}{\overset{K_{1a}}{\rightleftharpoons}} \text{AM}^+\text{S} \underset{k_{-1b}}{\overset{k_{1b}}{\rightleftharpoons}} \text{AM}^*\text{S} \underset{k_{-2}}{\overset{k_2}{\rightleftharpoons}} \left[\begin{matrix} \text{AMS} \leftrightarrow \text{AMDP} \\ \updownarrow \qquad\quad \updownarrow \\ \text{MS} \leftrightarrow \text{MDP} \end{matrix} \right] \underset{k_{-4}}{\overset{k_4}{\rightleftharpoons}} \text{AM}^*\text{DP} \overset{P}{\underset{K_5}{\rightleftharpoons}} \text{AM}^*\text{D} \overset{k_6}{\rightarrow}
$$

$$
X_0 \qquad X_{1a} \qquad X_{1b} \qquad X_2 \qquad X_3 \qquad X_4 \qquad X_5 \qquad X_6
$$

where A = actin, M = myosin head, S = MgATP, D = MgADP, and P = phosphate. An asterisk(*) or a dagger(+) mark is placed where the second and the third conformational states are identified. Det includes all the detached states (MS, MDP) and weakly attached states (AMS, AMDP);[12][13]. We found that the apparent rate constant $2\pi d$ reflects the transition between AM^+S and AM^*S states (ATP-isomerization step 1b); the apparent rate constant $2\pi c$ reflects the transition between AM^*S and Det states (detachment step 2), and both are sensitive to MgATP and MgADP concentrations. We also found the apparent rate constant $2\pi b$ reflects the transition between Det and AM^*DP states (cross-bridge attachment step 4), it is sensitive to the Pi concentration, and corresponds to the power stroke step[11]. We further found that the binding of nucleotides (MgATP, MgADP) and phosphate (Pi) are faster than our speed of observation, hence steps 0, 1a, and 5 can be approximated by equilibria.

Of seven elementary steps, we were able to determine three association constants and six rate constants by the sinusoidal analysis method, and one rate constant (k_6) by the ATP hydrolysis rate measurement. We are in a position, therefore, to identify the specific elementary step(s) that is altered by a specific change in experimental conditions. In this study, we are interested in the effect of the lattice spacing change, how this change affects the elementary steps of the cross-bridge cycle, and whether the particular effect on the elementary steps is consistent with the change in isometric tension.

MATERIALS AND METHODS

Chemicals and solutions, experimental procedures, and the method of calculating complex modulus and its fitting equations are generally the same as in our published work[10][11] and preceding paper, hence not repeated here. Dextran concentration is expressed in per cent (%), which represents the gram amount of dextran T-500 added to 100ml solution (g/dl). All the tension data and the magnitude of exponential processes including Y_∞ were normalized by the control tension (T_c) generated by the control activating solution (millimolar): 6 CaEGTA (pCa 4.82), 5.3 MgATP, 4.7 free ATP, 15 CP, 8 Pi, 35 KProp, 28 NaProp 10 MOPS, and 160 units/ml CK. Average T_c was 207 ± 8 kN/m^2 (n = 112, $\pm$ S.E.M.). All the other solutions were variations of the control activating solutions. Ionic strength was maintained at 200mM, pH was adjusted to 7.00, and all experiments were performed at 20°C.

RESULTS

The Effect of Compression on the Fiber Width

Dextran T-500 was added to the experimental solution, and the fiber width was compared in the relaxing and activating solutions in Fig. 1A, and in the relaxing solution and after rigor induction in Fig. 1B. The fiber width was measured at a

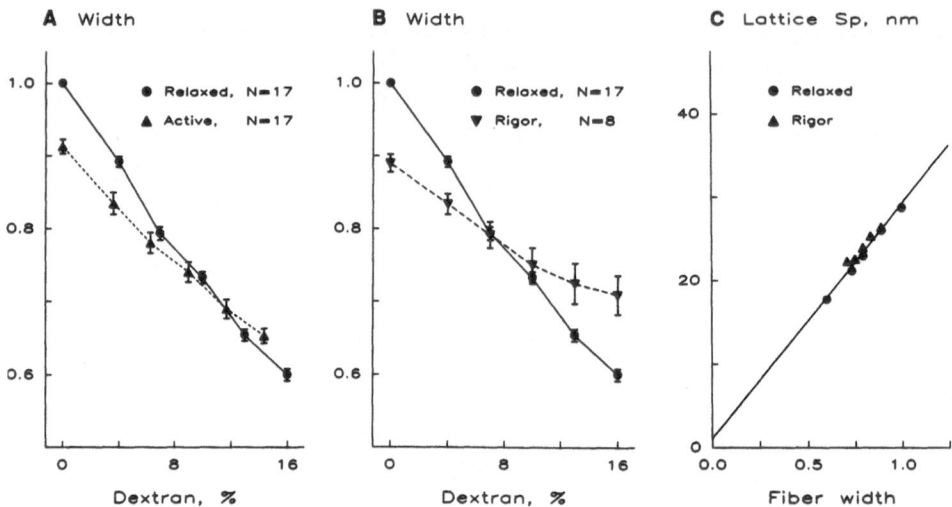

Fig. 1. Dependence of fiber width on dextran concentration (A, B) and relation between fiber width and filament lattice spacing (C).

fixed point along the preparation. As shown in Fig. 1, the width decreased linearly with an increase in dextran concentration in all conditions. The relaxed fibers were compressed by 40% with 16% dextran, the active fibers were compressed by 30% with 14.4% dextran, and the rigor fibers were compressed by 20% with 16% dextran. Interestingly, the plots of active and relaxed fiber widths cross in the high range of dextran concentration (10-12%, Fig. 1A): The fiber shrank on Ca-activation at low compression levels, whereas the fiber slightly expanded on Ca-activation at high compression levels. Similarly, the plots of rigor and relaxed fiber widths cross in the mid range (~7% dextran, Fig. 1B). The observation on the width of relaxed and rigor fibers is consistent with the lattice spacing measurement using equatorial x-ray diffraction studies. In Fig. 1C, the lattice spacing (center-to-center distance between thick and thin filaments) is plotted against the fiber width for the corresponding dextran concentrations. This plot demonstrates a near proportional relationship between the lattice spacing and the width measurements in the entire range of dextran concentrations under both relaxed and rigor conditions ($r = 0.986$). The observed effect of dextran on the fiber width is generally comparable to other reports[5)7)14-17].

Effect of Compression on Isometric Tension and Stiffness

The effects of compression on isometric tension and stiffness (Y_∞) are shown in Fig. 2. At low compression levels, the isometric tension increased and peaked at 6.3% dextran (Fig. 2A). The peak tension corresponded to $1.23T_C$. With further compression, the isometric tension decreased, and it became $0.49T_C$ at 14.4%

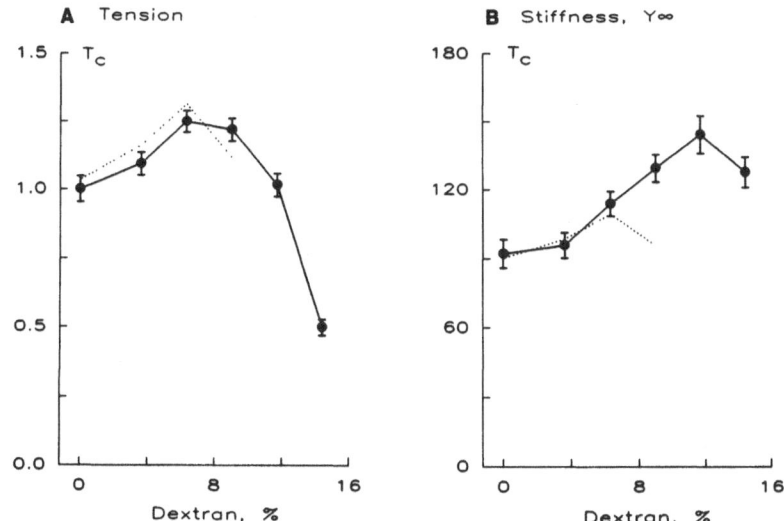

Fig. 2. Effect of fiber compression on tension (A) and stiffness (B).

dextran. The stiffness increased with compression and peaked at 11.7% dextran concentration (Fig. 2B). At the highest compression (14.4% dextran), stiffness somewhat decreased. Results of the compression effect on isometric tension and stiffness are generally consistent with earlier reports[6][7][18].

Effect of MgATP on $2\pi c$ and $2\pi d$ at Different Compression Levels

To characterize the kinetic constants involved in the MgATP binding step 1a, isomerizatioin step 1b, and the cross-bridge detachment step 2 (scheme 1), MgATP concentration was changed and the apparent rate constants of processes (C) and (D) were determined. The effect of MgATP was studied in the concentration range from 0.1 to 10 mM in presence of 8mM Pi. The high Pi concentration was chosen, because more cross-bridges are distributed in the states AM, AM+S, AM*S, and Det, hence there is better resolution of processes (C) and (D). The MgATP study was repeated for four different dextran concentrations (0, 3.6, 6.3, 9%). The data were fitted to Eqs.2 and 3 of the preceding paper, and the kinetic constants were deduced. They were then normalized to those in the absence of dextran (Table 1), and the results were plotted in Fig. 3. This figure demonstrates that the association constant K_{1a} and the rate constants k_{1b}, k_{-1b}, k_2 and k_{-2} of the elementary steps were minimally affected by compression.

Effect of MgADP on $2\pi c$ and $2\pi d$ at Different Compression Levels

To characterize the association constant involved in the MgADP binding (step 0, scheme 1), MgADP concentration was changed from 0 to 8mM at the fixed MgATP

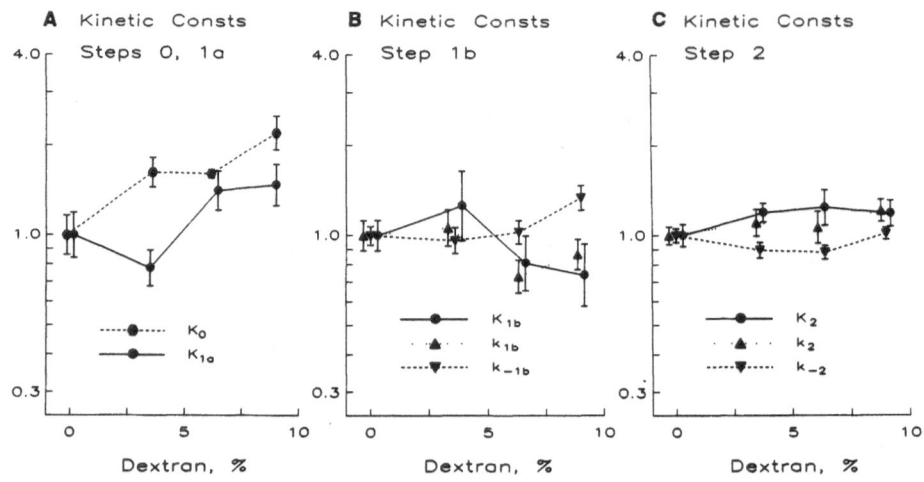

Fig. 3. Dependence of kinetic constants on dextran concentration.

concentration (5mM), and the rate constants of process (C) and process (D) were obtained. The MgADP study was repeated in four different dextran concentrations. The association constant K_0 thus obtained was normalized to that in the absence of dextran, and plotted in Fig. 3A against the dextran concentration. It is apparent in Fig. 3A that K_0 was minimally affected by compression for the dextran concentration in the range of 0 to 6.3%, but the effect was enhanced (2.2 times) at the 9% dextran concentration.

Effect of Phosphate on 2πb at Different Compression Levels

To characterize the effect of the lattice spacing change on force generation and Pi release steps, we studied the effect of Pi on exponential process (B) at four fixed dextran concentrations. The averaged 2πb as a function of Pi at different dextran concentrations is shown in Fig. 4A with discrete symbols and error bars (S.E.M.). This figure demonstrates that 2πb increased with an increase in Pi concentration, and it appeared to saturate at higher Pi concentrations. Such relationship can be fitted to Eq. 4 in the preceding paper to deduce k_4, k_{-4}, K_5. According to Eq. 4, which is based on scheme 1, the intercept to the ordinate (Pi = 0mM) represents the rate constant of the power stroke step (k_4) multiplied by the constant factor σ. The equilibrium constants (K_{1a}, K_{1b}, K_2) obtained from MgATP study at the corresponding dextran concentrations were used to calculate σ (Eq. 5, preceding paper). The increment to the large Pi concentration represents the rate constant of the reverse power stroke step (k_{-4}). The Pi concentration at half saturation point is the dissociation constant of Pi from the cross-bridge state AM*DP (step 5), and its reciprocal is the association constant K_5.

The relative change in the kinetic constants in response to compression is plotted in Fig. 4B,C. As shown in Fig. 4B, the rate constant of the power stroke step (k_4)

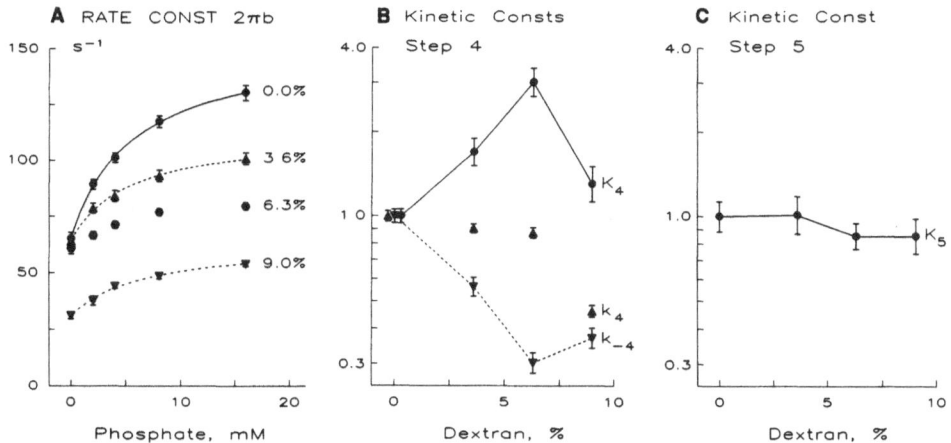

Fig. 4. Dependence of rate and kinetic constants on dextran concentration.

does not change much with 0-6.3% dextran, and it declines significantly with 9% dextran. In contrast, the rate constant of its reversal step (k_{-4}) decreases significantly at low compression levels, followed by a slight increase at high compression levels. These changes in the rate constants result in a 3 fold increase of the equilibrium constant ($K_4 = k_4/k_{-4}$) of step 4 at low compression levels, followed by a decrease of the equilibrium constant at high compression levels. Because step 4 leads into force generation[11] (preceding paper), this change in K_4 is consistent with the change in isometric tension shown in Fig. 2A. The association constant K_5 of Pi to cross-bridges in the AM*D state does not change significantly with compression (Fig. 4C).

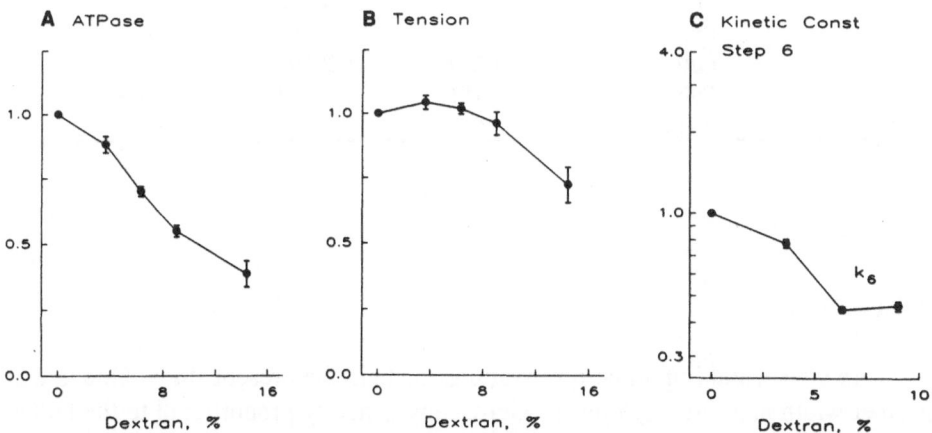

Fig. 5. Dependence of ATPase activity, tension and kinetic constant on dextran concentration.

ATP Hydrolysis Rate at Different Compression Levels

We recently showed that step 6 (AM*D isomerization, scheme 1) limits the ATP hydrolysis rate in the near isometric condition in rabbit psoas fibers[11]. To determine whether this rate-limiting step is altered by change in the lattice spacing, we measured the ATP hydrolysis rate at different dextran concentrations. The hydrolysis rate and the tension data were normalized to those of the control avtivation. The hydrolysis rate is shown in Fig. 5A, and isometric tension is shown in Fig. 5B. The ATP hydrolysis rate decreased almost linearly with the increase in dextran concentration. Isometric tension peaked at 3.6% dextran, and further increase in the dextran concentration caused a decline (Fig. 5B). To determine k_6, we assumed that the ATP hydrolysis rate to be $J = [AM*D]k_6$. The concentration of AM*D state was calculated based on the equlibrium constants shown in Table I. The rate constant k_6 is plotted aginst dextran concentrations in Fig. 5C: k_6 decreased progressively with an increase in the dextran concentration from 0 to 6.3%, and it remained the same for further compression.

Table 1. Kinetic constants that specify Scheme 1

Constant	Source	Best fit	Average(N)	Unit
K_0	ADP/c	0.530	0.58±0.09(7)	mM^{-1}
K_{1a}	ATP/c	0.262	0.23±0.04(10)	mM^{-1}
k_{1b}	ATP/d	1530	1880±220(10)	s^{-1}
k_{-1b}	ATP/d	1610	1510±110(10)	s^{-1}
K_{1b}	ATP/d	0.95	1.29±0.15(10)	
k_2	ATP/c	536	510±30(10)	s^{-1}
k_{-2}	ATP/c	130	132±7(10)	s^{-1}
K_2	ATP/c	4.12	3.9±0.3(10)	
k_4	Pi/b	105.7	106±4(11)	s^{-1}
k_{-4}	Pi/b	87.6	90±5(11)	s^{-1}
K_4	Pi/b	1.207	1.20±0.07(11)	
K_5	Pi/b	0.181	0.19±0.02(11)	mM^{-1}
k_6	ATPase	16.2	16.2±1.3(18)	s^{-1}

DISCUSSION

There are several important observations made in the present study. One is that the fiber width measured by optical microscopy is nearly proportional to the lattice spacing measured by the equatorial x-ray diffraction study (Fig. 1C). The proportionality is useful, because x-ray data are not always available and are difficult to obtain at the fast time scale during contraction. However, this proportionality may be limited to the chemically skinned rabbit psoas fibers used in the present study. The sarcolemma in this preparation becomes porous after skinning, but its mechanical constraint is present[19].

With the deduction of the kinetic constants of elementary steps, we were able to specify which elementary step is affected by the lattice spacing change. We found that the nucleotide binding steps and subsequent cross-bridge detachment step are minimally affected by the change in the actin-myosin lattice spacing (Fig. 3). We infer from these observations that the shape of the nucleotide binding site is not altered by compression.

The effect of compression is best seen on process (B). This process represents cross-bridge attachment (power stroke) and the subsequent phosphate (Pi) release steps[11]. Our results (Fig. 4) show that, with low level compression (dextran 0-6.3%), the rate of reverse power stroke (k_{-4}) decreases, and with high level compression (dextran 6.3-9%), the rate of power stroke (k_4) decreases. This results in a K_4 ($= k_4/k_{-4}$) to increase at low compression, and to decrease at high compression (Fig. 4B). Since K_4 is directly related to force production[11], an increase of force is predicted at low compression levels, whereas a decrease of force is predicted at high compression levels. The theoretical projection of isometric tension is included in Fig. 2A (broken line). The prediction is in accord with our observation of the isometric tension change with compression. On the other hand, stiffness is proportional to the number of attached cross-bridges. The theoretical projection of stiffness is displayed in Fig 2B as a broken line. As it is seen in Fig. 2B, the observed data and the theoretical prediction of stiffness agree well up to 6.3% dextran, then the predicted value and observed data diverge. It is apparent from this comparison that large stiffness at highly compressed state is not caused by the usual cross-bridge interaction with actin. We infer from these observations that, when the lattice spacing is highly compressed, the myosin heads are squeezed between two sets of filaments resulting in a large stiffness value but without effecting the normal contact with actin to transduce energy. From the Pi effect, we deduced the association constant K_5 of Pi to cross-bridges in the AM*D state. Our results (Fig. 4C) show that K_5 is unchanged in the entire dextran concentration range studied. We conclude from this observation that the Pi binding site on the myosin head is not altered by the compression of the lattice spacing.

We calculated the available spacing between thick and thin filaments based on x-ray diffraction studies and compared it with the approximate size of the myosin head (Fig. 6). We assumed the diameter of the thin filament to be 9.5 nm[20] and the thick filament to be 15 nm. We also assumed that the myosin head has a tadpole-shape with the long axis measuring 16.5 nm[21]. As demonstrated in Fig. 6A, the spacing between thick and thin filaments is slightly larger than the myosin head without compression. In this case, the myosin head may attach to actin to generate force; however, the head may come off easily because of the thermal vibration, and this will result in a large k_{-4}. It is likely that the cross-bridge force is generated diagonally and can be divided into longitudinal and radial elements. We then can visualize that the radial element helps the fiber shrink on Ca activation as observed at low dextran concentrations (Fig. 1A,). Our finding that the forward rate constant

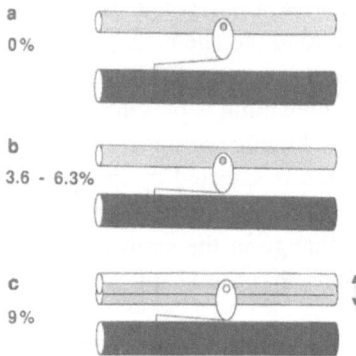

Fig. 6. Diagrams showing the effect of filament lattice spacing on the mode of attachment of a myosin head to the thin filament.

(k_4) does not change significantly at low compression level is consistent with the observation that the neck portion of the myosin head is flexible; when there is an adequate spacing between the head and the thin filaments, the distance change seems not to hinder the attachment (step 4) of the head to the thin filament as long as the thin filament is within reach. At low compression (Fig. 6B), which corresponds to the lattice spacing of the intact preparation, the thermal motion of the myosin head may be limited because the head is now trapped between the two set of filaments resulting in a smaller k_{-4} (Fig. 4B). A smaller k_{-4} in turn will result in a larger K_4, hence larger force as we have observed (Fig. 2A). This distance may be optimal for the cross-bridges to function.

The spacing between thick and thin filaments becomes smaller than the myosin head (Fig. 6C) when the fibers are highly compressed by dextran (9% or more). Under these conditions, it may be difficult for the myosin head to make the correct contact with actin to perform the power stroke reaction (step 4). This will result in a lower k_4 as we have observed (Fig. 4B). The critical power stroke reaction may occur only when adequate spacing becomes available. This is possible because the thin filaments are flexible in the radial direction, and they exhibit thermal fluctuations[22]. The degree of fluctuation even increases when Ca is bound to the thin filaments[23]. If this transition occurs for some cross-bridges, then the lattice spacing is fixed at this larger value, resulting in an apparent expansion of the lattice spacing. Consequently, it appears as if the myosin head pushes the actin filament away to make more space available to perform the power stroke reaction, resulting in an expansion of the fiber width on Ca^{2+} activation (Fig. 1A). The expansion of the width of highly compressed fibers is even larger on rigor induction (Fig. 1B). This may be caused by the loss of the radial force after rigor induction. With our experiments, the rigor tension declined nearly to zero by the time the width measurement was carried out, hence it is not likely that the radial force remained at a high level. We therefore conclude that the highly compressed state during Ca^{2+} activation is caused by the continuous presence of the radial force element. The shrinkage of the fiber width without compression and the expansion of the width at high compression levels were previously reported in rabbit psoas fibers on Ca^{2+} activation[7]) and in frog muscle fibers on rigor induction[9)15)24)].

The rate-limting step is characterized by ATP hydrolysis rate measurement. Our results show that the hydrolysis rate during Ca activation is significantly reduced as the lattice spacing is compressed by dextran (Fig. 5A). This indicates that the slowest step of the cross-bridge cycle, which is either the isomerization of AM*D state to form the AMD state or the MgADP desorption step to form the AM state, is senstive to the lattice spacing change. This might be possible when the isomerization step requires large scale rearrangement in the shape of myosin head, hence the spacing between thin and thick filament become a critical factor for the rearrangement process to proceed. This possibility is strengthened when we realize that the myosin head spends the most time at rate-limiting step 6 to enable filiment sliding and muscle shortening to occur. Evidently, such motion would require large scale molecular rearrangement, hence it is reasonable to be sensitive to the available spacing between thick and thin filaments.

In summary, our investigation demonstrates that the critical force generating transition (step 4) and the rate-limiting step (step 6) are modified by compression. Thus, it can be generalized that two slowest reaction steps in the cross-bridge cycle are sensitive to a change in the geometrical factors in the cross-bridges. These two steps are important in the cycle, because force is generated at step 4, and work is performed at step 6. Based on the facts that these steps interfere with the available spacing between thick and thin filaments, we conclude that both steps require large scale rearrangements in the conformation of the contractile proteins, most likely in the shape of the myosin head.

ACKNOWLEDGEMENTS

The present work was supported by grants from NIH (AR21530), NSF (DCB90-18096), and Iowa Affiliate of AHA (IA-91-G-13).

REFERENCES

1. Huxley, H.E. *Science* **164**, 1356-1366 (1969).
2. Gordon, A.M., Huxley, A.F. & Julian, F.J. *J. Physiol. (Lond.)* **184**, 170-192 (1966).
3. Edman, K.A.P. & Reggiani, C. *J. Physiol. (Lond.)* **385**, 709-732 (1987).
4. Granzier, H.L.M. & Pollack, G.H. *J. Physiol. (Lond.)* **421**, 595-615 (1990).
5. Godt, R.E. & Maughan, D.W. *Pflügers Arch.* **391**, 334-337 (1981).
6. Krasner, B. & Maughan, D.W. *Pflügers Arch.* **400**, 160-16 (1984).
7. Kawai, M. & Schulman, M.I. *J. Muscle Res. Cell Motility* **6**, 313-332 (1985).
8. Maughan, D.W. & Godt, R.E. *Biophys. J.* **28**, 391-402 (1979).

9. Maughan, D.W. & Godt, R.E. *J. Gen. Physiol.* **77**, 49-64 (1981).

10. Kawai, M. & Halvorson, H.R. *Biophys J.* **55**, 595-603 (1989).

11. Kawai, M. & Halvorson, H.R. *Biophys. J.* **59**, 329-342 (1991).

12. Greene, L.E. & Eisenberg, E. *Proc. Natl. Acad. Sci. USA* **77**, 2616-20 (1980).

13. Schoenberg, M. *Biophys. J.* **54**, 135-148 (1988).

14. Matsubara, I., Umazume, Y. & Yagi, N. *J. Physiol. (Lond.)* **360**,135-84 (1985).

15. Tsuchiya, T. *Biophys. J.* **53**, 415-423 (1988).

16. Podolsky, R.J., Naylor, G.R.S. & Arata, T. *Basic Biology of muscle: A Comparative Approach* (eds Twarog B.M., Levine R.J.C. & Dewey M.M.) 79-89 (Reven Press, New York., 1982).

17. Millman, B.M., Wakabayashi, K. & Racey, T.J. *Biophys. J.* **41**, 259-267 (1983).

18. Gulati, J. & Babu, A. *Biophys. J.* **48**, 781-787 (1985).

19. Eastwood, A.B., Wood, D.S., Bock, K.L. & Sorenson, M.M. *Tissue & Cell* **11**, 553-566 (1979).

20. Egelman, E.H., Francis, N. & DeRosier, D.J. *J. Mol. Biol.* **116**, 605-629 (1983).

21. Winkelmann, D.A., Baker, T.S. & Rayment, I. *J. Cell Biol.* **114**, 701-713 (1991).

22. Fujime, S. & Ishiwata, S. *J. Mol. Biol.* **62**, 251-265 (1971).

23. Ishiwata, S. & Fujime, S. *J. Mol. Biol.* **68**, 511-522 (1972).

24. Matsubara, I., Goldman, Y.E. & Simmons, R.M. *J. Mol. Biol.* **173**, 15-33 (1984).

MOLECULAR BIOLOGY OF THE LENGTH-TENSION RELATION IN CARDIAC MUSCLE

Jagdish Gulati

The Molecular Physiology Laboratory
Departments of Medicine and Physiology/Biophysics
Albert Einstein College of Medicine
Bronx, NY 10461

ABSTRACT

TnC in the thin filaent is found to act as the putative length-sensor in cardiac muscle. Techniques of genetic engineering and TnC-exchange in skinned fibers are combined to delineate the molecular principles governing the length induced modulations in TnC. Cardiac TnC is a better length-sensor than skeletal TnC, and the differences in the two moieties are used as a guide to explore the length-sensing mechanism. Studies are described with mutants of both cardiac and skeletal TnCs: (i) a mutant of cardiac TnC in which Ca^{2+}-binding to site 1 was restored (made by John Putkey), and (ii) two mutants of skeletal TnC, (a) with site 1 inactivated, and (b) a chimera, in which 41 residues in the N-terminus are entirely of the cardiac type and the remaining residues are all of the skeletal type (made at Albert Einstein). The results indicate that Ca^{2+}-binding property of site 1 is only partly responsible for the length-sensitivity of TnC, and that the residues outside loop-1 are also critical in both the on-off switching mechanism as well as the length-sensing mechanism. Further, we have also found that thin filaments are polarized such that Ca^{2+}-affinity may be high at the free end of the filament and decreases towards the Z-line, reaching its lowest value at the fixed end. This could serve as the basis for the conversion of sarcomere length change into a signal that can be transduced from the Z-lines to the TnC moiety in actin filaments.

INTRODUCTION

We have previously shown that cardiac TnC performs a length-sensing function in heart muscle[1], similar to that essential in the operation of Starling's law[2]. This provides a long sought molecular handle in the investigations of Starling mechanism. The original evidence consisted of the effect of length on Ca-sensitivity of skinned trabeculae containing either cardiac or skeletal TnC and the finding that length dependence was diminished with skeletal TnC. More recently, the converse

Mechanism of Myofilament Sliding in Muscle Contraction, Edited by
H. Sugi and G.H Pollack, Plenum Press, New York, 1993

593

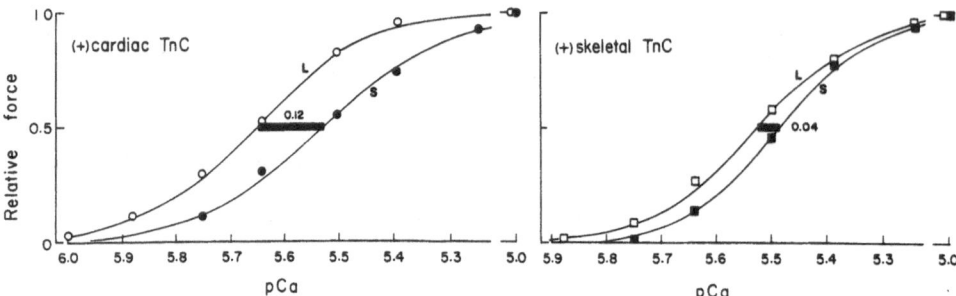

Fig. 1. The effect of TnC exchange on the length-dependence of Ca^{2+}-sensitivity in hamster trabeculae. L and S indicate sarcomere lengths of 2.4 μm and 1.9 μm respectively. (Reproduced with permission from ref.1)

experiments on fast twitch fibers were also attempted, both by ourselves and by others. Our own results are in agreement with the findings on myocardium, but the other group has come up with conclusions different from ours. These results are reviewed briefly before we discuss the new findings with engineered proteins.

Length-induced Modulation of Ca^{2+}-Sensitivity in Cardiac Muscle

The pCa-force relations in cardiac muscle at 1.9 and 2.4 μm sarcomere lengths reveal that Ca^{2+}-sensitivity is increased with length[3][4]. This is also an indication that the length-tension relation (between 1.9-2.4 μm) is affected by the degree of activation: the length-tension relation is steeper at lower activation levels[3]. With TnC exchange, Babu et al[1] found that the length-induced shift in pCa-force curve (and hence, steepness of the length-tension relation) diminished after cardiac TnC was replaced by skeletal TnC—from a rightward shift of 0.12 pCa unit with cardiac TnC to 0.04 pCa unit with skeletal TnC (Fig. 1).

Fig. 2 shows silver-stained gels of trabeculae from hamster hearts extracted and replenished with bovine cardiac TnC or rabbit fast skeletal muscle TnC. Table 1 indicates quantitative analysis of the various lanes: Normalized to LC1, cardiac TnC values were 0.31 and 0.09 in native (unextracted) and extracted trabeculae, respectively. Cardiac TnC was 0.33 in reconstituted trabecula—a slightly higher

Table 1. (from ref. 1)

Fiber treatment	CTnC	LC2	LC1
Native	0.31	0.81	1.00
(+) STnC	0.09	0.96	1.00
(+) CTnC	0.33	0.84	1.00

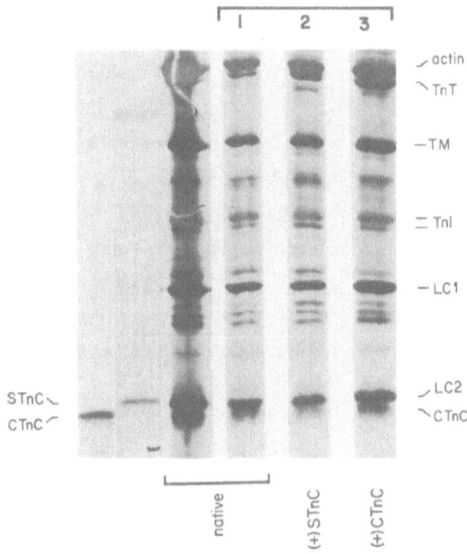

Fig. 2. Analysis of hamster trabeculae by SDS-PAGE (15% polyacrylamide, silver-stained). The first two lane are the purified cardiac and skeletal TnCs, to mark the displacements of the bands. The third lane is overloaded with native cardiac preparation, which serves to indicate the TnC band in cardiac muscle. The next three lanes (marked 1, 2, & 3) are experimental trabeculae: native, STnC-loaded, & CTnC-loaded, respectively. (Reproduced from ref.1)

number may be an artifact of the heavier gel loading in this case. Since the skeletal TnC band overlapped with the cardiac LC2 band in the present case, the amount of skeletal TnC uptake in the extracted trabecula could be estimated by subtraction—a difference of 0.15 was observed (in similarly loaded gels), which when adjusted for the lighter staining of skeletal TnC compared with that of cardiac TnC (1.00:1.44), accounted for a density ratio of 0.22. Thus, the density ratio of the total TnC content in the extracted trabeculae reloaded with skeletal TnC was 0.22 plus 0.09, or 0.31— indicating stoichiometric exchange of the TnC moieties in cardiac muscle.

In a recent review of the structure, function and regulation of the TnC gene expression, Parmacek and Leiden[5] questioned (i) whether the addition of skeletal TnC to the extracted cardiac muscle was stoichiometric in Babu et al[1], and (ii) claimed that the results were obtained from a "small number of preparations". The detailed analysis of the silver-stained gels discussed above should rest the first concern. The results similar to above were noted on four preparations by Babu et al[1]. Further, Gulati et al[5] separately reported additional 5 preparations with the same results on stoichiometry, which would invalidate the concern regarding the "number of preparations" as well.

Parmacek and Leiden[5] further noted (iii) that in contrast to the results of Babu et al[1] in cardiac muscle, cardiac TnC replacing skeletal TnC in fast twitch fibers had been demonstrated at the time in an abstract by a second group[7] to have no effect on the length dependence of Ca^{2+}-activated tension. Parmacek and Leiden were

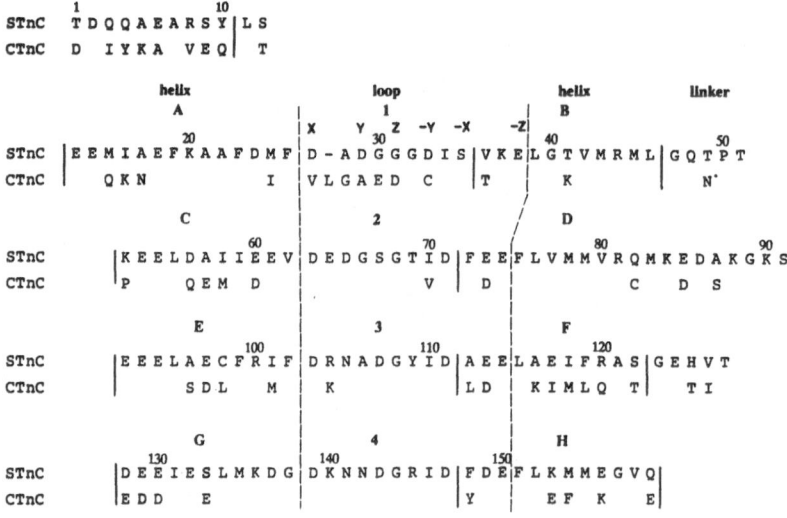

Fig. 3. The effect of TnC exchange on the length-dependence of Ca^{2+}-sensitivity in rabbit psoas fiber. Long length is 2.2 μm, and short length is 1.7 μm. The inset indicates the results on 13 preparations: 9 for STnC & 4 for CTnC. #ipK measures the shift in the pCa50 with length. (Reproduced from ref.11)

apparently unacquainted with our communications[8-10]), indicating that substitution of cardiac TnC for skeletal TnC in fast fibers does indeed enhance their length-dependence of contractility, under specified conditions. These results are discussed next.

TnC-Exchange in Fast Muscle

The diminished length-dependence in cardiac muscle with skeletal TnC indicated that TnC moiety was an important component of the length-sensing mechanism. However, in order to determine whether the length-sensing function of TnC was specific in myocardium or whether cardiac TnC would perform similarly in different milieus, it seemed worthwhile to check this with TnC exchange in fast fiber. The results in Fig. 3 show that cardiac TnC enhances the length dependence of rabbit psoas fiber[11]). These experiments on fast fibers were carried out in lower salt solution than the cardiac muscle experiments of Babu et al. (ionic strength was 100mM, compared to 180-200mM in the earlier experiments), because cardiac TnC was found to be more effective under these conditions in skeletal milieu[12)13]. The combined results of the length-dependence of Ca^{2+}-activated tension in psoas fibers[11]) and myocardium[1]) indicate that the property of TnC itself is subject to altered sarcomere length.

In contrast, on the basis of a short comunication by Moss et al.[7]), Parmacek and Leiden[5]) concluded that cardiac TnC has no effect on the length dependence of Ca^{2+}-activated tension in fast fibers. But careful examination of the results now published[14]), indicates that there were other possibilities. For instance, of the 15 fast-twitch fibers tested by Moss et al.12 did in fact indicate increased length-dependence following cardiac TnC exchange, but 3 did not (see Table 3 in ref.14). The extraction levels in two out of the 3 was incomplete and consequently any cardiac TnC effect in these fibers may not be expected to be detectable. In these 2 fibers P0 values of 0.44 and 0.47 were reported after extraction, indicating (on the basis of their own earlier finding, ref.15)) residual TnC values of 62% and 66%, respectively. Indeed, in general the less effective extraction documented, the less the effect of cardiac TnC substitution on the length-dependence of contractility.

Because a number of the 12 fibers indicating increased length-dependence after TnC-exchange generated incomplete force development—the basis of rejection by Moss et al—the agreement of these with our results may well be fortuitous. A number of possibilities for this discrepancy are currently being investigated in my lab, but time does not permit to discuss them here (see Cheng et al., *Biophys. J.* **61**, 294a).

Length-Sensing Domain of TnC

Since our results indicated that cardiac TnC may be a better length-sensor than skeletal TnC, we next proceeded to investigate the molecular basis of the functional distinctions between the two isoforms. The amino acid sequences are compared in Fig. 4. There is about 70% homology between the two, and although both contain 4 potential Ca^{2+}-binding EF-hands, and probably have a similar dumbbell crystal-

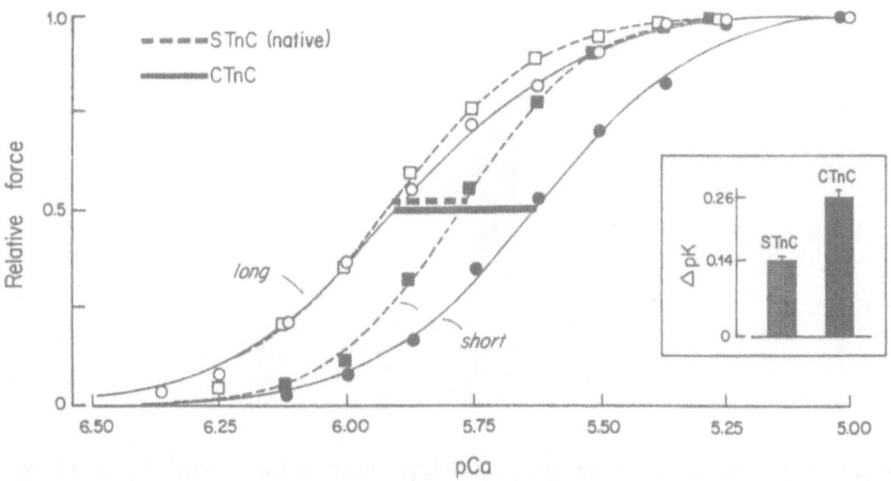

Fig. 4. The amino acid sequences of skeletal and cardiac TnCs. For the latter, only the residues different from STnC are indicated. The four helix-loop-helix cluster are marked.

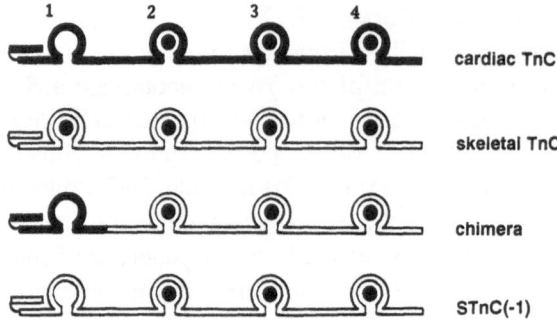

Fig. 5. Linear representation of the various TnC isoforms.

structure, site 1 in cardiac TnC lacks Ca^{2+}-binding ability. Furthermore, even though the majority of the differences in the two sequences are also confined to the first EF-hand and the N-terminal overhang, there are several other amino acid replacements throughout the molecules as well. To find which part of the molecule has the length-sensing function, the use of molecular genetics appeared to be the most promising. In one of the first experiments, we used a mutant of cardiac TnC in which the loop of site 1 was modified so that it could now bind Ca^{2+} like site 1 in skeletal TnC (Fig. 5). This mutant, termed CBM1 by John Putkey[17], was exchanged for endogenous TnC in skinned cardiac trabeculae and the force response was measured.

Fig. 6 summarizes the results of the length-dependence of pCa-force relations in trabeculae with CBM1. CBM1-loaded myocardium gave response intermediate between skeletal TnC and cardiac TnC. Thus, in addition to the metal-ion binding property of site 1, other residues/domains of the isoforms contribute to the length mechanism.

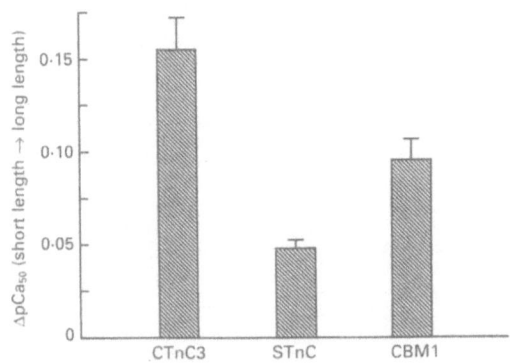

Fig. 6. The effect of CBM1 on the length-dependence of Ca^{2+}-sensitivity of hamster trabeculae. (Reproduced from ref.11)

Cassette Mutagenesis

Therefore, in efforts to identify these domains, we are now taking another approach to the problem. We have synthesized a skeletal-TnC-encoding DNA with multiple restriction sites[18)19)], and made mutants to imitate cardiac TnC functions. We made a mutant of the loop of site 1 in which the first two coordinates (both aspartate) in Ca^{2+}-binding in the loop of site 1 were each replaced with alanine, to inactivate the site (Fig. 5). Thus, we have a new analog of cardiac TnC, converse of CBM1—the new mutant (termed STnC minus 1, or STnC-1) binds only 3 Ca^{2+} ions like cardiac TnC, but in rest of the sequence it resembles skeletal TnC.

In physiological tests on skinned fibers, we find that STnC-1 is inactive in the on-off switching function in both skeletal and cardiac muscles—i.e. force regulation could not be restored when STnC-1 was replaced for the endogenous proteins. This is a further indication that residues outside the loop region can affect the structure of TnC and thereby modify both its switching function and the length-sensing mechanism.

Cardiac-Skeletal TnC Chimera

Next, we are now making chimeras, by mixing the domains of cardiac TnC with skeletal TnC. In one of the first such mutants (C1/S, Fig. 5)), the N-terminal 41 residues were made cardiac; the others were retained from the skeletal isoform. Our preliminary results with this chimera indicate that it too (like STnC-1) is unable to regulate in skeletal fiber. However, in contrast, the chimera could regulate the force development in cardiac muscle. This is an indication that, when site 1 is inactive, residues in the N-terminus part of the molecule, possibly including all of the first EF-hand (the loop plus the flanking helices), are critical for regulation in cardiac muscle. We are now in the process of measuring the length-dependence in cardiac muscle with the C1/S chimera. This should give a firm indication whether the cardiac-type residues in the N-terminal domain are sufficient for the phenotypic length response.

The Mechanism of Length-Transduction

For TnC moiety to act as the length-sensor there must be a mechanism that informs that the adjacent Z-lines have parted. A number of length associated structural alterations have been previously considered as possible candidates for such a mechanism: For example, inter-filament separation, cross-bridge number, and titin conformation. Another promising possibility is shown in Fig. 7.

This hypothesis proposes that (i) Ca^{2+}-affinity decreases along the length of thin filaments: Ca^{2+}-affinity is highest at the free end, and lowest near the Z-line. (ii) The pCa-force relation at a given sarcomere length reflects the mean Ca^{2+}-affinity of the filament only in the overlap zone. Therefore, since the new zone recruited at the shorter sarcomere length is of lower affinity, overall mean Ca^{2+}-affinity is also reduced. This would shift the force curve to the right.

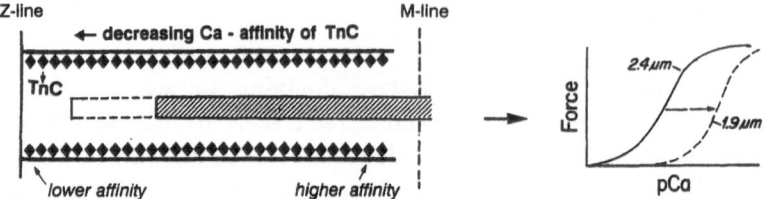

Fig. 7. The hypothesis that the TnC Ca^{2+}-affinity decreases progressively along the length of the thin filament, with highest affinity at the free end and lowest near the Z-line.

In the tests of this hypothesis, a helpful development was the technical ability to extract TnC either (i) selectively from the Z-line domain of the thin filament (asymmetric extraction), or (ii) uniformly from throughout the thin filament (symmetric extraction). These two types of extractions could be checked by loading the extracted fibers with biotinylated TnC, followed by rhodamine labelled avidin which conjugates selectively with biotin. Visualization of these fibers under the fluorescence microscope have indicated the placement of biotinylated TnC to be limited to a narrow strip for the asymmetric extraction, whereas fluorescence signal for symmetric extraction was much broader, spanning the entire thin filament length (Cheng and Gulati, unpublished).

A possible prediction is that the length dependence for asymmetrically extracted fibers should be increased above that for normal fibers. This prediction is the result of counting the vacated TnC sites near the Z-line as zero Ca^{2+}-affinity sites, which lowers the mean Ca-affinity of the TnC ensemble within the overlap domain at the short sarcomere length, below normal. Symmetrically extracted fibers, on the other hand, should indicate little net increase in their length dependence. Preliminary results do indicate an over two-fold greater length dependence for asymmetric fibers in comparison with the symmetric fibers (Babu and Gulati, unpublished).

Conclusions

Our experiments of TnC exchange in myocardium have indicated a new role for cardiac TnC: as the Starling's length-sensor. Cardiac TnC is a more efficient length sensor than skeletal TnC. We get the same results on fast fibers, but there is presently some controversy on this that may be explained by asymmetric extraction and exchange of TnC. The activation in low salt solutions (100mM vs 200mM physiological ionic strength) was found helpful in exposing the cardiac TnC effect in fast fibers. The molecular biological approach combined with skinned fiber studies have indicated that the metal-ion binding property of site 1 in the N-terminus may play a part of the role in the length-sensing mechanism. Other domains of the first EF-hand are important as well. The studies also have exposed another unexpected property of the thin filament—The interaction of TnC moiety with other proteins on the thin filament alters its Ca^{2+}-affinity such that the thin filament is polarized such that the affinity decreases progressively from its highest value near the M-line to its lowest near the Z-line.

ACKNOWLEDGEMENTS

Grant support for the studies was from NIH (AR 33736 & HL 37412), NY Heart Association, and the Blumkin fund.

REFERENCES

1. Babu, A., Sonnenblick, E. & Gulati, J. *Science* **240**, 74-76 (1988).
2. Starling, E.H. in *Starling Law of the Heart* (Longmans, Green, London, 1918).
3. Hibberd, M.G. & Jewell, B.R. *J. Physiol. (Lond)* **382**, 527-540 (1982).
4. Gulati, J. in *Molecular Biology of the Cardiovascular System* (eds, Roberts, R. & Schneider, M) 249-259 (Wiley-Liss, New York, 1990).
5. Parmacek, M.S. & Leiden, J.M. *Circulation* **84**, 991-1003 (1991).
6. Gulati, J., Scordilis, S. & Babu, A. *FEBS Lett.* **236**, 441-444 (1988).
7. Moss, R.L., Nwoye, L.O. & Greaser, M.L. *Biophys. J.* **57**, 146a (1990).
8. Babu, A. & Gulati, J. *J. Physiol. (Lond)* **410**, 71P (1989).
9. Gulati, J. & Babu, A. *J. Physiol. (Lond)* **429**, 22P (1990).
10. Gulati, J. & Babu, A. *Biophys. J.* **59**, 36a (1991).
11. Gulati, J., Sonnenblick, E. & Babu, A. *J. Physiol. (Lond).* **441**, 305-324 (1991).
12. Babu, A., Scordilis, S., Sonnenblick, E. & Gulati, J. *J. Biol. Chem.* **262**, 5815-5822 (1987).
13. Babu, A., Lehman, W. & Gulati, J. *FEBS Lett* **251**, 177-182 (1989).
14. Moss, R.L., Nwoye, L.O. & Greaser, M.L. *J. Physiol. (Lond)* **440**, 273-289 (1991).
15. Moss, R.L., Giulian, G.G. & Greaser, M.L. *J. Gen. Physiol.* **86**, 585-600 (1985).
16. Gulati, J, Babu, A., Cheng, R. & Su, H. in *Modulation of Cardiac Calcium Sensitivity* (eds. Allen, D & Lee, J) (In the press).
17. Putkey, J.A., Sweeney, H.L. & Campbell, S.T. *J. Biol. Chem.* **264**, 12370-12378 (1989).
18. Khorana, H.G. *J. Biol. Chem.* **263**, 7439-7442 (1988).
19. Babu, A., Su, H., Ryu, Y. & Gulati, J. *J. Biol. Chem.* **267** (1992, in press).

Discussion

Godt: Can you tell me how you achieved the asymmetric extraction versus the symmetric extraction? Can you give us a quick *précis* of how you did that, and how you tested that it was really the case?

Gulati: To get the different types of extractions, we used fibers of either 1.6 μm sarcomere length or 2.6 μm. We found that the extractions at the longer length were primarily from the Z-line domain. The fibers were next stretched to 3.4 μm and loaded with biotinylated TnC. So, what you see is a white band wherever the biotinylated TnC goes. In the case of symmetric fibers, the entire thin filament should light up. In the asymmetric case, there is a smaller light band and a longer dark band. The interpretation is fairly obvious. In the asymmetric case, the light band is shorter. The degree of extraction was similar in both cases.

Gordon: In reference to the idea that TnC is the sole determinant of the length sensitivity: Don Martyn in my lab has also extracted endogeneous TnC in the skeletal case and replaced it with cardiac TnC. In rabbit psoas muscles, we had no problems getting rabbit cardiac TnC back in and we had full regulation with cardiac TnC. We see no change in length-dependence with cardiac TnC. So, we conclude that in the skeletal case, the cardiac TnC may not be imparting enhanced length-dependence. We did this experiment with skeletal muscle because cardiac muscle is more problematic in our hands. We thought the skeletal case was probably a more critical test of your hypothesis.

Gulati: I think that is true. But the skeletal case is also trickier, as I said. I suggest that to resolve the discrepancy, Don has to really look at the quality of extraction. If it is partly asymmetric, it would counterbalance the effect and produce results opposite from the ones you would expect.

Gergely: What is the basis of this inhomogeneity or this gradient of troponin-C along the thin filament? Why is there a difference in troponin-C in transmitting this effect?

Gulati: I don't know the exact answer, but the following considerations were important in the hypothesis: (1) We thought the affinity of TnC binding to the thin filament may vary intrinsically along its length; (2) thin-thick filament overlap may affect the TnC binding. These points would explain asymmetric extraction at the long length. Now, what gives rise to the affinity gradient for the Ca-TnC reaction along the thin filament is an open question in our minds.

EFFECT OF CROSS-LINKING ON THE CONTRACTILE BEHAVIOR OF MYOFIBRILS

William F. Harrington, Trudy Karr and William B. Busa

The Johns Hopkins University
Department of Biology
Baltimore, MD 21218

ABSTRACT

When rabbit psoas myofibrils in rigor are cross-linked with DMS (dimethyl suberimidate) for various periods of time, they contract on activation to a final sarcomere spacing of 1.3-1.5 μm. This behavior is observed out to 100 min cross-linking time (2 mg/ml DMS; 10°C). Over the next 100 min of cross-linking, the sarcomere spacing, following activation and contraction, gradually increases and finally plateaus near its initial (rigor) value. We also determined the unloaded shortening velocity of the cross-linked myofibrils using an inverted microscope equipped with a video camera. Following photo-activation of caged ATP, the fast contracting process observed in control (untreated) myofibrils decreases in rate and magnitude with increasing cross-linking time. When taken together with earlier cross-linking studies, our present results suggest that the suppression of contraction may result from two distinct cross-linking reactions: (1)Cross-linking of myosin rods in the filament core which immobilizes the S-2 subunit and acts to decrease the isometric force (~90% at 100 min). (2)Cross-linking within the S-1 subunit. This latter reaction is believed to account for the continuous decay in the rate and magnitude of the unloaded shortening process.

INTRODUCTION

It seems clear from *in vitro* model studies that the myosin head alone (S-1 or HMM) can produce force[1)2)] and slide actin filaments[3)4)] at speeds approaching those obtained with muscle fibers under no-load conditions. Recent experiments also demonstrate that the movement is quantized[2)5)] suggesting, along with measurements of the sliding distance in muscle[6-8)] that the force-generating process

Mechanism of Myofilament Sliding in Muscle Contraction, Edited by
H. Sugi and G.H Pollack, Plenum Press, New York, 1993

603

involves a small (10-12 nm) axial displacement of the actin-attached head for each ATP molecule hydrolyzed near isometric tension. However it is not certain at the moment if the vectorial displacement observed in the *in vitro* system is an accurate reflection of the complete force-producing process in working muscle. A number of studies have also implicated the S-2 region of myosin and, particularly, the hinge domain of S-2 in the tension generating event. These include proteolytic digestion experiments of myosin rod[9], the S-2 subunit[10], glycerinated rigor[11], and activated[12] fibers, electron microscope studies of myosin and myosin rod at various temperatures[13][14] and laser temperature-jump investigations of rigor and activated fibers[15]. Molecular genetic studies have also provided evidence that alternative myosin hinges may play an essential role in the generation of different levels of contractile force or shortening speed[16-19] in activated muscle.

Recently we reported that polyclonal antibodies directed against the S-2 region of myosin depressed the isometric force of skinned psoas fibers, but had little effect on the actin-activated ATPase of these fibers or the shortening velocity of myofibrils[20][21]. These experiments suggest that the S-2 segment as well as the S-1 subunit contributes to active force in working muscle, but S-1 alone may be sufficient to support the unloaded shortening of myofibrils. The studies of Margossian et al.[22] also implicate the S-2 domain in the force-generating process. Their work demonstrates that a polyclonal antibody, directed against a 20 residue peptide of the cardiac myosin hinge, suppresses vectorial motion of actin filaments by monomeric myosin in *in vitro* motility assays. This antibody also reduces active shortening of skinned myocytes without affecting the actin-activated ATPase activity.

In view of the results and implications of the antibody studies, it was of interest to examine the effect of cross-linking within the thick filament core on the rate and extent of activated shortening of myofibrils. In earlier work[23] the isometric force at various stages of cross-linking glycerinated psoas fibers (with dimethyl suberimidate (DMS)) was observed to fall in parallel with the formation of high molecular weight myosin rod species, whereas the myofibrillar ATPase remains virtually unaffected by the cross-linking reaction. One interpretation of these results is that cross-linking between neighboring myosin rods within the core prevents release of S-2 from the thick filament surface and thereby inhibits this element from participating in the force-generating process. In the work described below, we examine the contractile properties of cross-linked myofibrils to obtain further insight into this question.

RESULTS

Contraction of Activated Myofibrils Following Cross-linking with DMS (Dimethyl Suberimidate)

Figures 1 and 2 show the average sarcomere spacing following contraction of activated myofibrils at various times of cross-linking with DMS. Myofibrils were cross-linked under rigor conditions and sarcomere lengths determined in a light

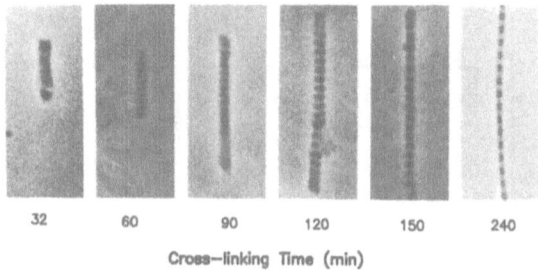

Fig. 1. Contraction of glycerinated psoas myofibrils after cross-linking with dimethyl suberimidate. Myofibrils in rigor buffer were cross-linked with DMS (2 mg/ml; 10°C) for various periods of time[23]. Cross-linking was terminated by addition of an equal volume of 0.1 M ethanolamine in 40 mM cacodylate, pH 7.0, followed by two low-speed centrifugation cycles in 40 mM cacodylate, 80 mM NaCl, 2 mM EGTA, pH 7.2. Contraction was initiated by addition of an equal volume of activating buffer (temp. 21-22°C), and after 2-4 min samples were transferred to a microscope slide and photographed using a phase contrast microscope equipped with a 35-mm camera. Separate tests showed that the cross-linked myofibrils are fully contracted within 1 min in activating buffer. Rigor and activating solutions were identical to those employed in our earlier cross-linking study[23].

microscope after completion of the shortening process. In the early stages (< 30 min reaction time) of cross-linking, the modified myofibrils undergo supercontraction on activation into globules or very short rods with diffuse and indeterminate sarcomere spacings. At cross-linking times between 30-100 min, the activated myofibrils contract rapidly to rod-like structures with similar and well-defined sarcomere spacings (1.3-1.5 µm). Beyond this stage of cross-linking, the final average sarcomere spacing of activated myofibrils gradually increases and levels off after ~200 min cross-linking time at 2.6-2.7 µm, which approximates the initial rigor spacing (2.8 µm).

We also observed a similar three-stage profile in the apparent width of the dark, A-band region vs. cross-linking time following activation (Fig. 2). During the first 30 min of cross-linking, activation of the rigor myofibrils resulted in a diffuse and indeterminate A-band region. Over the next stage of cross-linking (30-100 min) the dark "A-band" region shortens following activation from a width of 1.75 µm (initial rigor state) to about 1 µm independent of the cross-linking time. At cross-linking times above 100 min, the final A-band width following activation gradually increases and plateaus at 200 min near the A-band length (~1.75 µm) observed before activation. This observation suggests that the contractile force developed following activation is sufficient over the first 100 minutes of cross-linking to contract the Z-discs and associated actin and titin filaments and crush the tapered ends of thick filaments or, possibly, to disorient those filaments about the fiber axis. Beyond this time, the declining residual force results in a gradual elongation (or ordering) of the A-band filaments.

In our earlier cross-linking studies[23], the relative isometric force (f/f_0) of skinned glycerinated psoas fibers was determined as a function of cross-linking time

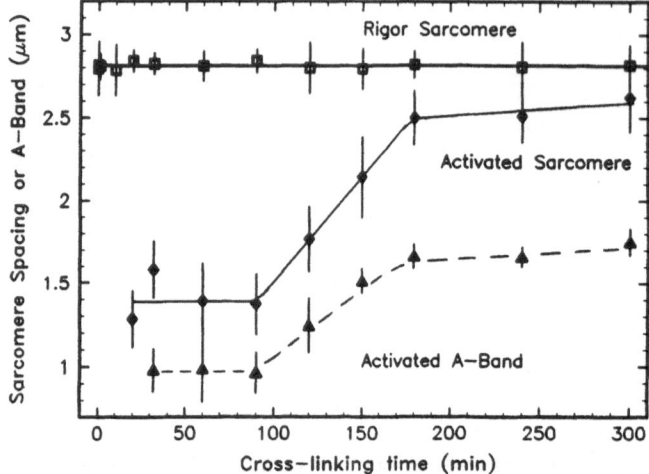

Fig. 2. Contraction of glycerinated psoas myofibrils after cross-linking with dimethyl suberimidate (2 mg/ml; 10°C). (□), average sarcomere spacing after cross-linking of myofibrils in rigor buffer with DMS at pH 7.2 for various periods of time; (♦), average sarcomere spacing following contraction of activated, cross-linked myofibrils. Data points were obtained after termination of active contraction. (▲), average A-band lengths following termination of active contraction. Average sarcomere spacings were obtained using a phase contrast microscope equipped with a 35 mm camera. Magnification was determined by photographing a lined, calibration reticle. Spacings were determined from measurements of photographic negatives in an X-Y microcomparator (Nikon Model 6C). Average spacings were obtained by measurement of 8-24 myofibrils at each time point.

under identical cross-linking conditions to those employed in the present work (2 mg/ml DMS; pH 7.2; 10°C). Following a distinct lag time of about 20 minutes, the normalized force decreased monotonically to about 10% of its initial value (f_0) at 100 min cross-linking time (Fig. 3). The time course of cross-linking within the thick filament was monitored by transferring single fibers at various stages of the reaction into SDS-containing solutions followed by SDS-gel electrophoresis of each sample. The time dependence of the heavy chain band during the cross-linking reaction (normalized to the actin band) followed a single exponential decay with half-time of about 15-20 min. In the present study, we observed a closely similar time-dependent profile and half-time of decay of the myosin heavy chain on cross-linking *myofibrils* with DMS (Fig. 3).

At the end of the second phase of the cross-linking reaction (~100 min), Ueno found intermolecular cross-linking within the thick filament core to be ~90% complete, judging by the formation of higher order (n ≥ 4) cross-linked rod species (see(ref. 23), Figs. 3 and 6). In these experiments, the S-1/rod junction was digested by chymotrypsin before electrophoresis of the cross-linked myofibrils and the time dependence of cross-linking rod segments within the thick filament core monitored by a disulfide oxidation procedure to distinguish between intermolecular and intramolecular cross-linking. Based on the results of these studies, we expect the S-2

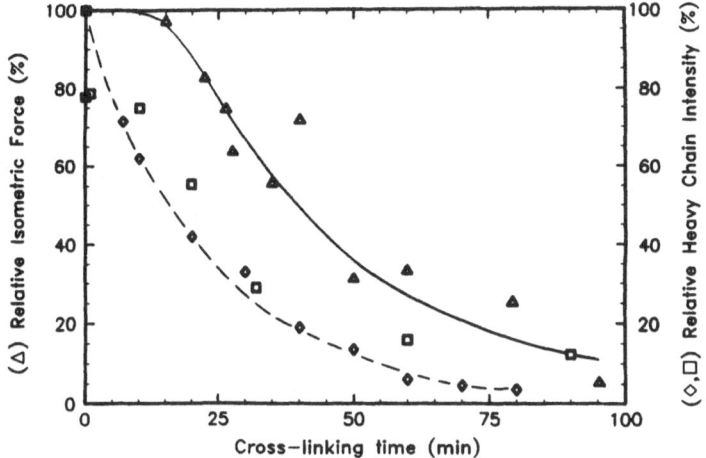

Fig. 3. Effect of DMS cross-linking on isometric force in glycerinated psoas fibers (taken from ref. 23). Glycerinated fibers were cross-linked (2 mg/ml, 10°C) in rigor buffer. (Δ), relative isometric force (f/f$_o$) vs. cross-linking time. (✧), myosin heavy chain intensity vs. cross-linking time of glycerinated fibers; (□), myosin heavy chain intensity vs. cross-linking time of *myofibrils* (present study; DMS, 2 mg/ml, 10°C). Band intensities of SDS-gel electrophoresis experiments (✧, □) were normalized to the actin band. The rate constant of cross-linking the myosin heavy chain in myofibrils is slightly larger than the rate constant observed for cross-linking the myosin rod (see ref. 23, Fig. 2A).

segment of the myosin rod to be locked down on the thick filament surface in myofibrils at this stage of the cross-linking reaction.

Velocity of Contraction of Myofibrils Following Cross-linking with DMS

We employed an inverted microscope equipped with a video camera to investigate the unloaded shortening velocity of cross-linked myofibrils. Contraction was initiated by photo-hydrolysis of caged ATP and the average sarcomere spacing at each time point determined by measuring the end-to-end length of the myofibrils[21]. Figure 4 compares the active shortening of rabbit psoas myofibrils which had been cross-linked in rigor at various times with that of control (untreated myofibrils). Solid lines are all single exponential fits to all of the data using a non-linear fitting routine[24]. Native myofibrils exhibit a rapid phase of shortening, following an 8 msec UV flash from a computer-controlled light shutter, which is complete in about 500 msec (21°C)[21]. This is followed by a much slower phase in which the myofibrils undergo further shortening and, in a few seconds, condense into the supercontracted state. The rate and magnitude of the fast shortening process decreases with increasing cross-linking time. From an inspection of the time-dependent profiles of Fig. 4, it is clear that this fast shortening process is still present at cross-linking times well beyond the end of the second stage of the cross-linking reaction (~100 min) where the isometric force has dropped to a small fraction of its

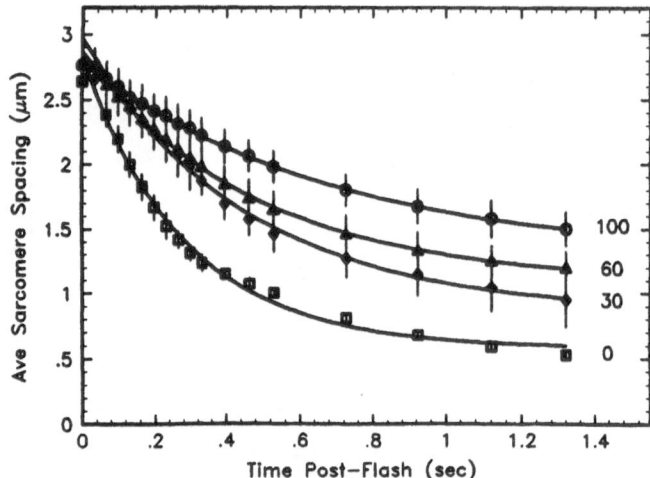

Fig. 4. Average sarcomere spacing vs. time of cross-linked, activated rabbit psoas myofibrils. Activation was initiated by photohydrolysis of caged ATP (8 msec per flash). (□), control (untreated) myofibrils; (◆), (△) and (O), activated after 30, 60 and 100 min cross-linking, respectively (2 mg/ml DMS, 10°C, pH 7.2) in rigor buffer. Solid lines are single exponential fits to all the data using the fitting routine of Johnson and Frazier[24]. Number of runs averaged for each data set: control, 4; 30 min, 9; 60 min, 12; 100 min, 10. Temperature 21-22°C. See ref. 21 for experimental details.

initial, f_0, value. This conclusion is supported by plots of the amplitude of the fast phase reaction and its initial rate vs. cross-linking time shown in Fig. 5. The amplitude of the fast shortening phase (A_1) decreases linearly with cross-linking time over the first 100 min. Assuming a similar dependence on cross-linking time beyond this point, the amplitude would extrapolate to zero well above 200 min. The initial rate exhibits a non-linear profile decreasing from 8 μm/sec (control) to about 2.2 μm/sec at 100 min cross-linking time (Fig. 5, insert). The behavior of both of these parameters, amplitude and initial rate suggest, therefore, that the fast process is completed near the time where shortening of the myofibrils following activation is no longer observed. The marked decrease in the rate of myofibrillar contraction following cross-linking stands in contrast to the effect of antibodies directed against the subfragment-2 region of myosin. As noted earlier, the polyclonal antibodies suppress contractile force in fibers, but have no significant effect on the shortening velocity of myofibrils.

DISCUSSION

In the earlier experiments of Sutoh in our laboratory[25] DTBP cross-linking between components of the thin filament complex (actin, tropomyosin and troponin) was shown to be a much slower process than the cross-linking reaction between

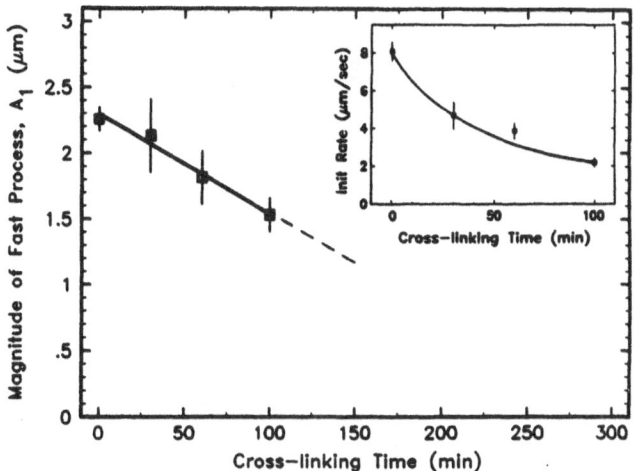

Fig. 5. Magnitude (A_1) and initial rates of unloaded shortening vs. DMS cross-linking time. Values for the magnitude (A_1) and initial rate (insert) were obtained from parameters derived from the single exponential plots of Fig. 4.

adjacent myosin rods in the thick filament core. He also found that the rate of cross-linking conditions. In the more recent experiments of Ueno[23] no measurable change in the intensity of the actin band was detected on SDS polyacrylamide gels following cross-linking of rigor myofibrils with DMS which could be attributed to bridging between the myosin heads and actin. Kimura and Tawada reported[26] similar conclusions based on DMS cross-linking studies of HMM-actin-tropomyosin complexes. Thus it appears that the cross-linking reactions responsible for the changes in contractile properties of myofibrils observed in the present work are directed primarily, if not exclusively, to cross-linking within the thick filaments themselves.

Two cross-linking processes have been detected and identified within myosin molecules of the thick filaments: (1) Intermolecular cross-linking between individual myosin rods[23][25][26][28] within the thick filament core, and (2) Cross-linking between the light and heavy chains of the S-1 subunits[23][27]. This latter reaction occurs at about one-third the rate observed for cross-linking myosin rods. This is based on the time course of the S-1 band intensity following proteolytic digestion at the S-1/rod junction of DMS cross-linked, glycerinated rigor psoas fibers[26] and myofibrils[23][25].

A similar correspondence in cross-linking rates was observed by Sutoh[25] in the DTBP cross-linking of myosin heads and myosin rods in rigor myofibrils and synthetic thick filaments.

Figure 6, taken from the paper of Ueno and Harrington[23], illustrates the internal cross-linking process. Following chemical cross-linking of myofibrils with

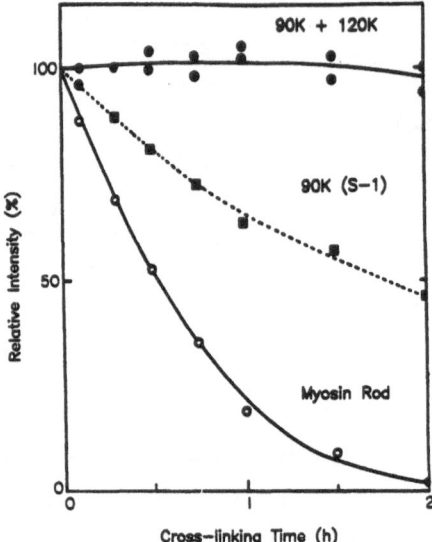

Fig. 6. Chemical cross-linking of rabbit psoas myofibrils with DMS (taken from ref. 23). Myofibrils (2 mg protein/ml) were cross-linked with DMS (1 mg/ml; 22°C) in 40 mM imidazole hydrochloride and 80 mM NaCl (pH 7.2). The S-1/rod junction in each sample was digested by chymotrypsin and the time course of cross-linking analyzed (following oxidation of thiol residues) by SDS-gel electrophoresis. (O), rod (260 K); (■), 90 K S-1 and (●), S-1 (90 K + 120 K) bands are plotted vs. cross-linking time. The decay of the 90 K band of S-1 results from cross-linking of the S-1 heavy chain with its associated light chains to form the 90 K + 120 K species.

DMS, the S-1 rod junction was digested at various cross-linking times and the time course of cross-linking monitored by SDS gel electrophoresis. The intensity vs. time of the 90K S-1 heavy chain species (dotted line) exhibits the characteristic time-dependent decay observed in earlier studies (e.g. ref.25). This band disappears at about one-third the rate of decay of the rod species (260 kDa), which is shown in the lower curve (open circles). Cross-linking between the S-1 heavy chain and its associated light chains forms a series of bands of Mr 90 to 120 kDa. The sum of the intensities of these bands remains invariant with cross-linking time (upper curve, closed circles).

When our present results are taken together with these earlier studies, it seems likely that two parallel but distinct cross-linking processes are responsible for the overall contractile behavior of the cross-linked myofibrils. Bridging of the S-2 region of the cross-bridge to neighboring myosin rods in the thick filament core appears to be responsible for the striking drop in the isometric force observed during the first 100 minutes of the cross-linking reaction. During this phase, the cross-linked myofibrils are capable of undergoing maximal shortening following activation and the system exhibits a high actin-activated ATPase[23][29].

Hence, immobilizing the S-2 subunit onto the thick filament surface may act to decouple this segment from the S-1 subunit in the overall force-generating mechanism. Concurrent with this process, cross-linking *within* the S-1 subunit acts to gradually inhibit the velocity and shortening amplitude of the myofibrils. This interpretation is based on the time-dependent decrease in the rate and magnitude of the fast sliding process observed in Figs. 4 and 5 which appear to correlate with the time-dependence of the internal cross-linking in the myosin head. This reaction has a half-time of about 60-75 min under our present experimental conditions (2 mg/ml DMS; 10°C) and would therefore be complete at about the time when active shortening of cross-linked myofibrils is fully suppressed.

CONCLUDING REMARKS

The present results provide support for the idea that there are two force-generating sites in vertebrate skeletal muscle. According to this view, the S-2 region contributes a large fraction of the isometric force—perhaps as much as 90%. This would be consistent with the proposal[21] that the S-1 subunit contributes the residual sliding force. Immobilization of the S-2 region on the thick filament surface by cross-linking or binding of anti-S-2 antibodies to this segment decouples this element from the S-1 subunit. This process has virtually no effect on the ability of S-1 to generate power in the presence of ATP and slide the thin filaments relative to thick filaments. When the myosin heads are modified, e.g. through internal cross-linking as in the present study, the sliding force is gradually suppressed. Beyond 100 min of cross-linking time under our experimental conditions, this leads to a gradual elongation of the average sarcomere spacing of activated myofibrils.

ACKNOWLEDGEMENTS

We are grateful to Drs. Emil Reisler and Stephen Lovell for helpful discussions.This work was supported by National Institutes of Health Research Grants AR-04349 to W.F.H. and HD-22879 to W.B.B.

REFERENCES

1. Kishino, A. & Yanagida, T. *Nature* **334**, 74-76 (1988).
2. Ishijima, A., Doi, T., Sakurada, K., & Yanagida, T. *Nature* **352**, 301-306 (1991).
3. Toyoshima, Y.Y., Kron, S.J., McNally, E.M., Niebling, K.R., Toyoshima, C. & Spudich, J.A. *Nature* **328**, 536-539 (1987).
4. Hynes, T.R., Block, S.M., White, B.T. & Spudich, J.A. *Cell* **48**, 953-963 (1987).
5. Uyeda, T.Q.P., Warrick, H.M., Kron, S.J. & Spudich, J.A. *Nature* **352**, 307-311 (1991).
6. Huxley, A.F. & Simmons, R.M. *Nature* **233**, 533-538 (1971).
7. Ford, L.E., Huxley, A.F. & Simmons, R.M. *J. Physiol. (Lond.)* **269**,441-515 (1977).

8. Higuchi, H., & Goldman, Y.E. *Nature* **352**, 352-354 (1991).

9. Ueno, H. & Harrington, W.F. *J. Mol. Biol.* **173**, 35-61 (1984).

10. Ueno, H. & Harrington, W.F. *J. Mol. Biol.* **180**, 667-701 (1984).

11. Ueno, H. & Harrington, W.F. *J. Mol. Biol.* **190**, 59-68 (1986).

12. Ueno, H. & Harrington, W.F. *J. Mol. Biol.* **190**, 69-82 (1986).

13. Walker, M. & Trinick, J. *J. Mol. Biol.* **192**, 661-667 (1986).

14. Walzthony, D., Eppenberger, H.M., Ueno, H., Harrington, W.F. & Wallimann, T. *Eur. J. Cell Biol.* **41**, 38-43 (1986).

15. Davis, J.S. & Harrington, W.F. *Proc. Natl. Acad. Sci. USA* **84**, 975-979 (1987).

16. George, E.L., Ober, M.B. & Emerson, C.P.Jr. *Mol. Cell Biol.* **9**, 2957-2974 (1989).

17. Collier, V.L., Kronert, W.A., O'Donnell, P.T., Edwards, K.A. & Bernstein, S.I. *Genes Dev.* **4**, 885-895 (1990).

18. McNally, E.M., Kraft, R., Bravo-Zehnder, M., Taylor, D.A. & Leinwand, L.A. *J. Mol. Biol.* **210**, 665-671 (1989).

19. Nyitray, L., Goodwin, E.B. & Szent-Györgyi, A.G. *Biophys. J.* **59**, 228a (1991).

20. Lovell, S., Karr, T. & Harrington, W.F. *Proc. Natl. Acad. Sci. USA* **85**, 1849-1853 (1988).

21. Harrington, W.F., Karr, T., Busa, W.B. & Lovell, S.J. *Proc. Natl. Acad. Sci. USA* **87**, 7453-7456 (1990).

22. Margossian, S.S., Krueger, J.W., Sellers, J.R., Cuda, G., Caulfield, J.B.,Norton, P. & Slayter, H. S. *Proc. Natl. Acad. Sci. USA* **88**, 4941-4945 (1991).

23. Ueno, H. & Harrington, W.F. *Biochemistry* **26**, 3589-3596 (1987).

24. Johnson, M.L. & Frazier, S.G. *Methods Enzymol.* **117**, 301-342 (1985).

25. Sutoh, K. & Harrington, W.F. *Biochemistry* **16**, 2441-2449(1977).

26. Kimura, M. & Tawada, K. *Biophys. J.* **45**, 603-610 (1984).

27. Labbe, J.P., Mornet, D., Roseau, G. & Kassab, R. *Biochemistry* **21**, 6897-6902 (1982).

28. Chiao, Y.C. & Harrington, W.F. *Biochemistry* **18**, 959 (1979).

29. Baker, A.J. & Cooke, R. *Biophys. J.* **49**, 7a (1986).

Discussion

Katayama: What would happen if you used the much longer or shorter cross-linkers, or if you used monofunctional reagents?

Harrington: We haven't tried longer or shorter cross-linkers. We have tried DTBP (Dithio-bis-Propionimidate), which is a reducible cross-linker. We find that after cross-linking with DTBP, we get the same general phenomenon that you saw here with DMS. But you can break the cross-links of DTBP. When you reduce the cross-links, the system will begin to slide again. Since by breaking the disulfide cross-links you can regenerate the sliding process within the myofibril, it doesn't appear as if the cross-linking reaction itself were responsible for this behavior.

Morales: Since the myosin head has lysine clusters that are very important in the interaction with actin, and it is conceivable that you could cross-link one cluster to another with a suberimidate, would it be worthwhile—perhaps with radioactive suberimidate—to see the effect on S-1 also?

Harrington: Very much so.

Pollack: Dr. Harrington, you measured what appeared to be differences of A-band width and there appeared to be some correlation with the degree of cross-linking, if I understand your data. Do you see a causal relation between the two?

Harrington: We assume that there is enough sliding force to crush the A-band region down to about a micron, or else just move the filaments out of their normal orientation.

Huxley: Did you check with polarized light whether the high refractive index region really was a birefringence region? This might be reversed striations like the diagram of Engelmann that I showed, where the striation reversal occurred.

Harrington: We did not check that.

ESSENTIAL ROLE OF MYOSIN S-2 REGION IN MUSCLE CONTRACTION

T. Kobayashi, K. Noguchi, T. Gross and H. Sugi

Department of Physiology
School of Medicine
Teikyo University
Tokyo 173, Japan

ABSTRACT

We studied the contraction characteristics and Mg-ATPase activity of glycerinated rabbit psoas muscle fibers in the presence and absence of polyclonal antibody directed against the subfragment-2 (S-2) region of myosin, to give information about the role of myosin hinge region in muscle contraction. The antibody was kindly supplied to us from Professor Harrington's laboratory. The antibody-induced decrease of Ca^{2+}-activated isometric force development was always accompanied by a parallel decrease of muscle fiber stiffness, so that the stiffness versus force relation remained the same by the antibody treatment. Force-velocity curves, obtained by applying ramp decreases in load from steady isometric force to zero, indicated that the antibody had no effect on the maximum shortening velocity and the shape of the force-velocity curve. Simultaneous measurements of Mg-ATPase activity and Ca^{2+}-activated isometric force showed that Mg-ATPase activity of the fibers remained unchanged despite the antibody-induced decrease of isometric force even to zero. These results indicate that, if the antibody attaches to the S-2 region of myosin molecules, their heads still hydrolize ATP without contributing to both muscle force generation and muscle fiber stiffness.

INTRODUCTION

Muscle contraction is believed to result from cyclic formation and breaking of cross-links between the myosin head (subfragment-1; S-1), extending from the thick filament, and a neighboring thin filament, the energy for contraction being supplied by ATP hydrolysis. Since the ATPase activity and actin binding site are localized in the S-1 region of myosin, S-1 is normally believed to play a major role in muscle contraction. This view has been strengthened by recent *in vitro* motility assays that

Mechanism of Myofilament Sliding in Muscle Contraction, Edited by
H. Sugi and G.H Pollack, Plenum Press, New York, 1993

615

S-1 alone is sufficient to produce force and move actin filaments[1][2]. From the standpoint of muscle physiology, however, it is not clear whether the ATP-dependent actin-myosin sliding observed in the assay systems is the same as that actually taking place in living muscle.

Contrary to the above general view, it has been proposed by Harrington and his coworkers that melting and shortening in the proteoritically sensitive myosin hinge region lying between the subfragment-2 (S-2) and the myosin tail contributes to muscle force generation[3-6]. In support of this hypothesis, polyclonal antibody directed against the myosin S-2 region has been shown to reduce Ca^{2+}-activated isometric force in glycerinated fibers, while ATPase activity of the fibers and the unloaded shortening velocity of isolated myofibrils exhibit little change[7][8]. In addition, it has been shown that, in the presence of antibody directed against a short peptide segment within the hinge region of cardiac myosin, actin filament movement in an *in vitro* motility assay is suppressed, while ATPase of cardiac myofibrils and S-1 remains unchanged[9].

In the reports cited above, however, measurements of force, shortening velocity, ATPase activity and motility assay have been made separately with different types of preparation. To obtain more conclusive results on the effect of anti-S-2 antibody on muscle contraction, we examined the contraction characteristics and ATPase activity of single glycerinated muscle fibers in the presence and absence of anti-S-2 antibody.

MATERIALS AND METHODS

Single glycerinated muscle fibers prepared from rabbit psoas were mounted horizontally at their slack length (sarcomere length, 2.2-2.3 μm) in an experimental chamber filled with relaxing solution (125 mM KCl, 20 mM Pipes, 4 mM MgCl$_2$, 4 mM ATP, 4 mM EGTA, pH 7.0); one end was connected to a force transducer (Akers, AE801) while the other end was attached to a servo-motor (General Scanning, G100PD with JCCX101 conrol unit). The fibers were maximally activated to contract isometrically with contracting solution, prepared by adding 4 mM CaCl$_2$ to relaxing solution. Muscle fiber stiffness was continuously measured by applying sinusoidal length changes (1 kHz, peak-to-peak amplitude 0.1% of fiber length) and recording the resulting force changes. Force-velocity (P-V) relation was determined by applying ramp decreases in force from the steady isometric force to zero. The resulting fiber shortening was recorded with a displacement transducer (differential capacitor) incorporated in the servo-motor. MgATPase of the fibers were measured by the decrease of NADH during cleavage of ATP using a dual wavelength spectrophotometer (Hitachi, 156). Further details of the methods used have been described elsewhere[10].

Polyclonal antibody directed against the long S-2 region of rabbit psoas muscle myosin was produced in goat[7] in Harrington's laboratory. All experiments were made at room temperature (18-20°C).

RESULTS

Effect of Anti-S-2 Antibody on Muscle Fiber Stiffness and Ca^{2+}-activated Isometric Force

Single glycerinated muscle fibers were first made to contract maximally in contracting solution without antibody, and then were repeatedly relaxed and activated by alternate application of relaxing and contracting solutions, both containing the same concentration of antibody (1.5 mg/ml). The results are summarized in Fig. 1. After each application of contracting solution, the fiber could be made to relax completely with relaxing solution, so that both stiffness and force started to change from "zero" baselines on the subsequent application of contracting solution. Though steady isometric force attained in contracting solution decreased markedly with time after administration of antibody, muscle fiber stiffness and isometric force always increased in parallel with each other, so that the stiffness versus force relation during isometric force development was linear and did not change appreciably in the presence of antibody.

Effect of Anti-S-2 Antibody on Force-Velocity Relation

Fig. 2A shows the P-V curves obtained by applying ramp decreases in load from steady isometric force to zero in the absence and presence of antibody. In spite of the

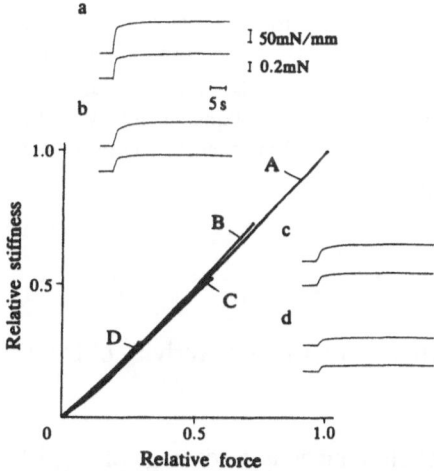

Fig. 1. Relation between muscle fiber stiffness and isometric force during Ca^{2+}-activated force development of a single muscle fiber before (curve A) and 30 (curve B), 60 (curve C), and 90 (curve D) min after administration of anti-S-2 antibody (1.5 mg/ml). Both stiffness and force are expressed relative to control values in the absence of the antibody. (*Insets*) Stiffness versus force curves A, B, C, and D were obtained from stiffness (upper traces) and force (lower traces) records a,b,c, and d, respectively. From Sugi et al.[10].

marked decrease of Ca^{2+}-activated isometric force with time after administration of antibody, the maximum shortening velocity remained unchanged. The *P-V* curves were found to be identical in shape when velocities were replotted against forces expressed relative to their respective steady isometric forces (Fig. 2B), indicating that the *P-V* curves were scaled in proportion to steady isometric forces at which ramp decreases in load were applied.

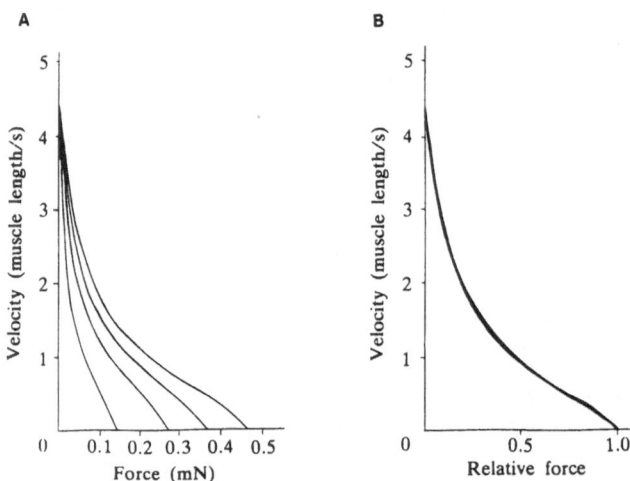

Fig. 2. Effect of anti-S-2 antibody on *P-V* relation in a Ca^{2+}-activated single fiber. (*A*) *P-V* curves obtained before (control) and 30, 60, and 90 min after administration of anti-S-2 antibody. Both velocities and forces are expressed in absolute values. Note that maximum shortening velocity remains unchanged in spite of marked reduction of steady isometric force. (*B*) The same *P-V* curves in which forces are expressed relative to their respective steady initial forces. Note that the curves are identical in shape. The curves were obtained from the results shown in Fig. 3. From Sugi et al.[10].

Effect of Anti-S-2 Antibody on ATPase Activity of the Fibers during Isometric Force Development

Typical examples of simultaneous recordings of MgATPase activity and Ca^{2+}-activated isometric force are presented in Fig. 3. In both the absence and presence of antibody, MgATPase activity of relaxed fibers was very small, and was not significantly different from the rate of spontaneous decrease of NADH concentration without fibers. Therefore, the result that the slope of ATPase records did not change appreciably when Ca^{2+}-activated force development was reduced even to zero indicates no appreciable effect of anti-S-2 antibody on MgATPase activity of Ca^{2+}-activated fibers.

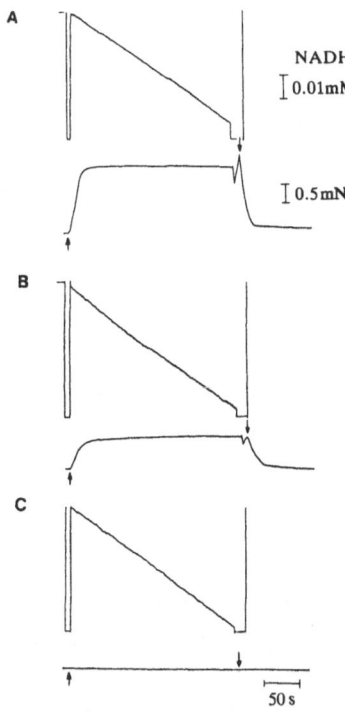

Fig. 3. Simultaneous recordings of Mg-ATPase activity (upper traces) and Ca^{2+}-activated isometric force development (lower traces) of a small fiber bundle consisting of three fibers before (*A*), and 100 (*B*) and 150 (*C*) min after administration of anti-S-2 antibody (1.5 mg/ml). Note that the slope of ATPase records does not change appreciably, even when Ca^{2+}-activated force development is reduced to zero (*C*). Decrease of ATPase trace shows decrease of NADH absorbance. Times of application of contracting solution and relaxing solution are indicated by upward and downward arrows, respectively. From Sugi et al.[10].

DISCUSSION

The parallel decrease of muscle fiber stiffness and Ca^{2+}-activated isometric force by anti-S-2 antibody (Fig. 1) indicates that the antibody-induced decrease of Ca^{2+}-activated isometric force development is due to a progressive decrease in the number of myosin heads (cross-bridges) involved in isometric force generation, since muscle fiber stiffness is believed to be a measure of the number of myosin heads attached to the thin filament[11]. This implies that, when anti-S-2 antibody attaches to the S-2 region of myosin molecule, its heads no longer contribute to both force generation and muscle fiber stiffness, precluding the possibility that the antibody-induced decrease of Ca^{2+}-activated force development might be due to internal "friction" between myosin heads and anti-S-2 antibodies with a size comparable to that of myosin heads.

The above view is entirely consistent with the result that the maximum shortening velocity remained unchanged despite the marked antibody-induced decrease of Ca^{2+}-activated isometric force development (Fig. 2A); if the decrease of force development is due to the internal friction, the maximum shortening velocity should certainly be reduced. Since the shape of P-V curve reflects kinetic properties of myosin heads interacting with the thin filament, the P-V curves scaled in proportion to steady isometric forces (Fig. 2B) may also be taken to indicate that only "negative" myosin molecules are involved in isometric force generation and fiber shortening. This implies that the heads of myosin molecules, with anti-S-2 antibodies attached to their S-2 region, neither contribute to muscle fiber stiffness nor provide any internal resistance against Ca^{2+}-activated actin-myosin sliding.

The present experiments have shown that MgATPase activity of the fibers remains unchanged even when Ca^{2+}-activated force development is decreased to zero in the presence of antibody (Fig. 3). This result is in accord with the previous reports that anti-S-2[8] and anti-hinge[9] antibodies have little or no effect on ATPase activity of Ca^{2+}-activated skeletal and cardiac muscles. The antibody-induced dissociation of Ca^{2+}-activated force generation from MgATPase activity, implies, together with the stiffness and P-V data, that the heads of myosin molecules, with anti-S-2 antibody attached to the S-2 region, still hydrolyze ATP but no longer contribute to force generation, fiber shortening and muscle fiber stiffness. According to current biochemical schemes [12][13], muscle contraction is associated with actin-myosin interaction, which includes both weak and strong binding states. A question arises how antibody attached to the S-2 region of a myosin molecule can exert an allosteric influence on its two heads, so that their ATPase reaction proceeds without involving strong actin-myosin binding to be detected by the stiffness measurement.

In summary, the present work clearly indicates the essential role of myosin S-2 region in muscle force generation as well as the danger of obtaining information about the mechanism of muscle contraction based on the simplified motility assays.

REFERENCES

1. Toyoshima, Y.Y., Kron, S.J., McNally, E.M., Niebling, K.R., Toyoshima, C. & Spudich, J.A. *Nature* **328**, 536-539 (1987).
2. Kishino, A. & Yanagida, T. *Nature* **334**, 74-76 (1988).
3. Ueno, H. & Harrington, W.F. *J. Mol. Biol.* **190**, 69-82 (1986).
4. Ueno, H. & Harrington, W.F. *Biochemistry* **26**, 3589-3596 (1987).
5. Applegate, D. & Reisler, E. *J. Mol. Biol.* **169**, 455-468 (1983).
6. Tsong, T.Y., Karr, T. & Harrington, W.F. *Proc. Natl. Acad. Sci. USA* **76**, 1109-1113 (1979).
7. Lovell, S., Karr, T. & Harrington, W.F. *Proc. Natl. Acad. Sci. USA* **85**, 1849-1953 (1988).
8. Harrington, W.F., Karr, T., Busa, W.B. & Lovell, S.J. *Proc. Natl. Acad. Sci. USA* **87**, 7453-7456 (1990).
9. Margossian, S.S., Krueger, J.W., Sellers, J.R. Cuba, G., Gaufield, J.B. Norton, P. & Slayter, H.S. *Proc. Natl. Acad. Sci. USA* **88**, 4941-4945 (1991).
10. Sugi, H., Kobayashi, T. Gross, T., Noguchi, K., Karr, T. & Harrington, W.F. *Proc. Natl. Acad. Sci. USA* **89**, 6134-6137 (1992).

11. Huxley, A.F. *Prog. Biophys. Biophys. Chem.* **7**, 255-318 (1957).

12. Lymn, R.W. & Taylor, E.W. *Biochemistry* **10**, 4617-4624 (1971).

13. Stein, L.A., Schwarz, R.P.Jr., Chock, P.B. & Eisenberg, E. *Biochemistry* **18**, 3895-3909 (1979).

Discussion

Yu: Do you know the value of lattice spacing?

Sugi: I have never measured the lattice spacing. We just started this experiment.

Yu: Could it be that there is some requirement for swinging out of S-2 to reach the thin filament? If you put in some dextran, maybe the force would not be affected so much.

Sugi: I appreciate your suggestion.

IX. KINETIC PROPERTIES OF ACTIN-MYOSIN SLIDING WITH INTACT MUSCLE FIBERS

INTRODUCTION

The mechanical response of contracting intact muscle or muscle fibers to externally applied perturbations, no doubt, reflect the properties of events taking place within muscle fibers. To obtain useful information about contraction mechanisms at the molecular level, however, various experimental techniques are required to remove complications arising mainly from nonuniform response among a number of sarcomeres connected in series along the entire fiber length.

The first three papers in this chapter deal with cardiac muscle. Brandt and others studied the force responses of electrically stimulated intact myocytes or Ca^{2+}-activated skinned myocytes of frog heart to rapid and slow length changes with the results analogous to those from intact skeletal muscle fibers. Saeki and coworkers explored the relationships among myoplasmic Ca^{2+}, tension and length in ferret papillary muscles steadily activated with ouabain or with repetitive electrical stimuli in the presence of ryanodine. By comparing the tension transients in response to step length changes and the aequorin light transients, they suggest that the Ca^{2+}-affinity of cardiac troponin C is dependent on tension. ter Kerus and Tombe compared twitch force (T_0) and unloaded shortening velocities (V_0) of rat ventricular traveculae under various levels of activation, while visco-elastic properties of unstimulated preparations were studied to estimate internal viscous resistance to shortening. The T_0 versus V_0 relation was explained in terms of number of active cross-bridges and viscous resistance.

The next five papers are all concerned with frog skeletal muscle. Edman showed that the double-hyperbolic shape of force-velocity relation in single frog muscle fibers did not change after reducing isometric force with dantrolene, suggesting the breakpoint of the force-velocity curve is determined by shortening velocity. The stiffness versus force relation was also biphasic, resembling closely the force-velocity curve. These results suggest the kinetics of cross-bridge function changes if shortening velocity is reduced below a critical level. Pollack and others made extensive discussion on the classification of sarcomere ;length-tension relation into "classical" and "flat" types, based on their observation that, when isometrically contracting sarcomeres shorten to a slightly shorter length (by about 40 nm), they produce isometric tension much greater than the ordinary isometric tension at the shorter length.

Evidence has been presented that, during muscle shortening, cross-bridges might interact with actin at a rate much faster than the rate of ATP splitting.

Piazzesi, Linari and Lombardi explored the above possibility by examining tension transients in response to conditioning and test release at various intervals. The results obtained indicate that the rate of detachment and reattachment of cross-bridges can take place at a rate much faster than the ATPase rate. To detect the presence of weakly binding cross-bridges in relaxed frog muscle fibers, Bagni and others analyzed force responses of relaxed fibers to fast ramp stretches with the conclusion that weakly binding cross-bridges may not be present in a significant amount in relaxed intact fibers.

Finally, Tsuchiya and others measured changes of the transverse stiffness of frog muscle during isometric contraction using a scanning laser acoustic microscope, and confirmed their previous observation that the transverse stiffness decreases during contraction.

TAKING THE FIRST STEPS IN CONTRACTION MECHANICS OF SINGLE MYOCYTES FROM FROG HEART

· P.W. Brandt*, F. Colomo, C. Poggesi and C. Tesi

*From the Dipartimento di Scienze Fisiologiche
Università degli Studi di Firenze
Viale G.B. Morgagni 63
I-50134 Firenze, Italy*

ABSTRACT

Intact or skinned atrial and ventricular myocytes from frog heart were mounted horizontally between the lever arms of a force transducer and a servo-controlled electromagnetic loud-speaker "motor" in a trough filled with Ringer or relaxing solution.

The myocyte length-sarcomere length relation for intact preparations at rest is linear at least in the range from l_0 (sarcomere length about 2.1 μm, resting force zero) to 1.6 l_0 (resting force about 100 nN).

The peak force value for control twitches (21-23°C, stimulus interval 10 s, $[Ca^{2+}]_o$ 1 mM) varies from 20 to 100 nN in atrial and ventricular intact myocytes. The effects induced by isoprenaline or changes in $[Ca^{2+}]_o$, stimulation pattern and bath temperature on twitch characteristics are comparable to those observed in multicellular preparations.

The steady force produced by maximally Ca^{2+}-activated skinned myocytes is much greater than that developed in control twitches and varies from 0.5 to 3.5 μN in different cells. The saturating pCa in the activating solution is around 5.50.

The force response of a resting myocyte to slow ramp stretches shows an initial velocity- and length-dependent component during the stretch itself and, after completion of the length change, a gradual recovery towards a steady level which only depends on the stretch extent. The force response of a stimulated myocyte to length steps complete in 2 ms consists of an apparently elastic change during the step itself and then of a rapid partial recovery followed by slowering of recovery. Whether or not the force recovery includes different phases as reported for skeletal muscle remains unclear.

---

*Present address: Department of Anatomy and Cell Biology, Columbia University, College of Physicians and Surgeons, 630 West 168st, New York, NY 10032.

Mechanism of Myofilament Sliding in Muscle Contraction, Edited by
H. Sugi and G.H Pollack, Plenum Press, New York, 1993

INTRODUCTION

During the last twenty years several attempts have been made to extend the study of myocardium mechanics to cell level[1]. The balance of this work is however unsatisfactory. The only definitive results are that enzymatic dissociation of myocytes does not alter the excitation-contraction coupling[2][3] and that the stimulus interval-force relation of the heart is a basic property of its cells[3].

The present paper reports the results of various classical mechanical experiments performed on intact or chemically-skinned frog-heart myocytes. The mechanical set-up recently described by Cecchi et al.[3][4] was used. The resolution of the force transducer is 0.5-1.0 nN, with a frequency response of at least 800 Hz in saline. The resolution of the "motor" by which length changes were imposed on the preparations is higher than 8.0 nm.

PRINCIPLES OF THE METHOD

Atrial and ventricular myocytes of frog heart (*Rana esculenta*) were isolated using the enzymatic-dissociation method described by Mitra and Morad[5]. The cells were mounted horizontally under Ringer solution ($[Ca^{2+}]_o$ 1 mM) by sucking their ends into glass micropipettes with tip bores of 1-3 µm, in a temperature controlled trough. The mounting procedure has been already described[3].

Before skinning, the Ringer solution in the trough was replaced with a "relaxing solution" (pCa 8). Skinning and activation of a mounted myocyte were obtained by highly localized applications of the appropriate solutions by means of micropipettes with tip bores of about 20 µm. The injection pipettes containing the skinning and the activating solutions were connected, via electrovalves, to pressure reservoires. They were positioned at a distance of about one cell length from the myocyte central region, perpendicularly to the longitudinal cell axis. The skinning solution was made of relaxing solution added with 0.05% Triton X-100 (Sigma, M.O., U.S.A.). It was applied to myocytes until to obtain the complete disruption of the membrane (usually for less then 10 s), which was signalled by an increase of cell diameters and improvement of striation pattern. The relaxing and the activating solutions were prepared as described by Brandt et al.[6].

One of the myocyte-holding micropipette acted as a calibrated-compliance cantilever force probe. The other one was stiff and was mounted on a servo-controlled electromagnetic "motor"[4][7] by which length changes (controlled-velocity ramps or steps complete in 2 ms) of different size and in either direction were imposed on a mounted myocyte. The preparation of the motor and force-probe micropipettes has been previously described[3].

The beam of a He-Ne laser was focused on the blackened tip of the force probe, whose image was magnified by means of a microscope and projected onto a photosensor (total magnification 125X). The deflection of the probe, caused by the force generated by a myocyte, changed the photosensor current output which was

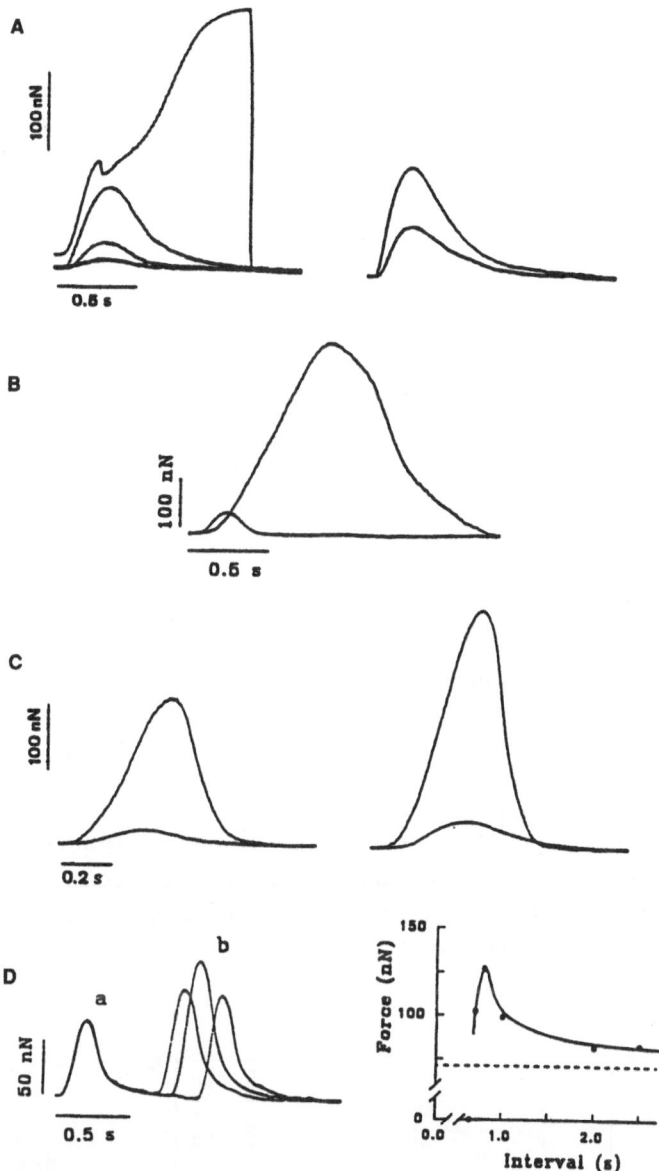

Fig. 1. Effects of various interventions on twitch characteristics. *A:* progressive increasing of
$[Ca^{2+}]_o$ from 1 to 4 mM and from 1 to 2 mM, respectively for an atrial (left) and a ventricular
(right) cell at 22°C produced proportional increases of peak twitch force; in the presence of 4
mM external Ca^{2+} the surface membrane of the atrial cell was presumably injured and this
resulted in contracture and detachment of the myocyte from one of the holding micropipettes.
B: effects of decreasing the bath temperature from 24 to 13°C in a ventricular myocyte. *C:*
effects of isoprenaline treatment (5 μM) in an atrial (left) and a ventricular (right) myocyte at
23°C. *D:* effects of changes in stimulation pattern in a ventricular myocyte at 22°C; *a:* control
twitches, *b:* premature test twitches. Apart from *A* the $[Ca^{2+}]$ in the Ringer solution was 1 mM.

electronically converted to voltage and recorded. The whole transducer system and the calibration procedures have been already described[3)4]. The force probe to use in a particular experiment was selected for the compliance, so that the myocyte shortening during a "fixed-end isometric" contraction did never exceed 1% of the slack length (l_0), independent of the force produced. The over-all sensitivity of the transducer system was 0.5-1 mV/nm, with a frequency response of 800-1,000 Hz in saline.

RESULTS

Characterization of the Preparation

The average value of peak force for control twitches (temperature 20-23°C, $[Ca^{2+}]_o$ 1 mM, resting sarcomere length 2.1-2.2 μm, stimulus interval 10s) was about 40 and 70 nN, in atrial and ventricular myocytes respectively.

Fig. 1A shows the effects of varying $[Ca^{2+}]_o$ on the twitch characteristics of an atrial and a ventricular myocyte. It can be seen that increase in $[Ca^{2+}]_o$ increased peak twitch force without affecting the contraction and relaxation times.

The effects of changing bath temperature on the twitch characteristics of a ventricular myocyte are illustrated in Fig. 1B. In this cell and in the other atrial and ventricular cells tested a decrease in temperature by about 10°C produced a 10 times potentiation of the peak twitch force and a conspicuous prolongation of the contraction and relaxation times.

Fig. 1C shows the effects of treatment with 5 μM-isoprenaline on twitch characteristics of an atrial and a ventricular myocyte. β-adrenergic stimulation led to prominent potentiation of the peak-twitch tension, prolongation of contraction time and shortening of relaxation.

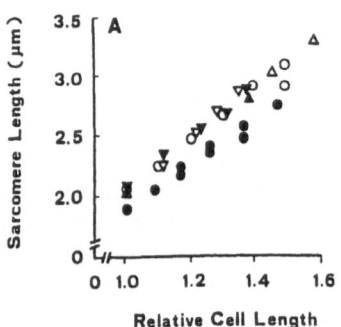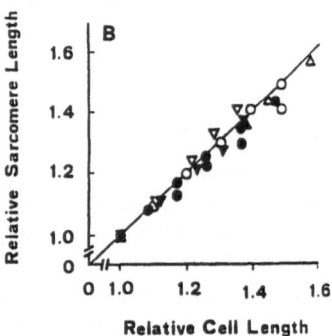

Fig. 2. *A*: absolute sarcomere length-relative myocyte length relations for six resting cells. *B*: Normalization for sarcomere length reduced the scatter between data-points from different myocytes.

The stimulus-interval dependence of twitch characteristics for a ventricular myocyte is shown in Fig. 1D. Similar results were also obtained in atrial cells. In almost all the cells tested the earliest premature twitch after the end of the mechanical refractoriness was potentiated relative to the controls. Only in a few cases was the earliest premature twitch depressed.

The Myocyte Length-Sarcomere Length Relation at Rest

The myocyte length (the distance between the ends of the holding micropipettes) was measured at a microscope magnification of 600X and was changed by displacing the motor micropipette with respect to the force probe. For sarcomere-length measurements a suitable region was selected along the myocytes in the third closest to the force probe, and the sarcomere length was determined by averaging the measurements of sequences of 10-20 striations in photomicrographs taken through a 40X objective and a 25X eyepiece.

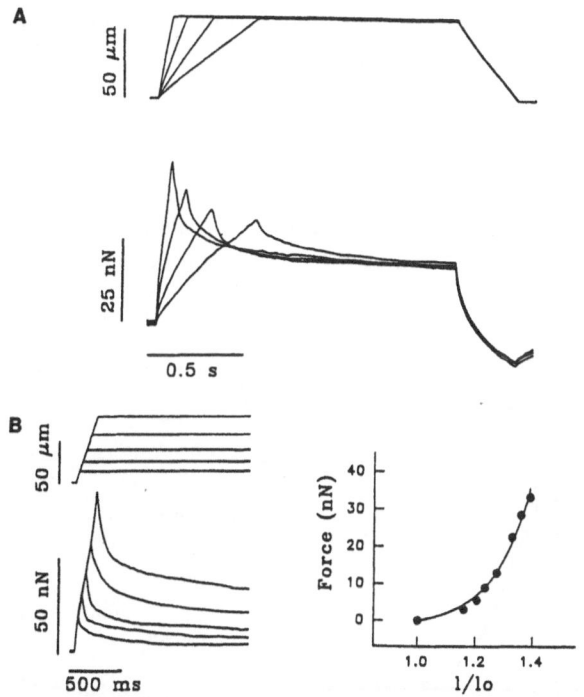

Fig. 3. Force responses (bottom traces) of two atrial myocytes (A, lo 345 μm; B, lo 362 μm; resting forces around zero) to ramp stretches (top traces). Temperature 21°C. In A the velocity of the lengthenings varied from 0.3 to 2.3 lo/s, while the extent remained constant, from 1.09 to 1.26 lo. In B the extent of the lengthenings varied from 1.16 to 1.39 lo, while the velocity was constant, 1.12 lo/s. The resting force at 1.16 lo was 3 nN. The graph refers to the length-resting force relation. Forces were measured at a time when they had settled to a steady level.

The myocyte length-sarcomere length relation for three resting atrial and three ventricular myocytes is shown in Fig. 2. The linearity of the relation, as well as the finding that the straight line fitted to data-points extrapolated to the axis origin, excludes the presence of regions along a myocyte with significantly different compliances. This also implies that the length of the cellular ends sucked into the holding micropipettes remains unvaried up to at least 100 nN.

The Force Response of Intact Myocytes to Length Changes

Fig. 3 shows the force-transient response of a resting atrial myocyte to slow ramp lengthenings. In panel A only the velocity of lengthening was varied. Conversely, in panel B the velocity of lengthening remained constant and the extent of the stretch was varied. The response is characterized by an initial increase in force during the lengthening, followed after the stretch was completed, by a gradual recovery to a steady level, according to the length-resting tension relation. The initial phase consists of two components: a former one, whose amplitude increases with the velocity independent of the extent of the lengthening, and a later component which increases with the extent of the lengthening, being only scarcely dependent on the velocity.

Fig. 4 shows the force-transient response of an electrically-stimulated ventricular myocyte to rapid releases and stretches complete in 2 ms. The transient shows an initial component during the length step itself, which increases as the step size is increased. This apparently elastic response, comparable to phase 1 of the force transient described by Huxley & Simmons[8] for skeletal muscle fibres, is followed by a rapid partial recovery and then by slowering of the recovery. Whether or not

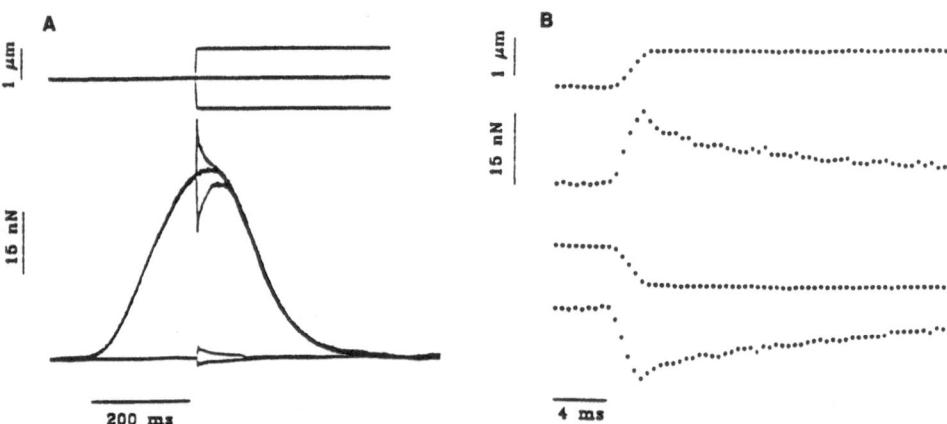

Fig. 4. *A*: superimposed records of length (top) and passive- and active-force (bottom) traces for an intact ventricular myocyte. Step releases and stretches were applied to the resting cell at *lo* (260 μm, passive force around zero) or during a twitch contraction (active force 45 nN). *B*: time-resolved records of the active-force transients in response to a step stretch (top pair of traces) or to a step release (bottom pair of traces).

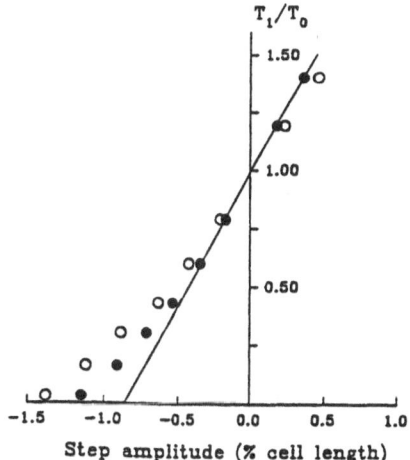

Fig. 5. Stress-strain relation at 20 nN during a twitch contraction of an intact ventricular myocyte. Data from the same experiment as in Fig. 4. T_1/T_0 is the ratio of the force at the length-step end to the force reached just before the step. Empty and filled circles refer to data-points before and after correction for force probe compliance (30 nm/nN).

the whole recovery of force includes different phases, as reported for skeletal muscle fibres[8)9)], remains unclear.

The relation of the size of the length steps imposed on a ventricular myocyte just before the twitch peak against the change in force at the step end (stress-strain relation) is shown in Fig. 5. It can be seen that, after correction for the force-probe compliance, the observed extent of shortening required to drop the active force to zero is about 1.12 %lo and that the straight line fitted to data points in the linear part of the stress-strain relation intercepts the length axis at 0.8 %lo.

The stress-strain relations obtained at 20 and 45 nN in different twitches of a ventricular cell are compared in Fig. 6. It is clear that the observed extents of length steps required to produce a given change in absolute force are smaller at 45 nN than at 20 nN. Moreover, the correction for force-probe compliance is more effective on data-points obtained at 45 nN than on those obtained at 20 nN, so that, after correction, the two stress-strain relations become practically superimposable, at least in the region of the shortest length steps. In other words, the extrapolated value of step release required to drop the twitch force to zero is independent of the force developed. The small differences still present in the region of the largest releases will be considered in "Discussion".

Ca^{2+}-activated Contractions in Skinned Myocytes

The experiments described in this section were aimed at determining the force developed by maximally Ca^{2+}-activated skinned myocytes.

Fig. 7*A* shows representative records of two contractions elicited in a skinned atrial myocyte at a pCa of 4.75. It can be seen that the speed of rise and the total amount of force are the same in either trial. Furthermore, in this as in the other atrial or ventricular myocytes tested, a decrease in either the Ca^{2+}-buffering capacity of the relaxing solution in the experimental trough or the pCa of the activating solution inside the injecting micropipette never increased the force produced.

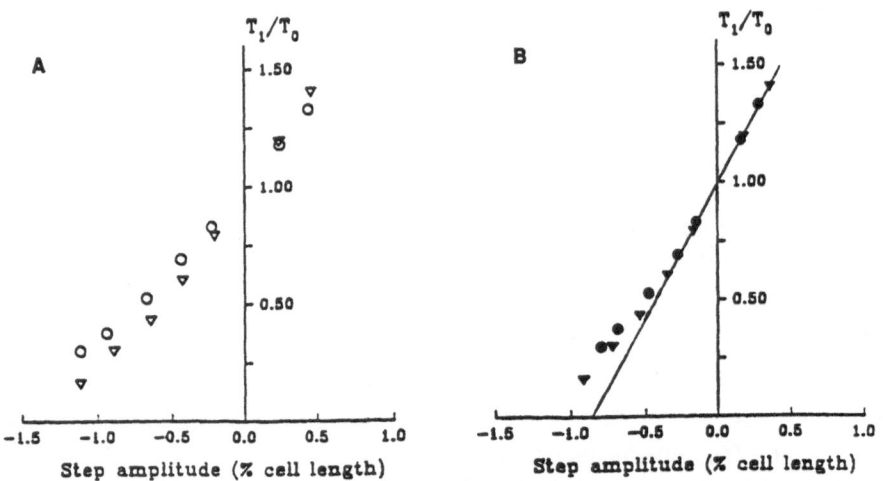

Fig. 6. Stress-strain relations at 20 nN (triangles) and at 45 nN (circles) during a twitch contraction of the same myocyte as in Figs. 4 and 5. T_1/T_0 as defined in Fig. 5. Empty (*A*) and filled (*B*) symbols refer to data-points before and after correction for force-probe compliance.

In the myocytes tested (twenty-eight from the atrium and seven from the ventricle) at a pCa of 4.75 the force reached the plateau level in 1-5 s. The average value for maximum absolute force was 1.49 ± 0.13 μN (mean $\pm$ S.E.M.; range 0.5-3.5 μN). The mean value for tension after normalization for myocyte cross-sectional area was 118.30 ± 7 kN/m^2 (mean $\pm$ S.E.M.; range 75-180 kN/m^2).

In a few experiments contractions were elicited using different-pCa activating solutions. Fig. 7*B* shows sample records from an experiment on an atrial myocyte. It can be seen that for this cell the saturating pCa is ≤ 5.50. At pCa 4.75 the speed of rise of the force is higher than at pCa 5.50, but the plateau forces are practically the same in either case. In contrast, the force developed at pCa 6.0 is considerably lower.

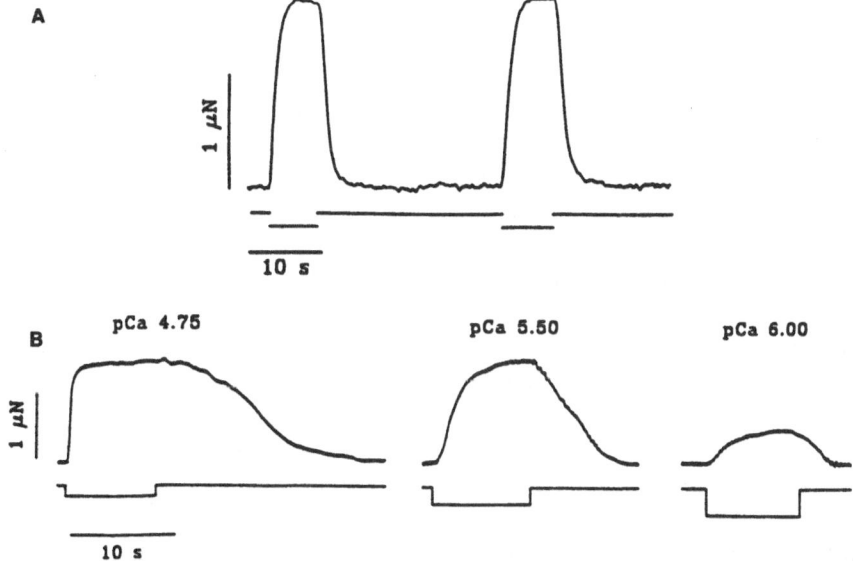

Fig. 7. *A*: contractions of a skinned atrial cell. The pCa of the activating solution in the injecting pipette was 4.75. *B*: contractions of another skinned myocyte by different pCa-activating solutions. In both panels the deflection of bottom traces are the injection signals of the activating solution. The different amplitude of the deflections identified the injecting pipette and the pCa of the activating solution.

DISCUSSION

Isolated Frog-Heart Myocytes as Preparations for Mechanical Investigations

The effects of inotropic interventions on twitch characteristics of intact myocytes, as reported in this paper, are comparable to those described by previous investigators for multicellular preparations from frog heart under similar experimental conditions[10-14]. This finding implies that the enzymatic isolation procedure does not alter the physiological properties of the cells which represent a reliable preparation for the study of the mechanics of myocardium.

The findings that the force produced by intact myocytes during a control twitch is potentiated by various inotropic interventions and that this force is 15-30 times smaller than that produced in a maximally Ca^{2+}-activated skinned cell imply that the level of activation of intact preparations is very low in the presence of 1 mM external Ca^{2+}.

Another point to note is that the maximum tension produced by skinned heart myocytes (about 120 kN/m^2) is lower than that produced by skeletal muscle fibres during a fused tetanus (200-300 kN/m^2), but this is attributable to a lower density of myofibrils in cardiac than in skeletal myocytes[15].

The Force Response of Intact Myocytes to Length Changes

The force response of a resting intact myocyte to slow ramp lengthenings is comparable to that of a complex mechanical system including a viscous element in parallel with a damped and an undamped elastic element. The identification of which component of a resting intact myocyte may be responsible for this force response is impossible at present. The result (not shown here) that in skinned myocytes, incubated in the relaxing solution, the velocity-dependent component of the force transient is greatly reduced suggests that most of the viscous properties of an intact frog-heart myocyte are determined by either membrane deformation or slippage of thin on thick filaments.

As to the extent of shortening required to bring to zero the twitch force developed by a single heart cell (y_0), it must be noted here that the value of 1.2 % lo (observed) and 0.8% lo (extrapolated), as reported in the present paper, are the smallest values among those reported in the specific literature on the myocardium. Moreover, the finding that, after correction for force-probe compliance, the extrapolated values of y_0 were the same for different force levels implies that a change in the active force developed by a myocyte is associated with a proportional change in stiffness or, in terms of cross-bridge theories of contraction, with a proportional change of the number of force-generating units. It must be also noted that, as it may occur in skeletal muscle fibres, the above values of y_0 may be overestimated because the length steps imposed on the myocytes could be not fast enough to avoid truncation of the instantaneous force change by the simultaneous quick recovery.

Finally, the differences between the stress-strain relations at 20 and 45 nN, still present in the region of the largest releases even after correction for force-probe compliance, may be attributed, as observed for skeletal muscle fibres, to the presence of a series compliance (e.g. the compliance of the thin filament[16]) as well as to the fact that (i) the initial force recovery is faster with large step releases[8][9]; (ii) the stiffness to force ratio and the speed of the initial quick recovery of force after a length step could be greater at low than at high force[17][18].

REFERENCES

1. Brady, A.J. *Physiol. Rev.* **71**, 413-428 (1991).
2. Tung, L. & Morad, M. *Am. J. Physiol.* **255**, H111-120 (1988).
3. Cecchi, G., Colomo, F., Poggesi C. & Tesi, C. *J. Physiol. (Lond.)* **448**, 275-291 (1992).
4. Cecchi, G., Colomo, F., Poggesi C. & Tesi, C. *Pflügers Arch.* (1992, in the press).
5. Mitra, R. & Morad, M. *Am. J. Physiol.* **249**, H1056-1060 (1985).
6 Brandt, P.W., Diamond, M.S., Rutchik, J.S. & Schachat, F.H. *J. Mol. Biol.* **195**, 885-896 (1987).
7. Cecchi, G., Colomo, F. & Lombardi, V. *Boll. Soc. Ital. Biol. Sper.* LI, 733-736 (1976).
8. Huxley, A.F. & Simmons, R.M. *Nature* **233**, 533-538 (1971)
9. Ford, L.E., Huxley, A.F. & Simmons, R.M. *J. Physiol. (Lond.)* **269**, 441-517 (1977).
10. Blinks, J.R. & Koch-Weser, *J. Pharm. Rev.* **15**, 601-652 (1963).

11. Chapman, R.A. & Niedergerke, R. *J. Physiol. (Lond.)* **211**, 389-421 (1970).
12. Allen, D.G. & Blinks, J. R. *Nature* **273**, 509-513 (1978).
13. Tung, L. & Morad, M. *Pflügers Arch.*, **398**, 189-198 (1983).
14. Niedergerke, R. & Page, S. *Quart. J. Exp. Physiol.* **74**, 987-1002 (1989).
15. Page, S.G. & Niedergerke, R. *J. Cell. Sci.* **11**, 179-203 (1972).
16. Bagni, M.A., Cecchi, G., Colomo, F. & Poggesi, C. *J. Muscle Res. Cell Motility* **11**, 371-377 (1990).
17. Cecchi, G., Griffiths, P.J. & Taylor, S.R. *Science*, **217**, 70-72 (1982).
18. Bagni M.A., Cecchi, G., Colomo F. & Tesi, C. in *Molecular Mechanism of Muscle Contraction* (eds. Sugi, H. & Pollack, G.H.) 473-488 (Plenum Publishing Corp, New York, 1987).

Discussion

ter Keurs: The T1 curve that you obtained suggests a large series elasticity. Have you done the experiments at room temperature, where it is known that T1 has a larger intercept?

Colomo: The value of 0.8% L_0 for the extrapolated intercept of the linear part of the T1 curve is certainly larger than that found in skeletal muscle fibers at low temperature. It is, however, the lowest value among those reported for myocardium. I agree with you that a decrease of temperature is expected to reduce the value of the T1 intercept because of the step time of 2 ms.

Edman: Was your measurement from the whole preparation or did you clamp the sarcomeres?

Colomo: The measurement was from the whole preparation, but I would like to point out that frog cardiac myocytes have no tendons and are 200-300 μm long, so they can be regarded as very short segments.

ter Keurs: Regarding the basis of the Frank-Starling Law, I have the following comment. Frank showed that pressure development in the isovolumic frog heart rises with an increase in volume of the heart, and then falls very clearly (Frank, O. *Z. für Biol.* **32**, 370-447, 1895). Nassar et al. showed that it is probably due to the fact that frog cardiac cells also work over the descending part of the force-length relationship (Nassar, R., Ranking, A., and Johnson, E. A. *The Physiological Basis of Starling's Law of the Heart,* Ciba Found. Symp. 24, 1974). Hence, my question: did you measure a force-length relation in these isolated cells, and did you find a descending limb?

Colomo: Not yet. The sarcomere length at which the contraction was produced was just above the slack length, about 2.1 μm.

Edman: I wonder whether it is possible to estimate the series compliance at the ends of the preparation. Do you have any problems with damaged ends?

Colomo: Yes, we sometimes had problems. Damage of the cell ends sucked inside the micropipettes was infrequent and led to contracture. Loose attachment was associated with detachment of the cell during contraction. In either case, that terminated the experiment.

TENSION AND INTRACELLULAR CALCIUM TRANSIENTS OF ACTIVATED FERRET VENTRICULAR MUSCLE IN RESPONSE TO STEP LENGTH CHANGES

Yasutake Saeki, Satoshi Kurihara*, Kenichi Hongo* and Etsuko Tanaka*

Department of Physiology
Tsurumi University School of Dental Medicine
2-1-3 Tsurumi, Tsurumi-ku
Yokohama 230 Japan
*Department of Physiology
The Jikei University School of Medicine
3-25-8 Nishishinbashi
Minato-ku, Tokyo 105 Japan

ABSTRACT

To elucidate the effects of mechanical constraints on the (Ca^{2+}) affinity of cardiac troponin C, we studied the relationships among the myoplasmic Ca^{2+} concentration ($[Ca^{2+}]_i$), tension and length in steadily activated intact cardiac muscle. The Ca^{2+} sensitive photoprotein, aequorin, was micro-injected into cells of ferret right ventricular papillary muscles to monitor the $[Ca^{2+}]_i$. The muscle was then steadily activated with ouabain (10^{-4} M)(ouabain contracture) or high frequency stimuli in the presence of ryanodine (5 μM)(tetanic contraction); the tension and aequorin light (AL) transients in response to a step length change were then analyzed. The tension transient response to either the stretch or release in length was oscillatory: tension decreased rapidly during the release and then increased, after which it lapsed into a new steady level in a series of damped oscillations. The opposite was true for the stretch. The oscillatory responses were conspicuous and less damped in the ouabain contracture. The transient AL response was also oscillatory, the time course of which corresponded exactly to that of the tension transient response, though no detectable changes in AL were observed at the initial phase of the stretch response. The increase in AL corresponded exactly to the decrease in tension, likewise the decrease in AL to the increase in tension. The steady level of AL after release was decreased in ouabain contracture, but was increased in tetanic contraction. These results suggest that the Ca^{2+} affinity of cardiac troponin C is increased with an increase in tension (i.e., the cross-bridge attachment) and decreased with a decrease in tension (i.e., the cross-bridge detachment), and that the myoplasmic calcium concentration is lowered by release, at least in a Ca^{2+}-overloaded condition, mainly through the sarcoplasmic reticulum.

Mechanism of Myofilament Sliding in Muscle Contraction, Edited by
H. Sugi and G.H Pollack, Plenum Press, New York, 1993

INTRODUCTION

It is well established that cardiac muscle contraction occurs through the binding of Ca^{2+} to a component of the thin filament, troponin C[1]. The Ca^{2+} binding allows the actin and myosin to interact and leads to the production of tension and shortening (i.e., mechanical events). There is now considerable evidence that the resulting mechanical events affect the Ca^{2+} affinity of troponin C[2-4]. Allen and Kurihara reported that in rat and cat ventricular muscle an increase in the muscle length increases the tension but decreases the duration of the aequorin light emission (AL, a function of intracellular Ca^{2+}, $[Ca^{2+}]_i$) without increasing its amplitude[5]. They also observed a transient increase in AL (i.e., extra Ca^{2+}) after a quick release in length. Housmans et al. reported a higher $[Ca^{2+}]_i$ during active shortening of cat papillary muscles than at a comparable time during isometric twitches[6]. These observations were attributed to a decrease in the affinity of troponin C for Ca^{2+} resulting from a decrease in muscle length and active tension development, since the extra Ca^{2+} can be explained by the release of Ca^{2+} from troponin C into the cytoplasm. However, the interpretation of these experiments is complicated by the possibility that Ca^{2+} release from the sarcoplasmic reticulum (SR) and/or Ca^{2+} entry across the surface membrane may be affected by muscle length. In addition, their results were obtained from twitching preparations which had periodically altered electrical and mechanical properties, and are thus difficult to directly interpret.

Therefore, for a more precise analysis, we studied the tension and AL transients in response to step length changes in steadily activated (ouabain contracture) cardiac muscle so that the complication due to the time-dependent nature of twitch contraction might be avoided. The same length perturbation was also applied to the quasi-steadily activated (ryanodine-treated tetanized) cardiac preparation in which the SR is thought to be inactive in the handling of Ca^{2+}[7][8], to exclusively discern the contribution of the SR to extra Ca^{2+}.

METHODS

Male ferrets (about 800 g body weight) were anesthetized with sodium pentobarbital (80 mg/kg, i.p.). Hearts were quickly removed and thin right ventricular papillary muscles were excised and placed in a petri dish that contained oxygenated normal Tyrode solution. Each end of the excised muscle was tied with a silk thread to a small tungsten wire (125 μm in diameter) hook in the dissecting dish. The preparation was transferred to an experimental chamber and mounted horizontally between al length driver (JOCX-101A, General Scanning Inc. Co., CA, U.S.A.), which is capable of imposing step length changes within 4 ms, and a tension transducer (Kulite semiconductor: BG-25, compliance 2.5 μm/g, unloaded resonant frequency 1 kHz).The experimental chamber was equipped with a pair of platinum-black wire electrodes for stimulation. The temperature of the solution was maintained at either 22°C or 30°C with a thermoelectric device (Haake, D-3, Germany) with an accuracy of ± 0.5°C.

Aequorin, purchased from Dr. J.R. Blinks, was dissolved in the 150 mM KCl and 5 mM HEPES solution at pH 7.0, with a final aequorin concentration of 50 - 100 μM. A glass micropipette with a resistance of 30-80 MΩ, measured after filling with aequorin solution, was used for the injection of aequorin. Aequorin was injected into 50 - 100 superficial cells of each preparation with 5 - 10 kg/cm^2 pressure while monitoring the membrane potential. Aequorin light signals detected with a photomultiplier (EMI 9789A, Ruislip) which was mounted in a small housing. A 10 mm diameter quartz light guide, which was attached to the photomultiplier, was placed just above the preparation. Light detection was further improved by placing a concave mirror under the preparation. Details of the method were previously described by Allen and Kurihara[5]. All data were stored on a tape (NFR-3515W, Sony Magnescale Inc., Tokyo, Japan) and a computer for later analyses. In order to improve the signal to noise ratio, the light signals were recorded through a low-pass active filter with an appropriate time constant. Signals were averaged with the computer or a signal processor (7T07A, NEC San-ei Co., Ltd., Tokyo, Japan). In the figures in the present study, the aequorin light signals (AL) are expressed as photomultiplier current.

The muscle was initially stimulated to contract at 0.2 Hz with a square pulse of 5 ms duration and a voltage of 50% above the threshold level. After a stabilizing period of 30 minutes, Lmax was determined as the length of maximum isometric twitch tension. At this point, the length and diameter of the preparation were measured with an ocular micrometer. The length ranged from 3.5 to 6.0 mm, and the diameter ranged from 500 to 900 μm. These values were used as a reference for normalizing absolute muscle length and tension. then, the electrical stimuli were stopped and aequorin was micro-injected into the superficial cells. After completion of the injection, the muscle was stimulated again in the same manner as before, and the normal Tyrode solution was changed to HEPES-Tyrode solution and the muscle was stabilized for about 60 minutes until the twitch tension and AL became stable. Immediately after the stabilizing period, the solution was changed to HEPES-Tyrode solution containing a high concentration of Ca^{2+} (8 - 20 mM) and either 10^{-4} M ouabain or 5 μM ryanodine, to obtain a steadily activated contractile state.

As seen in Fig. 1, ouabain produced a transient increase and then a continuous decrease in the systolic tension during which time the diastolic tension increased progressively. The twitch tension and AL became oscillatory (Fig. 1A), as has been reported previously[9]. Oscillatory AL changes always led the changes of oscillatory after-contraction, indicating that the AL changes were the cause of the after-contraction. Finally, the muscle lapsed into a state of stable contracture with a negligible amount of twitch tension, which could be eliminated by stopping electrical stimuli.

To obtain the tetanic contraction, the preparation was treated with ryanodine for at least 40 minutes, and then stimulated with square pulses with a 40 ms duration at 10 Hz and a voltage of about three-fold the threshold[10][11]. After the steady active

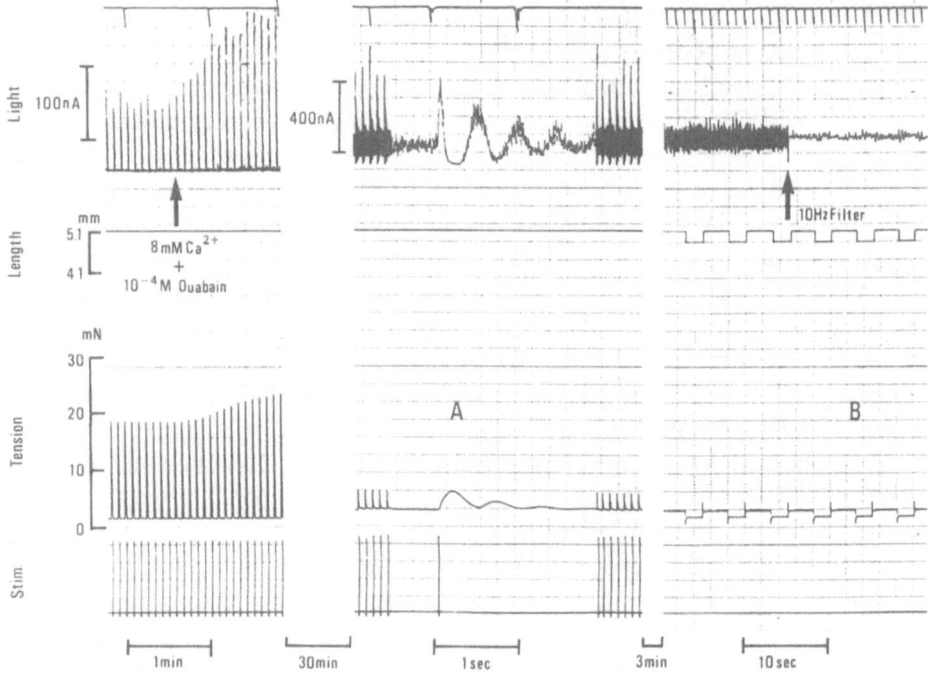

Fig. 1. The experimental protocol in ouabain-activated preparations. See text for explanation.

tension was obtained, the muscle length was released and stretched stepwise by amounts of 2 - 12% of the muscle length (Fig. 1B and Fig. 4), as in an earlier study[12].

RESULTS

Figure 2 shows the averaged transient response of AL (middle trace) and tension (lower trace) to a step change in length (5.2% of Lmax, upper trace) in the preparation activated with the solution containing 10^{-4} M ouabain and 8 mM $[Ca^{2+}]_o$ at 22°C, obtained from thirty-two records shown in Fig. 1B. As seen in this figure, the tension transient response to either release or stretch was oscillatory: tension decreased rapidly during the release and then increased, after which it lapsed into a new lower steady level in a series of damped oscillations. The opposite was true for the stretch back to the original length position, though not exactly symmetrical. The AL was also oscillatory, the time course of which corresponded exactly to that of the tension transient, though no detectable changes in AL were observed at the initial phase of the stretch response (i.e., the phase of the rapid increase in tension concurrent with the stretch). The increase in AL corresponded exactly to the decrease in tension, and likewise the decrease in AL to the increase in tension, as

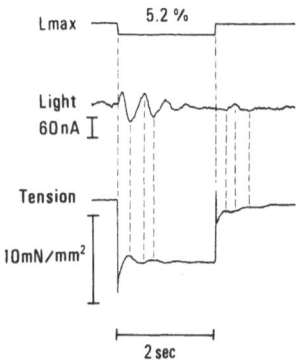

Fig. 2. Aequorin light (middle trace) and tension (lower trace) changes in response to step change in length (upper trace) during ouabain contracture at 22°C. Averaged records (n = 32).

indicated by the vertical dashed lines in Fig. 2. The amplitude of AL and tension transient were increased with the increase in the amplitude of step length change.

Figure 3 shows the averaged (thirty-two records) responses of AL (middle trace) and tension (lower trace) to the same amount (but different percent of initial muscle length) of step change in length (upper trace) at two different initial muscle lengths, Lmax(left) and 1.06Lmax (right) in the preparation activated with the solution containing 10^{-4} M ouabain and 8 mM $[Ca^{2+}]_0$ at 30°C. These records show that increasing the temperature markedly shortened the oscillatory time courses of AL and tension transients, but did not alter the characteristic relationships between AL and tension transients observed at 22°C (Fig. 2). The oscillatory frequency was quite independent of the initial muscle length (compare the left and right panel in Fig. 3) and the amplitude of step length changes; it was 2.1 to 2.3 Hz at 22°C and 4.3

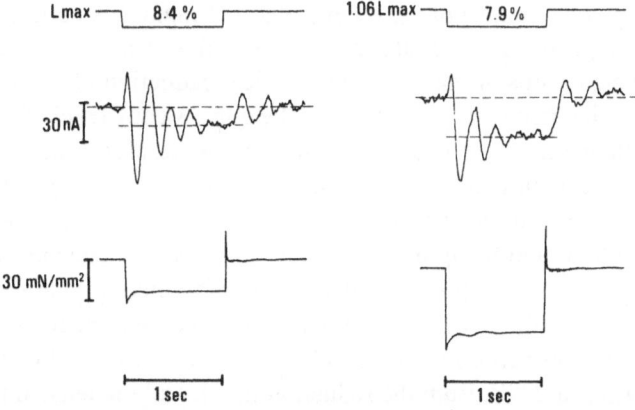

Fig. 3. Aequorin light (middle trace) and tension (lower trace) changes in response to step change in length (upper trace) at two different initial muscle length, Lmax (left panel) and 1.06Lmax (right panel) during ouabain contracture at 30°C. Averaged records (n = 32).

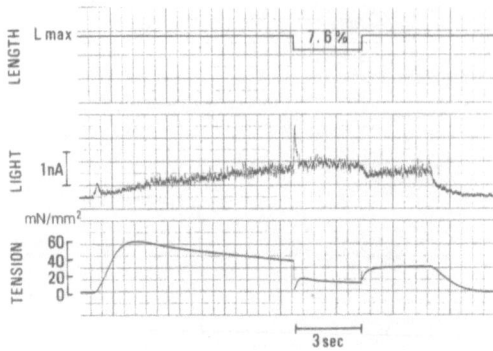

Fig. 4. Aequorin light (middle trace) and tension (lower trace) changes in response to step change in length (upper trace) during ryanodine-induced tetanic contraction at 30°C.

to 4.7 Hz at 30°C in each of the 4 preparations studied. The quasi-steady level of AL after release was lower than the level just prior to the release. This characteristic was much more clearly seen when the amplitude of length change was greater and the initial length was longer than Lmax, as indicated by the horizontal dashed lines in Fig. 3.

Figure 4 shows the AL and tension transient in response to a step length change (7.6% of Lmax, upper trace) during the tetanic contraction in the preparation activated with the solution containing 5 μM ryanodine and 20 mM [Ca^{2+}]o at 30°C. In most cases, the transient tension response comprised of three different phases. The first phase was a prompt and large tension change during the length change; the second phase, a quick recovery of tension to the value before the length change and the third a delayed tension change in the same direction as the first phase. In some preparations, the third phase was followed by a damped oscillation in tension, as seen in ouabain-activated preparations. The AL change in response to a step release in length was the initial rapid increase followed by a rapid decrease, the time course of which was in phase with that of the second phase of rapid tension rise. In contrast, no detectable change in AL was observed during the step stretch, as observed in ouabain-activated preparations. There was a clear reduction of AL concurrent with the delayed tension rise in the third phase. The quasi-steady level of AL after release was higher than the level observed without the step release, contrary to the observation in ouabain-activated preparations. The rate of the transient changes in both AL and tension became faster with the later step changes after the onset of tetanic contraction, and being markedly decreased with a decrease in temperature from 30°C to 22°C, as observed in ouabain-activated preparations. Both the reduction in AL and the tension rise in the second phase after release (9 s after the onset of tetanic contraction) were completed in about 200 to 400 ms at 30°C and about 1.1 to 1.3 s at 22°C. Both the reduction in AL and the tension rise in the third phase after stretch (12 s after the onset of tetanic contraction) were completed in about 400 to 800 ms at 30°C and about 2.0 to 2.5 s at 22°C.

DISCUSSION

The most important finding in the present study is that after the step change in length, the transient AL responses changed in phase with the tension responses regardless of the temperature in both ouabain-activated and ryanodine-treated tetanized preparations. The increase in AL corresponded exactly to the decrease in tension and the decrease in AL to the increase in tension. This phenomenon indicates that the AL changes are the consequence of the tension changes, suggesting, in agreement with earlier studies [6][12-16], the the Ca^{2+} affinity of cardiac troponin C is increased with an increase in tension (i.e., cross-bridge attachment) and decreased with a decrease in tension (i.e., cross-bridge detachment). If the AL changes are the cause of the tension changes, the increase in AL would not change in phase with the reduction in tension and would always lead the tension development, as observed previously[9] and seen in Fig. 1A. Allen and Kentish[12], recently, performed a similar length perturbation experiment on skinned ferret cardiac muscle in which all membranes were destroyed by Triton X-100, leaving troponin C as the major site of Ca^{2+} binding. They clearly showed that the transient AL changes are the consequence of the tension transients in response to step length changes by observing that increasing the EGTA (at constant $[Ca^{2+}]_i$) produces a considerable reduction of the AL response, while the magnitude and time course of tension changes were hardly affected. Except for one aspect, the relationships between the AL and tension transients in their skinned preparations are quite similar to those in our ryanodine-treated preparations, in which the SR is thought to be inactive in handling Ca^{2+}[7][8]. They did not observe the rapid AL decrease associated with the rapid tension rise in the second phase of the release response. If the Ca^{2+} affinity of troponin C is more closely correlated to the developed tension than to the muscle length, the AL should decrease in phase with the tension development, as observed in our ryanodine-treated tetanized preparations. The difference might be the greater tension rise in the second phase and/or the higher AL response in the first phase concurrent with the release in our preparations compared with those in their Triton X-100 skinned preparations.

The amplitudes of AL and and tension changes in the first phase of the release response increased in parallel as the size of the step release was increased from 2 to 12% of the initial muscle length (above the length change expected for cross-bridge detachment) in both ouabain-activated and ryanodine-treated tetanized preparations. This finding is consistent with that in the Triton X-100 skinned preparations[12], and suggests that the amount of Ca^{2+} released was determined chiefly by the magnitude of the change in tension (or length) and that it was affected only slightly by the sudden forcible detachment of cross-bridges, as proposed by Allen and Kentish[12]. If this is true, the fact that no detectable change in AL was observed at the first phase of the stretch response in the present study and in the Triton X-100 skinned cardiac preparations[12] can be partly explained by the quick tension recovery (reduction) in the second phase following the rapid tension rise in the first phase. In the Triton X-100 skinned slow-twitch rat soleus loaded with the Ca^{2+}-sensitive photoprotein, obelin, the quick stretch in length has been reported to produce a clear decrease in obelin light[2]. This might be related to the lower tension changes in the second phase compared with the that in cardiac muscles. Voltage-clamped barnacle muscle fibers have been reported to produce an increase in AL concurrent with both a stretch and release in length[16]. Although there is no direct evidence to date which explains the different finding in the first phase of the stretch response between the cardiac and barnacle muscles, one possible difference is the slower detachment rate of Ca^{2+} from the binding site of troponin C in cardiac muscle relative to that in barnacle muscle. Supporting evidence in the present study is that the initial rapid increase in AL lagged behind the stretch in length. In cardiac muscle, therefore, no change in AL concurrent with stretch may be due to the slow detachment rate of Ca^{2+} from the binding site of troponin C and/or the quick recovery of tension in the second phase as mentioned above. The length change per se might also contribute to the initial phase of AL transients so as to increase the AL for release and to decrease the AL for stretch, since the Ca^{2+} of troponin C is thought to be increased with an increase in the muscle length[2-5]. This may be an additional factor in the production of the asymmetrical AL responses immediately following the release and stretch.

The other interesting finding is that in the ryanodine-treated tetanized preparations, the quasi-steady level of AL after release was higher than the level just prior to the step release. This is consistent with the finding in the Triton X-100 skinned preparations[12]. This phenomenon can be reasonable explained by the lower affinity of Ca^{2+} to troponin C at shorter muscle length[2-5], since in these preparations the SR are thought to be inactive in the handling of Ca^{2+} and troponin C is the major site of Ca^{2+} binding. From this point of view, the intriguing fact is that the quasi-steady level of AL after release was lower than the level present just prior to the step release in the ouabain-activated preparations which have intact SR. this may indicate that the $[Ca^{2+}]_i$ is lowered by the release in length, at least in a $[Ca^{2+}]_i$-overloaded condition, through the SR. The quasi-steady level might be determined by the net effects of length release on the Ca^{2+}-restoring capacity of the SR and the Ca^{2+} affinity of troponin C.

REFERENCES

1.Ebashi, S, & Endo, M. *Prog. Biophys. Mol. Biol.* **18**, 123-183 (1968).

2. Stephenson, D.G. & Wendt, I.R. *J. Musc. Res. Cell Motility* **5**, 243-272 (1984).

3. Allen, D.G. & Kentish, J.C. *J. Mol. Cell. Cardiol.* **17**, 821-840 (1985).

4. Cooper, G. *Ann. Rev. Physiol.* **52**, 505-522 (1990).

5. Allen, D.G. & Kurihara, S. *J. Physiol. (Lond.)* **327**, 79-94 (1982).

6. Housmans, P.R., Lee, N.K.M. & Blinks, J.R. *Science* **221**, 159-161 (1983).

7. Sutko, J.L., Ito, K. & Kenyon, J.L. *Fedn. Proc.* **44**, 2984-2988 (1985).

8. Wier, W.J., Yue, D.T. & Marban, E. *Fedn. Proc.* **44**, 2989-2993 (19485).

9. Orchard, C.H., Eisner, D.A. & Allen, D.G. *Nature* **304**, 735-738 (1983).

10. Yue, D.T., Marban, E. & Wier, W.G. *J. Gen. Physiol.* **87**, 223-242 (1986).

11. Okazaki, O., Suda, N., Hongo, K., Konishi, M. & Kurihara, S. *J. Physiol. (Lond.)* **423**, 221-240 (1990).

12. Allen, D.G. & Kentish, J.C. *J. Physiol. (Lond.)* **407**, 489-503 (1988).
13. Gordon, A.M. & Ridgway, E.B. *J. Gen. Physiol.* **90**, 321-340.
14. Hofmann, P.A. & Fuchs, F. *Am. J. Physiol.* **253**, C541-C546 (1987).
15. Sweitzer, N.K. & Moss, R.L. *J. Gen. Physiol.* **96**, 1221-1245 (1990).
16. Gordon, A.M. & Ridgway, E.B. *J. Gen. Physiol.* **96**, 1013-1035 (1990).

Discussion

Sys: We previously showed that, during the later phases of relaxation, the rate of force decline is mainly determined by the amount of active force that is still in the muscle. Could that be attributed solely to the calcium affinity of troponin-C, or do you have to include force dependence of actomyosin detachment also?

Saeki: That result is obtained from twitching muscle, so it is difficult to answer the question because of the time-dependent nature. What this result really indicates is that the attachment itself affects troponin-C affinity. But it is very difficult to distinguish the two issues you just raised.

ter Keurs: I recognized that each of your stimulated twitches was followed by an aftercontraction. Is that a correct observation?

Saeki: Yes, right. A natural twitch contraction with an aftercontraction.

ter Keurs: Well, in that respect it is very interesting to note what you have observed in terms of calcium transients, because the aftercontraction, as we have described, always starts in the area near the clamps of the muscle. That area is stretched during the normal twitch and is nearly always quickly released during the relaxation phase of the twitch. So you have a more or less natural experiment on quick release there, causing the same effect as your intentional quick release. The interest of that observation is that those aftercontractions are the basic mechanism underlying triggered arrhythmias.

Pollack: Just to expand that a bit, are you familiar with the work of Nobu Ishide who has measured calcium waves travelling along the cardiac cell (Ishide, N. et al. *Am. J. Physiol.* **259**, H940-H950, 1990)? Do you think that perhaps some of your observations are related to these oscillatory waves?

Saeki: Calcium waves may be associated with tension or length changes.

Gordon: How did you distinguish between the oscillations and the calcium signal? Is that an oscillation in calcium release and uptake? Or, is it the mechanical response which is then feeding back to the calcium signal? Have you done experiments to try to distinguish between those two hypotheses?

Saeki: We always observed that the transient Aequorin light responses change in phase with the tension responses. This phenomenon seems to indicate that the Aequorin-light changes are the consequence of the tension changes. If the Aequorin light changes are the cause of the tension changes, the increase in Aequorin light would not change in phase with the reduction in tension, and would always precede the tension development, as observed in twitch contractions.

Kurihara: We did two kinds of relevant experiments: one was with the intact SR; in the other we treated the preparation with ryanodine so that the SR function was inhibited. We also did another experiment where we used BDM, so tension development was quite depressed. In that case we couldn't see any oscillatory change in tension or in the calcium signal.

DETERMINANTS OF VELOCITY OF SARCOMERE SHORTENING IN MAMMALIAN MYOCARDIUM

Henk EDJ ter Keurs and Pieter P de Tombe

The University of Calgary
Health Sciences Centre
3330 Hospital Drive NW
Calgary, Alberta, CANADA T2N 4N1

ABSTRACT

Maximal unloaded velocity of shortening of cardiac muscle (V_0) depends on the level of activation of the contractile filaments. We have tested the hypothesis that this dependence may be caused by viscous resistance of the muscle to length changes.

Twitch force (F_0) and sarcomere shortening were studied in trabeculae dissected from the right ventricle of rat myocardium, superfused with modified Krebs-Henseleit solution at 25°C. Sarcomere length (SL) was measured by laser diffraction techniques; force was measured by a silicon strain gauge; velocity of sarcomere shortening was measured using the "isovelocity release" technique.

V_0 and F_0 at slack SL were a sigmoidal function of $[Ca^{2+}]_0$, but V_0 was more sensitive to $[Ca^{2+}]_0$ (Km: 0.44 ± 0.04 mM) than isometric twitch force (Km: 0.68 ± 0.03 mM). At $[Ca^{2+}]_0 = 1.5$ mM, V_0 was independent of SL above 1.9 μm, but depended on SL at lower $[Ca^{2+}]_0$ and always depended on SL < 1.9 μm. A constant relation was observed between V_0 and F_0, irrespective whether F_0 was altered by variation of $[Ca^{2+}]_0$ or SL above slack length.

Visco-elastic properties of unstimulated muscles were studied at SL = 2.0 μm by small linear length changes at varied velocities up to 40 μm/s. The force response to stretch, after correction for the contribution of static parallel elasticity, consisted of an exponential increase of force ($\tau = 4$ ms) and an exponential decline after the stretch. This response would be expected from an arrangement of a viscous element in series with an elastic element. Viscous force increased in proportion to stretch velocity by 0.2- 0.5% of F_0/μm/s up to 15 μm/s, while the calculated stiffness of the elastic component was 25-45 N.mm^{-3}, suggesting that the most likely structural candidate for this visco-elastic element is titin. Dynamic stiffness at 500 Hz was proportional to instantaneous force during shortening and was 12% of stiffness at maximal twitch force when shortening occurred at V_0. This suggests that the number of active force generators, even at maximal activation, is strongly reduced during shortening at V_0.

Mechanism of Myofilament Sliding in Muscle Contraction, Edited by
H. Sugi and G.H Pollack, Plenum Press, New York, 1993

The observed relation between V_o and F_o could be explained by a model in which shortening velocity of the cardiac sarcomere depends on the level of activation and hence on the number of cross bridges supporting the viscous load.

INTRODUCTION

The effect of activation of the contractile apparatus by Ca^{2+} ions on the unloaded velocity of muscle shortening of striated muscle is still unclear[1]. A previous study of the central uniform region of intact right ventricular trabeculae of rat revealed that the unloaded velocity of sarcomere shortening (V_o) depends on $[Ca^{2+}]_o$ up to about 1 mM and then saturates, while isometric twitch force (F_o) saturated at about that the unloaded velocity of sarcomere shortening (V_o) depends on $[Ca^{2+}]_o$ up to about 1 mM and then saturates, while isometric twitch force (F_o) saturated at about $[Ca^{2+}]_o = 2.5$ mM[2]. That study also revealed that, at $[Ca^{2+}]_o = 2.5$ mM, V_o is independent of SL above slack length and that V_o diminishes with decreasing SL evaluated whether viscous force during shortening of the cardiac sarcomere can account for the effect of contractile activation on V_o. We also analyzed the possible basis for visco-elastic behaviour of the unstimulated muscle.

MATERIAL AND METHODS

Muscle Preparation and Electrical Stimulation

Sprague-Dawley rats were anaesthetized with diethyl-ether, and the heart was rapidly excised. Right ventricular trabeculae were dissected under a binocular microscope and mounted in an experimental chamber, as described before[6][7]. The preparations, 50-250 μm in width, 40-80 μm in thickness, and 2-5 mm in length, were stimulated via platinum electrodes positioned in the wall of the muscle bath. Stimulation strength was adjusted to 50% above threshold; Stimulus frequency was 0.5 Hz and duration was 2 ms at all times.

Solutions

The standard solution used was a modified Krebs-Henseleit solution with the following composition (mM): Na^+ 140.5, K^+ 5.0, Cl^- 127.5, Mg^{2+}1.2, $H_2PO_4^-$ 2.0, SO_4^{2-} 1.2, HCO_3^- 19, d-glucose, 10, $[Ca^{2+}]_o$ as indicated. All chemicals were of the highest purity (BDH Chemical Company, Toronto, Ont, Canada). The solutions were in equilibrium with 95% O_2/5% CO_2; pH was 7.40. Temperature was measured in the bath and maintained at 25.0-25.5°C within 0.1°C by circulating water bath and glass heat exchanger at the inflow of the muscle bath. Perfusion rate was approximately 2.5 ml/min.

Force, Sarcomere Length and Sarcomere Velocity

The methods employed to measure force-sarcomere velocity relation, F_0, and V_0 have been described in detail before[2)6)7)]. Briefly, force was measured by a silicon strain gauge, SL was measured by laser diffraction techniques, and muscle length was controlled by servo motor (Model 300 S, Cambridge Technology, Cambridge, Ma, USA). SL and force were displayed on a storage oscilloscope, on a pen recorder (Gould model 2800, Cleveland, Oh, USA), and sampled via an A/D converter (sampling rate 9 kHz) installed in an IBM PC-AT. A fixed 580 μs delay on

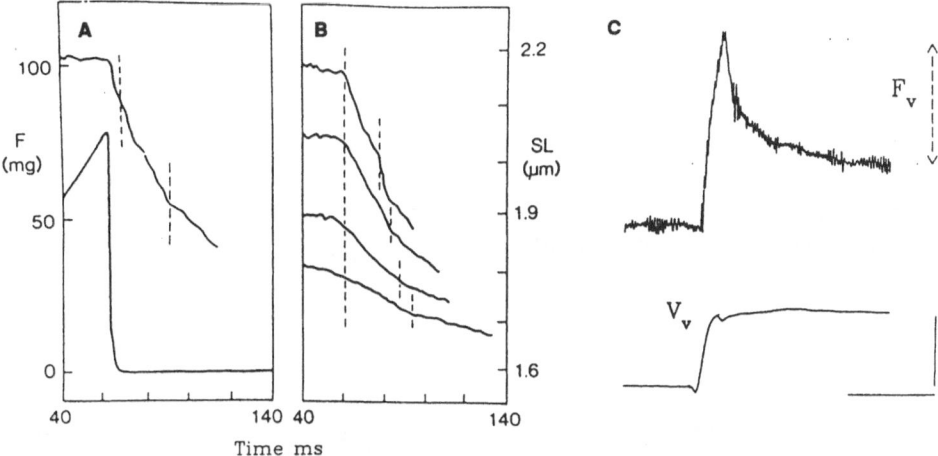

Fig. 1. Measurement of V_0 and passive viscosity. Force was measured by silicon strain gauge and SL by laser diffraction techniques. Panels A&B show 100 ms recording, starting 40 ms after the muscle was stimulated. Temperature was 25.2°C; Trabecula 2.5 mm in length, 150 μm in width, 60 μm in thickness. A) Shows a release from SL = 2.19 μm. SL top trace, F bottom trace. $[Ca^{2+}]_o$ = 0.2 mM. SL was kept constant during the initial phase of the twitch up to the quick release at 25 ms (65 ms after the time of stimulation). The dashed lines indicate the time window during which the velocity of sarcomere shortening was measured (about 8 μm/s). B) Illustrates quick releases from a SL of 2.18, 2.04, 1.89 and 1.81 μm respectively. The latter length, which is below slack length (1.85 μm) was obtained by allowing the trabecula to shorten actively early during the twitch. The dashed lines indicate time window during which the velocity of sarcomere shortening was measured: 8.85, 6.5, 3.7 and 2.8 μm/s respectively. $[Ca^{2+}]_o$ = 0.5 mM. Recording of force has been omitted for clarity. C) Illustrates the measurement of passive viscous forces in an unstimulated rat trabecula ($[Ca^{2+}]_o$ = 0.5 mM). Top panel force, bottom panel SL. The muscle was stretched (Vv = 18.2 μm/s) from a SL slightly above slack length for about 7 ms. Viscous force was measured as the difference between peak force and the final force level that was attained after the stretch was completed. It is clear from the tracing that force declined exponentially after the stretch; further analysis showed that the rise of Fv was also exponential. The time constant of this rise and fall of F_v are denoted in the text as τ_{rise} and τ_{decay}. In this muscle and at this velocity, F_v was 3.5 mg, equivalent to 3% of the twitch force that was measured at $[Ca^{2+}]_o$ = 1.5 mM and 1.9 μm SL. Calibrations: 20 ms; 2 mg; 0.1 μm. Trabecula 2.6 mm in length, 250 μm in width, and 50 μm in thickness.

the SL signal that was introduced by the scanning photodiode array system and analog peak detection circuit was corrected digitally.

SL was kept constant by stretching the muscle during the initial phase of the twitch until 70% of peak sarcomere isometric twitch force was reached. Next, the muscle was rapidly released to a new force level (Fig. 1A). This was followed by a slower linear sarcomere release to keep force constant. During the shortening phase, the velocity of sarcomere shortening was measured by computer (IBM PC-AT)[6][7] over a range of 0.2 µm. In order to measure V_O below slack length (1.9 µm), the trabecula was first allowed to shorten to a shorter SL early during the twitch, after which a similar protocol was used (Fig. 1B).

Measurement of Passive Viscous Force

For the measurement of passive viscous force, the muscle was perfused with medium containing $[Ca^{2+}]_0 = 0.5$ mM, but not stimulated. Visual inspection revealed no spontaneous activity under these conditions of low $[Ca^{2+}]_0$. Linear sarcomere stretches of 0.1 µm amplitude starting at a SL of 2.0 µm and at varied speeds (V_v; 2-40 µm/s) were applied to the preparation which caused a rapid increase in force. The parallel elastic force was calculated from the passive force-SL relation and

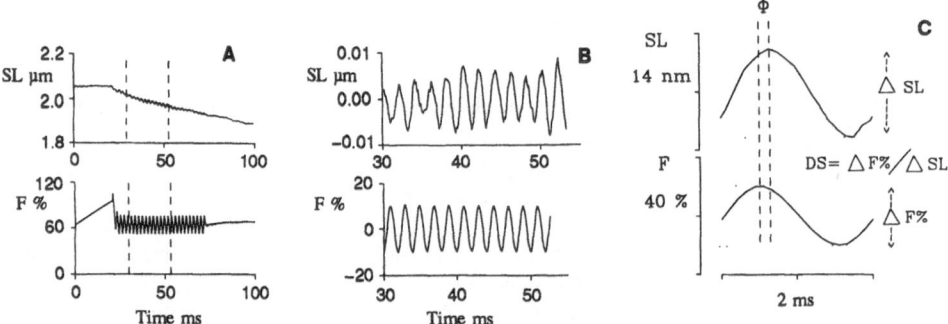

Fig. 2. Method used to measure dynamic stiffness during shortening. Dynamic stiffness was measured during the isovelocity release (Fig.1) by sinusoidal SL perturbations. Shortening ramps were synchronized to the length perturbations, recorded 10 times, and averaged. A) Shows a record of force (top panel) and SL (bottom panel) of the release at a slow time base. The velocity of sarcomere shortening was 2.2 µm/s. Note that the perturbations (500 Hz) were applied for about 20 ms, starting at the time of quick release. B) Shows SL (top panel) and force (bottom panel) after processing by digital high pass filter on a faster time base and at higher gain. This procedure allowed subtraction of the shortening ramp on the SL record. C) Shows one average cycle, calculated from the records in panel C by cross-correlation technique in order to further increase the signal to noise ratio. The methods used to measure dynamic stiffness (DS) and phase shift are illustrated. Sarcomere velocity = 2.2 µm/s; Phase angle = 27.6 degrees; Stiffness 1.9 %/nm; Trabecula 3.2 mm in length, 100 µm in width, and 60 µm in thickness; $[Ca^{2+}]_0 = 1.5$ mM. 25°C.

subtracted from the force transient 4, in order to obtain the visco-elastic force (F_V) (Fig. 1C).

Dynamic Stiffness

Dynamic stiffness was measured by sinusoidal length perturbation of the muscle during shortening (Fig. 2). We used 500 Hz perturbation frequency in seven muscles and 100 Hz in one muscle. Care was taken that the maximum SL perturbation did not exceed 15 nm (the average amplitude was 11 ± 0.7 nm). The digitised and averaged force and SL signal, acquired during a time window of about 20 ms following the quick release, were processed by a digital high pass filter. One average cycle of the sinusoidal SL and force changes, obtained by cross-correlatiom, was used to measure phase shift and dynamic stiffness (i.e. the ratio between force amplitude and SL amplitude). Force and stiffness were normalized to their maximum values during an isometric twitch.

Statistical Analysis

The data were fit via a non-linear least squares Marquardt fitting procedure or by linear regression. Force-sarcomere velocity relations were fit to Hill's hyperbola[8]): $V = (a*V_0-b*F)/(F + a)$. V_0-$[Ca^{2+}]_0$ and F-$[Ca^{2+}]_0$ relations were fit to: $Y = [Ca^{2+}]_0^h/([Ca^{2+}]_0^h + K^h)$, in which Y stands for F_0 or V_0, EC50 the latter stands for a compound affinity constant, and 'h' is the slope coefficient[5)7)]. Force-SL relations were fit to a power function : F, 'a', 'SL_0' and 'c' are constants[5)]. Data are presented as means $\pm$ S.E.M.

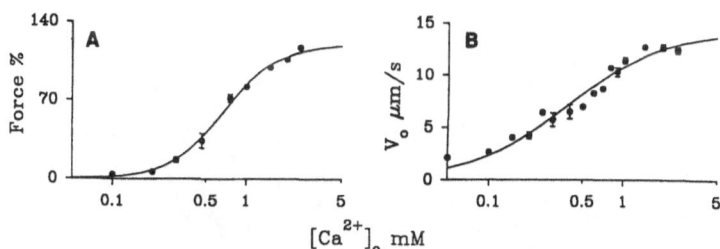

Fig. 3. Effect of $[Ca^{2+}]_0$ on V_0 and F_0 at slack length in rat myocardium. V_0 and F_0 were measured as described in the text and as illustrated in Fig. 1. Contractile activation was modified by varying $[Ca^{2+}]_0$ between 0.05 and 3.0 mM. The data were fit (solid line) to a modified Hill equation which had the form : $Y = Y_{max}([Ca^{2+}]_0^h/EC50^n+[Ca^{2+}]_0^h)$, in which EC50 is a compound affinity constant, Y stands for F_0 or V_0, and Y_{max} represents the saturation level of F_0 or V_0. Force is expressed as the percentage of force measured at SL = 1.9 μm, $[Ca^{2+}]_0 = 1.5$ mM. (note logarithmic scale on the abscissa in both panels). Data are shown as mean $\pm$ S.E.M. The average fit parameters were : (panel A) $Y_{max} = 120 + 4.7$ %; EC50 = $0.68 + 0.03$ mM; h = $2.6 + 0.6$; r > 0.99; 4 muscles. (panel B) $Y_{max} = 14.3 + 0.35$ μm/s; EC50 = $0.44 + 0.04$ mM (P < 0.01); h = $1.4 + 0.2$ (p < 0.01); r > 0.93; 6 muscles.

RESULTS

Fig. 3 shows the effect of $[Ca^{2+}]_o$ on F_o and V_o at slack length. $[Ca^{2+}]_o$ was varied between 0.05 and 3.0 mM; the data were fit to a modified Hill equation for a sigmoid. Maximal V_o was 14.3 ± 0.35 μm/s; F_o was 50 % of maximum F_o at $[Ca^{2+}]_o$ = 0.68 ± 0.03 mM. V_o was 50% of maximum V_o at $[Ca^{2+}]_o$ = 0.44 ± 0.04 mM (p < 0.01), indicating that V_o was more sensitive to $[Ca^{2+}]_o$ than F_o[2)7)10)]. We have recently shown a similar interrelationship in the myocardium of the cat[11)] and in myocardium of hypothyroid rats[9)]. The isomyosin composition of the myocardium of both cat and hypothyroid rat is in the V_3 form, as opposed to V_1 in myocardium of the euthyroid rat.

Hence, the sigmoidal relation between F_o or V_o and $[Ca^{2+}]_o$, and the higher sensitivity of V_o than F_o for $[Ca^{2+}]_o$, is not unique to myocardium of the euthyroid rat. An alternative way to alter contractile activation is via changes in SL[4)5)12)]. The change in contractile activation is believed to be caused by a change in the amount of activator calcium bound to the contractile filaments[13)]. Fig. 4 shows the effect of SL on F_o and on V_o at 3 levels of $[Ca^{2+}]_o$. The effect of $[Ca^{2+}]_o$ on the F_o-SL relation is similar to our previous results in rat myocardium[4)5)] and cat myocardium[11)]. Note that F_o increased by about 20% above slack length at the highest $[Ca^{2+}]_o$ studied. V_o on the other hand, was independent of SL above slack length at this level of $[Ca^{2+}]_o$. This result is consistent with our previous observations on rat[2)] and cat[11)] myocardium. At lower $[Ca^{2+}]_o$, however, both V_o and F_o varied significantly with SL over the full range of SLs studied. We have reported a similar dependence of V_o on SL at sub-maximal levels of activation in cat myocardium[11)]. Development of a large passive force by the parallel elastic element prevented stretch of the cardiac

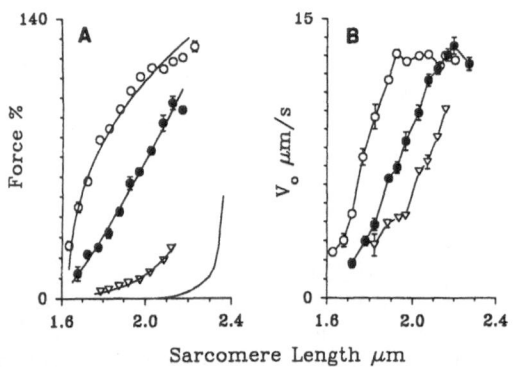

Fig. 4. The effect of SL on F_o and V_o. Shows F_o (A panel) and V_o (B) panel as function of SL. Data were collected in bins and are presented as mean ± S.E.M. F_o is expressed as percentage of F_o at 1.9 μm, $[Ca^{2+}]_o$ = 1.5 mM. $[Ca^{2+}]_o$ was 1.5 mM (open circles; 8 muscles), 0.5 mM (closed circles; 8 muscles), and 0.2 mM (triangles; 4 muscles). F_o-SL relations were fit to a power function in the form $F_o = a*(SL-SLo)^c$. The solid line without symbols indicates the passive force-SL relation, which was always independent of $[Ca^{2+}]_o$.

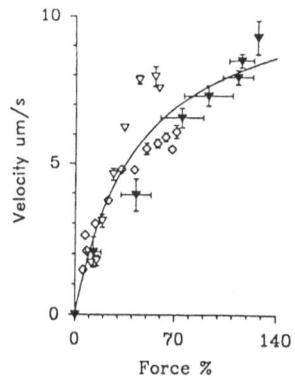

Fig. 5. V_o as function of F_o in myocardium. Shows V_o as function of F_o in cardiac trabeculae from cat (de Tombe & ter Keurs, 1991b, with permission). Data were collected in bins and are presented as mean ± S.E.M. F_o-V_o coordinates were collected from experiments in which : 1) $[Ca^{2+}]_o$ was varied at slack SL (filled triangles). 2) SL was varied at $[Ca^{2+}]_o = 1.5$ mM (open triangles). Note that only data at SL > 1.9 µm were selected. 3) Time into the twitch up to peak force (diamonds). Solid line reflects the fit of the data to $V_o = (V_{intr} * F_o)/(K+F_o)$ (see Discussion for details on this equation). The fit parameters were : $V_{intr} = 11.3$ µm/s; $K = 44.9$; r = 0.94.

sarcomere above about 2.3 µm. This precluded study of V_o and F_o over the range of SLs at which partial overlap between actin and myosin occurs.

The relation between V_o and F_o, compiled from the data as in Figs. 3 & 4 is shown in Fig. 5 for cat myocardium. F_o-V_o coordinates were only included for data obtained at or above slack SL; hence, V_o was not limited by internal restoring forces under these conditions. Similar relations between V_o and F_o were also observed in the myocardium of euthyroid rat[20], and myocardium of hypothyroid rats[9]. Myocardial myosin in the cat and hypothyroid rat is composed almost predominantly of the V_3 isomyosin. It is clear, therefore, that V_o is a unique function of the level of contractile activation independent of the iso-enzyme composition and of the manner in which the activation level is attained (i.e. variation of $[Ca^{2+}]_o$, SL, or time into the twitch).

The observed relation between F_o and V_o shown in Fig. 5 raises the question whether not only F_o, but also V_o is controlled by the amount of calcium bound to the contractile filaments, as we have concluded previously[2]. One possibility is the existence of an internal load, which would prevent maximal shortening of the myofilaments. Internal loads of both fixed magnitude[14-16] and viscous, i.e. velocity dependent magnitude[17], have been proposed. Such an internal load would play a larger role at lower levels of activation, and thus an apparent activation dependence of V_o would ensue. The V_o-F_o coordinates in Fig. 5 were carefully selected from a range of SL where the elastic load must have been negligible (1.9-2.2 µm). Hence we consider an internal load of fixed magnitude an unlikely source of the limiting factor for V_o.

Fig. 1C shows that a linear sarcomere stretch of 0.1 μm amplitude starting at a SL of 2.0 μm and at varied speeds (v = 2-30 μm/s) caused a rapid increase in force with a complex time course, with an overshoot over the static elastic force. With shortening, a similar response occurred but with opposite sign, i.e. during shortening an undershoot of force was observed under the static force level. Analysis of the response to shortening was limited by the low levels of static parallel elastic force levels in the range of SLs of interest (SL = 1.9 - 2.1 μm; $F_{passive}$ = 0 - 2 % F_0). Release of the sarcomeres studied in two muscles showed similar responses as the response to stretch, but even at modest velocity the release led to buckling of the muscle which precluded accurate measurement of SL.

In order to perform a detailed analysis of the response to stretch, force and SL signals of 10 consecutive stretches were averaged by computer before storage to magnetic disk for off-line analysis. Sarcomere velocity during the stretch was calculated by linear regression. Force returned to a steady level, which was higher than that before the stretch, in another 50 ms after completion of the stretch. The difference in force before and after the stretch was independent of stretch velocity, as would be expected from an elastic structure in parallel with the cardiac sarcomere. The F-SL relation measured after the force had stabilized was approximately linear, with a slope of 10 N.mm^{-3} between SL = 1.9 and 2.15 μm. The latter range is the range over which cardiac muscle is compliant[4)5)]. We used this relationship to calculate the force generated during the stretch by an assumedly "pure" and linear parallel elastic element. Next, we subtracted this force from the actually measured force in order to calculate the additional response of the muscle to stretch. Force increased, during the stretch, exponentially to a maximum (F_v) which depended on the rate of stretch, and decreased after the stretch also with an exponential time course; this behaviour would be expected from the arrangement of an elastic component and a viscous component in series. Fig. 6 shows the average of F_v in 6 trabeculae as a function of the velocity of stretch of the sarcomeres in the range SL = 2.0-2.1 μm, measured during the interval between twitches. It is clear that passive myocardium does not behave as a 'pure' Newtonian viscous element, since the slope of the relation between viscous force and stretch velocity decreased at higher velocities. The slope of the relation at SL < 2.2 μm was 0.15-0.35%F_0•s•μm^{-1}, which is similar to that reported in cat myocardium[17)18)], and skeletal muscle[19)].

In order to exclude the possibility that the apparent viscous force measured during the interval between twitches was caused by residual attached cross-bridges, the measurements were repeated in the absence of external Ca^{2+} and after the addition of EGTA. In Ca^{2+} free solution the viscous force increased slightly[20)].

The time course of the process, responsible for t_{rise} and t_{decay} was substantially faster (3 to 8 ms) than the rate of stress relaxation of the parallel elastic element (70 to 100 ms) that was observed upon stretch of the sarcomeres into a range of SLs at which passive force increases more steeply with SL (> 2.3 μm). The time constant of the force response to stretch (τ_{rise}) was slightly faster than τ_{decay} after the stretch: τ_{rise} = 6.6 ms at v $\simeq$ 4.0 μm/s, and τ_{rise} = 4.4 ms at v $\simeq$ 8 μm/s (τ_{rise} = 1.2 +

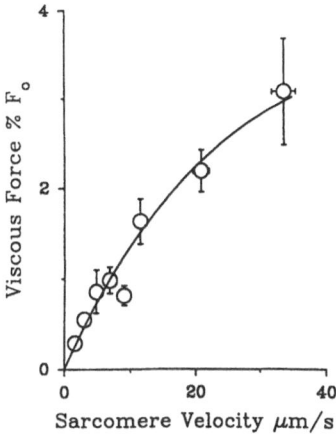

Fig. 6. Viscous force in unstimulated trabeculae. Passive trabeculae (5 muscles) were stretched 0.1 μm from a SL of 1.9-2.0 μm at varied velocities. Viscous force was measured by the method described in the text and as outlined in Fig. 1C. Data are presented as mean ± S.E.M. Viscous force is expressed as percentage of F_0 at 1.9 μm, $[Ca^{2+}]_0 = 1.5$ mM. The data were fit to an arbitrarily chosen single exponential function : $F_v = a(1-exp(-kV_v))$. Average fit parameters : a = 3.9 ± 0.84 %; k = 0.04 ±1.4 μm/s; n = 47; r = 0.818.

$0.7 \cdot \tau_{decay}$ (r = 0.6; p < 0.001)). τ_{rise} could not be assessed accurately at higher stretch velocities due to a limited number of data points available for the exponential fit at these large v. τ_{decay} decreased with increases in stretch velocity up to 30 μm/s. Furthermore, t_{decay} was similar in the presence of Ca^{2+} to that measured in the absence of Ca^{2+} over the SL range between 2.0 and 2.2 μm. The ratio of F_v over τ_{decay} of the combination of a viscous element and an elastic element in series provides an measure of the magnitude of the elasticity of the elastic element. $F/^\circ\eta/\tau_{decay}$ did not vary significantly with the velocity of the stretch (< 20 μm/s) and was 34 to 76 %F_0/μm in the muscles at $[Ca^{2+}]_0 = 1.5$ mM and at SL between 2.0 and 2.1 μm and 71 to 110 %F_0/μm in Ca^{2+} free solution, respectively.

In order to assess to what extent the shortening velocity is affected by this internal viscous load, it is necessary to know what fraction of the cross-bridges would support the load at a given shortening velocity. The relative number of attached cross-bridges during shortening can be assessed by measurement of muscle stiffness relative to the stiffness under isometric conditions[21-25]. Dynamic stiffness (DS) was a linear function of load in euthyroid rat cardiac trabeculae: DS% = 0.084*F + 12.0 (r = 0.84)(Fig. 7A). In contrast to dynamic stiffness, the phase shift Φ between force and SL was independent of force during shortening (Φ degrees = 0.05*F + 51.8 (r = 0.01). The simultaneously measured force-sarcomere velocity relation was hyperbolic (r = 0.89) with $V_0 = 9.2$ μm/s, b = 0.52 and a = 19.6, respectively (Fig. 7B). Assuming that dynamic stiffness reflects the number of attached cross-bridges[21-25] we calculated the average force-sarcomere velocity relation at the level of one cross-bridge. First, passive viscosity was corrected for by adding the appropriate viscous force at each sarcomere velocity to the load (Fig. 7),

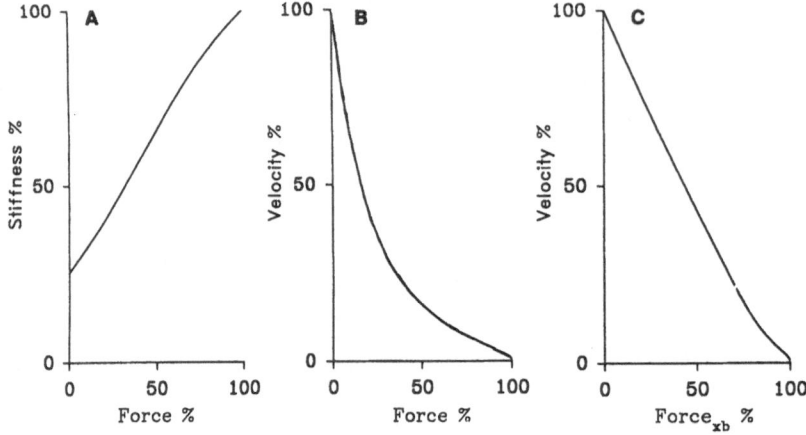

Fig. 7. Average force-velocity per single cross-bridge predicted from Huxley's 1957 model Calculations according to the model proposed by AF Huxley (1957). The following values were used for the constants: h = 15 nm; SL 2.5 μm; f1 = 13; g1 = 3; g2 = 63. A) Shows relative stiffness, calculated as the number of attached cross-bridges, as function of relative force during shortening. B) Shows relative sarcomere velocity as function of the calculated relative force during shortening. C) Shows relative sarcomere shortening velocity as function of the average relative force per attached cross-bridge as function. The latter was obtained by dividing the force at each velocity by the stiffness at that velocity.

an approach that has also been used by Chiu et al[26]. Next, the average force per cross-bridge was obtained by dividing the corrected force by the dynamic stiffness at each load level. This resulted in a linear force-sarcomere velocity relation which, in euthyroid rat, fitted[20]: V = 10.4 μm/s - 0.09*Force% (r = 0.97).

DISCUSSION

Both F_0 and V_0 were a sigmoidal function of $[Ca^{2+}]_0$, confirming our previous observations on rat myocardium[2][9] and cat myocardium[11]. Increase of $[Ca^{2+}]_0$ has also been shown to increase the maximal velocity of shortening (commonly referred to as V_{max}) of papillary muscles[27][28]. However, lack of segment length control in those studies on isolated papillary muscle preparations make it likely that V_0 was underestimated[7]. More recently, Martyn et al[29] have reported the effect of $[Ca^{2+}]_0$ on velocity of shortening under a light load in ferret papillary muscle. In that study, a magnetic sensing coil technique was used to measure and control the length of a central segment of the muscle. They also observed a reduction of segment velocity under a light load upon a reduction in $[Ca^{2+}]_0$.

SL had a prominent effect on V_0, which could be modulated by $[Ca^{2+}]_0$. At lengths below slack length, V_0 decreased in proportion to SL at all levels of contractile activation. At close to maximal levels of activation (i.e. $[Ca^{2+}]_0 = 1.5$ mM), V_0 was independent of SL above slack length, as has been observed previously in rat[2] and cat myocardium[11]. The decline of V_0 below slack SL is probably the result of restoring forces, which would act as internal elastic load on the cardiac sarcomere.

At sub-maximal levels of contractile activation, V_0 declined in proportion to SL over the full range of SLs studied in the present study on rat myocardium and in a recent study by us on cat myocardium[11]. This result is consistent with a previous study by Martyn et al[29], who used segment length control and observed that the velocity of shortening under a light load depends on segment length, including segment lengths above slack length.

Passive Visco-elastic Properties of Cardiac Trabeculae

The exponential increase of force during a linear stretch of the sarcomeres and the exponential decrease of force after the stretch of the unstimulated trabeculae is characteristic of an arrangement of an elastic (stiffness = k_{series}) and a viscous element (viscosity = η) in series. The steady force during the length change evidently reflects the viscosity of the viscous component ($F_v = \eta \cdot v$), while the time constant τ of increase of force to F_v reflects the time course of the length change of the elastic element in series with the viscous element. It can be shown that $\tau = \eta / k_{series}$. In the following we will discuss the possible structural counterparts of the visco-elastic behaviour and its potential implications for shortening of the active muscle.

Noble[18] and Chiu et al[17] have studied the diastolic viscous properties of cat papillary muscle using overall muscle length. The results of those studies are comparable to those of the present study, i.e. viscous force of a few percent and an apparently non-Newtonian[18] viscosity at higher stretch velocities. Using a value of 70 mN/mm^2 for F_0[9], η varied between 0.15 and 0.35 mN.s/mm^3 for the range SL < 2.2 μm. These values are comparable to the values the coefficient of viscosity obtained by Ford et al[19][21] in frog skeletal muscle. It is unlikely that the viscous behaviour is explained by properties of the sarcolemma or of the mitochondria, because the density of sarcolemma and mitochondria is approximately tenfold greater in cardiac muscle than in fast twitch skeletal muscle fibres. Neither is it likely that the passive viscosity measured in the present study is due to 'weakly attached' cross-bridges[30], which have been postulated on the basis of the force response to length changes at velocities far exceeding those used in the present study. Furthermore, it is unlikely that the viscous properties of the trabeculae were caused by residual attached Ca^{2+} dependent cross-bridges, since reduction of $[Ca^{2+}]_0$ to very low levels (pCa $\sim$ 9) did not affect the viscosity[20]. The magnitude of k_{series} is further evidence that it is unlikely that the visco-elastic behaviour is due to residual cross-bridges. We have estimated the series elasticity of the cross-bridge to be 100 %F_0/9.4 nm in cardiac muscle at 26°C[30]. The observed kseries (34 to 76 %F_0/μm) is at least 175 times lower than that of the cross-bridges.

Exclusion of the contribution by other structures leaves titin as the remaining possible structural basis for the passive visco-elasticity of the passive muscle. Further evaluation of the mechanical properties of titin is needed, but several of the characteristics of the behaviour of the visco-elastic element support its candidacy. Although kseries is much lower than that of the cross-bridges it is still substantial, e.g. it is approximately 5 times larger than the stiffness of the parallel elastic element of the sarcomere at lengths between 1.9 and 2.1 μm. This suggest the presence of a moderately stiff—probably protein—element in series with the viscous component. Titin would be a plausible candidate for this arrangement; the molecule is arranged in the form of a series of large macromolecular domains which can be extended enormously by unfolding[32], and which are coupled to each other by shorter protein domains. The shorter protein domains may provide the basis for k_{series}, while

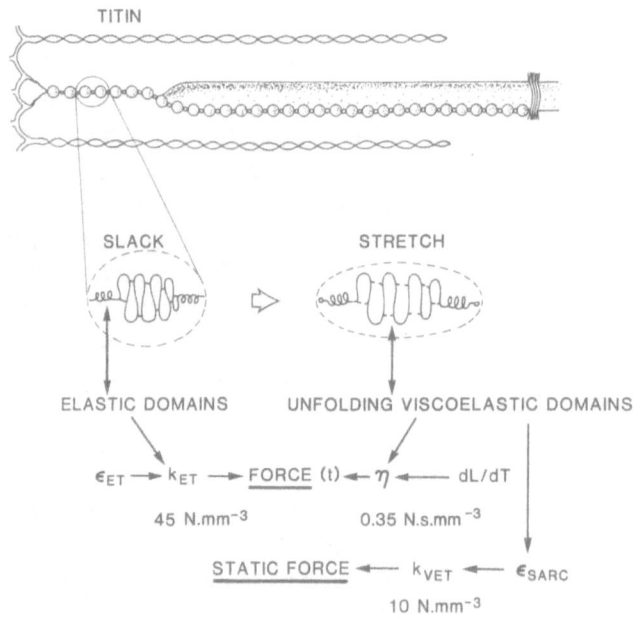

Fig. 8. The titin molecule as a basis for visco-elastic behaviour of unstimulated cardiac muscle. The proposed properties of titin are based on the visco-elastic characteristics of cardiac muscle observed during rapid stretches. The viscosity of the muscle (η) was 0.1-0.4 mN.s.mm^{-3}. The calculated stiffness of the elastic element in series (k_{et}) with the viscous element is 25-45 N.mm^{-3}; the strain (ε_{et}) of this element depends on the visco elastic force. We assume that these properties reside in the titin filament. We propose that the large macromolecular domains unfold during stretch and thereby provide a basis for the viscous behaviour. The links between the large macromolecular domains are assumed to be relatively stiff (k_{et}). The continuity of the chain thus formed implies, that the large domains may be responsible for the static parallel elastic behaviour of cardiac muscle after they have unfolded to their new length (depending on the sarcomere strain ε_{sarc}); this requires that the latter domains are compliant ($k_{vet} \sim 10$ N.mm^{-3}). For further explanation see text.

unfolding of the large domains could underlie the viscous behaviour. The unfolded domains could provide a basis for: i) the parallel elastic behaviour that is observed in isolated myocytes[33] and; ii) the compliant part of the parallel elastic structures that respond with a small increase of force with stretch of cardiac muscle above slack length (Fig. 8).

The hypothesis that titin may provide the basis for the visco-elastic behaviour of the muscle upon stretch as observed in this study deserves further testing. In particular, the visco-elastic properties of the muscle have to be evaluated in more detail during shortening. If titin is responsible for the viscous load during shortening we have to assume that titin does not buckle in the process. This assumption requires that titin is incorporated in lateral connections in between the filaments in the I-band. We have not performed a detailed quantitative analysis of the effect of shortening because of the low levels of static parallel elastic force in the range of SLs of interest which caused buckling of the whole muscle already at low velocities of shortening. Still, the results that we have obtained during shortening suggest that the behaviour of the muscle during stretch and during shortening is qualitatively similar. Dynamic stiffness at 500 Hz, which decreased with increase in velocity of active shortening probably reflects the T_2 component of short term stiffness. The T_2 component is assumed to be related to the redistribution of cross-bridge states after a rapid length change[19]. The T_2 component has been shown to be relatively linear over a short range both in skeletal muscle[19] and in cardiac rat trabeculae (Backx & ter Keurs, in preparation). Of importance for the present study, is the notion that the T_2 component of the force response to a rapid length change, and therefore the stiffness measured at 500 Hz, is proportional to the number of attached cross-bridges. In support of this conclusion is the observation that the relationship between dynamic stiffness and force during shortening at 100 Hz and at 500 Hz were almost identical.

Force-Velocity Relation of the Cardiac Crossbridge

We consider it unlikely that other factors such as deactivation, resonance of the muscle fibre and passive viscosity have contributed significantly to the measured DS20. Dynamic stiffness decreased as a function of force during shortening, as has been observed in skeletal muscle[21-25]. A finite stiffness was observed at zero external force, which is also consistent with the results from skeletal muscle. The magnitude (about 12%) of this finite stiffness at V_0 however, is smaller than that observed in the studies on skeletal muscle, probably due to the 50% lower level of activation of cardiac muscle[34][35]. Thus, in order to compare our results on myocardium to those of Ford et al.[19], a correction of about 50% is required. This analysis suggests that the stiffness at zero load is a function of the level of contractile activation. The average force-velocity relation calculated from the load and the corresponding stiffness at the level of a single cross-bridge was close to linear. A similar result has recently been reported for frog skeletal muscle[25]. A close to linear average force-velocity relation at the level of a single cross-bridge is in fact predicted by the 1957 Huxley model[36], as shown in Fig. 7.

Effect of Activation on Unloaded Velocity of Shortening

Our results indicate that V_0 of intact myocardium depends on the level of contractile activation. This raises the question of how the overall amount of calcium binding to troponin-C[37)38)] can regulate the actomyosin interaction of individual sites, i.e affect cross-bridge cycling rate, and thus regulate V_0. One possible explanation is the presence of an additional calcium regulatory site on myosin, as has been proposed[39-41)]. Another possibility is the existence of an internal load[14-17,26)]. We developed a model incorporating our results on passive viscosity (Fig. 5) and dynamic stiffness during shortening (Fig. 6). The model is based on the conclusion from the above mentioned observations that, during shortening, the velocity of shortening is linearly proportional to the average load per attached cross-bridge (slope coefficient: k_2). Second, the number of cross bridges at zero external load is assumed to be proportional (k_1) to the level of activation. Third, the viscous force generated by the shortening sarcomere is proportional to the velocity of shortening (η). Then it can be shown[20)] that:

$$V_0 = (V_{intr}*F_0)/(K + F_0) \tag{1}$$

in which $K = k_2*\eta/k_1$ and V_{intr} is the intrinsic maximal velocity. Fig. 6 shows that an adequate fit could be obtained for the data from cat myocardium. The numerical value of the constant K was 36.8, 44.9, 23.5 for euthyroid rat, cat and hypothyroid rat myocardium, respectively. The expected value for K, estimated from the values for k_1 k_2 and for viscosity of euthyroid rat myocardium varied from 23 to 38 and matched the value calculated from the V_0-F_0 relation (26.8) well. Hence the fit of the model to the actual data (Fig. 6) is surprisingly good, considering the widely varying conditions under which the data have been obtained. Thus, a passive viscosity together with a reduction of the number of cross bridges at V_0 seemed sufficient to account for the limiting effect of contractile activation on V_0.

It should be stressed however, that this conclusion is based on several assumptions. First, we assumed that the finite stiffness at zero external load is a function of the level of activation (see above). Even though this assumption appears to be reasonable, further experiments are necessary to confirm this assumption (i.e. measurement of dynamic stiffness during shortening at different levels of activation). Second, we have measured passive viscosity during a stretch in unstimulated muscles. We then assumed that this property also applied in the active muscle during shortening. The reports[19)31)], that in the analysis of instantaneous elasticity (F_1) of active muscle a viscous response remained undetected, seem to be at odds with the above assumption. The absence of a viscous response seems puzzling at first glance, but is readily explained if one realizes that during a rapid release (e.g. 18nm in 200 μs) to zero force the viscous force amounts to only 15% F_0 (See Fig. 6). The contribution of viscosity to the F_1 curve is proportionally less if smaller quick releases are applied.

Conclusion

We have found that the unloaded velocity of sarcomere shortening depends on $[Ca^{2+}]_o$ and SL. This dependence can be explained by the existence of a viscous force set up during shortening of the sarcomere. At low levels of activation, the number of cross bridges is reduced and the viscous load per cross bridge is substantial; hence, the cycling rate of the bridge slows down. It is evident that, if this hypothesis can only be valid if the measured viscosity resides in an element outside the contractile apparatus. Analysis of the visco-elastic response to length change suggests that this is indeed the case: the most likely element to be responsible for the visco-elastic behaviour of unstimulated cardiac muscle is titin.

ACKNOWLEDGEMENTS

This work was supported in part by a grant from the Canadian Medical Research council. H.E.D.J. ter Keurs is a Medical Scientist of the Alberta Heritage Foundation for Medical Research (AHFMR). P.P. de Tombe held a studentship from the AHFMR.

REFERENCES

1. Podolin, R.A. & Ford, L.E. *J. Muscle Res. Cell Motility* 4, 263-282 (1983).
2. Daniels, M., Noble, M.I.M., ter Keurs, H.E.D.J. & Wohlfart, B. *J. Physiol. (Lond.)* 355, 367-381 (1984).
3. Edman, K.A.P. *J. Physiol. (Lond.)* 291, 143-159 (1979).
4. ter Keurs, H.E.D.J., Rijnsburger, W.H., van Heuningen, R. & Nagelsmit, M.J. *Circ. Res.* 46, 703-714 (1980).
5. Kentish, J.C., ter Kerus, H.E.D.J., Ricciardi, L., Bucx, J.J.J. & Noble, M.I.M. *Circ. Res.* 580, 755-768 (1986).
6. de Tombe, P.P., Backx, P.H.M. & ter Keurs, H.E.D.J. *Circulation* 78, II- 68, (Abstract) (1988).
7. de Tombe, P.P. & ter Keurs, H.E.D.J. *Circ. Res.* 66, 1239-1254 (1990c).
8. Hill, A.V. *Proc. R. Soc. B* 126, 136-195 (1938).
9. de Tombe, P.P. & ter Keurs, H.E.D.J. *Circ. Res.* 68, 382-391 (1991a).
10. ter Keurs, H.E.D.J., de Tombe, P.P., Backx, P.H.M. & Iwazumi, T. *Biorheology* 28, 161-170 (1991).
11. de Tombe, P.P. & ter Keurs, H.E.D.J. *Circ. Res.* 68, 588-596 (1991b).
12. Allen, D.G. & Kentish, J.C. *J. Mol. Cell. Cardiol.* 17, 821-840 (1985).
13. Hofmann, P.A. & Fuchs, F. *J. Mol. Cell. Cardiol.* 20, 667-677 (1988).
14. Gulati, J. & Babu, A. *J. Gen. Physiol.* 86, 479-500 (1985).
15. Chiu, Y.L., Ballou, E.W. & Ford, L.E. *Circ. Res.* 60, 446-458 (1987).
16. Tsuchiya, T. *Biophys. J.* 53, 415-423 (1988).
17. Chiu, Y.L., Ballou, E.W. & Ford, L.E. *Biophys. J.* 40, 109-120 (1982a).

18. Noble, M.I.M. *Circ. Res.* **40**, 288-292 (1977).
19. Ford, L.E., Huxley, A.F. & Simmons, R.M. *J. Physiol. (Lond.)* **269**, 441-515 (1977).
20. de Tombe, P.P. & ter Keurs, H.E.D.J. *J. Physiol. (Lond.)* (submitted).
21. Ford, L.E., Huxley, A.F. & Simmons, R.M. *J. Physiol. (Lond.)* **361**, 131-150 (1985).
22. Julian, F.J. & Sollins, M.R. *J. Gen. Physiol.* **66**, 287-302 (1975).
23. Julian, F.J. & Morgan, D.L. *J. Physiol. (Lond.)* **319**, 193-203 (1981).
24. Ford, L.E., Huxley, A.F. & Simmons, R.M. *J. Physiol. (Lond.)* **311**, 219-249 (1981).
25. Haugen, P. In: *Molecular Mechanism of Muscle Contraction.* (eds. Sugi, H. & Pollack, G.H.) 461-469 (Plenum Press, 1988).
26. Chiu, Y.L., Ballou, E.W. & Ford, L.E. *Biophys. J.* **40**, 121-128 (1982b).
27. Sonnenblick, E.H. *Circ. Res.* **16**, 441-451 (1965).
28. Brutsaert, D.L., Claes, V.A. & Sonnenblick, E.H. *Circ. Res.* **29**, 63-75 (1971).
29. Martyn, D.A., Rondinone, J.F. & Huntsman, L.L. *Am. J. Physiol.* **244**, H708-H714 (1983).
30. Brenner, B.M., Schoenberg, M., Chalovich, J.M., Greene, L.E. & Eisenberg, E. *Proc. Natl. Acad. Sci.* **79**, 7288-7291 (1982).
31. Backx P.H. & ter Keurs H.E.D.J. *Circulation* **78II**, 68 (1988).
32. Wang, K. & Wright, J. The *J. Cell Biol.* **107**, 2199-2212 (1988).
33. Brady, A.J. *Physiol. Rev.* **71**, 413-428 (1991).
34. Fabiato, A. *J. Gen. Physiol.* **78**, 457-497 (1981).
35. Schouten, V.J.A., Bucx, J.J.J., de Tombe, P.P. & ter Keurs, H.E.D.J. *Circ. Res.* **67**, 913-922 (1990).
36. Huxley, A.F. *Prog. Biophys. Biophys. Chem.* **7**, 255-318 (1957).
37. Weber, A. & Murray, J.M. *Physiol. Rev.* **53**, 612-673 (1973).
38. Woledge, R.C., Curtin, N.A. & Homsher, E. *Energetic Aspects of Muscle Contraction.* (Academic Press, London, 1985).
39. Morimoto, K. & Harrington, W.F. *J. Mol. Biol.* **88**, 693-709 (1974).
40. Haselgrove, J.C. *J. Mol. Biol.* **92**, 113-143 (1975).
41. Lehman, W. *Nature* **274**, 80-81 (1978).

Discussion

Edman: I remember that you showed earlier that the mechanical V_{max} is independent of sarcomere length in cardiac muscle if you have a high degree of activation. Does that fit with your present ideas?

ter Keurs: A set of data points on the relationship between V_0 and F_0 in the upper right-hand corner of the saturating relationship of Fig. 5 is taken from the virtually constant velocity of shortening at high sarcomere length (Fig. 4B), where we are above the level of 80% force. The force still changes, but the velocity of shortening saturates, which is consistent with our previous observation.

Suga: Since you said that there is an internal viscous drag element, I think it must consume energy and generate some heat. I wonder whether this heat is a significant fraction when the cardiac muscle contracts under fairly normal conditions. It might be very small because you said the force is about 4% to 8% of the peak force. Do you have any idea about the quantitative value?

ter Keurs: I have no measurements of the energetics of this element. I would project from our data that force and the amount of shortening involved in this process are small and that the energy involved in this process is not more than a few per cent of the total energy involved in cardiac muscle contraction.

Suga: I am glad to hear that, because in a whole heart experiment, I actually stretched to a large pre-load and let the contraction go without any load. I found that excess oxygen consumption above the oxygen consumption of the unloaded condition is very small, so your statement matches with our result.

ter Keurs: I would expect a larger effect of internal opposing forces on energetics of the heart when it contracts below slack volume. When we allow the muscle to shorten below 1.9 μm, a precipitous fall in the maximal velocity of shortening is observed, which we assign to opposing forces that drive the muscle back to slack length.

MECHANISM UNDERLYING DOUBLE-HYPERBOLIC FORCE-VELOCITY RELATION IN VERTEBRATE SKELETAL MUSCLE

K.A.P. Edman

Department of Pharmacology
Univeristy of Lund, Sweden

ABSTRACT

The force-velocity relation of frog striated muscle exhibits two distinct curvatures located on either side of a breakpoint that occurs near 80 % of maximum isometric force (P_0) where the shortening velocity is approximately 1/10 of V_{max}. The present experiments have been performed to further elucidate the high-force deviation of the force-velocity curve in frog single muscle fibres.

The biphasic shape of the force-velocity curve appears at the same relative values of P_0 and V_{max} also after depressing the isometric force to 80 % of the control value by dantrolene, a substance known to reduce the release of activator calcium from the sarcoplasmic reticulum. This finding suggests that the breakpoint of the force-velocity curve is not related to the force level *per se* but rather to the speed of shortening of the contractile system. Thus as the speed of shortening goes below 1/10 of V_{max}, the performance of the myofilament system is changed such that less force and less motion are produced than expected from the main part of the force-velocity curve.

In a series of experiments active force and fibre stiffness were simultaneously recorded while the fibre shortened at various speeds during tetanus. Stiffness was measured as the change in force that occurred in response to a 4 kHz sinusoidal length oscillation of the fibre. A plotting of stiffness against force recorded under these conditions provides a biphasic relationship with a distinct transition between the two phases near 80 % of P_0, i.e. at the same relative force at which the breakpoint occurs in the force-velocity curve. Above 0.8 P_0 stiffness increases more steeply with force than below this point. This means that while more crossbridges than expected attach to the thin filaments when the load is raised above 0.8 P_0, the force output and the speed of shortening become *lower* than predicted from measurements at low and intermediate loads. The results suggest that the kinetics of crossbridge function is changed as the speed of filament sliding is reduced below a critical level, 1/10 of V_{max}. Beyond this point a greater portion of myosin crossbridges would seem to accumulate in a state where less force is being produced.

Mechanism of Myofilament Sliding in Muscle Contraction, Edited by
H. Sugi and G.H Pollack, Plenum Press, New York, 1993

Data are also presented to further elucidate the force-velocity relation at negative loads. In these experiments the passive tension at long sarcomere lengths has been utilized to produce a longitudinal compressive force on the sarcomeres during unloaded shortening (force-clamp recording). The results confirm [cf. Edman, K.A.P., *J. Physiol. (Lond.)* 1979: **291**, 143-159.] that the force-velocity relation at negative loads forms a smooth continuation of the positive force-velocity curve.

The relation between active force and velocity of shortening contains information of relevance to the understanding of the crossbridge mechanism of muscle contraction[1]. This relationship is not a simple hyperbolic function as was originally believed[2] but can be shown to contain two distinct curvatures on either side of a breakpoint near 78 % of the isometric force, P_o [3][4]. The two portions of the force-velocity curve are closely related to one another, and this interrelation between the two curvatures can be formulated by a modified version of Hill's hyperbolic equation[4]. Any change in curvature at low and intermediate loads, caused by a change in resting sarcomere length or by altered tonicity of the extracellular medium, is thus found to be associated with a specific, and predictable, change of the curvature in the high-force range.

The double hyperbolic shape of the force-velocity relation does seem to reflect the contractile behaviour at sarcomere level. This is suggested by the finding that the same biphasic shape is observed when measurements are made from the fibre as a whole and from short segments of the same intact fibre[4]. The double hyperbolic shape of the force-velocity relation can also be demonstrated in skinned muscle fibres provided that the sarcomere length is maintained reasonably uniform during contraction (Edman, unpublished data).

The purpose of the present experiments was to further elucidate the nature of the high-force deviation of the force-velocity curve during tetanic contraction of frog single muscle fibres. Results are also presented to characterize the force-velocity relation at negative loads.

METHODS

Single fibres isolated from the anterior tibialis muscle of *Rana temporaria* were mounted horizontally in a thermostatically controlled bath (1-3°C) between a force transducer and an electromagnetic puller as previously described[5]. Ringer solution was continously perfused through the muscle chamber at a rate of approximately 5 ml min[-1]. The bath temperature was constant to within 0.2°C throughout an experiment. The fibres were mounted to the hooks of the force transducer and puller by means of aluminium clips that were attached to each tendon as previously described[5]. The clips were folded around the hooks in a way that ensured that there was no change in position of the fibre during the experiment. With the approach used it was possible to adjust the attachment of the fibre so that any vertical, horizontal and twisting movements of the fibre were reduced to a negligible level

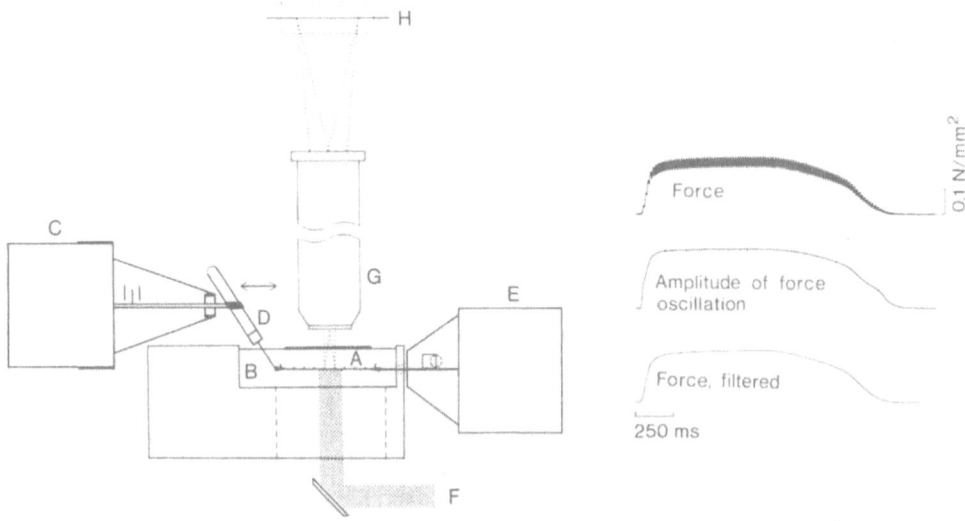

Fig. 1. Schematic illustration of the experimental set-up for force and stiffness measurements. A, muscle fibre with opaque markers on its upper surface. B, Trough containing Ringer solution. C, puller for shortening ramps. D, Force transducer (resonant frequency when submerged in Ringer fluid: 19 kHz). E, Puller for producing 4 kHz length oscillation of the fibre. F, Laser beam. G, Microscope tube. H, horizontal stage onto which an enlarged image of the fibre is projected. A photodiode array placed in the image plane records the relative position of adjacent markers on the fibre surface. Oscilloscope records from top: 1. isometric tetanus with superimposed 4 kHz force oscillation (only a fraction of the cycles displayed), 2. stiffness (=amplitude of force oscillation shown in upper record) and 3. tetanic force signal with oscillatory force response filtered off.

during contraction. The fibre was stimulated supramaximally to produce a fused isometric tetanus of 0.5-1 s duration at regular 2 min intervals. This was achieved by passing a train of rectangular current pulses (pulse duration: 0.2 ms, pulse frequency: 18-22 sec^{-1}) between two platinum plate electrodes that were placed on either side of the fibre. The pulse frequency was just sufficient to cause mechanical fusion in the respective fibre.

The approach used for simultaneous recording of force and instantaneous stiffness during contractile activity is illustrated schematically in Fig. 1. For these measurements the fibre was mounted in a special chamber (1-3°C) between two electromagnetic pullers. The left-hand puller (the same as that referred to above) was provided with a force transducer (see below) and was used to produce shortening ramps of different velocities during tetanus. The right-hand puller was actuated immediately before the stimulation volley to produce a 4 kHz sinusoidal length oscillation of constant amplitude (ca. 1.5 nm/half sarcomere) throughout the contraction period. Example records from an isometric tetanus are shown by the insets in Fig. 1. The upper record illustrates the isometric force myogram with the superimposed changes in force caused by the 4 kHz length oscillation. The amplitude of the force oscillation is an index of the fibre stiffness. This amplitude, i.e. the instantaneous stiffness, was measured electronically, on line (see further Edman & Lou[6]), and is illustrated by the middle record in Fig. 1. The force signal was also recorded without the superimposed force oscillation by passing the original signal through a notch filter with maximum damping at 4.0 kHz (bottom record in Fig. 1). The amplitude of the 4 kHz length oscillation (not illustrated) was also recorded throughout the contraction period. For a more detailed description of the stiffness measurement, see ref. 6.

As the above measurement of stiffness was based on length perturbation of the whole fibre, it was essential to minimize the tendon compliance. Care was therefore taken to attach the aluminium foils to the tendons (see above) as close to the fibre-tendon junction as possible without touching the fibre end. It was established in a separate experiment that the length change imposed on the whole fibre caused a constant change in length of a small "tendon-free" segment as tension was varied between 40 and 95 % of maximum tetanic force. The stiffness recorded from the whole fibre thus reflects sarcomere stiffness as long as the tension in the fibre exceeds 40 % of maximum tetanic force. The measurement of the distribution of the length change within the fibre (Edman, unpublished) was accomplished by means of an updated version of the surface marker technique that has previously been described[5][6].

The resting sarcomere length (measured by laser diffraction technique) was set to 2.10 or 2.25 μm unless otherwise stated. Care was taken to keep the intervals between tetani constant (2 min) throughout the experiment. At least 15 tetani were produced before the actual recordings were started. The signals from the force and displacement transducers were recorded and measured in a Nicolet digital oscilloscope.

The bathing solution had the following composition (mM): NaCl 115.5, KCl 2.0, $CaCl_2$ 1.8, NaH_2PO_4-Na_2HPO_4 2.0, pH 7.0.

RESULTS AND DISCUSSION

Double-Hyperbolic Force-Velocity Relation

The transition between the two curvatures of the force-velocity relation occurs at a load near 78 % of maximum tetanic force (P_o) at which point the the speed of shortening is approximately 1/10 of its maximum value, V_{max} (See further ref. 4). It was of interest to find out if the the high-force deviation of the force-velocity curve is related to the increase in force *per se* or to the decrease in speed of shortening. It could be argued that by going into the high-force range the kinetics of crossbridge function might change due to increased density of attached crossbridges. If this were the case, the high-force curvature would disappear after reducing the state of activation appropriately. Fig. 2 shows an experiment to test this point.

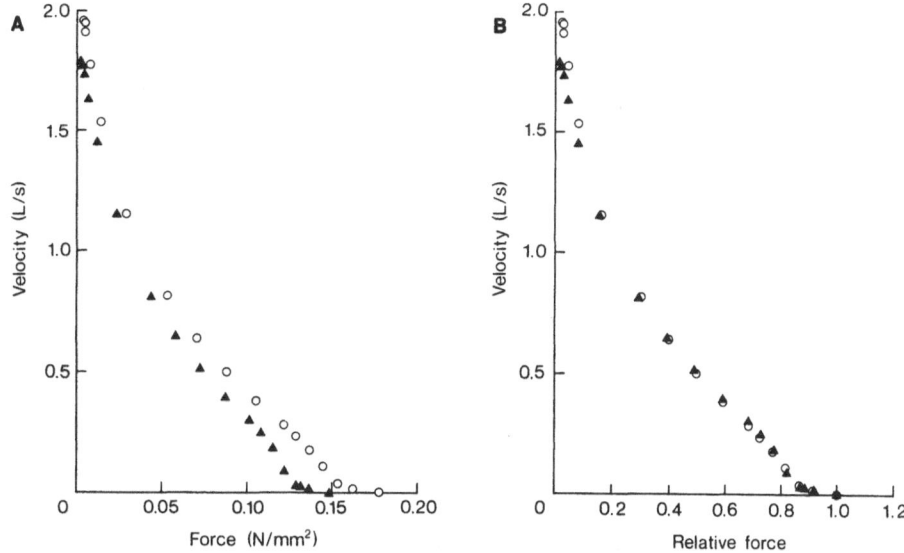

Fig. 2. Force-velocity relations recorded in a single muscle fibre at 2.2 μm sarcomere length in ordinary Ringer solution (open circles) and in the presence of 15 μM dantrolene (filled triangles). A. Force expressed in absolute values, N/mm^2. B. Data normalized to maximum isometric force (P_0) recorded in ordinary Ringer solution and dantrolene, respectively. Note that the biphasic shape of the force-velocity relation is maintained after dantrolene and that the high-force curvature covers the same range of relative loads as in the control. Unpublished observations by C. Caputo and K.A.P. Edman.

Illustrated are two sets of force-velocity data from the same muscle fibre. One series of measurements was performed in ordinary Ringer solution (open circles), the other in the presence of 15 μM dantrolene added to the external medium (filled triangles). Dantrolene, which is known to reduce the release of activator calcium from the sarcoplasmic reticulum [7-10], depressed maximum tetanic force to 80-85 % of the control value in the different experiments. The results clearly show that even after depression of the tetanic force the biphasic shape of the force-velocity curve was well maintained with the breakpoint of the curve appearing at the same relative values of P_0 and V_{max} as in the control (Fig. 2B). This finding suggests strongly that the high-force curvature is not related to the force level as such but rather to the speed of shortening of the contractile system. The breakpoint of the force-velocity curve thus occurs as a result of decreasing the speed of shortening to approximately 1/10 of V_{max}. Below this point the myofilament system produces less force and also shortens at a lower speed than expected from the main part of the force-velocity curve.

This behaviour of the contractile system could mean that there is a change in crossbridge function as the velocity of filament sliding is reduced below the critical level. It was therefore of interest to test if the change in force-velocity behaviour is associated with a corresponding change in fibre stiffness. To this end a series of

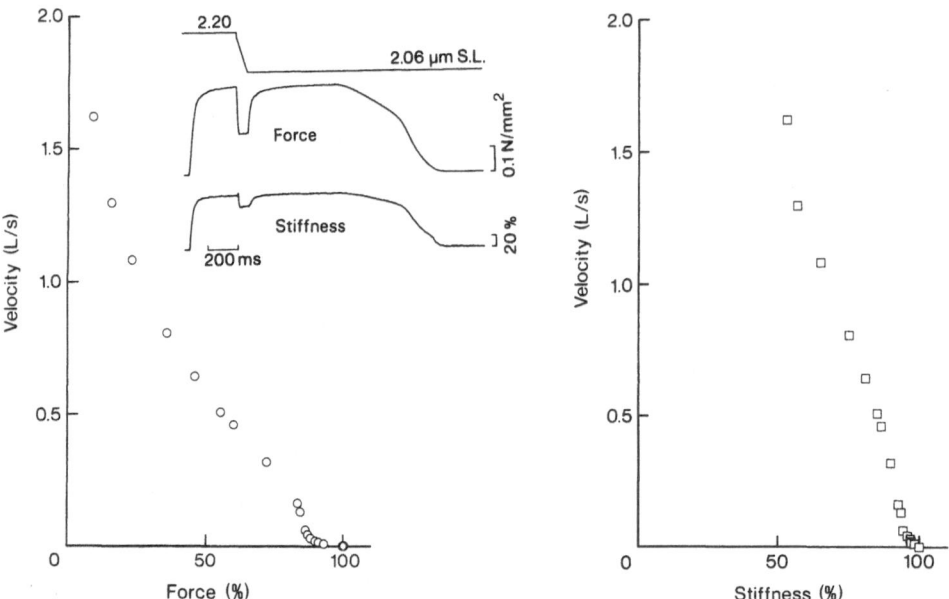

Fig. 3. Force-velocity (A) and stiffness-velocity (B) relationships determined in a single muscle fibre at 2.2 μm sarcomere length. Force and stiffness were determined simultaneously during shortening ramps of various speeds as described in text. Inset: Records of force and stiffness during shortening ramp induced during tetanus. Records from top: puller movement (downward, fibre shortening), force and stiffness.

experiments was performed in which force and stiffness were simultaneously recorded while the fibre shortened at different speeds during tetanus. Results from such an experiment are illustrated in Fig. 3, in which the force-velocity relation (A) and the corresponding stiffness-velocity relation (B) are shown together for comparison. The two relationships can be seen to differ substantially, but this difference becomes more informative when stiffness and force (measured at corresponding velocities) are plotted against one another as is shown in Fig. 4.

Stiffness can be seen to increase steadily as the force is raised to approximately 80 % of P_0, i.e. to the force level at which the breakpoint occurs in the force-velocity curve (Fig. 4). The stiffness-force relation below $0.8 P_0$ is slightly curved and can be fitted well by a rectangular hyperbola. Above $0.8 P_0$, however, there is a marked steepening of the stiffness-force relation. This would seem to make clear that as we go into the high-force range, and the speed of shortening is reduced below the critical level, the number of attached crossbridges increases more than predicted from measurements at low and intermediate loads. Nevertheless, as is seen in Fig. 3 A, this relative increase in crossbridge number is associated with *lower* force output and *lower* speed of shortening than expected from the main part of the force-velocity curve. These results would seem to make clear that some aspect of the crossbridge function is changed as the speed of filament sliding goes below the critical level. More bridges may accumulate in a state with low force production

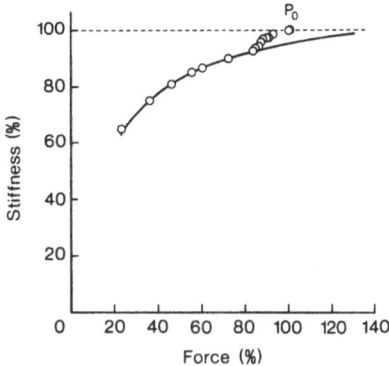

Fig. 4. Relation between force and stiffness during shortening replotted from data shown in Fig. 3. Note the biphasic shape of the force-stiffness relationship with a point of transition near 80 % of maximum force, P_0. Data below 0.8 P_0 fitted by a rectangular hyperbola (solid line) that is extrapolated to higher force values. Note that stiffness increases more steeply with force above 0.8 P_0 than predicted from the hyperbola.

under these conditions leading to a reduced driving force for the filament movement. Another possibility worth considering would be that a population of crossbridges with lower turnover rate and lower active force comes into function as the speed of filament sliding is reduced below the critical level, 1/10 of V_{max}.

Force-Velocity Relation at Negative Loads.

Using the slack-test method it was possible to demonstrate that maximum speed of shortening becomes progressively higher as a muscle fibre is prestretched to sarcomere lengths above approximately 2.7 μm[11]). The higher velocities attained under these conditions can be accounted for by the passive tension that exists at long sarcomere lengths. The passive tension will act as a longitudinal compressive force on the sarcomeres and will therefore increase the velocity above that observed at shorter lengths. By plotting the V_{max} data obtained at long sarcomere length against the compressive force (resting tension) recorded in each case, it was possible to estimate the force-velocity relation at loads lower than zero, i.e. at negative loads. The results so obtained indicated that the force-velocity relation does not undergo any significant change as the load goes from positive to negative values.

Experiments in apparent conflict with the intact-fibre results have recently been reported by Oiwa et al.[12]). The latter authors used an *in-vitro* assay system to study the speed of movement of myosin-coated polystyrene beads along isolated actin cables. The beads were subjected to external forces that counteracted or facilitated, respectively, the ATP-dependent movement along the actin cables. The results were interpreted to show that the speed at which thick and thin filaments are able to slide relative to one another during activity reaches a "ceiling" at zero load, i.e. the

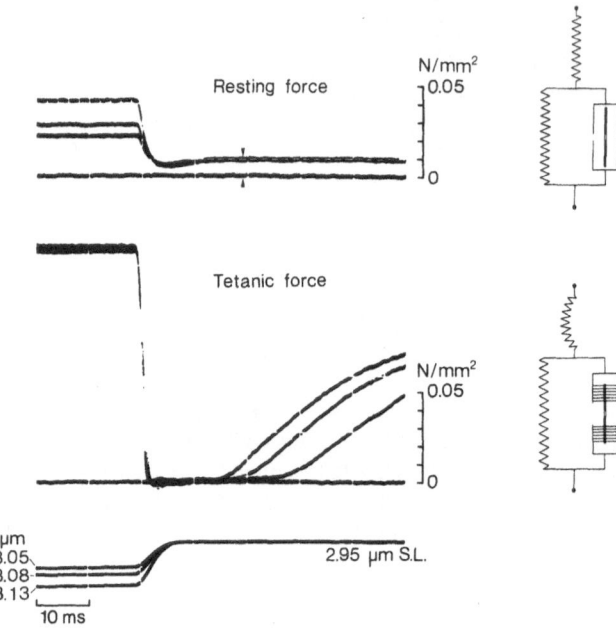

Fig. 5. Superimposed oscilloscope records of quick release performed at rest (upper panel) and during plateau of tetanus (middle panel) in a fibre that was prestretched to three different sarcomere lengths as indicated by bottom traces (puller signal). The release movement ended at 2.95 μm sarcomere length in all cases. Note that the force was reduced to the same finite value when release was performed at rest, whereas force dropped to zero when the same releases were made during activity. From Edman (1979)[11]. The schematic drawings to the right illustrate that as the fibre is released at rest (upper) there is not slackening of the fibre, the length change being distributed to both series and parallel elastic elements. When release is produced during activity (lower), on the other hand, the cross-bridges hamper the recoil of the parallel elastic element. The imposed length change therefore goes primarily into the series elastic element which slackens resulting in a drop in tension to zero. Tesion remains at zero level until the myofilament system has shortened sufficiently to take up the slack.

velocity of shortening at negative loads would not, according to the *in-vitro* assay[12], exceed V_{max}.

As information about the force-velocity relation at negative loads is essential to the understanding of the crossbridge mechanism, this question has been taken up anew in experiments on intact muscle fibres. The following two questions are addressed: 1. do crossbridges resist shortening when there is a longitudinal compressive force acting upon the sarcomeres? 2. Is there a smooth continuation of the force-velocity relation as the zero-load point is passed and the load goes negative?

The answer to the first question is given in Fig. 5. A single fibre is here prestretched to three different sarcomere lengths, 3.05, 3.08 and 3.13 μm, i.e. to lengths at which there is an increasing amount of resting tension. In each case the fibre is released to shorten to a given sarcomere length, 2.95 μm, in repeated runs.

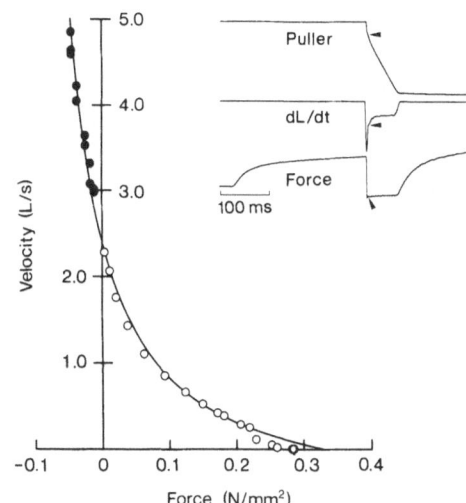

Fig. 6. Force-velocity relation in a single muscle fibre including measurements at both positive and negative loads. Filled circles: Velocities recorded at long sarcomere lengths plotted against the passive compressive force that acted upon the sarcomeres at these lengths (see text for further information). Open circles: load-clamp data derived at 2.10 μm sarcomere length. Hyperbola fitted to data points below 0.8 P_0 also including values at negative loads. Inset: Oscilloscope records of load-clamp recording during tetanus at long sarcomere length, 3.10 μm. Clamp level set close to zero load. Middle trace: first derivative of puller signal. Arrows indicate time at which velocity was measured. The difference between the force level at the onset of tetanus and the clamp level indicates the resting force.

When the release is made with the fibre at rest (upper panel), the tension drops to a finite value that corresponds to the resting tension at 2.95 μm sarcomere length. By contrast, when the same releases are performed during the tetanus plateau (middle panel), the fibre slackens and tension drops to *zero*. Tension thereafter remains at zero level until the fibre has shortened actively to take up the slack. The fact that tension is reduced to zero when the fibre is released during tetanus shows that the passive compressive force, shown in the upper panel, is absorbed by the contractile system during activity. This finding clearly demonstrates that the myosin crossbridges do have an ability to produce a braking force during muscle shortening.

Fig. 6 illustrates the force-velocity relation at negative loads. In this experiment the fibre was prestretched to sarcomere lengths (2.90-3.25 μm) where there was passive tension. During tetanus the fibre was released to shorten under *force-clamp conditions* at an external load that was set close to zero. During this shortening the parallel elastic elements that are responsible for the passive tension will exert a longitudinal compressive force on the sarcomeres, i.e. they will produce a negative load. The velocity of shortening was measured soon after the external force had dropped to zero and these values are plotted in Fig. 6 against the passive tension that existed at the point of measurement. In the same diagram are shown velocity data derived at positive loads in the same fibre near slack length. It was established in separate experiments, in which fibres were quickly frozen for electron microscopical examination (Edman, unpublished data), that shortening under a longitudinal compressive force did not involve buckling, or waviness, of the myofibrils or lead to disordering of the sarcomere pattern. The measured velocities at negative loads would thus seem to represent the speed of shortening at sarcomere level.

The results shown in Fig. 6 confirm (cf. ref. 11) that there is *no* discontinuity of the force-velocity relation as the load changes sign. Clearly the speed of shortening at zero load is not the upper limit of shortening velocity in active muscle as suggested by the *in-vitro* assay study of Oiwa et al.[12]. Thus, as is demonstrated in Fig. 6, the velocity of shortening can be raised far above V_{max} by applying a longitudinal compressive force on the sarcomeres. The present results support the view[1] that V_{max} represents the situation when bridges in pulling (force producing) position just match the bridges in braking position. When the load becomes negative and the speed of shortening exceeds V_{max}, a greater proportion of the attached crossbridges will end up in a braking position and counteract filament sliding.

REFERENCES

1. Huxley, A.F. *Prog. Biophys. Biophys. Chem.* **7**, 255-318 (1957).
2. Hill, A.V. *Proc. Roy. Soc. B.* **126**, 136-195 (1938).
3. Edman, K.A.P., Mulieri, L.A. & Scubon-Mulieri, B. *Acta Physiol. Scand.* **98**, 143-156 (1976).
4. Edman, K.A.P. *J. Physiol. (Lond.)* **404**, 301-321 (1988).
5. Edman, K.A.P. & Reggiani, C. *J. Physiol. (Lond.)* **351**, 169-198 (1984).
6. Edman, K.A.P. & Lou, F. *J. Physiol. (Lond.)* **424**, 133-149 (1990).
7. Putney Jr., J.W. & Bianchi, C.P. *J. Pharmacol. Exp. Ther.* **189**, 202-212 (1974).
8. Van Winkle, W.B. *Science, N.Y.* **193**, 1130-1131 (1976).
9. Desmedt, J.E. & Hainaut, K. *J. Physiol. (Lond.)* **265**, 565-585 (1977).
10. Morgan, K.G. & Bryant, S.H. *J. Pharmacol. Exp. Ther.* **201**, 138-147 (1977).
11. Edman, K.A.P. *J. Physiol. (Lond.)* **291**, 143-159 (1979).
12. Oiwa, K., Chaen, S., Kamitsubo, E., Shimmen, T. & Sugi, H. *Proc. Natl. Acad. Sci. USA* **87**, 7893-7897 (1990).

Discussion

Huxley: Did you also look at the region of negative velocities? I noticed that by the time you get down to the abscissa, the velocity curve has become almost horizontal, and if you project that to the right it will be very like what Bernard Katz saw in 1939 [*J. Physiol. (Lond.)* **96**, 45-64, 1939], which then appeared to be a discontinuity at zero speed. But it looks from this as though it might be continuous.

Edman: I actually published data on this particular point in 1988 in the *J. Physiol. (Lond.)* (**404**, 301-321, 1988). The curve is indeed continuous. I would like to take this opportunity to say that this is a mechanism we have to rely on extremely *in vivo*. For example, we would be badly off walking down the stairs or taking a jump if this flat region of the force-velocity relation did not exist, for it confers an enormous stability on the system. As I demonstrated in that paper, there is virtual isometry over a range of tensions between 90% P_0 and 120% P_0. The velocity varies there by merely 1.8% of V_{max}. This will give the individual segments in a fiber very good stability. The redistribution of sarcomere length between stronger and weaker segments will be moderate for this reason. The same mechanism will confer stability on opposing muscles, say, in the leg during body movements.

Kushmerick: I'm wondering, does the possibility exist of a slower myosin isoform in these frog fibers, which would then become manifest—possibly at lower velocities—and might then have higher populations of attached, high force-generating bridges?

Edman: Again, we actually published some data on that, also in 1988 [*J. Physiol. (Lond.)* **395**, 679-694, 1988] in collaboration with Reggiani, Schiaffino, and te Kronnie. There seem to be two different heavy chain isoforms in these frog muscle fibers, and they exist in different proportions along the length of the fiber. There is a correlation between the proportion of the two isoforms and the velocity of shortening in the different segments. However, I wouldn't think that the high-force curvature has any relation to the myosin isoform composition.

Lombardi: I agree with your comment on the stability of the force-velocity relation in the region of very slow negative velocities. I want to remind everyone that at the last meeting here in Hakone, Professor Colomo and I presented experiments on the variability of sarcomere length changes of different segments along the fiber during steady lengthening, as measured by using the striation follower. There are homogeneous fibers and fibers that develop inhomogeneity in sarcomere length in the different segments. Actually in all the fibers we found intrinsic stability in the region of lengthening velocity between isometric and say, 0.1 µm/s per half-sarcomere. This was measured directly on the different consecutive segments.

ter Keurs: I have a simple technical question regarding your Fig. 6. It seemed to me that velocity was a continuous function after the release, with a high peak which then subsides to the moment where you have drawn an arrow. Then it was shallower. How do you decide which value of velocity to plot?

Edman: I took the first possible opportunity to measure the velocity after tension had dropped to zero, because I'm interested here in correlating the velocity of shortening with the compressive force exerted by the parallel elastic elements. The magnitude of the compressive force was estimated from the passive tension at the sarcomere length considered. Of course, we are not dealing with steady-state conditions, so the velocities at negative forces are determined with less precision than are force-velocity data at positive loads. However, the measurements are precise enough to allow us to conclude that the speed of shortening continues to rise smoothly as the load changes sign and we move into the region of negative loads.

Huxley: You speak of this as a double hyperbolic but you gave that beautiful logarithmic plot with two straight lines. [see Fig. 2, *J. Physiol. (Lond.)* **404**, 301-321, 1988] Of course, the first force-velocity curve was published by Fenn and Marsh in about 1930. They fitted it with a straight line, but with an exponential decay coming down to the straight line. I wonder whether it would be better, in fact, to think of log velocity rather than putting it on a hyperbolic?

Edman: Well, I find it easier to work with the hyperbola. But you could equally well use an exponential function. As I tried to point out in my paper in 1988 (*J. Physiol. (Lond.)* **404**, 301-321, 1988), neither of these functions give a perfect fit. Of course, the logarithmic plotting certainly emphasizes the break-point very nicely.

SHORTENING-INDUCED TENSION ENHANCEMENT: IMPLICATION FOR LENGTH-TENSION RELATIONS

Gerald H. Pollack, Arie Horowitz, Manfred Wussling and Károly Trombitás

Bioengineering WD-12
University of Washington
Seattle WA 98195

ABSTRACT

Length-tension relations come in two types—the classical type with linearly descending limb, and the "flat," higher type. The classical type, now confirmed in several laboratories, is obtained when sarcomeres are servo-controlled to maintain constant length. The flat type, also confirmed in several laboratories, is obtained in fixed-end contractions, where some sarcomeres have the opportunity to shorten at least slightly. We find that the higher tensions seen in the flat type are indeed the result of very small shortening: when isometric sarcomeres shorten to a slightly shorter (*e.g.*, by 40 nm) length, they go on to produce considerably more isometric tension than if they had remained at the shorter length throughout contraction. We term this phenomenon "shortening-induced tension enhancement." The phenomenon accounts not only for the higher, flatter length-tension relation seen in fixed-end contractions but can explain the creep of tension that occurs in extended tetani. Thus, several issues concerning the interpretation of length-tension relations are resolved by this newly discovered phenomenon.

INTRODUCTION

The characteristic shape of the length-tension curve is generally taken as strong evidence in favor of the cross-bridge theory. In recent years, however, it has become clear that the classical shape is not unique: many authors have reported a rather "flat" curve that descends only at highly extended sarcomere lengths[1]. The flat curves, along with the classical linearly descending curves, are shown in Figure 1, which summarizes the results of virtually all systematic length-tension studies published up to the past year or two.

Why do the length-tension curves have two different shapes? Why is one flatter and higher than the other? Are these diverse results consistent with predictions of the cross-bridge theory? Or, is there another model that better explains the observations?

Mechanism of Myofilament Sliding in Muscle Contraction, Edited by
H. Sugi and G.H Pollack, Plenum Press, New York, 1993

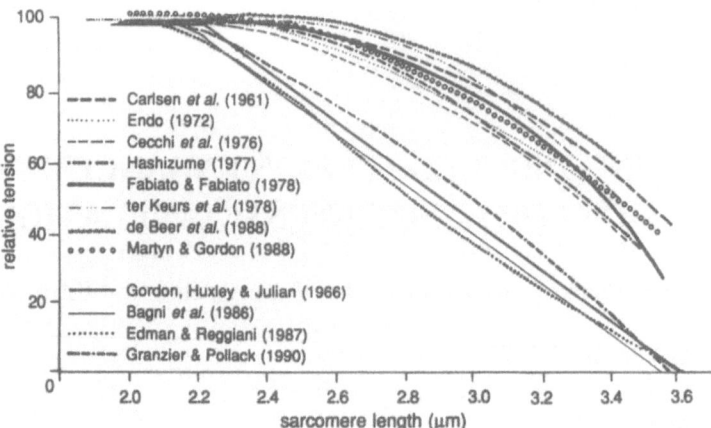

Fig. 1. Summary of length-tension relations obtained with local sarcomere length held constant using feedback control (bottom), and with the ends of the specimen fixed (top).

In this paper (see also ref. 11) we attempt to answer these questions. The paper highlights a phenomenon that has only just recently come to light, and that accounts for the discrepant length-tension relations—"shortening-induced-tension enhancement." Several features of the enhancement phenomenon do not fit the cross-bridge theory. A recent model[2], however, explicitly predicts the phenomenon, and goes on to explain how the phenomenon is the key element underlying the discrepant length-tension relations.

Predictions of the Cross-bridge Theory

The cross-bridge theory predicts a diminution of tension with decreasing filament overlap. The classical studies of Gordon, Huxley and Julian[3] appeared to have confirmed this prediction. More recently, the intercept has been found to differ in different muscles. In frog tibialis anterior muscle fibers, the intercept was found to lie at 3.53 μm[4], a shift to the left relative to semitendinosus muscle of more than 0.1 μm. In another study in which the length-tension relation was measured during shortening, the differences were confirmed: in semitendinosus, the intercept was at 3.65 μm, while in tibialis anterior it was at 3.53 μm[5].

Although the authors of these latter papers assumed that the shift could be explained by a putative difference of thin filament length, we found that there was no difference[6]: in tibialis anterior and semitendinosus muscle the I-segment length was 1.95 to 1.98 μm, depending on the method. Nor was there a difference in thick filament lengths. In tibialis anterior, thick filament length was 1.61 μm, while in semitendinosus it was 1.62 μm. The lateral shift of the length-tension relation is not therefore explainable in terms of differences of filament length. According to the cross-bridge theory, filament length alone is the determinant of the curve's shape.

An issue more serious, perhaps, than this shift, is the existence of the group of flat curves (Figure 1). Here the tension differs greatly from the theory's prediction.

At half-maximal overlap, approximately 2.8 μm, the theory predicts 50% of maximal tension, whereas many of the measurements show 80 - 90%.

These deviations from classical theory are often attributed to inhomogeneity. As tension rises, if some sarcomeres shorten (while others in series are stretched), the shortening sarcomeres "climb up the descending limb," and account for the "excess" tension that is observed. However, the large theorized shortenings—which can and do occur in some preparations[7]—are not seen consistently[8]. Even in diminutive preparations, where the absence of gross shortening is confirmed by direct observation[9][10] the "excess" tension is still observed. The flat curves cannot therefore be attributed to inhomogeneity. They remain to be explained by classical theories.

Predictions of the Filament-Shortening Theory

In this theory[2], contraction is brought about by filament shortening. All filaments have the capacity to shorten—thick, thin and connecting filaments. We do not detail the evidence in support of the theory in this paper (nor is it implied that all authors of this paper necessarily agree with all aspects of the theory). The reader is instead referred to the book, where this evidence is presented *in extenso*. Although filament shortening is not widely acknowledged, the evidence in its favor is surprisingly abundant.

One key feature of the theory is that functionally speaking, there are two types of sarcomere. A "generator" is a sarcomere that is actively generating tension. A "sustainer" is a sarcomere that is merely sustaining the tension generated by generators. Thus, sarcomeres that are stretched are sustainers. Sarcomeres that shorten are generators: by shortening a small amount (20 nm per half sarcomere), these sarcomeres' filaments are themselves able to shorten, to undergo a structural transition, and to generate tension.

The theory predicts that the length-tension curve of sustainers is different from that of generators. Sustainers merely support tension: if the tension on an isometric sustainer is raised only slightly, the sarcomere will stretch. Thus, isometric tension is a measure of holding capacity. This should vary directly with filament overlap. In other words, the length-tension curve for sustainers is the classical one—the linearly descending limb.

Generators, on the other hand, have a different length-tension curve. The main site of tension generation is argued to lie near the mid-portion of the thick filament: it is at this region that the thick filament begins shortening. Thin filaments are not required to mediate this process. Thus, whether there is full overlap or only moderate overlap, the tension-development process is the same, so the length-tension curve will be flat. At highly extended lengths the number of links between thick and thin filaments drops sufficiently that the generated tension is no longer sustainable, and the curve drops.

The filament shortening theory, then, explains the two types of length-tension curves of Figure 1, provided the classical curve characterizes sustainers, the flat curve generators. Further, the theory explicitly predicts (*cf.* ref. 2, p. 258) that sustaining sarcomeres that shorten by a small amount will undergo transition from sustainer to generator; that in so doing, the sarcomere can move from the lower curve to the upper.

Can Length-Tension Relations Be "Typed?"

What is striking about Figure 1 is that the curves fall rather distinctly into two types, linearly descending and flat. The linearly descending type is measured by those who employ sarcomere-length clamps. The flat type is measured by those who use fixed-end contractions. In the sarcomere-length clamp experiments, sarcomeres are not permitted to shorten, even by 20 nm per half sarcomere; these sarcomeres are therefore sustainers. Their length-tension curve descends linearly, as predicted. In the fixed-end contractions, on the other hand, it is inevitable that some sarcomeres along the fiber shorten at least slightly. These sarcomeres are generators. They generate the tension that is sustained by other sarcomeres in series (sustainers) and that is ultimately transmitted to the tension transducer. Their length-tension relation is predicted to be flat—and indeed it is.

Thus, the first prediction is validated. Sustainers—sarcomeres that do not shorten—are confirmed to have length-tension relation that is of the classical shape. Generators are confirmed to have a length-tension relation that is flat, at least until extended sarcomere lengths where tension ultimately falls off.

The second prediction, one that we will spend some time on, is that by shortening slightly, a sarcomere can be converted from sustainer to generator.

Motivated by the prediction that a small amount of shortening should give rise to a substantial increase of tension, we first combed the literature to see whether the relevant experiment had been carried out. The experiment requires that sarcomeres be clamped to precisely constant length during the period of tension rise to plateau, then allowed to shorten to a slightly shorter length, and kept there for the remainder of the tetanus. Surprisingly, this experiment had apparently not been carried out. Although shortening had been imposed on fibers, the absence of any sarcomere-length control vitiated the sarcomere-length constancy required prior to the imposed shortening. We thus initiated a series of experiments.

The experiments were carried out on single fibers of frog semitendinosus muscle. Sarcomere-length control was achieved by feeding back either an optical diffraction signal or a segment-length signal, the latter obtained by measuring the spacing between surface markers. Details of methods are given elsewhere[11].

A representative result is shown in Figure 2. Sarcomere length is initially clamped to a constant value as the fiber is stimulated and tension begins to rise. After tension reaches a plateau, sarcomeres are constrained to shorten in ramp-like fashion, and are subsequently held at a slightly shorter length. During this period tension undergoes a transient drop from its plateau, and then recovers. The recovery level is substantially higher than the pre-release level.

Figure 3 shows a summary of results of experiments carried out over the physiological range of sarcomere lengths. The lower curve shows the tensions measured during the period prior to release. In essence, these results should

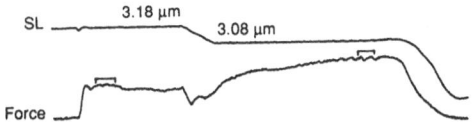

Fig. 2. Effect of shortening on isometric force. Sarcomere-length is clamped at 3.18 μm while tension rises to a plateau. Sarcomeres are then programmed to shorten to 3.08 μm. Tension following shortening is substantially higher than that prior to shortening.

resemble those obtained in other length-clamp experiments, and indeed they do. The linear descending limb is clearly evident. The upper curve represents the tensions measured subsequent to release. At shorter sarcomere lengths these tensions are no different from clamped tensions. At longer sarcomere lengths post-shortening tensions are substantially higher.

Potential artifacts were considered in detail. In particular, we were interested in whether sarcomere-length inhomogeneity might have played a role in generating the enhancement. Dispersion of sarcomere length within the sampled region was measured by two independent methods—striation imaging, and analysis of the intensity profile of the first diffraction order. With both methods, standard deviation of mean post-shortening sarcomere length was ± 0.03 μm, which was too small to account for the size of the force increase.

Experiments were carried out to determine the extent of release required to elicit the force increase. Force enhancement increased sharply with extent of shortening over the first 20 nm per half-sarcomere. Progressively larger releases had a more modest effect: releases of approximately five times this magnitude

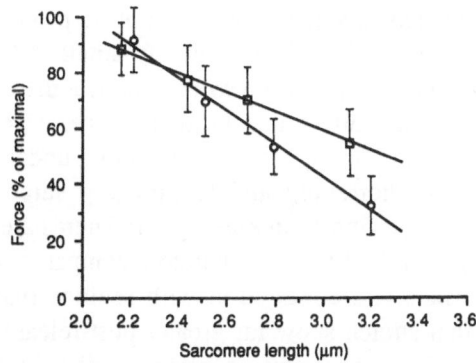

Fig. 3. Relation between pre-release (open circles) and post-release (open squares) sarcomere-length and force. Force is normalized relative to fiber tetanic force level at slack length, the latter given a value of 100%.

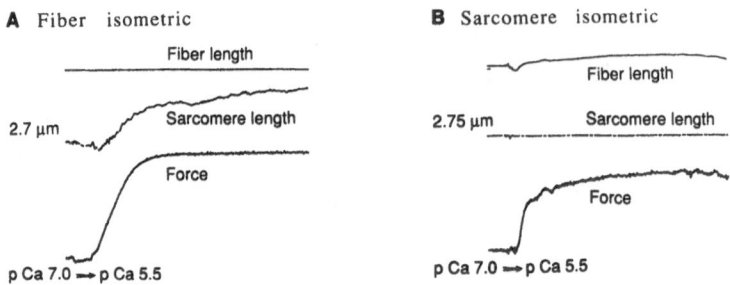

Fig. 4. Chemically skinned rabbit psoas fibers. Fiber-isometric (i.e., fixed-end) contraction produces more force than sarcomere isometric.

resulted in one-fifth the increase obtained with the first 20 nm per half-sarcomere of release. Thus, the effect occurs mainly over the first 20 nm per half sarcomere of release.

It appears, therefore, that the expected increase of force is realized. The prediction that a small amount of sarcomere shortening converts a sustainer to a generator, and therefore results in increased force, is confirmed.

A correlative prediction involves comparison of central and end regions of the fiber. The central region is ordinarily stretched during fixed-end contraction, a dynamic that is prevented by imposition of the sarcomere-length clamp. These central sarcomeres are sustainers. As shown above, small shortening converts these sustainers to generators.

End sarcomeres are different. During fixed-end contractions, these sarcomeres ordinarily shorten; they are therefore generators. As such, imposed shortening is not expected to bring about any further enhancement. As a test of this hypothesis we repeated the above protocols on end sarcomeres. No force increase was found.

Some of these experiments were carried out on skinned fibers, with mixed results[12]. Small releases were imposed on chemically skinned rabbit psoas muscle fibers. Figure 4 shows that, as in the intact fibers, fixed-end contractions produced more tension than those produced during sarcomere-isometric contraction. However, the enhancement phenomenon was not as pronounced. At 2.8 μm, a sarcomere release of 80 nm produced relatively small (13%) increase of force, barely above the amount (10%) predicted by the classical theory.

On the other hand, the difference between generators and sustainers was unmistakenly evident. In the skinned fibers contracting under fixed-end conditions, the distinction between shortening and lengthening sarcomeres is not tied to architecture, as it is in intact fibers. Shortening sarcomeres are apparently randomly located along the length of the fiber. We therefore compared responses of segments that shorten during fixed-end contraction with regions that stretch. We found, indeed that regions that stretch show far greater post-release tension than regions that shorten ($p < 0.01$). The results again confirm the distinction between generators and sustainers. Further, because such regions are randomly distributed along the fiber rather than confined to a particular region (*e.g.* ends vs. middle), the results show that the distinction is functional, not anatomical.

Interpretation of Length-Tension Relations

Armed with the shortening-induced tension enhancement phenomenon, we now return to the length-tension relation(s) and attempt to interpret them. The distinction between sustainers and generators is central.

In the sarcomere-length clamp experiment (generally carried out in the fiber's central region), the sarcomeres under study are sustainers. The clamp ensures that these sarcomeres do not shorten. The length-tension relation is that of the sustainer. It is a measure of the holding capacity of the actomyosin bridge, and as such, should vary linearly with overlap—as it does.

In the fixed-end contraction, by contrast, some sarcomeres along the preparation inevitably shorten. We found that almost maximum tension can be produced by as little as 20 nm per half-sarcomere of shortening. Thus, shortening of any sarcomere group by 0.02 μm should be sufficient to bring about maximal tension, the remaining sarcomeres in series sustaining the tension as they stretch or remain isometric. Thus, one anticipates a length-tension relation that is higher than that measured under sarcomere-isometric conditions. Since the enhancement phenomenon is virtually absent at 2.0 - 2.5 μm, and increases progressively with increasing sarcomere length, we anticipate the difference between fixed-end and sarcomere isometric length-tension curves to increase with increasing sarcomere length—as indeed it does (Figure 1).

This interpretation provides an alternative to the usual explanation of the discrepancy between the two sets of curves. The upper curves are generally thought to arise out of gross inhomogeneity, even though, as mentioned, the results of several studies contradict this interpretation. The present interpretation, in a sense also invokes inhomogeneity, but of an altogether different magnitude. To obtain the larger tension, all that is required is for some sarcomeres to shorten by at least 40 nm. Additional shortening may take place, and it will elevate the curve slightly, but the major element underlying the higher curve is minute shortening.

Creep

Closely related to this phenomenon is the phenomenon of creep. At long sarcomere lengths tension rises rapidly at first, and then "creeps" to a progressively higher value[3], until it reaches a plateau[8]. The usual explanation of creep is that it arises out of sarcomeres shortening and "climbing up the descending limb" of the length-tension relation to a progressively higher tension value.

While deceptively simple, this explanation has one serious fallacy. The data points along the length-tension curve are obtained in separate contractions; the curve is constructed by connecting these points. *A priori*, there is no assurance that such a "static" curve can be used to describe dynamic events. In fact, it is well known that when fibers are stretched slowly to longer lengths, their tension does not diminish, as the curve implies; it increases[13]. Likewise, when fibers shorten, there is no guarantee that their final tension will increase in accordance with the respective point along the length-tension curve. The length-tension curve is purely static. It takes no account of dynamic history.

Furthermore, the large degree of inhomogeneity demanded by the inhomogeneity hypothesis is sometimes not realized[8].

The current findings offer an alternative interpretation of creep. Creep occurs as generators are created out of sustainers. Consider, for example, a tetanus at long length under sarcomere-length control. If sarcomere length is extremely uniform, and the segment is short, sarcomeres may remain in almost perfect order. With no sarcomere shortening, no enhancement is predicted, and creep should be absent. If the sarcomeres are not perfectly homogeneous, however, some tension enhancement (creep) will occur as the tetanus proceeds. The important feature of this mechanism is that only minute amounts of inhomogeneity are required, not the gross inhomogeneity required of the "climbing up the descending limb" mechanism. Note that the presence / absence of creep should be rather unpredictable, depending on the degree of uniformity of the segment. This feature is confirmed[14].

Instability

A. V. Hill[15] was apparently the first to argue that the length-tension curve's descending limb implies sarcomeric instability: as sarcomeres shorten they grow stronger, pulling out the initially weaker ones, which get progressively weaker. The predicted result is ultimately disastrous, with sarcomeres eventually winding up very long or very short.

The instability argument suffers from the same fallacy as the creep argument: the length-tension curve describes static behavior; it does not necessarily reflect dynamic behavior during contraction. Thus, a negative slope does not necessarily reflect instability.

On the other hand, an instability with predictable consequence is implied by the enhancement phenomenon. Stability is assured if shortening diminishes tension while stretch increases tension (like a spring). Stretch is well known to increase tension in activated sarcomeres. And while shortening generally diminishes active tension, the special case in which sustainers shorten slightly to become generators results in a tension increase. This process is, in a sense, unstable. However, the range of instability is small: it takes place over shortening of less than 0.1 μm. Once the sarcomere becomes a generator, its behavior becomes stable: stretch increases tension, while shortening decreases tension.

The presence of local instabilities can lead to oscillations. To see why, consider a sarcomere group such as the one in Figure 2. Suppose during the course of the tetanus this sarcomere shortens as in the figure. Tension will decrease, then increase. Because the final tension is higher than that prior to the release, this sarcomere is now stronger, and able to shorten once again, so the cycle can repeat (though the effect was found to be progressively reduced in the second and subsequent releases). Thus, tension cycling can result from shortening. With different sarcomere groups along the fiber shortening at different times and speeds, one may readily anticipate tension oscillations of more-or-less unpredictable nature.

Such tension oscillations are present in some of the records we have obtained, particularly at the longer sarcomere lengths, where the enhancement effect is more pronounced. Oscillations of this sort are rather common among experimenters, and generally thought to arise out of quirks in the feedback system. However, the argument above implies that the oscillations may be an intrinsic feature of the process, and one that does not require gross inhomogeneity.

Conclusions

Small shortening causes a large increase of tension. This surprising phenomenon has implications not only for mechanics, but for the mechanism of contraction as well.

In terms of the cross-bridge theory, the phenomenon poses some difficulty. According to the theory, tension should be related directly to the degree of overlap. However, small shortening to a given sarcomere length results in more isometric tension than if the sarcomere had remained at the shorter length throughout the tetanus. Thus, overlap is not the sole determinant of tension.

On the other hand, the phenomenon is predicted explicitly by a recently proposed theory of contraction[2]. The tension increase results because the filaments themselves can shorten and generate active tension. The theory also predicts the finding that the phenomenon is absent in regions that shorten during fixed-end contraction, as sarcomeres in those regions are already generators and do not require shortening to achieve that status.

In terms of the length-tension relation, the phenomenon explains why there are two curves, the sarcomere-isometric curve and the fixed-end curve. The former characterizes sustainers; the latter characterizes generators. To move from the former to the latter, all that is necessary is a small amount of shortening.

REFERENCES

1. Pollack, G.H. *Physiol. Rev.* **63**, 1049-1113 (1983).
2. Pollack, G.H. *Muscles and Molecules: Uncovering the Principles of Biological Motion* (Ebner & Sons Publishers, Seattle, 1990.)
3. Gordon, A.M., Huxley, A.F. & Julian, F.J. *J. Physiol. (Lond.)* **184**, 143-169 (1966).
4. Bagni, M.A., Cecchi, G., Colomo, F. & Tesi, C. *J. Physiol. (Lond.)* **401**, 581-595 (1988).
5. Claflin, D.R., Morgan, D.L. & Julian, F.J. *Biophys. J.* **59**, 48a (1991).
6. Trombitás, K., Frey, L. & Pollack, G.H. (submitted).
7. Huxley, A.F. & Peachey, L.D. *J. Physiol. (Lond.)* **156**, 150-165 (1961).
8. ter Keurs, H.E.D.J., Iwazumi, T. & Pollack, G.H. *J. Gen. Physiol.* **72**, 565-592 (1978).
9. Sugi, H. Ohta, T. & Tameyasu, T. *Experientia* **39**, 147-148 (1983).
10. Fabiato, A. & Fabiato, F. *Eur. J. Cardiol.* **4** (*Suppl.*), 13-27 (1976).

11. Horowitz, A., Wussling, M.H.P. & Pollack, G.H. *Biophys, J.*, **63**, 3-17 (1992).
12. Wussling, M.H.P., Jonas, M., Horowitz, A. & Pollack, G.H. (submitted 1991).
13. Edman, K A.P., Elzinga, G. & Noble, M I.M. *J. Gen. Physiol.* **80**, 769-784 (1982).
14. Edman, K.A.P. & Reggiani, C. *J. Physiol. (Lond.)* **385**, 709-732 (1987).
15. Hill, A.V. *Proc R. Soc. B.* **141**, 104-117 (1953).

Discussion

Huxley: The suggestion that the extra tension with fixed ends is due to small amounts of shortening in other places was disproved before any of those experiments you describe, in the paper by Peachey and myself in 1961 [*J. Physiol. (Lond.)* **156**, 150-165, 1961], which I believe you were unaware of when you started these experiments. We had ciné photographs of the middle and the end of an isolated fiber and fixed-end-contractions, and the fiber ends shortened tremendously. The sarcomere length went down from 3.5 μm or so, right on the borderline of any contraction, down to something like 2 μm. In my mind, the rise of tension produced by stretching by an equivalent amount was closely equivalent to the tension produced in that situation, where isometrically, the fiber gave nothing. I think there is no doubt whatever that it's that extreme shortening that created the rise of tension. I think I'm right in saying that you have told me on previous occasions that you don't get this phenomenon if you operate near the peak in the plateau of the full length-tension curve. That is to say it's only in the region where it has negative slope and is grossly unstable—as pointed out particularly by Paul Edman in this symposium—that this phenomenon occurs. You would have to search the whole of the region that you have clamped in order to be sure that no region underwent strong shortening.

Pollack: I have certainly been aware of the experiments by yourself and Peachey. In those experiments it's possible to draw a correlation between extra tension and shortening of the ends. There are, however, nine or ten other published length-tension experiments (in Pollack, G.H. *Muscles and Molecules: Uncovering the Principles of Biological Motion.* Seattle, Ebner & Sons, 1990) that give similar results but didn't contain fiber ends adjacent to the tendon (*e.g.* skinned fiber segments). So, while shortening of the ends may have occurred in your experiments with Peachey, it's not the general explanation for the high tensions at long sarcomere length. We used every method that we could think of to look for sarcomere length inhomogeneity, and all the results of the experiments that we've done have proved negative.

Edman: May I add that Reggiani and I have published data on the redistribution of sarcomere length during isometric contraction [*J. Physiol. (Lond.)* **351**, 169-198, 1984]. We demonstrated that the contractile behavior is non-uniform along the entire fiber, so even if the ends are excluded, tension creep will occur due to redistribution of sarcomere length. I would also like to mention my recent experiment with Caputo [*J. Physiol. (Lond.)* **438**, 148P, 1991], where I had loaded shortening from a long sarcomere length down close to slack length, and compared the force with an isometric tetanus at the short length. If you do this under segmental length-clamp conditions, you have no shortening-induced force deficit.

Pollack: The experiments that you're talking about have to do with large shortenings. We have also done experiments with sizable shortening. Here I'm discussing a very specific protocol to keep the sarcomere length exactly constant during the rise of tension and only then impose a small release. If you don't take those precautions, what happens is that before the release, the sarcomeres begin to be inhomogenous, and the ones that shorten already give you the effect. So I don't think that there is any disagreement between our experimental results since the experiments were different.

Gordon: Do you see this extra force in the skinned fibers when you activate under length control?

Pollack: The amount of extra force that we get with skinned fibers is extremely small. One interpretation is that the extra tension doesn't exist in the skinned fibers. The effect in intact fibers would then be attributed to some aspect of activation. Another possibility is that you already have the effect before you impose the shortening. If you get some sarcomere shortening before you impose the shortening, then you already have this extra tension. If you then impose the shortening ramp, you won't get the effect. I think that is what's happening in the skinned fibers, because if you look at the diffraction pattern before the release, the width of the first order shows some increase. This means you are already getting some shortening and lengthening, and you therefore don't expect much additional effect. That could be the explanation.

Taylor: Dr. Sys showed in his poster that there is quite a bit of difference depending on what region of the muscle fiber you choose. I was just curious: how do you decide that the sarcomere length you are keeping constant is really representative of the sample? Do you try to look at several different positions on the fiber?

Pollack: That's a good question. The experiments I've shown you relate only to the central region of the fiber. That is, the region that ordinarily under fixed-end conditions is stretched. In such regions we get the increase of tension on a regular basis. On the other hand, if we look at the end sarcomeres in the fibers, the ones that ordinarily shorten during fixed-end contraction, we don't get the effect. So the effect is specific to sarcomeres that undergo stretch during fixed-end contraction. This distinction is predicted in the book that I recently wrote (Pollack, G.H. *Muscles and Molecules: Uncovering the Principles of Biological Motion*. Seattle, Ebner & Sons, 1990).

ter Keurs: Jerry, have you done an experiment in which the tetanus is followed by rapid freezing and then subjected to electron microscopical procedures? Have you looked at the sarcomere-length distribution across the region in which you were interested and at the alignment of the A-bands in the sarcomeres?

Pollack: No, we haven't. As the depth of quality freezing is limited to 10 or 20 μm at most, you could certainly sample only superficial sarcomeres. It's for that reason that we haven't done those particular experiments.

Huxley: When you clamp by light diffraction, how long is the piece of fiber that you clamp?

Pollack: It's about 150-200 μm—on that order.

Huxley: Well, as a way of checking homogeneity I would suggest looking at that area, perhaps with a horizontal microscope set up at the same time using polarized light at high aperture illumination so that you get a good optical section. You ought to be able to photograph it so that the whole of the length is in focus and you can examine it afterwards. This would need to be done at several planes of focus through the fiber, but this would be a way of examining the whole of the area within which you're keeping average sarcomere length constant.

Pollack: Thank you.

KINETICS OF REGENERATION OF CROSS-BRIDGE POWER STROKE IN SHORTENING MUSCLE

Gabriella Piazzesi, Marco Linari and Vincenzo Lombardi

Dipartimento di Scienze Fisiologiche
Università degli Studi di Firenze
Viale G.B. Morgagni 63, I-50134 Firenze, Italy

ABSTRACT

The force developed by a muscle during steady shortening is due to cyclic interactions between the cross-bridges extending from the thick myosin filament to the thin actin filament. Each interaction consists of a power stroke of the myosin molecule that accounts for a limited amount of sliding between the two sets of filaments (about 12 nm according to quick release experiments[1][2]), and is widely believed to be coupled to the hydrolysis of one ATP molecule[3][4]. On the other hand both energetics studies in muscle[5] and *in vitro* motility assays[6], indicating that shortening per ATP split is much larger than 12 nm, postulate that during shortening cross-bridges interact at a rate much faster than the ATP splitting rate. In the experiments reported here, made on intact fibres from frog skeletal muscle, the rate of regeneration of the power stroke was determined. Tension transients were elicited by imposing test step releases at different times (2-20 ms) after a conditioning release of about 5 nm. When the test step was imposed at 2 ms after the conditioning step, the tension attained at the end of the quick phase of recovery (T_2, due to the force generating stroke of the attached cross-bridges) was depressed and the T_2 curve (the plot of T_2 tension versus size of the test step) intercepted the length axis to the right, with respect to the intercept of the control T_2 curve, by an amount similar to the size of the conditioning step. By increasing the interval between conditioning and test step the T_2 tension increased progressively and the T_2 curve intercept approached the intercept of the control curve with a time constant of 6-7 ms. These results indicate that the force generating stroke elicited by a shortening step is followed by a relatively rapid process of detachment and reattachment by most of the cross-bridges, allowing for the generation of another power stroke. The rate of this process, 150/s, is one order of magnitude higher than that expected from the ATPase rate, suggesting that several actomyosin interactions occur in shortening muscle by the time one ATP is split. The results are simulated with a mechanical kinetic model of contraction[7], in which, for a critical amount of shortening, cross-bridges can detach, rapidly reattach and generate force before the completion of the "normal" isometric cycle.

Mechanism of Myofilament Sliding in Muscle Contraction, Edited by
H. Sugi and G.H Pollack, Plenum Press, New York, 1993

691

INTRODUCTION

Synchronization of cross-bridge action by means of step perturbations in length allows the mechanical characteristics of the cross-bridge cycle to be determined in intact muscle fibre preparation. Such experiments[1)2)] indicate that the length of the power stroke (the sliding distance between the two sets of filaments over which a cross-bridge remains attached to actin, while producing force and shortening) is about 12 nm. The energy that appears as work done by the cross-bridge during each cycle of attachment, generation of force and/or shortening and detachment is believed to be provided by the coupled hydrolysis of one molecule of ATP[3)4)].

On the other hand, energetics studies in both frog muscle[5)] and skinned fibres[8)], or *in vitro* motility assays[6)] indicate that the interacting distance of a myosin head per ATP split during rapid shortening is much larger (five times at least) than the cross-bridge throw as determined in Huxley & Simmons experiments. The discrepancy can be resolved if several power strokes per ATP split can be generated in shortening muscle, at a rate much faster than previously thought[9)10)]. The experiments reported here were aimed at determining the rate at which a cross-bridge can generate the power stroke. We found that the quick recovery of force after a shortening step (the energy releasing stroke of the attached cross-bridges) is followed by a relatively rapid process of detachment and reattachment by most of the cross-bridges, allowing for the generation of another power stroke within 10-15 ms.

METHODS

Single fibres, dissected from the lateral head of the tibialis anterior muscle of the frog (*Rana esculenta*), were mounted on the experimental trough between a loudspeaker motor[11)12)] and a capacitance force transducer with natural frequency 50 kHz[13)]. The fibres were stimulated at the sarcomere length of about 2.1 µm with pulse trains of current of suitable amplitude and frequency to produce fused tetani of 0.5-1 s duration. Temperature in the trough was kept at about 4 °C by means of a system consisting of a thermoelectric module stuck to the bottom of the experimental chamber, a thermistor probe and a suitable control circuit. The actual length change of a short segment selected along the fibre close to the force transducer end was controlled by means of a striation follower[14)], an optoelectronic device that detects the longitudinal translation of the sarcomeres at two points along the fibre bounding the selected segment. The loudspeaker motor was servo-controlled using as feedback signal either the output of the position sensor of the loudspeaker lever (fixed-end mode), or the output of the striation follower (length-clamp mode)[12)15)]. Fibre stimulation started in fixed-end mode and the control was shifted to length-clamp mode at a preset time during the rise of tetanic tension. When tension had attained the plateau value, single or double step length changes of different size were imposed on the selected segment. Segment releases of 5 nm per half-sarcomere (h.s.) or less

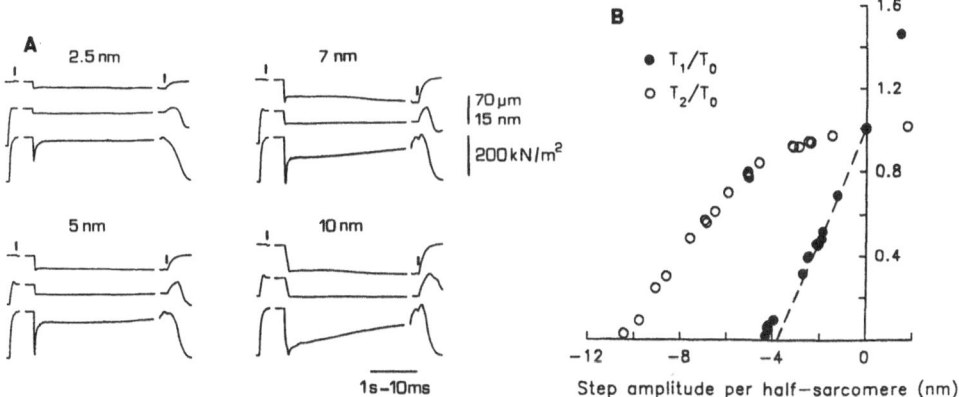

Fig. 1 *A*. Tension responses to step releases of different sizes applied at the plateau of an isometric tetanus. In each frame, from top to bottom traces are: motor position, sarcomere length change, tension response. Small bars mark the time of switch from fixed-end to length-clamp mode. Interruptions on the traces mark the time the trace speed increases one hundred times, in order to provide the appropriate resolution for the tension transient. The figure above each frame indicates the size of the step change of segment length (nm per h.s.). The step was complete within 120 μs for releases < 5 nm. It was made slower for larger releases, in order to avoid the segment length clamp failing caused by the fibre remaining slack. Fibre length 4.49 mm; segment length 1.6 mm; sarcomere length 2.07 μm; temperature 4 °C. *B*. T_1 and T_2 curves determined from the same experiment as in *A*. T_1 and T_2 are relative to the isometric tetanic tension (T_0). The intercept of T_1 curve on the length axis (Y_0) is estimated by extrapolating the line drawn by eye through the T_1 points obtained with small steps (dashed line), for which the relation is approximately linear[2)7)]. T_2 tension was estimated by extrapolating the tangent fitted to the later, slower recovery back to the time of the step[2)] (see also Fig. 6 *A* of Ref. 7).

were complete in about 110 μs; for larger amplitudes the release was suitably slowed to reduce the possibility of failure of the segment length-clamp, caused by the fibre remaining slack.

RESULTS

Fig. 1 *A* shows sample records of tension transients elicited when single steps of different size are imposed on the tetanized fibre. In agreement with previous work[1)2)] a step change in length produces a simultaneous drop in tension (phase 1), followed by a quick recovery (phase 2) that for releases of moderate size attains values close to the tension developed before the step. The larger the size of step release, the faster the time course of recovery during this phase. Phase 2 is complete in 1, 2 ms and is followed by a pause or even a reversal of tension recovery (phase 3, lasting 5-20 ms) that is more marked for step releases of 4-8 nm, and is briefer for larger releases. Eventually the tension approaches the value before the step within about 100 ms (phase 4).

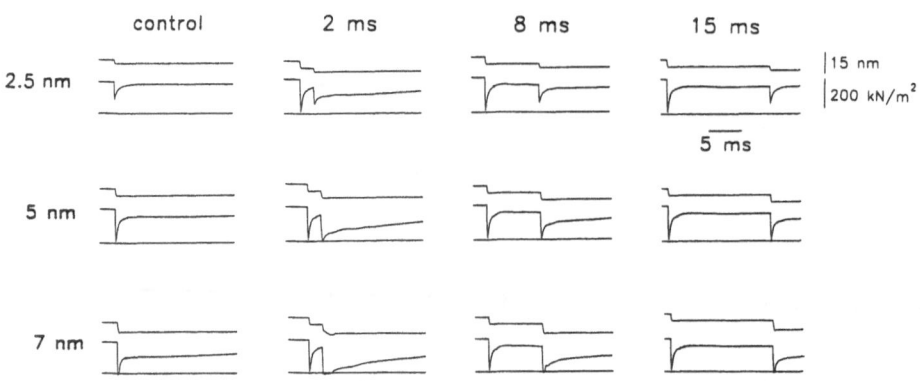

Fig. 2. Tension responses to length steps imposed either at the isometric tetanus plateau (control, first column) or 2 ms (second column), 8 ms (third column) and 15 ms (fourth column) after a conditioning step release of 5 nm. In each frame the upper trace is segment length change, the middle trace is tension response, the lower trace shows the resting tension. The figure on the left of each row indicates the size of the test step in nm per h.s. Following the test step release of 7 nm applied 2 ms after the conditioning release, the fibre went slack for more than 1 ms, because the effects of the two steps add up when they are close enough. Same fibre as in Fig. 1.

The extreme tension attained during the step (T_1) and the value of tension recovered at the end of quick phase (T_2) are plotted against the imposed length change in Fig. 1 B. When allowance is made for the effect of quick recovery during the step itself, the T_1 curve can be considered linear; it represents an estimate of instantaneous elasticity mainly residing in the cross-bridges themselves[1][2][16]. The intercept on the length axis (3.8 nm) of the straight line drawn across T_1 points obtained for small steps gives an estimate of the extension of the elastic element of the cross-bridge before the step. The slope of the straight line is a measure of the stiffness of the fibre segment under inspection, which, under the experimental condition used, represents an estimate of the number of the attached cross-bridges. The characteristics of T_2 curve are much more non-linear. The curve shows a downward concavity in the region of small steps, as a consequence of the fact that, after releases of small size, tension returns to being very close to the original value. For releases larger than 6 nm there is an almost linear fall in tension with the size of the length step, and the relation approaches zero at about 11 nm. Apart from the smaller value of the intercept of T_2 curve, these results agree very well with those of previous work[1][2], which gave the experimental basis to the Huxley and Simmons theory[1] for force generation.

According to this theory, attached cross-bridges are distributed between different force generating states, and the equilibrium as well as the rate of transition between states depend on the extension of cross-bridges. A step release reduces the average extension of the attached cross-bridges, producing a redistribution towards the higher force generating state: this redistribution is the energy releasing stroke that underlies the elementary force generating step. Detachment and reattachment

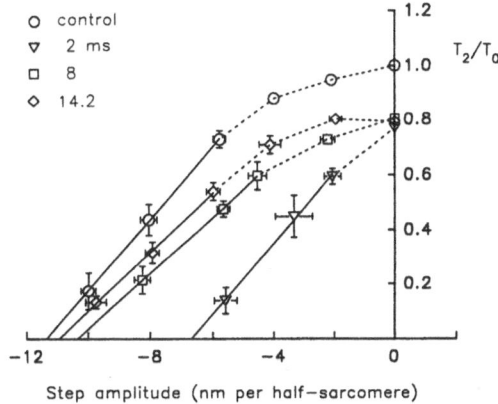

Fig. 3. T_2 curves obtained at the isometric tetanus plateau (control) and at different times after a conditioning release (test curves, identified by the symbols listed in the graph). Individual points are mean values ($\pm$ S.E.M.) of data pooled from three experiments and grouped in classes of length 2 nm wide. The dashed lines are drawn by joining the experimental points. The continuous lines are the regression lines on the T_2 points corresponding to the three largest steps. The intercepts of the lines on the length axis are:○, 11.33 nm; ◇, 10.92 nm; ▢, 10.33 nm;▽, 6.68 nm. Size of the conditioning step release: 5.20 $\pm$ 0.13 nm; temperature 3.65 $\pm$ 0.38 °C; sarcomere length: 2.09 $\pm$ 0.02 μm.

are assumed not to take place in the time scale of the quick tension recovery (a few ms). As regards the successive phases of the tension transient, phase 3 has not yet been given a satisfactory explanation, while phase 4 can very likely be attributed[9][10] to cross-bridge detachment and reattachment further along the actin filament.

In Fig. 2, tension responses to step releases of various sizes imposed at different times after a conditioning release of 5 nm per h.s. are compared with those obtained with single step releases (control). At 2 ms after the conditioning step, when the quick phase of tension recovery is almost over, the test step elicits a transient characterized by a strongly depressed quick recovery. During the next 13 ms, while tension before the test step changes very little, the level of tension recovered during the quick phase increases in a very striking manner as the time when the test step is applied increases. During the whole period, the segment stiffness, a measure of the number of attached cross-bridges, maintains a constant value, about 0.9 of the isometric plateau value[17]. Similar results were obtained in the three fibres used for these experiments. The data referring to the T_2 curve were pooled and means were calculated for T_2 values grouped in classes according to the size of the length step.

In Fig. 3 the test T_2 curves obtained at different times after the conditioning step are compared with the control T_2 curve. At 2 ms the T_2 curve shows almost no concavity. The intercept on the length axis of the straight part of the curve, estimated by linear regression analysis, was reduced from 11.3 nm (control) to 6.7 nm. This value is slightly larger than the difference between the intercept of the control curve and the size of the conditioning step (11.3 nm - 5.2 nm = 6.1 nm). In effect, at 2 ms, the test T_2 curve, when shifted to the left by the amount of the conditioning step, lies

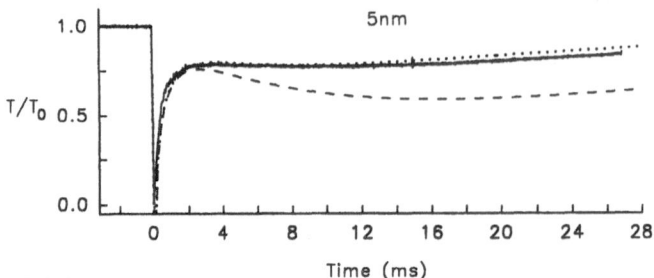

Fig. 4. Comparison of the experimental tension response to a step release of 5 nm (continuous line) with the tension response simulated by the model described in Piazzesi et al.[7] (dotted line). In the model, A_2 cross-bridges can detach through step 6 with a rate constant increasing with the degree of compression. Cross-bridges detached in this way can reattach (through step 5) at a rate 200 times faster than cross-bridges detached at the end of the normal cycle. The dashed line shows the model response when k_6 is made zero for values of $x < -1$ nm, so that there is no possibility of detachment once A_2 cross-bridges become compressed. The step in the model was completed in zero time; for reason of clarity the corresponding tension change takes 100 μs in the drawing.

only slightly above the control one. Both the concavity of the T_2 curve and its intercept on the length axis increase with the time elapsed after the conditioning step, so that within 10-15 ms the curve reattains the same shape as the control one. The time course of the recovery of T_2 curve can be estimated by plotting the T_2 curve intercepts against the time elapsed after the conditioning step. The experimental points are satisfactorily fitted with the exponential $l_t = l_c - (l_c - l_0) * \exp(-t/\tau)$, where l_t is the T_2 curve intercept at a given time t, l_c is the intercept of the control T_2 curve, l_0 is the difference between the intercept of the control T_2 curve and the size of the conditioning step. For the pooled data of Fig. 3 the estimated τ was 6.3 ms.

DISCUSSION

The double step experiments described in the present paper show that in the few tens of ms after a conditioning release of about 5 nm, although tension and stiffness change very little, very striking events occur at the cross-bridge level. At 2 ms, after the conditioning release, when the quick phase of tension recovery is almost over, a test step elicits a quick recovery that roughly corresponds to that expected after a step with the same amplitude as the sum of the conditioning and test step. This result is in agreement with previous work[2] and indicates that the dominant process occurring by the time phase 2 is complete is a redistribution between different force generating state of the attached cross-bridges without substantial detachment and reattachment. As expected, in accordance with Huxley & Simmons' theory[1], a test step at this time elicits the portion of the power stroke not consumed by the first step. Successively, within 10-15 ms, a time corresponding to phase 3, the effects of the conditioning step progressively disappear and another power stroke can be generated, indicating that most of the cross-bridges detach and reattach further along the actin filament. The rate limiting step in the process of repriming of the power stroke is likely to be cross-bridge detachment, since the stiffness (and therefore the number of attached cross-bridges) is not reduced significantly during this process[17].

An attempt was made to simulate the results of double step experiments with a mechanical kinetic model proposed to explain tension transients both in isometric conditions and during steady lengthening[7]. The model incorporates Huxley's 1957[9] theory and Huxley & Simmons' 1971[1] theory.

The scheme of cross-bridge reactions is

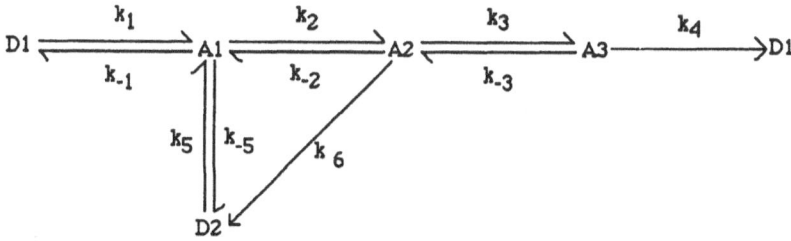

where D_1 and D_2 are detached states and A_1, A_2 and A_3 are attached states. The elementary force generating step is represented by each of the transitions $A_1 \rightarrow A_2 \rightarrow A_3$, since in each transition cross-bridge spring is extended by 4.5 nm. The rate constants for the transitions between attached states and from attached to detached states have been given ad hoc values in relation to cross-bridge extension. The functions expressing the values of k_3 and k_{-3} are shifted, with respect to those for k_2 and k_{-2} by 4.5 nm along the x axis (where x is the relative position between the myosin head and the actin site, taken as zero when, with cross-bridge in A_1 state, extension of cross-bridge elastic component is zero). Consequently in isometric conditions step 3 is rate limiting and attached cross-bridges are distributed between A_1 and A_2.

Following a shortening step, the reduction in cross-bridge extension causes a redistribution of cross-bridges toward A_3 and therefore an increase in rate of detachment through step 4. Since k_1 is moderate (in order to fit the development of tension in an isometric tetanus), the number of attached cross-bridges will reduce and the rate of attachment will increase. In this way a possible interpretation for the pause or inversion in tension recovery during phase 3 is provided. Actually, to maintain the depression of tension in phase 3 within realistic limits (Fig. 4) it is necessary to assume that shortening cross-bridges detach without completion of the normal cycle (step 6) and reattach at high rate through step 5. The rate constant k_6 is given values increasing with the increase in cross-bridge shortening[7]. If ATP splitting is associated with detachment at the end of the cycle, the possibility of detachment from an early stage provides a way of maintaining the ATP splitting rate lower than the rate of cross-bridge interaction.

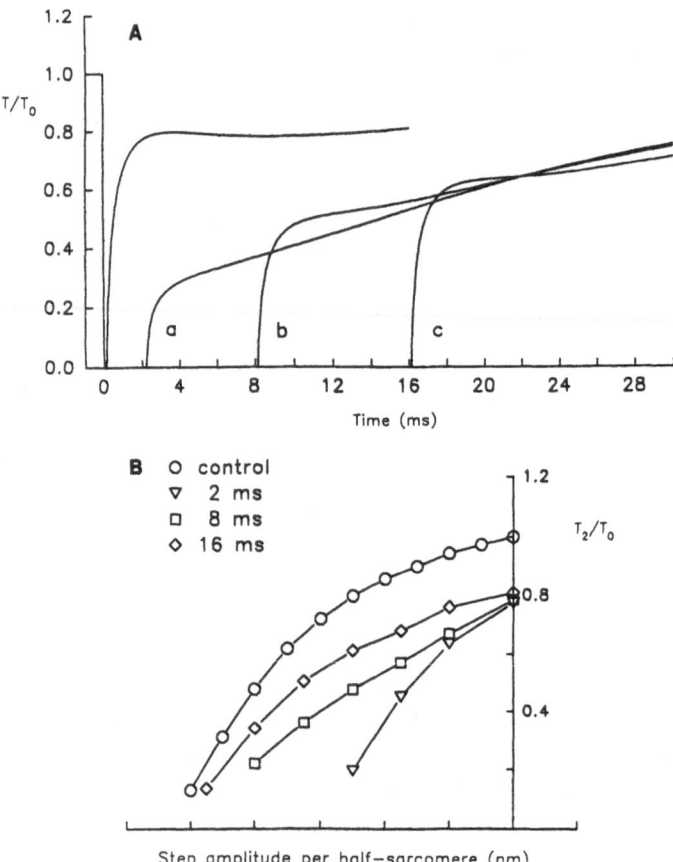

Fig. 5. *A.* Superposed simulated tension responses to a double step. A test step of 5 nm was imposed 2 ms (a), 8 ms (b) and 16 ms (c) after a conditioning step of the same size. Tension (*T*) is relative to the steady state isometric value (T_0). Negative values of T/T_0 are omitted. *B.* T_2 curves determined on the simulated responses obtained either with single (circle) or with double steps. Different symbols refer to different times between the conditioning and the test step as listed in the graph. The T_2 level was estimated by using the same method as for experimental records (see legend of Fig. 1). Lines are drawn by joining the points.

The response of the model to a double step protocol similar to that in Fig. 2, second row, is shown in Fig. 5, *A*. It can be seen that the simulated responses give a satisfactory representation of the increase of T_2 tension with the time elapsed after the conditioning step. The results for different sizes of the test step are summarized in Fig. 5 *B* by plotting the T_2 curves obtained at three different times after the conditioning step. The time dependent change in shape of the T_2 curve reproduces satisfactorily that which occurs in the experiment.

The change in fractional occupancy of attached states of cross-bridges accompanying the conditioning release of 5 nm are shown in Fig. 6. In isometric

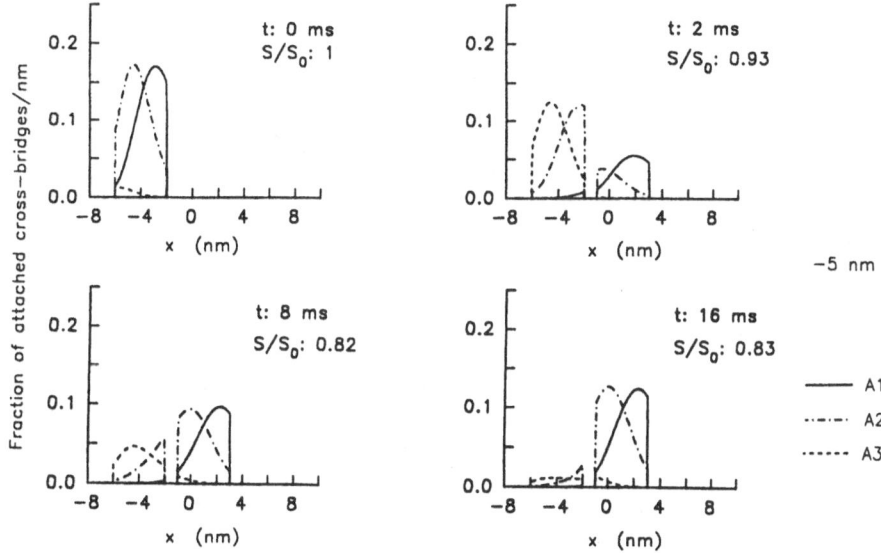

Fig. 6. x-distribution of attached cross-bridges at various times (*t*) after a shortening step of 5 nm imposed on the steady state isometric distribution. Continuous line, A_1 state; dot-dashed line, A_2 state; dashed line, A_3 state. The distribution at *t*= 0 ms is obtained by shifting the steady state isometric distribution 5 nm to the left. The ordinate gives the number of attached cross-bridges per nm as fraction of the total number. The areas under each curve, representing the fraction of attached cross-bridges in each state, are summed to give S/S_0, the stiffness at any time relative to that at the isometric steady state.

conditions only states A_1 and A_2 are populated. 2 ms after the release, when most of the quick recovery has occurred, there has been mainly a redistribution toward A_3, but a new population of A_1 (and A_2) is beginning to appear as a consequence of detachment and rapid reattachment to the next actin site through steps 6 and 5. During this phase, complete within 20 ms, tension shows almost no change, while the total number of cross-bridges has a small but significant reduction because some detachment, not followed by rapid reattachment, is occurring through step 4 at the end of the cycle. The complete recovery of tension and stiffness will occur at the slower rate that is characteristic of phase 4.

The possibility that, in shortening muscle, cross-bridges detach at an early stage of the cycle (without hydrolysis of ATP) and rapidly reattach and generate force, gives an explanation to the long interacting distance per ATP split found in muscle fibres[5)8)] and in myofilaments[6)]. This hypothesis implies that in each interaction only part of the energy associated with ATP splitting is consumed. In our model, in order to fit the rate of phase 2 recovery found experimentally[7)12)], the energy implied in the elementary force generating step represented by each of the transitions $A_1 \rightarrow A_2 \rightarrow A_3$ cannot be assumed to be much larger than $7*10^{-21}$ J. The energy made available by ATP hydrolysis, $56*10^{-21}$ J assuming an efficiency of 70 %, is 8 times greater than the energy for the elementary force generating step.

REFERENCES

 1. Huxley, A.F. & Simmons, R.M. *Nature* **233**, 533-538 (1971).
 2. Ford, L.E., Huxley, A.F. & Simmons, R.M. *J. Physiol. (Lond.)* **269**, 441-515 (1977).
 3. Kushmerick, M.J. & Davies, R.E. *Proc. R. Soc. B* **174**, 315-353 (1969).
 4. Lymn, R.W. & Taylor, E.W. *Biochemistry* **10**, 4617-4624 (1971).
 5. Homsher, E., Irving, M. & Wallner, A. *J. Physiol. (Lond.)* **321**, 423-436 (1981).
 6. Yanagida, T., Arata, T. & Oosawa, F. *Nature* **316**, 366-369 (1985).
 7. Piazzesi, G., Francini, F., Linari, M. & Lombardi, V. *J. Physiol. (Lond.)* **445**, 659-711 (1992).
 8. Higuchi, H. & Goldman, Y.E. *Nature* **352**, 352-354 (1991).
 9. Huxley, A.F. *Prog. Biophys. Biophys. Chem.* **7**, 255-318 (1957).
10. Huxley, A.F. & Simmons, R.M. *Cold Spring Harbor Symp. Quant. Biol.* **37**, 669-680 (1973).
11. Cecchi, G., Colomo, F. & Lombardi, V. *Boll. Soc. Ital. Biol. Sper.* **52**, 733-736 (1976).
12. Lombardi, V. & Piazzesi, G. *J. Physiol. (Lond.)* **431**, 141-171 (1990).
13. Huxley, A.F. & Lombardi, V. *J. Physiol. (Lond.)* **305**, 15-16P (1980).
14. Huxley, A.F., Lombardi, V. & Peachey, L.D. *J. Physiol. (Lond.)* **317**, 12-13P (1981).
15. Cecchi, G., Colomo, F., Lombardi, V. & Piazzesi, G. *Pflügers Arch.* **409**, 39-46 (1987).
16. Ford, L.E., Huxley, A.F. & Simmons, R.M. *J. Physiol. (Lond.)* **311**, 219-249 (1981).
17. Lombardi, V., Piazzesi, G. & Linari, M. *Nature* **355**, 638-641 (1991).

Discussion

Brenner: What is the difference between your model and a model that was published by Cooke and Bialek some years ago in the Biophysical Journal (*Biophys. J.* **28**, 214-258, 1979) They had exactly the same thing: based on other experimental evidence in their model, at -14 nm, cross-bridges are forced to come off and rapidly reattach to subsequent actin sites during release or during shortening.

Lombardi: You are referring to the one in which they don't account for the tension transient. They assume as in the Huxley '57 model that as cross-bridges attach they are generating force. I don't think they made a model in which a transition between different force-generating states is implied. So their model cannot fit our data because I was giving evidence to imply that detachment-reattachment occurs at an intermediate state in the cross-bridge cycle after part of the force-generating process.

Brenner: I am not really convinced about that, because what you are saying is that the recovery of phase 2 comes with a rate constant of something like 100/s. You say that there is some detachment going on with fast reattachment and the detachment persists after some negative strain. That's exactly what Cooke and Bialek said, but based on other experimental evidence.

Lombardi: In my model, the detachment from the intermediate state of the cross-bridge cycle starts to play a role for very slightly compressed cross-bridges. There is a 4.5 nm extension in the spring from each of the transitions toward higher force-generating states (A1→A2→A3). Now, detachment through step 6 becomes significant with 4-5 nm release, just when cross-bridges in the A2 state start to become compressed. Actually, it is not true that they all go through step 6. The fact is that the probability of cross-bridges ending the power stroke (through step 3), or just detaching and quickly reattaching (through steps 6 and 5) is comparable at this point. At -14 nm I would expect a very high detachment rate for negatively-compressed cross-bridges (through step 3 and 4) without fast reattachment.

Brenner: I'm not sure I heard it correctly, but didn't you say that it is impossible from your experiment to assume that, in the very early part of the tension recovery, or maybe even during the truncated part of the recovery, that there is rapid detachment-reattachment taking place?

Lombardi: Yes, indeed. The double step experiment I showed indicates that at 2 ms after a 5 nm step release, when phase two is complete, you find a stiffness that is of the order of 90% of original stiffness. We are now trying to do the experiments to see when this 10% of cross-bridges is lost. In any case, I think that even Huxley and coworkers [Ford et al. *J. Physiol. (Lond.)* **269**, 441-515, 1977] agree with the idea that after a step release some of the cross-bridges detach and this influences the T2 curve to some degree. Complete recovery occurs with phase 4. So, after allowing for this, you have recovery in phase 2 occurring without substantial detachment-reattachment. In the example of the 5 nm step, the only evidence for substantial detachment-reattachment, with the kinetics as shown, is after two milliseconds; that is, when phase 2 is well over.

Brenner: I didn't mean to say that you couldn't see very fast detachment-reattachment or that you ruled it out. I just wondered if there could have been very fast detachment-reattachment during the truncation the way there is in the rabbit.

Lombardi: If you are referring to the lost 10%, the 10% that we don't find at the end of phase 2, we have not characterized their kinetics. It could be that they are lost between zero and 100 μs, or between zero and 200 μs or later.

FORCE RESPONSE OF UNSTIMULATED INTACT FROG MUSCLE FIBRES TO RAMP STRETCHES

M.A. Bagni, G. Cecchi, F. Colomo and P. Garzella

Dipartimento di Scienze Fisiologiche
Università degli Studi di Firenze
Viale G.B. Morgagni 63, I-50134 Firenze, Italy

ABSTRACT

The possibility that weakly binding bridges are attached to actin in the absence of Ca^{2+} under physiological conditions was investigated by studying the force response of unstimulated intact muscle fibres of the frog to fast ramp stretches. The force response during the stretching period is divided into two phases: phase 1, coincident with the acceleration period of the sarcomere length change and phase 2, syncronous with sarcomere elongation at constant speed. The phase 1 amplitude increases linearly with the stretching speed in all the range tested, while phase 2 increases with the speed but reaches a plateau level at about 50×10^3 nm/half sarcomere per second. The analysis of data shows that phase 1, which corresponds to the initial 5-10 nm/half sarcomere of elongation, is very likely a pure viscous response; its amplitude increases with sarcomere length and it is not affected by the electrical stimulation of the fibre. Phase 2 is a viscoelastic response with a relaxation time of the order of 1 ms; its amplitude increases with sarcomere lengths and with the stimulation. These data suggest that weakly binding bridges are not present in a significant amount in unstimulated intact fibres. .

INTRODUCTION

Force responses of skinned rabbit psoas fibres to stretches of various speeds, have shown[1][2] that the stiffness of relaxed fibres at low ionic strength was strongly dependent upon the velocity of stretch. Stiffness was almost negligible at low stretching speeds, but it increased up to 1/2 the stiffness of a fibre in rigor with the fastest stretch used. In agreement with biochemical findings[3][4], this result was taken as an indication that at low ionic strength an important fraction of cross-bridges is attached to actin even in the absence of Ca^{2+}. The speed dependence of stiffness suggested that these bridges were not statically attached but in a "weakly binding state" characterized by a rapid equilibrium between the attached and detached states[5]. Though fibre stiffness was greatly reduced at normal ionic strength, it

Mechanism of Myofilament Sliding in Muscle Contraction, Edited by
H. Sugi and G.H Pollack, Plenum Press, New York, 1993

703

showed the same speed dependence as at low ionic strength indicating that a small population of weakly binding bridges was attached to actin also under physiological conditions[5]. This has led to the suggestion that, in contrast to the steric blocking hypothesis, the major role of Ca^{2+} in the regulation process is to induce the transition of crossbridges from the non force-generating weakly bound states[5][6], populated in relaxed muscle, to more strongly bound force-generating states. To test whether the presence of weakly bound bridges can be demonstrated also in intact muscle fibres under physiological conditions, we studied the force response of unstimulated intact frog muscle fibre to stretches of various amplitudes and speeds.

METHODS

Single intact fibres, 2.0-2.5 mm long, isolated from the lumbricalis digiti IV muscle of the frog (Rana esculenta) were mounted by means of aluminium foil clips, between the lever arms of a force transducer (35-60 kHz resonance frequency) and a fast displacement generator (minimun ramp time 50 μs) in a chamber fitted with a glass floor for ordinary and laser light illumination. The sarcomere length was usually adjiusted to about 2.15 μm. Ramp stretches at speed between 5 and 250 sarcomere length per second (l_0/s), amplitude up to 6% l_0 were applied to one end of an unstimulated fibre and force responses were measured at the other. Sarcomere length changes were measured in a fibre region (about 200 μm long) located at about 500 μm from the force transducer, by means of a laser diffractometer[7]. This procedure virtually eliminated measurement artifacts due to the delay line behaviour of the fibre. A low coherence solid state laser (power 3mW, wavelength 0.778 μm) was used as a light source for the diffractometer. The rise time of the system was less than 3 μs and the peak to peak noise corresponded to about 0.2 nm. The incident angle of the laser beam on the fibre was adjusted so as to maximize the intensity of the first right diffraction order which was utilized to compute the sarcomere length. The output from the diffractometer was electronically differentiated (rise time of the differentiator, 3 μs) to obtain the instantaneous sarcomere lengthening speed. To check fibre viability and measure active tension, fibres were stimulated at regular intervals with brief (0.6-0.8 s duration) tetanic volleys. Experiments were made at 5 and 15°C.

Force, fibre length (l_f), sarcomere length and sarcomere lengthening speed were measured on a digital oscilloscope (4094 Nicolet U.S.A.) and stored in floppy disks for further analysis. To improve the signal to noise ratio, especially for the sarcomere lengthening speed signal, up to 10 responses were often averaged.

RESULTS

Fig. 1 shows the typical force response of an unstimulated intact fibre subjected to a constant velocity stretch. As shown previously[9], the force transient can be separated into at least four different phases: 1) a fast initial tension rise coincident

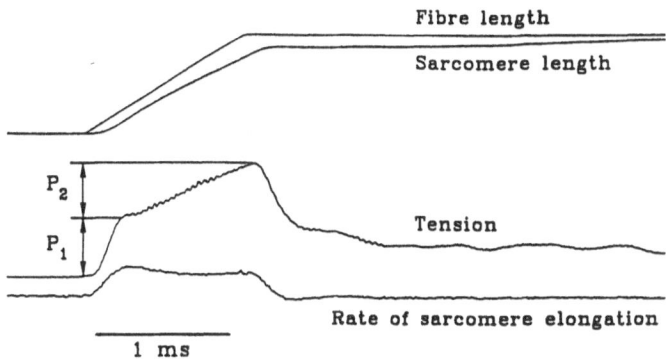

Fig. 1. Sample records of force response in an unstimulated fibre subjected to ramp lengthening at speed of 22 l_0/s. From top to bottom: fibre length, sarcomere length, tension and rate of sarcomere elongation. The initial force response is clearly divided into two phases indicated in the figure as P_1 and P_2. The sharpness of the P_1 -P_2 transition depends on the shape of the sarcomere length change and this in turn depends on the mechanical properties of the fibre and on the compliance of the fibre connections to the transducers. Fibre length, 2,190 µm; sarcomere length 2.18 µm; temperature 5°C. Experiment of 23-X-1990.

with the beginning of the stretch (P_1); 2) a slower tension rise (P_2); 3) a fast drop of tension at the end of the stretch and 4) a slow return to the new steady state level. Note that in spite of the fast acceleration imposed on the fibre, the lengthening speed at the sarcomere level required at least 200 µs of acceleration to reach a steady constant value in this fibre. This effect is due to the delay line behaviour of the fibre that delays and distorts the sarcomere length changes with respect to the fibre length

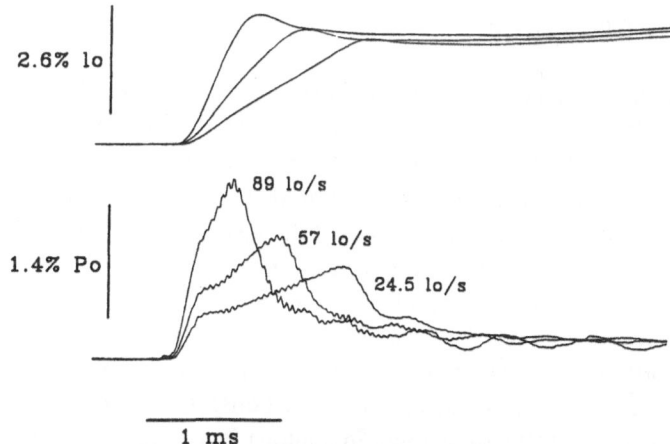

Fig. 2. Effects of stretching speed (indicated by the figures close to the records) on the force response. Upper traces: sarcomere length; lower traces: force response. Stretching amplitude: 2.77% l_0. Sarcomere length traces show clearly the distortion due to the delay line behaviour of the fibre. Same fibre as in Fig. 1; temperature 15°C.

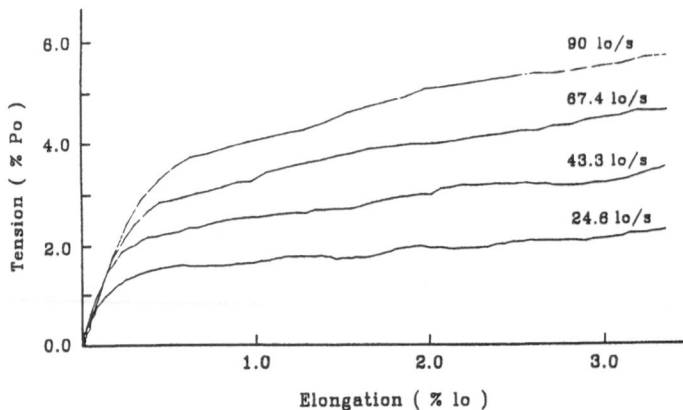

Fig. 3. Instantaneous length-tension relation in an unstimulated fibre, at 4 different shortening speeds. Figures close to the traces indicate the steady state speed that is reached only at the end of the initial fast rise of tension. Fibre length: 2,000 μm; sarcomere length: 2.13 μm; temperature 15°C. Expt. of 20-VI-1991.

changes. The acceleration period corresponds to phase 1, while phase 2 represents the force response of the sarcomeres stretched at constant speed. The comparison between sarcomere lenghtening speed and force traces, shows that the peak of phase 1 is reached exactly at the end of the acceleration period when the stretching speed has become constant (see also Fig. 6). Phase 3 is equivalent to phase 1 and it is synchronous with the deceleration period. Phase 4 (equivalent to phase 2) is the return to the resting tension relative to the new sarcomere length. The last two phases will not be further considered in this paper.

Effects of Stretching Speed

Records in Fig. 2 show the effects of stretching speed on force response. It can be seen that P_1 force increases with the speed while P_2 response remains almost constant in this speed range. Note that the peak of P_1 is attained at same time independent of the stretching speed and therefore it corresponds to different amounts of sarcomere elongation. In Fig. 3 the instantaneous relations between force and length at 4 different shortening speeds are reported. Again phases 1 and 2 are clearly evident. Note that during phase 1 the tension does not increase linearly with the elongation.

The relationships between P_1 and P_2 amplitudes and stretching speed are reported in Fig. 4. It can be seen that P_1 response is linearly correlated with the speed even at the fastest speeds used. In contrast P_2 response increases initially with the speed, but reaches a plateau as the speed increases. The linear speed dependence of P_1 is also evident in Fig. 5 that shows the instantaneous relationship between sarcomere-stretching speed and force response measured during the acceleration period (140 μs) that corresponds in this fibre to the initial 12 nm/half sarcomere of

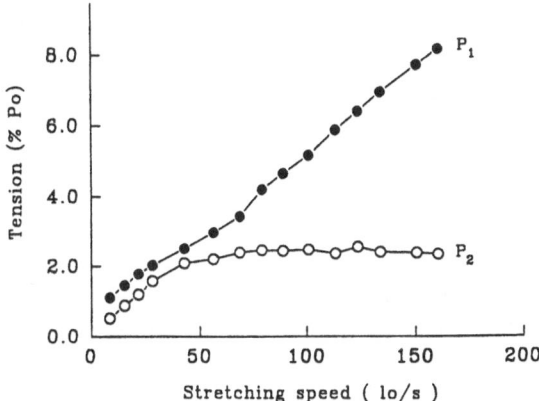

Fig. 4. P_1 (●) and P_2 (○) responses versus stretching speed. P_1 increases linearly with the speed, resembling a viscous response, while P_2 reaches a plateau at high stretching speeds resembling a viscoelastic response. The extrapolated value of P_1 at zero speed is not zero. This is very likely due to the superimposition on the P_1 response of the tension developed by the short range elastic component [8] that at low speeds is higher than the P_1 response. P_2 values are corrected for the resting tension measured 50 ms after the end of the stretch. Stretching amplitude: 3.7% l_0; fibre length: 2,160 μm; sarcomere length: 2.13 μm; temperature: 15°C. Expt. of 3-VII-1991.

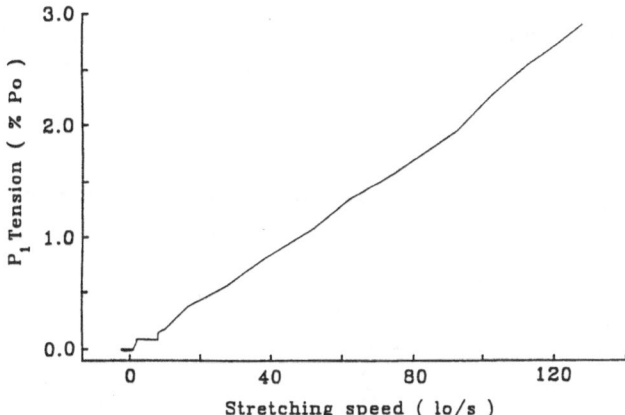

Fig. 5. Instantaneous relation between P_1 tension and stretching speed measured during the acceleration period (140 μs) corresponding, in this fibre, to the initial 12 nm/half sarcomere of elongation. Same fibre as in Fig. 1.

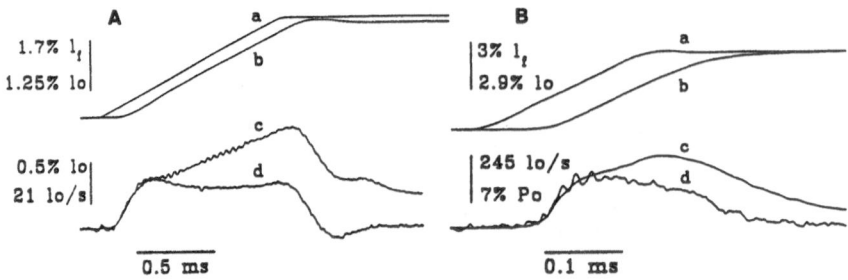

Fig. 6. Fast time base records of force response to ramp stretching in two unstimulated fibres. Stretching speeds were 24.4 l_0/s (A) and 245 l_0/s (B). Upper pair traces: fibre length (a) and sarcomere length (b). Lower pair traces: tension (c) and sarcomere lengthening speed (d). The delay between speed and force during the acceleration period, if present, is no longer than few microseconds. *A*, fibre length: 2,190 µm; sarcomere length: 2.18 µm; temperature 15°C. Expt. of 23-X-1990. *B*, fibre length: 2,250 µm; sarcomere length: 2.17 µm; temperature 15°C. Expt. of 5-XII-1990.

elongation. These results indicate that P_1 could be explained by the response of viscosity in the fibre, while P_2 is a viscoelastic response. The viscous coefficient estimated from the ratio of force per unit of cross-sectional area and rate of stretching in each sarcomere, is of the order of 0.5×10^8 N.s.m^{-3}, a value somewhat smaller than that found previously in tibialis anterior fibres[9].

The linear relationship between speed and P_1 response of Fig. 4 is not exclusive of a pure viscous response but it is also consistent with that of a simple viscoelastic system having a relaxation time much shorter than the minimum stretch duration used for measurements. Since it was not possible to decrease the stretch duration

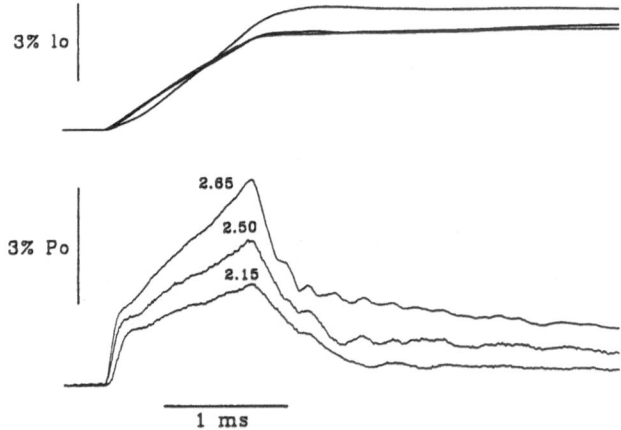

Fig. 7. Effects of sarcomere length on the force response to ramp stretch in an unstimulated fibre. Upper traces: sarcomere length; lower traces: force response. Sarcomere length (µm) is indicated by the figures close to the traces. Fibre length: 1,935 µm at sarcomere length of 2.15 µm; temperature 15°C. Expt. of 21-VI-1991.

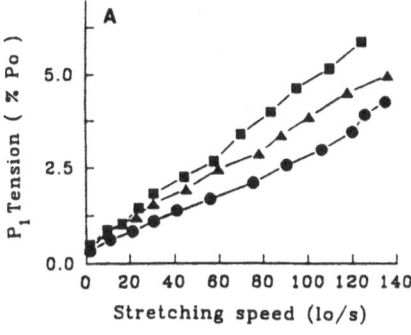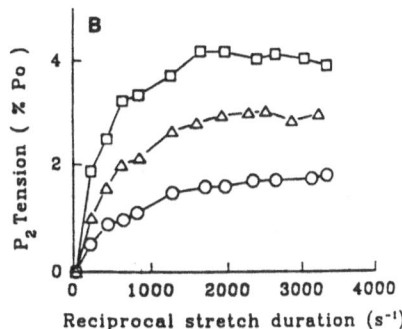

Fig. 8. P_1 response versus stretching speed (A) and P_2 versus reciprocal stretch duration (B) at 2.15 μm (circles), 2.50 μm (triangles) and 2.65 μm (squares) sarcomere length. In spite of decreasing of myofilament overlap, P_1 and P_2 responses increases with sarcomere length. P_2 values corrected as in Fig. 4. The relaxation time (τ) associated with P_2 response, can be calculated according to the equation: $P_r = (\tau/t_d) [1-\exp(-t_d/\tau)]$ where: P_r= force relative to the plateau and t_d=stretch duration. Same experiment as in Fig. 7.

below 200 μs while keeping a fair ramp shaped sarcomere elongation, a different approach was used to improve the resolution of the relaxation time measurement. It can be shown that in a simple viscoelastic system stretched at constant acceleration, the force response lags the speed by a time corresponding to the relaxation time. Records reported in Fig. 6, show that the delay between force and speed during the acceleration period and therefore the relaxation time, cannot be longer than 5 μs.

Effects of Sarcomere Lengths

The nature of the passive force response was further investigated by studying the effects of sarcomere length. In fact, if P_1 and P_2 responses are due to cross-bridges, their amplitudes are expected to scale with the degree of overlap between thick and thin filaments. Fig. 7 shows sample records at sarcomere length of 2.15, 2.50 and 2.65 μm, while the relationships between P_1 and P_2 responses and stretching speed at these sarcomere lengths are reported in Fig. 8. It is seen that increasing the sarcomere length increases both P_1 and P_2 amplitudes without affecting the relaxation time associated with P_2.

Effects of Activation

Fig. 9 shows the effects of stimulation on the force response to elongation ramps. In *A*, stretches were applied at different times after a single shock stimulation during the latent period when the tension developed was negligible. It can be seen that P_1 tension is not affected by the stimulation while P_2 response increases progressively as the stretch is delayed with respect to the stimulus. Note however, that also the

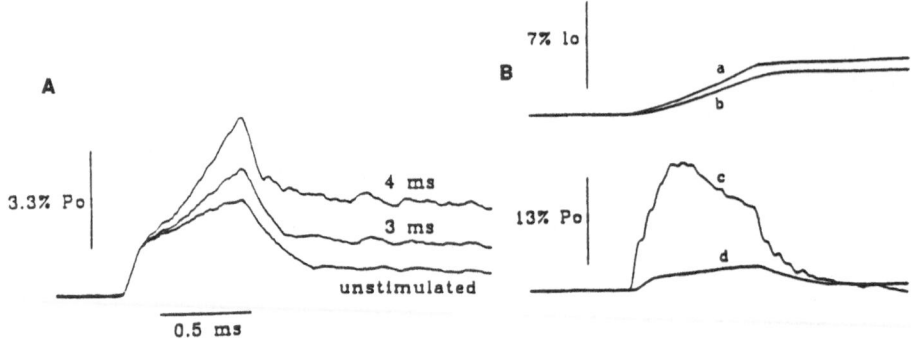

Fig. 9. Effects of activation on force response to stretching at constant speed. A: stretch applied at progressively increasing delays (indicated by the figures close to the traces) after a single shock stimulation. Bottom trace is the response of unstimulated fibre. The active tension developed at the beginning of the stretch was zero in all the records. B: superimposition of force responses (lower pair traces) to constant stretching speed in an unstimulated fibre (d) and in an active fibre (c) at peak of twitch tension. Upper pair traces are sarcomere length changes (a, active; b, unstimulated fibre). Note that P_1 response is evident also in the activated fibre. The second phase of force response is instead partially truncated by the "give" phenomena. Fibre length: 2,770 μm; sarcomere length: 2.13 μm; temperature 15°C. Expt. of 14-V-1991.

"steady" level at which the tension decays after the completion of the force transient increases by about the same amount. In Fig. 9B the stretch was applied at peak of a twitch response, when the active tension was 0.14 P_0. P_1 response is now higher than in passive fibre, however, sarcomere length traces show that the stretching speed is also higher so that the calculated viscous coefficient resulted about the same in both cases. P_2 response is clearly truncated by the "give"[10] phenomena which occurs at 12 nm/half sarcomere of elongation.

DISCUSSION

The force response of unstimulated intact fibres to ramp stretches shows two initial phases clearly distinct: a fast rise of tension at the beginning of the stretch, P_1, and a much slower rise during the stretch, P_2. The initial slope of the instantaneous length-tension relation has been taken as a measure of the fibre stiffness and consequently as a measure of the number of attached bridges[1][6]. However our results show that this procedure is not applicable to intact unstimulated fibres since the initial phase of the force response is not elastic but viscous in nature. Owing to viscosity in the fibre, the slope of the length-tension relation corresponds to stiffness only if the stretching speed is strictly constant during all the length change. In our preparation some 100-200 μs of acceleration were usually nedeed for the sarcomere stretching speed to attain a constant value, and this means that with fast stretches speed was not constant during the initial 5-10 nm/half sarcomere of elongation.

As shown by Schoenberg[11] the force response of a simple cross-bridges model with only one attached state is equivalent to that of a simple viscoelastic system whose relaxation time is equal to the reciprocal detachment rate constant. The detachment rate constant of the weakly binding bridges at normal ionic strength was found to be in the range of 3×10^3 s^{-1} with a possible maximum value of 2×10^4 s^{-1}[5] that corresponds to a minimum relaxation time of 50 μs. Results reported in Figs. 4, 5 and 6 show that P_1 force is either a pure viscous response giving a resistive force proportional to the stretching speed, or a viscoelastic response with a relaxation time of about 5 μs. A pure viscous response would be inconsistent with weakly binding bridges, while a relaxation time of 5 μs, in terms of weakly binding bridge kinetics, would mean a detachment rate constant of 2×10^5 s^{-1} or more, one order of magnitude higher than the maximum value proposed for skinned fibres. In addition, if P_1 response is due to cross-bridges its amplitude is expected to scale with the amount of overlap between actin and myosin filaments. However results reported in Fig. 8A show exactly the opposite: P_1 amplitude increases with decreasing of overlap. Finally, if as suggested weakly bridges following activation are converted to strongly binding bridges a decrease[5][12] or a transient change[6] in their number is expected upon stimulation. However, Fig. 9 shows that activation and force generation do not affect P_1 amplitude. All the above findings indicate that P_1 response cannot be attributed to weakly binding bridges.

One interesting question is whether P_1 force can be attributed to the viscous resistance to the relative sliding motion of the thick and thin filaments. A rough calculation of this resistance, made according to A.F. Huxley[13] but using the myosin dimension (including S1) resulting from recent studies[14-16], gives values of the same order of magnitude as the P_1 response. In addition the calculation would also explain the finding that P_1 response increases with sarcomere length: the reduction of viscous resistance due to the reduced overlap could be in fact smaller than the increase due to the reduction of filament spacing resulting from sarcomere length increase. It should be also pointed out that owing to the constant myofilament lattice volume behaviour[17][18], the elongation of the passive fibre produces a synchronous reduction of the lateral separation between thick and thin filaments. With fast stretches this involves rapid movements of filaments towards the center of the fibre and this could contribute to the viscous resistance during elongation.

As shown in the results section P_2 behaviour corresponds to a viscoelasticity in the fibre and is therefore consistent in principle with the presence of weakly binding bridges. The relaxation time associated with P_2 (assuming a simple viscoelastic system) is about 1 ms. This would correspond to a detachment rate constant of 10^3 s^{-1}, one order of magnitude smaller than that found in skinned fibres. Although it is quite possible that the kinetics of weakly binding cross-bridge is slower in intact fibre, it is more difficult to explain in terms of weakly bridges the observation that P_2 response increases with decreasing the overlap between myofilaments. This result is, in fact, just the opposite of that expected on the basis of a direct proportionality between cross-bridge number and extent of overlap. In addition, P_2 tension, measured at high speeds, was found to increase uniformly with the stretch amplitude with no sign of "give" even when the sarcomere elongation was bigger than 6% l_0, corresponding to more than 64 nm/half sarcomere. This again suggests that P_2 response is not due to cross-bridges since it is very unlikely that they could account for such a large scale elastic response.

P_2 response could be related to the presence of the elastic titin filaments that join thick filaments to Z discs[19]. These filaments, that account for most of the resting tension, could have in fact a viscoelastic behaviour. Data reported in Fig. 9A show that upon stimulation P_2 amplitude increases even during the latent period when the active tension is zero. This finding can be accounted for by assuming that cross-bridges formed initially upon activation, are in a non-force generating state as suggested previously[20]. These bridges would be strained by the relative sliding motion of the filaments produced by the stretch and tension would be generated. The observation that tension produced by the stretch persists for a relatively long time indicates that these bridges would have a low detachment rate.

REFERENCES

1. Brenner, B., Schoenberg, M., Chalovich, J.M., Greene, L.E. & Eisenberg, E. *Proc. Natl. Acad. Sci. USA.*, 7288-7291 (1982).
2. Yanagida, T., Kuranaga, I. & Inoue, A. *J. Biochem.* **92**, 407-412 (1982).
3. Stein, L.A., Schwarz, R.P., Chock, P.B. & Eisenberg, E. *Biochemistry* **18**, 3895-3909 (1979).
4. Hibberd, M.G. & Trentham, D.R. *Ann. Rev. Biophys. Chem.* **15**, 119-161 (1986).
5. Schoenberg, M. in : *Molecular Mechanism of Muscle Contraction* (eds. Sugi, H. & Pollack, G.H.) 189-202, (Plenum Publishing Corp., New York, 1988).
6. Brenner, B. in : *Molecular Mechanisms in Muscular Contraction* (ed. Squire, J.M.) 77-149 (The Macmillan Press LTD, Southampton, G.B., 1990).
7. Bagni, M.A., Cecchi, G. & Colomo, F. *J. Muscle Res. Cell Motility* **6**, 102 (1985).
8. Hill, D.K. *J.Physiol. Lond.* **199**, 673-684 (1968).
9. Ford, L.E., Huxley, A.F. & Simmons, R.M. *J. Physiol. (Lond.)* **269**, 441-515 (1977).
10. Katz, B. *J. Physiol. (Lond.)* **96**, 45-64 (1939).
11. Schoenberg, M. *Biophys. J.* **48**, 467-475 (1985).
12. Schoenberg, M. *Biophys. J.* **54**, 135-148 (1988).
13. Huxley, A.F. in : *Reflections on muscle.* (Liverpool University Press, 1980).
14. Harford, J. & Squire, J.M. in : *Molecular Mechanisms in Muscular Contraction* (ed. Squire J.M.) 287-320 (The Macmillan Press LTD, Southampton, G.B., 1990).
15. Yu, L.C. & Podolsky, R. in : *Molecular Mechanisms in Muscular Contraction* (ed. Squire, J.M.) 265-286 (The Macmillan Press LTD, Southampton, G.B, 1990).
16. Umazume, Y., Higuchi, H. & Takemori, S. *J. Muscle Res. Cell Motility* **12**, 466-471 (1991).
17. Matsubara, I & Elliott, G.F. *J. Mol. Biol.* **72**, 657-669 (1972).
18. Cecchi, G., Griffiths, P.J., Bagni, M.A., Ashley, C.C. & Maeda, Y. *Science* **250**, 1409-1411 (1990).
19. Horowits, R. & Podolsky, R.J. *J. Cell Biol.* **105**, 2217-2223 (1987).
20. Cecchi, G., Griffiths. P.J. & Taylor, S.R. *Science* **217**, 70-72 (1982).

Discussion

Sugi: Your experiment reminds me of D.K. Hill's old experiment done on whole muscle [*J. Physiol.* (*Lond.*)**199**, 637-684, 1968]. It's my impression that your result is more or less the same as his results. What point is different from D.K. Hill?

Cecchi: It's true they are the same type of experiments, but with an important difference. D.K. Hill used a maximum stretching speed of less than $0.4 \, L_0/s$ while we are using much greater speeds in the range of $5-250 \, L_0/s$. At a speed slower than one length per second, our force response is equivalent to that of D.K. Hill; however, as we increase the speed, a viscous component (P_1), not present in Hill's response, becomes prominent.

Huxley: Jan Lannergren showed that this short-range elastic component shown by D. K. Hill was greatly increased in hypotonic solutions. Have your tried your responses in hypotonic solution?

Cecchi: Yes, we have tried hypotonic solutions and many other solutions. The "viscous phase" (P_1) increases in hypertonic solution and decreases in hypotonic solution, similar to D.K. Hill's response.

Kawai: How much of the phenomenon that you are seeing is due to titin, or connecting filaments?

Cecchi: I don't know. The first part of the force response I've been looking at is a viscous response, so I think it is difficult to attribute this response to titin.

Kawai: If you have a rubber band you can have a viscous response.

Cecchi: A pure viscous response? I don't believe so. I think you should get some elasticity somewhere. Anyway, the second part of the response I showed, the P_2, is due to some viscoelasticity in the fiber and may be equivalent to the one showed this morning by ter Keurs, attributable to titin.

Kawai: But you ruled out the contribution from the cross-bridges entirely, didn't you?

Cecchi: Yes. This is not surprising since even in skinned fibers at high ionic strength only a few cross-bridges attach in the absence of Ca^{2+}.

Kawai: But your effect is larger at the longer sarcomere lengths.

Cecchi: Yes. This is one of the proofs that our force response cannot be attributed to the weakly binding bridges.

ter Keurs: You showed that the P_1 response is a viscous response that occurs even in the presence of calcium in the muscle. If so, why don't I see it in the T1 curve? I should see a different intercept on the ordinate for stretch compared to shortening.

Cecchi: The situation is the following: the viscous response P_1 is present unchanged during the latent period at time when the internal Ca^{2+} concentration is already increased; it seems unchanged also during the development of active tension in a twitch, although I would like to investigate this point a little further. Regarding your question, I think that the viscous response does not affect the T1 curve very much for the following reasons: 1) the viscous force during the small length steps used to determine the T1 curve is relatively small due to the low stretching (or releasing) speed (about 10 L_0/s) associated with these length changes; 2) the major contribution of viscosity to force response will appear at the mid-point of the step corresponding to the maximum speed, while the effect on the tension reached at the end of the step, usually plotted on the T1 curve, will be negligible since at this time the speed is close to zero.

Huxley: I just want to point out that D.K. Hill did attribute his short-range elastic component to interactions between actin and myosin. So that was weakly binding cross-bridges more than twenty years ago.

Brenner: Well, I think it was already pointed out by Cecchi that the speed of stretches in Hill's experiments was at least one or two—perhaps up to four—orders of magnitude slower than had been used by Cecchi, and maybe two or three orders slower than we have been using. I think Hill's short-range elasticity was not due to the type of weak-binding interaction we are talking about, because the detachment-rate constants we find are much too fast to be picked up in Hill's experiment. We think there might have been a few of the real strong interactions left. So, done some twenty years ago, his experiment does not imply weak interactions.

ter Keurs: I missed your interpretation of the P_2 response, and would like to ask you to repeat it.

Cecchi: Actually, you missed the interpretation because I didn't say anything about it. I wanted to see if this was attributable to cross-bridge interaction, and I think that the kind of explanation you gave for your viscoelasticity would probably be okay. In the titin-filament P_2 phase, are you talking about the viscous elasticity?

ter Keurs: Yes, definitely.

MEASUREMENT OF TRANSVERSE STIFFNESS CHANGE DURING CONTRACTION IN FROG SKELETAL MUSCLE BY SCANNING LASER ACOUSTIC MICROSCOPE

T. Tsuchiya, H. Iwamoto, Y. Tamura* and H. Sugi

Department of Physiology
School of Medicine
Teikyo University
Itabashi-ku, Tokyo 173, Japan
*Department of Physics
Suzuka College of Technology
Suzuka, Mie 510-02, Japan

ABSTRACT

The scanning laser acoustic microscope (SLAM) was utilized to measure the change in the propagation velocity in the transverse direction during contraction in living skeletal muscles of the frog. The SLAM was operated at 100 MHz and interferograms were produced on a CRT in real time. The images of the interferogram were processed by image-analyzer and the propagation velocity was calculated from the shift of the interference line at rest and during contraction. In all the measurements (n = 15), the velocities during contraction were clearly slower (-7.6 m/s) than at rest and this means that the transverse stiffness decreased during contraction (-2.4x10^7 N/m^2). The decrease in the propagation velocity preceded the increase in force by 30-40 ms after stimulation, suggesting that the decrease in the transverse stiffness reflects the basic molecular change in muscle contraction.

INTRODUCTION

It is generally believed that the longitudinal stiffness of muscle during contraction reflects the relative number of the attached cross-bridges and the increase in the stiffness during contraction was proved by applying step or sinusoidal length changes[1-3]. In these experiments, however, the perturbations of applied length changes were not negligibly small and the length change itself might affect the stiffness. To avoid this effect, we developed a technique in which the stiffness changes could be continuously recorded by measuring the propagation velocity of ultrasonic waves applied to frog skeletal muscle both in the longitudinal and in the

Mechanism of Myofilament Sliding in Muscle Contraction, Edited by
H. Sugi and G.H Pollack, Plenum Press, New York, 1993

715

transverse directions[4][5] and found unexpectedly that the stiffness decreased in the transverse direction during isometric contraction in frog skeletal muscle, though the stiffness increased in the longitudinal direction. Present experiment was conducted to study the phenomenon of the decrease in the transverse stiffness during contraction two dimensionally by the scanning laser acoustic microscope (SLAM).

METHODS

Scanning Laser Acoustic Microscope

The scanning laser acoustic microscope (Sono-microscope System 2140, Sonoscan Inc., Bensenville, IL, USA) was employed to determine the propagation velocity in the transverse direction of muscle and the schematical diagram of the microscope is illustrated in Fig. 1. The operational details of the instrument may be found elsewhere[6] and are only summarized here. The electrical oscillation signal (100 MHz) from ultrasonic driver is converted into continuous plane waves of ultrasound by piezo-electric transducer and they are transmitted to a sample on a sonically activated fused silica stage and a gold-mirrored plastic coverslip is placed on a sample. Mechanical perturbations at the coverslip surface (dynamic ripple) due to the acoustic energy transmitted through a sample are sensed by a focused, scanning laser beam (30 frames /s) and the reflected light on the gold layer is detected by a photodiode probe. The reflected light contains the information of the

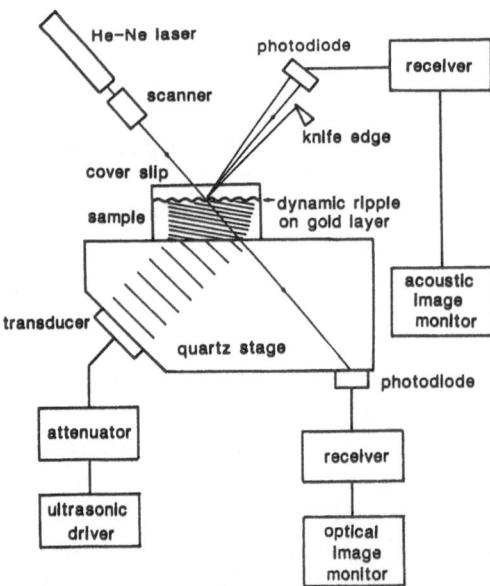

Fig. 1. A schematical diagram of the scanning laser acoustic microscope. Read the text for detail.

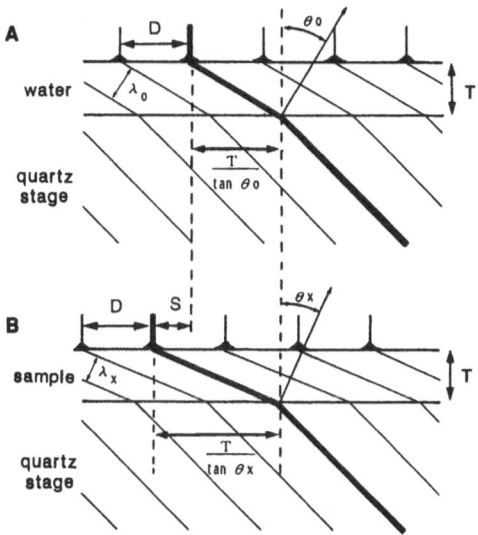

Fig. 2. Schematical representation to show how the acoustic wave fringes take place at the surface of a cover slip on water (A) and a sample (B). Read the text for detail.

propagation velocity of a sample and is converted electrically into acoustic images, which are stored in video tape.

Principle of the Measurement of Propagation Velocity

Fig. 2 schematically depicts the constant phase wavefronts and shows how the acoustic wave fringe (dynamic ripple) on the surface of a cover slip takes place. In Fig. 2A and B, reference substance (water) and a sample (muscle) is on a stage respectively and they have the same thickness (T). Sound waves are transmitted from the quartz stage to water at an angle θ_0 and to a sample at an angle θ_x, relationship being described by Snells' Law, namely,

$$\frac{V_0}{\sin \theta_0} = \frac{V_x}{\sin \theta_x} \tag{1},$$

where V_0 and V_x is propagation velocity of sound in water and a sample. From the Eq (1), the distance between adjacent fringes (D) in water and in a sample is known to be equal, that is,

$$D = \frac{\lambda_0}{\sin \theta_0} = \frac{\lambda_x}{\sin \theta_x} \tag{2},$$

where λ_O and λ_x, is wave length in water and in a sample. Shift of interference fringe lines (S) takes place between in water and in a sample because of the difference in acoustic index of refraction and

$$S = \frac{T}{\tan \theta_O} = \frac{T}{\tan \theta_x} \qquad (3)$$

The normalized shift of fringe (N) is defined as the shift from the known fringe in water to the unknown fringe in a sample, divided by the distance between two adjacent known fringes (D), that is,

$$N = S/D. \qquad (4)$$

From the Eqs. (2), (3) and (4), the propagation velocity in a sample (Vx) is formulated, that is,

$$V_x = \frac{V_O}{\sin \theta_O} \cdot \sin \left\{ \tan^{-1} \left(\frac{1}{(1/\tan \theta_O) - (N\lambda_O/T\sin \theta_O)} \right) \right\}. \qquad (5)$$

In this equation, $V_0 = 1514.0$ m/s, $\theta_0 = 10.29$, $\lambda_0 = 15.14$ μ, $T = 0.9$ mm, and the normalized shift of fringe (N) is obtained experimentally (Fig. 6). Details of the theory of velocity measurement appeared elsewhere[7].

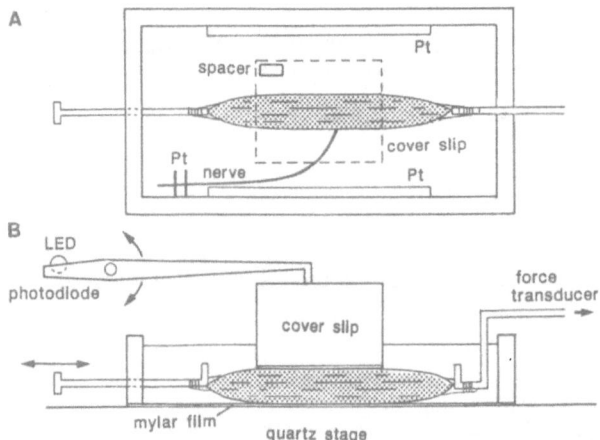

Fig. 3. Schematical representation of the experimental arrangement for the measurement of the propagation velocity in the transverse direction in the muscle. A; top view of the bath containing a muscle preparation. B; side view of the bath.

Muscle Preparation and Experimental Arrangement

Fig. 3 shows the experimental arrangement for the measurement of the propagation velocity on the quartz stage of the acoustic microscope. One end of a whole muscle of the ileofiburalis of the frog (*Rana japonica*) was connected to the force transducer (U-gage, Shinkoh, Tokyo) and the other end to a micro-manipulator to adjust muscle length. The space between the floor of the bath and the quartz stage was filled with water. A plastic cover slip, underside of which was mirrored by gold, was placed in contact with a muscle preparation and fixed firmly so that the thickness of a preparation could be kept constant (0.9 mm). Furthermore, a photodiode-LED device was used to check the thickness during contraction. Repetitive stimulation (50-100 Hz) was made directly with a pair of platinum electrodes attached on the side walls of the bath or indirectly through nerve with the other pair of electrodes. The muscle was stimulated isometrically at rest length in every 10 min at room temperature (23°C) and was constantly soaked in Ringer solution, which had the following composition (mM): NaCl, 115; KCl, 2.0; $CaCl_2$, 1.8; Na_2HPO_4-NaH_2PO_4, 2; pH 7.0.

RESULTS

Typical interferograms at rest and during isometric contraction are shown in Fig. 4A, B, in which a muscle preparation is placed in the middle of each frame. The distance between two adjacent fringes is constant (theoretically 84.5 μm) both in the region of water and in the muscle as explained in Methods and the wave fringes are

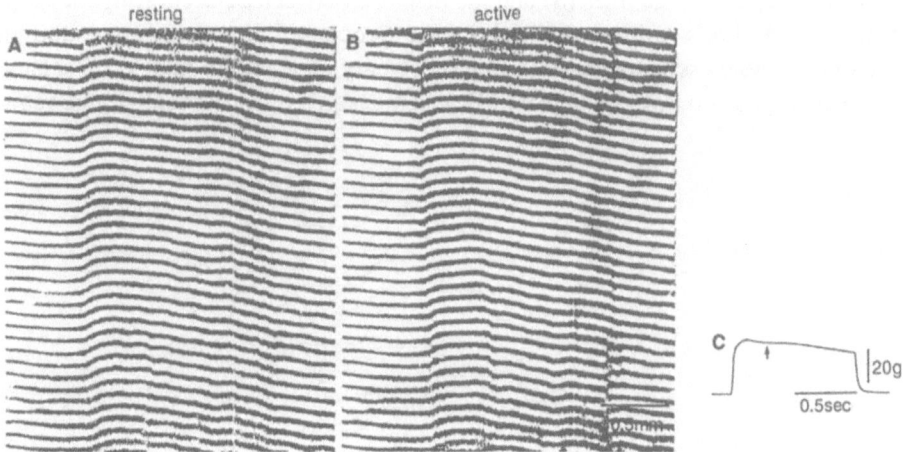

Fig. 4. Original interferograms of a resting muscle (A) and an active one (B). A muscle preparation is in the middle of each frame and the fringe lines are shifted upwards, suggesting that the propagation velocity in the muscle is faster than in water in the both states of the muscle. The interferogram (B) was recorded at the time of the arrow in (C).

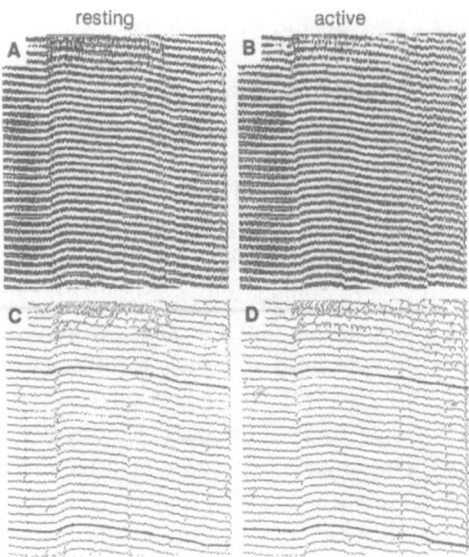

Fig. 5. The image-processed interferograms of the muscle by binarization (A; resting muscle, B; active one) and by thinning (C; resting muscle , D; active one).

shifted upwards in the muscle, indicating that the propagation velocity is faster in the muscle than in water. The interferogram during contraction was taken at the time of an arrow in the force record (Fig. 4C). The fringe lines of original images are too thick to measure exactly the separation of lines for the calculation of the velocity, therefore, they were processed first by binarization (Fig. 5A, B) and then by thinning (Fig. 5C, D) with a computer (PC9801, NEC, Tokyo).

The shift of the fringe lines (S) and the distance between adjacent fringe lines (D) defined in Methods were measured in the processed interferograms and the procedure is shown schematically in Fig. 6. Twenty clear fringe lines were selected in one frame of a processed image and D was measured at forty points in the region

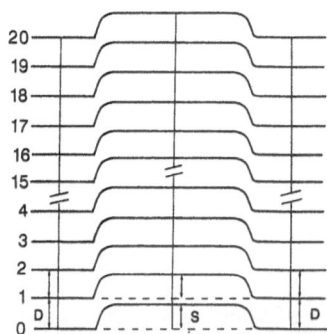

Fig. 6. Schematic illustration to show the shift of the fringe lines (S) and the distance between adjacent fringe lines (D). Normalized shift of fringe lines (N) is S/D.

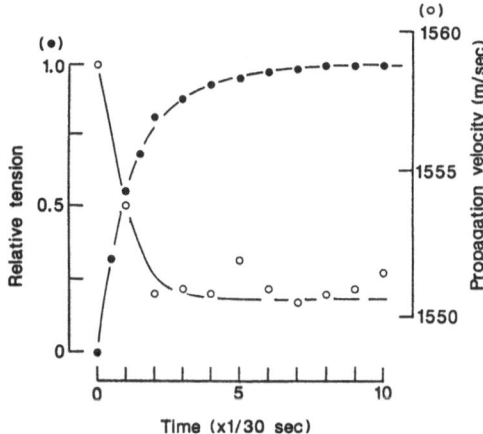

Fig. 7. The time courses of the decrease in the transverse propagation velocity (O) and of the increase in force (●) after stimulation of muscle.

of water at the both sides of the muscle and S at twenty points in the region of the muscle. The normalized shift of fringe lines (N) were obtained from the averaged S and D and the velocity in the transverse direction was calculated by the Eq. (5).

The changes in force and in the transverse velocity after stimulation are illustrated in Fig. 7. The decrease in the velocity reached a plateau faster by 30-40 ms than the increase in isometric force did. The decrease in the velocity was around 8 m/s in this measurement.

The data obtained in 15 preparations are summarized in Table 1. "d" and "n" represent direct or indirect (nerve) electrical stimulation respectively. In all the measurements, the transverse propagation velocity is slower in active muscle and the mean velocity at rest and during isometric contraction is 1545.4 m/s and 1537.8 m/s respectively, the amount of the decrease being 7.6 m/s. The decrease in the stiffness by activation is 2.4×10^7 N/m^2, calculated by the Eq. of $E = (V_x) \times \rho$, where E is stiffness and ρ is the density of the muscle, 1.035 g/cm^3 (8).

DISCUSSION

Characteristics of Stiffness Measured by the Acoustic Microscope

Interpretation of the stiffness measured with the propagation velocity of sound depends on the wave length of the sound used and the diameter of a sample measured. If the wave length is much larger than the diameter, the value of stiffness is nearly equal to so-call Young's modulus, on the other hand, if the wave length is much smaller than the diameter, the velocity is regarded as the phase velocity and the stiffness means the elastic modulus, which is larger than Young's modulus. In this experiment, the frequency of sound and the wave length is 100 MHz and 15 μm

Table 1. Changes in the transverse propagation velocity in resting muscle and in active one in 15 preparations. "d" and "n" denote the contraction induced by direct and indirect (nerve) electrical stimulation respectively. Note that the velocity decreased during contraction in all cases.

No.	stimulation	propagation velocity at rest (m/s)	propagation velocity during contraction (m/s)
1.	n	1560.1 ± 3.9	1552.4 ± 8.6
2.	n	1557.3 ± 2.7	1550.3 ± 2.3
3.	n	1557.9 ± 2.8	1549.1 ± 1.9
4.	d	1551.9 ± 3.0	1546.1 ± 2.3
5.	d	1537.9 ± 2.7	1532.9 ± 2.2
6.	d	1542.9 ± 5.0	1534.8 ± 5.0
7.	d	1544.1 ± 3.3	1536.5 ± 3.2
8.	d	1530.3 ± 2.5	1525.3 ± 2.5
9.	n	1545.6 ± 3.3	1533.7 ± 3.0
10.	n	1542.7 ± 3.0	1528.5 ± 3.8
11.	n	1547.2 ± 4.2	1543.9 ± 2.6
12.	n	1541.3 ± 2.8	1535.3 ± 1.8
13.	n	1548.5 ± 1.6	1544.3 ± 3.1
14.	d	1537.2 ± 2.3	1526.3 ± 2.6
15.	d	1535.7 ± 3.6	1527.1 ± 4.4
mean ± S.E.M.		1545.4 ± 2.2	1537.8 ± 2.3

respectively and the width of a muscle preparation is 1.25×0.9 mm, therefore, the interpretation of the stiffness is thought to be the latter case in the above theoretical consideration. If the stiffness of resting muscle is measured with step length changes or with sound of kilohertz region, it is negligibly small compared with the stiffness during contraction in the longitudinal direction. In this experiment, however, the resting stiffness in the transverse direction largely depends on the bulk modulus of elasticity of myoplasm and the increase in the stiffness by activation can be known by subtracting the resting stiffness from the active one

Possible Mechanism of the Decrease in Stiffness in the Transverse Direction during Contraction

The transverse propagation velocities in resting muscle measured by an acoustic microscope by Vinson et al.[9] and in the techniques of ultrasonic waves by Hatta et al.[5] are 1569-1587 m/s at 25°C and 1530 at 20°C respectively. These values agree with ours in the present experiment, 1545 m/s at 23°C.

The decrease in the propagation velocity in the transverse direction during contraction was found by Tamura et al.[4] using new techniques of ultrasound and studied in detail by Hatta et al.[5]. According to their results, the decrease in the stiffness by activation is 5.6×10^7 N/m^2 at 1 - 2°C and 6.4×10^7 N/m^2 at 19 - 20°C. These values are larger than the present result (2.4×10^7 N/m^2 at 23°C). In the previous experiments[4)5)], the direction of the propagation of ultrasound is just perpendicular to the long axis of the muscle and it is a little oblique (cf. Fig. 2, $\theta_0 =$ 10.29) in this experiment. In the longitudinal direction, the stiffness is known to increase during contraction[2)5)] and it is provable that the longitudinal component of the stiffness change affects the the measurement in the transverse direction.

One possibility of the reason for the decrease in the stiffness in the transverse direction may be the change in the density of a muscle preparation during contraction. If water is squeezed out of the space between fibers by contraction, the density of a whole muscle and then the total stiffness may increase. As a matter of fact, however, the stiffness decreased and furthermore, as shown in Fig. 7, the decrease in the stiffness preceded the increase in force, therefore, the possibility of the change in the total muscle density can be ruled out.

The other possibility is the change in the state of water molecules between myofilaments. Ogata[10)] indicated that there are two different levels of water structures in muscle corresponding to the different resonance frequencies in dynamic ^{1}H-NMR study, major signal being related to a free water group and minor signal to a highly structured one and that the former changes its state depending on relaxation and contraction cycle, being well structured in relaxation and free as pure water in contraction. In addition, it is known that the compressibility of proteins in water may change depending on hydration-dehydration states, because the compressibility of water in dehydration is larger than that of water in hydration, suggesting that hydrated protein is stiffer than dehydrated one[11)]. If this kind of change takes place in muscular contraction, the decrease of the transverse velocity of ultrasound during contraction may be possible.

In the previous experiment[5)], it was proved that the amount of the decrease in the transverse stiffness during contraction decreased linearly with the increasing sarcomere length. This fact indicates that the transverse stiffness change reflects the active process during contraction and an unknown mechanism, e.g. hydration-dehydration change as mentioned above, may be responsible for the phenomenon of the decrease in the transverse stiffness.

REFERENCES

1. Julian, F.J. & Sollins, M R. *J. Physiol. (Lond.)* **66**, 287-302 (1975).
2. Ford, L.E., Huxley, A F. & Simmons, R.M. *J. Physiol. (Lond.)* **269**, 441-515 (1977).
3. Cecchi, G., Griffith, P.J. & Taylor, S. *Science* **217**, 70-72 (1982)
4. Tamura, Y., Hatta, I., Matsuda, T., Sugi, H. & Tsuchiya, T. *Nature* **299**, 631-633 (1982)
5. Hatta, I., Sugi, H. & Tamura, Y. *J. Physiol. (Lond.)* **403**, 193-209 (1988)
6. Korpel, L., Kessler, L.W. & Palermo, P.R. *Nature* **232**, 110-111 (1971)
7. Goss, S.A. & O'Brien,Jr.W.D. *J. Acoust. Soc. Am.* **65**, 507-511 (1979)
8. Truong, X.T. *Am. J. Physiol.* **226**, 256-264 (1974)
9. Vinson, F.S., Eggleton, R.C. & Meiss, R.A. *Ultrasound in Medicine* **4**, 519-534 (1978)

10. Ogata, M. *J. Muscle Res. Cell Motility* **12**, 311 (1991)
11. Gekko, K. & Noguchi, N. *J. Phys. Chem.* **83**, 2706 (1979).

Discussion

Goldman: The transverse stiffness of the filament lattice seems to increase during contraction, according to Yu and Brenner's results [*J. Physiol.* (*Lond.*) **441**, 703-718, 1991]. But you show that the stiffness of the whole muscle, including the water, decreases during contraction. Can you relate these two observations? Why do you think it decreases during contraction?

Tsuchiya: In Brenner and Yu's experiment, radial stiffness was measured by the relation between lattice spacing and osmotic radial force and the interpretation of this kind of stiffness is different from ours, which was based on ultrasonic waves. In our method the bulk modulus of elasticity of the myoplasm can be measured in addition to that of the filament lattice. Furthermore, whole muscle preparations were used in our experiment and we should consider the effects of water between fibers. If water is squeezed out of a muscle during contraction, the density may increase and the propagation velocity should increase. As a matter of fact, it decreased; therefore, we can rule out the possibility of the water between fibers being squeezed out during contraction.

Huxley: Is there any possibility of a change simply in thickness in the muscle if it became narrower and thicker? How would that influence the measurement?

Tsuchiya: The constant thickness of the sample is very important, so we kept the distance between the plastic slip and the quartz stage constant by using a spacer. In addition, we continued to check the thickness by an LED-photodiode device and found that the thickness of the preparation was very much constant during contraction.

Pollack: Professor Ogata (this symposium) finds a change in the structure of water, a collapse of water structure during contraction. Do you think that observation could explain your result?

Tsuchiya: Yes, I am aware of that possibility. Generally, a hydrated protein is stiffer than a dehydrated protein. We must take Professor Ogata's finding into consideration to interpret our results.

Yu: Did you measure the transverse transmission in rigor muscle?

Sugi: We already measured rigor muscle transverse and longitudinal stiffness and found that the values are almost the same. Our work is already published in *Nature* and in the *Journal of Physiology* [Tamura et al. *Nature* **299**, 631-633, 1982; Hatta et al. *J. Physiol.* (*Lond.*) **403**, 193-209, 1988]. The stiffness value doesn't change appreciably, irrespective of the direction of the ultrasonic wave measurement. So there is a considerable difference between contracting and rigor states.

X. KINETIC PROPERTIES OF ACTIN-MYOSIN SLIDING STUDIED BY ENERGETICS EXPERIMENTS

INTRODUCTION

In the field of muscle energetics, the main object of study is how a muscle converts chemical energy derived from ATP hydrolysis into mechanical work under a variety of conditions. All theories of muscle contraction hitherto presented have been primarily evaluated by the way to account for a phenomena obtained in this field, such as the Fenn effect and the force-velocity are force-energy relations established by A.V. Hill and coworkers.

The first two papers by Curtin and Woledge are concerned with power output and efficiency in white muscle fibers of the dogfish when they are subjected to sinusoidal length changes with varying timing of stimulation and frequency of length changes. The results are compared with those obtained from ramp fiber shortening and tetanic stimulation. They put forward a hypothesis which accounts for the results in terms of the behaviour of strained cross-bridges. Using the technique of phosphorous NMR spectroscopy and HPLC analysis, Kushmerick and coworkers established a criterion to distinguish mammalian skeletal muscle fiber types based on the content of energetically important metabolites.

Godt and others made extensive studies on the effect of ionic strength on maximal Ca^{2+}-activated force and ATPase activity in skinned rabbit and lobster muscle fibers, and found that, in rabbit fibers, high ionic strength decreased Ca^{2+}-activated force, but had little effect of ATPase activity, indicating an increased tension cost of contraction. They explained the effect of high ionic strength to be partly due to destablization of actomyosin complex. Taylor and Suga finally discussed the mode of cardiac muscle performance based on their contraction model consisting of parallel and series elastic elements and cross-bridge contractile element with appropriate cycling characteristics.

POWER AND EFFICIENCY: HOW TO GET THE MOST OUT OF STRIATED MUSCLE

Nancy A. Curtin* and Roger C. Woledge**+

*,**Marine Biological Association UK
Plymouth
*Charing Cross & Westminster Med. Sch.
London W6 8RF
**University College London
London WC1E 6BT

ABSTRACT

White muscle fibres from dogfish were stimulated during sinusoidal cycles of shortening and lengthening that mimic the *in vivo* pattern of movement. The results show that the timing of the stimulation relative to movement and also the frequency of the movement affect mechanical power output and efficiency. Maximum mechanical power is produced at a higher frequency of movement than maximum efficiency. The value of maximum power for a cycle of movement is less than half that produced during ramp shortening of fully active fibres. Higher efficiency is achieved during cyclic movement than during ramps, probably because work can be done with lower "overhead" costs for activation and because little ATP is used during lengthening.

INTRODUCTION

In vivo locomotion is produced by repeated cycles of shortening and lengthening by skeletal muscle fibres and the blood is pumped by the heart during similar cycles of the myocardial cells. Another feature of *in vivo* function is that activation occurs during only a fraction of each cycle of movement.

Although this is how muscles work *in vivo*, this type of performance has not been studied much in experiments on isolated muscle. The aim of our experiments was to measure the muscle capabilites when stimulated during cycles of sinusoidal

---

+Correspondence to: Dr. N.A. Curtin Dept. Physiology Charing Cross & Westminster Med. Sch. Fulham Palace Rd. London W6 8RF UK

Mechanism of Myofilament Sliding in Muscle Contraction, Edited by
H. Sugi and G.H Pollack, Plenum Press, New York, 1993

movement. We focused on the muscle's ability to produce mechanical power and on its efficiency at converting chemical energy into work.

From earlier studies we know that shortening affects the ability to produce force both during and after shortening[1-4]. These effects of shortening depend on the velocity of movement and on the pattern of stimulation. Thus we expected that the results for tetani during sinusoidal movements might be different from the results of experiments using the more usual design, ramp shortening after a period of isometric contraction.

METHODS

All of the results are from the fast, white fibres from the body of the dogfish, Scyliorhinus canicula, contracting at normal body temperature, 12°C. We have chosen to use muscle from this small shark, because during swimming the muscles move in a simple sinusoidal way. In addition, the dogfish has been a particularly useful experimental animal for studies of motor control and from these we have some information about the temporal pattern activation of the muscle fibres by motor neurons during swimming [5] and references therein). The muscle fibres are suitable for study as isolated preparations because they are easy to dissect, have good tendons, and the fibre types are well segregated in different parts of the fish body.

Small bundles of muscle fibres were attached at one end to a motor that controlled fibre length and at the other to a force transducer. They were stimulated tetanically by end-to-end electrical stimulation. The temperature of the fibre preparation was measured with a thermopile consisting of antimony-bismuth thermocouples made by vacuum deposition of the metals onto a mica substrate [6] supplied to us by G. Elzinga). The heat production by the muscle fibres was calculated from the temperature change using the heat capacity of muscle, etc. determined from control heating by the Peltier effect.

RESULTS AND CONCLUSIONS

Fig. 1 shows a record of length change and force during sinusoidal movement with stimulation during 25 % of the mechanical cycle. The amplitude of the length change was 9 % of L_0, the fibre length giving maximum isometric force. The net work was found by multiplying force by the rate of change of length and then integrating the result. The net work is the difference between the work done *by* the muscle during shortening and the work done *on* it (by the motor) during lengthening. The average mechanical power during one complete cycle of movement was found by dividing the net work in the cycle by the duration of the cycle.

The efficiency was found by dividing the work by the total energy output, net work + heat, during the cycle. All of this energy is provided by chemical reactions, ultimately ATP splitting.

We varied the the frequency of the movement in the range 1.33 to 5 Hz and the timing of stimulation relative to the movement (stimulus phase) to find the combination that gave the maximum mechanical power output and the maximum efficiency.

For faster mechanical cycles, we found that it was necessary to start stimulation earlier in the cycle, before shortening, in order to get the maximum mechanical power and efficiency. So, the faster the mechanical cycle, the more of the stimulation occurred during lengthening. Fig. 1 shows records for conditions that

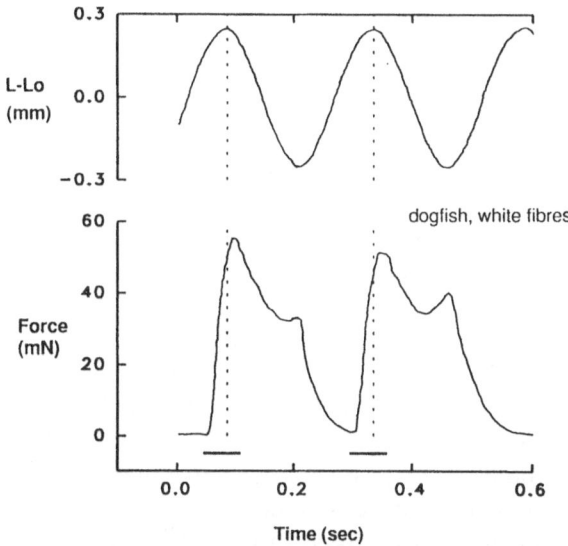

Fig. 1. Example records of length change and force for one preparation of muscle fibres. The vertical dotted lines mark the start and end of the cycle from which measurements were made. The heavy lines below the force record show the two periods of stimulation. Downward movement of the length record indicates shortening. Movement was centered around L_0, the length giving maximum isometric force.

gave the maximum power for this fibre preparation; movement was at 4 Hz. Note that the stimulation occurred largely during the lengthening part of the cycle, before shortening started.

Fig. 2 summarizes maximum values of mechanical power and efficiency found by optimizing the timing of the stimulation for each mechanical frequency. The results for this preparation of white fibres are typical in showing that maximum power is produced at a higher mechanical frequency (faster movement) than maximum efficiency. In this case the fibre preparation had to move twice as fast, 4 Hz, to produce maximum power as it did for maximum efficiency, 2 Hz.

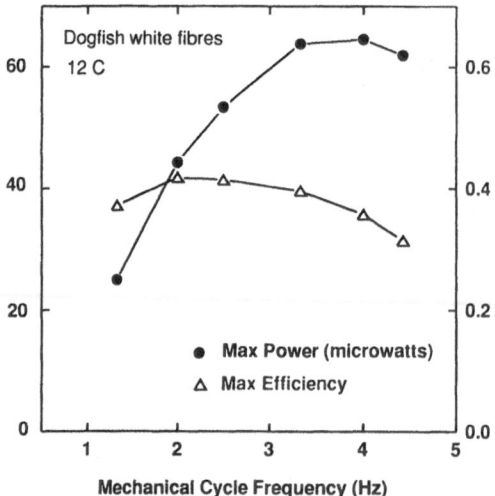

Fig. 2. Values of maximum power and maximum efficiency at each of the frequencies of sinusoidal movement tested. The maximum power and maximum efficiency values were found by varying the timing of the stimulation relative to movement. Results from the same fibre preparation as Fig. 1. Left vertical axis = POWER. Right vertical axis = efficiency.

It is interesting to compare the maximum power we found during sinusoidal movement with the maximum power produced during ramp shortening of fully active fibres. For one complete sinusoidal cycle of shortening and lengthening under optimum conditions the mechanical power was 0.144 ± S.E.M. 0.018 W/g dry weight (n = 11 fibre preparations). In previous experiments we found that during ramp shortening the maximum power output from white fibres is 0.308 ± S.E.M. 0.019 W/g dry weight (n = 6 fibre preparations)[7)8)]. Thus, the power output is lower for sinusoidal cycles. This must be due, at least in part, to the fact that during half of the sinusoidal cycle the muscle is lengthening and cannot do useful mechanical work.

How do the velocities of movement compare for conditions giving maximum power? During the sinusoidal movement, velocity continuously changes, but the average velocity was 0.62 L_0/s (= 0.09 L_0 shortening / [0.5 * 0.29 s], where 0.29 s is the average cycle period giving maximum power). In our previous experiments with ramp shortening maximum power was produced at shortening velocity 1.07 ± S.E.M. 0.03 L_0/s. So it seems that the velocity of movement optimal for power output is slower for the sinusoidal movement than for ramp shortening.

Maximum efficiencies for sinusoidal movement and for ramp shortening can also be compared. Sinusoidal cycles give a maximum efficiency of 0.42 ± S.E.M. 0.02 (n = 11 fibre preparations), which is somewhat higher than that found during ramp shortening, 0.33 ± S.E.M. 0.01 (n = 23 from 7 fibre preparations)[8)]. Why might cycles of movement give a higher efficiency? During the cyclic movements, the muscle is only stimulated for part of the time, whereas stimulation is continuous during ramp shortening. Thus there is some saving in the cost of the Ca^{2+} reactions that use ATP, but do not directly contribute work. These "overheads" are probably less during cycles than ramps. It is also relevant when considering cycles that ATP splitting by the fibres during lengthening is small, even when high force is produced[9][10].

ACKNOWLEDGEMENTS

We thank the Wellcome Trust for financial support.

REFERENCES

1. Hill, A.V. *Proc. R. Soc. Lond. B* **126**, 136-195 (1938).
2. Edman, K.A.P. *J. Physiol. (Lond.)* **246**, 225-275 (1975).
3. Marechal, G. & Plaghki, L. *J. Gen. Physiol.* **73**, 453-467 (1979).
4. Edman, K.A.P. *Acta Physiol. Scand.* **109**, 15-26 (1980).
5. Mos, W., Roberts, B.L. & Williamson, R. *Phil. Trans. R. Soc. Lond. B* **330**, 329-339 (1990).
6. Mulieri, L.A., Luhr, G., Trefry, J. & Alpert, N.R. *Am. J. Physiol.* **233**, C146-C156 (1977).
7. Curtin, N.A. & Woledge, R.C. *J. Exp. Biol.* **140**, 187-197 (1988).
8. Curtin, N.A. & Woledge, R.C. *J. Exp. Biol.* **158**, 343-353 (1991).
9. Curtin, N.A. & Davies, R.E. *Cold Spring Harbor Symp. Quant. Biol.* **37**, 619-626 (1973).
10. Curtin, N.A. & Davies, R.E. *J. Mechanochem. Cell Motility* **3**, 147-154 (1975).

Discussion

Suga: Two questions. You showed the power and efficiency curves. If you draw the work curve, does the work peak come between the power peak and efficiency peak? The second question is: you compared ramp and sinusoidal curves, but if you chose rectangular external work curves would the efficiency be higher?

Curtin: The answer to your first question is that the peak work (the net work for a complete cycle of movement) is produced during movement at 2 Hz; so, the work curve is shaped like the efficiency curve. As to your second question, we have done experiments in which we imposed a ramp-and-hold pattern instead of a sinusoidal movement. We were surprised at how similar the results were, at least for power, to the sinusoidal movement. We haven't measured efficiency under these conditions. But I suspect that it will be similar to what we found with sinusoidal movement.

Huxley: Do you think that in the sinusoidal movement, some ATP is being reconstituted from ADP and P_i during the stretch phase, or is that what Roger Woledge is going to tell us in the next communication? This was discussed a great deal more than thirty years ago by A.V. Hill and his collaborators (e.g., Hill and Howarth, *Proc. R. Soc. Lond. B* **151**, 169-193, 1959).

Curtin: I don't think we have any reconstitution. Certainly, in the experiments on frog muscle done in Bob Davies's lab to follow up on A.V. Hill's ideas, the results showed that when a tetanus included an isometric phase followed by a stretch phase, more ATP was used than in the isometric phase alone; there was no net gain in ATP (Infante et al. *Science* **144**, 1577-1578, 1964). In my experiments, there was always a net breakdown of ATP during stretch (Curtin, N.A. and Davis, J.S. *Cold Spring Harbor Symp. Quant. Biol.* **37**, 619-626, 1973).

Woledge: I'll just add that there is no endothermic phase during the stretch such as one might expect from the reconstitution of ATP. The extra efficiency in the cycles is not due to the presence of an endothermic process during the stretching phase.

THE EFFICIENCY OF ENERGY CONVERSION BY SWIMMING MUSCLES OF FISH

Roger C. Woledge and Nancy A. Curtin*

Department of Physiology
University College London
Gower St., London WC1E 6BT, U.K.
*Department of Physiology
Charing Cross & WestiminsterMedical School
London W6 8RF.

ABSTRACT

The efficiency of energy conversion by fish myotomal muscle undergoing sinusoidal length changes with brief periods of stimulation in each cycle has been found to be greater than the efficiency of the same type of muscle when tetanically stimulated. This finding is reviewed in relation to the well-known energetic properties of muscle undergoing shortening. It is suggested that an adaptation of the ideas of Lombardi & Piazzesi[18] which were proposed to explain the behaviour of muscle during lengthening could be used to provide one possible explanation of this finding. Their theory proposes that crossbridges, when they have sufficient energy because they are strained, can detach to a crossbridge state which can rapidly re-attach without having to split ATP first. This type of theory might also provide an explanation of the findings on ventricular energetics which are expressed in the time varying elastance model of Suga[11].

Introduction

The efficiency of muscle in performing mechanical work is interesting from several points of view. Firstly it is a constraint on theories of muscle contraction. For this purpose the relevant definition of efficiency would be the ratio of the work done to free energy provided by ATP splitting by the myofibrils. Secondly high muscular efficiency must confer significant evolutionary advantages on an animal because it tends to increase the maximum steady running speed, the migratory range, the maximum duration of a sprint and so on. In fact, muscle efficiency is very variable across different species[1][2] and even between red and white muscle fibre types in one species (Curtin and Woledge, unpublished experiments on dogfish).

Mechanism of Myofilament Sliding in Muscle Contraction, Edited by
H. Sugi and G.H Pollack, Plenum Press, New York, 1993

This raises the question of what compensating advantages can be obtained by animals having relatively inefficient muscles. The relevant measures of efficiency for the purpose of such a discussion is the ratio of work done to the free energy provided by all the ATP split during the contraction-relaxation cycle, including that split by the SR as well as by the myofibrils. For some purposes one might prefer to compare the work done with the total free energy from the oxidation of substrate required to resynthesise the ATP used. There are thus a number of useful definitions of efficiency, each appropriate to its own purpose. In studying fish muscle we have been expressing the efficiency as the ratio of work done to the total energy given out as heat and work during either the whole or part of a contraction relaxation cycle. In discussing the results we shall assume that the energy comes from the splitting of ATP and the concomitant resynthesis of ATP from PCr. This ATP splitting will include that split by the SR calcium pump.

We have reported previously[3] that the efficiency of white myotomal muscle, measured during a period of isovelocity shortening by a tetanised muscle, has a maximum value of 33 ± 1% (S.E.M.)). This is a low value compared to that in comparable experiments on frog (39%)[4] or tortoise (72%)[5] muscle. These differences are probably due to differences in the efficiency of energy conversion by the myofibrils rather than to differences in the proportion of ATP used by non-myofibrillar ATPases. In the study reported here[6] we measured the overall efficiency for a single cycle of sinusoidal movement of the muscle during which the muscle is activated and then relaxes. We expected the value to be below the efficiency we had measured for a period of steady shortening, because the muscle can be shortening under optimum conditions for only a part of the cycle. During the rest of the cycle the muscle is either shortening under conditions less than optimal, or lengthening, i.e. absorbing work, while the rate of energy output would be expected to remain positive throughout the cycle. It is surprising then that the value obtained for the overall efficiency was not less, but greater than that from steady shortening (in fact 42 ± 5% (S.E.M.)). This paper attempts to relate this finding to other reported facts and theories about the energetics of shortening and muscle efficiency.

The Historical Background

Ideas about the energetics of shortening have been dominated for many years by the observations of Hill in 1938[4] and the explanations given of them by Huxley in 1957[7]. As this work is so well known I will summarise it here extremely briefly. Hill discovered that the rate of energy output, i.e. rate of work production plus rate of heat production, increased markedly above the isometric level when muscle was made to shorten. Fig. 1 illustrates this phenomenon with some of our experiments with dogfish muscle. Huxley's explanation is that the energy from crossbridge turnover and consequent ATP splitting is increased by shortening. In the isometric state the turnover is limited by the low rate at which the attached crossbridges break off. Because the crossbridge has quite a short working stroke in the model, the rate

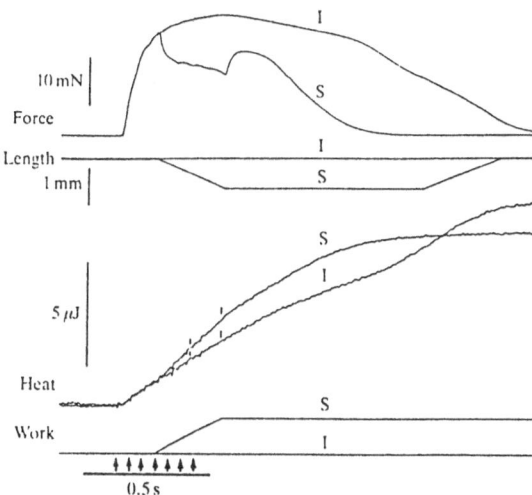

Fig. 1. Experimental records obtained from a bundle of white myotomal muscle fibres from the dogfish. The records show force length, heat and work during (I) a tetanus under isometric conditions and (S) a tetanus that includes a period of shortening at constant velocity. The amplitude of shortening was 0.115 fibre lengths and the velocity was 0.46 fibre lengths per second. Dry mass of fibres was 199 μg. Reproduced from reference 3.

of breaking has to be supposed to increase markedly when shortening occurs. As crossbridge breaking is the rate limiting step this causes an increase in energy output, part of which becomes work, and the remainder heat. This behaviour has been found in every kind of skeletal muscle that has been investigated, although the relative quantities of work and heat vary[3)5)8)]. It is important to recognise that all these experiments are made while the muscle is continuously tetanically stimulated, and thus probably fully, or almost fully, activated. As far as I know it is an open question whether cardiac muscle would also behave in this way if a high level of activation could be sustained.

The Fenn effect, as originally described in 1924[9)], referred to the way in which the total energy output in a twitch varied with how much shortening occurred. Fenn's experiments on frog muscle showed that shortening caused an increase in the total energy output above the isometric level. This result was in contrast to the then current theoretical idea that the amount of energy that would be produced was determined by the initial length of the muscle and was independent of what subsequently happened mechanically during the contraction. Seemingly Hill's 1938 discoveries encompass Fenn's results. If a twitch is regarded simply as a brief period of muscle activity during which the rate of energy output is dependent on shortening velocity, then clearly the total energy output will depend on how much shortening has occurred. This view which was proposed for example by Hill[10)] is however much too simple; muscles actually behave in a way which is more complex, more efficient energetically and more interesting than suggested by this notion. Much of the evidence for this is quite old, but more recent work, particularly that of Dr

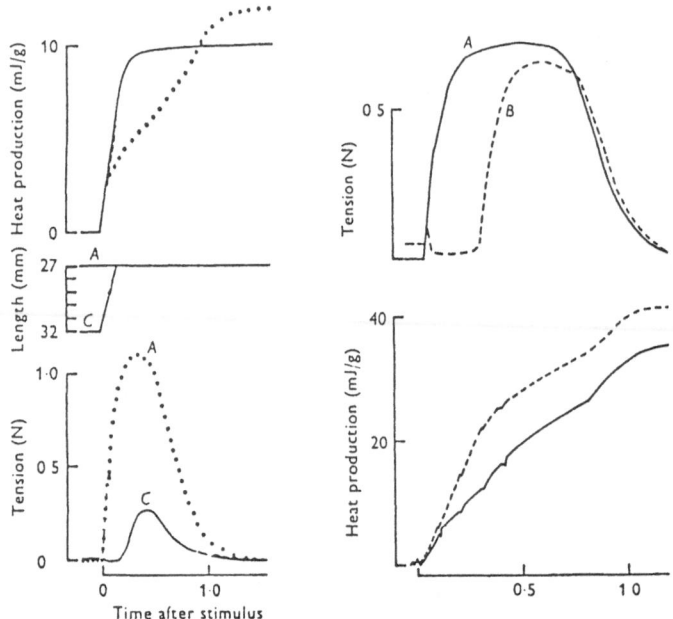

Fig. 2. Experimental records from frog sartorius muscle. *Left*, The records show heat, length and tension during twitches. In A the muscle was isometric, in C it shortened 5 mm against a very small load. *Right*, Heat and tension records are shown for two contractions. In A the muscle is isometric. In B it shortens 8 mm under light load at the start of the contraction. The muscle was stimulated in each case by five stimuli at 100 msec intervals. Reproduced from reference 19.

H. Suga[11] and his colleaques on cardiac muscle, have highlighted the energetic significance of these phenomena, which as yet are not fully explained in crossbridge terms.

Unlike Hill's observations on rates of energy output in well activated muscle, Fenn's observations did not seem to be equally applicable to all forms of muscle. Even skeletal muscles did not always show any increase above the isometric level of energy ouput when shortening occurred in a twitch[12] and such an increase is generally absent in cardiac muscle. These facts have been reviewed repeatedly[13-15] and two ideas have emerged to explain them.

The first of these ideas is that shortening has two distinct energetic consequences: (1) a increase in crossbridge turnover and therefore of energy output; (2) an increase in the rate of relaxation, a deactivating effect, which spares energy output. Fig. 2 shows observations which illustrate the existence of both these effects during a twitch of frog skeletal muscle. The heat produced in a twitch with rapid shortening under a light load is at a very high rate during the early part of the twitch, well above the isometric rate. But then heat production slows to well below the isometric rate and the muscle is unable to exert appreciable isometric force. The shortening has induced an earlier relaxation than in the isometric contraction and the energy spared in this way is greater than the extra energy expended in shortening. Thus the total energy output in the twitch with shortening is less than in the isometric twitch. This tendency of the shortening to deactivate is of course well known from mechanical studies[16)17)] and is greatly diminished by giving further stimuli, with the result that the total energy output in the contraction is then greater than it is in the isometric control. On this view the explanation that would be given of the variable results of experiments like Fenn's is that the balance between the energy sparing and energy dissipating effects of shortening differs in different muscle preparations. In particular, in cardiac muscle, with its generally lower level of activation the energy sparing effects are much more marked and therefore the total energy falls as the amount of shortening is increased, rather that rising as it did in Fenn's experiments.

The second idea to be raised, to account for the failure of cardiac muscle to show an increase in energetic cost above the isometric level when it shortens, is that the energy should be partitioned into that required for tension maintenance and that required to perform work[14)]. This approach does not lead readily to a crossbridge analysis of the phenomena because although each crossbridge cycle contributes both work and tension-time integral (impulse) there seems to be no constraint that increase in impulse would be accompanied by decreased work or vice-versa.

Recent Work

Two types of experiment have suggested recently that the available explanations of the energetics of shortening are not by any means complete. The first is the observations by Suga[11)] on the energetics of cardiac muscle. These have been summarised in the remarkably simple empirical description named the "time varying elastance model". What has been found is that the energetic cost of a beat is determined at an early stage in its development (not as early as the moment of stimulation) and is then independent of whether or not shortening occurs. The ideas in the previous paragraph could in principle explain this sort of result, but it would be very unlikely that the positive and negative effects of shortening would frequently just cancel each other. It appears rather that the cardiac muscle is behaving late in the twitch in a visco-elastic manner! This certainly challenges theories based on turnover of crossbridges which can act elastically only over a short range.

The second recent experiment is our own observations on fish muscle undergoing brief contractions with cyclic motion simulating the normal locomotor behaviour of the fish. As mentioned above we find that overall efficiency in this type of contraction is certainly as high as, and probably higher than the maximum efficiency during a continuing tetanus. Deactivation by shortening should certainly help to improve efficiency in these experiments, for if the muscle remains active after shortening is over, then it not only incurs an energetic cost but also absorbs work during the subsequent lengthening, which would be largely dissipated as heat. However, this phenomenon itself cannot expain how the efficiency for a whole cycle could exceed the maximum during shortening in a tetanus. There has to be a period within the cycle when the efficiency exceeds that maximum during a continuing

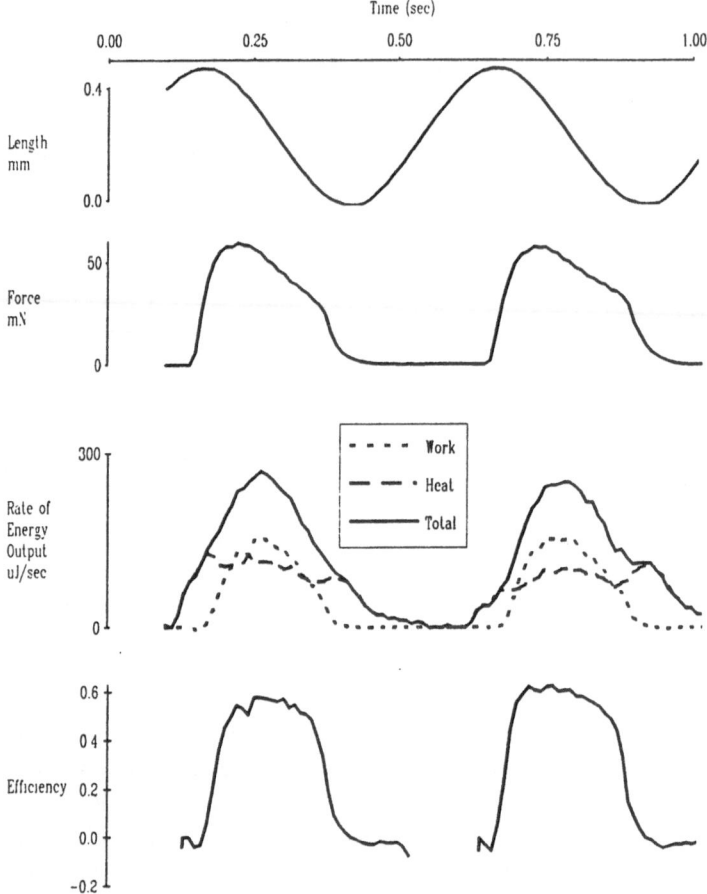

Fig. 3. Experimental records from a bundle of white myotomal muscle fibres from the dogfish. The records show length, force, rate of energy output and efficiency. The rate of energy output as work and heat are shown as well as the total. The efficiency is the rate of work output divided by the total rate of energy output. The efficiency record is omitted when the rate of energy output is below 3% of its maximum value. The overall efficiency for the complete cycle shown between the arrows was 0.432. The timing of the stimulus and the rate of the oscillations were chosen to give the largest efficiency that could be obtained from this preparation.

tetanus with steady shortening. That this is indeed the case is illustrated by the experimental records in Fig. 3. Two ideas occur to us which might be able to provide an explanation. Firstly there could be a large saving in the energetic cost of activating the muscle because in the cycle experiments it is not fully active during the whole time when it is shortening. Secondly that the process of energy conversion by the myofibrils might be intrinsically more efficient when the muscle is not fully activated. While we have not at present investigated these two possibilities experimentally, it is interesting to note that following the second could lead to ideas applicable to explaining the results of Suga on cardiac muscle.

These two experimental situations, the beating heart and the fish muscle undergoing sinusoidal length changes, have in common that the number of crossbridge cycles by each crossbridge per complete sequence of activation and relaxation is small. We can estimate this number for our experiments by dividing the total energy output per gram of muscle per cycle of movement by the molar enthalpy for PCr splitting to get the number of moles of ATP split by 1g of muscle (dry wt) in one cycle of movement. The answer is about 3 µmoles. Since a gram of dry muscle contains about 1.5 µmol of myosin heads, the number of ATP molecules split by an average site is about 2. A similar calculation for the beating heart (Dr.H.Suga, personal communication) gives a value of about 1 ATP per myosin head per beat.

An Hypothesis

According to crossbridge models of muscle contraction the formation of a crossbridge is a reaction unlikely to be reversed; the crossbridge once formed is committed to go through a cycle involving ATP splitting before it can return to its initial state. As a muscle becomes progressively less active after responding to one or a few stimuli it might well lose the ability for new crossbridges to be formed at a stage when attached crossbridges are still exerting force, and doing work. When the number of crossbridge cyles in the whole contraction is as low as 1 or 2, this stage, at which new crossbridge cycles can no longer be committed, should be considered to start relatively early, and represent a significant fraction of the whole contraction-relaxation cycle. Since the total number of crossbridge cycles is fixed by the onset of this phase of the contraction, total energy output from the muscle is then determined. If the muscle was isometric during this phase of contraction all the energy would appear as heat. If shortening occurred there would be some work done, and a corresponding reduction in the heat production. It is suggested that this type of behaviour underlies the success of the time-varying elastance model of Suga in explaining cardiac energetics.

If, as is generally supposed, crossbridges can only exert force over some 15 nm displacement of the filaments which they connect, then a shortening of about 1 % of muscle length would reduce force to zero, and no recovery would be possible unless new crossbridges could form. How can we explain a continued ability to shorten and do work while larger length changes occur, within the general framework of a crossbridge type of model? A possible method is suggested by the theoretical work of Lombardi & Piazzesi[18]. Their work is concerned with finding an explanation for the behaviour of active muscle while it is being stretched. In their successful model, which is summarised in Fig. 4, a crossbridge that has been highly stretched can detach into a state (D2, see Fig. 4) capable of immediate re-attachment (without having to split ATP). These crossbridges are capable of much more rapid attachment than a normal detached bridge which has not been subjected to stretch. Thus

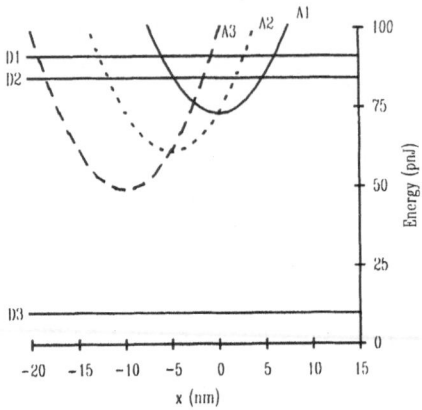

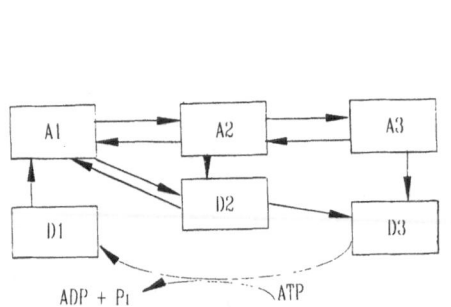

Fig. 4. *Top,* The crossbridge states hypothesised in the theory of Lombardi and Piazzesi[18], A1, A2 and A3 denote attached states of the crossbridge which exert progressively more force. D1 denotes the normal detached state from which crossbridges can form attachments. D3 (not included in Lombardi & Piazzesi's treatment) denotes a crossbridge which, having detached, must split ATP before another cycle of attachments can occur. D2 is the special detached state which can only be formed by breaking of a crossbridge under considerable strain. It is supposed that re-attachment of D2 to form A1 occurs more rapidly than attachment from the D1 state. The arrow connecting D1 to D3 shows a possible reaction not considered by Lombardi & Piazzesi. *Bottom,* The free energy of the different crossbridge states considered in this model as a function of the crossbridge distortion x. As usual the attached states are represented by parabolas because of the work that can be stored in the bridges as they are distorted. The important feature of this model is the detached state D2 which has a high energy level. Because of this, bridges can only enter this state when they are considerably distorted and bridges in this state are able to attach and do work without splitting ATP.

crossbridges are capable of going through more than one cycle per ATP split. There is no energetic problem inherent in this for muscles that are being stretched because the energy is provided by the work done on the muscles.

A possible extension of their model, which would be applicable to the situations we are discussing here, is to suppose that highly shortened crossbridges as well as highly stretched crossbridges are capable of detaching to state D2. As is illustrated in Fig. 4 this can only happen if these bridges acquire enough energy to reach the point at which the A3 curve crosses the D2 line. This energy would have to come from other crossbridges which were descending the energy curves, so only a proportion of the bridges can reach the D2 state. A crossbridge that does do so will have gone through a cycle without contributing net work. It will be able to re-attach rapidly and then contribute work in subsequent shortening. The effect of such cycles is to allow the same amount of work to be done per ATP but over a greater length range than on the traditional view. We do not know whether a self consistent kinetic scheme using this idea could be worked out that would explain the mechanics and energetics of shortening in muscle which is not fully active. Because of the success of this type of model in explaining the previously puzzling phenomena during lengthening it would seem worthwhile to attempt to extend it to the behaviour during shortening.

What would the effect of the mechanism proposed above on the efficiency during shortening ? We can get some idea about this by considering the reasons for the efficiency being less than its maximum value. The maximum would be attained only if each crossbridge became attached at the x value at which the A1 curve and D1 line cross, and then remained attached until it reaches the minimum energy point on the A3 curve. The causes of a lesser efficiency would be (1) attachment to the left of the optimum x-value (2) detachment before reaching the minimum energy (3) remaining attached after passing this point. In the absence of the D2 state any crossbridge that passes the minimum on the A3 curve is destined to waste energy. The D2 state allows some bridges that remain attached to complete a cycle without this waste, and to contribute work in a subsequently. Thus the addition of a route to this state (D2) for shortening muscle will tend to increase efficiency. What we have observed in fish muscle is a higher efficiency when the muscle is not fully active. Perhaps we could suppose that the action of the D2 state in enhancing efficiency is less effective in fully active muscle. This could be explained by a competition for attachment sites between the D1 and D2 states. When the muscle was fully active the D2 state might fail to find an attachment site and therefore "decay" to the D3 state. Clearly these ideas are highly speculative, but by analogy with the work of Lombardi and Piazzesi on which they are based, they perhaps warrent further development.

REFERENCES

1. Hegland, N.C. & Cavagna, G.A. *Am. J. Physiol.* **253**, C22-C29 (1987).
2. Woledge, R.C., In: *Energy Transformation in Cells and Organisms; Proc. 10th Conf. Eur. Soc. for Comp. Physiol. & Biochem*, pp 36-45. eds: Wieser, W. & Gnaiger, E. (1989).
3. Curtin, N.A. & Woledge, R.C. *J. Exp. Biol.* **158**, 343-353 (1991).
4. Hill, A.V. *Proc. Roy. Soc. Lond. Ser.B* **126**, 136-195 (1938).
5. Woledge, R.C. *J. Physiol. (Lond.)* **197**, 685-707 (1968).
6. Curtin, N.A. & Woledge, R.C. This volume, 729-734.
7. Huxley, A.F. *Prog. Biophys. Biophys. Chem.* **7**, 255-318 (1957).
8. Gibbs, C.L. & Gibson, W.R. *Am. J. Physiol.* **223**, 864-871 (1972).
9. Fenn, W.O. *J. Physiol. (Lond.)* **58**, 175-203 (1923).
10. Hill, A.V. *Proc. Roy. Soc. Lond. Ser.B.* **136**, 195-211 (1949).
11. Suga, H. *Physiol. Rev.* **70**, 247-277 (1990).
12. Hill, A.V. *Proc. Roy. Soc. Lond. Ser. B.* **109**, 267-303 (1930).
13. Woledge, R.C. *Prog. Biophys. Mol. Biol.* **22**, 37-74 (1971).
14. Mommaerts, W.F.H.M. *Naturwissenschaften* **57**, 326-330 (1970).
15. Rall, J.A. *Am. J. Physiol.* **242**, H1-H6 (1982).
16. Edman, K.A.P. *J. Physiol. (Lond.)* **246**, 225-275 (1975).
17. Edman, K.A.P. *Acta Physiol. Scand.* **109**, 15-26 (1980).
18. Lombardi, V. & Piazzesi, G. *J. Physiol. (Lond.)* **431**, 141-171 (1990).
19. Dickinson, V.W. & Woledge, R.C. *J. Physiol. (Lond.)* **233**, 659-671 (1973).

Discussion

Suga: You first showed that among the different animal species—tortoise, frog and dogfish—the efficiency changes between 40% and 80%. But when you chose the figure of dogfish and calculated the efficiency in red and white muscles, they were about the same. Probably the heart contains two different types of myosin and ATPase activity, and the difference is a factor of about three. Is it reasonable to have similar efficiency values for those two types of myosin isoforms?

Woledge: Yes. I don't think everything you do that alters the power of the muscle will alter its efficiency. I think rather that it might be the other way about: the things you do to alter the efficiency alter the power. It is not difficult to think of ways to change the power output while leaving the efficiency unchanged. One way would be to change the rates of all the reactions in proportion. That would leave the efficiency unaltered, but make the muscle more powerful. So, I would suggest that in the case of the cardiac isoenzymes this kind of change is what takes place.

Gergely: I wonder how you relate your efficiencies to thermodynamic efficiencies? What would the relationship of ΔG and ΔH be under the prevailing concentration conditions?

Woledge: 1.35 is the answer to your question. The ΔH value is 34 kcal/mole. The ΔG value is 46. The ratio is 1.35. So the efficiencies should be divided by 1.35 if one wants to find the thermodynamic efficiency, assuming that ATP is driving it. However, if you want that, you must also dissect off that part of the energy used by calcium pumping, which is irrelevant. That means you must then increase it by about 25%. So, the net correction to be made is on the order of 10% or 15% rather than 30% or 40%.

Brenner: This concept that during shortening cross-bridges are moved above the detached state, *i.e.*, are forced to detach with rapid reattachment to a subsequent binding site on actin is exactly the concept proposed by Cooke and Bialek some years ago (*Biophys. J.* **28**, 214-258, 1979), at least in the part of the paper concerned with isotonic shortening.

Woledge: Good. I'm glad that this is a reasonably well-accepted mechanism. To me it sounded rather revolutionary. I'm not sure whether Pate and Cooke ever discussed the energy level of the detached state. It seems to me that this is a critical point.

Huxley: I think you showed that the high efficiency of the tortoise muscle is linked to the highly curved force-velocity curve. Does that correlation exist in any of these other differences? I mean, comparing dogfish with frog, does the dogfish have a very straight force-velocity curve?

Woledge: The curve is just a tiny bit straighter than that for frog muscle. It isn't very remarkable. If I could just amplify Professor Huxley's point briefly: if you make all the bridges break off at the minimum energy point, then V_{max} is infinite. This is because there is no force opposing the force pushing the bridges on. In tortoise muscle, V_{max} is very high compared to the other speeds at which the muscle can work.

Goldman: With regard to the idea mentioned by Roger Woledge that free energy might be recovered if a cross-bridge remains attached while filament sliding carries it past the minimum of its free energy curve, a similar idea was mentioned before in a paper by Cooke and Bialek (*Biophys. J.* **28**, 241-258, 1979). The curve shape represents the component of free energy stored in the compliant part of the cross-bridge that is under tension when the cross-bridge is exerting a normal sliding force. The compliant region is compressed when the cross-bridge has slid beyond the minimum of the free energy curve and is exerting negative force. If the energy recovered when the cross-bridge detaches from the compressed position is to be available to produce another increment of sliding, the structure must return to the tension configuration. Do you have a physical or structural idea of how this rearrangement of the cross-bridge compliance can occur?

Woledge: Actually, I don't. But it seems that the energy is no longer in the spring once you've detached it. It has to be spread out into some other places in the molecule—but where that is I have no clue. Maybe someone could throw some light on that.

Huxley: I've always drawn the spring somewhere in the S2, but that doesn't mean I positively believe it's there. One of the diagrams I had long ago put the spring inside the myosin head. You need to have some trapping mechanism for holding it in the compressed state. That's complicated, but not inconceivable.

Morales: In response to Dr. Goldman's comment, I think if you have actin and the S1 moiety bound with an external force and you pull on that, then you distort the interface, or something of that sort. Then if you go the other way, you distort the proteins too, but almost surely it would be asymmetrical. So, it wouldn't be a parabola, but something homologous to it. However, the physical part of it would be quite understandable.

Kushmerick: It would require storing that energy in some other molecule once the cross-bridge is attached.

Morales: Well, no. If you're just speaking of actin and S1 and with some external force you force it to go in the opposite direction, you can't say what parts of the molecule will receive energy that way, but the system will.

Suga: According to my experience in cardiac muscle energetics, series elasticity is not the only source of energy storage because, as Tad Taylor explained, we can convert almost 95% of the mechanical potential energy into external work if we appropriately reduce the load during relaxation, without increasing any oxygen consumption. This means there must be some element that can store the mechanical potential energy. There isn't any external work during relaxation in normal contraction. The existence or extractability of external work means there must be some energy-storing element in cardiac muscle.

Woledge: What is so striking about these phenomena in cardiac muscle that Professor Suga is referring to is that, once you're past the peak of contraction, the amount of energy that is going to be released seems to be determined. It may come out as heat, if there is no shortening and the muscle is held isometric, or it may come out partly or wholly as work, as Professor Suga has described. So it behaves as if it contained a certain amount of energy that can be collected as work and that collection occurs over a shortening distance much larger than the throw of one cross-bridge. So one wants to find a mechanism by which this energy already committed into the cross-bridge cycle, presumably by bridge-attachment, will now be liberated over a number of subsequent cross-bridge throws as the muscle is shortened down.

Pollack: A question about Professor Suga's observation: does it refer to papillary muscle or to the intact ventricle? In papillary muscle the possible storage mechanism is clear. Because of clamping, the ends of the preparation have a tendency to be elastic and can easily store energy, which can then be released.

Suga: Series elasticity has been observed in both whole heart intact ventricle and excised papillary muscle. George Cooper has shown that force-length area (a linear version of PVA) is linearly correlated with oxygen consumption. In that case, the energy storage place might be in the damaged ends. But in the whole heart, there are no damaged ends. This means that the damaged ends are beside the point. Energy storing might be distributed throughout the myocardial muscle.

ter Keurs: To that response, I would like to add that, however indirect our measurement of ATP consumption may have been, the same result as Suga's PVA appears to apply to the force-sarcomere length area, in which case all the energy of the series elastic element would have to be stored in the Z-band.

Gergely: Isn't it a problem if one uses the simple two-state model to discuss the possibility of going up the force-energy curve and then is still able to reverse the reaction? If it is a force-generating state, then some chemical change has occurred in the substrates bound to the myosin bridge. For instance, if the phosphate is lost and it is in the stage where this is irreversible, can one still go back and generate force? Maybe one would have to consider more complicated models where the various chemical species are considered.

Woledge: I think Earl Homsher has shown that the loss of phosphate is a reversible process and it will quite quickly go back on. Your suggestion is interesting to me, though, because if the reattachment to phosphate is required, then here's something we can vary—which might make a difference to these energetic properties. So, that does suggest some kind of experiment.

Homsher: On the other hand, too, it may well be that if the force-generation process is an isomerization, you can have a whole variety of different isomers or changes in the structure of the protein—like some sort of ratchet mechanism, for example. You wouldn't have to have a change in chemical state. Once we admit the possibility of isomers, many possibilities open up.

Goldman: I would like to make two further points: if the cross-bridge travels all the way up the negative-force section of the free energy curve (point D on Roger Woledge's diagram) and regains all of the free energy it had before it attached, then no work production occurred. The muscle can't shorten solely on that basis. But if there are several detached states (D_1, D_2, etc.), with decreasing free energies, then net work can be obtained. The second point is that an experiment bearing on the biochemical identity of the cross-bridge state undergoing multiple attachment-detachment cycles is the effect of inorganic phosphate on the amount of sliding per ATP molecule used. Hideo Higuchi and I reported some experiments (*Nature* **352**, 352-354, 1991; *Biophys, J.* **61**, A140, 1992) similar to those shown by Dr. Takenori Yamada at this meeting, indicating that in fibers, the sliding distance per ATP molecule utilized by each myosin head is greater than 60 nm. In further experiments, we didn't find any change when we added 10 mM P_i.

TWO CLASSES OF MAMMALIAN SKELETAL MUSCLE FIBERS DISTINGUISHED BY METABOLITE CONTENT

Martin J. Kushmerick*,**, Timothy S. Moerland† and Robert W. Wiseman*

*Department of Radiology
**Department of Physiology and Biophysics
University of Washington
Seattle, WA
†Department of Biological Science
Florida State University
Tallahassee, FL

ABSTRACT

Phosphorus NMR spectroscopy and HPLC analyses were made on isolated rat and mouse muscles selected for different volume fractions of the major known fiber types. We tested the hypothesis that muscle cell types at rest have intrinsically different contents of PCr, ATP and Pi. The Pi content was low and the PCr and ATP contents were high in muscles with large contents of type 2b and 2a fibers, and vice versa in muscles with large volume fraction of types 1 and 2x fibers. From the profile of these metabolites we could distinguish only two classes of fibers in the murine muscles and predict well the composition of cat muscles. For the first class, types 2a and 2b fibers, the intracellular concentrations were: ATP 8 mM; total Cr 39 mM; PCr 32 mM; Pi 0.8 mM; ADP 8 μM. For the second class, type 1 and 2x fibers, these quantities are: ATP 5 mM; TCr 23 mM; PCr 16 mM; Pi 6 mM; ADP 11 μM. Thus our results establish a new and apparently general criterion upon which to distinguish skeletal muscle cells, one based on the resting content of bioenergetically important metabolites.

INTRODUCTION

Individual muscle fibers have been classified by a number of anatomical, physiological and biochemical methods[1] into various categories: i) fatigue-sensitivity and recruitment order of individual motor units[2-4]; ii) oxidative, glycolytic, and mixed oxidative-glycolytic types based on histochemical staining of characteristic enzymes in the cells[5]; iii) fast- and slow-twitch types, or types 1 and 2 (and further subtypes), based on myosin ATPase staining intensity[6,7], mechanical

Mechanism of Myofilament Sliding in Muscle Contraction, Edited by
H. Sugi and G.H Pollack, Plenum Press, New York, 1993

749

properties[8-10] and myosin heavy and light chain isoform composition by immunochemistry and by gel electrophoresis[11-16]. Although the number of possible cell types distinguishable by these and additional criteria is very large[1)17], individual muscles with predominantly one physiologic characteristic can be obtained and their energy metabolic and other functions can be studied. Thus murine *soleus* muscles were described as slow-twitch and highly oxidative, and the *extensor digitorum longus* muscles as fast-twitch and glycolytic[18]. The rate of ATP splitting during an isometric tetanus was higher in the fast-twitch murine *extensor digitorum longus* muscle than in the slow-twitch *soleus*, as was the speed of shortening; the rate of aerobic ATP resynthesis in recovery was faster in the *soleus* [18].

The following evidence suggests that muscle fibers can differ concerning their contents of phosphocreatine (PCr), creatine, ATP and inorganic phosphate (Pi). Measurements of metabolite composition by ^{31}P NMR demonstrated higher PCr and lower Pi contents in the feline *biceps* muscle than in the *soleus*. This compositional difference corresponded to the predominance of slow-twitch fibers in the *soleus* and their absence in the *biceps* [19]. Chemical analyses of single fiber segments dissected from resting and stimulated rat *plantaris* and *soleus* muscles[20] showed correlation among fiber types with metabolite content at rest as well as with the extent of PCr and ATP splitting during stimulation. The mechanistic basis for the differences in metabolite concentrations was attributed to the greater average rate of neural activation of type 1 fibers in the animal compared with type 2 fibers. Thus the reduced content of PCr and ATP was explained by increased muscle activity, not to a phenotypic characteristic. An alternative possibility, viz. that there are characteristic differences among fiber types in their content of PCr, ATP and other metabolites at rest, was not considered. We therefore designed experiments to test the hypothesis that there are significant differences in bioenergetically important metabolites in the major classes of fiber types. This report is derived from a full paper published elsewhere[21].

METHODS

The strategy taken was to analyze small muscles or strips of muscles whose fiber composition is known. We employed *in vitro* conditions under which the muscles are well supplied with oxygen in an unstimulated state. A selection of muscles from rat and mouse with a range of compositions of fiber types was obtained for this purpose. We obtained ^{31}P spectra and HPLC analyses of phosphorus-containing metabolites from the individual muscles. If different fiber types contain different levels of phosphorus metabolites at rest, then these values, when measured in the whole muscle, will vary systematically depending on the volume fraction of each fiber type in that muscle.

Muscle Preparation:

We studied the *extensor digitorum longus* (EDL), the *soleus* (SOL), the *tensor fascia lata* (TFL), and the *diaphragm* (DPH) dissected from 150 - 250 g Sprague-Dawley rats anesthetized with pentobarbital (60 mg/kg i.p.). We also studied EDL and SOL muscles from 25 - 40 g Swiss-Webster mice. The component fiber types in the adult muscles that we chose for study have been identified and classified into type 2b, 2a, 2x or 1; the volume fraction of each muscle studied is listed in Table 1. Fiber type 2x has been recently identified[13-16]. These fibers are mechanically in the fast-twitch category[8], have a high aerobic capacity and therefore have sometimes been confused with type 2a fibers[13]. Type 2x is most likely same fiber type as 2d identified in Pette's laboratory[17]; we had earlier identified the 2x fibers as 2a based on conventional histochemical ATPase staining[18].

Table 1.

Volume Fraction of fiber types

Fractional volume were obtained data from the references cited.

MUSCLE	2b	2a	2x	1	Ref
Rat TFL	1.0	0	0	0	11
Rat EDL	0.85	0.13	0	0.02	11
Rat DPH	0.07	0.35	0.33	0.25	30
Rat SOL	0	0.1	0	0.9	11
Mouse EDL	0.63	0	0.36	0.01	13, 18
Mouse SOL	0	0	0.63	0.37	13, 26

Note that the Gorza paper[13] shows that in the mouse fibers previously classified as 2b are really 2x whereas those previously classified as 2a are really 2b so the data from Crow and Kushmerick[18] and Moerland et al[26] were altered accordingly.

Because the wet weight of some of the rat muscles exceeded the range of 30 mg that can be kept in good physiological condition by superfusion with oxygenated physiological salt solution (PSS) and because their shape was inappropriate for our NMR methods, the rat DPH and TFL muscles were split by blunt dissection along the axis of the fibers. Only the middle portions of the muscle lay in the sensitive volume of the coil such that attachment regions did not contribute to the spectral signal. All muscles were incubated in PSS at their *in vivo* length for one hour at 22°C to allow recovery from any decrement in PCr caused by the dissection. The PSS was equilibrated with 100% O_2 and contained (mM) 116 NaCl, 4.6 KCl, 26.2 MOPS (titrated to pH 7.4 with NaOH), 2.5 CaCl$_2$, 1.2 MgSO$_4$ and gentamycin (10 mg/L) at pH 7.4. Muscles were frozen between brass blocks cooled to -196°C after the completion of spectral acquisition. As a control, muscles from the opposite limb were prepared similarly and frozen at a time when its mate was put into the NMR probe. These preparations gave reproducible ^{31}P NMR spectra for up to three hours, so we concluded that the preparations were in a metabolic steady state during the duration of our experiment.

Nuclear Magnetic Resonance Spectroscopy:

Spectroscopy of the rat muscles was performed both on a 8.5T high-field spectrometer built by the Francis Bitter National Magnet Laboratory (MIT) and studies of the mouse muscles were made on a 7T GN300 General Electric Omega spectrometer (University of Washington). The superfusion temperature was controlled by a heat exchanger through which thermostatic water was pumped to maintain the temperature of the probe (25°C $\pm$ 2°). PSS flow was usually 0.5 ml/min; the velocity of superfusate over the surface of the muscle was approximately 4 cm/sec. The important factors that enabled the NMR measurements was the optimized filling factor and the realization that, for conductive biological samples in PSS, the sample contributes predominantly to the noise at the frequencies used[22]. With a 140 mM Na$_2$HPO$_4$ standard in the smaller capillary (as used for the mouse muscle experiments), the signal-to-noise ratio for a single acquisition was 12:1. The integrated area of each of the five peaks (Pi, PCr, γATP, αATP and βATP) in the spectrum was expressed as the fraction of the total phosphorus integral in the spectrum. Relative spectral areas were converted into chemical content based on the HPLC-measured ATP content as described in Figure 2. Figure 3 displays the cellular concentrations derived from chemical content per gram wet weight using cellular water fractions measured previously[19].

High Performance Liquid Chromatography:

Stable neutralized perchloric acid extracts were prepared from frozen tissues as previously described[18][19]. Tendon ends were cut from the frozen muscle before the muscle's weight was measured. The creatine contents of the muscle extracts were measured with a Waters Amino Acid column (cation exchange column #80002; Waters, Milford, MA) run isocratically with 25 mM phosphate buffer (pH 7.8). Eluted peaks were quantified by optical absorbency at 210 nm in comparison with a known standard. Nucleotide and phosphocreatine analyses used an anion exchange column (#303NT405; Video, Hesperia, CA) under a phosphate gradient (25 mM KH$_2$PO$_4$, pH 4.5 to 400 mM KH$_2$PO$_4$ + 100 mM H$_3$PO$_4$, pH 2.7, in 20 min); creatine elutes with the void volume with this method. Eluted peaks were quantified by optical absorbance at 210 nm using pure standards for ATP and PCr. Total creatine content was the sum of creatine and phosphocreatine.

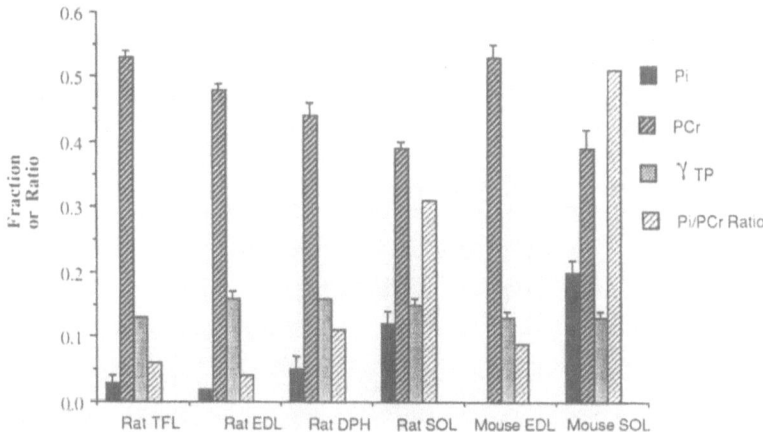

Fig. 1. Quantities are the relative peak area of each species as a fraction of the summed areas of all detected peaks. Error bars represent one standard error of the mean for N = 5. γATP is the area for the γATP peak only; the other ATP peaks were similar and are not shown.

Statistical Analyses:

Regression analyses were performed with the SYSTAT computer program[23].

RESULTS

Composition of Individual Muscles

The fractional peak areas for all the muscles studied (peak area for each component of the spectrum divided by the sum of all peak areas) is displayed in Figure 1. The spectral peak areas for each muscle were then divided by the average of the γATP and βATP areas (or in the case of the mouse muscle by the βATP peak) and multiplied by the ATP content measured by HPLC; the resultant quantities for each metabolite are displayed in Figure 2 in units of μmoles metabolite content per gram wet weight.

Inspection of the data in Figures 1 and 2 with the muscles' fiber type composition (Table 1) demonstrates that the content of inorganic phosphate of muscles containing predominantly type 2b and 2a fibers was lower than in those containing primarily type 1 and 2x fibers; rat TFL ~ EDL < DPH << SOL and mouse EDL << SOL. The content of PCr in muscles containing predominantly 2a and 2b fibers was higher than in those containing types 1 and 2x: rat TFL > EDL > DPH >> SOL and mouse EDL >> SOL. The ATP content was also clearly different in *soleus* muscles compared to the others. The intracellular pH was 7.0 to 7.1 by calculation from the chemical shift of Pi with respect to PCr (data not shown); there were no systematic differences in the muscles studied.

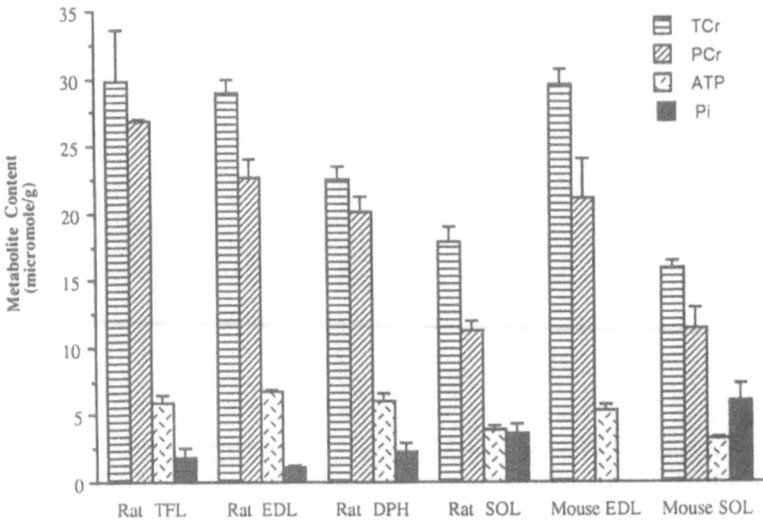

Fig. 2. ATP and total creatine contents are HPLC analyses of perchloric acid extracts of rapidly frozen samples normalized to frozen weight. This measure of ATP was used to calibrate the spectral analyses summarized in Figure 1 to obtain the contents of Pi and PCr in terms of content per gram wet weight. Error bars represent one standard error of the mean for N = 5. Note that no Pi was detected in 4 of the 5 mouse EDL muscles.

Quantitative Analysis of Fiber Type Composition and Metabolite Content

The metabolite composition characteristic of each of the fiber types given in Table 1 can be obtained by combining the data used to construct Figure 1 into a matrix of simultaneous linear equations of the form:

$$[S]_i = \alpha_i (vf_{2b}) + \beta_i (vf_{2a}) + \gamma_i(vf_{2x}) + \delta_i(vf_1) \qquad \text{Equation 1}$$

where the subscript i refers to each of the four metabolites (total creatine, ATP, Pi and PCr), [S] to the measured whole muscle metabolite content, and the terms vf_{2b}, vf_{2a}, vf_{2x} and vf_1 refer to the volume fractions respectively of types 2b, 2a, 2x and 1 fibers in each muscle. The coefficients α, β, γ and δ represent the fiber-specific content of each ith metabolite respectively for type 2b, 2a, 2x and 1 fibers. The solution of this matrix of equations was conveniently solved by treating Equation 1 (and others that follow) as a problem of multiple regression analysis. The results of this analysis for total creatine (PCr + Cr) yielded the following coefficients (and their standard errors): α = 30.8 (2.0) μmole/gram wet weight of fiber for fiber type 2b; β = 28.2 (8.9) μmole/gram wet weight of fiber for type 2a, γ = 18.9 (4.1) μmole/gram wet weight of fiber for type 2x and δ = 16.0 (3.4) μmole/gram wet weight of fiber for type 1. Similarly the analysis for ATP yielded: α = 6.1 (0.1), β = 10.7 (0.6), γ = 3.5 (0.3) and δ = 3.2 (0.2) μmole/gram wet weight of fiber. The total creatine and ATP contents of fiber type 1 were clearly smaller than that in type 2 fibers. The composition of type 2x appeared more similar to the composition of type 1 fibers than to the metabolite contents of types 2b and 2a; this hypothesis is tested below. Thus the null hypothesis (that all fiber types have the same resting composition) is excluded by the results of the first regression model.

We tested a second hypothesis, that the metabolite compositions of type 2b, 2a and 2x fibers are similar to each other but are different from type 1, by a second multiple regression model that grouped all the type 2 fibers into one class:

$$[S]_i = \theta_i \ (vf_{2b} + vf_{2a} + vf_{2x}) + \eta_i(vf_1) \qquad\qquad \text{Equation 2}$$

where the coefficient θ represents the ith metabolite content of fiber types 2b, 2a and 2x postulated to be common to those three types and the coefficient η represents the ith metabolite content of type 1 fibers. The third hypothesis tested was that fiber types 2x and 1 have common metabolite contents, but that composition is different from that of fiber types 2a and 2b. The regression model for this hypothesis is:

$$[S]_i = \kappa_i \ (vf_{2b} + vf_{2a}) + \lambda_i(vf_{2x} + vf_1) \qquad\qquad \text{Equation 3}$$

Table 2.

Regression analysis predicting compositions of specific fiber types

Values given were obtained from the second and third multiple regression models. Units are μmole/gram wet weight with the standard error of the parameter value in parenthesis.

Fiber Type	Total Creatine ATP		Pi	PCr

Regression model 2: the composition of types 2b, 2a and 2x fibers are postulated to be similar to each other but different from type 1

	Total Creatine ATP		Pi	PCr
θ: Types 2b, 2a and 2x	27.8 (2.1)	5.9 (0.6)	1.5 (1.0)	22.8 (2.0)
η: Type 1	13.3 (4.1)	3.3 (1.1)	5.2 (1.9)	7.6 (3.9)
F ratio	111	72	6.4	80

Regression model 3: fiber types 2x and 1 are postulated to have common metabolite contents, but different from that of fiber types 2a and 2b.

	Total Creatine ATP		Pi	PCr
κ: Types 2b and 2a	30.7 (1.4)	6.5 (0.5)	0.6 (0.9)	25.6 (1.4)
λ: Types 2x and 1	16.9 (1.5)	3.8 (0.5)	4.5 (1.0)	11.6 (1.5)
F ratio	373	163	11	245

where the coefficient κ represents the ith metabolite content of fiber types 2b and 2a, postulated to be common for both types, and the coefficient λ represents the ith metabolite content of type 1 and 2x fibers postulated to be common for both types but different from that of types 2b and 2a. The results of these analyses showed that the third hypothesis correlated better with data matrix than did the second

Table 3.

Comparison of observed and predicted chemical composition of skeletal muscles

Metabolite	<<-------------------Predicted---------------------->>		<<---Observed--->>
	Mouse SOL (hypothesis 2)	SOL (hypothesis 3)	Mouse SOL
Total Creatine	22.4	16.9	15.9 (0.5)
PCr	17.2	11.6	11.4 (1.6)
ATP	4.9	3.8	3.3 (0.1)
Pi	2.9	4.5	6.0 (1.3)
	Cat Biceps (hypothesis 2)	Biceps (hypothesis 3)	Cat Biceps
Total Creatine	27.8	30.7	27.7
PCr	22.8	25.6	27.6
ATP	5.9	6.5	7.0
Pi	1.5	0.6	2.4
	Cat Soleus (hypothesis 2)	Soleus (hypothesis 3)	Cat Soleus
Total Creatine	13.3	16.9	17.8
PCr	7.6	11.6	12.1
ATP	3.3	3.8	3.7
Pi	5.2	4.5	7.4

Observed data for rat and mouse muscles and from ref. 19 for cat muscles; it is not known whether type 2x fibers are found in cat biceps and soleus muscles.
Hypothesis 2 is given by regression equation 2; type 1 different from type 2b = 2a = 2x.
Hypothesis 3 is given by regression equation 3; type 1 = type 2x and type 2b = 2a.

hypothesis, based on F test and displayed in Table 2. Finally note that we used regression models 2 and 3, which obviously were derived from the rat and mouse data in the present work, to predict the composition of cat biceps and soleus muscles previously measured[19]; the composition of the cat muscles was better predicted by regression model 3 than by model 2, Table 3.

DISCUSSION

The main conclusion relates to important differences between the composition of the specific fiber types. Our analyses indicate categories based upon the intracellular content of bioenergetically important metabolites, PCr, ATP, Pi and total creatine, do not necessarily correspond to other schemes of classifying fiber types which were summarized in the Introduction. Types 2a and type 2b fibers appear to have little differences in chemical content and therefore in their spectral characteristics at rest, yet these fibers differ in their oxidative capacity and shortening speed. The content of type 1 fibers were clearly different from types 2b and 2a. The composition of type 2x fibers was more similar to type 1 fibers than to the other type 2 fibers. The conclusion grouping 2x and type 1 fibers into the same compositional class was unexpected because 2x fibers are mechanically fast[8]. The Pi content of whole muscles increased with increasing content of type 1 and 2x fibers. The PCr content decreased with increasing content of type 1 and 2x fibers. These results were seen in the rat *diaphragm* and are especially prominent in the mouse *soleus* muscle. Thus, there is about a ten-fold range in the Pi/PCr ratios in normal resting muscles, from 0.05 (or lower) in fast-twitch fibers (types 2a and 2b) to 0.5 in muscles containing predominantly type 1 and 2x fibers.

It is possible that the results from the muscles *in vitro* systematically differ from their composition in the intact animal because of incubation conditions or damage during preparation. The following comparison shows no evidence for this possibility. Rat lower limb musculature was sampled by a surface coil[24] in a way that sampled mixed fast muscle which are types 2a and 2b[25]. The PCr/ATP ratio from Table 1 of that paper (calculated the ATP value in the same way as in the present manuscript) was 3.6 and the Pi/PCr ratio was 0.10. Both values are similar to the present data in Figure 2 for muscles composed predominantly of those muscle types. The Pi/PCr ratios in the present data is not higher nor are the PCr/ATP ratios lower as would have been expected if some unsuspected damage to the *in vitro* muscles produced an artifactual result.

The results obtained from murine muscles can be extended to other mammalian skeletal muscles. For example, in Table 3 our analysis and conclusions concerning fiber type composition observed in the rat and mouse predicted accurately the composition of cat *biceps* and *soleus* muscles previously measured. Therefore the results obtained herein appear to be representative of mammalian fibers generally. On the basis of the conclusion that our regression model 3 is the best explanation of our data, we recalculated the data in terms of intracellular metabolite concentration to yield our best estimate of the fiber type-specific metabolite contents. These estimates of the total creatine, ATP, PCr and Pi concentrations in cell water in the two classes of mammalian fibers is displayed in Figure 3. Despite the large differences in composition between the two classes of fibers, the ratios of the metabolites are related such that their resting concentration of ADP and free energy available for hydrolysis of ATP were not distinguished. ADP contents were calculated from the creatine kinase equilibrium relation as reported previously[19];

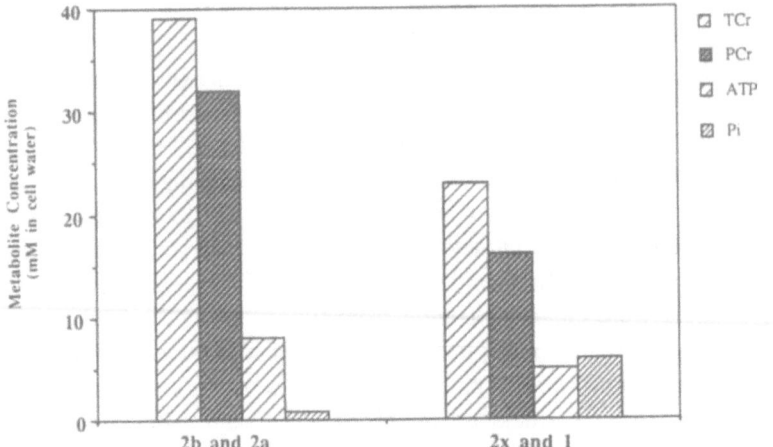

Fig. 3. These estimates were obtained as explained in the text from multiple regression analyses. The chemical content expressed elsewhere in this paper are in units of content per gram wet weight. Here these were converted into cellular concentrations on the assumptions that the metabolites observed in the muscle were entirely intracellular and that the fraction of cell water in total tissue water was 0.79 for types 2a and 2b muscle and 0.73 for types 1 and 2x muscle[19].

the values obtained were similar in all fibers, being 8 μM in type 2b and 2a fibers and 11 μM in the others. The chemical potential of ATP, μATP, was calculated as reported previously[19], and was similar in all fibers, the values being -68 kJ/mole in type 2b and 2a fibers and -61 kJ/mole in the others.

It remains an open question why the content of phosphate-containing metabolites differs in these fiber types. It is possible the content of metabolites observed is related to the fiber's capacity for oxidative metabolism. However, this is not the complete explanation because fiber types 2a and 2b, which differ in their oxidative capacity, were not distinguishable. These cells are subject to the same extrinsic hormonal regulation and extracellular fluid composition in the animal, so the differences in cellular content appears to be a genuine phenotypic distinction between muscle cells. We have speculated that differing steady-state composition of bioenergetically important phosphate compounds (which can be altered by exercise and in experimental animals by chronic stimulation and by uptake of creatine analogs) may be an important factor influencing the muscle's phenotype[26]. However a causal relationship has not been established. Our interpretation of fiber type specific composition is clearly different from that of Hintz *et al.*[20]. Types 1 and 2x fibers have substantially lower PCr and ATP and higher Pi content at rest than do types of 2a and 2b fibers. Thus it is not necessary to have different degrees of muscle activity to observe differences in PCr, Pi and ADP contents. Our results are consistent with the findings that the maximal isometric force per cross sectional area when measured in skinned fibers (with comparable composition of the incubation medium) show little fiber-specific differences, whereas the force generating capacity of slow motor units is less than fast ones[1][3].

What might be the functional significance of these different metabolite levels? Inorganic phosphate inhibits isometric force in permeabilized single fiber preparations[27-29] with a half maximal inhibition in the several millimolar range. Thus type 1 and 2x fibers have a Pi content that would be inhibitory of actomyosin interactions producing force from the beginning of contractile activity. Fiber-specific effects of Pi on actomyosin interactions in skinned skeletal muscle fibers are poorly known. We have preliminary data that some Pi inhibition is present (RWW, SK Phillips, RC Woledge and MJK, unpublished data) because addition of pyruvate to the incubation medium of a mouse *soleus* decreases [Pi] as it increases isometric force. It could be a useful strategy to arrange a cellular milieu such that the concentration of this inhibitory metabolite changes relatively little with respect to its inhibitory constant, as would be the case for type 1 and 2x fibers at the concentration of Pi we measured. Metabolic fatigue attributable to changes in [Pi] would then be quite small in type 1 and 2x compared to that in the other fiber types. Although speculative, we envision a possible survival benefit to this type of strategy wherein the primary consideration is the maintenance of mechanical output even at a cost of partial inhibition. The mechanical power output of these fibers whenever activated would be diminished relative to their full potential, but their function would be maintained during prolonged activity by two mechanisms: i) energy balance is more easily achieved by a higher ATP synthesis rate relative to the ATP demand; ii) our speculation that any further change in [Pi] would result in little reduction in mechanical power output.

ACKNOWLEDGEMENTS

P. Bryant Chase, Kevin E. Conley, Richard D. Hedges and Christopher D. Hardin provided me with helpful criticism during the writing of this report, as did William LaFramboise who also shared his gel electrophoresis data prior to publication. Rudolph Stuppard's assistance with the HPLC and Elizabeth Egan's editorial assistance were also significant. This work was supported by grants from NIH (AR36281 and AR38782 to MJK; F32 AR07763 to TSM; F32 AR08105 to RWW), the University of Washington Department of Radiology and the Massachusetts Institute of Technology Research Resource Grant (RR00995). This paper is a summary of the original publication of this experimental work[21].

REFERENCES

1. Saltin, B. & Gollnick, P.D. in *Handbook of Physiology: Section 10 Skeletal Muscle* (eds. Peachey, L.D.) 555-631 (American Physiological Society, Bethesda, 1983).

2. Henneman, E., Somjen, G. & Carpenter, D.O. *J. Neurophysiol.* **28**, 560-580 (1965).

3. Burke, R.E. & Eggerton, V.R. *Exercise and Sports Sci Revs* **3**, 31-81 (1975).

4. Lewis, D.M., Parry, D.J. & Rowlerson, A. *J. Physiol. (Lond.)* **325**, 393-401 (1982).

5. Peter, J.B., Barnard, R.J., Edgerton, V.R., Gillespie, C.A. & Stempel, K.E. *Biochem.* **11**, 2627-2633 (1972).

6. Ariano, M.A., Armstrong, R.H. & Edgerton, V.R. *J. Histochem. & Cytochem.* **21**, 51-55 (1973).

7. Burke, R.E., Levine, D.N., Tsairis, P. & Zajac III, F.E. *J. Physiol. (Lond.)* **234**, 723-748 (1973).

8. Bottinelli, R., Schiaffino, S. & Reggiani, C. *J. Physiol. (Lond.)* **437**, 655-672 (1991).

9. Sweeney, H.L., Kushmerick, M.J., Mabuchi, K., Sreter, F.A. & Gergely, J. *J. Biol. Chem.* **263**, 9034-9039 (1988).

10. Metzger, J.M. & Moss, R.L. *Biophys. J.* **52**, 127-8131 (1987).

11. Tsika, R.W., Herrick, R.E. & Baldwin, K.M. *J. Appl. Physiol.* **63**, 2101-2110 (1987).

12. Ausoni, S., Gorza, L., Schiaffino, S., Gundersen, K. & Lomo, T. *J Neurosci* **10**, 153-160 (1990).

13. Gorza, L. *J. Histochem. Cytochem.* **38**, 257-265 (1990).

14. Maier, A., Gorza, L., Schiaffino, S. & Pette, D. *Cell Tissue Res* **254**, 59-68 (1988).

15. LaFramboise, W.A., et al. *Biochem. Biophys. Acta* **1035**, 109-112 (1990).

16. LaFramboise, W.A., et al. *Develop. Biol.* **114**, 1-15 (1991).

17. Pette, D. & Staron, R.S. *Rev. Physiol. Biochem. Pharmacol* **116**, 1-76 (1990).

18. Crow, M.T. & Kushmerick, M.J. *J. Gen. Physiol.* **79**, 147-166 (1982).

19. Meyer, R.A., Brown, T.R. & Kushmerick, M.J. *Am. J. Physiol.* **248**, C279-C287 (1985).

20. Hintz, C.S. et al. *Am. J. Physiol.* **242**, C218-C228 (1982).

21. Kushmerick, M.J., Moerland, T.S. & Wiseman, R.W. *Proc. Natl. Acad. Sci. USA* **80**, 7521-7525 (1992).

22. Hoult, D.I. & Richards, R.E. *Proc. R. Soc. Lond*, A**344**, 311-340 (1975).

23. Wilkinson, L. *SYSTAT: The System for Statistics.* (SYSTAT, Inc., Evanston, IL, 1989).

24. Kushmerick, M.J. & Meyer, R.A. *Am. J. Physiol.* **248**, C279-C287 (1985).

25. Armstrong, R.B. & Phelps, R.O. *Am. J. Anat.* **171**, 259-272 (1984).

26. Moerland, T.S., Wolf, N.G. & Kushmerick, M.J. *Am. J. Physiol.* **257**, C810-C816 (1989).

27. Cooke, R. & Pate, E. *Biophys. J.* **48**, 789-798 (1985).

28. Godt, R.E. & Nosek, T.M. *J. Physiol. (Lond.)* **412**, 155-180 (1989).

29. Millar, N.C. & Homsher, E. *J. Biol. Chem.* **265**, 20234-20240 (1990).

30. LaFramboise, W.A., Watchko, J.F., Brozanski, B.S., Daood, M.J. & Guthrie, R.D. *Am. J. Respir. Cell Mol. Biol.* **6**, 335-339 (1992).

Discussion

Rall: Do you have a sense about why nature would create this difference in metabolic conditions? Is there some functional consequence of this?

Kushmerick: I don't know what it is yet. I imagine there is some sort of function. One function that comes to mind immediately is that, independent of the extent of chemical change, a slow twitch muscle will not demonstrate much fatigue due to inorganic phosphate changes. We also know that the pH inhibition of force is less. So this might be a mechanism to prevent fatigue.

Morales: I think that the expectation based on your result—that the control could be a simple mass law result of ADP accumulation—was an early guide but that cannot be the case now. An important reason the control becomes more complicated is because the enzymes involved in this are often in assemblies. Dr. Maughan and Paul Srere at the University of Texas–Dallas, for example, have shown kinetic differences between assemblies and solutions. So when you have the enzymes in an assembly, the nature of the control can be much more complicated and includes allosteric and other effects.

An intermediate situation between the completely solubilized and the whole muscle would be to make artificial systems in which you put the assemblies together and see how the control behaves—for example, mitochondria and fibers. Has that been successful?

Kushmerick: No, it hasn't. There have been some natural extracted enzymes that bind together—let's say two, three, or more enzymes. Srere has done this with mitochondrial assemblies. It has also been done with some of the glycolytic intermediates. But there has been nothing larger than a few enzymes together for the glycolytic enzymes. Chris Hardin in my lab actually tried to make up some models of this using Sephadex beads and putting enzymes on them, but we couldn't make anything as finely graded as the cell seems to be able to do.

INFLUENCE OF IONIC STRENGTH ON CONTRACTILE FORCE AND ENERGY CONSUMPTION OF SKINNED FIBERS FROM MAMMALIAN AND CRUSTACEAN STRIATED MUSCLE

R.E. Godt, R.T.H. Fogaça, M.A.W. Andrews* and T.M. Nosek

Dept. of Physiology & Endocrinology
Medical College of Georgia
Augusta GA 30912 U.S.A.

ABSTRACT

Increased ionic strength decreases maximal calcium-activated force (F_{max}) of skinned muscle fibers via mechanisms that are incompletely understood. In detergent-skinned fibers from either rabbit (psoas) or lobster (leg or abdomen), F_{max} in KCl-containing solutions was less than in potassium methanesulfonate ($KMeSO_3$), which we showed previously was the least deleterious salt for adjusting ionic strength. In either salt, lobster fibers were considerably less sensitive to elevated ionic strength than rabbit fibers. Trimethylamine N-oxide (TMAO, a zwitterionic osmolyte found in high concentration in cells of salt-tolerant animals) increased F_{max}, especially in high KCl solutions. In this regard, TMAO was more effective than a variety of other natural or synthetic zwitterions. In rabbit fibers, increasing ionic strength decreases F_{max} but has little effect on contractile ATPase rate measured simultaneously using a linked-enzyme assay. Thus high salt increases the tension-cost of contraction (i.e. ratio ATPase/F_{max}). At both high and low salt, TMAO decreases tension-cost. Given a simple two-state model of the cross-bridge cycle, these data indicate that ionic strength and TMAO affect the apparent detachment rate constant. High ionic strength KCl solutions extract myosin heavy- and light-chains, and troponin C from rabbit fibers. This extraction is virtually abolished by TMAO. Natural zwitterions, such as TMAO, have been shown to protect proteins against destabilization by high salt or other denaturatants. Our data indicate that, even in the best of salts, destabilization of the actomyosin complex may play a role in the effect of high ionic strength on the contractile process.

---

*Current address: Dept. of Physiology & Biophysics; Univ. of Vermont; Burlington VT 05405 U.S.A.

Mechanism of Myofilament Sliding in Muscle Contraction, Edited by
H. Sugi and G.H Pollack, Plenum Press, New York, 1993

INTRODUCTION

Changes in ionic strength have been used to probe the nature of the acto-myosin interaction. For example, the so-called weakly-binding states are thought to be important steps in the cross-bridge cycle. These states have been studied at low ionic strength which favors their formation[1]. Furthermore, comparative studies with intact and skinned muscle indicate that the decrease in force of intact skeletal mucle in hypertonic solutions[2][3] is due in large part to the osmotically-induced elevation of intracellular ionic strength [4][5].

It has been suggested that certain natural osmolytes can counteract the destabilizing effects of high salt on isolated proteins[6]. These osmolytes are found in high concentration (hundreds of mM) in cells of euryhaline organisms, i.e. those able to live in environments of widely varying salinity. Such compounds are thought to play an essential role in stabilizing the structure and function of intracellular proteins[7]. We have used certain of these stabilizing osmolytes as tools to better understand how changes in ionic strength affect the contractile apparatus in skinned muscle fibers.

Some of these data have been reported previously in abstract form[8][9].

METHODS

Rabbit Psoas

Small bundles of muscle from rabbit psoas were excised and skinned at 22°C in a solution containing (in mM): 1 Mg^{2+}, 2 MgATP, 5 EGTA, 20 imidazole, and 60 potassium methanesulfonate (KMeSO$_3$), ionic strength 150, pCa = -log [Ca^{2+}] > 8, pH 7, and 0.5% vol/vol purified Triton X-100, a non-ionic detergent. After skinning, the fibers were stored at -20°C in a similar solution which also contained 0.1 mM leupeptin, 0.1 mM phenylmethylsulfonylfluoride (PMSF), cytidine-5'-triphosphate (CTP) instead of ATP, and 50% vol/vol glycerol. Fibers were used within 2 mo after dissection. Immediately before experimentation, single fibers were mounted between a force transducer and a moveable arm and stretched to a sarcomere length of 2.6 μm as determined using He-Ne laser diffraction. Further details are given in Andrews et al.[10].

Lobster Fibers

Single muscle fibers from walking leg (extensor carpopodite) or abdomen (extensor abdominal superficial) were excised and skinned in a solution similar to that given above, except it contained (in mM): 50 EGTA, 5 Mg^{2+}, and a cocktail of protease inhibitors (0.1 leupeptin, 0.1 PMSF, 1 benzamidine, and 0.01 aprotinin). Dissection and skinning were carried out at ca. 5°C. After skinning, fibers were stored at -20°C in the same solution as for rabbit psoas fibers but also containing the protease inhibitors listed above. Most lobster muscle fibers are larger than 300 μm in diameter. To decrease diffusion distances, we pared the fibers to 150-200 μm diameter before attachment to the force transducer. After attachment, fibers were stretched 15-20% beyond slack length to maximize Ca^{2+}-activated force.

Procedures for Experiments Where Only Force Was Measured

Solutions were contained in 2.5 ml troughs milled into a plexiglas base and were stirred continuously. To change solutions, fibers were transferred from trough to trough. Before each experiment, the pH of the solution in each trough was adjusted to 7.00. This was especially important when adding high concentrations of zwitterions since these altered the solution pH slightly. The solutions used in all experiments were formulated using a computer program which describes the multiple binding equilibria of ions in the solutions, using binding constants given previously[10)11)] The basic relaxing solution for force measurements contained (in mM): 1 Mg^{2+}, 1 MgATP, 15 Na_2phosphocreatine, 5 EGTA, 20 imidazole, 85 KCl or $KMeSO_3$, 100 U/ml creatine kinase, pH 7.00. The calculated ionic strength of this solution was 165 mM. The basic activating solution was similar but contained $CaCl_2$ to adjust the concentration of free calcium (pCa 4), and had only 75 mM KCl or $KMeSO_3$ to keep ionic strength equal to 165 mM. Ionic strength was increased by adding either KCl or $KMeSO_3$ to these basic solutions. All force experiments were conducted at 22°C.

Procedures for Combined Force and ATPase Experiments

Simultaneous measurements of fiber force and ATPase rate were made using a specially designed microspectrofluorometer (Scientific Instruments G.M.B.H., Heidelberg, Germany). One or two fibers were mounted in a small cuvette (25 μl) between a force transducer and a moveable arm. Solutions were exchanged by perfusing the cuvette. The ATP hydrolysis rate was estimated using a linked-enzyme assay:

$$PEP + ADP \Leftrightarrow ATP + pyruvate \qquad (1)$$
$$pyruvate + NADH \Leftrightarrow NAD^+ + lactate \qquad (2)$$

where PEP is phosphoenolpyruvate, and reaction 1 is catalyzed by pyruvate kinase (PK) and 2 is catalyzed by lactate dehydrogenase (LDH)[12)]. The ATPase rate is proportional to the decrease in NADH fluorescence (excitation: 340 nm; emission: 470 nm). After each experiment, the system was calibrated using ATP-free solutions with known concentrations of NADH.

The basic relaxing solution for the force/ATPase experiments contained (in mM): 1 Mg^{2+}, 1 MgATP, 15 Na_3PEP, 5 EGTA, 50 N,N-Bis(2-hydroxyethyl)-2-aminoethanesulfonic acid (BES), 50 KCl or $KMeSO_3$, 0.5 NADH, 0.1 P1,P5-di(adenosine-5') pentaphosphate, 100U/ml PK, and 140 U/ml LDH. Calculations assumed a binding constant of 147 M^{-1} for Mg^{2+} binding to PEP[13)], and pKa of 3.4

and 6.35 for proton binding to PEP[14]). The calculated ionic strength of this solution was 165 mM. The basic activating solution (pCa 4) had only 40 mM KCl or $KMeSO_3$ to maintain ionic strength at 165 mM. Ionic strength was increased by adding either KCl or $KMeSO_3$ to these basic solutions. The force/ATPase experiments were conducted at 23-29°C.

Protein Extraction from Psoas Fibers

To determine the extraction of protein from fibers, single fibers were immersed, under mineral oil, in a series of small (5 µl) droplets of relaxing solution at 315 mM ionic strength (KCl) for 1-, 9-, and 20-min intervals (30 min total). The droplets containing any extracted protein were placed in 10 ul of sample buffer containing 62.5 mM Tris (pH 6.8), 1% wt/vol SDS, 0.01% wt/vol bromophenol blue, 15% glycerol, and 15 mM dithiothreitol. The remaining fiber was placed in 10 µl of sample buffer and sonicated for 2 min. Using the volume of the fiber estimated optically, additional sample buffer was then added so that the fiber was diluted 1:1000. Aliquots of sample buffer containing droplets or fiber were loaded onto polyacrylamide gels (7.5% acrylamide). After electrophoresis, gels were stained using an ultrasensitive silver technique[9)15).

RESULTS

Ionic Strength and Maximal Ca^{2+}-Activated Force (F_{max})

The relation between F_{max} and ionic strength for rabbit psoas fibers is shown in Fig. 1A. The data for KCl and $KMeSO_3$ are similar to that reported recently[10) In

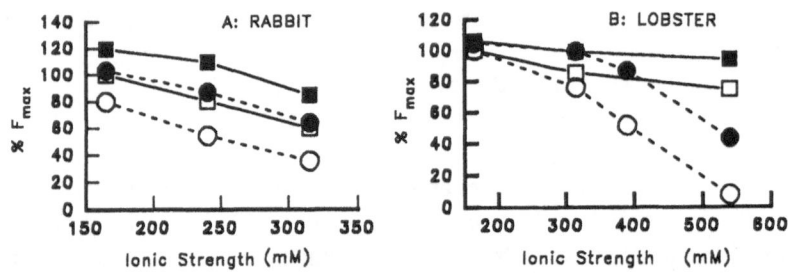

Fig. 1. Relation between F_{max} and ionic strength for (A) rabbit psoas fibers and (B) lobster ext. abdominal superficial fibers in solutions with $KMeSO_3$ (squares & solid lines) or KCl (circles & dashed lines). F_{max} was attained at pCa 4. Filled symbols indicate F_{max} in solutions containing 300 mM TMAO. All data are expressed relative to F_{max} at 165 mM ionic strength ($KMeSO_3$) control solution without TMAO. Standard error bars are visible when larger than symbol. Note difference in axis scales in A and B. Under all conditions, treatments were reversible since F_{max} in control solution decreased no more than 10% over the course of the experiment.

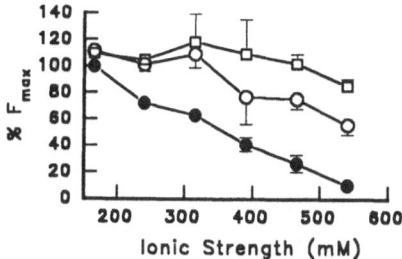

Fig. 2. Relation between F_{max} and ionic strength for lobster walking leg fibers in KCl-containing solutions. Closed circles are control solutions without TMAO, open circles are for 300 mM TMAO and open squares for 500 mM TMAO. All data are expressed relative to F_{max} at 165 mM ionic strength without TMAO. Standard error bars are visible when larger than symbol.

prior studies, we determined that, of the salts tested, $KMeSO_3$ had the least deleterious effect on skinned psoas fibers. In addition, we found that maximal Ca^{2+}-activated force was achieved at pCa 4 in all salts[10]. As can be seen in Fig. 1, addition of 300 mM trimethyamine N-oxide (TMAO), a zwitterionic osmolyte found in high concentration in muscles of euryhaline animals[6], increased F_{max} under all conditions.

Fig. 1B shows the relation between F_{max} and ionic strength for lobster abdominal muscle fibers. As with rabbit psoas, i) F_{max} in KCl is lower than in $KMeSO_3$ and ii) TMAO increases F_{max} under all conditions. Lobster fibers are less sensitive to increased ionic strength than rabbit fibers (note change in scale between Figs. 1A & 1B). In either KCl or $KMeSO_3$, 300 mM TMAO counteracts the depressant effect of high ionic strength on F_{max}.

The effect of TMAO on F_{max} is dependent on concentration. Fig. 2 shows the effect of two concentrations of TMAO on F_{max} of fibers from lobster walking leg in

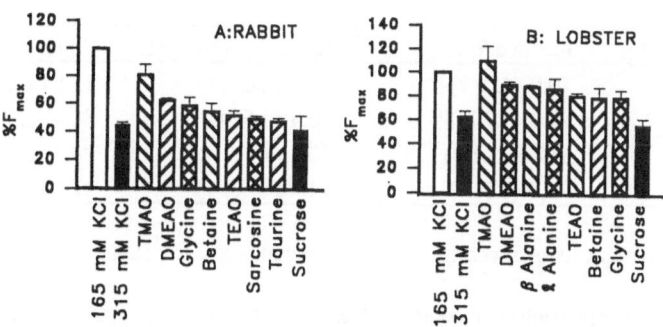

Fig. 3. Influence of zwitterions and sucrose on F_{max} in 315 mM ionic strength KCl solutions. The concentration of all compounds was 300 mM. Data expressed relative to F_{max} in 165 mM ionic strength KCl solution without zwitterions. DMEAO and TEAO are synthetic analogs of TMAO (see text). Panel A is data for rabbit psoas and panel B is for lobster walking leg.

KCl-containing solutions. As can be seen, 500 mM TMAO is more effective than 300 mM in counteracting the depressant effect of elevated ionic strength. In rabbit psoas fibers, on the other hand, F_{max} at 240 mM ionic strength (KCl or $KMeSO_3$) is higher with 300 mM TMAO than with 500 mM TMAO (data not shown).

These effects of TMAO are shared by other natural and synthetic zwitterions, but not by sucrose. Fig. 3 A & B show that a variety of zwitterions can reverse the depressant effect of high ionic strength KCl solutions. Of all compounds studied, TMAO has the greatest ameliorative effect. Dr. M. Gary Newton (Dept. of Chemistry, Univ. of Georgia, Athens GA) synthesized two TMAO analogs, dimethylethylamine N-oxide (DMEAO) and triethylamine N-oxide (TEAO). Both of these compounds were less effective than TMAO, perhaps due to the increased size of the alkyl groups. We feel the effects shown in Fig. 3 are not simply due to osmotic or other non-specific effects on water activity since: i) at a fixed concentration, zwitterions are not equally effective, and ii) addition of 300 mM sucrose had no significant effect on F_{max}.

Effect of Ionic Strength on Fiber ATP Hydrolysis Rate

We studied the effect of ionic strength on force and fiber ATPase rate in rabbit psoas fibers using the linked-enzyme assay. As seen in the Table 1, increased ionic strength ($KMeSO_3$) has a more marked depressant effect on force than on ATPase rate, thus the tension-cost (i.e., the ratio ATPase/force) is increased by elevation of ionic strength. At low salt (165 mM ionic strength), 300 mM TMAO decreases ATPase with little effect on F_{max}, whereas at high salt (315 mM ionic strength), TMAO increases F_{max} with little effect on ATPase. Therefore, at both ionic strengths, TMAO decreased relative tension-cost. TMAO does not eliminate the

Table 1. Effect of ionic strength and 300 mM TMAO on force, ATPase rate, and tension-cost in rabbit psoas fibers. Values are percentage of control.

Ionic strength (mM)	Relative force	Relative ATPase rate	Relative Tension-cost
315*	70.9 (4.0, 7)	92.3 (2.6, 7)	132.4 (8.3, 7)
165 + TMAO*	98.3 (4.7, 4)[#]	86.5 (4.5, 4)	88.4 (5.3, 4)
315 + TMAO**	114.5 (3.1, 4)	103.5 (4.7, 4)	90.4 (3.2, 4)
315 + TMAO*	92.8 (1.8, 7)	118.2 (5.3, 7)	127.3 (5.2, 7)

Each fiber served as its own control. Values in parentheses are: (Standard error, number of fibers).

* Data expressed as percentage of control values in 165 mM ionic strength without TMAO.

** Data expressed as percentage of control values in 315 mM ionic strength without TMAO.

Note that in these experiments TMAO had no significant effect on F_{max}, unlike data shown in Fig. 1A. This may be due to differences in experimental conditions (see Methods).

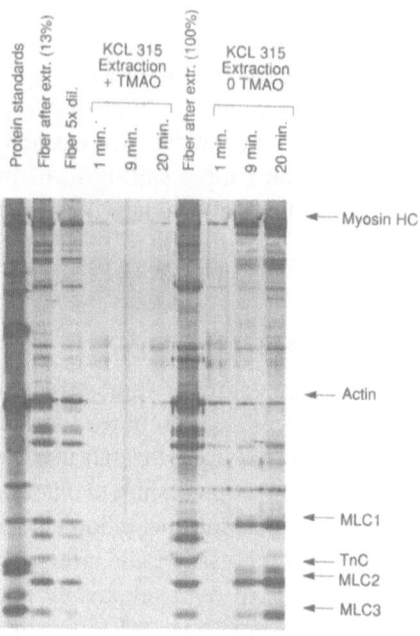

Fig. 4. Extraction of contractile proteins by high ionic strength KCl solutions and prevention by 300 mM TMAO. As described in text, two halves of a single fiber are exposed to droplets of solution for times given. Right-most four lanes are entire fiber (after exposure to droplets) and droplets after times indicated. On left of gel is 13% of fiber (after exposure to droplets with TMAO), 13% of fiber diluted a further 5-fold, and entire droplets after times indicated.

effect of increased ionic strength on tension-cost, however, since tension-cost at high salt with TMAO had an elevated tension-cost relative to the low-salt control (without TMAO).

Extraction of Contractile Proteins at High Ionic Strength (KCl) and Prevention by TMAO

We have previously reported that high ionic strength solutions made with KCl extract myosin heavy- and light-chains, as well as troponin C from skinned psoas fibers. Such extraction was not seen in similar solutions with $KMeSO_3$[10] If this is due to a destabilization of the contractile apparatus by high concentrations of KCl, one would expect that addition of TMAO would prevent this extraction. To test this prediction, we cut a long psoas fiber into two pieces of equal length. One piece was immersed in small droplets of 315 mM ionic strength (KCl) for varying periods, The other piece was exposed to similar solutions which contained 300 mM TMAO. As can be seen in Fig. 4, prolonged exposure to 315 mM ionic strength relaxing solutions with KCl extracted myosin heavy- and light-chains, and TnC (right four lanes of gel). These data are similar to those shown previously at 390 mM ionic strength (KCl)[10]. Note that addition of TMAO virtually eliminated extraction even at long times (30 min total exposure time). Quantification of the amount of protein extracted at 390 mM ionic strength (KCl) showed that proportionally more than twice as much of each of the three myosin light-chains were extracted than of heavy-chain (Table VII in ref. 10). This may indicate that extraction by high concentrations of KCl is more complex than the simple salting-in of myosin, where one would expect evolution of equal proportions of heavy-and light-chains into solution.

DISCUSSION

It is well documented that elevation of ionic strength decreases the isometric force of skinned muscle fibers[5)16-19]. Recently we showed that the relation between force and ionic strength is dependent upon the salt used to alter ionic strength, and we found that the least deleterious was potassium methanesulfonate[10].

What is the mechanism by which changes in ionic strength affect the contractile process? Others have shown that increased ionic strength leads to a decrease in the number of attached cross-bridges in both relaxed[20] and fully activated skinned fibers[21]. In accord with these observations, we found that increased ionic strength decreased both force and fiber stiffness proportionately[8]. Simultaneous measurements of force and ATPase in skinned rabbit psoas fibers (ref. 21 and Table 1) showed that increased ionic strength decreased force more sharply than ATPase, i.e. the tension-cost was increased, which in a simple two-state model of the cross-bridge cycle, is proportional to the apparent detachment rate constant[22]. From tension-cost estimates and sinusoidal analysis Kawai et al[21] concluded that ionic strength affects force primarily by modifying the rapid equilibrium between the detached and the weakly-attached state, with little effect on the other steps of the cross-bridge cycle. In skinned fibers from rabbit and frog, Schoenberg[20] proposed that the cross-bridge attachment rate is also affected by ionic strength, although it should be noted that his experiments were conducted in relaxed fibers at lower ionic strengths than used by Kawai or us.

From solution measurements, Geeves & Goldmann[23] found that the overall affinity of rabbit myosin S-1 for actin is decreased when ionic strength is increased from 0.1 to 0.5 M with K acetate. They suggest that ionic strength decreases the rate constant for the isomerization from an "attached" to a "rigor-like" cross-bridge state, analogous to the weakly attached (low-force) and strongly attached (high-force) cross-bridge states[23)24]. Geeves & Goldmann[23] also found that chloride (but not acetate) destabilized the attached state by accelerating the detachment of cross-bridges in the low-force state. This accords with our observation that force is lower in solutions containing KCl (see Fig. 1). Our observation that lobster skinned fibers are less sensitive to elevated ionic strength than rabbit psoas (Figs. 1A & B) is consistent with other observations from crustacean skinned fibers (crayfish[16]; barnacle[25]). This suggests that the sensitivity to ionic strength of key steps in the cross-bridge cycle is different in crustacean muscle.

It is known that increasing ionic strength causes swelling of the myofilament lattice in skinned fibers[21][26-28]. This could affect the cross-bridge cycle. However, lattice swelling probably does not play a major role since force was depressed by increased ionic strength even when swelling was prevented by addition of Dextran T-500 (a high molecular weight polymer) to the bathing solutions[10]. From X-ray diffraction evidence, Kawai et al[21] also argue that changes in lattice spacing cannot account for the effects of ionic strength on force production.

If the interaction between myosin and actin is "strongly ionic in character"[29][30], elevation of ionic strength could decrease the number of attached cross-bridge by simple electrostatic screening. If this were the case, one would expect a priori the same relation between force and ionic strength regardless of salt. While this is clearly not observed (ref. 10, and our Fig. 1), this prediction ignores known salt-specific effects on force production by skinned fibers[5][31]. A stronger argument is that if electrostatic screening were the sole mechanism, how could TMAO and the other zwitterions counteract such screening and thereby increase force (Figs. 1 & 2, Table 1)?

Biologically ocurring zwitterions, especially TMAO, are known to protect proteins from destabilization of structure and function under a variety of conditions, e.g. elevated salt concentrations or urea[6][7]. For example, Altringham et al[32] showed in skinned dogfish muscle fibers that TMAO prevented the depressant effect of urea on isometric force. We suggest that TMAO and the other zwitterions (Fig. 3) counteract the depressant effects of high salt by "stabilizing" cross-bridges. Although at present we can only speculate, this could include effects on the isomerization step[23] and/or the attachment or detachment steps of the cross-bridge cycle[20][21][23]. Our observation that TMAO decreases tension-cost (Table 1) suggests that it is likely to decrease the apparent detachment rate. This should lead to an increase in the number of attached cross-bridges, which seems to be the case since TMAO increases both force and stiffness of skinned rabbit psoas fibers in proportion at all ionic strengths tested (90-390 mM)[8]. A further indication of the role in muscle of protein destabilization by high salt is the extraction of myosin heavy- and light chains, as well as TnC in high ionic strength KCl solutions[10], and Fig. 4). As seen in Fig. 4, this extraction was virtually abolished by TMAO.

How do TMAO and similar osmolytes work? Timasheff and Arakawa[33] have proposed that certain osmolytes can stabilize protein structure against denaturation by being preferentially excluded from the surface of the protein. This leads to a preferential hydration of the protein which makes denaturation thermodynamically unfavorable. Using lysozyme as a typical protein, Arakawa & Timasheff[34] found a variety of osmolytes that could stabilize protein structure against thermal denaturation and which were strongly excluded from the protein domain. All of the zwitterions (except those we synthesized) that counteracted the depressant effects of high KCl solutions on force of skinned fibers (Fig. 3) were among those stabilizing osmolytes.

ACKNOWLEDGEMENTS

We wish to thank Dr. M. Gary Newton, Univ. of Georgia, for synthesizing

TMAO analogs, and Jane Chu and K.Y. Fender for their excellent technical assistance. Rabbit muscle was kindly supplied by Dr. Keith Green, Medical College of Georgia. This work supported by N.I.H. grant AR 31636 and Navy contract #N00014-91-C-0044 Office of Naval Research.

REFERENCES

1. Brenner, B., Schoenberg, M., Chalovich, J.M. & Greene, L.E. *Proc. Natl. Acad. Sci. USA* **79**, 7288-7291 (1982).
2. Hodgkin, A.L. & Horowicz, P. *J. Physiol. (Lond.)* **136**, 17p (1957).
3. Gordon, A.M. & Godt, R.E. *J. Gen. Physiol.* **55**, 254-275 (1970).
4. Howarth, J.V. *J. Physiol. (Lond.)* **144**, 167-175 (1958).
5. Gordon, A.M., Godt, R.E., Donaldson, S.K.B. & Harris, C.E. *J. Gen. Physiol.* **62**, 550-574 (1973).
6. Yancey, P.H., Clark, M.E., Hand, S.C., Bowlus, R.D. & Somero, G.N. *Science* **217**, 1214-1222 (1982).
7. Hochachka, P.W. & Somero, G.N. *Biochemical Adaptation* pp. 305-354 (Princeton University Press, Princeton, 1984).
8. Andrews, M.A., Martyn, D.A., Fogaca, R.T.H. & Godt, R.E. *Biophys. J.* **59**, 45a (1991).(Abstract)
9. Fogaca, R.T.H., Andrews, M.A. & Godt, R.E. *Biophys. J.* **57**, 546a (1990).(Abstract)
10. Andrews, M.A.W., Maughan, D.W., Nosek, T.M. & Godt, R.E. *J. Gen. Physiol.* **98**, 1-21 (1991).
11. Godt, R E. & Lindley, B.D. *J. Gen. Physiol.* **80**, 279-297 (1982).
12. Güth, K. & Wojciechowski, R. *Pflügers Arch.* **407**, 552-557 (1986).
13. Manchester, K.L. *Biochem. Biophys. Acta* **630**, 225-231 (1980).
14. Wold, F., & Ballou, C.E. *J. Biol. Chem.* **227**, 301-328 (1957).
15. Switzer, R.C. III, Merril, C.R. & Shifrin, S. *Analyt. Biochem.* **98**, 231-237 (1979).
16. April, E.W. & Brandt, P.W. *J. Gen. Physiol.* **61**, 490-508 (1973).
17. Thames, M.D., Teichholz, L.E. & Podolsky, R.J. *J. Gen. Physiol.* **63**, 509-530 (1974).
18. Gulati, J. & Podolsky, R.J. *J. Gen. Physiol.* **72**, 701-716 (1978).
19. Kawai, M. in *Basic Biology of Muscles: A Comparative Approach* (eds. Twarog, B.M., Levine, R.J.C. & Dewey, M.M.) 109-130 (Raven Press, New York, 1982).
20. Schoenberg, M. *Biophys. J.* **54**, 135-148 (1988).
21. Kawai, M., Wray, J.S. & Güth, K. *J. Muscle. Res. Cell Motility.* **11**, 392-402 (1990).
22. Brenner, B. in *Molecular Mechanisms in Muscular Contraction* (ed. Squire, J.M.), pp. 77-149 (CRC Press, Inc., Boca Raton, 1990).
23. Geeves, M.A. & Goldmann, W.H. *Biochem. Soc. Trans.* **18**, 584-585 (1990).
24. Geeves, M.A. *Biochem. J.* **274**, 1-14 (1991).
25. Ashley, C.C. & Moisescu, D.G. *J. Physiol. (Lond.)* **270**, 627-652 (1977).
26. April, E.W., Brandt, P.W. & Elliott, G.F. *J. Cell Biol.* **53**, 53-65 (1972).
27. Godt, R.E. & Maughan, D.W. *Biophys. J.* **19**, 103-116 (1977).
28. Matsubara, I., Umazume, Y. & Yagi, N. *J. Physiol. (Lond.)* **360**, 135-184 (1985).
29. Highsmith, S. *Arch. Biochem. Biophys.* **180**, 404-408 (1977).
30. Moos, C. *Cold Spring Harbor Symposium of Quantitative Biology* **37**, 137-143 (1973).
31. Jacobs, H.K. & Guthe, K.F. *Arch. Biochem. Biophys.* **136**, 36-40 (1970).
32. Altringham, J.D., Yancey, P.H. & Johnston, I.A. *J. Exp. Biol.* **96**, 443-445 (1982).

33. Timasheff, S. & Arakawa, T. in *Protein Structure & Function: A Practical Approach* (ed. Creighton, T.E.) 331-345 (IRL Press, Oxford, 1989).
34. Arakawa, T. & Timasheff, S.N. *Biophys. J.* **47,** 411-414 (1985).

Discussion

Brenner: How do you explain the change in ATPase on your scheme, which is cut from Geeves's kinetic scheme? (Geeves, M. and Goldman, Y. *Biochem. Soc. Trans.* **18**, 584, 1990; Geeves M. *Biochem. J.* **274**, 1, 1991).

Godt: I don't think that can be explained. In this meeting, Earl Homsher and I talked about the fact that there are likely to be changes at other steps in the cycle. But that was the only biochemical scheme in which they frankly talked about changes in ionic strength and looked at different ions.

Edman: An argument against the idea that ionic strength changes the dissociation rate constant of bridges is that mechanical V_{max} is quite insensitive to ionic strength. This has been shown for ionic strength changes in the range of 50-190 μM.

Godt: Are you speaking of intact fibers or skinned fibers?

Edman: The experiments I refer to are those published by Gulati and Podolsky (*J. Gen. Physiol.* **79**, 233-257, 1981) and Julian and Moss [*J. Physiol. (Lond.)* **311**, 179-199, 1981]. In both cases, the experiments were carried out on skinned fibers.

Godt: We haven't checked that, but certainly in our hands there is very little effect on the ATPase rate. So if V_{max} and ATPase go together, as some have suggested, that would make sense.

Edman: V_{max} is reduced only when ionic strength is decreased to a very low level, to 20 μM or so (see reference to Gulati and Podolsky above).

Godt: These experiments only went from physiological up to 315 mM ionic strength, so we've only really looked at the two extremes.

Ando: You changed ionic strength from 0.09 to 0.54 M. The effectiveness of ionic strength works according to the square root. But this square root rule holds only for ionic strengths up to 50 mM. If you increase the ionic strength more than that, its effectiveness is much less than the square root. Also, I studied the energy transfer between the charged donors and the acceptors. I measured energy transfer efficiency by changing the ionic strength up to 0.6. Between 0.1 and 0.6 of ionic strength there was almost no significant change of the energy transfer. But below 50 mM of ionic strength you can see a very big change. I also studied energy transfer measurement with fiber systems with no consistent results. With positively charged acceptors, energy-transfer efficiency decreased with increasing ionic strength, but with negatively charged acceptors, again, energy-transfer efficiency decreased with increasing ionic strength. So this means that, within the fiber, we cannot study the ionic strength effect in a purely physical way.

Godt: It's a pity that you can't study these effects of ionic strength in fibers, because that is what is of interest to me. I'm sorry I can't get any information from your kinds of experiments. But it is interesting that you say ionic strength above the physiological doesn't have any more effect on your accumulation, but it certainly does on fiber force and ATPase and other contractile properties.

VARIABLE CROSSBRIDGE CYCLING-ATP COUPLING ACCOUNTS FOR CARDIAC MECHANOENERGETICS

Tad W. Taylor and Hiroyuki Suga*

National Cardiovascular Center Research Institute
Suita, Osaka 565 Japan
*The 2nd Department of Physiology
Okayama University Medical School
Okayama 700 Japan

ABSTRACT

Cardiac twitch contractions were simulated by Huxley's sliding filament crossbridge muscle model. Huxley's model was extended to include cardiac twitch contractions with a model structure having parallel and series elastic components with a crossbridge contractile element. The appropriate crossbridge energetics were added based on the crossbridge cycling rate and the energy of ATP hydrolysis. The force-length area (FLA) as a measure of the total mechanical energy was computed for both isometric and isotonic contractions in a manner similar to the pressure-volume area (PVA), (Suga, H. *Physiol. Rev.*, **70**, 247-277, 1990). Experimental studies have demonstrated that the pressure-volume area (PVA) correlates linearly with cardiac oxygen consumption and hence with the energy expenditure of a cardiac contraction. PVA correlates linearly with cardiac oxygen consumption, and since FLA is analogous to PVA, FLA should correlate with the ATP expended. Simulations comparing FLA with the crossbridge cycling ATP usage showed that at lower muscle fiber activation levels (shorter initial fiber lengths and lower preload levels) FLA decreased more rapidly than the number of muscle fiber crossbridge cycles. This could imply that one ATP can cause more than one crossbridge cycle at lower fiber activation levels as was proposed by Yanagida et al. (*Nature*, **316**, 366-369, 1985). If the number of crossbridge cycles to ATP ratio is allowed to increase at lower activation levels, Huxley's model agrees with the experimental findings on FLA and PVA.

INTRODUCTION

It has been shown from years of experimentation that pressure-volume area (PVA) is linearly correlated with cardiac oxygen consumption and therefore with the energy expended in cardiac contractions[1]. Suga[2] originally postulated that the

Mechanism of Myofilament Sliding in Muscle Contraction, Edited by
H. Sugi and G.H Pollack, Plenum Press, New York, 1993

775

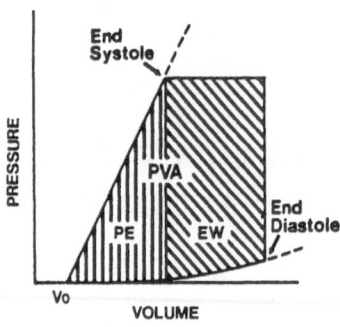

Fig. 1. Pressure-volume area.

total mechanical energy generated in a cardiac contraction is correlated with cardiac oxygen consumption. The total mechanical energy generated by each left ventricular contraction was measured by the PVA. As is shown in Figure 1, PVA consists of the areas corresponding to external work (EW) and potential energy (PE). Suga et al.[3] consistently observed a linear correlation between the magnitude of the PVA and the oxygen consumption in the canine left ventricle with a stable contractility.

The force-length area (FLA) has also been shown to correlate linearly with energy expenditure[4][5], analogously to PVA. From papillary muscle preparations, Hisano and Cooper[4] found that FLA is correlated with oxygen consumption. Mast and Elzinga[5] found that heat liberated in relaxation and FLA are linearly correlated. These results show that both PVA and FLA represent the total mechanical energy generated by a myocardial contraction.

To try to relate the PVA or in this case FLA to actual crossbridge kinetics, we used Huxley's crossbridge model since the mechanical energy of a myocardial contraction is known to be generated by crossbridge cycles. Similar to Wong[6][7], a Maxwell-type three-component Hill model was used to represent the papillary

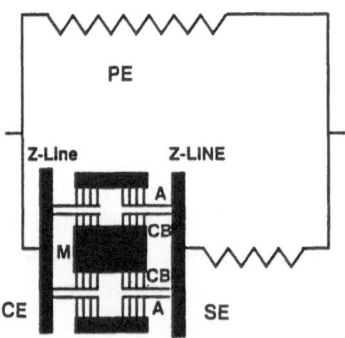

Fig. 2. Three-component cardiac muscle model.

muscle which is illustrated in Figure 2 with parallel and series elasticity elements (PE, SE) along with a contractile element (CE). The sliding filament muscle sarcomere model which was proposed by Huxley[8] represents the CE. This model is still useful for predicting muscle fiber mechanics and energetics when the proper applications and parameters are implemented, as was mentioned by Woledge[9]: "The points of agreement between this theory and experimental observations are remarkable for a theory which is so simple and contains rather few adjustable parameters." For this reason, we applied Huxley's model to this investigation.

THEORETICAL DEVELOPMENT

Basic Model Formulation

The interaction of actin and myosin molecules, through the myosin crossbridge, generate muscle fiber tension with the reaction[8]:

$$A + M \underset{g}{\overset{f}{\rightleftharpoons}} A\text{-}M \tag{1}$$

where A and M are actin and myosin, A-M represents attached crossbridges, f is the association rate constant, and g represents the dissociation rate constant. Huxley's expressions for $f(x)$ and $g(x)$ were implemented. Exponential relations were used to describe the behavior of the PE and SE as was done by Wong[6][7].

The number of attached crossbridges in a sarcomere at position x and time t can be written as[10-12]:

$$\partial n(x,t)/\partial t - v(t)\partial n(x,t)/\partial x$$
$$= [A(t)-n(x,t)]f(x) - n(x,t)g(x) \tag{2}$$

where $A(t)$ represents the number of actin molecules which can react with myosin molecules. An inhibitory effect on the actin-myosin reaction is exerted by tropomyosin. After stimulation, calcium is released around the myofilaments, troponin reacts with calcium and the actin-myosin reaction proceeds. Calcium was assumed to react with the actin-troponin complex with one calcium ion being important, as was experimentally verified by Holroyde et al.[13]:

$$A\text{-}T + Ca^{2+} \underset{c_2}{\overset{c_1}{\rightleftharpoons}} A + CaT \tag{3}$$

where A-T is the concentration of the actin-tropomyosin complex, the maximum number of crossbridges that can be formed for a given myosin concentration is given by A, and the reaction binding and dissociation rate constants are c_1 and c_2. The rate of change of A(t) can be written as:

$$dA(t)/dt = c_1(Ca)[AT_0-A(t)] - c_2A(t) \tag{4}$$

with:

$$Ca = \gamma Ca_m \tag{5}$$

$$c_2 = c_{20}\exp\{k_iDL_{pe}[(L_{max} - L)/(L_{max} - L_{min})]^q\} \tag{6}$$

where AT_0 is the total amount of actin which can react in the muscle fiber and γ is the time-varying calcium stimulus function multiplying the maximum calcium concentration Ca_m. The calcium stimulus relation was implemented as:

$$\gamma(t) = C_{fac} \left[\frac{\alpha}{\beta-\alpha}\right] [\exp(-\alpha t) - \exp(-\beta t)] \tag{7}$$

The cardiac muscle fiber tension developed from the CE crossbridges was determined by integrating over all x the number of attached crossbridges multiplied by the crossbridge distance and force constant.

The crossbridge cycle rate was assumed to be proportional to the crossbridge energy utilization rate found from:

$$CB = \int_0^1 \int_{-\infty}^{\infty} g(x)n(x,t)dxdt \tag{8}$$

Assuming one ATP per crossbridge cycle, the ATP utilization rate and hence oxygen consumption rate should be proportional to the crossbridge energy utilization rate.

RESULTS

Isometric Contractions

Figure 3 shows CE force versus time and crossbridge cycles versus time in isometric contractions performed at various preload values and corresponding muscle fiber lengths. Panel A shows that most of the contraction is completed by 0.35 seconds. As the initial fiber length is increased (which corresponds to increasing the preload), the peak contraction force increases. In the isometric case, the total fiber length is fixed, but the CE and SE will still move. As the preload increases, the time to peak force also increases. Crossbridge cycles versus time at different initial muscle fiber lengths for an isometric contraction are illustrated in Figure 3, Panel B. Analogous to the force versus time curves, increasing the fiber length increases the number of crossbridge cycles. Initially the crossbridge cycles rapidly increase, but the rate levels off at approximately 0.35 seconds. These results

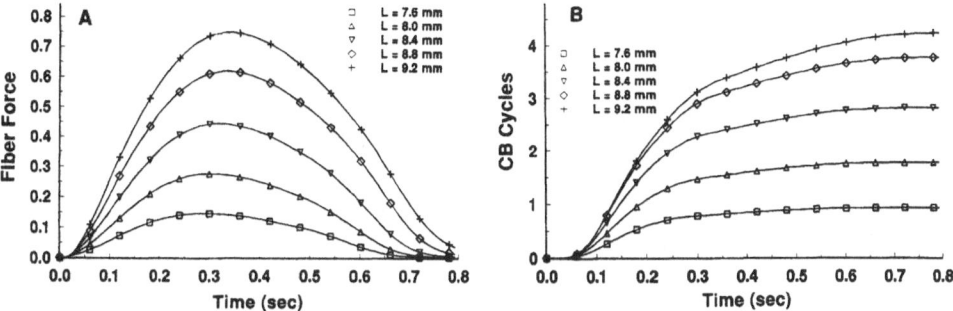

Fig. 3. (A), CE force versus time. Preload varied in isometric contractions. (B), Crossbridge cycles versus time. Preload varied in isometric contractions.

imply that after the first part of the cardiac muscle fiber contraction most of the crossbridge cycle energy has been expended.

Isotonic Contractions

Figure 4 shows muscle fiber force and crossbridge cycles versus time at different fiber afterloads. In Panel A, as the afterload is increased the fiber approaches isometric behavior. Since the maximum level of the FLA is produced with isometric contractions, as the afterload is increased the FLA will increase which directly correlates with energy consumption. However, in Panel B as the afterload is decreased the model predicts that the crossbridge energy will increase, contrary to what is expected from FLA predictions.

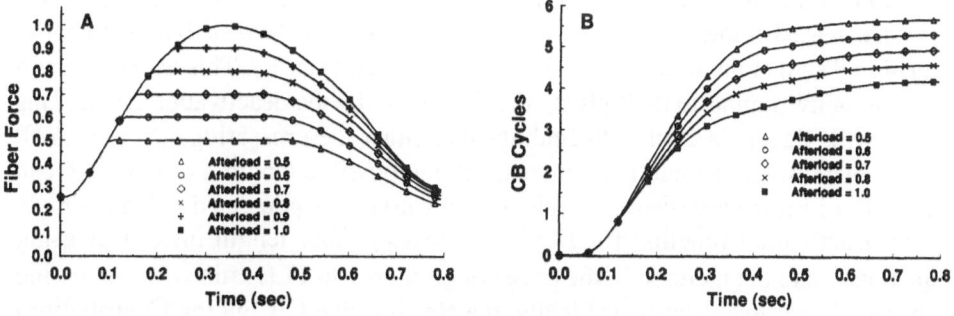

Fig. 4. (A), Total force versus time. Afterload varied in isotonic contractions. (B), Crossbridge cycles versus time. Afterload varied in isotonic contractions.

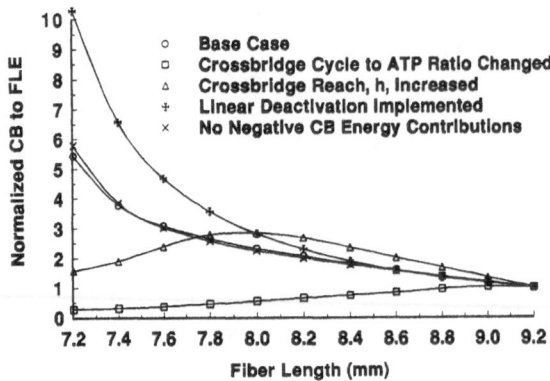

Fig. 5. Normalized crossbridge to force-length energy ratio, isometric.

Crossbridge to Force-Length Energy Ratio

Figure 5 illustrates the relative normalized CB cycles to force-length energy (FLE, proportional to FLA) ratios in the isometric case; one represents the ratio at a fiber length of 9.2 mm. The final value (at 9.2 mm fiber length) of the CB to FLE ratio was used to normalize all other values on the same curve (absolute values are also calculated in the model but relative values are shown here for clarity). The Base Case Curve reflects the crossbridge to FLE ratio when the initial muscle fiber length is decreased; this causes the fiber force also to decrease. Since FLE is known to correlate with cardiac fiber energy expenditure, the ratio of crossbridge energy to FLE should be constant. In the Crossbridge Cycle to ATP Ratio Changed Curve, the crossbridge cycle rate was multiplied by a deactivation factor which is the relative time-independent amount of muscle fiber force deactivation that has occurred (dividing $A(t)$ by γ gives a relatively time-independent value):

$$Dact(L) = A(t)/\gamma \tag{9}$$

This empirical measure effectively decreases the ATP to CB cycle ratio at shorter fiber lengths and has the effect of keeping the normalized CB to FLE ratio close to unity. In the Crossbridge Reach, h, Increased Curve, the crossbridge reach, h, was increased at lower activation levels by dividing h by the deactivation factor in Equation 9 so that the new crossbridge reach in was h/Dact(L). This keeps the ratio closer to unity, similar to multiplying the CB energy by the deactivation factor. This agrees with Yanagida et al.'s[14] findings that either the crossbridge cycle to ATP ratio is increasing, or there is an increase in the possible length a crossbridge can reach. The Linear Deactivation Implemented Curve was generated using a simple linear deactivation function for f_1 with decreasing fiber length instead of using Equation 4, and the increase in the crossbridge energy to FLE ratio with decreasing fiber length was even stronger. Finally, the No Negative CB Energy Contributions

Curve was generated assuming no ATP was hydrolyzed at negative crossbridge reaches, and it is similar to the Base Curve.

DISCUSSION

Yanagida et al.[14] experimentally found that more than one crossbridge cycle occurred per ATP, six or more, or that the range of movement of a crossbridge during one ATP hydrolysis under an unloaded condition was at least 600 Å. Therefore, the sliding distance of the thin filament in a crossbridge with one ATP hydrolysis cycle without a load could be much longer than the distance in the maximum loaded isometric tension case.

Yasumura et al.[15] experimentally showed that the oxygen consumption of an almost unloaded contraction from a relatively large end-diastolic volume only slightly exceeded the oxygen consumption of an isovolumic contraction. This implies that at low cardiac muscle fiber force levels little oxygen and consequently little energy is used for shortening; hence several crossbridge cycles per ATP hydrolysis may be occurring.

Table 1 shows some comments from the literature regarding both the cardiac PVA relationship and tight versus loose coupling in crossbridge formation. The PVA concept has been well studied and proven to be true, while the tight versus

Table 1. Comments in the literature about muscle behavior.

Gibbs and Chapman[16] (1985): "Therefore, the experimental finding of a constant thermodynamic efficiency of potential energy (pressure-volume) generation in cardiac muscle suggests that some kind of optimally interactive autoregulatory mechanism is at work in the myofilaments, previously undiscovered but required to explain the otherwise puzzling findings of Suga and co-workers."

Gibbs and Chapman[16] (1985): 'In summary, however, we appear to be faced by a paradox. Although many features of the cardiac results imply that there is a stoichiometric relation between ATP consumption and the manifestation of pressure-volume (or force-length) potential energy, the stoichiometric slippage essentially built into all models of muscle contraction"...."predicts that this should not be the case."

Yanagida[14] (1985): "If there are six or more crossbridge cycles per ATP cycle, the sliding distance obtained here would be similar to that deduced from mechanical studies, 80 Å."

Huxley[17] (1990): "It is clear that the evidence is not yet complete on this fascinating and important question of whether or not there is tight coupling between crossbridge movement and ATPase activity, and there are many very interesting experiments still in progress or in the course of publication."

Gibbs[18] (1991): "It may even be necessary to consider the possibility that during shortening myosin crossbridges can interact with several actin sites, each time donating part of their stored energy but not having to be re-primed by ATP hydrolysis after each contact."

loose coupling in crossbridge function question is controversial as is mentioned by Huxley[17] in the table.

Both decreasing of the ATP to crossbridge cycle ratio and increasing the crossbridge reach using Equation 9 are empirical measures, but they do allow better agreement between the crossbridge energy and FLE. The behavior of the simple Huxley model used here definitely does not prove or disprove whether there is tight or loose coupling in crossbridge function, but it does suggest some paths for future investigation.

ACKNOWLEDGEMENTS

We wish to thank the Japan Science and Technology Agency for an STA fellowship supporting the first author. We also want to thank Professor Roger C. Woledge for his suggestions on the relationship between crossbridge kinetics and PVA.

REFERENCES

1. Suga, H. *Physiol. Rev.* **70**, 247-277 (1990).
2. Suga, H. *Am. J. Physiol.* **236**, H498-H505 (1979).
3. Suga, H., Hayashi, T. & Shirahata, M. *Am. J. Physiol.* **240**, H39-H44 (1981).
4. Hisano, R. & Cooper, G. IV. *Circ. Res.* **61**, 318-328 (1987).
5. Mast, F. and Elzinga, G. *Circ. Res.* **67**, 893-901 (1990).
6. Wong, A.Y.K. *J. Biomech.* **4**, 529-540 (1971).
7. Wong, A.Y.K. *J. Biomech.* **5**, 107-117 (1972).
8. Huxley, A.F. *Prog. Biophys. Biophys. Chem.* **7**, 255-318 (1957).
9. Woledge, R.C., Curtin, N.A. & Homsher, E. in *Energetic Aspects of Muscle Contraction* (Academic Press, London, Soc. Monogr., No. 41, 1985).
10. Panerai, R.B. *J. Biomech.* **13**, 929-940 (1980).
11. Taylor, T.W., Goto, Y. & Suga, H. (submitted).
12. Taylor, T.W., Goto, Y., Hata, K., Takasago, T., Saeki, A., Nishioka, T. & Suga, H. (submitted).
13. Holroyde, M.J., Robertson, S.P., Johnson, J.D., Solaro, R.J. & Potter, J.D. *J. Biol. Chem.* **255**, 11688-11693 (1980).
14. Yanagida, T., Arata, T., & Oosawa, F. *Nature* **316**, 366-369 (1985).
15. Yasumura, Y., Takashi, N., Futaki, S., Tanaka, N. & Suga, H. *Am. J. Physiol.* **256**, H1289-H1294 (1989).
16. Gibbs, C.L. & Chapman, J.B. *Am. J. Physiol.* **249**, H199-206 (1985).
17. Huxley, H.E. *J. Biol. Chem.* **265**, 8347-8350 (1990).
18. Gibbs, C.L. *Proceedings of the Australian Physiological and Pharmacological Society* **22**, 1-21, 1991

Discussion

Pollack: In your model you have three elements: the contractile, the series elastic, and the parallel elastic. Did you use values for series elasticity that are the sort of large values characteristic of papillary muscle, or the small values that are characteristic of skeletal muscle?

Taylor: That's a good question. I started out with a pretty large value of about 8% from some values that Sonnenblick had come up with (Sonnenblick, E.H. *Am. J. Physiol.* **202**, 931-939, 1962). But after reading some of your research in which you state that this is probably too large a value, I lowered it to about 2% and it really didn't make much difference. The point is that a lot of the Huxley-type models predict that the force basically decreases almost linearly with length. The problem is that the force-length area decreases almost quadratically with length and so there has got to be some other mechanism at work there from force-length area.

SUMMARY AND DISCUSSION OF POSTER PRESENTATIONS

SUMMARY AND DISCUSSION OF POSTER PRESENTATIONS

Uniformity and Nonuniformity of Cardiac Muscle Performance

Sys: It may be clear from previous discussions that most people consider nonuniformity as an artefact or a nuisance for molecular or subcellular interpretation of cardiac muscle performance. In our interpretation of the control of cardiac muscle contraction and relaxation by load and (in) activation, the need is felt to consider nonuniformity at different levels of integration, from the crossbridge, the single cell, the multicellular preparation, to the intact heart. A principle of uniformity through nonuniformity was proposed to be activate at each level of cardiac mechanics. I would like to present briefly some results to illustrate this principle at the multicellular level in isolated cat papillary muscle.

To analyse patterns of longitudinal nonuniformity, rather than 'internal shortening' only, muscles were partitioned into four longitudinal segments by insertion of three microelectrode tips in the core (Fig. 1). Segment length measurements were obtained during isometric twitches at different calcium concentrations and muscle lengths. The upper panel shows force versus time traces of an isometric twitch obtained under the same conditions in three different muscles labelled a, b, and c; notice the nearly identical time course of contraction and the similar time course of relaxation in these three muscles. Nonuniformity of segmental kinetics is qualitatively illustrated in the lower right panel. In a given muscle, substantial differences in the pattern of length change of different segments were observed. Moreover, when one particular segment was considered, its pattern of length change appeared not to be predictable from muscle to muscle. These two types of differences in pattern of segment length change demonstrated the wide variation or nonuniformity in segmental kinetics despite the remarkable parallelism or uniformity of the force traces. The reversed proposition of uniformity through nonuniformity (or of homogeneity through heterogeneity) was further supported by observations at other levels of cardiac muscle performance, ranging from the subcellular level to the intact heart as a whole.

In a given muscle, S_1, S_2 and S_3 segment length measurements at peak twitch force allowed calculation of their variation coefficient, which could be used as a measure of nonuniformity within the muscle. Averaged results for segment length variation coefficient at rest and at peak twitch force—at given external calcium concentration and at given muscle length—are shown in Fig. 2. Due to segment length normalization, nonuniformity at rest increased when muscle length was deviating from optimal length l_{max}. Nonuniformity increased with isometric contraction. The increase in nonuniformity from rest to peak, although more

787

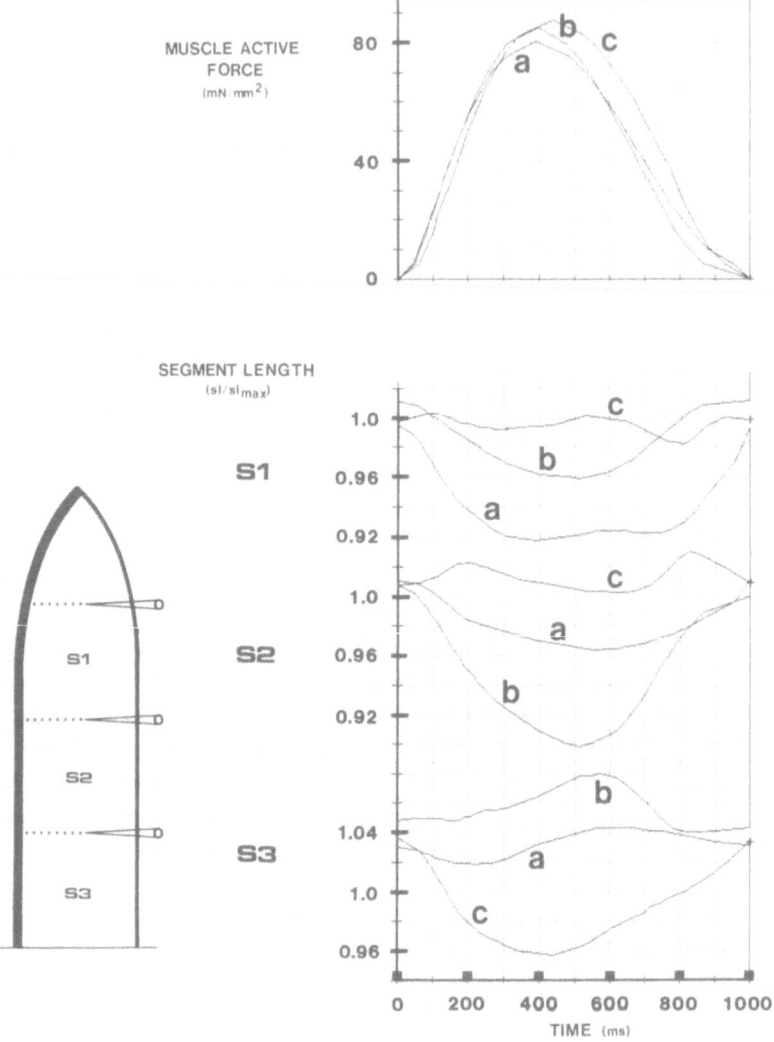

Fig. 1.

pronounced at higher external calcium, was independent from muscle length during the isometric twitch; this is in contrast to segment shortening by itself which was described to be more marked during isometric twitches at shorter muscle length. The invariance of this increase in segmental nonuniformity for changes in muscle length again illustrated the principle of uniformity at the muscular level through nonuniformity at the segmental level.

Although nonuniformity may hamper a quantitative evaluation of cardiac performance at a lower level, the described findings brought the following message. Isolated cardiac muscle displays an important degree of physiologic nonuniformity in performance of longitudinal segments. This physiologic nonuniformity appears to

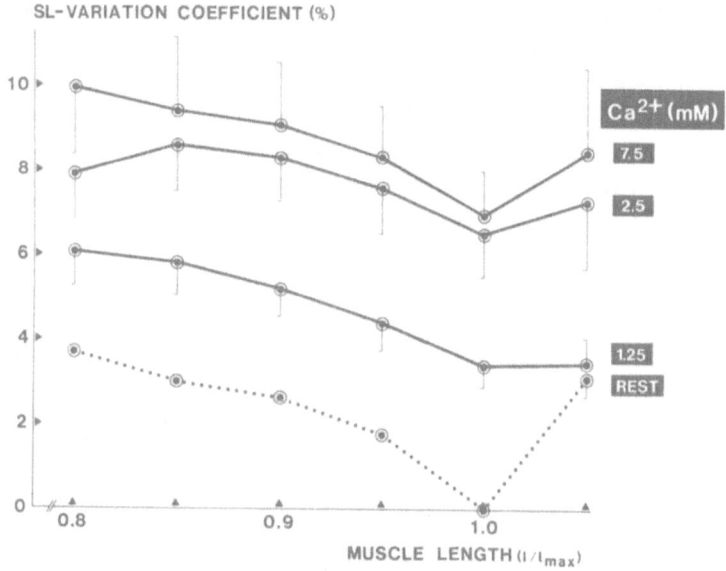

SL-VARIATION COEFFICIENT (%)

MUSCLE LENGTH (l/l_{max})

Fig. 2.

serve a regulatory or stabilizing role in order to induce uniform performance of the muscle as a whole. Understanding the role of nonuniformity in normal conditions is a prerequisite to interpret an inappropriately increased nonuniformity in several pathologic conditions.

ter Keurs: I have a problem with your conclusion. Don't the properties of the contractile system in the muscle allow stability of its performance despite the presence of non-uniform behavior of the segments? That leads me to the observation that was made by us at least eleven years ago; that is, if one takes a muscle with a uniform distribution of the cross-sectional area along a long stretch of its center— let's say two or three millimeters—then one finds no variation in the behavior of the sarcomeres along that length of the muscle. That leads to the question of whether the cross-sectional areas of these papillary muscles are uniformly distributed along the length.

Sys: Indeed, with respect to the central segments S1 and S2, no differences in cross-sectional area could be detected. We are presently also looking at the distribution of muscle *vs.* non-muscle tissue and we do not see any consistent differences that could be correlated to mechanical performance of the different segments. Moreover, if the tapered end would extend beyond the tendon-end segment into the upper central segment S1, it would emphasize the importance of nonuniformity of strength between S1 and S2 (per unit cross-sectional area).

Huxley: Is the non-uniformity in a particular preparation the same in successive contractions? What I have in mind is a paper by A.V. Hill and J.V. Howarth [*J. Physiol. (Lond.)* **139**, 466-473, 1957] describing a relaxation heat which was a dissipation of mechanical energy when a part of the muscle elongates during relaxation. They found in successive twitches that this relaxation heat appeared alternately at one end of the muscle and the other.

Sys: I don't know about the stability of Hill's preparation, but these preparations of isolated cat papillary muscles are very stable. They can easily display the same performance over several hours, up to maybe six or seven hours. We also do not see differences in performance of a given segment from contraction to contraction, unless of course conditions are changed.

ter Keurs: Could you please answer my earlier question? Your statement was that the non-uniformity of segment-length behavior leads to uniformity of overall behavior. Isn't it the other way around? The properties of the contractile apparatus in the cells in various segments of the muscle allow uniformity of the force output, thanks to their force-length relation and force-velocity relations, despite non-uniformity of the composition of the segments that you have observed.

Sys: We recently proposed several arguments for a system-dynamics approach to cardiac muscle performance (*Circulation* **83**, 1444-1445, 1981). Uniformity through nonuniformity is an example of the concept that the interaction of subsystems may add new properties to the composite system. We do agree that at present this approach is still a matter of interpretation.

Interaction of Nonpolymerizable Actins with Myosin

Arata: I would like to talk about monomeric MBS-actin. Polymerization of G-actin in the presence of salt was blocked by treatment of G-actin with *m*-

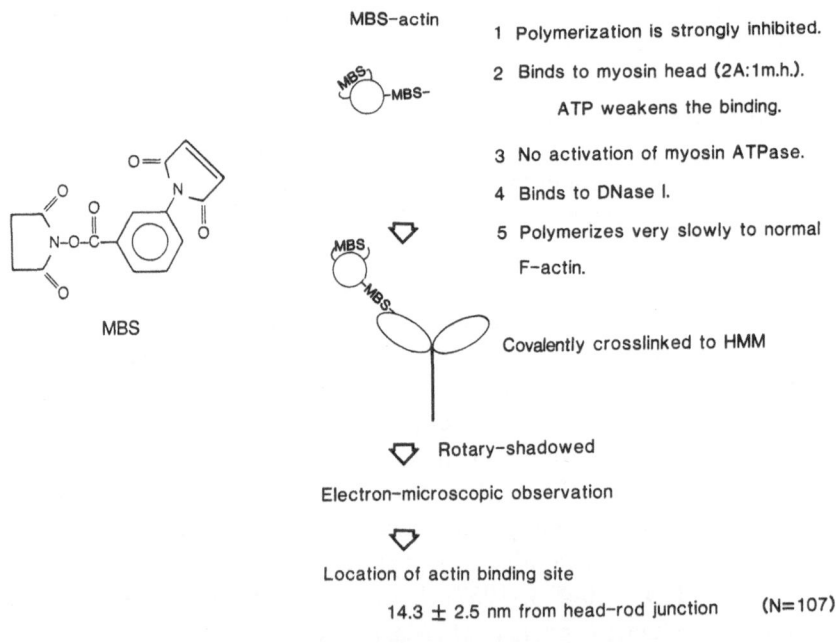

MBS–actin

MBS

1 Polymerization is strongly inhibited.

2 Binds to myosin head (2A:1m.h.).
   ATP weakens the binding.

3 No activation of myosin ATPase.

4 Binds to DNase I.

5 Polymerizes very slowly to normal
   F–actin.

Covalently crosslinked to HMM

Rotary–shadowed

Electron–microscopic observation

Location of actin binding site

14.3 ± 2.5 nm from head–rod junction   (N=107)

Fig. 1.

maleimidebenzoic acid *N*-hydroxysuccinimide ester (MBS; Fig. 1). However, MBS-actin polymerized during long incubation with phalloidin and stimulated myosin S1 ATPase as native F-actin, suggesting no serious damage. This actin retained the ability to bind DNase I. This actin also retained the ability to bind to myosin heads with apparent dissociation constants of 3-8 x 10^{-6} M and formed a *2:1* actin-head complex, suggesting at least two latent actin-binding sites on a myosin head. ATP weakens only 2- to 6-fold the binding of this complex. Native F-actin blocked the binding of MBS-actin on myosin heads in the presence and absence of ATP.

Monomeric MBS-actin was covalently crosslinked to S1 or HMM (by MBS). The resulting complex (180 kDa on SDS gels) did not bind to F-actin and showed no F-actin-activated ATPase, suggesting that monomeric actin- and F-actin-binding sites overlap or interact with each other on myosin head. Two complexes (180 and 140 kDa) were produced to the less extent in the presence of ATP. The electron-microscopic examination of these monomeric acto-HMM complex is now in

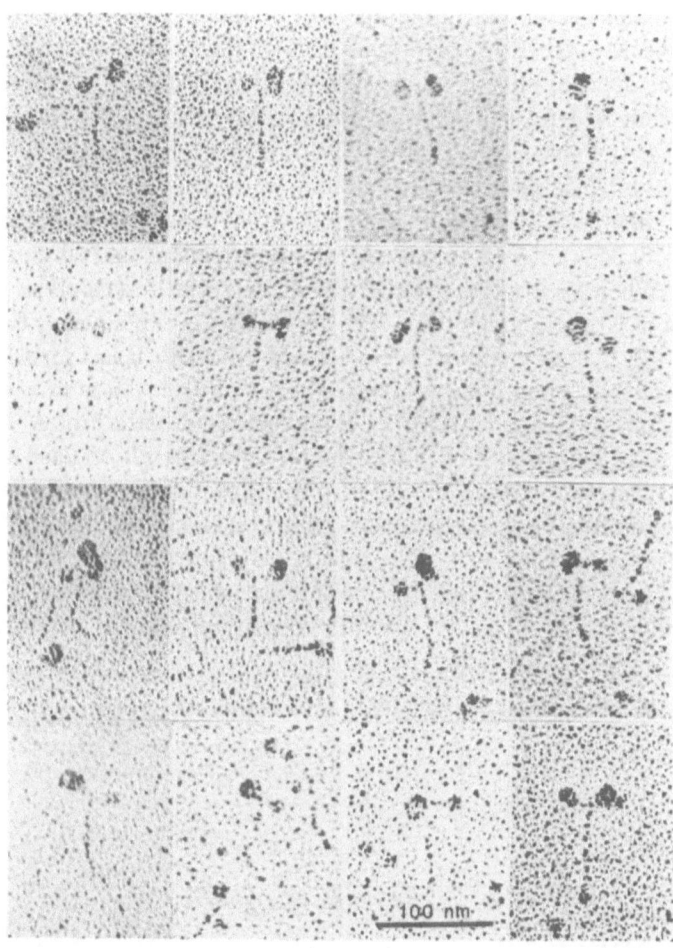

Fig. 2.

progress to explore the spatial location of actin binding site on a myosin head. The preliminary experiments showed that the location of the MBS-actin binding site of the complex produced in the absence of ATP is near the distal end of the head and 14.3 nm away from the head-rod junction (Fig. 2).

A Model for the Molecular Mechanism of Sliding Motion of Actin-Myosin System.

Ando: I present here a phenomenological model for the molecular mechanism of muscle contraction, in which conformational change in actin plays an important role. This model can account for well known basic phenomena about muscle contraction such as the P-V relation and the Fenn effect. The model also harmonizes and integrates two observations which seem to contradict each other at first glance (Osaka group's observation: a long step size of actin sliding per an ATP hydrolysis; Stanford group's observation: very slow sliding of actin under a low density of myosin). The motif of this model is depicted in Fig. 1. [1] The high-energy state of M*•ADP•Pi excites actin. [2] Sliding force is produced between the excited state of actin (A*) and M•ADP•Pi. [3] A* goes back to the ground state (A) with a rate constant which depends on how much work was done. [4] the ATP-splitting products are release from AM•ADP•Pi, but not from A*M•ADP•Pi. [5] The interaction of myosin (M, M•ADP, M•ADP•Pi, M*•ADP•Pi) with A generates frictional force that is proportional to the sliding velocity, Vs.

The P-V relation is accounted for as follows. When Vs is small, actin which is interacting with M•ADP•Pi is mainly A*. So, positive tension is high and frictional force is low. When Vs is large, M•ADP•Pi interacts with A as well as A*, resulting in generation of large frictional force. The Fenn effect is explained as follows. When Vs is small, the ATPase activity is low because M•ADP•Pi has to wait in order to release the ADP and Pi until the A* relaxes to A. (The relaxation rate is small, since the work performed is small.) With Vs being increased, the A* of A*M•ADP•Pi is replaced by A as actin filaments slide, which results in high ATPase activity. With Vs near the maximum, before the ATP-splitting products are released from AM• ADP•Pi, A will often be replaced by A*, which results in saturation of the ATPase activity. When actin filaments slide, one A* is available to several M•ADP•Pis before the A* goes back to A (see Fig. 2). This accounts for a long step size of sliding per an ATP hydrolysis. however, when the density of myosin is low, A* does not

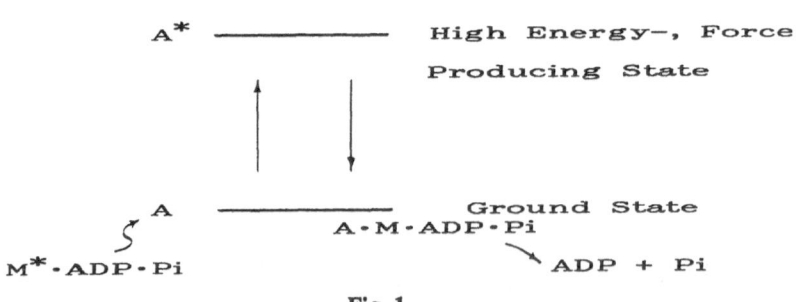

Fig. 1.

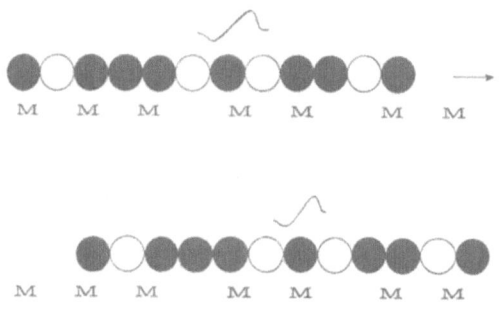

Fig. 2.

have many chances to interact with several M•ATP•Pis. In this situation, the energy stored in A* is not effectively used. So, one ATP hydrolysis causes just one actin-monomer-size of dislocation, i.e., very slow sliding as has been observed by the Stanford group.

After this Hakone symposium, we made mathematical analyses of the present model. We obtained several results which quantitatively agreed with the corresponding experimental observations.

Effect of Isozyme Redistribution on the Kinetic Property of Cardiac Myosin Studied Using an *in vitro* Force-Movement Assay System

Sugiura: In mammalian heart three myosin isoforms (Vl, V2, and V3) have been identified depending on the mobility on the pyrophosphate gel electrophoresis. Of these isoforms, Vl and V3 are the homodimers of α and β myosin heavy chains respectively and V2 is the heterodimer of α and β. It has also been shown that the distribution of these isoforms changes under various physiological and pathophysiological conditions. Since Vl has the highest ATPase activity and V3 the lowest, functional significance of isoforms redistribution has often been studied using papillary muscle preparations. In this study we studied the effect of isoform redistribution on the kinetic property by comparing the force-velocity relations of the sliding movement of bead coated with different cardiac myosin isoforms using an *in vitro* force-movement assay system.

We obtained Vl isoform from the cardiac muscle of 4 week-old rats and V3 from that of hypothyroid rats. Purity of each isoform was confirmed by pyrophosphate gel electrophoresis. Ca^{2+}-activated ATPase activity was 1.11 (Vl) and 0.46 (V3) μmole Pi/mg/min. Small polystyrene beads (2.8 μm in diameter) were coated with either of these myosin isoforms and made to slide on actin cables of an internodal cell of an alga *Nitellopsis obtusa* in the presence of Mg-ATP. Then, these internodal cell preparations were placed on the rotating stage of centrifuge microscope. By revolving this stage, we could observe the sliding movement of myosin coated beads under various centrifugal force applied to them as a load.

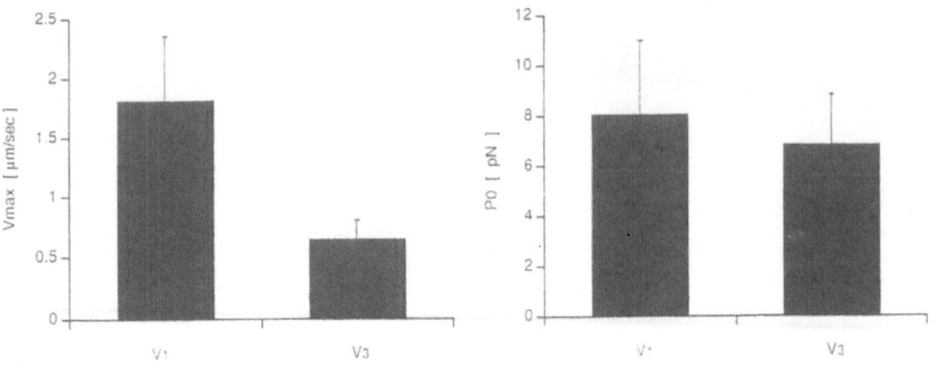

Fig. 1.

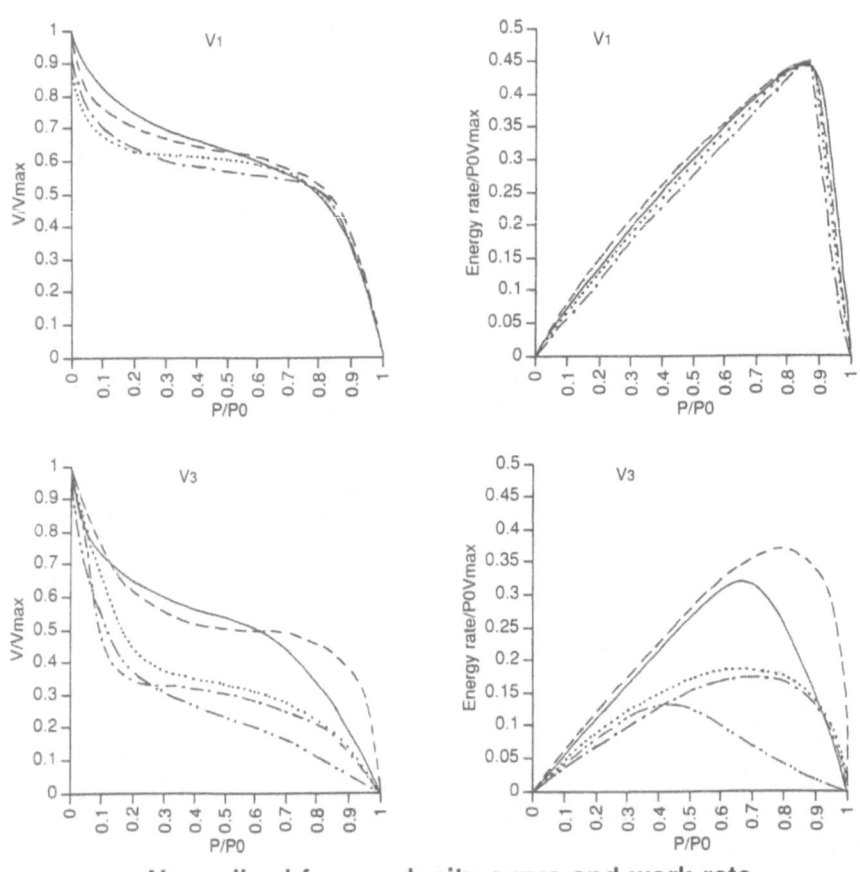

Normalized force-velocity curve and work rate

Fig. 2.

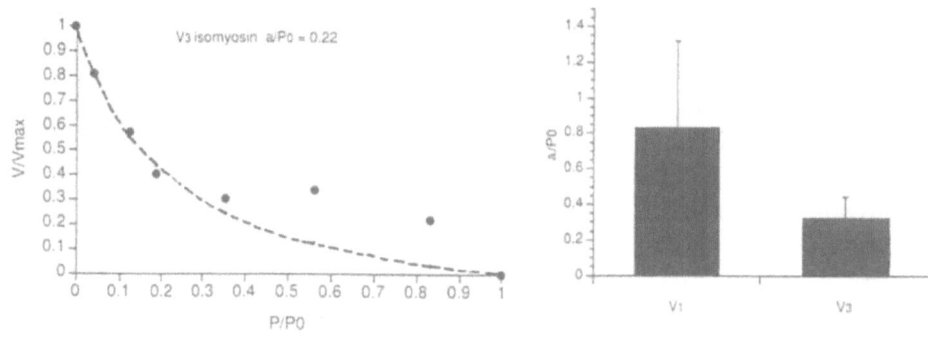

Fig. 3.

Under constant load the myosin-coated beads or bead aggregate moved smoothly along the actin cables over a long distance and the velocity decreased as the load was increased

The obtained force-velocity relation was double hyperbolic in shape similar to what was observed with skeletal muscle myosin. Maximal unloaded sliding velocity (V_O) was higher in Vl isoform (1.8 vs. 0.6 µm/sec) but we found no significant difference in the maximum force (P_0) between the two isoforms (8.0 vs. 6.8 pN) (Fig. 1). When both force and velocity were normalized to their maximum values, the curves of Vl isoform were shifted upward. The normalized (work rate)-load relations had higher peak values in the higher load region in Vl isoform (Fig. 2). To further characterize the shape change, the force-velocity data in the low load region were fit to Hill's equation and the a/P_0 value was calculated. The obtained a/P_0 value was higher for Vl (0.83 vs. 0.32) suggesting a difference in kinetic property (Fig. 3).

We found difference in kinetic property of actin-myosin sliding between the two cardiac myosin isoforms. Kinetic property observed in cardiac muscle may reflect the characteristics of each myosin molecules.

Velocity Fluctuation in Unloaded Shortening of Sarcomeres in Striated Muscle Myofibrils

Tameyasu: Though analyses of the shortening of a short segment in intact, single frog skeletal and cardiac muscle fibers [Pollack, G.H., *Physiol. Rev.*, **63**, 1049-1113 (1983); Tameyasu, T. et al, *Biophys. J.*, **48**, 461-465 (1985); Toride, M. & Sugi, H., *Proc. Jpn. Acad.*, **65B**, 49-52 (1989)] have suggested that the shortening is interrupted by short phases in which the velocity is close to zero, it has been generally thought that the shortening velocity of the sarcomere is constant at a steady load [Huxley, A.F., *Circ. Res.*, **59**, 9-14 (1986)]. Using a new method for rapid activation, I studied the time course of an unloaded shortening of sarcomeres in myofibrils isolated from glycerinated scallop striated muscle.

A single scallop striated muscle fiber consists of a single layer of 1 μm thick myofibril with 1.8 μm long-thick and 1.0 μm long-thin filaments without M-line in the A-band, resembling the slow skeletal muscle fiber of frog. The myofibril preparations were made by gently homogenizing the glycerinated scallop striated muscle in the rigor solution (in mM: K acetate 100, Mg acetate 5, EGTA 5, imidazole 10, $CaCl_2$ 5, pH 7.0). Then they were treated with 0.5 % Brij for 20 min. The myofibril which was usually 4-5 μm wide and less than 50 μm long was placed between two thin glass plates spaced by about 30 J m and filled with the rigor solution, the upper of the two plates being covered by 1 mm layer of an activation solution made by adding 5 mM Na_2ATP to the rigor solution. The solution around the myofibril was changed from the rigor to the activating one by rapidly withdrawing the upper thin glass plate, the solution change being completed within

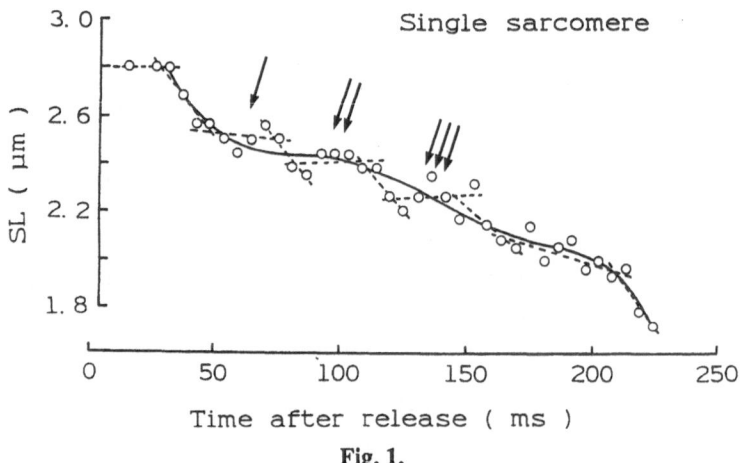

Fig. 1.

several ms. The central, 2 μm wide- and 8 μm long-segment in the myofibril was scanned at < 200 Hz by projecting the phase contrast image of the sarcomeres onto a photomultiplier with a slit *via* an optical scanner. The length of a single sarcomere, SL, was defined as the distance between the centroids of the adjacent I-bands. The space resolution was about 0.04 μm and the time resolution 5.4 ms at best.

After the rapid solution change, the myofibril commenced to shorten from the initial SL = 2.8-3.2 μm with a latent period of about 40 ms. When the data points were fitted with up to tenth order polynominal using the least square method, the resulting shortening curve exhibited a pause, in which the shortening velocity was close to zero. It was noticed in some sarcomeres, as shown in Fig. 1, that the pause appeared more than once during a shortening of the sarcomere with a phase of rapid shortening at a velocity of 20-30 μm/sarcomere/s (20°C) between them. The extent of the rapid shortening from the resting SL to the first pause and between the pauses was 0.26 ± 0.13 μm/sarcomere (mean $\pm$ S.D., n = 14). The pause lasted for 0.05 ± 0.03 s (mean $\pm$ S.D.). These observations suggest that the unloaded shortening of the sarcomere in the scallop striated muscle essentially consists of the alternation of the rapid shortening phase and the pause.

It seemed difficult to ascribe the fluctuation of the shortening velocity to a noise inherent to the photomultiplier, since such a noise caused only a random fluctuation of the data points as examined using a 2 μm grating. Neither the mechanical vibration of the setup nor the electrostatic interaction between the specimen and the glass surface might be the sources of the velocity fluctuation, since the pause appeared to occur asynchronously among the longitudinally contiguous sarcomeres. So it is likely that the shortening velocity of the sarcomere really fluctuates under the steady, zero external load in the glycerinated scallop striated muscle, being inconsistent with the widely accepted view that the contraction velocity of a contractile unit is steady at a constant load.

Pollack: I would like to know whether you have measured the size of the steps between the pauses. If so, do your measurements agree with the measurements we published earlier?

Tameyasu: Yes, I measured them. So far, we have obtained over 2.6 μm on average, so our measurement is much larger than yours.

Huxley: Can this be related to the spontaneous oscillatory contractions that Dr. Ishiwata told us about earlier today?

Tameyasu: In his case, a solution which is much different from ours was used and this solution contained a large amount of magnesium ADP. In our case, the solution does not contain magnesium ADP at all. So the situation is different.

Huxley: You used a very different muscle.

Tameyasu: Yes.

The Relationships between Force and Velocity of Shortening or Lengthening in Cardiac Muscle Obtained by Ramp-Load Release

Okuyama: Force-velocity relationship is usually obtained by the quick-release or the afterload method which demands muscle to contract several times for drawing a whole curve. During the measurement of velocities by these classical methods, the contractile state of living muscle is sometimes changed mainly due to the time lag, so that the curves might be modified by the experimental protocols.

A new method, a ramp-load release, is proposed to measure velocities under the stable conditions. When the muscle generates the maximal tension during an isometric tetanic contraction, the afterload is released at a constant rate (dF/dt = constant) to a resting value. Then the muscle starts shortening with various velocities (Fig. 1). The velocities are time variant, which is the result of time-variant force. The instantaneous tension and the simultaneous velocity of shortening show the force-velocity relationship, so that the force-velocity curve can quickly and simply

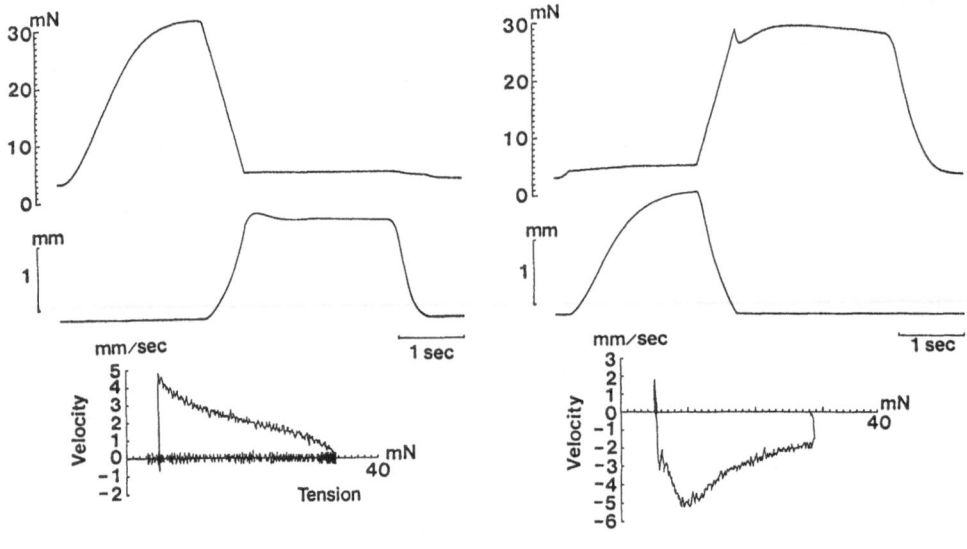

Fig. 1.

be obtained during one contraction. On the contrary, when the muscle shortens to a minimal length after the isotonic shortening, the afterload is increased at a constant rate to a maximal tension development, then the muscle is stretched with various velocities. This shows another force-velocity relationship, that is, a relationship between tension and lengthening velocity.

Eight ventricles of frog (*Rana nigromaculata*) are cut in strip and connected to a tension-length transducer which can generate force and detect tension and length signal. The tension or the afterload can be controled externally as is desired. The muscle strips are mounted vertically in the bath which solution contains 9 mM Ca^{2+} so that tetanic contraction may occur by electric stimulation. The temperature of the bath is controlled at 20°C or 25°C.

The force-velocity relationship obtained during only one contraction is quite similar to Hill's, but is not identical (Fig. 2). The curves depend on the ramp rate of tension. The greater is the decreasing ramp rate after the minimal tension development during isometric contraction, the shortening velocity is the larger, too. In contrast, the greater is the increasing ramp rate after the maximal shortening during isotonic contraction, the lengthening velocity is the larger, too. This rate-dependent relationship too complicated to explain from Hill's hyperbolic equation, $(T + a)(v + b) = (To + a)b$, where T is tension, v is shortening velocity and a, b, T_0 are constants.

To simplify the equation, it is assumed that the virtual velocity might include internal shortening or lengthening of the muscle. Then the two-component analog of the muscle is employed in order to calculate the real velocity of the contractile component. In this analog, if the series elastic component should follow Hooke law, the velocity v should be only the term that is to be modified by the ramp rate of

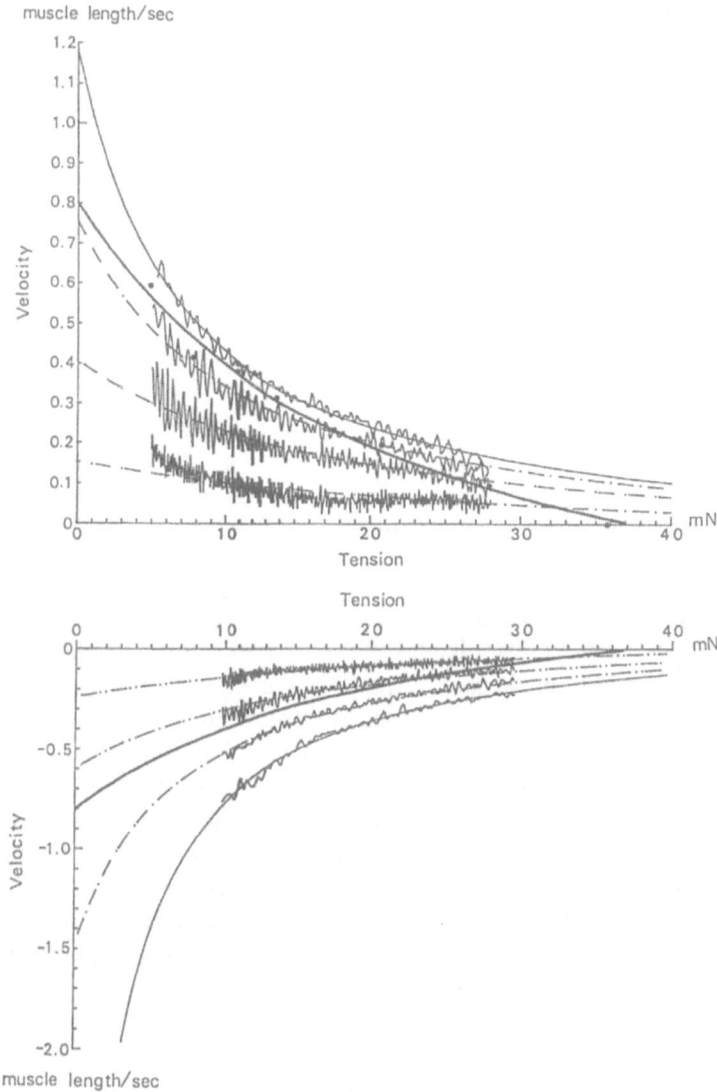

Fig. 2.

tension. In this case the curve should shift in parallel with the ramp rate. As is shown in Fig. 2, the velocity is more enhanced at smaller load with the ramp rate, so that this hypothesis should be dismissed, Second, if it is assumed that the virtual constant b might be affected by the ramp rate, it well follows the ramp rate using the curve-fitting procedure. The virtual b versus dF/dt is quite linear even when the sign of dF/dt changes. Further, this linearity does not change by the temperature (Table 1).

To explain the force-velocity relationship in this method from Hill's is not so simple as is expected. There will be other descriptions, one of which is: if the nonlinearity of SEC is taken into account, the shortening or lengthening velocity of

Table 1.

$$(T+a)\left\{v+b\beta\left(\alpha\cdot\Delta\frac{dF_n}{dt}+\frac{\gamma}{\beta}\right)\right\}=(T_0+a)b\left(\alpha\cdot\Delta\frac{dF_n}{dt}+1\right)$$

n = 8		20 ℃	25 ℃
α	$(mm^2\cdot sec/mN)$	-0.0124 ± 0.0015	-0.00939 ± 0.0014
β		$0.766\ \pm0.118$	$0.836\ \ \pm0.095$
γ		$0.787\ \pm0.114$	$0.856\ \ \pm0.094$
γ/β		$1.029\ \pm0.021$	$1.026\ \ \pm0.011$
dF_{n0}/dt	$(mN/sec\cdot mm^2)$	$-81.4\ \ \ \pm10.7$	$-105.3\ \ \pm16.2$
muscle length	(mm)	6.4 ± 1.2	
preload	(mN/mm^2)	5.2 ± 1.3	
total Tension	(mN/mm^2)	38.9 ± 5.5	

SEC becomes greater in smaller load. this is one of the qualitative explanation of virtual velocity for the greater load. It does not look simple to take the nonlinearity into account. In addition, the nonlinearity is not always steady.

As the Hill equation of hyperbola is composed of the theory of of energy liberation, the amount of energy liberation for unit time is proportional to the difference between the maximal tension development and the current load, that is, $(T + a)v = b(T_0 - T)$. When the absolute value of dF/dt increases, the b should increase. This is the reason the b is linear to the ramp rate. Moreover, when the ramp rate increase, decreases the duration of the ramp load applied because the dF/dt multiplied by the duration is equal to the constant T_0 (in case preload is zero). If the ramp duration decreases, the duration of energy liberation decreases, which explains the higher velocity for the same ramp rate.

Table 2.

$$(T + a)(v + b\cdot\gamma) = (T_0 + a)\cdot b$$

$$\text{in case}\quad\Delta\frac{dF_n}{dt} = 0$$

$$\text{that is}\quad\frac{dF}{dt} = CSA\cdot\frac{\alpha\cdot\frac{dF_{n0}}{dt}}{\alpha}$$

$$= \frac{CSA}{\alpha}\cdot const$$

	20°C	25°C
$\alpha\cdot\frac{dF_{n0}}{dt}$	$0.9911\ \pm0.0375$	$0.9672\ \pm0.0348$

The present formula is lightly complicated. Taking the particular ramp rate, the equation between tension and velocity is simplified to $(T + a)(v + b\gamma) = (To + a)b$, where γ is constant (0.86 at 25°C frog cardiac muscle)(Table 2.). Therefore, the force-velocity curve is quickly and easily obtained in one contraction by this new method of ramp-load release.

Double Pulse Photolysis of Caged ATP in Skinned Fibers from Rat Psoas Muscle

Horiuti: The rapid force recovery (phase 2) of the isometric force transient (Huxley & Simmons, 1971) is equivalent to the high-frequency phase advance (process C) in the sinusoidal analysis (Kawai & Brandt, 1980), so that they probably represent the same chemo-mechanical process in the cross-bridge reaction. However, while Huxley & Simmons (1971) assumed that phase 2 represents structural changes of the attached corss-bridges, Kawai & Halvorson (1989 & 1991) assigned process C to the transition of cross-bridges from the last attached state to the detached states.

The 'quadrature stiffness' at a high frequency is an index of the magnitude of process C and phase 2 (Goldman et al., 1984). In this study, following Goldman et al. (1984) but at a low temperature (10°C, $\Gamma/2 = 0.2$ M), we photo-released ATP in rigor skinned fibers to induce an initial relaxation and a subsequent contraction (10 mM CaEGTA). The quadratrue component of 1 kHz-stiffness (low-pass-filtered at 100 Hz) rose shortly behind the initial relaxation but earlier than the following contraction. In the absence of Ca^{2+} (10 mM EGTA), the increase in the quadrature stiffness was small and only transient.

Our results sem to suggest that the rigor cross-bridges, after binding ATP and before generating force, enter a 'weakly-bound' state which is different from the detached state that appears in the resting muscle fiber.

Measurement of the State of Water in Muscle at Rest and during Contraction

Ogata: I will talk on the NMR behavior of water in muscle as studied with 500 Hz high resolution FT-NMR. Fig. 1 shows three water signals taken before (A), during (B) and after (C) tetanic stimulation of a frog skeletal muscle fiber bundle. The signal from relaxed muscle was not always symmetrical but often exhibited a minor peak (A). As this minor peak disappeared by freezing the fibers at -15°C and warming them again, it seems to reflect some biological phenomena. The minor peak became much less pronounced during stimulation (B), with some recovery after relaxation of the fibers (C). These signals suggest the presence of two different groups of water arising from two different levels of water structure in muscle. There might be exchange of proteins between the two water structures (Wf and Wa). I could calculate parameters to fit the observed results with the following characteristics:

Protein is more mobile in Wf than in Wa. P_{wf} (proportion of Wf) does not change appreciably during and after tetanic stimulation. Twf (transverse relaxation

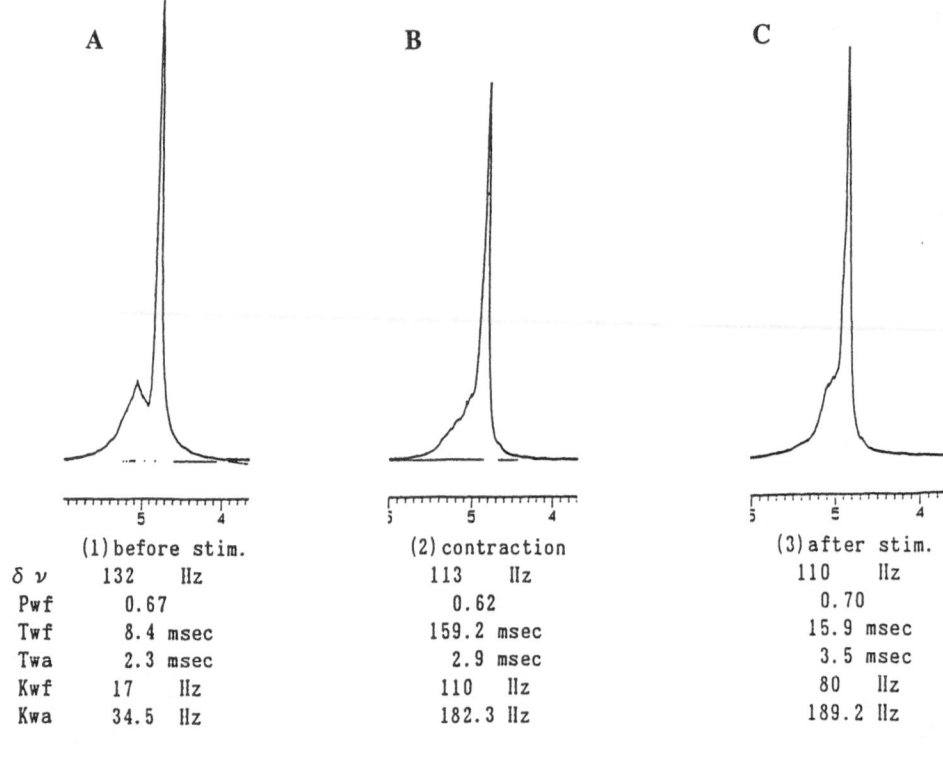

	A	B	C
	(1) before stim.	(2) contraction	(3) after stim.
$\delta\nu$	132 Hz	113 Hz	110 Hz
Pwf	0.67	0.62	0.70
Twf	8.4 msec	159.2 msec	15.9 msec
Twa	2.3 msec	2.9 msec	3.5 msec
Kwf	17 Hz	110 Hz	80 Hz
Kwa	34.5 Hz	182.3 Hz	189.2 Hz

Fig. 1.

time of Wf) increases from 8.4 to 159 ms during tetanus, a value very close to the value for free water, and then decreases down to 1.59 ms after tetanus. On the other hand, T_{wa} does not change markedly. K_{wf} and K_{wa} (rate constants of the water groups) increase during tetanus, At present, I would like to say that water structure is well organized in relaxed muscle, and collapses during contraction.

Pollack: It wasn't clear to me how you were able to assign one fraction to intramyofibrillar water and the other to extramyofibrillar water. What was the basis for this distinction? How did you know which was which?

Ogata: I jumped to that conclusion. I can only say that from the NMR result, there are two water groups that have different resonant frequencies. Our intramyofibrillar story is a bit premature.

Kushmerick: NMR studies of the state of water in muscles, nerves, and other tissues have been made for a long time. I don't recall anybody demonstrating a frequency difference. Usually, one talks about this in terms of NMR spin-relaxation rates, which you also observed. I think the unique feature here is that you found a second resonance.

CONCLUDING DISCUSSION

CONCLUDING DISCUSSION

Chaired by G.H. Pollack, H. Sugi, J.M. Gillis, E. Homsher and R.C. Woledge

STRUCTURAL CHANGES DURING MUSCLE CONTRACTION

Pollack: Two structural approaches that have been used increasingly over the past years have been especially well represented at these meetings. One is the use of X-ray diffraction, and the other is the use of EM methods. With the advent of synchrotron radiation, the ability to obtain information from X-ray diffraction has increased enormously. One of several examples is the use of the equatorials to determine the number of cross-bridges attached. The EM approach has also taken a step forward with the use of quick-freeze techniques. One application is the study of cross-bridge angles.

At this meeting, several people are represented in each of these categories. In X-ray diffraction, for example, we heard papers by Drs. Wakabayashi, Iwamoto, Squire, Griffiths, Yagi, Yu, and Brenner. With regard to the EM approach to infer cross-bridge angle, we have heard Drs. Katayama, Suzuki, Sugi, Reedy, and Goldman. I'm sure I haven't named everyone for both categories. So we have a lot of information that has been presented.

What I would like to do—and this discussion is not a rehash of all of those papers because I think we've done enough of that—is draw more general conclusions. The central question is this: given these two approaches, what unambiguous information have we gained during the past twenty years about the mechanism of contraction? I stress the word, "unambiguous."

Let me suggest that there is very *little* unambiguous information that we have learned in the past twenty years. With that provocative statement, perhaps we can get started.

Probes to Detect Myosin Head Movement

Morales: This is not in reply to your question, but I think that of the people who have studied probes as a way to determine myosh head rotation—and it is really in many ways the ideal way to detect the head rotation—probably Dr. Burghardt has the most information and analysis, and it seems to me you ought to ask him to participate in this. His work is not EM or X-ray.

Burghardt: From the probe work labelling SH-1, we have learned that the cross-bridge rotates when it binds actin and nucleotide. I think that's unambiguous.

Mechanism of Myofilament Sliding in Muscle Contraction, Edited by
H. Sugi and G.H Pollack, Plenum Press, New York, 1993

Pollack: Can we take this as an unambiguous result?

Maughan: Well, I am a little bit puzzled because of the results from Cooke and Thomas (*e.g.,* Cooke, R. et al. *Nature* **300**, 776-778, 1982) which seem to contradict, to some extent, the rotation of the head. I wonder if you could comment on that?

Burghardt: Originally in 1982, there was some suggestion that the SH-1 maleimide spin label showed there was no rotation. But I think since that time we have shown that if you include that data with all the other probes—fluorescent and spin—that can label that site, it does rotate. I don't think there is any contradiction or controversy about that now.

Huxley: Can someone say whether the sort of millisecond or submillisecond fluctuations described a year or two ago by David Thomas and colleagues are unambiguously established, or are they still in doubt?

Morales: Before commenting on that, I would like to say that we also got rotational fluctuations some years before that. So the fact that the fluctuation approach indicates rotation is unambiguously established, I would like to think. The difference between what we saw and what Dr. Thomas saw is due to the great difference in the speed at which these go. Ours are much slower, not in the millisecond or microsecond range. I think ours were a lot more like Dr. Yanagida's fluctuations. But in a general way, there is rotation indicated from that approach also.

Reedy: I think there are two or three sources of ambiguity. One concerns the fact that the probes are attached mostly to thiol #1 in rabbit muscle, and rabbit muscle doesn't have well-oriented, relaxed cross-bridges under most of the conditions that have been studied. It's not even easy—except for a few people who have been working on it recently—to get well-ordered cross-bridges in the relaxed state for X-ray diffraction purposes. But it hasn't been done very often, that the conditions of temperature and ionic solution required to sustain that for X-ray diffraction have been transferred to the spin label work. So that's one source of a problem there.

Another one I think is that Meg Titus and some other people, Andrew Szent-Györgyi and so on, found that thiol #1, being labelled, gradually deprived rabbit muscle of its capacity to be regulated. The time came when a relaxed fiber, labelled with thiol #1, didn't care whether the calcium was there or not; it was going to behave like an activated fiber. So any disorder in the bridges found in a relaxed fiber that's been labelled on that reactive thiol could be ascribed to either the disordering that's due to temperature effects, or the disordering that might be due to some degree of activation of the bridges. Those are sources of ambiguity that have troubled me.

The third source doesn't apply to rabbit at all. Roger Cooke and I have worked on insect muscle where there is good evidence for a well-ordered orientation in relaxed muscle of the bulk of the myosin cross-bridges—these 90° bridges which may not be at 90° but they are well ordered. John Squire presented a model in which the heads may just lean partly into those 90° shelves. But, in any event, they are very well ordered. Roger Cooke and I used a spin-labelled nucleotide which avoided all problems with thiol labelling—something we couldn't achieve properly in the insect muscle anyway. Under those conditions, the relaxed bridges remained very well ordered by X-ray and EM criteria, but the spin label gave a highly disordered signal, indicating that it was not telling us very much about the ordering of the bulk of the cross-bridges.

So, I see those as three sources of ambiguity, two of them applying to rabbit and one to insect, that make it difficult to say that you know from what the probe tells you how the bridge was before it went into rigor.

Pollack: Can we restrict our comments to ambiguity *vs.* lack of ambiguity, as Mike has?

Goldman: I think most people who have used probes have found rotations of the probe, and this doesn't include just on SH1 but, as Mike mentioned, at the nucleotide binding site, and now in several studies of the probes on the light chains. A crucial point we haven't mentioned yet is whether those rotations occur during the transition in the cross-bridge that leads to the force-generating state. Are they associated directly with the power stroke? I don't think anyone has shown that in a situation where the cross-bridges are synchronized well enough to allow us to know that the power strokes are occurring during the rotation. During the synchronization just after a length change, for example, or if the fiber is activated rapidly, people have generally *not* observed probe rotation.

Interpretation of X-ray Diffraction Results

Pollack: I think it is a good time to move to X-ray diffraction, where the timing and tension can be measured. Can we have some comments about the equatorial reflections and to what extent they can determine, unambiguously, the degree of cross-bridge attachment?

Yu: As John Squire and I reported in this meeting, there are several factors that can affect equatorial intensities. We have done partial activation studies on frog and rabbit fibers, in which we activated the muscle in a graded level—i.e., we increased the number of cross-bridges in isometric condition. (Yu, L.C. et al. *J. Mol. Biol.* **132**, 53-67, 1978; Brenner, B., and Yu, L.C. *Biophys. J.* **48**, 829-834, 1985). For these cases, there is some kind of scale that shows proportionality between force-generating cross-bridges increasing with increasing $I(1,1)/I(1,0)$. It would however be dangerous to correlate a change in number with the change in $I(1,1)/I(1,0)$.

Pollack: Other comments on the intensity ratio?

Squire: Some people say that the $(1,1)$ is proportional to the number of attached heads, others that it is not proportional to number. But there is actually some theory behind this and I think it's worth showing what the theory is.

First of all, let's forget about cross-bridges and pretend there are just myosin filament backbones and actin filaments. We know where these are in the unit cell and we can calculate what we would expect the amplitude of the diffraction to be. For the $(1,0)$, the amplitude is the difference of the scattering factors of the myosin backbone and the actin. For the $(1,1)$, it is the myosin plus two scattering factors for actin. That's a very simple relationship, which was the original relationship found by Gerald Elliott et al. (*J. Mol. Biol.* **6**, 295-305, 1963; *J. Mol. Biol.* **25**, 31-45, 1967) and is the basis on which all of this apparent linearity between changes of the (1,1) and number is based. But that is just a very primitive theory. What you have to do of course is to add in cross-bridges. The question is, "What difference does it make?"

Theoretically, when the myosin heads are included as a separate structure, there is no obvious reason why the (1,0) and (1,1) intensities should be directly proportional to the number of heads on either. If you have a cross-bridge on actin in a particular configuration and it swings in the way we have been thinking about in the past, then there is a radial movement of mass, but we are talking about a head that is nevertheless always attached to actin. A transition that swings the head away from you so that there is more mass in towards actin will affect the intensities, even though it doesn't change the number. So, You mustn't make the too-simple assumption that the intensities are proportional to the number. That may be true in certain situations, but in general, it isn't. The important message is that equatorial X-ray diffraction patterns contain information about both the number of heads in each state and the structure of that state. It is a very powerful tool with which to study the cross-bridge cycle.

Pollack: Thank you, John, for these important admonitions about the interpretation of X-ray results. There is another worrisome issue about the interpretation. I wonder if somebody might be willing to say something about the transitions that Bill Harrington suggests may take place in the S2. Would such transitions affect the intensity ratio as well?

Yu: They would, if they change mass distribution on the equator, and certainly if they change the helical distribution.

Squire: Jeff Harford, Michael Chew, and I have done X-ray experiments looking at S2 by recording the high-angle diffraction pattern from muscle (Squire, J.M. et al. *Daresbury Ann. Rep.* **189/90**, 153). It gives a 5.1 Å reflection, which is a characteristic of α-helix reflection. There might be contributions there from other things apart from myosin, but we were able to rule out that possibility. For example, tropomyosin could contribute, but we don't think it does. The conclusions we have so far from a rather difficult experiment are that the intensity changes from relaxed to fully active are rather small, probably less than 2%. We want to pin it down a little further than that, but that is as far as we have gone so far. The Harrington model would, in fact, predict only a small per cent of the change.

Morales: There are two problems I think apply to all the methods. One is that S1, as an irregular solid, can rotate around any one of its principal axes. Looking at the pictures Ken Holmes showed (Holmes and Milligan, this symposium), one sees that S1 is not likely to rotate about the axis as so often shown in textbooks, but rather about its long axis ("torsional rotation"); this is precisely the axis that Burghardt has inferred from his probe work. The other problem is that S1 may have semi-autonomous "domains," so the observed rotation of an orientational probe may indicate only the rotation of the domain that bears it.

Tregear: I don't think things are really quite so bad. Just to take a point you had made specifically about the S2: changes in the S2 itself, I suppose, wouldn't change the equatorial very greatly. What Leepo Yu meant was that that might then cause movement of the S1, which would cause a change. That's just a clarification.

A second point is that, in fact, the way the equatorials have been used mostly is by correlation with other variables. What you can be sure of is that when you get a change in the equatorials, then something has moved. The overall assumption is that the only thing moving is the head of the myosin. That is just an axiom. Of course, if there is something else moving—titin or whatever—you're in trouble. But the correlation is either with mechanical measurement, as we know, or with some other part of the diagram. And now that X-ray diagrams are being obtained and will be obtained in two dimensions rapidly, as Dr. Yagi showed, for example, we think that correlative work will be much easier to do, and I think it will be a lot more credible; we will be able to connect a definite event that we call a weak-binding state or a strong-binding state or some form of state, with the change in the equatorial.

Lombardi: I want to comment on the combined X-ray and mechanical experiments on intact fibers, that I reported in this symposium. We found such a large change in the 145 Å meridional reflection following a length step, with the same time course of quick tension recovery. This is unambiguous. This cannot demonstrate rotation *per se*, but demonstrates unambiguously axial movement of the myosin head occurring together with the force-generating step. The other point is that, in the same records, the area detector did not show significant changes in reflections that have much larger intensity like the equatorial, supporting the conclusion that there is no change in mass distribution in the plane across the longitudinal axis during the tension transient following a step release.

Sugi: It is now time to finish discussion on the structural changes during contraction. It is a pitty that we did not have time to discuss the EM results.

MUSCLE CONTRACTION MECHANISM AS STUDIED WITH *IN VITRO* MOTILITY ASSAY SYSTEMS

Sugi: I want to discuss the results obtained with *in vitro* motility assay systems. The assay systems hitherto developed are listed below:

(1) Fluorescently labelled F-actin versus myosin system (Kron, S.J., and Spudich, J.A. *Proc. Natl. Acad. Sci. USA* **83**, 6272-6276, 1986);

(2) Myosin-coated bead versus actin cable system (Sheetz, M.P., and Spudich, J.A. *Nature* **303**, 31-35, 1983; Shimmen T., and Yano, *Protoplasma* **121**, 132-137, 1984);

(3) Myosin-coated needle versus actin cable system (Chaen, S. et al. *Proc. Natl. Acad. Sci. USA* **86**, 1510-1514, 1989);

(4) Myosin-coated bead versus actin cable system combined with the centrifuge microscope (Oiwa, K. et al. *Proc. Natl. Acad. Sci. USA* **87**, 7893-7899, 1990);

(5) Fluorescently labelled F-actin versus native myosin filament system [Sellers, J.R., and Kachar, B. *Science* **249,** 406-408, 1990; Yamada, A. et al. *J. Biochem. (Tokyo)* **108,** 341-343, 1990].

These systems enable us to observe the ATP-dependent actin-myosin sliding under a light microscope. The number of actin and myosin molecules involved in the sliding is very small compared to that in muscle fibers. In addition, we can characterize and chemically modify actin and myosin in the system. The results obtained with these systems are extremely useful in eliminating the gap between muscle physiology and muscle biochemistry.

Fluorescently Labelled F-actin versus Myosin System:

Sugi: I shall begin with the fluorescently labelled F-actin versus myosin system, which is now popular and used in many laboratories. The most remarkable finding brought about by this system is normally believed to be that S1 is essential for producing sliding motion. But to my understanding, it was first pointed out by Dr. Oplatka and his co-workers that continuous filamentous structures are not essential for producing muscle contraction. They showed that ghost myofibrils could again be made to shorten and generate force after irrigation with HMM or S1 (Oplatka, A. et al. *Biochem. Biophys. Res. Commun.* **58,** 905-912, 1974). However, people dealing with *in vitro* assay systems have totally ignored their pioneering work. Dr. Yanagida, why have you and others ignored Dr. Oplatka?

Yanagida: I have not ignored him. His ideas are always very good, but his experiment at that time was not enough to convince us of his conclusion.

Sugi: Did you repeat Dr. Oplatka's experiment?

Yanagida: Yes. For example, we added S1 to the myosin-free ghost fiber. Sometimes the muscle fiber contracted, but sometimes it did not. So, his results were difficult to reproduce.

Sugi: I have a fundamental question about the state of S1 on a glass surface, as pointed out by Yoko Toyoshima (Chapter IV). According to Dr. Yanagida and others, there should be some artifactual rigor-like linkages, which may produce enormous internal load against the actin-myosin sliding. Dr. Sellers, would you like to comment on how we can escape from such an artifactual internal load?

Sellers: One thing we routinely do when we set up a motility assay is that, before we add the rhodamine-phalloidin label to actin, we wash it through with actin that is not labelled with rhodamine-phalloidin so that it is invisible. This will improve the movement of the system. But I should say that we still have a much slower velocity of movement than does Dr. Yanagida's laboratory. He attributes this difference to the fact that we use nitrocellulose film while he uses silicone-coated surfaces. I just have one question before we leave S1: as far as I am aware, in all the reports, F-actins have always moved more slowly with S1 than with HMM or myosin. I wonder whether anyone would like to venture a guess as to whether that is because it is too close to the surface and therefore has problems interacting, or whether there may be mechanistic reasons why the S1 moves a little bit more slowly? Do we think S1 moves slowly because it is missing the S2, or because it only has one head, or is it due to surface interactions?

Gordon: Jim Spudich in a seminar reported that, in the genetic engineered case, where you put a very small amount of myosin rod onto S1, thin filaments move at the high velocities you would see for HMM.

Sugi: In this connection, I also want to ask Dr. Yanagida about the two-head or one-head story. At the previous Hakone meeting, there was some debate between our group and you concerning whether two-head cooperativity (Chaen, S. et al. *J. Biol. Chem.* **261**, 13632-13636, 1986) is essential for muscle contraction or not. At that time, Hugh Huxley judged that you took the better of the argument. But, I noticed that in your system, S1 or single-headed myosin can cause smooth F-actin movement only under low ionic strengths. At physiological ionic strengths, F-actin tends to detach from the glass surface. This is the reason I think that two heads may be necessary for physiological actin-myosin sliding.

Yanagida: I showed that the double-headed structure is not essential for movement and producing force (Harada, Y. et al. *Nature* **326**, 805-808, 1987). I don't know why the myosin molecule has a two-headed structure. Recently, we measured the force for actin-myosin sliding by using single-headed HMM, not S1. The force is almost zero. The velocity produced by single-headed HMM is almost the same as that produced by two-headed HMM, but the force produced by the one-headed HMM is very small.

Sugi: It's very strange and interesting. It should be examined by others.

Yanagida: S1 can produce almost the same force as that produced by each head of the two-headed HMM or myosin, so maybe this is why the affinity of the single-headed HMM for actin is very small compared to that of the two-headed HMM. In that sense, your argument is correct.

Sugi: I agree with you that S1 can induce sliding on a glass surface, but I am not sure whether S1 alone can produce sliding in muscle fibers at high ionic strengths.

Myosin-coated Needle and Myosin-coated Bead versus Actin Cable Systems

Sugi: Now I want to move to the myosin versus actin cable systems. Among these systems, the myosin-coated bead versus actin cable system was first developed independently by Drs. Sheetz and Spudich in the U.S., and by Drs. Shimmen and Yano in Japan. Drs. Shimmen and Yano developed it one year in advance of the U.S. group, but they published it only in 1984. These systems are not widely used, mainly because it is difficult to have a constant supply of good algal cells on which myosin can move smoothly and rapidly. It is my impression that the F-actin versus myosin system we have just discussed resembles a water solution system in which molecules collide spontaneously. But in the actin cable versus myosin system, although myosin is randomly oriented on the bead surface, the properties of this system closely resemble those of muscle fibers, since actin cable sonsists of actin filaments.

As we have reported in this symposium, the myosin-coated needle versus actin cable system on the centrifuge microscope exhibits a hyperbolic-shaped steady-state

P-V relation analogous to that of muscle fibers (Chaen et al., this volume) and the myosin-coated needle versus actin cable system shows a bell-shaped load versus work relation in response to a constant amount of ATP application (Sugi et al., this volume). The latter result reminds me of the Fenn effect, though we do not measure the actual amount of ATP used by the actin-myosin sliding. An important issue brought about from this result is that the amount of work done by the system with a constant amount of iontophoretically applied ATP increases with increasing initial baseline force to a certain value, indicating an increase in the apparent efficiency of the system. It is generally held that, in contracting muscle, the hyperbolic P-V curve, which yields the bell-shaped power versus load relation (and also the bell-shaped work versus load relation found by Fenn), is associated with the change in the number of cross-bridges interacting with actin. This is Professor Huxley's 1957 scheme.

As far as our *in vitro* system is concerned, however, the average velocity of actin-myosin sliding does not change appreciably, while we get the increase in the apparent efficiency of the system. Since the number of myosin molecules is limited to a few hundred in the myosin-coated needle versus actin cable system, and even less than ten in our myosin-coated bead versus actin cable system on the centrifugal microscope, I tend to think that these kinetic properties may stem from properties of the individual cross-bridges, and not from a change in the number of the cross-bridges interacting with actin.

Gillis: For those of us who are not using this very sophisticated assay, could you tell us why the first assay—F-actin versus myosin systems—is not as close to living systems as the second one? I don't understand.

Sugi: I mean that this system resembles water solution systems rather closely, since there is a close parallelism between the Michaelis-Menten V_{max} and the maximum F-actin sliding velocity. Michaelis-Menten's analysis is based on the spontaneous collision of molecules in water solutions, and can apply to this system in many cases.

Sellers: May I make a comment? There is one way you can make the sliding system much more equivalent to a muscle, and that is to use native thick filaments. In that case, you have an actin sliding over presumably a very properly oriented thick filament, and this would be more comparable to a muscle than either the random sliding system or the alga system.

Sugi: I would now like to mention some results obtained with the myosin-coated needle versus actin cable system developed in our laboratory. Using the myosin-coated meedle versus actin cable system, Dr. Oiwa and I decreased the amount of iontophoretically applied ATP bit by bit until needle movement was just barely also minimized, since the applied ATP around the needle is removed by the hexokinase-glucose system [Oiwa, K. et al. *J. Physiol. (Lond.)* **407**, 751-763, 1991]. In other words, if you minimize the total current passed through the ATP electrode, it would result in a very short, pulse-like application of ATP to myosin. Fig. 1 are histograms showing the distribution of actin-myosin sliding distance induced by small amounts of iontophoretically applied ATP. The amount of charge passed through the ATP

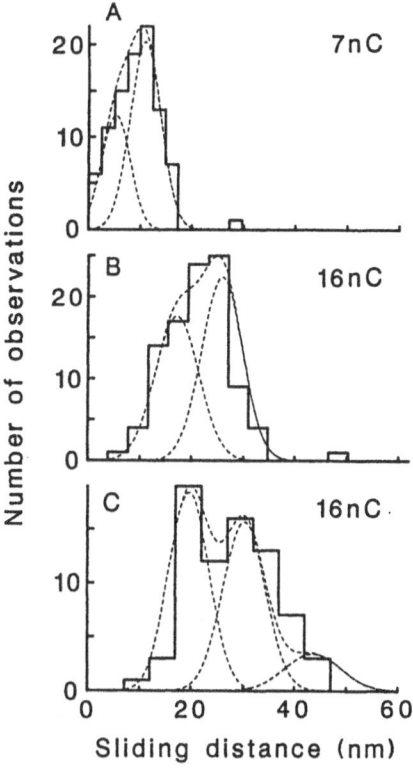

Fig. 1.

electrode was 7 nC in A and 16 nC in B and C. Assuming that the above distribution is statistically built up of unitary distance of actin-myosin sliding, a few Gaussian curves were fitted to the histograms. The above analysis indicates that the sliding distance distribution with 7 nC ATP consists of two Gaussian curves detectable. This means that the time in which myosin may utilize the applied ATP is with mean values of 5 and 10 nm respectively (A), while those with 16 nC ATP consists of two or three Gaussian curves with mean values equal to integral multiples of 10 nm (B and C). As the peak of 5 nm is likely to reflect mechanical vibration of the experimental apparatus, we think that the above results are consistent with the idea that the unitary distance of actin-myosin sliding is about 10 nm at least in our experimental conditions.

We have also obtained information about myosin step size from the experiments with laser flash photolysis of caged-ATP (Yamada et al., this volume); when one ATP molecule is supplied to each myosin molecule within a muscle fiber (i.e., only one of the two myosin heads can use ATP while the other does not), then we again have a minimum discrete myofilament sliding of 10 nm (Fig. 2). On the other hand, if an ATP molecule is applied to each of the two myosin heads in the muscle fiber, we get a myofilament sliding of 60 nm. The figure 10 nm resembles that reported by

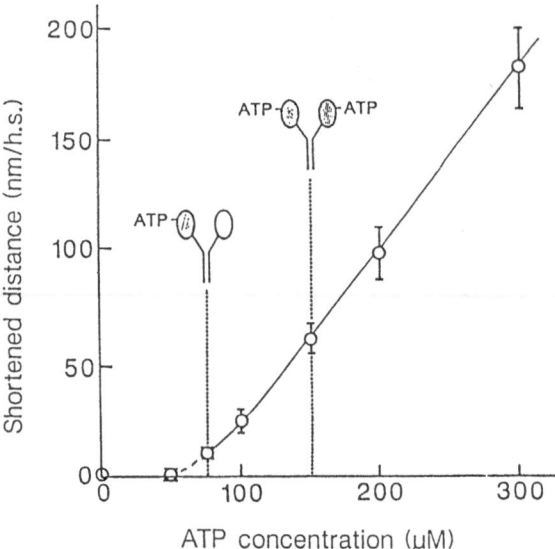

Fig. 2.

Yoko Toyoshima and Jim Spudich, while the figure 60 nm resembles that reported by Dr. Yanagida.

An interesting issue in our caged-ATP experiments is that, when one ATP molecule is applied to only one of the two myosin heads, what is the state of the other head? If it is a rigor state, it may produce enormous resistance to sliding movement. This fact seems to be consistent with Dr. Inoue's report at the previous Hakone Meeting that one ATP molecule per one of the two heads of myosin is enough to detach rigor of HMM from actin [Inoue, A. et al. *Molecular Mechanism of Muscle Contraction*, (ed. H. Sugi and G.H. Pollack) 227-235 (Plenum, New York, 1988)]. An alternative explanation may be that one head of a myosin molecules causes filament sliding of about 10 nm by pulling the other rigor head.

Goldman: We get a similar result to yours: there is a considerable amount of fiber shortening at about 100 μM ATP. However, I think it is very surprising that this happens, and I wonder what you think about it? You have drawn it so that one ATP molecule binds to one head of the two, but I think it is more likely that the ATP molecules would bind randomly and therefore about a quarter of the molecules would have no ATP—that's both heads without ATP, but sliding still occurs. Why?

Sugi: We start with a simple, optimistic assumption that I stated, because we always observe very uniform sarcomere shortening under the electron microscope whenever we have a discrete fiber shortening. There is enough time for ATP to diffuse uniformly among myofilaments.

Goldman: Do you actually have reason to believe that ATP liberated at half-stoichiometry will really bind as your diagram shows—the leftmost myosin molecule with one ATP bound to one head and no ATP bound to the other, rather than randomly?

Sugi: Of course, we cannot be entirely sure about that. But we tend to believe that the actual condition is close to what I stated.

Edman: You like to emphasize that your assay system based on myosin-coated beads has many similarities with the intact fiber system, but in fact there are very many differences. In the P-V relation of your system, there is some kind of curvature, which may be hyperbolic. Yet especially with beads of small P_0 values, you have an almost constant velocity over a large range of loads, and the velocity suddenly drops down with loads close to P_0 (see Chaen, S. et al., this volume; see also Oiwa et al. *Proc. Natl. Acad. Sci. USA* **87**, 7893-7897, 1990). Your V_{max} is lower than that in muscle fibers at corresponding temperatures, and then you have an unexplained finding of a decrease in sliding velocity at negative loads. Does the bead in your system rotate or is it just sliding?

Sugi: It doesn't rotate at all. It slides straight along actin cables.

Edman: How do you know that?

Sellers: You can sometimes observe clumps of beads. Where there are two or three beads clumped together, they slide uniformly.

Sugi: Yes, the beads slide together in groups. So, they do not rotate. I forgot to mention the discrepancy of the effect of negative load. Dr. Edman, in your fiber system, the compressive force increases the velocity of shortening very much [Edman, in this volume; see also Edman, K.A.P. *J. Physiol. (Lond.)* **291**, 143-159, 1979] while in our system, it slows the velocity down. My explanation for this is that if you suddenly slacken the fiber with resting tension, the cross-bridges may be deactivated so that the fiber rapidly becomes taut again.

Edman: I wouldn't agree with that, because that would be to deny the whole measurement of the P-V relationship.

Sugi: Normally, we mechanics people tend to escape from the complexities arising from the resting tension. In your experiments, you give a release to a resting muscle fiber (Fig. 7, Edman, K.A.P., 1979) the amount of which you still have some resting tension after the release. But, when the fiber is tetanized, you have no resting tension immediately after the same amount of release. This is a very strange phenomenon.

Edman: I could mention that if you apply the same release to the resting fiber with long sarcomere lengths, you will have much higher shortening velocities. So, the velocities you see in tetanized fibers at the same sarcomere lengths are certainly restricted by the presence of the cross-bridges.

Sellers: If you mix a fast myosin and a slow myosin in either the sliding F-actin assay or the actin cable assay, this may be regarded as putting a negative load arising from the slow myosin onto the fast myosin. You do indeed get graded sliding velocities above those of the fast myosin and the slow myosin. It seems to me that this situation is close to what you do when you centrifuge in the direction in which the bead is moving.

Fluorescently Labelled F-actin versus Native Myosin Filament System:

Sugi: We should move to the fluorescently labelled F-actin versus native myosin system. This has been developed independently by Dr. Jim Sellers at NIH, as well as Drs. Yamada, Ishii, and Takahashi at the University of Tokyo. To my knowledge, it has also been developed by Dr. Sleep at King's College, though he did not publish his results. The results of these groups agree in that actin moves unidirectionally across the M-line, but the speed—and probably the force—may change very much across the M-line. According to Yoko Toyoshima, however, there is no difference in velocity if actin moves across the M-line of some crustacean thick filaments. Unfortunately, we do not have her with us now. Dr. Sellers, do you have any opinion about this discrepancy?

Sellers: We have only tried molluscan thick filaments, in part because they are long and in part because it is much easier to make the native thick filaments—at least in our hands—from those than from other species. But with four different molluscan thick filaments—including both catch and non-catch-type thick filaments—we always see almost exactly a ratio of ten in the velocity between going the correct way and the incorrect way, even though the absolute values of velocity can vary by an order of magnitude. In this connection, I was very pleased to see that Dr. Yanagida also had a large difference in speed depending on the direction with synthetic rabbit thick filaments.

Sugi: If I believe Yoko (Toyoshima)'s results, then the difference might come from the arrangement of myosin heads.

Yanagida: Yoko Toyoshima reported that even if she used lobster muscle, the velocity was the same. But she did not report clear data. She did not show video records and micrographs. We also made experiments by using native thick filaments of crab muscle, which are almost the same as lobster, and we confirmed that the velocity produced by incorrectly oriented myosin heads is much smaller than that produced by properly oriented ones.

Sugi: As a final note, I would like to have Dr. Simmons' comments about the advantages and disadvantages of the application of optical tweezers for future research work with *in vitro* assay systems.

Simmons: I have never used any other kind of system, so I can't really give you any first-hand comparisons. I think the nice thing is that it is a very flexible system. It is much easier to manipulate objects than to use rigid mechanical needles and so on, so I think that's a big advantage for some experiments. If you'll excuse the pun, a laser beam is light. It doesn't cost very much effort to move it around. The major disadvantage that we have come across stems from limitations on the signal-to-noise ratio.

Some of those limitations can be overcome, but noise arising from the Brownian motion of the beads is endemic. There is also an upper limit to the strength of the trap, probably several hundred pN. In our experiment, the limit is set mainly by the maximum power limit of the acousto-optic modulator, but there is an ultimate limit set by what the objective will take. So far, we have not tested our objective to destruction.

REGULATORY MECHANISMS OF MUSCLE CONTRACTION

Gillis: The session on muscle regulation (Chapter II) encountered difficulty concerning the question of the increased affinity of troponin for calcium when myosin binds to thin filaments. I would like to try to clarify the situation by calling on the people who were involved in the problem.

Dr. Ashley presented a kinetic model and experimental evidence that the binding of S1 to the thin filament is not necessary to explain the results; his model completely excludes, or doesn't take into account, the fact that S1 could modify the situation (Ashley et al., this volume). But on the other side, Dr. Gordon and Dr. Saeki showed that, once the muscle is fully activated, a small variation of force can produce a variation of Ca^{2+} concentration (Gordon and Ridgway, this volume). I was very impressed by one of the conclusions of Albert Gordon's talk. Apparently, according to his results, the affinity of TnC can change by as much as twenty times when the force sustained is practically zero, or goes to 1 kg/cm^2, which is about one fifth of the maximum tension. So, this point sets the difficulty.

I would also like to mention what Dr. Gulati told us in this symposium about the sensitivity of TnC along the length of the thin filament. This may or may not be related to Ca^{2+} sensitivity. The length-sensing mechanism he presented may be somehow related to this question of the calcium sensitivity affected by S1 binding.

Effect of S-1 Binding to Actin on Ca^{2+} Affinity of TnC

First of all, I would like to ask Chris Ashley whether his model excludes the possibility that S1 binding would increase TnC affinity or if he simply doesn't need it and used the Occam's razor criterion.

Ashley: I think you actually misquoted me. I was very careful to say that I didn't exclude the possibility of S1 binding (Ashley, C.C. et al. *Q. Revs. Biophys.* **24**, 1, 1991). I made the point, I think in the modelling, that the end-to-end effect of tropomyosin in our hands was more important, perhaps the major effect. I was very impressed by two things at this meeting in this particular regard. First of all, by Dr. Saeki (Saeki et al., this volume) and Dr. Horiuti's (summary of poster, this volume) presented work showing that the extra Ca^{2+} appeared to come from the cross-bridges breaking, by simply showing BDM sensitivity and the ryanodine insensitivity, so that the effect was not something to do with enhanced Ca-induced Ca release during the period of elevated free Ca^{2+} (see Kurihara, S. et al. *Jpn. J. Physiol.* **40**, 915, 1990). That clarified in my mind where the calcium may well be coming from. Dr. Gordon also clarified a very important point: an affinity change of up to about 10% P_{max} force in barnacle. You quoted 30%—I think it was about 0.5 kg/cm^2 P_0—but since P_{max} in barnacle is 6 kg/cm^2, it's closer to 10% of the effect. Above this 10% value there was a saturation of the effect, if I'm right. We are thus putting an upper limit on the S1 effect in barnacle.

Therefore, I would restate the case. We felt that for intermediate forces (50% P_{max} or greater) in frog—way above the 10 % level—if we can make the extrapolation from barnacle to frog, we are in the force region of tropomyosin end-to-end interactions, and not S1 feedback, which is quite slow. I would also add as a final point that there are people in the audience here—Dr. Cecchi and Dr. Griffiths—who presented at a previous Hakone meeting the finding that if you release single frog fibers injected with aequorin from P_{max}, there is no increase in aequorin light; there is actually a slight decrease. So, the experimental data in frog is rather in favor of our modelling rather than against it. Perhaps I can stop at this point.

Gillis: Thank you, Chris. You clarified the situation. Dr. Gordon, do you have a comment on this?

Gordon: Yes, our data imply that the major effect of cross-bridges/force on TnC calcium affinity occurs at fairly low forces. In Chris Ashley's modelling, there are little data at low forces, with most forces above 25% of maximum. Our data from the barnacle muscle fibers show little change in calcium affinity in the range of forces above 25% of maximum. Thus, there may not be a major discrepancy. The physiological importance of the increase in affinity at very low forces is that it gives the muscle an extra boost at low calcium activation.

Chris Ashley mentioned David Allen's work on frog muscle [*J. Physiol. (Lond.)* **275**, 63P, 1978]. David showed that with shortening steps in tetanized muscle, he could get small decreases and small increases in free calcium, depending on the initial sarcomere length. These results were obtained during maximal activation, not under the submaximal activation conditions we investigated. The other point is that the magnitude of this cross-bridge-dependent effect may vary with the different muscle—for example, in measuring calcium binding in skinned muscle fibers. Franklin Fuchs has shown a strong cross-bridge-dependent Ca binding in cardiac muscle (Hoffmann, P.A. and Fuchs, F. *Am. J. Physiol.* **253**, C541-C546, 1987), but using the same experimental techniques, he has been unable to show such an effect in rabbit skeletal muscle. It may be that the difference lies in the different myosin isoforms and how they interact with the thin filament or in the different troponin isoforms and how they interact.

Additional information on cross-bridge effects on the thin filament come from the X-ray diffraction studies that I discussed. Hugh Huxley (Kress, M. et al. *J. Mol. Biol.* **188**, 325-342, 1986) showed some time ago that the decline in intensity of the second actin-layer line during relaxation from contraction at a sarcomere length with full filament overlap with cross-bridges attached was greatly prolonged compared to the decline in sarcomere length with no filament overlap. It is as if the attached cross-bridges somehow delay the tropomyosin return to the relaxed position, whether because of a direct mechanical effect on the tropomyosin or an effect to enhance calcium affinity, which would serve to do the same thing. Thus, there is some evidence in an intact muscle preparation for feedback from cross-bridges to thin-filament activation.

Ashley: As Dr. Gordon just mentioned, there is the other difficult point in the literature, which is the disparity between the equilibrium-radioactive-Ca^{2+}-binding experiments and the experiments with TnC DANZ, etc. Equilibrium-binding experiments show quite clearly in the hands of Franklin Fuchs (*J. Muscle Res. Cell Motility.* **6**, 477, 1985) that in rigor, calcium affinity goes up, at least in rabbit, while in actively cycling bridges this increase in calcium affinity isn't observable. Therefore, one might extrapolate to say well, in fact it's very small, and maybe Dr. Gordon hasresolved the issue by saying that *it is* relatively small. But certainly the equilibrium ^{45}Ca binding experiments are difficult to explain away by a vast change in affinity brought about by the cycling bridge state, anyway in rabbit skeletal muscle, because it's not observed.

Maéda: Just one comment. We reported that the extent of the intensity change of the second actin-layer line in the extremely stretched muscle is smaller than in the fully overlapped muscle. But, the problem is that it is very difficult to confirm experimentally whether it is overstretched to non-overlap or not. Therefore, we recently oriented the thin filaments in the capillary. We used actin plus tropomyosin. We changed the calcium by using caged calcium, and measured the second actin-layer line. This means that this actin-layer line intensity change is free from the influence of myosin. Then, this intensity change was compared with the intensity change observed from the muscle.

Gillis: At which overlap?

Maéda: At full overlap. Our conclusion is that the extent of the intensity change of the second actin-layer line due only to the calcium is 40%, compared with the intensity change induced by calcium plus the myosin head. Therefore, whatever the mechanism of the actin-layer line change is, this thin filament structural change caused by calcium alone is only a part of the story. An extra change is caused by S1 binding. This is our conclusion. Unfortunately, we were not able to introduce S1 to the capillary without disordering the gel orientation. Therefore, we were not able to do this type of experiment.

Faruqi: The experiments of Hugh Huxley (done in collaboration with the late Marcus Kress and myself: Kress, M. et al. *J. Mol. Biol.* **188**, 325-342, 1986) showed that the time course of the actin second layer line for both full overlap and non-overlap cases was ahead of the equatorials by about 12-17 milliseconds at 6°C; the implications of this observation are that the tropomyosin movement is a prerequisite for myosin binding.

Change in the Ca^{2+} Affinity of TnC along the Thin Filament

Gillis: At the beginning of this discussion, I mentioned that Dr. Gulati presented very interesting results about the change of affinity of TnC along the length of the thin filament. I think that's a very new and provocative idea. I would give the audience the opportunity to offer an explanation or ideas about this interesting hypothesis.

Rüegg: In the overlap region, you could have TnC-S1 feedback, and so increase the Ca^{2+} affinity of TnC.

Gillis: Yes, but Dr. Gulati, your scheme is that the affinity progressively diminishes, whatever the amount of overlap.

Gulati: That's right. I suggested that affinity is greatest at the tip of the thin filament and decreases toward the Z-line. So, when you increase the overlap at short lengths, and increase the possibility of cross-bridge interactions, the expected result would be the opposite of what you really see.

As to the specific comment by Dr, Rüegg, I don't really have any objection to cross-bridges being involved; in fact, in my view cross-bridges probably do affect the Ca^{2+} sensitivity of TnC. My hypothesis concerns mainly whether such a cross-bridge hypothesis can account for the length-sensing mechanism in cardiac muscle.

One objection that commonly arises concerns the descending limb. It is true that the cardiac muscle people are more interested in the ascending domain because that's where the heart operates, and there is a need to have increase in calcium sensitivity with increasing length. But, you run into major problems in the descending limb. If the cross-bridges were setting calcium sensitivity, the increasing length should produce lowered calcium sensitivity, which is the opposite of that observed. Our mechanism is attractive because it can explain the length-dependence of calcium sensitivity over the entire length-tension relationship without *ad hoc* assumptions for every region. I propose this as a working hypothesis for our future experiments.

Huxley: Something puzzled me in this symposium about Dr. Gulati's presentation. After his asymmetric extraction, where TnC was left only near the tip of the thin filament and he had obtained this increased effect of length on Ca^{2+} sensitivity, the difference between the greater length and the shorter length is in the part of the thick filament that is opposite the troponins. So, it seemed to me that it implied some difference between the cross-bridges in the different parts of the thick filament.

Gulati: There are two possibilities: one is to extract TnC so that in increasing the overlap you encounter additional TnC's at the lower affinity. The other is to extract TnC up to the very end of the thick filament, then no additional TnC is encountered with increasing overlap. Two types of responses are then possible: 1) no length-dependence of calcium sensitivity; or 2) an enhanced length dependence. The observed response was of the second kind. In our minds this raised the possibility that the length-dependence reflects either a) the intrinsic TnC-actin interaction or alternatively b) overlap has a feedback effect on TnC-actin interaction. These are questions that we will have to answer with future experimentation.

Gordon: That brings up another issue. Removing TnC does something else in addition to removing a Ca^{2+}-binding/activating site. It decreases cooperative interactions along the thin filament, so that the activation by rigor cross-bridges is not as cooperative, presumably not spreading to the next tropomyosin-seven actin unit past the region where TnC has been removed (Brandt, P. et al. *J. Mol. Biol.* **212**,

473-480, 1990). So, in your experiments where a given amount of TnC is removed asymmetrically and the calcium sensitivity compared with one where presumably the same amount of TnC is removed symmetrically, there may be a major difference in cooperativity.

Gulati: You are quite right. But I think, in addition, that the cooperativity issue has quite separate implications. TnC has an on-off switching mechanism, and I think it also has a length-sensing mechanism as well as a cooperativity mechanism. We have evidence suggesting that all three can be separately controlled. As far as cooperativity is concerned, the entire thin filament acts as a concerted transition mechanism; surely this would impact the asymmetric fibers. But whether this effect is different in symmetric fibers, I don't know.

Gordon: Your new hypothesis on the variation in Ca^{2+} affinity along this filament to explain sarcomere length effects on Ca^{2+} sensitivity is an interesting one and quite testable. However, I think that your argument to rule out a cross-bridge effect is a spurious one. You indicated that if cross-bridges enhanced calcium sensitivity, one would expect that decreasing filament overlap should decrease, not increase, calcium sensitivity. As you decrease the overlap, you indeed decrease the total number of cross-bridges, but if you think about the simplest regulatory unit— seven actins, one tropomyosin, one troponin—this troponin could be affected by cross-bridges near it and thus only the troponins on the thin filaments near the overlapped region would be affected. So I think the total number of cross-bridges is not as important as the number of cross-bridges near the regulatory unit. Of course, the question is what "near" means with cooperativity. You suggest that the whole filament may activate cooperativity, but if that is the case, the number and position of the cross-bridges should not be important.

Rüegg: I'll try to remind you that signal transduction from TnC to S1 is a two-way affair; one should offer a testable hypothesis concerning the TnC-S1 feedback. As is well-known, there is signal information flow in the forward direction: Ca^{2+} binds to TnC. This increased Ca^{2+} occupancy then increases interaction between TnC and TnI, probably involving residues 104-115 of TnI (Talbot, J.A. and Hodges, R.S. *J. Biol. Chem.* **256**, 12374-12378, 1981). This, then, has the consequence of decreasing interaction between TnI and actin, and this effect would increase the interaction between actin and S1, because the inhibitory effect of TnI is repressed. But, the reverse is also true, inasmuch as an increase in interaction between TnI and TnC increases the Ca^{2+} affinity. It is well-known that TnC alone has a much lower Ca^{2+} affinity in the presence of TnI. However, the addition of actin to the system would then decrease the Ca^{2+} affinity of TnC again. So, actin-TnI interaction decreases the Ca^{2+} affinity. This interaction may be inhibited by S1 actin-interaction attaching "rigor" cross-bridges or cycling cross-bridges (Güth, K. and Potter, J.D. *J. Biol. Chem.* **262**, 13627-13635, 1987), as S1 may compete with TnC for actin. Consequently, the Ca-affinity of TnC would be increased.

These predictions regarding S1-TnC feedback are testable. For instance, one might want to inhibit these interactions selectively, one by one, and see how the affinity of TnC for Ca^{2+} is affected and whether S1-TnC feedback is blocked. To target actin-myosin interaction sites, for instance, one could use anti-peptides of the type used by Chaussepied and Morales (*Proc. Natl. Acad. Sci. USA.* **85**, 7471-7474, 1988), or one could use peptide competition approaches (Keane, A.M. et al. *Nature* **344**, 265-268, 1990) or perhaps even drug mimetics of such peptides. We believe that, for instance, a peptide, S1 701-717, could be used to inhibit this interaction. This peptide would presumably compete with TnI for actin, because it binds to the same region of actin as TnI, namely actin 1-12. I have shown in my paper that, indeed, the apparent Ca^{2+} sensitivity is increased by the addition of such a peptide to skinned cardiac fiber systems. One would have to find out how the Ca^{2+} binding to TnC at submaximal Ca^{2+} levels is increased under these conditions.

Using DANZ-labelled TnC in the skinned fiber system (Güth and Potter, 1987), one has to be cautious: the fluorescent signal is simply an indicator of conformational change in the TnC, which may or may not be determined by the Ca^{2+}binding to TnC directly (Morano, I. and Rüegg, J.C. *Pflügers Arch.* **418**, 333-337, 1991). Anyway, using this method one could find out whether the conformation of TnC would be affected by certain parts of S1 interacting with actin, and how the effect of S1 is relayed to TnC within the thin filament system *via* several protein-protein interactions.

Winegrad: I want to make a comment on Jag Gulati's hypothesis, because I think the curves he drew for the Ca^{2+}-tension curves under his conditions may not have represented the situation completely accurately. Actually, there is a simple way of testing his hypothesis. If you have full overlap and the Ca^{2+} sites are acting independently and have different affinities, as he has described, then what you should get is not a shift of the curve from the left to the right with the same shape, but a combination of two curves. The curve takes off from the left one and reaches its peak at the right one because you are titrating different affinities and so you will get a broader curve. If your data actually show a shift to the same curve, that rules out your hypothesis.

Gulati: I can't say you're right or wrong. There are shifts in the slopes of the curve, if that's what you're referring to. But there is no question in my mind that with length-induced change in the affinity of TnC, the curves will have a rightward movement whether the slope is the same or not.

Winegrad: If they are acting independently and you have different affinities, you are going to have to go to the highest calcium concentration to see maximum force when you have full overlap. When you have half overlap, you are not going to have to go to that a high Ca^{2+} concentration to see maximum force at that particular sarcomere length. So, you won't get a parallel shift from left to right under those conditions.

Gulati: The cooperativity changes alone with length are not going to explain our results.

Godt: Jim Sellers, can you remind us about molluscan muscle? Do molluscan muscles that have long thick filaments also have long thin filaments?

Sellers: Yes, I think so.

Godt: So, that might be a really good place to test this kind of idea, where there would be a spatial difference between the tip of the I-band and the Z-line.

Sellers: The problem is that the question of the contribution of the troponin system in molluscan muscle has not yet been adequately resolved. I think Andrew Szent-Györgyi's lab still believes that there is not enough troponin there to be an effective regulatory system.

Godt: But what about the thin filament-regulated lobster muscles, for example?

Sellers: Yes, those are better. Those sarcomeres are never as long as the molluscan ones, but that might be a system you could do it in.

Godt: How would we go about testing to see if Ca^{2+} binding was different near the Z-line and at the tip of the I-band? Could we use some phosphorescent Ca^{2+}-like compound or something like that?

Gillis: Possibly yes. If we were to use a large sarcomere of crab or of lobster, it could be possible. However, the probes just sample the free calcium resulting from the equilibrium between the various binding sites and the probe itself, so it is not that easy.

ter Keurs: When John Kentish and I measured the force-Ca^{2+} relations at a number of different sarcomere lengths (Kentish, J.C. et al. *Circ. Res.* **58**, 755-768, 1986), we observed 1) a large shift in the half maximum of the curves with stretch of the sarcomeres, reflecting an increase in Ca^{2+} sensitivity of the filaments; and 2) a three-fold increase in cooperativity as reflected by the slope coefficient of the sigmoid fit to the relationship between force and Ca^{2+}. I did not see quite such an increase in the slope coefficient of the curves shown in Jag's presentation. The question to Jag is: how do you intend to evaluate cooperativity of Ca^{2+} binding to TnC? Technically, it means that you will have to measure and control sarcomere length, since during contractions at constant muscle length, shortening of the sarcomeres in the center of the muscle will occur, which in our experience causes force-pCa curves to collapse to a slope coefficient of approximately three; moreover, internal shortening will reduce the shift of the force-pCa curves.

A comment about Jag's interesting model of the regional distribution of TnC properties: I look forward to the results of your studies. You might want to compare your model with the model of regional distribution of the sensitivity of TnC to Ca^{2+} ions along the thin filaments that I proposed a few years ago [ter Keurs, H.E.D.J., in *Cardiac Metabolism*, (A.J. Drake-Holland and M.I.M. Noble, eds. Wiley and Sons, New York, 1983)].

Rüegg: To test Gulati's model, one would have to "map" TnC-Ca^{2+} affinity within the sarcomere. To do this, one could use the TnC-DANZ method (Güth and Potter, 1987), but combine it with a TnC-DANZ imaging technique using confocal microscopy, so that one sees how fluorescence at a given submaximal Ca^{2+}-level varies within the sarcomere (*e.g.*, in the I-band and A-band).

Gordon: Another way to detect differences in calcium binding within the sarcomere is to use electron probe microanalysis as Marie Cantino has been doing (Cantino, M. et al. *Biophys. J.* **57**, 337a, 1990). The experiments were designed to test whether there was an increased calcium binding in the region of overlap of thick and thin filaments in rigor. She was able to show that there is an enhanced Ca^{2+} binding localized to the region of overlap within the sarcomere. Her images, particularly at a pCa of 6, show that there was only an enhanced Ca^{2+} binding in the region of overlap on the thin filament and no change in Ca^{2+} binding along the thin filament toward the Z-line. To test Gulati's hypothesis, she needs to do more experiments at different Ca^{2+} levels and at sarcomere lengths where there is no filament overlap.

Gillis: So, what you said just confirms Franklin Fuchs's experiment.

Gordon: Absolutely. Rigor cross-bridges enhance calcium binding in the overlap region.

Ashley: Has this been done for cycling bridges?

Gordon: No, the electron probe microanalysis experiments have only been done for rigor.

THE CROSS-BRIDGE CYCLE IN MUSCLE CONTRACTION

Homsher: This discussion will focus on weakly bound and strongly bound cross-bridges. The basic topic we hope to address is the sorts of bonds that are involved in cross-bridge attachment. There are a variety of different bonds and bond energies associated with those specific types of bonds and I have asked Dr. Morales to consider this briefly. We will then move to a definition of the characteristics of weakly and strongly bound cross-bridges that Bernhard Brenner and Vincenzo Lombardi will speak about. Next, Professor Huxley will discuss the relationship between biochemical states and steps in the cycle.

Weakly-bound and Strongly-bound Cross-bridges

Morales: This was an honor for which I was not totally prepared, and so I've consulted Bill Harrington and John Gergely about my memory of these numbers. So if my numbers are wrong it's their fault! I'm doing this somewhat from memory. In the 1970's, several people including ourselves were concerned with the equilibrium of S1 with nucleotide, S1 with actin, and then the ternary complex of actin, S1, and nucleotide. This was prompted by Martonosi's discovery that you could get this ternary complex. So many of us spent time getting these numbers—in our group especially Stefan Highsmith. The idea is this: the stability constant of S1 with actin is about 10^7 M^{-1} in ionic strength of 50 mM, and we already knew about structure disrupting ions, so this was cesium acetate or some other "benign solvent." However, when the nucleotide is present, this stability constant is very much weakened. If the nucleotide happens to be ADP, this constant drops to about 10^5 M^{-1}; if it happens to be ATP—that is to say, if the occupancy is ADP•P_i—then the constant is about 10^4 M^{-1}. Now, that is really what I think Dr. Homsher wanted me to say, but I can't refrain from saying a couple of other things.

If you study just the rigor situation, you can do experiments in which you vary the ionic strength. It turns out that if you go to infinite ionic strength you drop -log K to about 4.5. So you can imagine what the ionic strength does: if it were acting ideally, you would be screening out the electro-static attraction so the residue would be "hydrophobic bonding." You could parcel out the free energy into a coulombic part and a hydrophobic—at least insensitive to ionic strength—part. Now that is rather interesting, because it turns out the rigor attachment isn't so sensitive to ionic strength, probably because only a small part of the attachment is "coulombic." But once you have nucleotides there, the hydrophobic portion "dissolves" and the residue is very sensitive to ionic strength.

My final remark is on the question of what bonds might be involved. Ken Holmes and I stress that there are clusters of positive and negative charge on the two partners. Actin has several negative clusters; S1 has at least two positive clusters and probably more. There is strong reason to think that these clusters match one another. In the case of two of the clusters the matching is certain. That is where EDC cross-links positive and negative clusters, as Kazuo Sutoh originally found. So I have put here two pluses to represent the clusters. With the rest, you have to assume there is a hydrophobic bonding, which hasn't really been determined, although Dr. Holmes showed me some material yesterday that suggests where the hydrophobic patch is on actin. We don't know for sure where it is on myosin. So you might say this is the rigor situation, in which all the forces contribute. It seems that when nucleotide binds, it somehow cancels the hydrophobic interaction. It creates some crazy distortion that cancels the hydrophobic interaction and leaves mainly the coulombic surfaces. That's how a strong dependence on ionic strength is acquired after nucleotide is added. Nucleotides lower the attraction, and ATP does it better than ADP. The NIH phrase-makers took over from there.

Homsher: So what do you think the bond strength is of a weakly attached cross-bridge, say AM•ADP•P$_i$?

Morales: Those are the constants that I cited; in the presence of ATP the affinity is about 10^4 M^{-1}, if our trio's memory is correct.

Homsher: So, it isn't really so weak.

Morales: Well it's a thousand times less than the rigor affinity. It's a considerable amount, but it isn't trivial, no.

Godt: While we are talking about hydrophobic forces, I didn't get a chance to talk about this, but the major theory on how TMAO and these other osmolites work is by preferentially being excluded from the region around the protein which leads to preferential hydration of the protein. So would that lead to a strengthening of hydrophobic bonds if there were more water around the protein? I'm trying to understand how TMAO increases force in high salt.

Morales: In the first place "hydrophobic" is a catch-all for something that is not completely understood, even in simple systems. I wouldn't hazard a guess about water. I think I understand how the water is clustered around charges; the water is electrostricted around the charges and then when the charge meets another charge, the water goes away and gives you a big entropy increment. I don't know, you may be right, but I can't say anything certain about how the water interacts with the hydrophobic bonds. So far as the electrolytes are concerned the malignant electrolytes like chloride, thiocyanate, and perchlorate polarize the bonds of the macromolecule itself and distort it. Those are really structure-destroying ions; you can replace them with benign ions and you stop the disruption—God knows what the ions do in addition to that.

Harrington: I just wanted to add a point. In the case of water, the pressure dependence of cross-bridge interaction is very important. I think Julien Davis may be able to tell us something about that.

Davis: Mike Geeves and colleagues (Coates, J.H. et al. *Biochem. J.* **232**, 351-356, 1984) examined the effect of pressure on the interaction of myosin and actin and found that the transition from the weakly bound to the strongly bound cross-bridge state is associated with a change in volume (ΔV) of about 100 cm^3 mol^{-1}. Charged group solvation has frequently been found to be the source of such large volume changes in protein-protein interactions. For example, the assembly of charge-rich myosin rods into the backbone of the thick filament is associated with a volume increase of 200-300 cm^3 per mol of monomer incorporated. The rupture of a typical salt bridge between carboxyllate and cationic acid ions causes a volume decrease of 30 cm^3 mol^{-1}. The change in volume arises from the electrostriction (condensation) of solvent water about the newly-exposed charged groups. The observed volume change therefore squares roughly with the breaking of the 2-3 salt bridges between actin and myosin mentioned earlier by Dr. Morales. Thus, several lines of evidence seem to support a significant role for charge-charge interactions in the formation of the strongly bound state of the cross-bridge.

Brenner: I was asked to say something about the definition of weak and strong-binding cross-bridge states. So let me first list and explain the different properties. Then I will mention some examples.

The first criterion is actin affinity. At the same ionic strength, the actin affinity of weak-binding states is much lower than that of strong-binding states. The second criterion is rapidly reversible actin interaction. This is found for both classes, but for strong-binding states the rate constant for dissociation appears much smaller. The third property is the Ca^{2+} effect on actin affinity. For weak-binding states the Ca^{2+} effect on actin affinity is small (up to some 2 to 5 fold) while for strong-binding states Ca^{2+} has a much larger effect on actin affinity. The fourth point, which is related to the third point, is the inability of weak cross-bridge interactions with actin to activate the regulated thin filament while strong cross-bridge interactions can activate the thin filament. This means that if we increase the number of cross-bridges that are strongly bound to regulated actin, we can reach a point where the actin filament becomes "activated" even in the absence of Ca^{2+}. Thus, if some ATP is present, some cross-bridges may now hydrolyse the present ATP and go through their cycle even in the absence of Ca^{2+}. This active cycling can be detected by force production or hydrolysis of ATP.

While the first two properties only represent quantitative differences, property four is a qualitative difference, i.e., even if we match actin affinity by lowering ionic strength for weak-binding states or raising ionic strength for strong-binding states, the last property still remains different. Thus, even at matched actin affinities the states of the two classes do not become identical, i.e., a strong-binding state at such high ionic strength that its actin affinity matches the weak-binding states still retains its ability to activate the contractile system. Consequently, to classify cross-bridge states, the ability/inability to activate regulated actin is the essential criterion.

Thus far, we have classified as weak-binding states the states present in relaxed fibers (presumably M•ADP•Pi and M•ATP) and cross-bridges in the presence of ATP-γ-S. Strong-binding states are considered to be cross-bridges without nucleotide, and cross-bridges in the presence of ADP, AMPPNP, and PPi.

Kawai: We found one state, AM*•ADP•Pi, which should be classified as a strongly bound state, because it has the largest tension. What do you think of that, Bernhard?

Brenner: In the cross-bridge cycle, at the transitions between the weak and strong-binding states, both actin affinity and the bound nucleotide or products change. We think that the two changes occur sequentially and not in a single step. For example, at the transition from the weak-binding AM•ADP•Pi state to the strong-binding AM•ADP state both actin affinity increases and Pi is released, but in two steps. The question is which one changes first. As a consequence, at the transitions between weak and strong-binding states, some states with the same nucleotide or products bound exist both in a strong and weak-binding configuration.

Kawai: According to our experimental results, isomerization happens first, then the phosphate is released. In other words, we positively identify a strong ADP•Pi state. This is designated by AM*•ADP•Pi.

Tregear: Would you agree, Bernhard, that the artificial states that can be classified somewhere in the weak-binding states—ADP vanadate is an obvious one, ADP aluminum flouride seems to be an added one now, and my own oddity which was AMPPNP ethelyne glycol—none of them is as well characterized as yours?

Brenner: I did not include them because I have to ask you whether, for example, your AMPPNP ethelyne glycol can activate the contractile system. This would be the property that allows us to classify it as a weak or strong-binding state.

Tregear: It certainly doesn't activate the contractile system; that one can surely say.

Brenner: So then it is a weak-binding state.

Tregear: But after all it's blocking the site, isn't it? So it could hardly activate the contractile system. You couldn't get ATP running through that.

Brenner: The question is whether it is possible to have a few cross-bridges as sensors to see whether you can activate the contractile system. This is the general approach we use to classify cross-bridge states as weak or strong-binding. For example, in a situation where we have just enough ATP to keep a skinned fiber relaxed (i.e., cross-bridges have ATP or no nucleotide), increasing concentrations of MgPPi added to the solution result, above a minimum MgPPi-level, in an increasing amount of active force (or ATP-hydrolysis), although Ca^{2+} is low. This "activation" of the contractile system is caused by the increasing number of cross-bridges with MgPP$_i$ at the cost of nucleotide free cross-bridges, or, to some extent, at the cost of cross-bridges with ATP that now, however, can go through the active cycle. Thus, cross-bridges with MgPP$_i$ apparently can activate the contractile system and therefore are classified as strong-binding cross-bridge states.

Inoue: We are studying this problem in solution, but this is my comment. If we want to study weakly binding states we have to make a very low ionic strength solution. But at very low ionic strength when we increase the temperature to 20°, even in the absence of Ca^{2+}, it can bind to the regulated actin. Yanagida and I showed previously that such conditions in muscle fibers can induce tension. This fiber has a very low ATPase but the tension is very high. So the situation is very complicated, and I have to think of other methods for studying it. In muscle fibers, the addition of ATP can induce tension. That state is not ADP•P$_i$. We can get ADP•P$_i$ only in solution. How do you explain this?

Brenner: Well, I can comment on at least one of your points with the observation that, in muscle fibers, when the ionic strength is really low, you observe active tension. This has to do with taking proper care of skinned fibers. In the very beginning, we also had some problems with possible activation at very low ionic strength or in the absence of Ca^{2+}. I don't really know what the reasons are. It appeared to us that we had to be careful about the regulatory system. With fibers no older than about a week, and with 2 mM of free Mg^{2+}, we could go up to 20° C at ionic strength of 20 mM without getting active force.

Homsher: I want Vincenzo Lombardi to come up and describe some characteristics of the strong cross-bridge from his point of view.

Lombardi: Thank you. I will try to show the "paradoxical" possibility of having strongly bound states of cross-bridges that have no force and, depending on the mechanical conditions, can detach and reattach quickly. It has been shown [Lombardi, V. and Piazzesi, G. *J. Physiol.* (*Lond.*) **431**, 141-171, 1990] that when you stretch a tetanized fiber, tension rises above the isometric value and then attains a steady level beyond a given amount of lengthening (just above 10 nm, Fig. 3A). Also stiffness, measured by small step releases, attains a constant value. So, during steady lengthening, steady-state relations between velocity of lengthening and

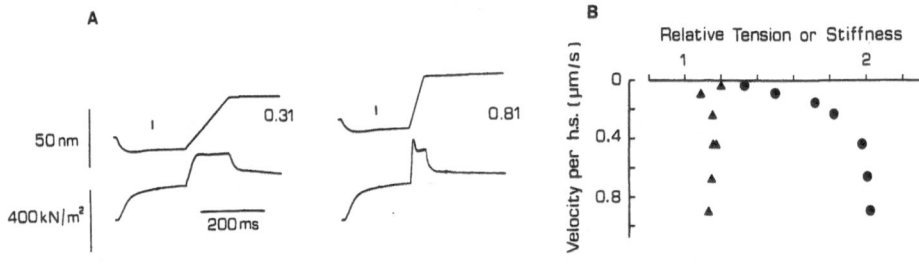

Fig. 3.

tension and stiffness can be drawn (Fig. 3B). First consider the region of very slow lengthening, for instance 0.1 µm/s. This means that for each cross-bridge just attached it takes 100 ms to attain 10 nm of lengthening. Under these conditions, when we perturb the system by imposing a step length change there is no problem of confusing the tension transient with cross-bridge detachment at the end of the interaction cycle. This is a crucial condition. You have very high tension (1.6-1.8 T_0) while stiffness is only about 10%-20% larger than in isometric conditions. Such high force with such slow lengthening with only a small increase in the number of attachments is evidently attained because the cross-bridges on average are more strained. So these are, from a mechanical point of view, strongly bound cross-bridges.

The kinetic characteristics of these cross-bridges can be investigated by eliciting the tension transient with step length changes of different size. Fig. 4 shows T1 and T2 curves obtained with these kinds of experiments. Compared with the isometric control, during slow lengthening (0.1 µm/s, tension before the step 1.8 T_0) the T1 curve has an intercept on the abscissa shifted to the left by about 1.5 nm. This means that the elastic component of the cross-bridge is more extended. On the other hand there is even some contribution to the increase in steady force by a recruitment of attachments, because the slope of the T1 curve and therefore the stiffness is a bit

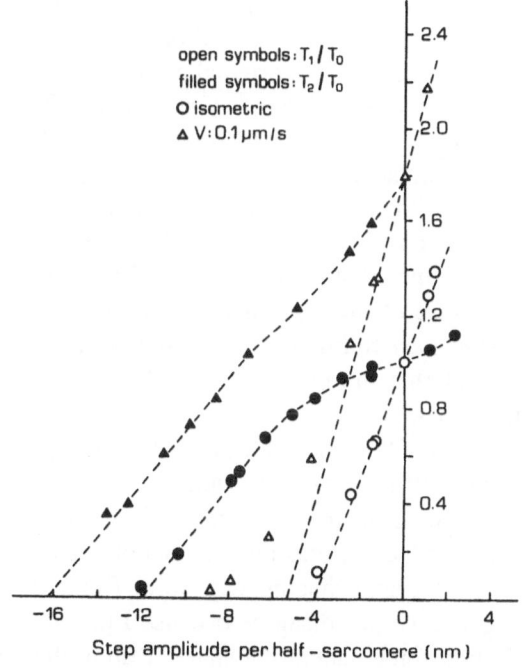

open symbols: T_1 / T_0
filled symbols: T_2 / T_0
O isometric
△ V: 0.1 µm/s

Step amplitude per half-sarcomere (nm)

Fig. 4.

larger than in the control. Note that the abscissa intercept of the T2 curve determined during steady lengthening is shifted by something like 4 nm (Fig. 4). According to Huxley and Simmons's theory (*Nature* **233**, 533-538, 1971), the intercept of the T2 curve is the sliding distance the attached cross-bridges can afford while they are exerting force. So this distance becomes 4 nm larger during lengthening, indicating a corresponding increase in cross-bridge extension.

So why is their spring extended by only 1.5 nm? Because they have redistributed toward the lower force-generating state. This is the second point: we have a kind of strongly bound low force-generating cross-bridge. Though it looks a bit misleading to have low force-generating states of a cross-bridge exerting a force of 1.8 T_O, we have to realize that the high force depends on the work we are doing on the muscle fiber. When the velocity of steady lengthening is increased above 0.3-0.4 µm/s you get into the region of the T-V relation where force does not increase very much with lengthening speed (Fig. 3B). At the same time stiffness does not change in the whole range of velocities used. So it seems that attachment is not rate limiting, because in this case stiffness would be reduced by increasing the lengthening speed, as it happens, by increasing shortening speed [Ford et al. *J. Physiol.* (*Lond.*) **361**, 131-150, 1985]. So, during steady lengthening a high-speed cross-bridge attachment is very fast. But, beyond a critical amount of strain, cross-bridges are rapidly detaching too, since force does not increase with lengthening speed. There should be a critical point they reach during lengthening, where the detachment rate becomes very quick. This is the way to explain the fact that stiffness and force remain constant and power increases linearly with the velocity of forcible lengthening. Note that even if cross-bridges rapidly detach, it does not occur where they attach; they are strongly bound. You bring the strongly bound bridges to a point where they rapidly detach and then rapidly reattach at the original point.

Holmes: Vincenzo, how fast is rapid on your timetable?

Lombardi: It can be 1000/second or more to fit the stiffness and force-velocity relation. Actually we simulated these things, providing a detachment rate constant that rises exponentially with cross-bridge extension. You have to think that this way of cycling has nothing to do with the normal cycle, because the characteristics of the tension transient show that cross-bridges are redistributed toward a low force-generating state. Because of the high mechanical energy barrier, they cannot go through the cycle. They have to find another way to detach before the completion of the normal cycle, and this implies another detached state responsible for the unusually fast reattachment.

Brenner: So, strong binding doesn't mean that they all generate the same amount of tension on average. In fact, as I showed in this symposium, to account for isometric transients according to the Huxley-Simmons model, we assume that strong-binding states produce increasing amounts of active force as we proceed in the cycle. So I don't see any discrepancy here. Detachment-reattachment is a different matter. I guess we just disagree because I provided evidence that it does occur in strong-binding force-generating states. I think the real matter is to find a proper model to account for all of them.

Lombardi: I'm sorry I missed your records. In any case I don't think you have to ignore the results of the double step experiments. Can you account for the finding that there is no detachment occurring during the force-generating step, the power-stroke?

Sugi: Interpreting the effect of stretch, I've already published a couple of papers in *Journal of Physiology* in 1988, [Amemiya, Y. et al. *J. Physiol. (Lond.)* **407**, 231-241, 1988; Sugi, H. and Tsuchiya, T. *J. Physiol. (Lond.)* **407**, 215-229, 1988] stating that the stretch-induced force increment is very complicated in nature and in my opinion is associated with distortion of hexagonal lattice of filaments. Also, you should be careful about how the turnover rate may change. You should measure heat production, or P_i production. I think it is the concept of stretch-induced locked-on cross-bridges that explain the result. So you should be careful about cycling.

Lombardi: Actually, during slow lengthening the tension transient looks as you would expect from Huxley and Simmons's theory. The other thing to account for is energetics. I think I have a bit of considering to do. The results I showed were expected from the finding that the energy balance becomes negative. Energy is stored in the muscle during high-speed lengthening, and it has been shown that it is not stored in a biochemical way (Woledge et al. *Energetic aspects of muscle contraction*. Academic Press, London, 1985). We have found a mechanical way of storing energy: an attached state of cross-bridge, which is an early state in the force-generating process, is a high potential energy state; by stretching a cross-bridge in this state, you even put mechanical energy into it. I think this is a way of storing energy. So I think it is the right answer to the energetics requirements.

Simmons: I was going to ask if the binding constant of this low force-producing state actually was strong.

Lombardi: You mean the A-1 state in my model?

Simmons: Yes. Obviously it's a bit difficult to translate that into a biochemical thing; you would have to assume an actin concentration. But by comparison with the other states, how much lower is it?

Lombardi: Maybe I can put it this way. I have to provide an attachment rate constant from this detached state that is 200 times faster than the isometric rate constant for attachment. Is that a good answer? (Note added in proof: According to our simulation [Piazzesi et al. *J. Physiol. (Lond.)* **445**, 659-711, 1992], $k_1/k_{-1} \approx 40$).

Simmons: No.

The Crossbridge Cycle

Huxley: Fig. 5 shows what I have thought for many years about cross-bridges. I showed this in a lecture in 1977 and it was published three years later (Huxley, A.F. *Reflections on Muscle* University Press, Liverpool, 1980).

This scheme is really Hugh Huxley's tilting cross-bridge (Huxley, H.E. *Science* **164**, 1356-1366,1969) adapted by A.F. Huxley and Simmons (*Nature* **233**, 533-538,1971) to provide step-wise rotation and an elastic element within the cross-bridge (diagrammatically in S2 connecting the head to the backbone of the thick

Non-
refractory
A1 A2 A3 A4
A1 A2 A3 A4
Refractory
Pi
I
II
III
ATP
ADP
1
2
3
4
5
6
7
8
9
10

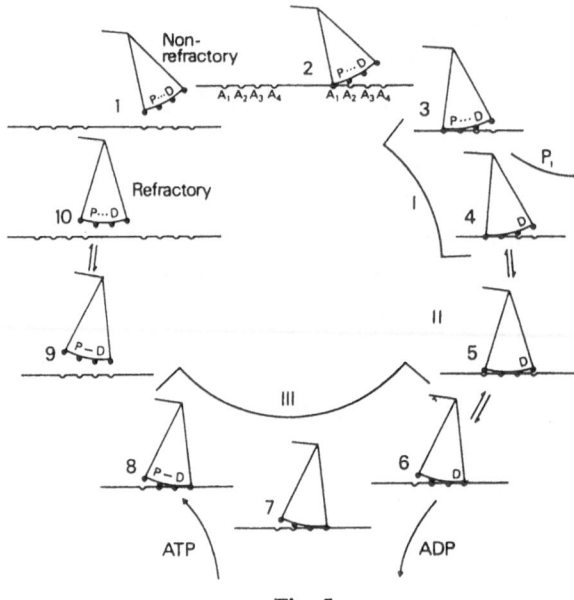

Fig. 5.

filament; not shown in Fig. 5). These postulates are the way in which we provided an explanation for the isometric transient. The chemical states associated with these mechanical states have been added in what I hope is a plausible way on the basis of the work of E.W. Taylor, D.R. Trentham, H.D. White, E. Eisenberg, and others (reviewed for example by Hibberd, M.G. and Trentham, D.R. *Ann. Rev. Biophys. Biophys. Chem.* **15**, 119-161, 1986).

One thing I would like to illustrate on the basis of this scheme is that more than once in this meeting, contributors *(e.g.*, Squire and Kawai) have emphasized that one must not identify one chemical state with one structural state. In particular, Dr. Kawai showed diagrams of force for his various states; the states immediately before and after phosphate release had the same force. The same was also true for later chemical steps. This is exactly what I had in mind when I constructed the scheme shown in Fig. 6. In state 2, there is a weak and very rapidly reversible attachment; if the system is activated through troponin, the myosin head is allowed to go forward one step to state 3, so that it is doubly bound and is therefore capable of showing stiffness. The fact of having moved into this position permits the phosphate to come

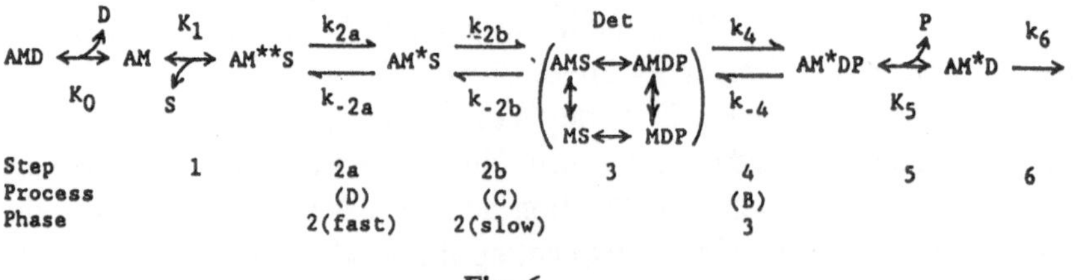

AMD
D
K0
AM
K1
S
AM**S
k2a
k-2a
AM*S
k2b
k-2b
Det
AMS AMDP
MS MDP
k4
k-4
AM*DP
P
K5
AM*D
k6
Step 1 2a 2b 3 4 5 6
Process (D) (C) (B)
Phase 2(fast) 2(slow) 3

Fig. 6.

off, which it does while the head is still in the same position and the force in the spring connecting to the thick filament is therefore unchanged. So states 3 and 4 are the two states that Dr. Kawai postulated with equal force, one with and one without phosphate. The same would be true for states 6 and 7, respectively, before and after the dissociation of ADP—the same structural situation, same stretch in the spring, and therefore the same force, but a different chemical situation.

Both Dr. Kawai and Drs. Millar and Homsher have emphasized that P_i release is a two-stage process. In the latters' abstract it is stated that "recent studies have shown that the release of P_i from the AM•ADP•P_i cross-bridge occurs in two steps—an initial force-generating isomerization which is followed by the release *per se* of P_i"; the first would correspond to the step from state 2 (which can exert no force because the head is pivoted about its one point of attachment) to state 3 which is doubly bound and therefore can exert force; the second is the step from state 3 to state 4 with no change of force.

So far, I have been emphasizing respects in which my scheme (Fig. 5) has been supported by what we have been told during this meeting. Dr. Kawai et al (this syposium), however, proposed the sequence of events shown in Fig. 6 which reproduces one of his slides. He put phase 2 of the isometric transient (rapid redevelopment of force after a shortening step) at Steps 2a and 2b, i.e., *after* ATP has bound to the cross-bridge (Step 1). In my thinking (Fig. 5), it is the transitions from state 4 to state 5 and from state 5 to state 6 that are responsible for the rapid force redevelopment. During the shortening step itself, the thick filament is shifted to the right, lowering the force in the spring, and the force is redeveloped by the clockwise rotation of the head which re-stretches the spring. We have put these steps before release of the diphosphate which gives the opportunity for ATP to come on, whereas Dr. Kawai has put it after the ATP has gone on. I do not know of any direct evidence for whether ATP is already on during this redistribution. We put it not coming on until state 8, causing dissociation, because of the long-standing evidence from rapid-reaction experiments that tells us that when ATP is added to an actomyosin preparation, dissociation is extremely rapid—within a millisecond or so. If one had ATP coming on at state 4, one would have to postulate something rather subtle that would prevent the ATP from causing dissociation before the head has gone through the force-generating steps down to state 6. I find it difficult to visualize how ATP could be on in states 4-6, but that does not mean it cannot be so; if there is evidence that it is on during the redistribution process, it would be an important difference from the scheme of Fig. 5. The other reason for liking to think that it is coming on at state 8, after the head has reached the presumed rigor state, is that energy is clearly needed to break the strong rigor bond.

Now as regards the detachment that Dr. Lombardi described during stretch, it was well established before the days of sliding filaments that during stretch, energy utilization is very low. The evidence came both from heat measurements (Abbot B.C. et al. *Proc. R. Soc.* **B139**, 86-104, 1951) and from measurements of ATP utilization (Curtin, N.A. and Davies, J.S, *J. Mechanochem. Cell Motility* **3**, 147-154, 1975). On this basis, it was natural to postulate long ago (Huxley, A.F. *Prog.*

Biophys. Biophys. Chem. **7**, 255-317, 1957, cf. p. 292) that a stretched cross-bridge could be detached without having gone through the complete chemical cycle. On the basis of that postulate alone, however, the number of myosin heads attached (and therefore stiffness) would fall with increasing speed of stretch, but Lombardi and Piazzesi [*J. Physiol.* (*Lond.*) **431**, 141-171, 1990] found a small increase of stiffness, implying very rapid reattachment of cross-bridges detached during stretch. What I would like to think in terms of Fig. 5 is that the cross-bridges that are forcibly detached have reached state 4 and are pulled off in that condition (i.e. having lost P_i) and are able to reattach very rapidly, 200 times faster than those that have gone through the complete cycle to state 1.

A.V. Hill (*Biochim. Biophys. Acta.* **4**, 4-11, 1950) issued a challenge to biochemists to find direct evidence for the utilization of ATP during muscle contraction. It is perhaps time now to issue a challenge to biochemists to identify the chemical state of cross-bridges forcibly detached by stretch while the muscle is activated. Obviously, it is a transient state because of its very rapid reattachment, and it will therefore be difficult to identify chemically. But something of this kind must exist, though whether it is the same as state 4 is highly speculative.

I am always a little worried about the classification of cross-bridge states into weakly binding and strongly binding—it's like chemists not liking to talk about high-energy and low-energy phosphates because there is a whole range. The many different states must have very different strengths of binding. In the same way, it is sometimes convenient to classify muscles into fast and slow, but Kushmerick's paper (this symposium) emphasized again that fast and slow do not always go with low oxidative and high oxidative—there are many permutations and combinations. I wonder how many permutations and combinations really may exist of the different properties that Dr. Brenner showed (this symposium) associated with these two categories of binding. Another possibility that arises out of this scheme is a conceivable explanation for the force-velocity curve with a discontinuity of slope that Dr. Edman [*J. Physiol.* (*Lond.*) **404**, 301-321, 1988; and this symposium] showed us, where at tensions just below the isometric, the speed of shortening is much less than one would expect by extrapolating the hyperbola that fits the points at lower tensions. One imagines that in the isometric state most of the cross-bridges are in states 4 and 5, with a rapid alternation between the two—going back and forth on a time scale that must be a little less than a millisecond in order to account for the speed of the transient when the equilibrium is altered by a stretch or release. If the muscle is shortened very slowly the equilibrium will be pushed toward state 5; there will be only a very small drop in tension. We put this in in order to explain the way that the T2 curve stays up. But the number of cross-bridges going on further in the cycle and being liberated by the ATP coming on might still be very low, although we have lowered tension a little. But shortening depends on cross-bridges being detached by ATP, ready to attach again, raise the tension, and pull, going around the cycle. This is speculative, and I have not yet thought it through thoroughly.

Davis: I would just like to make a comment concerning the assignment of phase-2_{slow} to cross-bridge dissociation. In T-jump experiments, that is the phase in which

I see a very large amplitude increase in tension above isometric on T-jumping, so it is hardly likely to be due to a dissociation process in the cross-bridge.

Huxley: Yes, I would agree with that. I would think that dissociation begins seriously in phase 3. There is very little change of stiffness at this stage.

Kawai: One thing I want to add to Professor Huxley's scheme is the rate-limiting step somewhere around step 5 or 6. The scheme is effectively open at that step. If force is generated at step 4, there is ample time for the myosin head to perform work at the rate-limiting step. Thereafter, the myosin head binds to ATP and goes through the next cycle. But I totally disagree that state 3 is a dissociation. I have many pieces of evidence against that.

Huxley: Phase 3 is the most puzzling of these phases.

Kawai: One thing we must not forget with tension transients is that if you stretch the muscle preparation, the transition must occur toward a lower-tension state, as shown in your model (Huxley, A.F. and Simmons, R.M. *Nature* **233**, 533-538, 1971). However, phase 3, in which the delayed tension rise takes place, is opposite to it. There have to be a mechanism and a scheme that explain the delayed tension rise. Our cross-bridge scheme explains the delayed tension if we assume the stretch induces more substrate binding. Can you explain the delayed tension rise?

Huxley: My ideas about phase 3, particularly in stretch, are very uncertain. I don't know whether to think it is closely related to the stretch activation of insect asynchronous muscle which, one supposes, has a rather special mechanism providing for it, or whether it is a universal property of muscle that is somewhat exaggerated in the asynchronous muscles.

Kawai: I believe that phase 3 is present in all skeletal muscles. Dr. Tawada and I cross-linked perhaps 20% of the myosin heads to the thin filament in rabbit psoas fibers. For the remaining 80% of the cross-bridges, we have seen phase 3, which was very much enhanced under the partially cross-linked condition (Tawada, K. and Kawai, M. *Biophys. J.* **57**, 643-647, 1990). Therefore, we believe that phase 3 is a universal phenomenon.

Brenner: This is a question for Ken Holmes. Is there any evidence for four distinct binding sites on actin?

Holmes: I don't think we will know that until we know what myosin looks like. There are certainly two that we can distinguish, but more than that, we don't know. We suppose there are more.

Huxley: I would like to emphasize that from my point of view, Fig. 5 is a purely schematic diagram indicating the existence of several steps with progressively increasing force onto the thick filament. I have emphasized [Huxley, A.F. *J. Physiol. (Lond.)* **243**, 1-43, 1974] that from the point of view of muscle mechanics, it might be something like my diagram with the myosin head rocking or it might be flexure within the head, or it might be flexure within the thin filament. From my point of view, I have to rely on structural people for evidence of that kind. I regard this in the same way that when one draws an electrical circuit, one draws a zig-zag for a resistance but this does not mean that it is a zig-zag structure. The diagram is meant only to imply that there are definable states with rapid equilibrium between adjacent ones, controlled by force. One has to have an elasticity somewhere, which I have customarily drawn as a spring in S2, but again, it might be flexibility in the head itself or in the attachment to actin.

HOW MUCH WORK FROM ONE CROSS-BRIDGE CYCLE?

Woledge: The question is, "How much work?" That means, how much work can we get from one cross-bridge head as it goes through one cycle splitting one ATP? I can conceive of three ways that an estimate of this number can be obtained. It would be nice for us, if time allowed, to compare these numbers and their uncertainties. First, the kinds of experiments that Nancy Curtin and I do lead to some estimate of how much work actually is delivered by a live muscle as it shortens. Our estimate would be 30-40 pnJ (J x 10^{-21}) per cross-bridge. That is, it would be within that range for fish or frog muscle. We will leave tortoise out of it, since the number would be greater, and because other people don't generally work on tortoise muscle. Perhaps they would like to, however, if they are interested in this question of how much work you can get.

Dr. Lombardi showed us estimates of how much work you can get out of the cross-bridge. We have always imagined that the area under the T2 curve gives us the amount of work obtained from cross-bridges without them going through a cycle and splitting ATP. He showed us that that is incorrect, if you allow the cross-bridges to rest for a short time—10 or 20 milliseconds—you can then, in a further release, get more work than otherwise would have been the case. We can conceive of an experiment in which we prevent the bridges from recycling, having split an ATP. We could then continue this process. By a succession of releases and rests or—the same thing—a slow release, we would collect as much work as we could. So, we could imagine the T2 curve extending to something like 20 nm. The area under it would represent the work we could get for one cross-bridge per one ATP. We can only put it in the right terms if we know what the force is. So, for the sake of argument, we can use the guessed figure of 2 pN. Then, the area under the extended T2 curve will come out to something between 30 and 40 pnJ. I find that coincidence slightly reassuring. I wonder if Dr. Lombardi wants to comment?

Lombardi: Three pnJ is the energy in each stroke elicited at the steady state with 5 nm step release (Lombardi et al. *Nature* **355**, 638-641, 1992). If the stroking rate is 10 times faster than the ATP split you end up with 30 pnJ and 50 nm total sliding. Is this what you said? I just want to be sure.

Woledge: I was suggesting that we take the area under the whole of the curve, which ought to be the work that you can get for one cross-bridge from one ATP. The number is in fact of the same order as the conclusion from our results.

Simmons: Would it be worthwhile for someone to do an energetics experiment of the kind that Dr. Lombardi has done with multiple releases just to check that there isn't an extra energy liberation?

Woledge: We have been discussing this with Vincenzo. If we can get the techniques right, we will do it. The last point I want to make is just to draw attention to the experiments that Dr. Suga made on the beating heart (Suga, H. *Physiol. Rev.* **40**, 247-277, 1990). I want to make the outrageous suggestion that in the heartbeat, in this situation I have been inviting you to imagine, a set of cross-bridges with no more allowed to join might actually exist. If we subsequently allow the heart—having reached the peak of its contraction—to shorten, we can obtain a certain amount of energy out of it. As I have already commented, this energy seems to be fixed: you can either obtain it as work, or if you don't do that, it comes out as heat. The challenge is to test the suggestion I am making that, after that point, no new cross-bridges form. If we knew this were true, it would be a very interesting physiological situation.

Suga: Yes. I think this is a very interesting point, and as Dr. Tad Taylor already showed in this symposium, if we plot cardiac oxygen consumption, either isovolumic or ejecting, against PVA, we have a linear relationship. This means that the efficiency of energy conversion from oxygen consumption to total mechanical energy measured by the PVA method is constant at about 40%. Dr. ter Keurs's graduate student showed (personal communication) very recently that if one measures ATP instead of the oxygen consumption and plots it against FLA, the slope becomes much smaller, and efficiency becomes about 60%. Since we know that the oxygen consumption to ATP energy conversion efficiency is about 60%-70% in myocardium, if the ATP to FLA efficiency is 40%, then 60% from both oxygen to ATP and also from ATP to mechanical energy accounts for the 40% from oxygen to FLA or PVA. What we are saying is that each cross-bridge step does not give any particular amount of mechanical energy, but as a whole, there is a tight stoichiometry between the energy consumption of cross-bridges and the mechanical energy generated by them. This does not mean that a single cross-bridge step gives the same amount of energy. There is some unknown autoregulatory mechanism, as Gibbs suggested (Gibbs, C.L. and Chapman J.B. *Am. J. Physiol.* **249**, H199-H206, 1985).

We recently performed a very interesting experiment: we tried to extract the mechanical energy from PVA in the beating heart and found that almost 95% of PVA comes out as actual mechanical work without increasing energy consumption (Hata, K. et al. *Am. J. Physiol.* **26**, H1778-H1784, 1991). This means that producing external work does not require extra energy once PVA is produced. It means that PVA energy has to stay at the end systole as some form of mechanical energy. This has also been supported by Mast and Elzinga's study (Mast F. and Elzinga G. *Circ. Res.* **67**, 893-901, 1990) in which they showed that FLA generates equivalent heat if the extra energy is not produced and is simply relaxed isometrically.

SUMMARY AND CONCLUSION

A.F. Huxley:

Professor Sugi told me that his original plan was to have Hugh Huxley summarize the Symposium, but Hugh was unable to attend this meeting. He has missed a great deal by not being here. So I am doing it instead.

The title I am provided with is, "Summary and Conclusion". We have had 68 communications and 16 posters, and if I were to summarize those in half an hour, you would find it insufferably dull. Further, as regards a conclusion, I don't think anyone here thinks that there is one well-defined conclusion that has been reached. We are all further on, but I think most of us are still confused. I am not sure who first used the phase, "We are still confused, but at a higher level".

Some of the topics I might have discussed in this summary session have been covered in the general discussion, and that will be a reason for not spending much time on them now. But I will try to discuss items that are closest to the topic of this Symposium — the mechanism of myofilament sliding in muscle. I will concentrate on skeletal muscle, partly because most of what we have heard directly relevant to the mechanism of sliding has come from studying skeletal muscle, both historically and in what we have heard in the last few days. Most of the things we heard about smooth muscle and cardiac muscle have had to do principally with activation processes, and are extremely interesting and important; I am not going to spend on them any of the little time I have but I have listed a few of the highlights in Table 1. To some extent, these features of smooth muscle and cardiac muscle activation are probably special to those types of muscle, but I was greatly impressed by several of

Table 1. Highlights on activation presented at this Symposium.

1. Rall, J.A., Hou, T-t. & Johnson, J.D. a. Quantitative evidence for importance of Ca-binding by parvalbumin in relaxation of frog skeletal muscle. b. Activation of a mechanically-skinned fibre by Ca release from a caged compound gave a faster rise of tension than electrical stimulation of an intact fibre.
2. Winegrad, S. Evidence for endothelial factors regulating contractility in rat heart.
3. Pfitzer, G. a. Electrical stimulation of chicken gizzard muscle elicits phasic contractions preceded by phosphorylation of the 20 kDa light chain of myosin. b. The actin-binding fragment of caldesmon inhibits the rise of tension elicited by myosin light-chain kinase.
4. Gailly, Ph. & Gillis, J.M. Brevin, a protein which severs F-actin filaments, is present in smooth muscle but is lost on skinning. Replacement causes a reduction in force of isometric contraction but an increase in speed of unloaded shortening.

Mechanism of Myofilament Sliding in Muscle Contraction, Edited by
H. Sugi and G.H Pollack, Plenum Press, New York, 1993

839

Table 2. New techniques presented at this Symposium (in order of presentation).

1. Scanning tunnelling microscope (Faruqi A.R., Kendrick-Jones, J. & Cross, R.A.)
2. Competing peptides (Rüegg, J.C., Zeugner, C., Keane, A. & Trayer, J.).
3. Mutated proteins. a. Troponin - C (Gulati, J.) b. Tropomyosin in asynchronous muscle (Molloy, J., Kreuz, A. & Maughan, D.) c. Actin in Dictyostelium (Sutoh, K.)
4. Luminescent/paramagnetic probes (Ajtai, K. & Burghardt, T.P.)
5. Optical traps (Simmons, R.M., Finer, J., Warrick, H., Kralik, B., Chu, S. & Spudich, J.)
6. Force measurement in single filaments, with high frequency- response (Ishijima, A., Saitoh, K., Harada, Y. & Yanagida, T.)
7. Force application to in vitro systems with centrifuge microscope. (Chaen, S., Oiwa, K., Kamitsubo, E., Shimmen, T. & Sugi, H.)
8. Imaging plate for recording X-ray diffraction patterns. (Yagi, N., Takemori, S. & Watanabe, M.)
9. "Flash and smash" followed by electron microscopy (Goldman, Y.E.)
10. Acoustic microscope for measuring ultrasonic velocity (Tsuchiya, T., Iwamoto, H., Tamura, Y. & Sugi, H.)

the things we have heard that were quite unexpected to me. Of course, we have all heard about phosphorylation for many years, but we heard it in much more detail this time. We have also heard about a great number of more or less new techniques. Again, I am not going to spend time on them, but I would like just to acknowledge how impressive and numerous they are by including a list (Table 2), which may well not be complete. But I want to spend most of my time on some of the long-standing questions about which we have heard new evidence.

For many years, people have used the phrase that the movement of one filament relative to the other is brought about by a conformational change. This is an ambiguous phrase. Is it meant only in the strict sense of a switch between two well-defined structures within one protein molecule, or is it used in an extended sense, which would include rotation of the S-1 relative to the thin filament, which might be called a conformational change in actomyosin? And should it be used for the changes that Dr. Morales described,i.e. for shifts within the myosin head connecting ATP attachment to actin dissociation? I will repeat a quotation that I have used before about the use of the word "conformational". Dean Inge, who was Dean of St Paul's Cathedral in London when I was a boy, is recorded as having said that the word "bloody" had become "simply a sort of notice that a noun may be expected to follow". In the same way, it seems to me that the word "conformational" has become merely a notice that the word "change" may be expected to follow. People often don't distinguish between conformational changes in the strict sense and other sorts of change that would not fall within the narrow definition of conformational.

We have had in these few days at least two indications of something that one would have to think of as a conformational change in the narrow sense. Dr. Katayama with his very elegant electron microscopy described a "kinked" configuration in S-l when attached to actin with the appropriate nucleotides bound, and Dr. Wakabayashi gave us differences in X-ray scattering. I am not clear to what extent those two observations are fully consistent with one another, but in any case there remains the question: assuming that the change is real, is it a change of shape that actually pulls the myosin relative to the actin, or is it a by-product of the allosteric action by which nucleotide binding influences the strength with which the myosin head attaches itself to actin? Another long-standing question is, Why does myosin have two heads? Twenty years ago Manuel Morales asked "Does a myosin cross-bridge progress arm-over-arm on the actin filament?"as if pulling on a rope[1]. And long ago, Professor Tonomura[2][3] who, sadly, died about ten years ago, had evidence that the two heads on one myosin molecule are chemically different from one another. This has not been generally accepted by biochemists. I can't add any view on that myself—I am not competent to evaluate the evidence. But it was exciting to hear the new evidence from Inoue et al for an actual amino acid difference, a difference of one residue between two types of head, and having antibodies against these two types. From the discussion, however, my impression was that those present and able to judge this kind of work—which I am not—did not feel that it had been shown conclusively whether these two types of head are really on the same myosin molecule, or whether the myosin was a mixture of two homodimers. It has been clear for many years that two heads are not necessary for sliding of some kind. I am glad that Professor Sugi reminded us of the experiment of putting S-1 back into a ghost preparation and getting motion[4], but I think most of us are more conscious of the *in vitro* filament assays, in which motility of an actin filament is readily obtained with S-1 or single-headed HMM. But it remains true—and I think it has been repeatedly found—that HMM gives more force or higher speeds of actin filament sliding than S-1, but as far as I have been able to understand, it is uncertain whether this is due to the presence of some part of the rod attached to the head, or whether it is due to two heads being present.

All the theorizing that I have done has been on the basis of one head operating independently of the other (Fig. 1), and this is not because it is excluded that the two interact in some way or cooperate, but simply in order to have a hypothesis that is manageably simple. Even now, anything involving the two heads in different ways would be so speculative that one could put almost anything into the theory.

The question of one versus two heads being active is also relevant to the question of the length of the step, one of the questions I think is uppermost in all our minds. We heard again from Dr. Yanagida about his evidence for the absence of substantial fluctuations of tension when the sliding speed gets above a rather low level[5][6]—that is impressive evidence suggesting that a high proportion of the heads are attached at any one moment, but we did not hear a counterattack by Jim Spudich, who, in the same issue of Nature, had elegant evidence of quantized sliding speed and calculated a step length that would fit with a single stroke of the myosin head[7] I am in no position to judge between them. Both arguments appear convincing to a reader, and I am sure there will be more evidence on this point.

The experiment that Dr. Lombardi[8][9] described—of repeated short steps with apparently complete re-priming from one to the next—also implies that several mechanical strokes must take place using the energy from a single ATP. I would

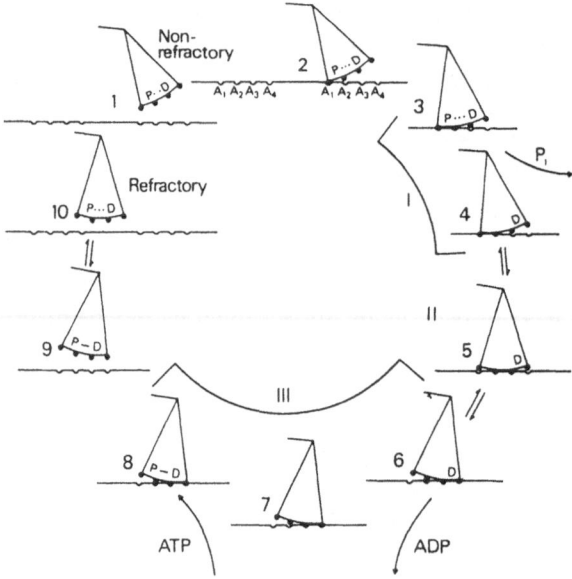

Fig. 1. Diagram of a cross-bridge cycle proposed in 1977 (fig.16 of ref. 27). The mechanical steps are based on A.F. Huxley & Simmons[19] but with three rigidly attached positions (I, II and III) in place of the two positions assumed in that article. The spring in S-2 connecting the top of the myosin head to the back- bone of the thick filament is not shown. P and D represent inorganic phosphate and ADP respectively; P-D is therefore ATP and P...D is ATP after hydrolysis but with the products still bound to the myosin head. The original legend included the following: "Attachment in position I promotes dissociation of Pi, which in turn permits the myosin head to move forward to attitude II and attitude III; these movements are restrained by tension in the S-2 component (assumed elastic) which connects the top of the myosin head to the thick filament. As a result of the head reaching position III, ADP dissociates from the myosin head, permitting ATP to become bound in its place; this promotes dissociation of the myosin head from the thin filament. There may well be more than one attached state in between I and III. Double arrows indicate rapid reversibility. In addition to the steps indicated it is possible that the initial attachment (1-2) and the first forward movement of the head with Pi still attached (2-3) are also readily reversible."

hope that this idea fits with the idea of the rapid reversibility of attached states suggested by Dr. Brenner[10]. This is a very interesting and important suggestion; I confess that I did not fully take in his evidence for it, and I am not clear whether it really excludes the possibility that I raised in discussion: he tests by applying a stretch, and we know from other experiments of Lombardi and his colleagues[11] that stretch can detach myosin heads in a state capable of reattaching with great rapidity. Were the states that Brenner observed pre-existent, or were they created by the stretch?

Another controversial result from *in vitro* preparations emerged from the contribution by Chaen et al.[12], who applied force to myosin-coated beads moving on actin cables by means of a centrifuge microscope. When the force opposed the motion, the speed of movement was reduced as was to be expected but it was not

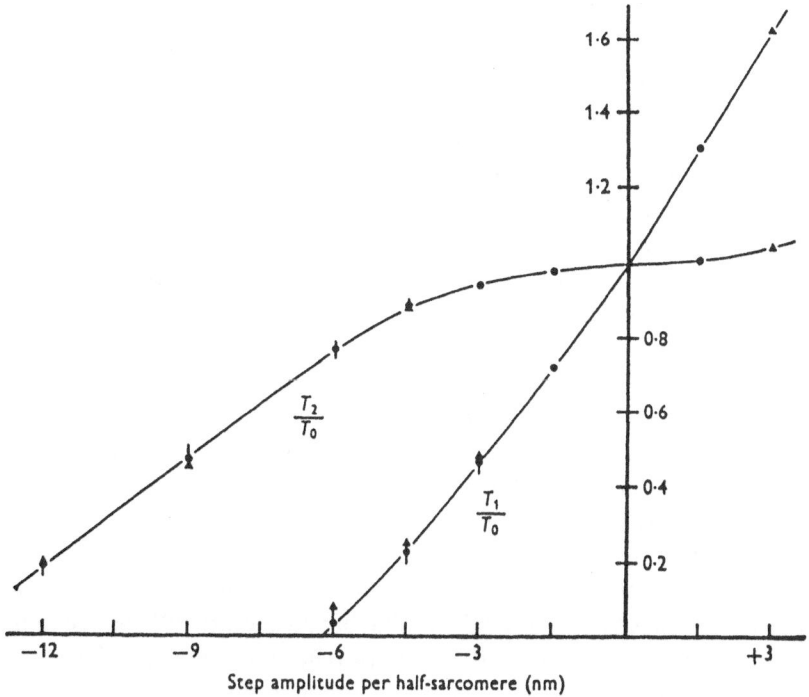

Fig. 2. Curves of T_1 (extreme tension) and T_2 (tension approached during early recovery phase) after step changes of length of various amplitudes, imposed during plateau of an isometric tetanus. Expressed as fractions of T_0, the tension immediately before the step. If there were negligible detachment of cross-bridges before T_2 reaches zero and very rapid thereafter, the area under the T_2 curve would represent the work done during shortening at a speed low enough for the early recovery phase to be complete; making T_0 equal to the isometric force per cross-bridge makes the area equal to the work per cross-bridge in a single cycle with utilisation of 1 ATP molecule. With $T_0 = 2$ pN, the area is 17 pnJ and with $T_0 = 4$ pN it is 34 pnJ. From fig.13 of ref. 28..

increased when the force was in the direction of motion. This disagrees with the conclusion of Edman[13] who found, in intact isolated frog muscle fibres, an increase in the speed of unloaded shortening when a fibre was stretched beyond about 2.7μm per sarcomere and attributed the increase to the effect of passive tension pushing the filaments in the shortening direction. A point that ought perhaps to be checked is whether the fibrils remain straight in this situation or whether they are thrown into waves as occurs when non-activated myofibrils in an intact fibre are passively shortened below their slack length[14].

Fig. 2 is a diagram equivalent to one that Roger Woledge showed. Putting two piconewtons as the isometric force, the area under the curve comes to 17 piconanojoules, which is too low to give the kind of efficiency that we heard about in this meeting from Curtin & Woledge[15]. The total free energy of ATP hydrolysis is about 80 of these units so that 40% efficiency would mean 32, and we would only have.about half that efficiency if the area in Fig. 2 is a real representation of the maximum work obtainable from a single stroke and only one stroke is performed per ATP used. One big uncertainty about that diagram is the figure of 2 pN for the isometric force per cross-bridge. You come to something like that for frog muscle if you assume that both heads of every myosin molecule are generating their full force at the same time during an isometric contraction. But if you assume that only one of the heads of each myosin molecule can be active at a time, you have half as many heads. The total isometric force is a measured quantity, so you would have to say that the force per head is 4 pN. The area under the curve in Fig. 2 would be doubled, and would be sufficient to give the efficiency that is observed.

So the question of one versus two heads interacts strongly with the question of the distance that a cross-bridge can shorten with the use of one ATP, in the sense that if we had good evidence that both heads are exerting full force independently in the isometric state, we should be forced to say that the complete working stroke per ATP is much larger than the 12 nanometres that one obtains from transients, which of course covers only what happens within a very few milliseconds.

It seems to be generally agreed that all heads are in some sense attached in rigor, and that stiffness in the isometric state is not very difference from that in rigor. I think this is sometimes taken as evidence that both heads are fully active in isometric contraction and are exerting full force. But in rigor, the binding strength of attachment of myosin to the thin filament, although much greater than for S-1 or other single-headed preparations, is not as great as would be expected if each of the two heads was independently bound with the strength of binding that is found with a single-headed preparation such as S-1. It makes one wonder whether the situation in rigor might be something like what is shown in Fig. 3, with one head properly attached in a rigid configuration. The other one might not be capable of swinging around into a situation of that kind, but is attached at only one point and gives less total binding affinity—bound, in a sense, but in a situation that does not correspond to force generation.

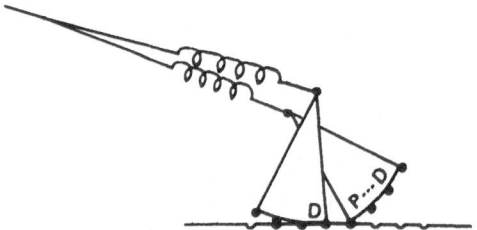

Fig. 3. Possible "arm-over-arm" action of the two heads of one myosin molecule (Botts et al, 1973)[1]. Even if steric factors restrict the range of positions in which the second head can bind[29], it is plausible to suggest that when head 1 has reached position III (Fig. 1), head 2 is placed so that it can enter the cycle with a higher rate constant than when the other head is not attached.

If there were a situation like Fig. 3 in which it is impossible for both heads to exert full force at the same time, a possibility that is then raised is that when one myosin head has completed its stroke and is very soon going to be detached by an ATP coming on, the other one is ready-placed to make an attachment and go through its stroke. Being held in the right position, the rate constant for attachment might be enormously higher than it is for a fully free and detached head. That is an interesting possibility, and one that would certainly upset such kinetics as one might calculate from the kinds of theories that I have produced. It does seem to me that it might perhaps be an alternative explanation for the results that Dr. Lombardi has presented; it might be the way in which the effective working stroke of a myosin molecule is extended by involving the two heads successively. This would really be Manuel Morales's arm-over-arm operation[1]. I have no idea whether this sort of idea would stand up if it were properly simulated. It may well turn out that it is just one of those attractive, but wrong, ideas.

Another aspect of this is that, supposing that the second head becomes attached 7 nm ahead of the first, it would be out of phase with the 14.5 nm spacing that gives rise to the meridional reflection that Lombardi so elegantly showed is decreased in strength in parallel with the force-generating operation by a myosin head. It just might be that it is the attachment of another head half-way along the spacing that causes the reduction of intensity. Again, I have no idea whether that idea would fit quantitatively.

Yesterday, Jerry Pollack challenged us to say whether any conclusion had been reached conclusively and unambiguously from X-ray diffraction, and I think that this experiment of Irving, Lombardi, Piazzesi & Ferenczi[16] showing that the decrease of the 14.5 intensity is closely parallel with the working stroke of the myosin head, is conclusive evidence that there is redistribution of mass along the fiber axis. A strong suggestion that the particular interpretation by Irving et al—that it is directly due to tilt, spreading out the mass along the length—is very likely but perhaps not absolutely unambiguous. However, I think it is quite unambiguous in showing that some redistribution of mass along the axis of the fiber takes place in parallel with the working stroke by the myosin head.

Then, another long-standing question: Does the head tilt? Dr. Suzuki and his colleagues[17] made elegant direct measurements under the electron microscope of the angle of the heads and presented changes in the proportion of heads within 10° of 90°, but he did not give us figures for the mean angle in the different states. This is a technique that is going to be extremely informative.

We have heard rather little about probes during this meeting. Dr. Burghardt told us about the luminescent and paramagnetic probe that he is developing, but we have heard rather little about previous evidence. But I think Dr. Burghardt did say that the early evidence that appeared to exclude change of tilt to the head is no longer valid. The thing that I have most wanted to know in the last twenty years, since the publication of A.F. Huxley & Simmons[18], is whether the tilt of heads—or any other parameter that can be measured—fluctuates on the time scale of a millisecond or a little less, corresponding to the equilibrium that we suppose must exist between adjacent states of the myosin head. It is very difficult to imagine a mechanism for the rapid recovery from either stretch or release that we see in phase 2 of the transient without postulating an equilibrium between the states from which and to which it is going. I don't say that it is impossible, but it is much easier to think of it in terms of an equilibrium which is displaced by the length change. The time scale in question is too long to be investigated by ordinary fluorescence depolarisation or electron paramagnetic resonance, but I was greatly encouraged by the observation by David Thomas and his colleagues using saturation-transfer EPR[19] and phosphorescence depolarisation[20] that myosin heads do indeed undergo substantial rotational motion on the right time scale during isometric contraction. This mobility is not shown by a label on F-actin during ATP hydrolysis with myosin S-1[21]. I shall look forward to further evidence in the future on this very important point.

Another question: is there a substantial change in the actin? Again, Dr. Suzuki reported large changes in his measured spacing on thin filaments, but according to my memory the X-ray experts in the audience told us that they had not seen anything comparable. A substantial change is predicted by the theory of Schutt & Lindberg[22], which postulates that contraction is brought about by actin alternating between the "helix" and "ribbon" states that they have observed in actin-profilin crystals[23]. Rotation of individual actin monomers is one of the possibilities that has to be kept in mind, but the very elegant pictures of the polarization of rhodamine fluorescence on actin filaments by Kinosita et al.[24] seem to exclude rotations in certain directions. There was also the astonishing video by Ishiwata, in which thin filaments twisted themselves up into little coils; I was relieved that he did not suggest that this happened *in vivo*. It was interesting that Reedy thought the torque that gives rise to this might be the basis of the change in the Z-lattice. But we must remind ourselves that Jarosch for many years has been putting forward the idea that the thin filaments screw their way along between the thick filaments. We mustn't dismiss this a priori.

I do not know how well known is the work of Borovikov and his colleagues in St Petersburg. He has looked at structural changes in thin filaments by recording the depolarization of intrinsic fluorescence or of fluorescence from probes, finding that the state of polarization is altered when S-1 binds, and the alteration is different according to the nucleotide that is on the S-1[25][26]. Very important work, and like so much good Russian work, it fails to get noticed in the West. There are plenty of other things that are conspicuously in need of explanation, but because of the time limit, I will have to leave them to your imaginations

Every speaker here has expressed the very sincere thanks that we all feel to Professor Sugi and his numerous collaborators for this absolutely splendid meeting. I want now not only to reinforce these thanks and again to express my own, but to congratulate Professor Sugi and his group on the extraordinary number of different and interesting techniques and experiments that he has under way in his laboratory. At least nine of the contributions in this meeting were from his laboratory, all of them using different techniques: deep-freeze etch; the glass microneedle for force measurement; injecting ATP near an actin filament in one of these *in vitro* preparations; his centrifuge microscope; caged ATP experiments; cardiac myosin isozymes; antibody to S-2; actin with different nucleotides bound and all behaving in the same way; the ultrasonic microscope; and so on. It is most impressive. I think I am expressing what all of us are feeling when I say that we congratulate Professor Sugi as well as thanking him for organizing this meeting, which has been absolutely splendid.

Sugi: I thank Professor Huxley for his extremely nice summary and conclusion. It is very true that we are still confused, but hopefully at a little higher level.

REFERENCES

1. Botts, J., Cooke, R., dos Remedios, C., Duke, J., Mendelson, R., Morales, M.F., Tokiwa, T., Viniegra, G. & Yount, R. *Cold Spring Harbor Symp. Quant. Biol.* **37**, 195-200 (1973).
2. Tonomura, Y. *Muscle proteins, muscle contraction and ion transport.* (University of Tokyo Press, 1972).
3. Inoue, A., Takenaka, H., Arata, T. & Tonomura, Y. *Adv. Biophys.* **13**, 1-194 (1979).
4. Borejdo, J. & Oplatka, A. *Biochim. Biophys. Acta,* **440**, 241-258 (1976).
5. Yanagida, T., Ishijima, A., Saitoh, K. & Harada, Y. This volume, 339-349.
6. Ishijima, A., Doi, T., Sakurada, K. & Yanagida, T. *Nature* **352**, 301-306 (1991).
7. Uyeda, T.Q.P., Warrick, H.M., Kron, S.J. & Spudich, J.A. *Nature* **352**, 307-311 (1991).
8. Lombardi, V., Piazzesi, G. & Linari, M. *Nature* **355**, 638-641 (1992a).
9. Piazzesi, G. & Linari, M. & Lombardi, V. This volume, 691-714.
10. Brenner, B. This volume, 531-543.
11. Piazzesi, G., Francini, F., Linari, M. & Lombardi, V. *J. Physiol. (Lond.)* **445**, 659-711 (1992).
12. Chaen, S., Oiwa, K., Kamitsubo, E., Shimmen, T. & Sugi, H. This volume, 351-360.
13. Edman, K.A.P. *J. Physiol. (Lond.)* **291**, 143-159 (1979).
14. Gonzalez-Serratos, H. *J. Physiol. (Lond.)* **212**, 777-799 (1971).
15. Curtin, N.A. & Woledge, R.C. This volume, 729-734.
16. Irving, M., Lombardi, V., Piazzesi, G. & Ferenczi, M. *Nature* **357**, 156-158 (1992).
17. Suzuki, S., Oshimi, Y. & Sugi, H. This volume, 57-70.
18. Huxley, A.F. & Simmons, R.M. *Nature* **233**, 533-538 (1971).
19. Barnett, V.A. & Thomas, D.D. *Biophys. J.* **56**, 517-523 (1989) .
20. Stein, R.A., Ludescher, R.D., Dahlberg, P.S., Fajer, P.G., Bennett, R.L.H. & Thomas, D.D. *Biochemistry* **29**, 10023-10031 (1990) .
21. Ostap, E.M. & Thomas, D.D. *Biophys. J.* **59**, 1235-1241 (1991).
22. Schutt, C.E. & Lindberg, U. *Proc. Natl. Acad. Sci. USA* **89**, 319-323. (1992)
23. Schutt, C.E., Lindberg, U., Myslik, J. & Strauss, N. *J. Mol. Biol .* **209**, 735-746 (1989) .
24. Kinosita, K. Jr, Suzuki, N., Ishiwata, S., Nishizaka,T., Itoh, H., Hakozaki, H. & Marriott, G. This volume, 321-329.
25. Borovikov, Yu.S. *Microscopica Acta,* **82**, 379-388 (1980) .
26. Borovikov, Yu.S., Wrotek, M., Aksenova, N.B., Lebedeva, N.N. & Kakol, I. *FEBS Lett.,* **223**, 409-412 (1987).
27. Huxley, A.F. *Reflections on Muscle.* (Liverpool: University Press, 1980).
28. Ford, L.E., Huxley, A.F. & Simmons, R.M. *J. Physiol. (Lond.)* **269**, 441-515 (1977).
29. Schoenberg, M. *Biophys. J.* **60**, 679-689 (1991).

HAKONE SYMPOSIUM Nov. 11~15 1991

(*First row*) N. Oishi, H. Suga, Mrs. Simmons, L.C. Yu, K. Tawada, K. Wakabayashi, T. Ando, S. Suzuki, T. Yamada, H. Iwamoto, K. Oiwa, K. Noguchi, H. Okuyama, J. Inoue (*Second row*) Y. Okamoto, T. Yanagida, E. Katayama, R.E. Godt, N.A. Curtin, Y. Zhao, M. Kawai, A.F. Huxley, H. Sugi, W.F. Harrington, M.F. Morales, G. Cecchi, I. Takagi, J.M. Gillis, J.C. Rüegg, N. Itagaki (*Third row*) S. Sugiura, S. Ishiwata, T. Tameyasu, N. Yagi, K. Horiuti, R.C. Woledge, A.R. Faruqi, Mrs. Winegrad, S. Winegrad, K.A.P. Edman, A. Inoue, B. Brenner, Y.E. Goldman, J. Gergely, R.M. Simmons, G. Pfitzer, H.E.D.J. ter Keurs, G.H. Pollack, Mrs. Homsher, E. Homsher, P.B. Chase, M.J. Kushmerick (*Fourth row*) T. Arata, J.M. Squire, K. Yamada, E. Kamitsubo, V. Lombardi, R.T. Tregear, J.R. Sellers, J. Gulati, S. Kurihara, T. Tsuchiya, J.A. Rall, Mrs. Maéda, S. Sys (*Fifth row*) M. Ogata, P.J. Griffiths, Mrs. Holmes, C. Poggesi, M.K. Reedy, A.M. Gordon, T.W.Taylor, F. Colomo, Y. Saeki, H. Onishi, S. Chaen, Y. Maéda (*Sixth row*) K.C. Holmes, M. Schoenberg, T.P. Burghardt, D.W. Maughan, C.C. Ashley, J.S. Davis, Y. Ogawa, T. Kobayashi

849

PARTICIPANTS

Ando, T.
  Department of Physics
  Faculty of Science
  Kanazawa University
  Kakumamachi, Kanazawa 920-11
  Japan
Arata, T.
  Department of Biology
  Faculty of Science
  Osaka University
  Toyonaka, Osaka 560
  Japan
Ashley, C.C.
  Department of Physiology
  University of Oxford
  Parks Road
  Oxford OX1 3PT
  U.K.
Brenner, B.
  Department of General Physiology
  University of Ulm
  Oberer Eselsberg
  D-7900 Ulm
  F.R.G.
Burghardt, T.P.
  Department of Biochemistry
  and Molecular Biology
  Guggenheim 14, Mayo Foundation
  200 First Street South West,
  Rochester, Minnesota 55905
  U.S.A.
Cecchi, G.
  University of Florence
  Dipartimento di Scienze Fisilogiche
  Viale Morgagni 63,

I-50134 Firenze
  Italy
Chaen, S.
  Department of Physiology
  School of Medicine
  Teikyo University
  Itabashi-ku, Tokyo 173
  Japan
Chase, P.B.
  Department of Radiology
  Uniersity of Washington
  Medical Center SB05
  Seattle, Washington 98195
  U.S.A.
Colomo, F.
  Dipartimento di Scienze Fisilogiche
  Università degli Studi di Firenze
  I-50134 Firenze
  Italy
Curtin, N.A.
  Department of Physiology
  Charing Cross & Westminster
  Medical School
  Fulham Palace Road
  London W6 8RF
  U.K.
Davis, J.S.
  Department of Biology
  The Johns Hopkins University
  34th and Charles Streets
  Baltimore, Maryland 21218
  U.S.A.
Edman, K.A.P.
  Department of Pharmacology
  University of Lund

Sölvegatan 10, S-223 62 Lund,
Sweden
Faruqi, A.R.
   Laboratory of Molecular Biology
   Medical Research Council
   Hills Road
   Cambridge CB2 2QH
   U.K.
Gergely, J.
   Department of Muscle Research
   Boston Biomedical Research
   Institute
   20 Staniford Street
   Boston, Massachusetts 02114
   U.S.A.
Gillis, J.M.
   Laboratory of Physiology
   Faculty of Medicine
   Université Catholique de Louvain
   U.C.L. 5540
   1200 Brussels
   Belgium
Godt, R.E.
   Department of Physiology &
   Endocrinology
   Medical College of Georgia
   Augusta, Georgia 30912-3000
   U.S.A.
Goldman, Y.E.
   Department of Physiology
   School of Medicine
   University of Pennsylvania
   Philadeophia, Pennsylvania 19104-
   6085
   U.S.A.
Gordon, A.M.
   Department of Physiology and
   Biophysics SJ-40
   University of Washington
   Seattle, Washington 98195
   U.S.A.
Griffiths, P.J.
   University Laboratory of
   Physiology
   Parks Road

Oxford OX1 3PT
U.K.
Gulati, J.
   Department of Medicine
   Division of Cardiology,
   Albert Einstein College of Medicine
   1300 Morris Park Ave.,
   Bronx, New York 10461
   U.S.A.
Harrington, W.F.
   Department of Biology
   Johns Hopkins University
   34th & Charles Streets
   Baltimore, Maryland 21218
   U.S.A.
Holmes, K.C.
   Max-Plank-Institute for Medical
   Research
   Department of Biophysics
   Jahnstraße 29
   6900 Heidelberg
   F.R.G.
Homsher, E.
   Department of Physiology
   School of Medicine
   UCLA
   Center for Health Sciences
   Los Angels, California 90024
   U.S.A.
Horiuti, K.
   Department of Physiology
   Medical College of Oita
   Oita-gun, Oita 879-56
   Japan
Huxley, A.F.
   Manor Field
   1 Vicarage Drive
   Grantchester, Cambridge
   CB3 9NG
   U.K.
Inoue, A.
   Department of Biology
   Faculty of Science
   Osaka University

Toyonaka, Osaka 560
Japan
Ishiwata, S.
  Department of Physics
  School of Science and Engineering
  Waseda University
  Shinjuku-ku, Tokyo 160
  Japan
Iwamoto, H.
  Department of Physiology
  School of Medicine
  Teikyo University
  Itabashi-ku, Tokyo 173
  Japan
Kamitsubo, E.
  Biological Laboratory
  Hitotsubashi University
  Kunitachi, Tokyo 186
  Japan
Katayama, E.
  Department of Fine Morphology
  Institute of Medical Science
  University of Tokyo
  Minato-ku, Tokyo 108
  Japan
Kawai, M.
  Department of Anatomy
  College of Medicine
  The University of Iowa
  Iowa City, Iowa 52242
  U.S.A.
Kinosita, K. Jr.
  Department of Physics
  Faculty of Science and Technology
  Keio University
  Kohoku-ku, Yokohama 223
  Japan
Kobayashi, T.
  Department of Physiology
  School of Medicine
  Teikyo University
  Itabashi-ku, Tokyo 173
  Japan
Kurihara, S.
  Department of Physiology

The Jikeikai Medical School
Minato-ku, Tokyo 105
Japan
Kushmerick, M.J.
  Department of Radiology,
  University of Washington
  Medical Center SB05
  Seattle, Washington 98195
  U.S.A.
Lombardi, V.
  Dipartimento di Scienze Fisilogiche
  Viale G.B. Morgagni 63
  50134 Firenze
  Italy
Maéda, Y.
  EMBL c/o DESY
  Notkestrasse 85
  W-2000 Hamburg 52
  F.R.G.
Maughan, D.W.
  Department of Physiology and
  Biophysics
  University of Vermont
  Burlington Vermont 05405
  U.S.A.
Mitsui, T.
  Department of Physics
  School of Science and Technology
  Meiji University
  Tama-ku, Kawasaki 214
  Japan
Morales, M.F.
  Laboratory of Physiology and
  Biophysics
  University of the Pacific
  2155 Webster Street
  San Francisco, California 94115
  U.S.A.
Ogata, M.
  Director
  Institute of Health Science
  Kyushu University
  Kasuga, Fukuoka 816
  Japan

Ogawa, Y.
  Department of Pharmacology
  Juntendo University
  School of Medicine
  Bunkyo-ku, Tokyo 113
  Japan
Oishi, N.
  Radioisotope Research Center
  Teikyo University
  Itabashi-ku, Tokyo 173
  Japan
Oiwa, K.
  Department of Physiology
  School of Medicine
  Teikyo University
  Itabashi-ku, Tokyo 173
  Japan
Okinaga, S.
  President
  Teikyo University
  Itabashi-ku, Tokyo 173
  Japan
Okuyama, H.
  Department of Physiology
  Kawasaki Medical School
  Okayama 701-01
  Japan
Pfitzer, G.
  II Physiologisches Institut
  Universität Heidelberg
  Im Neuenheimer Feld 326
  D6900 Heidelberg
  F.R.G.
Poggesi, C.
  Dipartimento di Scienze Fisilogiche
  Università degli Studi di Firenze
  I-50134 Firenze
  Italy
Pollack, G.H.
  Bioengineering WD-12
  University of Washington
  Seattle, Washington 98195
  U.S.A.
Rall, J.A.
  Department of Physiology

Ohio State University
333 West 10th Avenue
Columbus, Ohio 43210
U.S.A.
Reedy, M.K.
  Department of Cell Biology
  Duke University Medical Center
  Durham, North Carolina 27710
  U.S.A.
Rüegg, J.C.
  II. Physiologisches Institut
  Universität Heidelberg
  Im Neuenheimer Feld 326
  D-6900 Heidelberg
  F.R.G.
Saeki, Y.
  Department of Physiology
  Faculty of Dentistry
  Tsurumi University
  Tsurumi-ku, Yokohama 230
  Japan
Schoenberg, M.
  Bldg. 6, Room 108
  National Institutes of Health
  Bethesda, Maryland 20892
  U.S.A.
Sellers, J.R.
  Building 10, Room 8N202
  National Institute of Health
  Bethesda, Maryland 20892
  U.S.A.
Shimmen, T.
  Department of Life Science
  Faculty of Science
  Himeji Institute of Technology
  Ako-gun, Hyogo 678-12
  Japan
Simmons, R.M.
  Department of Biophysics
  King's College London
  26-29 Drury lane
  London WC2B 5RL
  U.K.
Squire, J.M.
  Biophysics Section

The Blackett Laboratory
Imperial College of Science,
Technology & Medicine
Prince Consort Road,
London SW7 2BZ
U.K.
Sutoh, K.
Department of Pure & Applied
Sciences
College of Arts & Sciences
University of Tokyo
Meguro-ku, Tokyo 153
Japan
Suga, H.
The 2nd Department of Physiology
Okayama University Medical
School
Okayama 700
Japan
Sugi, H.
Department of Physiology
School of Medicine
Teikyo University
Itabashi-ku, Tokyo 173
Japan
Sugiura, S.
The second Department of Internal
Medicine
School of Medicine
University of Tokyo
Bunkyo-ku, Tokyo 113
Japan
Suzuki, S.
Department of Physiology
School of Medicine
Teikyo University
Itabashi-ku, Tokyo 173
Japan
Sys, S.
Department of Physiology
University of Antweap
Groenen Borgerlaan 171,
B-2020 Antwerp
Belgium
Tameyasu, T.
Department of Physiology

St. Marianna University
School of Medicine
Miyamae-ku, Kawasaki 216
Japan
Taylor, T.W.
Department of Cardiovascular
Dynamics
National Cardiovascular Center
Research Institute
Suita, Osaka 565
Japan
ter Keurs, H.E.D.J.
University of Calgary
Department of Medicine and
Medical Physiology
3330 Hospital Drive N.W.,
Calgary T2N 4N1
Canada
Toyoshima, Y.
Department of Biuology
Ochanomizu University
Bunkyo-ku, Tokyo 112
Japan
Tregear, R.T.
AFRC Institute of Animal
Physiology
and Genetics Research
Babraham, Cambridge CB2 4AT
U.K.
Tawada, K.
Department of Biology
Faculty of Science
Kyushu University
Fukuoka, Fukuoka 812
Japan
Tsuchiya, T.
Department of Physiology
School of Medicine
Teikyo University
Itabashi-ku, Tokyo 173
Japan
Wakabayashi, K.
Department of Biophysical
Engineering
Faculty of Engineering Science
Osaka University

Toyonaka, Osaka 560
Japan

Winegrad, S.
Department of Physiology
School of Medicine
University of Pennsylvania
B307 Richards Building,
37th and Hamilton Walk
Philadelphia, Pennsylvania 19104-
6085
U.S.A.

Woledge, R.C.
Department of Physiology
University College London
Gower Street
London WC1E 6BT
U.K.

Yagi, N.
Department of Pharmacology
Faculty of Medicine
Tohoku University
Sendai 980
Japan

Yamada, K.
Department of Physiology
Oita Medical College

Oita-gun, Oita 879-56
Japan

Yamada, T.
Department of Physiology
School of Medicine
Teikyo University
Itabashi-ku, Tokyo 173
Japan

Yanagida, T.
Department of Biophysical
Engineering
Faculty of Engineering Science
Osaka University
Toyonaka, Osaka 560
Japan

Yu, L.C.
Building. 6, Room 114
National Institutes of Helth
Bethesda, Maryland 20892
U.S.A.

Zhao, Y.
Department of Anatomy
University of Iowa
Iowa City, Iowa 52242
U.S.A.

CONTRIBUTORS INDEX

Entries in bold types indicate papers, while other entries are contributions to discussions.

SUBJECT INDEX

Acoustic microscope 715-716, 719,
721-722
Actin
atomic model of 15-17, 21, 23-24,
371
—cable 303-305, 308-311, 313-
314, 351, 353-359, 377, 673,
793, 795, 809, 811-812, 815, 842
conformational changes of 21, 243
Cys-374 of 361, 372
electric field around 376
electric potential on 370, 373
—DNase I complex 15, 791
—genes 20, 241-242
MBS— 790-792
mutant— 241-243
myosin binding site on 15, 18-22,
118, 273, 368, 373
subdomains of 241
Actin filament
1st layer line of 445, 449, 451-453,
456, 458-460, 475, 482-484
2nd layer line of 432, 445, 459-
460, 819
5.9 nm layer line of 21, 439, 458
axial spacing of monomers in 57,
64, 846
fluorescent actin filament 217-218,
269, 279-281
S-1- decorated— 15, 17
stiffness of 57, 65, 68-69
target zone on 33, 40-41, 43, 46
Actomyosin ATPase 162, 243, 247,
268, 270
Aequorin 98, 107, 183, 185, 187,
189, 191, 193, 639-641, 647, 818

AMPPNP 33, 35, 37, 310, 393, 396-
397, 560, 563-565, 575, 827
Antibody
anti-actin— 402, 406
anti-hinge— 620
anti-projectin— 72
anti-S-2— 608, 611, 615-620, 846
anti-titin— 74, 76, 78
Antipyrylazo III 146
Atomic force microscope 90
Atropine 195, 197-199
ATP
—analogues 426, 563, 575
iontophoretic application of—
303-305, 307-308, 313-314, 316-
317, 812, 846
—splitting 291-292, 579, 691-692,
697, 699, 731, 733, 735-736,
741-742, 750, 836
ATP-γ-S 134, 393, 395, 398-400,
403, 827

BAPTA 97, 100, 107
Barnacle 109, 111, 183-185, 187-
188, 192-193, 202, 646, 770,
817-818
BDM(butanedione monoxime) 186,
395, 648, 817
Birefringence 1-4, 436, 476, 613
Brevin 205-211
Brownian motion 294-295, 321, 323,
335-336, 816

Caged ATP 100, 103, 318, 489-493,
498, 505-506, 511, 603, 607,
801, 813-814, 846